Lecture Notes in Electrical Engineering

Volume 748

The book series *Lecture Notes in Electrical Engineering* (LNEE) publishes the latest developments in Electrical Engineering - quickly, informally and in high quality. While original research reported in proceedings and monographs has traditionally formed the core of LNEE, we also encourage authors to submit books devoted to supporting student education and professional training in the various fields and applications areas of electrical engineering. The series cover classical and emerging topics concerning:

- Communication Engineering, Information Theory and Networks
- Electronics Engineering and Microelectronics
- Signal, Image and Speech Processing
- Wireless and Mobile Communication
- Circuits and Systems
- Energy Systems, Power Electronics and Electrical Machines
- Electro-optical Engineering
- Instrumentation Engineering
- Avionics Engineering
- Control Systems
- Internet-of-Things and Cybersecurity
- Biomedical Devices, MEMS and NEMS

For general information about this book series, comments or suggestions, please contact leontina.dicecco@springer.com.

To submit a proposal or request further information, please contact the Publishing Editor in your country:

China

Jasmine Dou, Editor (jasmine.dou@springer.com)

India, Japan, Rest of Asia

Swati Meherishi, Editorial Director (Swati.Meherishi@springer.com)

Southeast Asia, Australia, New Zealand

Ramesh Nath Premnath, Editor (ramesh.premnath@springernature.com)

USA, Canada:

Michael Luby, Senior Editor (michael.luby@springer.com)

All other Countries:

Leontina Di Cecco, Senior Editor (leontina.dicecco@springer.com)

**** This series is indexed by EI Compendex and Scopus databases. ****

More information about this series at http://www.springer.com/series/7818

Vijay Nath · J. K. Mandal
Editors

Proceeding of Fifth International Conference on Microelectronics, Computing and Communication Systems

MCCS 2020

Springer

Editors
Vijay Nath
Department of Electronics
and Communication Engineering
Birla Institute of Technology Mesra
Mesra, India

J. K. Mandal
Department of Computer Science
and Engineering
Kalyani University
Kolkata, India

ISSN 1876-1100 ISSN 1876-1119 (electronic)
Lecture Notes in Electrical Engineering
ISBN 978-981-16-0274-0 ISBN 978-981-16-0275-7 (eBook)
https://doi.org/10.1007/978-981-16-0275-7

This Springer imprint is published by the registered company Springer Nature Singapore Pte Ltd.
The registered company address is: 152 Beach Road, #21-01/04 Gateway East, Singapore 189721, Singapore

Preface

This volume contains research papers presented at the 5th International Conference on **Microelectronics Computing & Communication System (MCCS-2020)** organized by Indian Society for VLSI Education (ISVE) and IETE Ranchi Centre with support of BSNL at ARTTC BSNL Ranchi during 11–12 July 2020. It provided a great platform for researchers from across the world to report, deliberate, and review the latest progress in the cutting-edge research pertaining to smart computing and its applications to various engineering fields. The response to MCCS-2020 was overwhelming with a good number of submissions from different areas relating to artificial intelligence, machine learning, Internet of Things, VLSI design, micro and nanoelectronics circuits and devices, green energy, signal processing, smart computing, computational intelligence, and its applications in main tracks. After a rigorous peer-review process with the help of program committee members and external reviewers, only quality papers were accepted for publication in this volume of LNEE series of Springer. Several special sessions were offered by eminent professors in many cutting-edge technologies. Several eminent researchers and academicians delivered talks addressing the participants in their respective field of proficiency. Our thanks are due to Prof. P. S. Neelakanta, Florida Atlantic University USA; Prof. Ramjee Prasad, Aarhus University Denmark; Prof. Gopal Pathak, VC Jharkhand Technical University Ranchi; Prof. Nand Kunar Yadav 'Indu' VC, Central University of Jharkhand; Sh. Anurag Dubey, Xilinx USA; Sh. Samuel Tensingh, STMicroelectronics Singapore; Prof. Abhijit Biwas, University of Kolkata; Prof. A. A. Khan, Former VC Ranchi University; Sh. K. K. Thakur, CGMT BSNL Ranchi; Dr. Raj Kumar Singh, Ranchi University; Sh. Ajay Kumar, Governing Council Member IETE New Delhi; Sh. Viney Kakkar, Treasurer IETE New Delhi; Dr. P. R. Thakura, BIT Mesra; Dr. J. B. Sharma, Rajasthan Technical University Kota; Dr. Shylashree N. RV College of Engineering Bangalore; for their valuable talks. We would like to express our appreciation to the members of the program committee for their support and cooperation in this publication. We are also thankful to the team from Springer for providing a meticulous service for timely production of this volume.

Our heartfelt thanks to our loving founder Dr. J. W. Bakal, President IETE New Delhi, and Sh. K. K. Thakur, Chairman IETE Ranchi Centre; Prof. P. R. Thakura, President ISVE Ranchi and Executive Committee of IETE and ISVE Ranchi for extending excellent support to host this at ARTTC BSNL Campus. Professor P. S. Neelakanta, Florida Atlantic University USA, and Prof. Ramjee Prasad, Arhus University Denmark deserve a big round of applause from all of us for their continuous guidance and support from the beginning of the conference. Without their support, we could never have executed such a mega event.

Special thanks to all guests who have honoured us with their presence in the inaugural day of the conference. Our thanks to all special session chairs, track managers, and reviewers for their excellent support. Last but not least, our special thanks go to all the authors who submitted papers and all the attendees for their contributions and fruitful discussions that made this conference a great success.

Mesra, India
Kolkata, India

Vijay Nath
J. K. Mandal

Contents

About the Editors

Dr. Vijay Nath was born in Gorakhpur (U.P.), India, in 1976. He received his bachelor's degree in Physics and master's degree in Electronics from DDU Gorakhpur University, India, in 1998 and 2001. He received PGDCN (GM) from MMMUT Gorakhpur in 1999 and Ph.D. degree in VLSI Design and Technology from Dr. RML Avadh University Ayodhya in association with CEERI, Pilani, in 2008. From 2000 to 2001, he was a project trainee in IC Design Group, CEERI, Pilani. Presently, he is an associate professor in the Department of ECE, BIT Mesra Ranchi (JH), India, and joined this institute in 2006. His research interests include micro and nanoelectronics, analog and digital VLSI design, ASICs, embedded systems designs, Internet of things and machine learning. He has to his credit around 195 publications in reputed Scopus and SCI journals and conferences, and three scholars were awarded Ph.D. degree under his supervision. He has successfully completed two R&D projects funded by DST New Delhi with DRDL Hyderabad and MHRD New Delhi, and the third project is in ongoing stage funded by RESPOND SAC ISRO Ahmedabad. He has received 100 crore projects as Co-PI in NM-ICPS (TIH) in BIT Mesra (Spokes) for development of e-farming platform for enhancement of production marketing and income of Indian farmers in August 2020. He is the editor of Nanoelectronics, Circuits and Communication Systems, Proceeding of NCCS-2015, NCCS-2017, NCCS-2018, and NCCS-2019 published in 2017, 2018, 2019, and 2020, and the editor of Microelectronics, Computing and Communication Systems, Proceeding of MCCS-2015, MCCS-2017, MCCS-2018, and MCCS-2019 published in 2017, 2018, 2019, and 2020 in Scopus book series: LNEE, Springer. He is a member of several professional societies and academic bodies including IETE, ISTE, ISVE, and IEEE.

Prof. J. K. Mandal received his M.Sc. in Physics from Jadavpur University in 1986 and M.Tech. in Computer Science from the University of Calcutta and was awarded a Ph.D. in Computer Science and Engineering by Jadavpur University in 2000. Currently, he is a professor of Computer Science and Engineering and was the dean of the Faculty of Engineering, Technology, and Management, Kalyani University, West Bengal, for two consecutive terms. He started his career as a lecturer at NERIST, Arunachal Pradesh, in September 1988, and has 30 years' teaching and research experience. His areas of research include coding theory, data and network security, remote sensing and GIS-based applications, data compression, error correction, visual cryptography, steganography, security in MANET, wireless networks, and unify computing. He has to his credit more than 30 proceedings volumes, 6 books, 160 journals, and 300 other articles. He is one of the editors for the Springer AISC and CCIS Series, and 23 scholars were awarded Ph.D. degree under his supervision.

Strategic Design and Optimization of Vedic Low Pass FIR Filter for ECG Signals

Sudhanshu Janwadkar and Rasika Dhavse

Abstract ECG signals are contaminated by numerous noise sources during signal acquisition and need to be filtered before any sort of analysis can be made. Although numerous filter implementations have been proposed in literature, the considerations to be made in VLSI design and implementation of filter are scattered over numerous papers. This paper aims to provide readers a systematic procedure to design Finite Impulse Response (FIR) filter architecture to filter high frequency noise from ECG signals. Crucial parameters in design of FIR filters viz. Choice of Window Technique, Cut-off and Sampling Frequencies, Order of the Filter, Number of Bits for representation, Stop Band attenuation etc. are discussed at length and strategic decisions are made based on experimentation. VLSI architectures based on Vedic sutras result in low power design which is desirable in portable equipments. This paper therefore presents Vedic Sutra Urdhva Tiryagbhyam based FIR Filter. Software tools used for this study include MATLAB (and its FDA tool) and Xilinx ISE 14.7. The Vedic FIR filter has been implemented targeted for Spartan-3e starter FPGA board. The proposed 64th order FIR filter architecture occupies 869 slices, 1483 4-input LUTs and consumes dynamic power of 80 μW at 100 MHz clock frequency.

Keywords Digital filter · Electrocardiogram · Filter design · Vedic mathematics · VLSI architectures

---

The authors express their gratitude to Special Manpower Development Program for Chips to System Design (SMDP-C2SD) project under Ministry of Electronics and Information Technology, Government of India for providing access to tools required to conduct this research work.

S. Janwadkar (✉) · R. Dhavse
Electronics Engineering Department, Sardar Vallabhbhai National Institute of Technology, Surat, India

R. Dhavse
e-mail: rsk@eced.svnit.ac.in

V. Nath and J. K. Mandal (eds.), *Proceeding of Fifth International Conference on Microelectronics, Computing and Communication Systems*, Lecture Notes in Electrical Engineering 748,
https://doi.org/10.1007/978-981-16-0275-7_1

1 Introduction

Electrocardiogram (ECG) is the electrical activity of heart manifested by its contractions. It is measured by systematic placement of electrodes on chest and limbs [1]. The ECG signal is strongest among all biomedical signals with amplitude of the tone of few millivolts [2]. The bandwidth of ECG signal is from 0.5 to 150 Hz [1, 3]. ECG pattern consists of six identifiable sections called P-wave (atrial depolarization), Q-wave, R-Wave (initial positive deflection), S-wave (negative deflection, QRS together represent ventricular depolarization) and T-wave (ventricular re-polarization), each with non-overlapping frequency spectra [4, 5]. The wave pattern of a typical Electrocardiograph [6] is shown in Fig. 1. ECG signals extracted from patients are weak and need to be amplified using an instrumentation amplifier. However, throughout the process, ECG signals are contaminated by various sources of high frequency noise and interference. ECG signals need to be filtered before spectral analysis of the signals can be accomplished, followed by diagnosis.

Filtering of ECG signals using conventional analog low pass filters [7, 8] was reported to have considerable effects on the QRS complex, J-waves and epsilon. Also, they suffer from phase distortions [9, 10]. Therefore, ECG signals are converted to digital and filtering is performed in digital domain in modern ECG machines. Current quest of research is towards developing portable systems for acquisition and analysis of the biomedical systems, implying area and power consumption constraints in the filtering circuitry. Digital filters with linear filter response can be employed which do no effect the phase of the signals and hence do not contaminate QRS complex etc. The VLSI implementation of Vedic Mathematics sutra, Urdhva-Tiryagbhyam, is known to result in low power design [11–14]. Urdhva-Tiryagbhyam Sutra multiplier based digital low pass filter has been proposed in this paper.

Despite numerous filters reported in literature for ECG filtering, the design considerations and trade-offs to be made during design are not consolidated and literature on this is unorganized. The paper is aimed to describe the various trade-offs to be examined while designing FIR low pass filter (LPF). This paper aims to help readers to systematically decide the design parameters with justifications for a given set

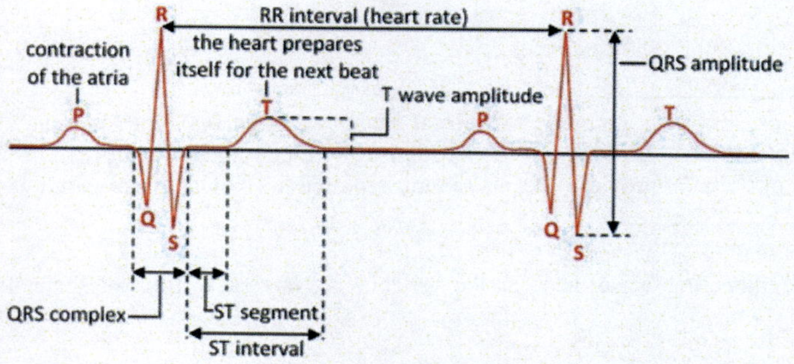

Fig. 1 Normal wave pattern of Electro-cardiogram [6]

of specifications. Similar approach may be followed for designing filter for other biomedical signals. Towards the end, this paper presents area-power efficient solution to implement FIR filter for eliminating high frequency noise from raw ECG signals.

The paper is organized in this manner: Sect. 2 describes the sources of noise in ECG signal extraction and role of LPF in signal filtering. Section 3 discusses the trade-off in design of FIR filter using MATLAB FDA tool. Urdhva Tiryagbhyam Sutra and its VLSI implementation to design 8 × 8 multiplier and design of 64th order FIR filter are described in Sect. 4. Section 5 summaries the experimentation, results and discussion based on it.

2 Sources of Noise in ECG Signal Extraction

The raw ECG signal is corrupted by several low-frequency and high-frequency noise sources [2, 7, 15, 16]. The low frequency noise sources include Motion Artifacts (accounted to poor electrode contacts, body movements during measurement etc.) [17], Baseline-Wander (caused due to activities like patients respiration) [18], Abdominal Artifacts (accounted to fetus heart activity in pregnant mother) [19] etc. The high frequency noise sources include Electromyographic (EMG) noise [20, 21], Powerline Interference [22], Electromagnetic (EM) interference and Noise coupled from other electronic equipments at high frequencies [2, 15, 16, 23]. They are typically eliminated using digital LPFs. In ECG filtering, LPF can be employed to [7]:

- Remove high frequency EMG noise
- Remove EM interference and high frequency coupling noise caused by other electronic equipments
- Remove noise due to artifacts such as coughing

3 Considerations in Design of Low Pass FIR Filter

A number of parameters are necessary to be decided for the design of a low pass filter. Experimentations performed using FDA tool of MATLAB are discussed in the current section.

3.1 Choice of Filter Design Method

FDA tool of MATLAB was used to design 80-order filters for each Rectangular, Hamming, Hanning, Blackman and Bohman window design techniques. Each filter was designed for cut-off frequency of 50 Hz and sampling frequency of 250 Hz. A

Table 1 Comparison of window design techniques for MSE and SNR improvement obtained from MATLAB

Window technique	MSE	SNR improvement (in dB)
Rectangular	0.4392	2.7697
Hamming	0.4191	2.9709
Hanning	0.4217	2.9881
Blackman	0.4174	2.9911
Bohman	0.4219	2.9896

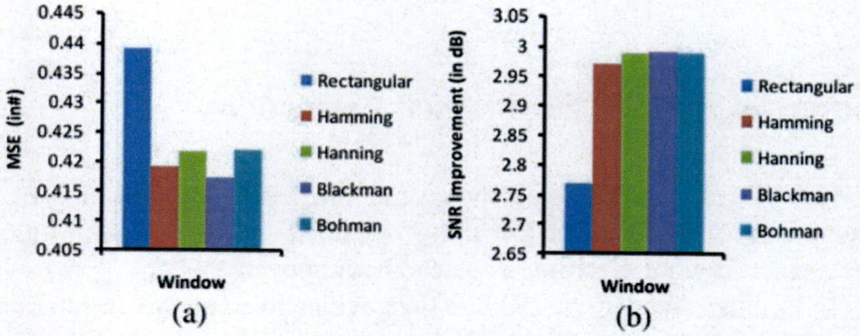

Fig. 2 MATLAB simulation comparison of a. MSE b. Improvement in SNR for window design techniques

high order was chosen to relatively ease other critical design considerations. To make a choice of filter, two parameters have been considered:

$$\text{Mean Square Error (MSE)} = \frac{1}{N}\sum_{i=1}^{n}(s - s_f)^2 \tag{1}$$

where s and s_f represent the signal and the filtered signal respectively.

$$\text{Signal to Noise Ratio (SNR) (indB)} = 10\log_{10}\frac{P_{signal}}{P_{noise}} \tag{2}$$

where P_{signal} and P_{noise} are signal power and noise power respectively.

A sine wave of 45 Hz frequency corrupted with synthetic noise of 80 Hz frequency was filtered using each of the windowing technique. The MSE between the input signal and the filtered signal has been captured using MATLAB function. The improvement in SNR and MSE values (calculated from MATLAB) are tabled in Table 1.

Values for MSE and the Improvement in SNR for various Window techniques are shown in Fig. 2. A lower value of MSE indicates good filtering. It is clear that the MSE is least for Blackman window. Also, Blackman Window offer very good improvement in SNR compared with other windowing techniques. This trend holds good for other high order FIR filters as well [24]. Hence, Blackman window is chosen as the Window Design technique.

3.2 Stop-Band Attenuation

For attenuating frequency components outside the stop-band, a minimum of −60 dB attenuation is recommended in the stop band [24]. This is ensured by monitoring the power of second lobe level while fixing the order of the filter.

3.3 Cut-Off Frequency

American Heart Association (AHA) in its guidelines released in 2007 [1] has recommended to use an upper cut-off of at least 150 Hz for monitoring ECG of adults. Further, it states that monitoring ECG upto 40 Hz will invalidate only the amplitude measurements and when such sub-optimal cut-off frequency is used; the ECGs must alert the user [1]. Jayarani et al. [15] and Li et al. [16] have argued that although most of the frequency spectrum of normal ECG signal ranges between 0.01 and 100 Hz, but 90% of the spectral energy of the signal is between 0.25 to 35 Hz. Rangayyan [2] recommends that clinical ECG may be filtered to a bandwidth of about 0.05–100 Hz while ECG for heart rate monitoring could use a reduced bandwidth 0.5–50 Hz. Christov et al. [25] use a high-pass filter of cut-off frequency 35 Hz to remove the ECG artifacts from the EMG signal, indicating that the frequency spectrum of ECG signals to be below 35 Hz. Drake and Callaghan [26] have used Hamming window and 40-order filters and cut-off frequencies at 20, 30, 40, 50 and 60 Hz. They conclude that 30 Hz cutoff frequency provided the optimal balance between ease of implementation, time investment, and performance.

The MIT-BIH database for ECG signals was referred. The frequency spectrum of the ECG signals acquired was plotted with help of MATLAB. Let f_{max} represent the frequency beyond which the amplitude of ECG signal in spectrum is below 1 mV. The components beyond this frequency do not contribute significantly to the spectrum. Therefore, for all signals in the MIT-BIH database, f_{max} frequency was noted from the spectra. A scatter plot representation of same is shown in Fig. 3. Based on spectrum analysis of multiple subjects, as represented in Fig. 3, it was found that the ECG signals occupied major spectrum below 50 Hz only. Based on literature and the study of database, the cut-off frequency for ECG signals is chosen to be 50 Hz.

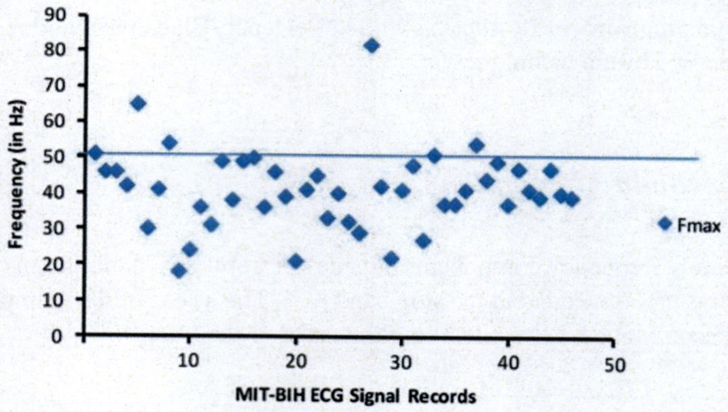

Fig. 3 Scatter plot of f_{max} of ECG Records in MIT-BIH database

Table 2 FDA tool simulation results for fixing filter order

Filter order N	Power of first side lobe (in dB)	Transition bandwidth (in Hz)
8	−80	185
16	−74	108.6
32	−76	44.5
64	−75	22.5
128	−75.5	13.5

3.4 Sampling Frequency

Nyquist criteria obligates sampling to be performed at twice the rate of the desired high-frequency cutoff [1, 2]. AHA recommends sampling rates of at least 2–3 times the theoretical maximum frequency considered [1]. Considering, cut-off frequency of 50 Hz, sampling rate of 250 Hz was considered sufficient.

3.5 Order of the Filter

The order of the filter was decided by monitoring the frequency response on FDA tool of MATLAB, for different filter order. Filter order was finalized by monitoring that power of the first side lobe is below −60 dB and transition band is narrow. Based on FDA tool simulations, the power of first side lobe and transition band width for $f_c = 50$ Hz for various filter order was determined. The observations are reported in Table 2.

From Table 2, it may be noted that with increase in filter order, the transition band decreases, implying a higher filter order as desirable. But, realizing high filter order dictates larger hardware. In Table 2, 64th order filter has transition bandwidth of 22.5 Hz while it reduces to 13.5 Hz for 128th order filter. However, 128th order filter will consume double hardware resources as 64th order filter, which is unnecessary considering a mere improvement of 9 Hz in transition band. Hence, a 64th order filter is finalized.

3.6 Number of Bits for Representation of Each Filter Coefficients

Choosing the number of bits for representation of each filter coefficient is important, as it not only affects the sensitivity of the filter but also has an effect on the power consumption of the design. The amplitude of ECG is in millivolts range [27, 28]. The number of bits for representing filter-coefficients depends on the number of bits used by the preceding ADC. An N-bit A/D converter is capable of converting infinite range of analog input values into a finite range of 2^{N-1} digital steps, where N is the resolution of the converter [33]. Resolution defines the smallest voltage change that can be distinguished by a single bit and is defined as:

$$\text{Resolution} = \frac{V_{DD}}{2^{N-1}} \tag{3}$$

For the supply voltage of 1.8 V and 3.3 V, the resolution values in mV, obtained from Eq. 3 are represented in Table 3.

Using 4-bit representation would not be sufficient, where as using 16-bit representation would give excellent resolution, but dictate requirement for more hardware and also consume more power. Therefore, a resolution of 8-bits was finalized.

The magnitude-phase response of the final design of the filter based according to aforementioned considerations is shown in Fig. 4.

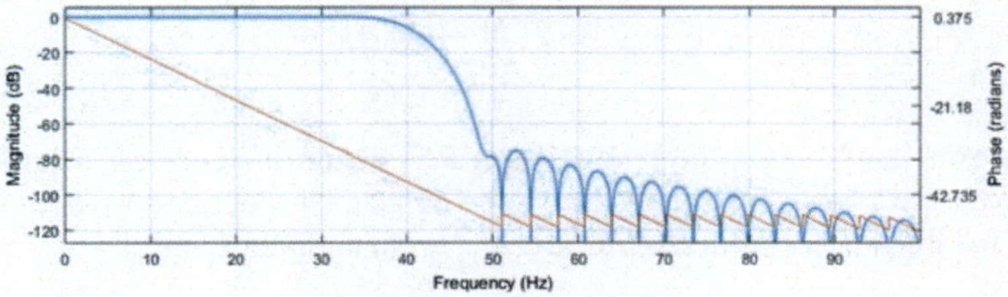

Fig. 4 Magnitude-phase response of filter designed using FDA tool

Table 3 Resolution versus number of bits

Number of bits (N)	Resolution (in mV)	
	$V_{DD} = 3.3$ V	$V_{DD} = 1.8$ V
4	220	120
8	12.94	7.06
16	0.05	0.028

4 Urdhva Tiryagbhyam (UT) Sutra

The Urdhva Tiryagbhyam (UT) Sutra is a Vedic mathematic algorithm for multiplication of two input numbers. The meaning of Urdhva Tiryagbhyam is vertically and crosswise. There is no restriction on range of the input numbers. Various architectures of binary multipliers, targeted for both ASIC and FPGA platforms, have been designed on basis of UT Sutra and reported in literature [11, 12, 14].

4.1 The Mathematical Analysis of UT Sutra

P. Saha et al. [14] have elaborated modern mathematics understanding of the UT Sutra. Assume two N-bit words described in binary as $A = \sum_{i=0}^{N-1} a_i 2^i$ and $B = \sum_{j=0}^{N-1} b_j 2^j$, where x_i, y_j are in range 0 to 1. The multiplication can be described as:

$$P = AB = \sum_{i=0}^{N-1} a_i 2^i \sum_{j=0}^{N-1} b_j 2^j = \sum_{i=0}^{N-1} \sum_{j=0}^{N-1} a_i b_j 2^{i+j} \tag{4}$$

Let $i + j = k$.

$$P = AB = \sum_{k=0}^{2N-1} a_i b_{k-i} 2^k \tag{5}$$

This essentially means the product term P is a sum of terms pk obtained by multiplying the kth parts of a and b in a crosswise manner. Consider two 8-bit inputs A[7:0] and B[7:0]. According to Eq. 5 for UT Sutra multiplication,

$$\begin{aligned} P = {} & A[7:4] X B[7:4] + A[7:4] X B[3:0] + \\ & A[3:0] X B[7:4] + A[3:0] X B[3:0] \end{aligned} \tag{6}$$

According to Eq. 6, Implementing 8 X 8 UT multiplier dictates the use of four 4 X 4 UT Sutra multipliers and three 8-bit ripple carry adders. The resulting architecture is shown in Fig. 5.

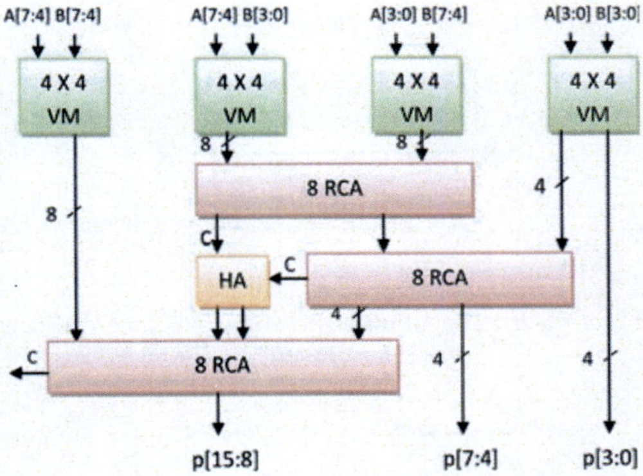

Fig. 5 Architecture of 8-bit multiplier using UT sutra

In Fig. 5, UT sutra multipliers are represented as VM, Ripple Carry Adders are represented as RCA and Half Adder is represented as HA. The prefixes indicate the size of these components. Each of the 4 × 4 UT Sutra multipliers are in turn composed from of 2 × 2 UT Sutra multipliers. Therefore, the design was extended from 2 × 2 multiplier to 8 × 8 multiplier in bottom-up approach based on Eq. 5. This essentially implies that UT Sutra based Vedic Multiplier is a modular structure. Ripple Carry adders were chosen as they are most area efficient when compared to other adder architectures.

4.2 Design of FIR Filter Using UT Sutra Multiplier

FIR filters are called so because their impulse response dies into zero after finite duration. They are provide linear phase shift to the input samples. If x(n) represent the input samples to an FIR filter and has impulse response h(k) for k^{th} instance (where $0 \leq k \leq N$), then for an Nth order filter, the output response y(n) is defined as [29]:

$$\mathrm{y(n)} = \sum_{k=0}^{N-1} h(k).x(n-k) \tag{7}$$

The terms x(n-k) obtained in Eq. 7 are referred as filter taps. The basic building blocks for an FIR filter are shifter, multiplier and adders. Among these, multipliers are the most crucial blocks in terms of area, speed and power consumption. Any optimization in multiplier can greatly improve the performance of the overall design [34]. The proposed architecture which realizes Eq. 7 consisting of a UT Sutra multiplier, shifter and adder blocks is shown in Fig. 6.

Synthesis of architecture shown in Fig. 6 reported usage of 1308 slices and 2208 4-input LUTs, which could be lowered. Power consumption was reported to be 80 μW at 100 MHz clock frequency. Hence, need to further optimize the design for

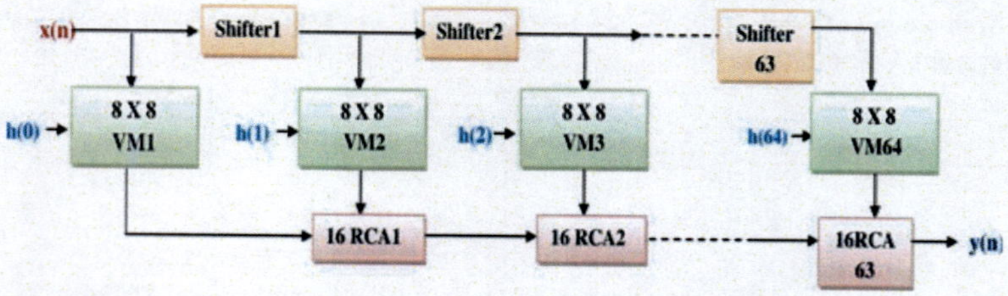

Fig. 6 Schematic of proposed vedic 64th order FIR filter

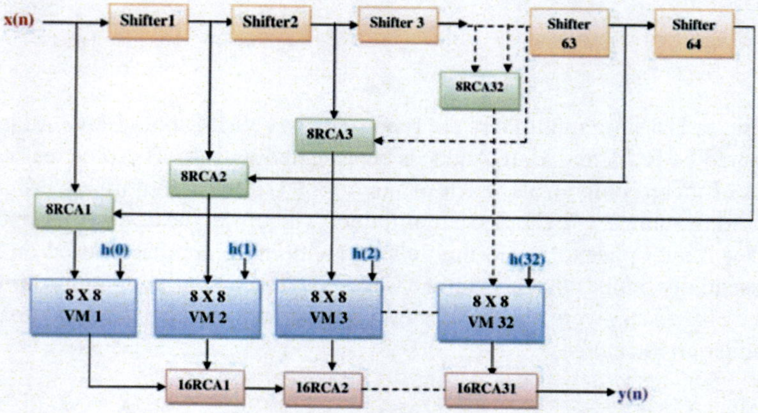

Fig. 7 Schematic of proposed and optimized vedic 64th order FIR filter

device utilization was deemed necessary. The design and number of slices required to implement can be further optimized if the symmetry between filter coefficients is exploited. If the filter has to give a linear phase response, the filter coefficients are required to be symmetrical, i.e.

$$h(n) = h(N - n) \tag{8}$$

Filter coefficients obtained for Blackman Window using FDA tool of MATLAB are also symmetrical. Hence, the property in Eq. 8 was used to optimize the design. The optimized version is represented in Fig. 7. The synthesis and post implementation results for the FIR architecture shown in Fig. 7 is reported in Table 4.

5 Experimentation, Results and Discussion

Software tools used for this study include MATLAB (and its FDA tool) and Xilinx ISE 14.7. Also, the MIT-BIH Arrhythmia Database [30] was considered for this study. The filter design was accomplished using FDA tool of MATLAB for specifications

Table 4 Comparison of post-implementation results for proposed vedic FIR filter

		Vedic FIR filter proposed in [32]	Proposed vedic FIR filter
Device utilization summary	Number of slices:	1095	869
	Number of slice flip flops:	312	312
	Number of 4 input LUTs:	Total: 1889 Logic: 1832 Registers: 57	Total: 1483 Logic: 1426 Registers: 57
Timing summary	Minimum clock period	3.706 ns	4.014 ns
	Maximum operating frequency	269.832 MHz	249.128 MHz
Power consumption summary	Dynamic (signal + control) power consumption	90 μW	80 μW
	Leakage power	270 μW	270 μW

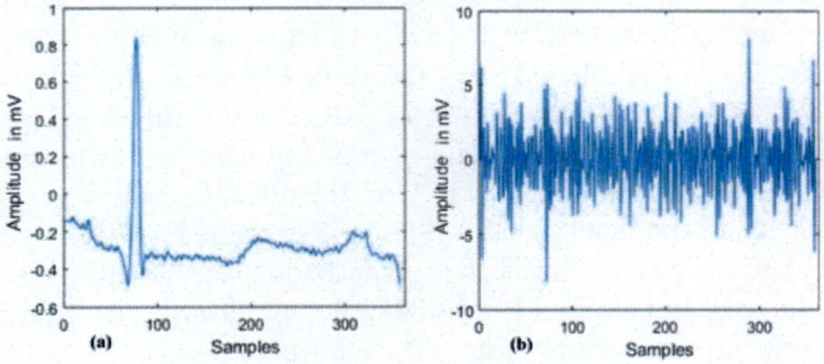

Fig. 8 (a) Raw ECG signal, MIT-BIH database (record 100) (b) Noise corrupted ECG signal

decided (and as mentioned in Sect. 2). Vedic based FIR Filter as shown in Fig. 7 and 8 was then implemented using VHDL coding. The target platform chosen was Spartan-3e starter board from Xilinx. The board features 500 K gate Spartan-3E FPGA (XC3s500e) [31]. Later, MATLAB was used to generate test vectors and verify the functionality of the implemented filter.

The ECG signals (in textual form) from MIT-BIH database were imported into MATLAB. The database doesn't mention if any filter were used in the recording instrument during ECG recording. This raw ECG signal was then corrupted using synthetic noise (sine wave) of high frequencies (f > 40 Hz). The noise corrupted signal was then converted into 8-bit binary values. This file containing 8-bit binary values of the ECG signal was then read using VHDL file IO. The filter response containing 8-bit binary values of the filtered signal was imported again in MATLAB and filtered signals were plotted.

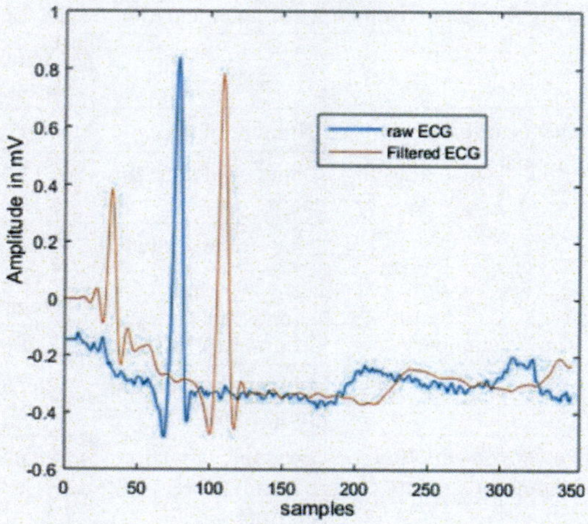

Fig. 9 Raw ECG signal versus vedic filtered ECG signal

Figure 8(a) shows the ECG signal of record 100. A single QRS complex has been shown. The signal is sampled at 334 Hz. The signal is corrupted with noise signal. The resulting Noisy ECG signal has been represented in Fig. 8(b). The noise corrupted signal was then filtered using the Vedic FIR filter, shown in Fig. 7. The filtered ECG signal along with the raw input ECG signal is shown in Fig. 9.

Designed Vedic FIR filter is capable of rejecting noise and is retaining all the crucial features of the ECG signal [3]. Also, the filtered signal is smoothened in comparison with the raw input signal. It is also evident from Fig. 9 that the amplitude of the filtered ECG signal is attenuated in comparison to the raw ECG signal. This is observed from peak amplitude of P-wave, R-wave and T-wave. Also, the samples in the filtered response are linearly shifted. The number of samples shifted equals half the filter order.

To establish the essence of our work, we have compared the performance of our proposed Vedic Low Pass FIR filter with previously existing architecture [32]. The comparison is done by implementing both the architectures for common Spartan-3e FPGA kit. Improvement in area aspect has been verified by reduction in number of slices and 4-input LUTs. Minimum clock period gives an indication of timing. The dynamic power is reported at 100 MHz clock frequency, 5pF output load, 3.3 V supply and 12.5% toggle rate, using Xpower Analyzer.

Comparison of Device Utilization of the proposed Vedic Low Pass FIR filter with previous proposed filter architecture [32], indicates reduction in number of slices (27.37%) and 4-input logic LUTs (26%). A major achievement is that improvement in device utilization doesn't affect much to the performance of our proposed FIR filter (maximum frequency of operation of 249.128 MHz compared to 269.832 MHz). The reason can be attributed to the fact that high speed operation is not obligatory in filtering of biomedical signals, as they are sampled at very low sampling rate (typically few hundreds of Hertz). Vedic Multiplier results in lower power consumption than all other conventional multiplier architectures. Although the leakage power remains same, there is improvement in dynamic power consumption of the proposed architecture. In pursuit of modern applications such as the portable battery operated measuring equipments, our proposed Vedic Low Pass FIR Filter architecture is clearly superior.

6 Conclusion

ECG signal is contaminated by numerous sources of noise during acquisition. Numerous filter implementations have been proposed in literature for filtering of ECG signals. However, the works are either application specific or lack micro-planning. Moreover, a designer needs to survey many references so as to understand strategic planning procedure for these architectures. We have consolidated and presented a simple systematic procedure to proceed with the design considerations for filtering ECG signals. Similar approach may be extended for filtering other biomedical signals.

Each parameter to be considered while designing FIR filter was systematically deduced based on simulations of FDA tool of MATLAB and Literature Survey. It was observed that Blackman Window is most efficient over other windowing techniques due to lower MSE (= 0.4974) and high signal to noise ratio (= 2.9911 dB) under the experimental conditions. Further, based on Literature Survey and observing frequency spectra of ECG signals of a numerous subjects from MIT-BIH database, the cut-off frequency was chosen to be 50 Hz. The filter order was then systematically designed in a way such that the stop-band is attenuated at least by −60 dB and transition bandwidth is narrow. Based on trade-offs, a 64th order filter was designed. Based on typical amplitude values of ECG signal, 8-bit representation throughout the system was finalized.

In this paper, we have discussed the design and implementation of a Vedic UT Sutra multiplier based low pass FIR filter. The architecture for 8 X 8 UT Sutra multiplier was then deduced based on the UT sutra. This UT Sutra multiplier was used for design of FIR filter and filtering operation was demonstrated. The filter implemented on Spartan-3e starter board occupies 869 slices and 1483 4-input LUTs and consumes 80 μW dynamic power at 100 MHz clock frequency.

References

1. Kligfield P, Gettes L, Bailey J, Childers R, Deal B, Hancock E, van Herpen G, Kors J, Macfarlane P, Mirvis D, Pahlm O, Rautaharju P, Wagner G (2007) Recommendations for the standardization and interpretation of the electrocardiogram. J Am College Cardiol 49(10):1109–1127
2. Rangayyan RM (2010) Biomedical signal analysis: a case study approach. Wiley, New Delhi, pp 73–175
3. Luo S, Johnston P (2010) A review of electrocardiogram filtering. J Electrocardiol 43(6):486–496

4. Lei WK, Dong MC, Shi J, Fu BB (2008) Automatic ECG interpretation via morphological feature extraction and SVM inference nets. In: APCCAS 2008 - 2008 IEEE Asia Pacific Conference on Circuits and Systems, pp 254–258
5. Panda R, Pati UC (2012) Removal of artifacts from electrocardiogram using digital filter. In: 2012 IEEE Students' Conference on Electrical, Electronics and Computer Science, pp 1–4
6. Martinek R, Dek J (2012) A system for improving the diagnostic quality of fetal electrocardiogram. Przeglad Elektrotechniczny 88:164–173
7. Hartmann E (2003) ECG front-end design simplified with microconverter. Analog Dial. 37. https://www.analog.com/en/analog-dialogue/articles/ecg-front-end-design-simplified.html
8. Dozio R, Burke MJ (2009) Second and third order analogue highpass filters for diagnostic quality ecg. In: IET Irish Signals and Systems Conference (ISSC 2009), pp 1–5
9. Watford C (2014) Understanding ECG filtering. EMS-topics. https://ems12lead.com/2014/03/10/understanding-ecg-filtering/
10. Ezenwa BN, Gupta MM, Nikforuk PN, Prasad K (1988) A baseline tracking algorithm for drift reduction in electrocardiography. In: Proceedings of the Annual International Conference of the IEEE Engineering in Medicine and Biology Society, vol 3, pp 1230–1232
11. Jie LS, Ruslan SH (2017) A 2 × 2 bit vedic multiplier with different adders in 90 nm CMOS technology. In: Proceedings of the International Conference on Electrical and Electronic Engineering (IC3E 2017), vol 1883, pp 020017
12. Sivanandam K, Kumar P (2019) Design and performance analysis of reconfigurable modified vedic multiplier with 3-1-1-2 compressor. Microprocess Microsyst 65:97–106
13. Kumar D, Saha P, Dandapat A (2017) Vedic algorithm for cubic computation and VLSI Implementation. Eng Sci Technol Int J 20(5):1494–1499
14. Saha P, Banerjee A, Dandapat A, Bhattacharyya P (2011) ASIC design of a high speed low power circuit for factorial calculation using ancient vedic mathematics. Microelectron J 42(12):1343–1352
15. Jeyarani AD, Jaya Singh T (2010) Analysis of noise reduction techniques on QRS ECG waveform - by applying different filters. In: Recent Advances in Space Technology Services and Climate Change 2010 (RSTS CC-2010), pp 149–152
16. Li J, Deng G, Wei W, Wang H, Ming Z (2017) Design of a real-time ECG filter for portable mobile medical systems. IEEE Access 5:696–704
17. Nagai S, Anzai D, Wang J (2017) Motion artifact removal for wearable ECG using stationary wavelet multi-resolution analysis. In: 2017 IEEE 5th International Symposium on Electromagnetic Compatibility (EMCBeijing), pp 1–5
18. Gupta P, Sharma KK, Joshi SD (2015) Baseline wander removal of electrocardiogram signals using multivariate empirical mode decomposition. Healthcare Technol Lett 2(6):164–166
19. Jaros R, Martinek R, Kahankova R (2018) Non-adaptive methods for fetal ECG signal processing: a review and appraisal. Sensors 18(10):36–48
20. Abbaspour S, Lindn M, Gholamhosseini H (2015) ECG artifact removal from surface EMG signal using an automated method based on wavelet-ICA. Study Health Technol Inf 211(6):91–97
21. Guohua Lu, Brittain J-S, Holland P, Yianni J, Green AL, Stein JF, Aziz TZ, Wang S (2009) Removing ECG noise from surface EMG signals using adaptive filtering. Neurosci Lett 462(1):14–19
22. Levkov C, Mihov G, Ivanov R, Daskalov I, Christov I, Dotsinsky I (2005) Removal of power-line interference from the ECG: a review of the subtraction procedure. Biomed. Eng. Online 4:50
23. Maggio A, Bonomini M, Laciar E, Arini P (2012) Quantification of ventricular repolarization dispersion using digital processing of the surface ECG. In: Advances in Electrocardiograms - Methods and Analysis
24. Characteristics of Different Smoothing Windows, National Instruments, NI LabWindows CVI 2012 Manual (2012)
25. Christov II, Daskalov IK (1999) Filtering of electromyogram artifacts from the electrocardiogram. Med. Eng. Phys 21(10):731–736

26. Drake JD, Callaghan JP (2006) Elimination of electrocardiogram contamination from electromyogram signals: an evaluation of currently used removal techniques. J Electromyogr Kinesiol 16(2):175–187
27. Dori G, Gershinsky M, Ben-Haim S, Lewis B, Bitterman H (2009) Evaluating rest ECG amplitude changes using the ECG variability contour method. In: 2009 36th Annual Computers in Cardiology Conference (CinC), Park City, UT, pp 837–840
28. Becker DE (2006) Fundamentals of electrocardiography interpretation. Anesthesia Progr 53(2):53–64
29. Gustafsson O, Johansson K, Johansson H, Wanhammar L (2006) Implementation of polyphase decomposed FIR filters for interpolation and decimation using multiple constant multiplication techniques. In: APCCAS 2006 - 2006 IEEE Asia Pacific Conference on Circuits and Systems, pp 924–927
30. Goldberger A, Amaral L, Glass L, Hausdorff J, Ivanov P, Mark R, Mietus J, Mark G, Moody GB, Peng C, Stanley H (2014) Physiobank, Physiotoolkit, and Physionet: components of a new research resource for complex physiologic signals. Circulation 101(23):215–220
31. Digilent, Digilent Spartan-3E Starter Board. Accessed 11 Nov2019, https://www.xilinx.com/products/boardsand-kits/1-elhacw.html
32. Sumalatha M, Naganjaneyulu PV, Satya Prasad K (2019) Low power and low area VLSI implementation of vedic design FIR filter for ECG signal de-noising. Microprocess Microsyst 71:102883
33. Principles of Data Acquisition and Conversion. Texas Instruments (2015). https://www.ti.com/lit/an/sbaa051a/sbaa051a.pdf
34. Keshab K (2007) Parhi, VLSI Digital Signal Processing Systems: Design and Implementation. Wiley India Pvt., New Delhi

Computing Transient Response of Drain Current for High Electron Mobility Transistor in Presence of Hot Electrons

Swarnav Mukhopadhyay, Anwita Nath, Purbasha Ray, and Arpan Deyasi

Abstract Transient response effect on drain current in HEMT is numerically computed considering presence of hot electrons inside the channel. Hot electrons are released from trap state to form 2DEG, which also increases device temperature. Effect of gate width and gate-to-drain length are computed on drain current, and leakage current is also evaluated for different temperatures. High peaks in drain current characteristics are observed, which is increased with higher temperature, larger gate width and lower gate-to-drain width. at the same instant, leakage current is reduced which speaks for effectiveness of the design. Current peak also appears earlier with increasing hot electron concentration, and diminishes if concentration becomes moderate. Care is taken so that width of the heated region should be less than barrier layer thickness, otherwise changing piezo-polarization can permanently damage the device.

Keywords Transient response · Drain current · Hot electrons · Leakage current · Structural parameters · Heated region width

1 Introduction

High electron mobility transistor becomes a choice of research due to the possibility of delivering large power [1] at lower operating temperature and also at lower supply voltage. Owing to high carrier mobility and formation of 2DEG [2], electrical output becomes very high; and therefore fits for wireless applications and RADAR systems. Its inherent noise reduction property originates due to the formation of quantum–mechanical electron gas layer [3], which effectively eliminates the scatterign probability between free carriers and impurity agencies. However, arise of

S. Mukhopadhyay · P. Ray
Department of Electronics and Telecommunication Engineering, Jadavpur University, Kolkata, India

A. Nath · A. Deyasi (✉)
Department of Electronics and Communication Engineering, RCC Institute of Information Technology, Kolkata, India

V. Nath and J. K. Mandal (eds.), *Proceeding of Fifth International Conference on Microelectronics, Computing and Communication Systems*, Lecture Notes in Electrical Engineering 748,
https://doi.org/10.1007/978-981-16-0275-7_2

small geometry effects is a major concern for all submicron devices, and henceforth gate control becomes essential, either by incorporating different materials [4] or by modified architectures [5]. A few well-known small geometry effects like mobility degradation, channel length modulation, self-heating etc. are added with the theoretical models [6] in order to explain the experimental results; and these proposed models are utilized in explaining the RF behaviour [7, 8] of HEMT. But the analysis is severely modified when effect of surface traps are considered.

Current collapse and drain current dispersion effects [9] are measured in presence of pulse inputs, and charging and discharging phenomena are observed [10] in presence of crystal defects. Different substrates are considered to characterize hole trapping [11]. Apart from surface traps, effect of deep level traps are also investigated to characterize drain current [12]. Works are later used to design high-frequency inverter circuit with low noise [13]. However, both drain and leakage currents are simultaneously not detected as far the knowledge of the authors. In this paer, these currents are calculated in presence of hot electrons, and shift of current peak is noted with tuning of several structural parameters. A reasonable amount of hot electron concentration is considered otherwise current response becomes flatter. Design is made considering the effect of piezo-polarization, and result shows the possibility of getting high current at very low voltages.

2 Device Structure and Design

The AlGaN/GaN HEMT structure is taken as shown in Fig. 1(a). GaN substrate layer of thickness3 μm is taken, followed by unintentionally doped AlGaN layer with 30% Al mole fraction and thickness of 20 nm and n-type doping concentration of $2 \times 10^{16}/m^3$. Moderate n-type doped AlGaN layer of doping concentration $2 \times 10^{18}/m^3$ and thickness of 10 nm is considered over UID-AlGaN layer. Ni/Au alloy is taken as a gate material with gate length of 0.5 μm, gate to drain distance (L_t) 2–4 μm, gate width (W_t) 20–50 μm. Gate metal thickness is taken as 1 μm.

When the transistor is in OFF mode ($V_i = -10$ V) in Fig. 1(a), then the channel is depleted and there is no 2DEG available at AlGaN/GaN interface. But due to available surface states the electrons are trapped at the surface between drain-gate and gate-source area. As there's a high voltage present between drain ($V_D = 10$ V) and gate ($V_G = -10$ V) and it is assumed that the transistor is in OFF state for a long time (t >> switching time), the trapped electrons gets heated up (shown in red) and gains energy, which causes leakage current to flow from drain to gate. But due to increased temperature, the mobility of hot electrons starts to reduce and they starts to accumulate, which recursively increases the temperate of the region. The magnitude of the leakage current increases since the time duration of the OFF state increases.

At the instant when $V_G = 0$ V that is transistor is ON condition, the hot electrons that are released from the trap causes a large leakage current to flow between drain and gate, at transient. The heated region starts to reduce as currents starts to flow between drain and source via 2DEG or channel. And it reduces exponentially with

Fig. 1 AlGaN/GaN HEMT device structure while transistor is in (a) OFF state; (b) ON state

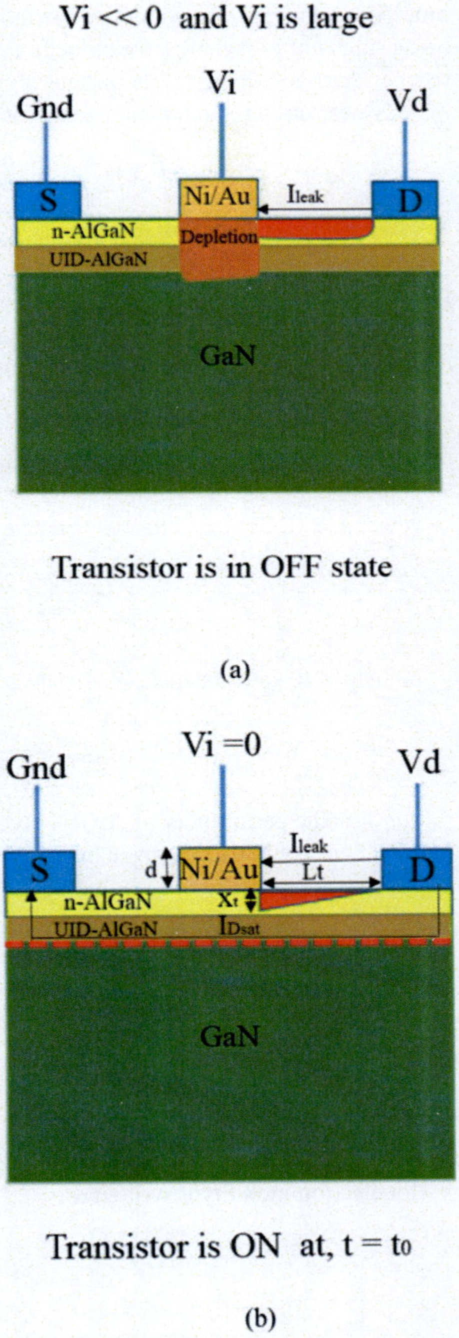

time. The temperature also reduces as the hot electrons starts to move, rather than being stagnant. In Fig. 1(b), the reduction of heated region is shown, as the electrons reaches drain region just at the instant of turning ON, so the heat is dispersed more quickly near drain region, than gate region.

3 Mathematical Modelling

Let, at $t < 0$ HEMT is OFF for a long time, generated heat is T_1. At $t = 0$, temperature is T_1, transistor is ON and total drain current $= I_{Dsat} + I_{leak}$, where, I_{Dsat} and I_{leak} are drain saturation current and leakage current respectively.

At, $t \rightarrow \infty$ Temperature is T_2, Drain current $= I_{Dsat}$.

$$I_D(t) = (I_{Dsat} + I_{leak})(1 - e^{(-t/\tau)}) \tag{1}$$

where $\tau = (RC)_{channel}$, channel is created by hot electrons released from trap.

$$R_{channel} = \frac{L_t}{qW_t n_h} \tag{2}$$

L_t is drain to gate distance, W_t is gate width, n_h is the hot electron concentration.

$$C_{channel} = \frac{\varepsilon_{AlGaN} x_t}{2} \tag{3}$$

ϵ_{AlGaN} is the permittivity of AlGaN layer, x_t is the depth of heated region inside AlGaN. Leakage current is given by.

$$I_{leak} = \mu_n C_{dg} \frac{W}{L_t} \frac{V_t^2}{2} \Theta \tag{4}$$

where Θ is the tunneling probability, C_{dg} is the gate-to-drain capacitance, may be written in the form.

$$C_{dg} = \frac{Q}{V_{dg}} = \frac{\varepsilon_{ox} A_{ch}}{d} \tag{5}$$

Hot electron mobility is written as.

$$\mu_n = \mu_0 \left(\frac{300}{T}\right)^{\theta_i} \tag{6}$$

θ_i depends on experimental results.

$$T = T_2 + (T_1 - T_2)e^{-nt} \tag{7}$$

Where T_1 and T_2 are initial and final temperature of the heated region and '*n*' is a constant depends on experimental results. Tunneling probability is.

$$\Theta = \exp\left(-\frac{4}{3}\frac{\sqrt{2qm^*}}{\hbar}\frac{\phi_B^{1.5}}{E}\right) \tag{8}$$

φ_B is the barrier height, $\varphi_B = (E_{C,\,AlGaN} - E_{F,\,metal}) = (\varphi_M - \chi)$, where χ is electron affinity of AlGaN.

4 Results and Discussion

The concentration of hot electron can affect the device performance in case of leakage as well as current degradation. This model shows that, as the hot carriers are released from trap at the instant of switching ON, the current slowly starts to flow from drain to source in Fig. 2. If the released carriers are more it causes less drain delay. But as the number of released carriers increased, then fast switching is observed. But increasing hot carrier also increases the transient leakage current from drain to gate. As the surface carriers flows from drain to gate at transient, the gate leakage current increases initially, but reduces as the total number of surface carriers get reduces after switching on the device. So it can be observed as the hot carrier concentration increases the peak current increases at first, then decays after certain time and finally becomes the drain saturation current.

Fig. 2 Transient response of drain current with different concentration of hot electrons

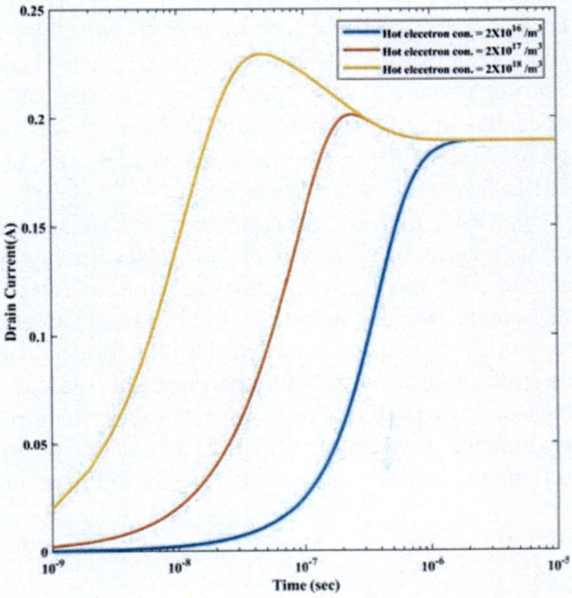

With increasing duration of the OFF state, the temperature of the device starts to increase. As the temperature increases the energy of the trapped electrons also starts to increase. Because of that, some electrons acquire enough energy to get released from the trap, which increases the leakage current, shown in Fig. 3(a). After turning on the device the leakage current decreases monotonically. The drain lag reduces little bit with increasing temperature (Fig. 3b), due to increased concentration of released electrons from trap. But the peak drain current increases with temperature as the overall leakage current increases. The leakage current dies out after 1 μsec, which is a great concern for the device operation. As sudden increased current can damage the device if it gets out of control. Device will become unstable and can cause increased power consumption.

Decreasing the gate to drain length increases the leakage current but in decreases the drain lag. As the distance becomes shorter, carriers take less time to transport from rain to gate initiallycausing transient leakage, after that due to increased electric field the drift velocity of the electrons increases which reduces delay (Fig. 4a). The leakage current increases due to reduction of distance and increase of tunneling probability.

In Fig. 4b, it is observed that, as the device/gate width increases, the transient leakage current increases, but drain saturation current increases also. It happens due to reduction of resistance of both hot carrier channel and 2DEG channel simultaneously. The drain lag also reduces due to increase of gate width, which eventually reduces RC_{delay} of the device.

The depth of heated region can play crucial role for the stability of the device. As the depth of heated region (X_t) increases but X_t < thickness of AlGaN, the mobility of the carriers near the channel gets degraded, as a result drain lag increases. But leakage current decreases as the heated carriers get dispersed from the surface area. So gate leakage decreases at transient, which is shown in Fig. 5. If the heated region becomes comparable with AlGaN thickness, it can cause permanent damage of the device and might change piezo-polarization. This can cause damage in crystal structure or change the crystal orientation. That will create interfacial traps and degrade the performance of the device, such as increased delay, reduced drive current, increased DIBL, increased leakage current etc.

Increasing device temperature will increase tunneling probability and associated leakage current. As temperature increases the highly energized electrons will be able to cross the Schottky barrier of gate and enter AlGaN barrier layerand cause thermionic emission as well as band to band tunneling current to flow from drain to gate. As the device is switched on the leakage current starts to decrease due to increase of effective Schottky barrier height, reduction of energized electrons and the total leakage current depends on the device temperature. If the overall device and environment temperature is high then tunneling probability increases. In Fig. 6, the variation of leakage current with time for different tunneling probability is shown.

Fig. 3 (a). Transient response of leakage current for different temperature, (b). Transient response of drain current for different temperature

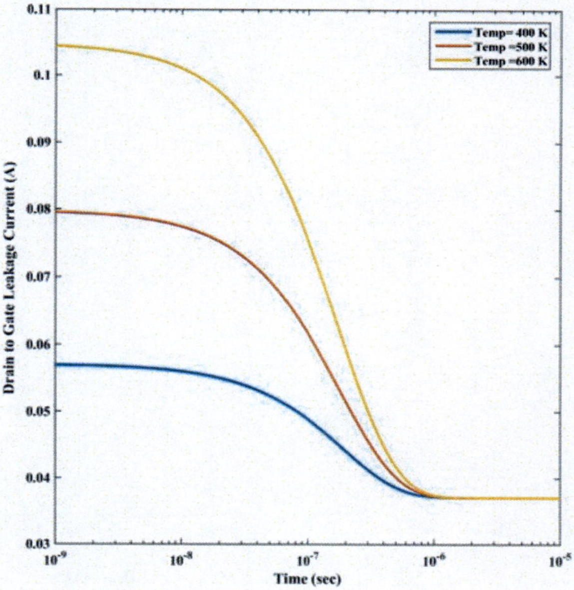

(a)

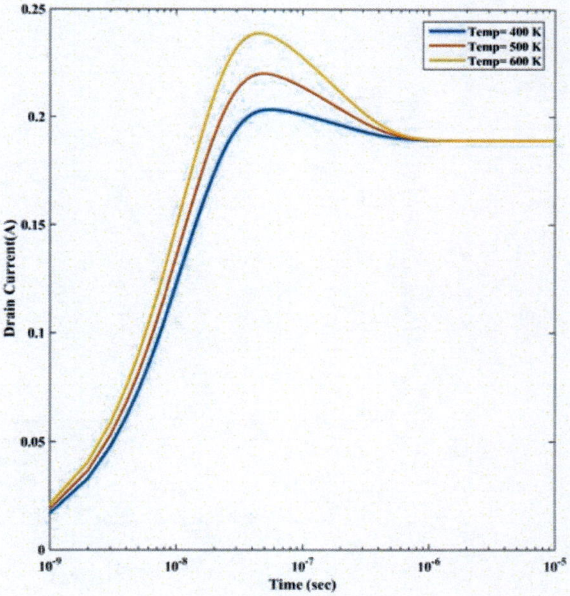

(b)

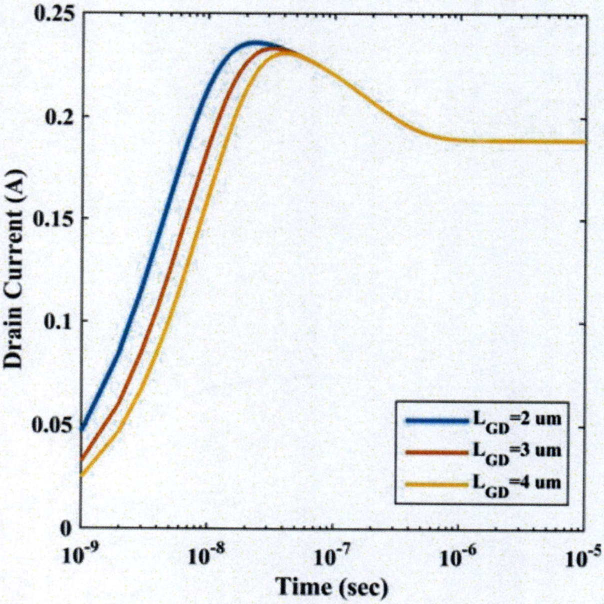

(a)

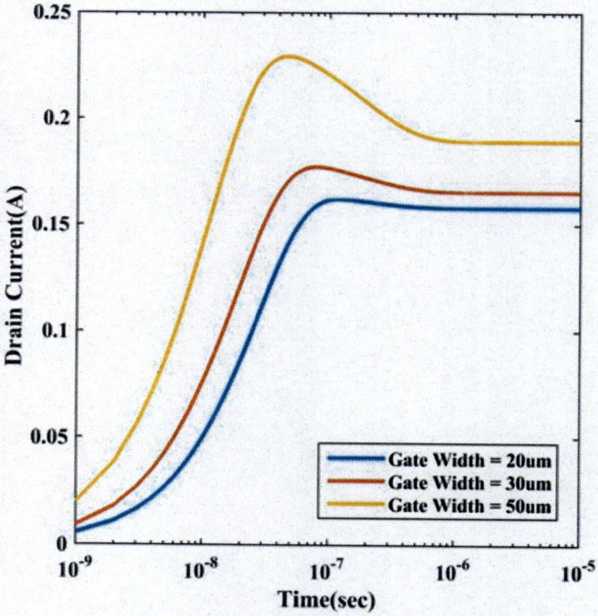

(b)

Fig. 4 (a) Transient response of drain current for different gate to drain length, (b): Transient response of drain current for different gate width

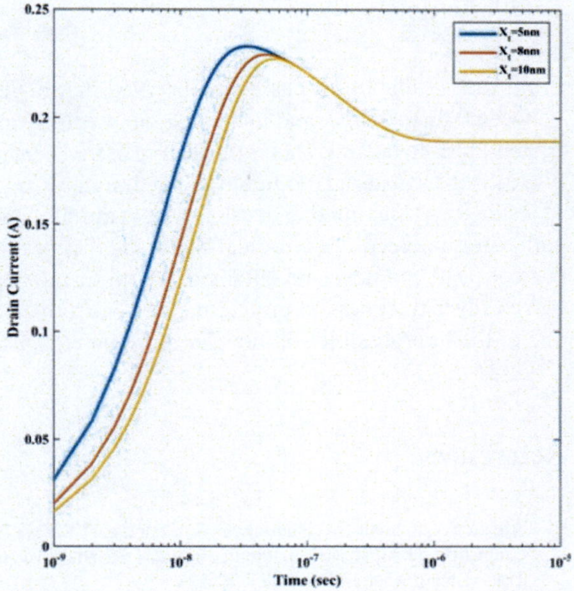

Fig. 5 Transient response of drain current for different depth of heated region from surface

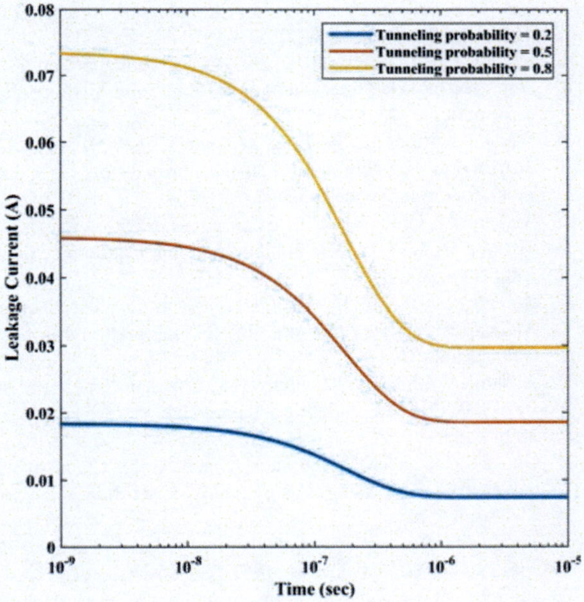

Fig. 6 Transient response of leakage current for different tunneling probability

5 Conclusion

From the simulation, it has been observed that higher drain current peak is found between 0.01 to 0.1 μs, and in that region, precisely in the vicinity of 0.1 μs; leakage current starts to fall. this shows the feature of transient response, and therefore suitable bias as well as material properties of the design can be said optimum as both lowering of leakage and maximizing of drain current are simultaneously obtained. Hot electron concentration becomes a critical factor, as if it becomes less than 10^{17} cm^{-3}, then current peak can never be obtained. Depth of the heated region is kept lower than barrier layer thickness to protect the device from self-heating effect. The design is one of the best possible solution for transient response on HEMT.

References

1. Ahsan SA, Ghosh S, Khandelwal S, Chauhan YS (2017) Analysis and modeling of cross-coupling and substrate capacitance in GaN HEMTs for power-electronic applications. IEEE Trans Electron Devices 64(3):816–823
2. Zanato D, Gokden S, Balkan N, Ridley BK, Schaff WJ (2004) The effect of interface-roughness and dislocation scattering on low temperature mobility of 2D electron gas in GaN/AlGaN. Semicond Sci Technol 19(3):427–432
3. Kumar A, Pathak NP (2014) Design and characterization of a wideband p-HEMT low noise amplifier. In: IEEE International Conference on Advances in Computing, Communications and Informatics
4. Wei J, Zhang M, Li B, Tang X, Chen KJ (2019) An analytical investigation on the charge distribution and gate control in the normally-off GaN double-channel MOS-HEMT. IEEE Trans Electron Devices 65(7):2757–2764
5. Alsharcf M, Granzner R, Ture E, Quay R, Racko J, Breza J, Schwierz F (2014) Design of GaN Tri-gate HEMTs. In: The 10th International Conference on Advanced Semiconductor Devices and Microsystems
6. Khandelwal S, Fjeldly TA (2012) A physics based compact model for drain current in AlGaN/GaN HEMT devices. In: 24th International Symposium on Power Semiconductor Devices and ICs
7. Ahsan SA, Ghosh S, Khandelwal S, Chauhan YS (2017) Physics-based multi-bias RF large-signal GaN HEMT modeling and parameter extraction flow. IEEE J Electron Dev Soc 5(5):310–319
8. Ghosh S, Ahsan SA, Dasgupta A, Khandelwal S, Chauhan YS (2016) GaN HEMT modeling for power and RF applications using ASM-HEMT. In: IEEE International Conference on Emerging Electronics, India
9. Tirado JM, Sanchez-Rojas JL, Izpura JI (2007) Trapping effects in the transient response of AlGaN/GaN HEMT devices. IEEE Trans Electron Devices 54(3):410–417
10. Taguchia H, Akahori K, Shimazu T, Tanabe H (2018) Analysis of drain current transient response of gate pulse voltage in AlGaN/GaN high electron mobility transistors 3(5):216–219
11. Gassoumi M, Salem MMB, Saadaoui S, Grimbert B, Fontaine J, Gaquiere C, Maaref H (2011) The effects of gate length variation and trapping effects on the transient response of AlGaN/GaN HEMT's on SiC substrates. Microelectron Eng 88(4):370–372

12. Mukherjee S, Patrick EE, Law ME (2017) Simulation of deep-level trap distributions in AlGaN/GaN HEMTs and its influence on transient analysis of drain current. ECS J Solid State Sci Technol 6(11):S3093–S3098
13. Paul S, Mondal S, Sarkar A (2019) Characterization and analysis of low-noise GaN-HEMT based inverter circuits. Microsyst Technol 153:1–9

An Approach for Alert Correlation Using ArcSight SIEM and Open Source NIDS

Chandan Singh Negi, Nisha Kumari, Pankaj Kumar, and Siddhant Kumar Sinha

Abstract In the era of Information Warfare, the malicious actors try to penetrate the network devices and delete their traces. In order to detect such network intrusion we need to do continuously monitor logs of each and every network connected devices. SIEM solution is used for centralized collection, consolidation and correlation for analyzing logs in real time. NIDS and NIPS inspect each & every packet travelling through the network, and generate alerts using pre-defined signatures or heuristics. This paper discusses an approach to design and implement, a solution by integrating the different IDS/ IPS like Snort, Suricata, and Bro with ArcSight SIEM solution in order to provide enhanced Information Security in an environment by reducing false positives/negatives and achieving better alert confidence. We show that the alerts generated on different NIDS for same types of attacks are different and this when integrated in the ArcSight SIEM solution gives a much higher degree of confidence of detection than each of the solution used individually.

Keywords ArcSight · SIEM · Snort · Suricata · Bro IDS · Open Source NIDS

1 Introduction

Since the evolution of Internet, cyber security has been one of the big concerns, more so in today's world, where millions of digital transactions take place per second. Nowadays, the terms data breaches, malware, identity theft, and hackers sound very common [1]. One of the main reasons why these risks occur is that organizations either do not have any picture of info security status of their environment or, in case they have it, there are gaps in it. Attackers often exploit these gaps using multiple attack vectors to gain unauthorized access of system. In order to fix these gaps, most modern Enterprises use a large number of network and security devices such as firewalls, desktops, web servers, VPNs, etc. and each of these generate security logs at different layer of the OSI stack. Taken together, these bits and pieces of data can

---

C. S. Negi · N. Kumari (✉) · P. Kumar · S. K. Sinha
Department of Computer Science, Birla Institute of Technology, Mesra, Ranchi, India

V. Nath and J. K. Mandal (eds.), *Proceeding of Fifth International Conference on Microelectronics, Computing and Communication Systems*, Lecture Notes in Electrical Engineering 748,
https://doi.org/10.1007/978-981-16-0275-7_3

paint a big picture of the current and potential risk. First off, security needs to be restructured considering the new age threat vectors. To conclude, an approach should carter defense in depth so the management has a holistic view of the organization being fully aware of the direct/indirect threats marching against them.

2 Literature Review

The SIEM System is based on Systems, applications, databases, networking devices and security products can generate millions of log records daily. While log records that are generated for each hardware device or software product can be stored independently on each device, there are mechanisms that can be configured to direct log record data that is generated by multiple hardware and software products to a single, centrally managed file system. Without specialized tools, it is very difficult to collect this information into a central location, analyze it, understand the results and take appropriate action. Security Information and Event Management (SIEM) [2] system provides a real-time analysis of security alerts generated by hosts, applications and network hardware.

2.1 The Architecture of SIEM

The SIEM comprises of generally following steps,

a) Collection: Log generating sources generate logs related to security activities within these devices, which can be further used for monitoring to give an idea of the activities happening in that system within that point of time. Data collection from the end devices and equipment can be done by two mechanisms i.e. PULL and PUSH [9]. In both the cases, the connector is placed closed to end devices to minimize the risk of losing any logs.

   - Pull Mechanism: In pull mechanism, the connector is responsible for the collection of logs from the end devices. The connector has the credentials to log into device and collect the logs specified as configured by administrator. For Example, logs of Window devices can be collected by connectors by adding the machines in the connector.
   - Push Mechanism: In push mechanism, the end devices are responsible for sending the logs to the connector. The end devices are configured to send the specified logs from the device to the remote connector on the specified port or by using the smart connector. For example: Linux devices send syslog events using RSYSLOG daemon to connector directly.

b) Normalization: After receiving raw logs from different sources this stage deals with unification of logs. Features from raw logs are extracted and mapping of logs from different vendor into a single format is done to get better understanding. This phase also consist of parsing, Data Compression, Data Deduplication, and Data Reduction. Logs from the log generating sources are collected at a common storage and can be used for monitoring as well as forensic analysis [4].

c) Correlation: The correlation engine, the brain of SIEM is a place where complex rules are formed in order to make proper utilization of logs and get actionable intelligence out of it. These rules help SIEM to detect malicious and suspicious events and generate alerts.

d) Monitoring: The main objective of deploying SIEM is continuous monitoring of logs and catching bad guy/ traffic/activity in our environment.

2.2 *The Intrusion Detection System (IDS)*

An IDS [3] is a device or software application that monitors a network or systems for malicious activity or policy violations. Any malicious activity or violation is typically reported either to an administrator or collected centrally using a security information and event management (SIEM) system. The IDS can be divided into two types [5],

a) Host based (HIDS): A HIDS detects malicious activity on host application and network activity and also covers activities of network, transport and application layer of TCP/IP.

b) Network Based (NIDS): A NIDS detects malicious activity on network, transport, and application layer of TCP/IP. There are a huge number of open source NIDS available but we are restricting ourselves to three of them Snort, Suricata, and Bro for our research.

   - Snort: It is a free and lightweight open source network Intrusion Prevention System (IPS) and network Intrusion Detection System (IDS) considered best choice to run on any Operating System [6]. Snort [7] NIDS has the ability to perform real-time traffic analysis and packet logging on Internet Protocol (IP) networks. Snort performs protocol analysis, content searching and matching. The program can also be used to detect probes or attacks, including, but not limited to, Operating System fingerprinting attempts, semantic URL attacks, buffer overflows, server message block probes, and stealth port scans. Snort can be configured in three main modes: sniffer, packet logger, and network intrusion detection. In sniffer mode, the program will read network packets and display them on the console. In packet logger mode, the program will log packets to the disk. In intrusion detection mode, the program will monitor network traffic and analyse it against a rule set defined by the user. The program will then perform a specific action based on what has been identified.
   - Suricata: It is a free and open source, mature, fast and robust network threat detection engine. The Suricata engine [5] is capable of real time Intrusion

Detection (IDS), Inline Intrusion Prevention System (IPS), Network Security Monitoring (NSM) and offline pcap (packet capture) processing. Suricata inspects the network traffic using a powerful and extensive rules and signature language, and has powerful Lua scripting support for detection of complex threats. With standard input and output formats like YAML and JSON integrations with tools like existing SIEMs, Arcsight, Splunk, Logstash/Elasticsearch, Kibana, and other database become effortless.

- Bro: It is a passive, open-source network traffic analyser [8] that inspects all traffic on a link in depth for signs of suspicious activity. Bro can be used to build a NIDS and can be compared to tcpdump, Snort, netflow, and Perl (or any other scripting language) all in one. Bro can be conceptualized in three layers: (i) Bro event engine which analyses live or recorded network traffic or trace files to generate neutral events, (ii) Bro policy scripts, which analyse events to create action policies, and (iii) Bro analysers. Most of the analysers are located in Bro's event engine with an accompanying policy script. Its scripting platform gives space for experimentation and customization and makes it suitable for high speed network as compared to snort.

3 Research Methodology

In order to increase the confidence level of detection and reduce the chances of false positives/ negatives, we tested three different IDS tools viz. Snort, Suricata and Bro. The tools were configured on the network which had normal legitimate user traffic. Known malware traffic was injected into the network and alerts from these IDSs were integrated and correlated. Specifically, we tested PCAPs of various malware's from Spider Labs [14] against our SIEM implementation. Each IDS performed differently according to its inbuilt signature/ rule set. The experimental setup is as indicated in Fig. 1.

The setup is a typical representation of a modern Enterprise. The DMZ (Demilitarized Zone) Network contains all the network facing devices that are accessible from

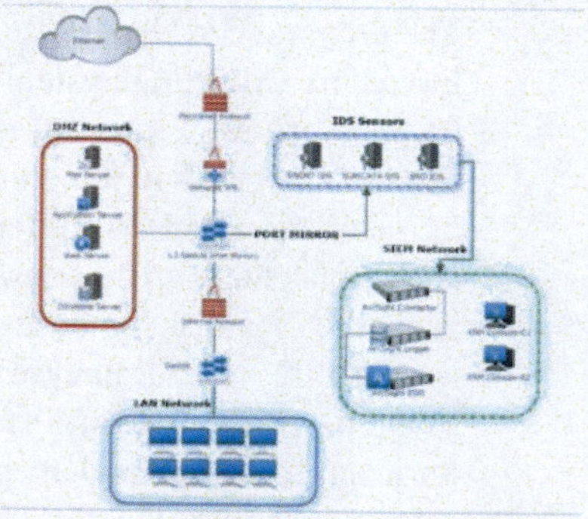

Fig. 1 Framework for intrusion detection

the untrusted network/ Internet. The DMZ Servers are placed between the two Firewalls with Ingress and Egress rules configured to allow traffic only on specific ports. The planning and implementation of NIDS at the right location in the network plays a significant role. The NIDS must be placed to collect the traffic from the critical network sources and devices. The configuration of SPAN (Switch Port Analyzer) in Switch [11] is required to get the port mirror traffic that is flowing in the network. The source ports of the SPAN must contain the collection of devices that need to be mirrored and sent to IDS Sensors.

3.1 Collection and Normalization of Logs from NIDS

Snort is capable of producing logs in various formats such as syslog and unified logs. However, we have to modify the configuration file of snort to produce unified logs. ArcSight connector [10] comes with SNORT Smart connector out of the box which is capable of reading the unified logs, converting it into CEF (Common Event Format) and sending it to Logger. Suricata is capable of generating Snort like fast.log which can be easily configured to push and pull logs to send to a centralized server. For our experiment we used ArcSight Snort Multiple Files Smart collector [11] to collect logs, parse it to CEF (Common Event Format) format and forward it to ArcSight Logger. Bro writes logs into Tab Separated Values [TSV] format suitable for post processing with external software. Users can however also choose from a set of alternative output formats and backend to interface directly with, e.g., external databases. We use ArcSight smart connector named as Bro IDS, NG files to collect from log directory, parse and forward logs to ArcSight Logger [12].

3.2 ArcSight Logger and ESM Configuration

- Logging: We use ArcSight logger as a centralized logging server for storing logs. ArcSight logger provides us UI for searching and viewing logs seamlessly from the participating devices. Analysing the field properties of the parsed logs helps us to make rules for correlation and visualization. The logs are then correlated by ArcSight ESM which is the brain of ArcSight SIEM. ArcSight ESM contains preconfigured rules as well as a platform to configure rules for compliance monitoring and audit.
- Correlation: Deriving actionable intelligence from the logs from different logging sources is the primary task of ArcSight Enterprise Security Manager (ESM). It provides various features like active channel, graphical representation, notification on alerts and many more. We make use of multiple Use Cases for correlating and raising an alert and notification. Correlation of the logs was done by matching Source IP addresses and Destination IP addresses coming from multiple NIDS based on filters.

Fig. 2 Filter1 for combining logs from multiple NIDS

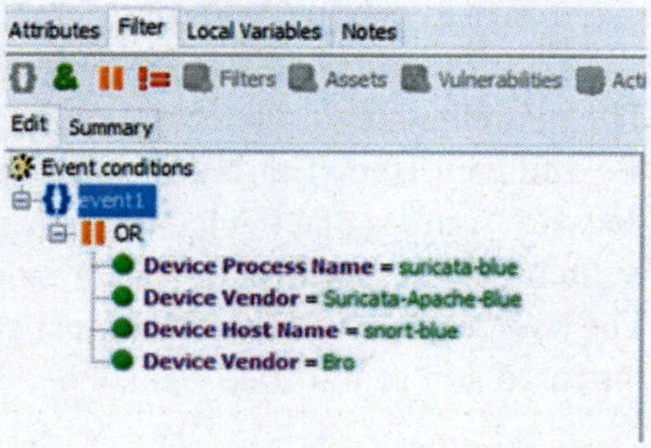

– Filters: Filters are the basic conditions on which logs are evaluated. It contains set of conditional operators like AND, OR and NOT etc., which we use to compare the field values of the logs. We make different filters for the events coming from different sources to merge into a single Event List. These are as given in succeeding paragraphs.

Filter 1: In this filter (Fig. 2) we combine events coming from the multiple IDS devices. This filter is used as base filter for creating further filters.

Filter 2: In this Filter (Fig. 3) we filtered out the events using Filter1 and further matched events only related to Network based Attacks using pattern matching using keywords like SCAN, WEB_SERVER.

Filter 3: In this Filter (Fig. 4), we filtered out the events using Filter1 and further matched events only related to System Based Attacks using pattern matching using keywords like POLICY, NETBIOS.

Fig. 3 Based on Filter1 & matching alerts related to network scan

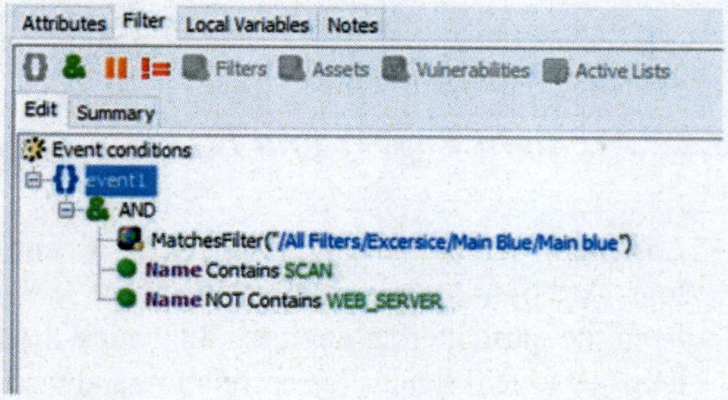

Fig. 4 Based on Filter1 & matching logs related to system changes

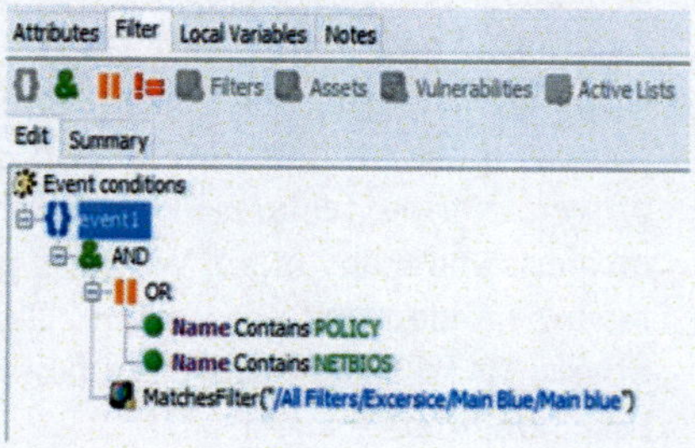

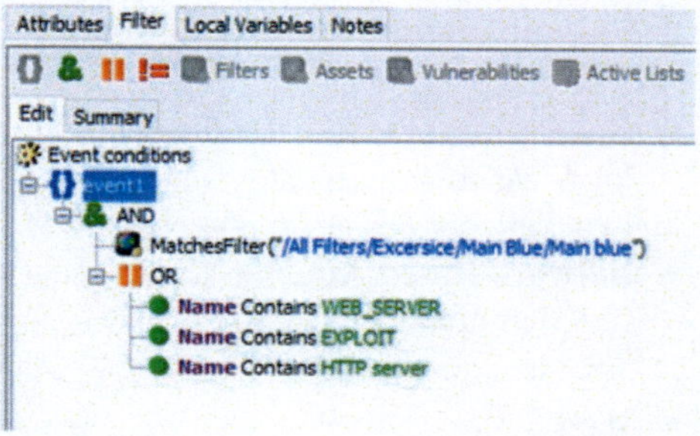

Fig. 5 Based on filter 1 & matching logs related to web server attacks

Filter 4: In this Filter (Fig. 5) we filtered out the events using Filter1 and further matched events only related to Web-Server Based Attacks using pattern matching using keywords like WEB_SERVER, EXPLOIT, and HTTP server.

3.3 SIEM Alerts and Notifications

A ‘rule’ is a programmed procedure that evaluates incoming events for specific conditions and patterns, and when a match is found, can initiate actions in response. Rules are the centerpiece of the ESM Correlation Engine [13], and are what reveals specific meaning out of the steady event stream. Rules are similar to IDS rules, except that they operate on an event stream instead of a bit stream. They are constructed using aggregation and Boolean pattern matching to evaluate objects, such as event fields, network models, and active lists: There are three types of rules:

- Standard Rules, are the normal rules that can operate with all type of fields and list.
- Lightweight Rules, are only for maintaining active and session lists. They do not create a correlation event whenever they are triggered.
- Pre-persistence Rules, are very similar to Lightweight rules, but can only set event fields.

The ArcSight SIEM comes with rule editor which can be used to create any rule on the specific field. In our testing environment, we create rules to correlate the event based on the Attacker IP Address and Target IP Address. We combine the events coming from all the three NIDS in an active channel and filter and see live events from all three IDS into an active channel. We use Standard rules which will be triggered when it matches conditions.

3.4 Attack Visualization and Dashboards

- **Data Monitors:** It is the small visualization of values of particular fields in the logs. We create separate data monitors for visualization of different attack types like Network based Attacks, Web Server Attacks & System based Attacks.
- **Network Based Attacks**: We combine the network related attacks from the three IDS and combined in a single Dashboard. Figure 6, demonstrates that NMAP Scan is in progress and Fig. 7 represent the IP-address of top network attacks.
- **Web Server Attacks**. The Web Server based attacks are collated within Web Server Attacks. Figure 8, demonstrates that the web server scan is in progress. That denotes DirBuster Web Scan & Nmap Scan is in top HTTP user agents. That reveals that Web Server is under attack.

Figure 9 represent the top web server attacker. The most IP address is listed and percentage of the attack is executed.

- **System Based Attacks:** The system based alerts are collated within System Based Attacks. Figure 10, depicts that SMB share access is in progress. Figure 11, shows the top system attackers.

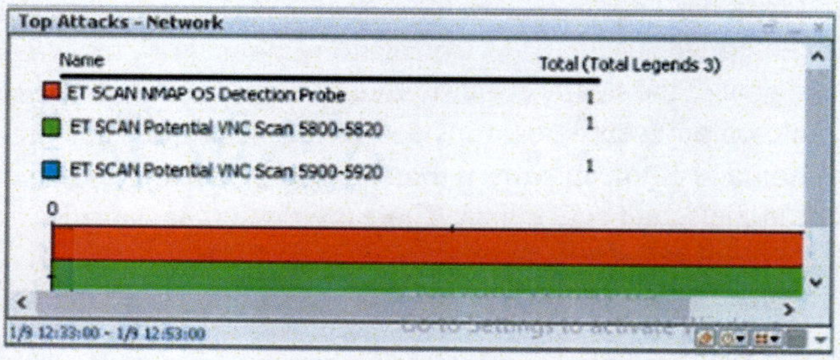

Fig. 6 Network alerts

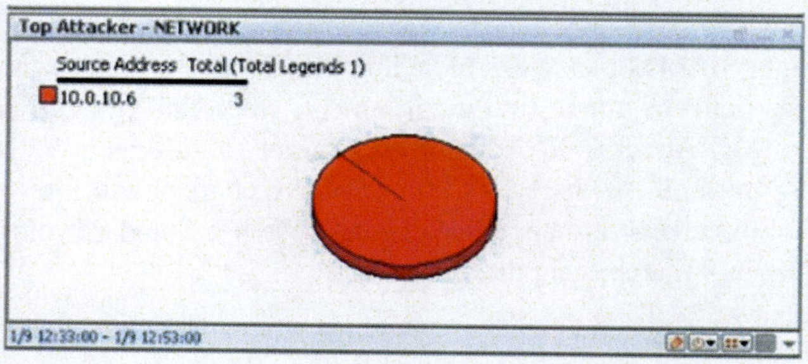

Fig. 7 Top network attackers

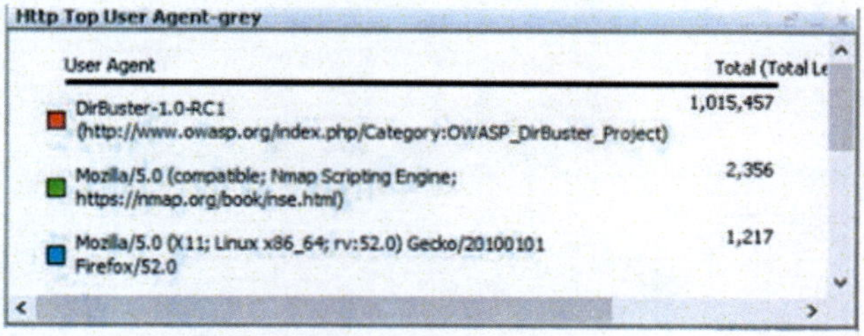

Fig. 8 Web server top HTTP user agents

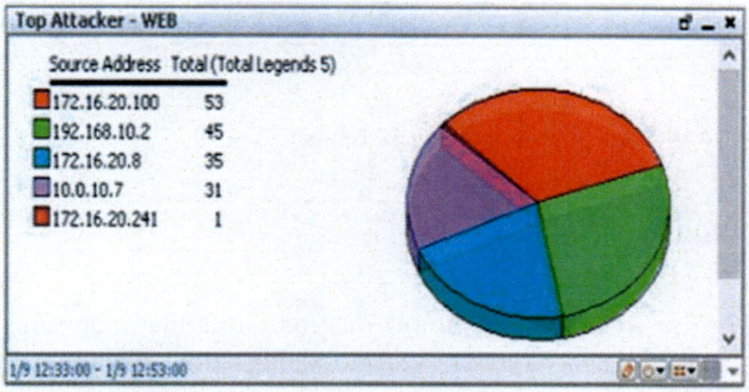

Fig. 9 Top web server attackers

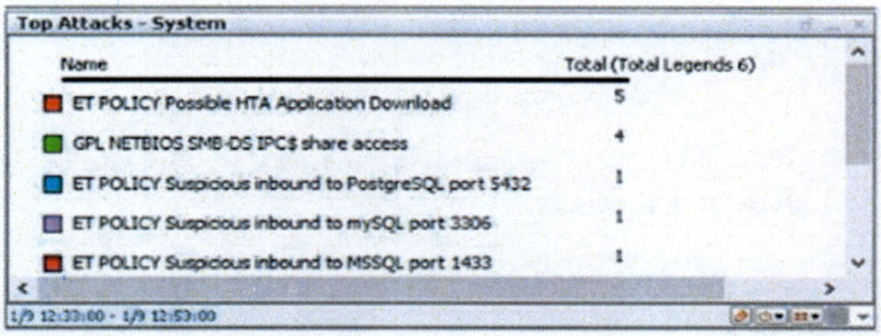

Fig. 10 Top system based alerts

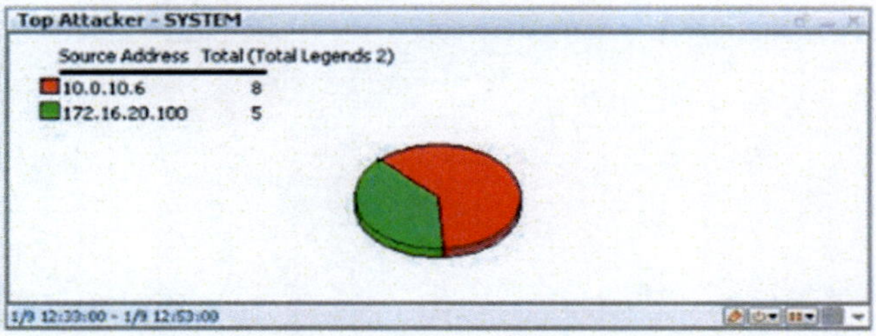

Fig. 11 Top system based attackers

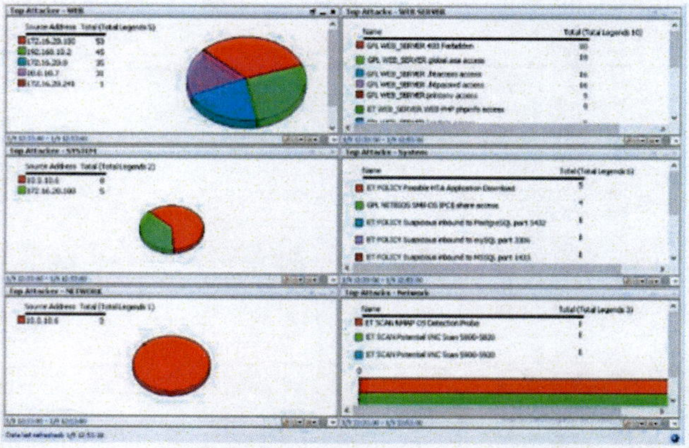

Fig. 12 Combined dashboard of various attack vectors

3.5 Dashboards

A Dashboard is a collection of multiple data monitors that represents a complete graphical view of alerts for a certain time period. Figure 12, represents the combined dashboard of various attack vectors.

3.6 Event Graphs

Figure 13 represents the event graph which contains the set of entities and relationships of alerts and attackers.

Fig. 13 Event graph of attack on web server

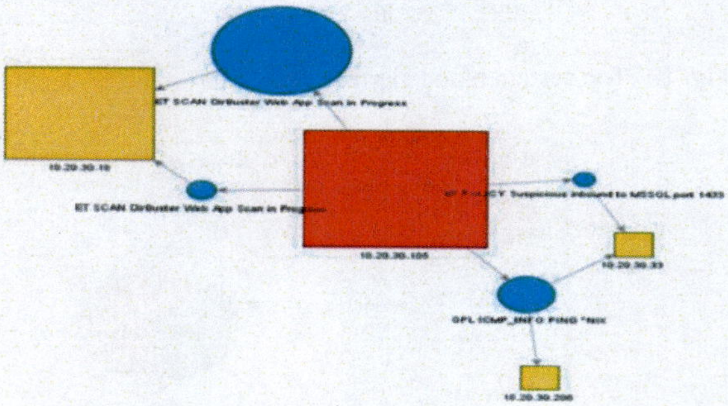

Table 1 Comparison matrix

Sr. No	Malware (Pcap)	Snort 3 (Community rules)	Suricata (Emerging threats)	Bro IDS (Security onion)	ArcSight (Snort)	ArcSight (Combined)
1	Adwind	1	0	0	1	1
2	Cerber	1	1	1	1	3
3	GlobelPoster	1	1	0	1	2
4	GPON	1	0	1	1	2
5	GrandCrab	1	1	0	1	2
6	Locky	1	1	1	1	3
7	Nemucod	1	1	1	1	3
8	Petya	0	1	1	0	2
9	Shade	0	1	0	0	1
10	Urnsnif	1	1	0	1	2
11	WebLogic	1	0	0	1	1

4 Result and Discussion

We've tested our setup against the malicious traffic stored in packet captures. The outcome helped us realize how different NIDS performed under diverse malware traffic. The packet captures (PCAPs) had traces of different ransomware, web application attacks and Trojan activity. This helped us benchmark our system against multiple attack vectors. Analysis helped in charting down the statistical chart which later worked as the pivot for building out our theory. The result is as shown in Table 1 above. As can be seen from the figure below, each of the malware inject is detected by one or more IDS, but the combined result in each case outperforms the result of the capability of the individual IDS. However, the result also depends on the network design and the architecture in order to ensure that all traffic logs are available to all the IDSs, the configuration of the rules at the individual IDS and proper correlation of the rules in the SIEM. ArcSight configured with Snort, had same detection rate as Snort. However, ArcSight with feeds from multiple NIDS sources performed way better than individual.

5 Conclusion

In our research, our findings reveal that each IDS engine has limitations which can result in many malware samples getting undetected and thus creating a false sense of security and reduced confidence level. Multiple IDS incorporate signatures/ rule sets for detection of malware and thus a malware which is undetected by one might be detected by the other. By implementing multiple IDS and correlating events from all of them, increases the probability of detection of the malware threat in the network. However, in order to achieve the same, proper network architecture and design is important so that it is ensured that all the logs are available to all the sensors without causing any adverse performance on the normal day to day operation of the network.

References

1. McHugh J (2001) Intrusion and intrusion detection. Int J Inf Secur 1(1):14–35
2. Miller D (2011) Security information and event management. In: SIEM implementation, vol xxxiv. McGraw-Hill, New York, pp 430. ISBN 00-717-0109-5
3. White JS, Fitzsimmons T, Matthews JN (2013) Quantitative analysis of intrusion detection systems: snort and Suricata. In: Cyber Sensing 2013
4. Pavlik J, Komarek A, Sobeslav V (2014) Security Information and event management in the cloud computing infrastructure
5. Pir RM (2014) Intrusion detection techniques and open source intrusion detection tools. IJEDR 2(3):3290–3295. ISSN 2321-9939
6. Park W, Ahn S (2016) Performance comparison and detection analysis in snort and suricata environment. Wirel Pers Commun 94(2):241–252
7. Zhou Z, Chen Z, Zhou T, Guan X (2010) The study on network intrusion detection system of Snort. In: International Conference on Networking and Digital Society (2010)
8. Sanders C, Smith J (2014) Applied Network Security Monitoring. Collection, Detection, and Analysis
9. Chuvakin A, Schmidt KJ, Phillips C, Moulder P (2013) Logging and log management: the authoritative guide to understanding the concepts surrounding logging and log management, p 431. ISBN 15-974-9635-9
10. ArcSight SmartConnector for Snort Syslog Configuration Guide Version 7.6.0.8009.0
11. ArcSight SmartConnector for Bro IDS NG File Configuration Guide
12. ArcSight logger Administrative guide Version: 6.5
13. ArcSight ESM 6.11.0 Administrator's Guide
14. SpiderLabs (2019) Repository holding malware pcap samples. https://github.com/SpiderLabs/IOCs-IDPS, Accessed 31 Mar 2019

High Impedance Fault Detection in Microgrids Using Cross-Alienation Coefficient

Bhatraj Anudeep, Prakriti Kumar Srivastava, Sauvik Biswas, and Paresh Kumar Nayak

Abstract High impedance fault (HIF) in modern power distribution system increases risk of fire and threats to people. In this paper, a cross-alienation coefficient based secure HIF detection technique is proposed for microgrids using current signals at both ends of feeder. Different HIF and non-HIF cases are simulated on a standard microgrid through PSCAD/EMTDC for performance evaluation of the proposed algorithm. Results of different cases justify the efficacy of the proposed technique.

Keywords Alienation coefficient · High impedance fault · Microgrid

1 Introduction

Faults in power system are undesirous events because, they interrupt the system operation, apart from this they produce various technical issues such as cascading outages, voltage sag propagation, power blackout [1]. With suitable protection arrangement, faults are detected and cleared quickly [2]. For these reasons new and appropriate methods for detection, identification and classification of faults in power system are being constantly developed [3]. However, low current and high impedance faults (HIFs) are difficult to identify and these faults pose significant threat to power systems if they are allowed to exist for longer time [4].

B. Anudeep (✉) · P. K. Srivastava · S. Biswas · P. K. Nayak
Department of Electrical Engineering, Indian Institute of Technology (ISM) Dhanbad, Dhanbad 826004, Jharkhand, India
e-mail: bhatraj.anudeep.2015dr33@ee.ism.ac.in

P. K. Srivastava
e-mail: prakriti.18mt0327@ee.iitism.ac.in

S. Biswas
e-mail: sauvik.iit.16dr000225@ee.ism.ac.in

P. K. Nayak
e-mail: paresh@iitism.ac.in

V. Nath and J. K. Mandal (eds.), *Proceeding of Fifth International Conference on Microelectronics, Computing and Communication Systems*, Lecture Notes in Electrical Engineering 748,
https://doi.org/10.1007/978-981-16-0275-7_4

In distribution systems, HIFs are one of those unwanted faults of current magnitudes 0–75 A that impose protection challenges in terms of detection and classification of fault [5]. The HIF takes place while energized conductor breaks and makes contact with high-impedance surfaces like concrete, dry ground, gravel, asphalt road, sand and soils or when a live conductor comes in contact with grounded objects like wood fences, tree limbs, vehicles etc. [6]. These faults can create hazardous situations and may lead to serious danger accident such as electric arc which results fire [7]. Due to low values of current, the HIFs may not damage the network components of distribution system, but it may be dangerous for human life if the live conductors come in contact [8].

As per authors knowledge various time-domain analysis, time-scale analysis, training based methods are proposed in the literature [9]. Frequency domain analysis methods include sequence components of current signals as in [10], some methods use phase angle between voltage phasor and harmonic current phasor [11, 12], some methods used high frequency components parameters for detecting HIFs [13]. Time domain analysis methods based on power dissipation factor is proposed in [14]. HIF detection through fractal technique is proposed in [15]. A time domain analysis approach based on signal superposition technique is proposed in [16] for detection and isolation of HIFs.

A discrete wavelet transform based scheme for HIF detection is proposed in [17]. However, multilevel decomposition, choice of basis function is a difficult issue. Methods based on support vector machine (SVM), decision tree (DT), probabilistic neural networks (PNN) are used for HIF detection [18–21]. However, larger training data is required for practical implementation of ANN techniques and also these methods have high computational burden. A phasor measurement units (PMU) based technique for HIF detection is proposed in [22]. However, PMU data transfer requires communication medium make the scheme improper for economical point of view. The communication medium which has been a revolution in modern distribution networks which gives a faster and secure protection in microgrids.

In this manuscript, a cross-alienation coefficient based secure HIF detection technique is proposed for microgrids using current signals at both ends of feeder. The accuracy and performance of the proposed technique is tested under different fault and non-fault transients such as capacitors switching, balanced and unbalanced load switching etc. The effects of HIFs depend on the connection model of the grounding of the distribution network [11]. The different HIF model viz. (i) isolated grounding, (ii) solidly grounding, (iii) impedance grounding and (iv) resonant grounding are studied and their performance is evaluated.

The paper is organized as follows: the proposed cross-alienation algorithm is described in detail in Sect. 2. The system modelling is provided in Sect. 3. The HIF modelling is provided in Sect. 4. The results of various faults and non-fault cases are provided in Sect. 5. Comparative result is provided in Sect. 6. Finally, the conclusion of the paper is provided in Sect. 7.

2 Proposed Technique

In this section, three-phase current signals from both ends of feeder on microgrid are monitored. To determine the cross-alienation coefficient two sets of synchronized data is required i.e. moving current signal at both the ends of feeder for each phase. Cross-alienation coefficient is usually positive and its value lie between zero and unity which represent the degree of similarity between the two variables.

If alienation coefficient is unity then the degree of similarity between the two measured sets of moving current signal at both the ends of feeder will be lowest. If it is zero then then the degree of similarity will be highest. Based on this alienation concept, internal or external faults are discriminated.

2.1 Calculation of Three-Phase Cross-Alienation Coefficients

The cross alienation coefficient (K) between the two measured sets of moving current signal at both the sides of feeder is expressed as,

$$\mathrm{K} = \left(1 - R^2\right) \tag{1}$$

Where, R is the correlation coefficient between the two measured line current samples at the feeder ends. If the correlated coefficient has n samples, then R has value equal to

$$R = \frac{\sum_{t=1}^{N}[i_s(\mathrm{t}) - i_s(\mathrm{t})_{am}]\left[i_r(\mathrm{t}) - i_r(\mathrm{t})_{am}\right]}{\sqrt[2]{\sum_{t=1}^{n}[i_s(\mathrm{t}) - i_s(\mathrm{t})_{am}]^2\left[i_r(\mathrm{t}) - i_r(\mathrm{t})_{am}\right]^2}} \tag{2}$$

Where, $N = 20$ represents the number of samples in one cycle; $i_s(t)$ and $i_r(t)$ are the measured current signals from the sending and receiving end of the feeder, respectively, while $i_s(t)_{am}$ and $i_r(t)_{am}$ determine the arithmetic means of current signals for m samples and are calculated as

$$i_s(\mathrm{t})\mathrm{am} = \frac{1}{m}\sum_{t=1}^{m} i_s(\mathrm{t}) \tag{3}$$

$$i_r(\mathrm{t})\mathrm{am} = \frac{1}{m}\sum_{t=1}^{m} i_r(\mathrm{t}) \tag{4}$$

2.2 Fault Detection and Classification Algorithm

The system operation is said to be in normal operation mode or external fault if the cross-alienation coefficient between the sending and receiving end current signals is zero or close to zero. Then, both the sampled current windows will be similar if there is an internal fault, the cross-alienation coefficient increases from zero to unity or close to unity. In this case, the sampled current windows at feeder ends are not similar. Due to discretisation error, the calculated values of the alienation coefficients will be compared with the threshold value $\varepsilon = 0.005$ in order to obtain good results.

1. Relay blocking- If three-phase alienation coefficients (K_a, K_b, K_c) < ε, then the system operation is said to be in normal operation mode or external fault. If fault is symmetrical, $K_a = K_b = K_c \leq \varepsilon$; and if it is non-symmetrical then $K_a \neq K_b \neq K_c \leq \varepsilon$.
2. Relay tripping- If three-phase alienation coefficients $\varepsilon <$ (K_a, K_b, K_c) ≤ 1 the fault will be internal. If fault is symmetrical, $\varepsilon < K_a = K_b = K_c \leq 1$; while if it is non-symmetrical internal fault ($K_a \neq K_b \neq K_c$) then at least one of K_a, K_b and $K_c \leq \varepsilon$ but less than unity.
3. For non-symmetrical internal faults, There are two cases: (i) During single ph-to-ground fault, the faulty ph alienation is greater than ε but less than 1, while the values of healthy phases are less than < ε. (ii) During two phases fault; the faulty phases alienation coefficients is greater than ε and ≤ 1, while the value of healthy phase is less than < ε.

Based on the measured cross-alienation coefficients shown in Fig. 1, the protective relays have different operating characteristics that can be operated locally or

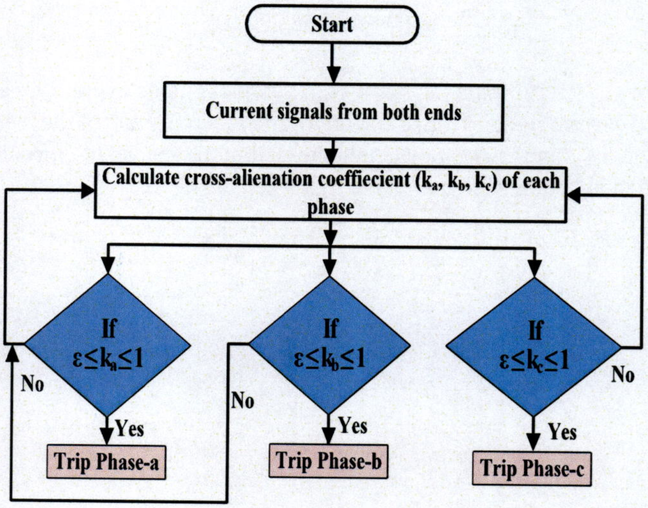

Fig. 1 Step by step flowchart of the proposed method

remotely via communication links. Therefore for monitoring relay operation, a wireless communications technique based on WiMAX technology is introduced. Through this communication system individual relays can share information with a central computer [23].

WiMAX technology has less interference than other technologies hence the rate of transferring a wireless signal is much faster. For measuring 3-ph current signals from both ends of the microgrid feeders the smart meters are situated at both the ends of the feeders.These 3-ph current signals are measured, sampled and then transmitted to the main protection centre from smart metres. In this centre, the transmitted data is analysed and then cross-alienation coefficients are calculated for relay actions.

3 System Modelling

For the assessment of the proposed algorithm a standard model (IEEE standard 399–1997) is simulated using PSCAD/EMTDC environment with few modifications as proposed in Fig. 2 [24]. The DG1 is a synchronous generator (SDG) of 15 MVA, 6.6 kV, 50 Hz is connected to PCC2 through a (delta/star) transformer T2 of rating 20 MVA, 6.6 kV, 50 Hz the DG2 of 9 MW is connected to PCC3 is power electronics interfaced PV plant (PVDG) through a (delta/star) transformer T3 of rating 15 MVA, 34.5 kV/4.40 kV, 50 Hz. A combination of linear and non-linear loads P1 = 2.4 MW + j1.3MVAr is connected at PCC, P2 = 6.6 MW + j2.6MVAr is connected at PCC1, P3 = 5.2 MW + j1.8MVAr is connected at PCC2, P4 = 6.6 MW + j2.2 MVAR is connected at PCC3, P5 = 6.4 MW + j1.62 MVAR is connected at PCC4 are

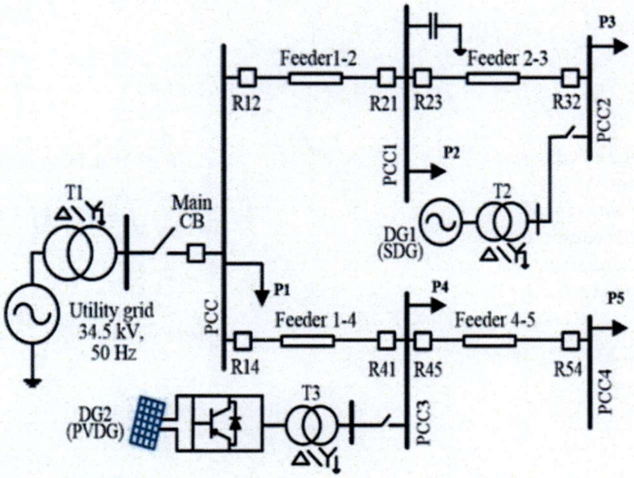

Fig. 2 Schematics of test system studied

connected. Utility grid of 48 MVA, 34.5 kV, 50 Hz is connected at PCC. Through a (delta/star) transformer T1 of rating 48 MVA, 66 kV/34.5 kV, 50 Hz.

4 HIF Modelling

A HIF with single-line-to-ground type is shown in Fig. 3. A phase voltage V_{ph} is taken and the HIF is connected to a single phase line. The construction of HIF has DC voltage sources V_x and V_y which are connected in series with diodes respectively D_x and D_y. the resistances R_x and R_y are connected in series with the diodes respectively [25]. The resistances represents arcing phenomena in the HIF. This HIF model gives fault current which is unsymmetrical in positive and negative half cycle. The graphs given in Fig. 4 show the faulted current and voltages inside the HIF model. Figure 4(a) shows the voltage characteristics form pre-fault to fault occurring stage for HIF, Fig. 4(b) shows the current characteristics through the HIF model which is random

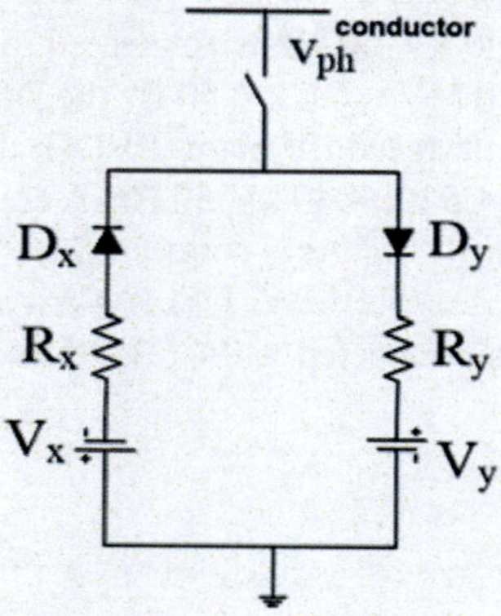

Fig. 3 Single line diagram of the HIF model

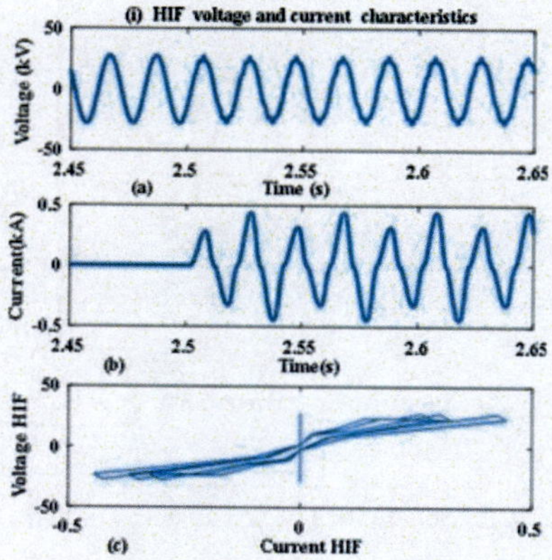

Fig. 4 Current and voltage characteristics of the HIF model, (a) HIF voltage in phase–a, (b) HIF current in phase–a (c) HIF voltage versus current characteristic graph.

with unsymmetrical in positive and negative half cycle. Figure 4(c) shows the v-i characteristics obtained at the HIF model which shows the randomness in the fault current.

5 Results and Discussions

The performance of the proposed technique is tested by creating HIFs at different locations, different mode of operations like grid connected mode and islanded mode. The tests are conducted in PSCAD/EMTDC simulation at various balanced and unbalanced switching in a microgrid shown in Fig. 2. The performances on the various typical cases are demonstrated below.

5.1 Performance of HIF Detection Techniques for Isolated Grounding

A high impedance A-G fault (isolated grounded) at 2.5 s is created on feeder 1–2, 5 km far from relay R_{12} while the SDG and the utility grid only (Fig. 2) are in operation but PVDG is not grid connected. The voltage and current signals retrieved at relay R_{12} and R_{21} are shown in Fig. 5(a) and (b), respectively. The 3-ph voltage and currents at both sides of feeder 1–2 are shown in Fig. 5. The cross-alienation coefficient is calculated from the Eq. (1) in Sect. 2. From the Fig. 5(a) it is clearly observed that there is a disturbance in the current signal of phase–a at 2.5 s in at both

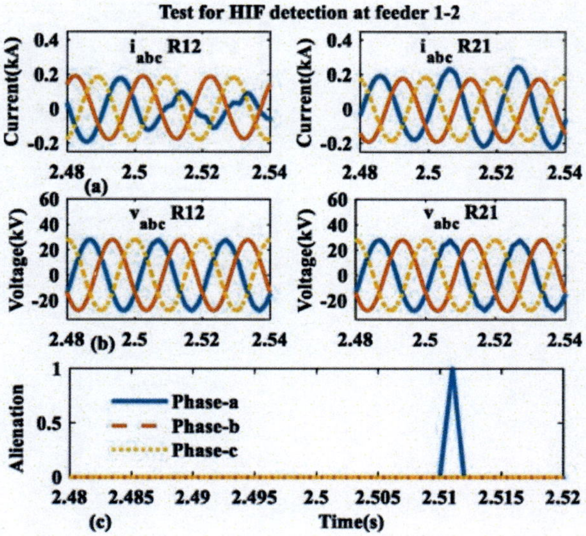

Fig. 5 Results of an HIF isolated grounding on feeder 1–2 initiated at 2.5 s (a) Current at relay R_{12} and R_{21}, (b) Voltage at relay R_{12} and R_{21} and (c) Cross-alienation coefficient at feeder 1–2

the sides of feeder 1–2. The cross-alienation coefficient is shown in Fig. 5(c) the coefficient of phase-a raises to "1" just after 10 ms after the occurrence of HIF fault. This concludes the proposed algorithm detects high impedance fault in any phase of the microgrid accurately.

5.2 *Performance of HIF Detection Techniques for Solid Grounding*

A high impedance A-G fault (solidly grounded) at 2.5 s is created on feeder 1–2, 8 km far from relay R_{12}while the the PVDG and utility grid only (Fig. 2) are in operation but SDG is not. The 3-ph voltage and current signals retrieved at relay R_{12} and R_{21} are given in Fig. 6(a) and (b), respectively. The 3-ph voltage and currents at both sides of feeder 1–2 are shown in Fig. 6. It is clearly observed that there is a disturbance in the current signal of phase–a at 2.5 s in Fig. 6(a) at both the sides of feeder 1–2. The cross-alienation coefficient is shown in Fig. 6(c) the coefficient of phase-a raises to "1" just after 11 ms after the occurrence of HIF fault. This concludes the proposed algorithm detects high impedance fault in any phase of the microgrid accurately.

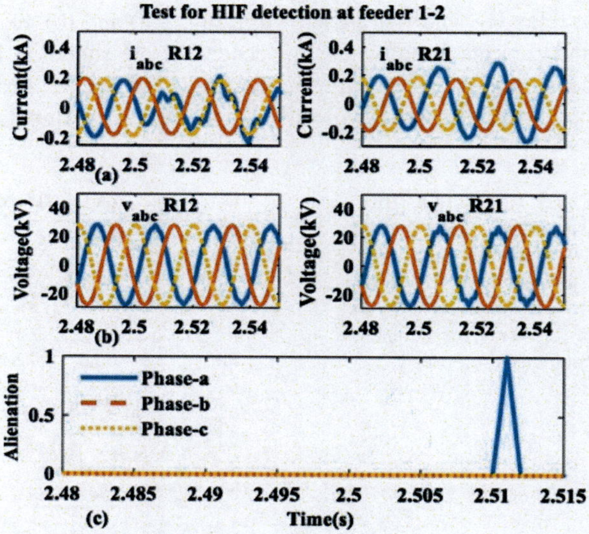

Fig. 6 Results of an HIF solidly grounding on feeder 1–2 initiated at 2.5 s (a) Current at relay R_{12} and R_{21}, (b) Voltage at relay R_{12} and R_{21} and (c) Cross-alienation coefficient at feeder 1–2

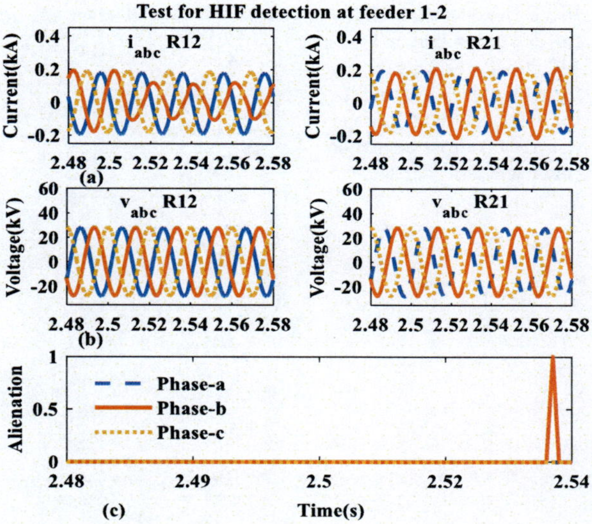

Fig. 7 Results of an HIF impedance grounding on feeder 1–2 initiated at 2.5 s (a) Current at relay R_{12} and R_{21}, (b) Voltage at relay R_{12} and R_{21} and (c) Cross-alienation coefficient at feeder 1–2

5.3 Performance of HIF Detection Techniques for Impedance Grounding

A high impedance B-G fault (impedance grounded) at 2.5 s is created on feeder 1–2, 2 km far from relay R_{12} while the PVDG and the SDG onlyare in operation in islanded mode (Fig. 2) and utility grid is not. The 3-ph voltage and current signals retrieved at relay R_{12} and R_{21} are shown in Fig. 7(a) and (b), respectively. From the Fig. 7(a) it is clearly observed that there is a disturbance in the current signal of phase –bat 2.5 s in Fig. 7(a) at both the sides of feeder 1–2. The cross-alienation coefficient is shown in Fig. 7(c) the co-efficient of phase-b raises to "1" just after 30 ms after the occurrence of HIF fault.

5.4 Performance of HIF Detection Techniques for Resonant Grounding

A high impedance C-G fault (resonant grounded) at 2.5 s is created on feeder 1–4, 2 km far from relay R12 while the PVDG and the SDG only are in operation (Fig. 2) and utility grid is not. The 3-ph voltage and current signals retrieved at relay R12 and R21 are given in Fig. 8(a) and (b), respectively. Accordingly, it is clearly seen that there is a disturbance in the current signal of phase –c at 2.5 s in Fig. 8(a) at the both sides of feeder 1–2. The cross-alienation coefficient is shown in Fig. 8(c) the co-efficient of phase-a raises to "1" just after 22 ms after the occurrence of HIF fault.

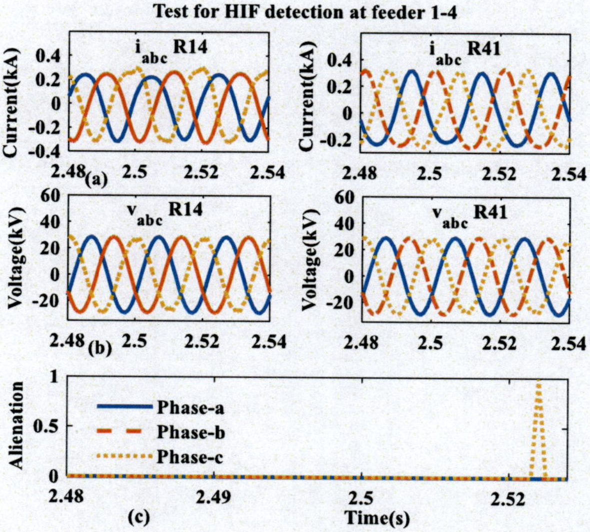

Fig. 8 Results of an HIF isolated grounding on feeder 1–2 initiated at 2.5 s (a) Current at relay R_{14} and R_{41}, (b) Voltage at relay R_{14} and R_{41} and (c) Cross-alienation coefficient at feeder 1–2

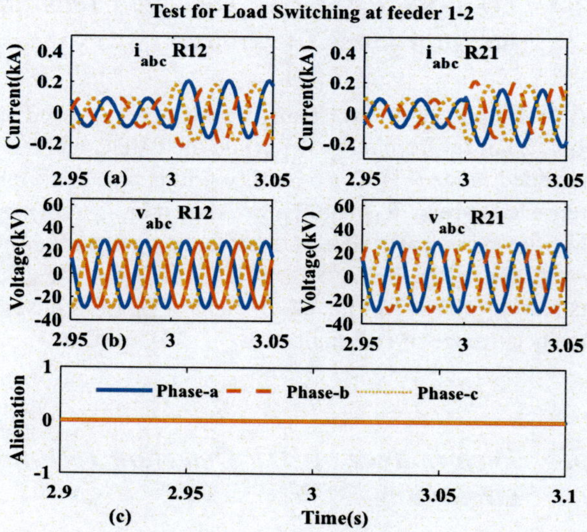

Fig. 9 Performance on the load switching on feeder 1–2 initiated at 3.0 s (a) Current at relay R_{12} and R_{21}, (b) Voltage at relay R_{12} and R_{21} and (c) Cross-alienation coefficient at feeder 1–2

5.5 Performance During High Load Switch on

A large load P2 of 140% (Fig. 2) is switched on suddenly at 3.0 s when the SDG and utility grid only are in operation but PVDG is not connected to grid. The current and voltage signals retrieved at relay R_{12} and R_{21} are given in Fig. 9(a) and (b), respectively. The 3-ph voltage and currents at both sides of feeder 1–2 are shown in Fig. 9. From the Fig. 9(a) it is clearly observed that there is a raise in current at R_{12} and R_{21}in the feeder1-2 at 3.0 s in Fig. 9(a). The cross alienation co-efficient is shown in Fig. 9(c).There is no change in the coefficient at the time of load switching which conclude a safe operation at any local disturbances in the microgrid.

5.6 Performance of HIF Detection Techniques for Capacitor switching

The capacitor switching at PCC1 is done suddenly of its rated value 0.5 MVAR at 2.6 s during the islanded operation of the PVDG and the SDG only (Fig. 2) where utility grid is not connected. The voltage and current signals retrieved at relay R_{12} and R_{21} are shown in Fig. 10(a) and (b), respectively. The 3-ph voltage and currents at both sides of feeder 1–2 are given in Fig. 10. From the Fig. 10(a) it is clearly observed that there is a disturbance in the 3-ph voltage and current signal of at 2.6 s at both the sides of feeder 1–2. The cross-alienation coefficient is shown in Fig. 10(c) the coefficient remains "0" after the capacitor switching in the feeder 1–2. This concludes the proposed algorithm is safe and secure for capacitor switching in the microgrid.

6 Comparative Analysis

The present technique is compared with one of the important existing method: time–frequency based differential energy method [26],

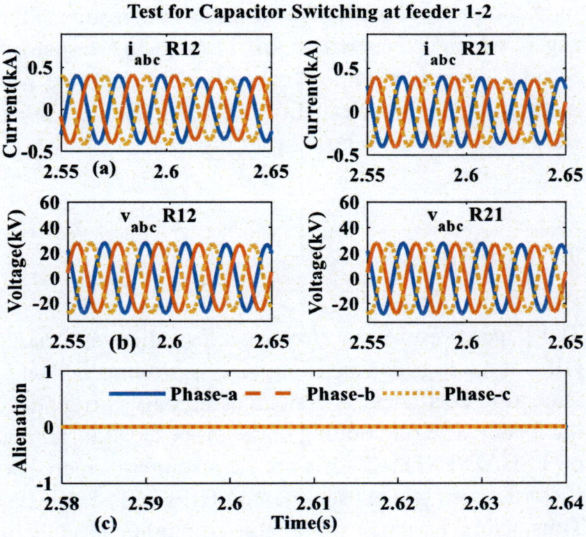

Fig. 10 Performance on capacitor switching on feeder 1–2 initiated at 2.6 s (a) Current at relay R_{12} and R_{21}, (b) Voltage at relay R_{12} and R_{21} and (c) Cross-alienation coefficient at feeder 1–2

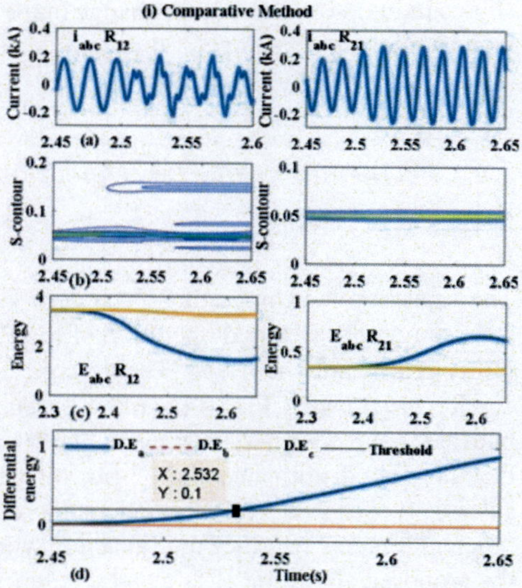

Fig. 11 Performance on HIF in phase-a (solid grounding) on feeder 1–2 initiated at 2.5 s (a) Current at relay R_{12} and R_{21}, (b) S-transform contour at relay R_{12} and R_{21} (c) Spectral energy of each phase and (d) Differential energy of each phase

6.1 *Performance for Comparison of HIF Detection for Isolated Grounding*

A high impedance A-G fault (solid grounding) at 2.5 s on feeder 1–2 is created, which is 5 km far from relay R_{12} during islanded mode of operation (Fig. 2). As shown in Fig. 11(d), the S-transform based differential energy scheme is able to detect faulted phase (phase-a) in 32 ms after the initiation of the fault. Considering the page limit only comparative method is shown and the results of the proposed method is given in Fig. 6, which detects HIF in just 11 ms. The S-transform based method has threshold setting issues [26] which may differ for different network configurations and the method has high computational burden. This proves the proposed method is faster easy and free from threshold setting issues.

7 Conclusion

The paper presents an effective cross-alienation based technique for detection of HIFs in microgrids consisting of synchronous DG and photovoltaic DG. The cross-alienation coefficient is derived using only current data retrieved from both ends of the feeder. The functioning of the proposed algorithm is assessed by generating data on PSCAD/EMTDC software on a standard microgrid system. The results clearly show that using proposed method HIFs can be detected securely in the microgrids. Thus, using the modern available communication facility, the proposed scheme can be used as an improved protection method for a smart microgrid.

References

1. Tengdin J, Westfall R, Stephan K (1996) Impedance fault detection technology. Report of PSRC Working Group D15. https://www.pespsrc.org/Reports/High_Impedance_Fault_Detection_Technology.pdf
2. Wang C, Wang S (2008) Study on some key problems related to distributed generation systems. Autom. Electr Power Syst 20:1–4
3. Wang S, Wang H, Cai S (2009) A review of optimization allocation of distributed generations embedded in power grid. Autom Electr Power Syst 18:110–115
4. Wang B, Geng J, Dong X (2018) High-impedance fault detection based on nonlinear voltage-current characteristic profile identification. IEEE Trans Smart Grid 9(4):3783–3791
5. Baqui I, Zamora I, Mazon J et al (2011) High impedance fault detection methodology using wavelet transform and artificial neural networks. Electr Power Syst Res 81(7):1325–1333
6. Ghaderi Amin, Ginn Herbert L, Mohammadpour Hossein Ali (2017) High impedance fault detection: a review. Electric Power Syst Res 143:376–388
7. Tripathi A, Kori AK (2013) High impedance fault detection on rural electric distribution and power quality control systems. Int J Emerg Technol 4(2):128–131
8. Torres V, Guardado JL, Ruiz HF et al (2014) Modeling and detection of high impedance faults. Electr. Power Syst. Res. 61:163–172
9. Wester CG (1998) High impedance fault detection on distribution systems. In: 1998 Rural Electric Power Conference Presented at 42nd Annual Conference, St. Louis, MO, USA, p c5-1
10. Yu D, Khan S (1994) An adaptive high and low impedance fault detection method. IEEE Trans Power Del 9(4):1812–1821
11. Lien K-Y, Chen S-L, Liao C-J, Guo T-Y, Lin T-M, Shen J-S (1999) Energy variance criterion and threshold tuning scheme for high impedance fault detection. IEEE Trans Power Del 14(3):810–817
12. Detection of Arcing Faults on Distribution Feeders (1982). Texas A & M University, EPRI Report EL-2757
13. Macedo JR, Resende JW, Bissochi CA Jr, Carvalho D, Castro FC (2015) Proposition of an interharmonic-based methodology for high impedance fault detection in distribution systems. IET Gener Transm Distrib 15(16):2293–2601
14. Li KK, Chan WL (2003) Novel methods for high impedance grounding fault protection in low voltage supply systems. Electric Power Compon Syst 31:1133–1150
15. Mamishev AV, Russell BD, Benner CL (1996) Analysis of high impedance faults using fractal techniques. IEEE Trans Power Syst 11(1):435–440
16. Zamora I, Majon AJ, Sagastabeitia KJ, Zamora JJ (2007) New method for detecting low current faults in electrical distribution systems. IEEE Trans Power Del 22(4):2072–2079
17. Elkalashy NI, Lehtonen M, Darwish HA, Taalab I, Izzularab MA (2008) DWT-based detection and transient power direction-based location of high-impedance faults due to leaning trees in unearthed MV networks. IEEE Trans Power Del 23(1):94–101
18. Samantaray SR, Panigrahi BK, Dash PK (2008) High impedance fault detection in power distribution networks using time-frequency transform and probabilistic neural network. IET Gener Transm Distrib 2(2):261–270
19. Michalik M, Lukowicz M, Rebizant W, Lee S-J, Kang S-H (2008) New ANN-based algorithms for detecting HIFs in multigrounded MV networks. IEEE Trans Power Del 23(1):58–66
20. Sheng Y, Rovnyak S (2004) Decision tree-based methodology for high impedance fault detection. IEEE Trans Power Del 19(2):533–536
21. Etemadi AH, Sanaye-Pasand M (2008) High-impedance fault detection using multi-resolution signal decomposition and adaptive neural fuzzy inference system. IET Gener Transm Distrib 2(1):110–118
22. Zanjani MGM, Kargar HK, Zanjani MGM (2012) High impedance fault detection of distribution network by phasor measurement units. In: Proceedings of 17th Conference on Electrical Power Distribution Networks (EPDC), Tehran, Iran, pp 1–5

23. Sedoyeka E, Hunaiti Z, Nabhan MA, et al (2008) WiMAX Mesh networks for underserved areas. In: The Sixth ACS/IEEE International Conference on Computer Systems and Applications, AICCSA 2008, Doha, Qatar, 31 March–4 April 2008, pp 1070–1075
24. Anudeep B, Nayak PK (2019) Transient energy-based combined fault detector and faulted phase selector for distribution networks with distributed generators. Int Trans Electric. Energy Syst 30:1–29
25. Sarwagya K, De S, Nayak PK (2017) High-impedance fault detection in electrical power distribution systems using moving sum approach. IET Sci Measur Technol 12(1):1–8
26. Kar S, Samantaray SR (2014) Time-frequency transform-based differential scheme for microgrid protection. IET Gener Transm Distrib 8(2):310–320

Robust Image Hashing Using Chromatic Channel

Abdul Subhani Shaik, Ram Kumar Karsh, and Mohiul Islam

Abstract Image security remains a challenging issue in modern age of multimedia communication. In recent days, image hashing technique has gained a lot attentions because of the availability of different multimedia editing tools. Color image hashing has been explored limited. In this work, chromatic channel information has been used to construct hashing. The ring partition techniques are generally designed on the intensity channel. Chromatic information of an images are also added along with intensity channel information to create an image hash that is sensitive to global as well as local color changes. Experiment has been conducted on different standard database and results illustrate capability of the proposed method for detection of color forgery, while maintaining robustness against operations which preserve content and good discrimination. The efficacy of the proposed method is shown using receiver operating characteristics.

Keywords Color forgery · Chromatic channel · Ring partition · Image hashing · Image authentication

1 Introduction

Image security is currently performed using watermarking. This kind of techniques do not provide enough flexibility for a number of practical applications. Thus, image hashing techniques may help in achieving the better flexibility and desired security. In image hashing approach, generally the source images are mapped to a short string based on visual content [1]. The short string should keep the following properties: robustness against content-preserving operations, and good discriminations. Good discrimination may be defined as: the hashes of a source image and counterfeit

A. S. Shaik (✉) · R. K. Karsh
Department of ECE, National Institute of Technology Silchar, Silchar 788710, Assam, India
e-mail: abdul_rs@ece.nits.ac.in

A. S. Shaik · M. Islam
Department of ECE, CMR College of Engineering and Technology, Hyderabad 501401, Telangana, India

V. Nath and J. K. Mandal (eds.), *Proceeding of Fifth International Conference on Microelectronics, Computing and Communication Systems*, Lecture Notes in Electrical Engineering 748,
https://doi.org/10.1007/978-981-16-0275-7_5

version of source one (i.e., content addition, content deletion, color modification etc.) should be sufficiently different [2]. Hashes of a source image and operated version by content-preserving operations (i.e., compression, rotations, and filtering etc.) of source one should approximately same [3]. Monga et al. [4] has developed image hash based on two-stage framework, in which first stage creates an intermediate hash from features and coding of intermediate hash gives the final hash. This technique has become the de-facto standard in different image hashing techniques.

In literature researchers have proposed various features for hashing. Some of the features are either global [5, 6] or local [7–10] in nature. The Fourier-Mellin transformation coefficients were used to form hash [5]. This approach is robust to moderate geometric attack, rotation of image within 20^0 and filtering operations, but misclassified the color forgery. In another work, Radon Transform is exploited to construct hash and applied for image authentication [6]. This method is limited for a large degree of rotations. The dictionary of feature vectors of an image, called as "word" constructed from different image blocks is developed for hashing [7]. This method is fragile to higher degree of rotations. A technique based on arbitrarily selected sub images and NMF has been proposed in [8]. The final hash is created by concatenating entries of the reduced dimension matrix. This approach is not robust to watermark embedding.

Virtual watermark has been used for hash generation [9, 26]. This algorithm shows good robustness for digital operations, geometric operations, and capable to detect content change in fairly large regions, but robust for rotation only up to 5^0. Fouad et al. [10] examined the security of NMF based hashing method and showed that if secret key has been used in many stages, then first stage secret key shows an vital part in safety of hash. A new hashing technique by combining scale space theory with radon transform has been proposed in the literature [11]. It has been successfully applied for image authentication. It can also be used for several other issues such as detecting image manipulation type and estimating geometrical operation parameters. The SIFT key points and dictionary-based approach [12] for image hashing is robust to geometric operation carried out on the original image. This method is unable to detect color forgery. Zernike moments and saliency-based hashing approach is proposed in [13]. Local feature detection mainly depends on good saliency detection method. This algorithm is limited in case of texture images.

Selective quaternion has been used to construct hash [14]. The authors proposed a method based on selective quaternion to create a hash. It is robust to some digital operation but brightness and contrast changes has not been addressed. The block truncation coding (BTC) based hashing approach discussed in [15]. This approach is robust, secure but limited to brightness changes. Ring partition-based hashing technique is introduced in [16]. In another work, NMF was combined with ring partition [17]. This algorithm constructs secondary image via mapping of source image into several rings and follows NMF to reduce the dimension of an image. This approach is robust to arbitrary rotation and has better discriminative capability than [16]. The main drawback of this approach is that approximately 20 percent of an image's corner is not considered during the process of hash creation. The authors extended the earlier work by removing the NMF and introducing a new concept of invariant vector distance along with ring partition. It has been shown that the invariant vector distance is insensitive to content preserving operation [18]. The key dependency of hash has also been included to make the hash secure. The main drawback of this approach is that around 20 percent of image's corner knowledge was not considered. This method works only on the CIE $L^*a^*b^*$ intensity channel and therefore unable to detect any global or local color changes [19–22].

The paper's contribution can be highlighted as: The ring partition algorithm are generally designed on the intensity channel. Chromatic information of an image are also added along with intensity channel information to create image hash that is sensitive to global as well as local color changes.

The article is set out as follows. Section 2, explains proposed image hashing algorithm. Section 3, discusses about the various experimental results of the work. Summary of the research and future scope is addressed in Sect. 4.

2 Proposed Image Hashing

The method proposed is composed of three phases: preprocessing, ring partitioning, and hash generation as shown in Fig. 1.

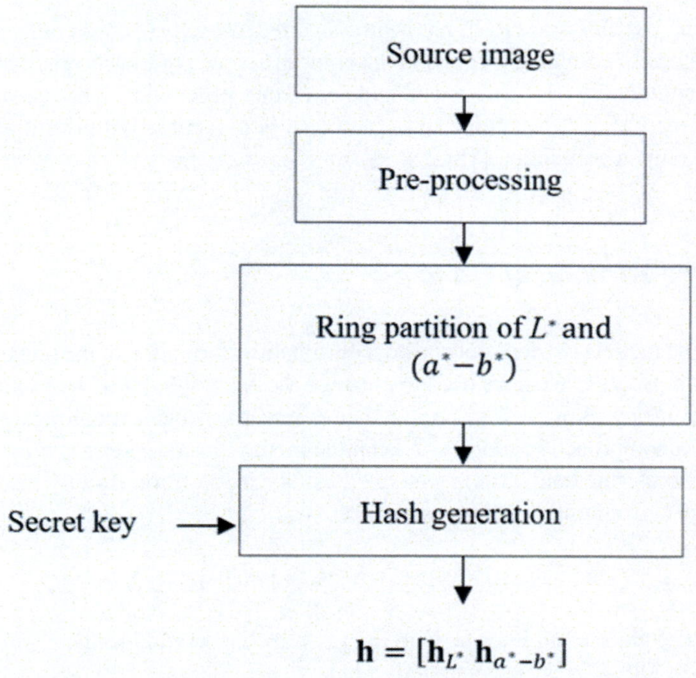

Fig. 1 Proposed method

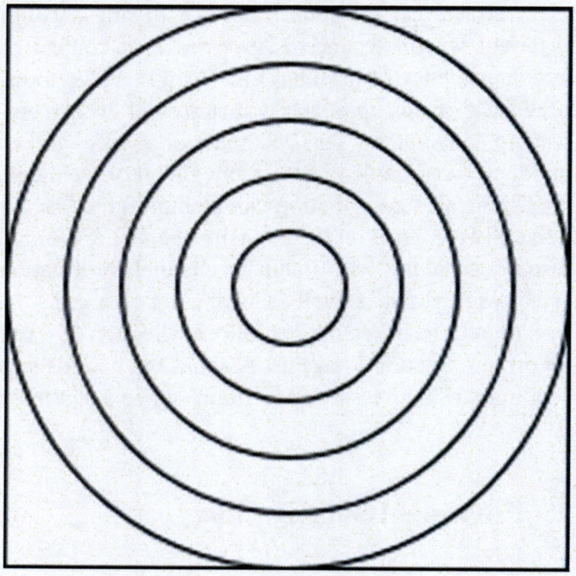

Fig. 2 Mapping from square to ring

2.1 Pre-Processing

In this step, the color space RGB has been converted into CIE $L^*a^*b^*$ color space [23] and the intensity component (L^*) and the color difference of the chromatic channels (a^*-b^*) component has been considered for further processing. The reason behind choosing the CIE $L^*a^*b^*$ color space is that it is perceptually uniform and fairly related to human perception [18].

2.2 Ring Partition of L and a-b

The existing model [18] does not consider chromatic information of the images during the hashing process. We have tried to enhance the existing method by including the chromatic information so as to provide better sensitivity against color forgery. The ring partition approach is adopted to keep the method invariant to rotations.

The first intermediate hash is generated using ring partition (briefly discussed in Algorithm 1) from only intensity channel (L^*) as

$$\mathbf{h}_{L^*} = [h(1), h(2), \ldots, h(n)] \tag{1}$$

Similarly, the second intermediate image hash (i.e. used chromatic information ($a^* - b^*$) of CIE $L^*a^*b^*$ is generated as below:

$$\mathbf{h}_{a^*-b^*} = \left[h'(1), h'(2), \ldots, h'(n)\right] \tag{2}$$

Now, the final image hash is obtained by concatenating the two intermediate image hash as

$$\mathbf{h} = [\mathbf{h}_{L^*}\mathbf{h}_{a^*-b^*}] \tag{3}$$

Therefore, the final length of the hash will be sum of the hash length of intensity channel (L^*) and the hash length of chromatic channel ($a^* - b^*$) i.e. $2n$.

2.3 Metric for Performance Comparisons

Let $\mathbf{h}_1$ and $\mathbf{h}_2$ are considered hashes of transmitted and received image. The hash distance is calculated using L2 norm as.

Algorithm 1. Mapping from a square to ring and generation of hash
Input: Preprocessed image, **I**
Output: Image hash, **h**

1. Convert arbitrary size image, I, into size $m \times m$, I'
2. Mapping I' into rings as follows (Shown in Fig. 2)
3. Find area of inscribed circle as $A' = \pi r_n^2$, Where $r_n = m/2$, n is number of rings
4. Area of other rings are $A = \lfloor A'/n \rfloor$
5. Radius of first ring, $r_1 = \sqrt{A/\pi}$
6. Radius of other rings are $r_k = \sqrt{(A + \pi r_{k-1}^2)/\pi}$; $k = 2,3,\ldots,n-1$
7. Create first mask, as $\mathbf{C}_1 : (x-a)^2 + (y-b)^2 = r_1^2$, $1 \le x \le m, 1 \le y \le m$, (a,b) is a image center coordinate of I', Inside ring 1 and outside zero values is called mask gives M_1.
8. Other masks, $\mathbf{C}_j - \mathbf{C}_{j-1}$ yields M_j; $j = 2,3,\ldots,n-1$
9. First column, $S_1 = I' \times M_1$
10. Other columns, $S_j = I' \times M_j$
11. Find the mean, variance, skewness, and kurtosis (ie., features) for each column from **S**.
12. Generate the average of each features. $\mathbf{u}_{\text{ref}} = [u_{\text{ref}}(1), u_{\text{ref}}(2), u_{\text{ref}}(3), u_{\text{ref}}(4)]^T$
13. Let $\mathbf{f}_k = [f_k(1), f_k(2), \ldots, f_k(l)]$ represents the k-th row of **S** ($1 \le k \le 4$). Then, $g_i(j) = (f_k(j) - \overline{f_k(j)})/\sigma_k^2$
$\overline{f_k(j)}$ and σ_k^2 are the mean and standard deviation of $\mathbf{f}_k$, respectively.
13. $e(i) = \sqrt{\sum_{j=1}^{4}(u_i(j) - u_{\text{ref}}(j))^2}$
14. The $e(i)$ is pseudorandom scrambled using a secret key to generate a hash **h**

$$\text{Hash distance} = \sqrt{\sum\nolimits_{i=1}^{2l} |h_1(i) - h_2(i)|^2} \quad (4)$$

If the hash distances are below the selected threshold (discussed in section III-A), T, then the image pairs are considered as similar or authentic, otherwise forged or different. True positive rate (TPR) and false positive rate are generated via hash distances of image pairs as follows

$$TRP = \frac{(Number\ of\ visually\ similar\ image\ truely\ \det ected)}{(Total\ number\ of\ visually\ similar\ images)} \quad (5)$$

$$FRP = \frac{(Number\ of\ wrongly\ \det ected\ similar)}{(Total\ number\ of\ different,\ forged,\ and\ color\ forged)} \quad (6)$$

3 Experimental Results

Based on the experimental results on large image pairs the parameters are selected for the proposed method are as follows $m \times m = 512 \times 512, n = 32$. Proposed image hashing approach is applied for image authentication. The proposed method efficacy has been evaluated in following sub-sections A, B, C, and D.

3.1 Robustness and Discriminative Capability

The proposed approach has been evaluated from USC-SIPI database [24] with 53 different images. There are total of 53 × 74 = 3,922 visually similar image pairs after various manipulation of digital operations using StirMark, Photoshop, and Matlab as shown in Table 1. Eventually, 199 separate internet-selected images used to produce 199 × (199–1)/2 = 19,701 pairs of visually different images. In order to determine generic threshold value (T), the hash distance from the original image to correspond visually similar image pairs and among different pairs are then measured using the model that is being suggested. The hash distance histogram is shown in Fig. 3. The range of hash distance [177 270] can discriminate between similar and different image pairs. Empirically, threshold has been selected T = 181 for authentication of image. Figure 3(a) shows that most of the similar image pairs are truly detected similar as distances as below the selected threshold, whereas for different greater than the T.

Table 1 CPOs

Software	Manipulation	Parameter	Parameter Values
Stir Mark	JPEG compression	Quality factor	30, 40, 50, 60, 70, 80, 90, 100
Stir Mark	Watermark embedding	Strength	10, 20, 30, 40, 50, 60, 70, 80, 90, 100
Stir Mark	Scaling	Ratio	0.5, 0.75, 0.9, 1.1, 1.5, 2.0
Stir Mark	Rotation	Rotation angle in degree	±5, ±15, ±30, ±45, ±90
Photoshop	Brightness adjustment	Photoshop Scale	±10, ±20
Photoshop	Contrast adjustment	Photoshop Scale	±10, ±20
Matlab	Gamma correction	γ^{1}	0.75, 0.9, 1.1, 1.5
Matlab	3 × 3 Gaussian‘ low pass filter‘	Standard‘_ deviation	0.3, 0.4, 0.5, 0.6, 0.7, 0.8, 0.9, 1‘
Matlab	Salt and peppernoise	Density	0.001–0.01
Matlab	Speckle noise	Variance	0.001–0.01
Total			74 parameter values

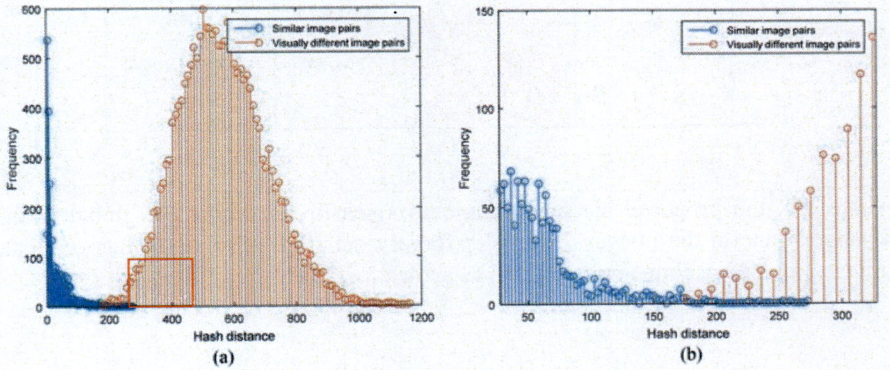

Fig. 3 (a) Hash distance distribution for perceptually similar and different image pairs, (b) expanded version of red rectangle from (a)

3.2 *Analysis for Global Color Changes*

The same 53 images from USC-SIPI database are also used for global color changes and is modified using image editing tool, few samples are shown in Table 2.

Now, there are 53 forged image pairs available for the global color forgery experiments. The inclusion of (a^*-b^*) component of the CIE $L^*a^*b^*$color model, makes the proposed hashing technique sensitive to the global color changes. The results displayed in Table 2 indicate that the hash distances are greater than the threshold

Table 2 Analysis of global color

Sl. No.	Original Image	Forged images using global Color changes	Proposed method	Ring and Vector distance based [18]
1			318.80	11.66
2			526.94	20.97
3			418.94	17.60
4			408.29	17.05
5			530.57	18.44

value. So, the proposed hashing technique is sensitive towards the global color changes done in the images. The color forgery pairs are misclassified as original image pairs in baseline approach [18] as shown in Table 2.

3.3 *Analysis for Local Color Changes*

The same 53 color images from USC-SIPI database used for global color changes are also used for the experiments performed for local color changes and is modified using Photoshop. So, there are 53 forged pairs of images available in the database to perform the experiment for local color changes. It has been observed that the baseline method is unable to detect the local color change forgery. In case of local color changes small area is being modified. From the Table 3, it is obvious that the entire hash distance is greater than the threshold pre-selected, i.e. 181, wrongly classified as similar in baseline [18].

Table 3 Sensitivity detection for local color changes

Sl. No.	Original images	Forged images using local color changes	Proposed method	Ring and Vector distance based [18]
1			183.08	9.21
2			183.51	10.14
3			280.45	4.24
4			200.04	4.12
5			190.40	16.15

3.4 Sensitivity Analysis

A set of 400 images and their tampered versions have been taken from CASIA database [25] for sensitivity analysis of the proposed hashing technique. Forged area has been highlighted by using red circle as shown third column of Table 4. It may be observed that the obtained hash distances are far greater than the threshold value of 181. It is also observed that all the manipulated images are correctly identified as forged images. Hence, the efficiency of the proposed method in the case of forgery detection is 100% for the selected dataset.

3.5 Key Dependent Analysis

Key dependency of any hashing technique is very necessary to make it secure. Therefore, it may be required to test the key dependency of the proposed hashing technique. So, in order to do this the hash for an image has been generated with a key value of

Table 4 Forgery (content addition/deletion) detection

Sl. No.	Original images	Forged images	Hash distance using proposed method
1			209.5524
2			291.1263

4. The different hash has been also generated with different key values for the same image keeping other parameters constant. Then, obtained the hash distances. For the same key value (i.e. 4), it is observed that hash distance is zero, shown in Fig. 4. This is because of the fact that for same key, the proposed model generates same hash each time. It may be observed that for different key values the hash distances are very large as compared to the threshold value of 181. So, it can be determined that the proposed hashing technique is highly dependent on the secret key, which is a good indication for the security of the proposed hashing technique.

3.6 ROC for Robustness and Discrimination Comparison

In addition to baseline method [18], the proposed method has compared with some more existing methods [6, 16, 27], shown in Table 5. Due to inclusion of color channel into the hash generation process, the proposed method has become sensitive to color forgery unlike the existing techniques [6, 16, 18]. The method [27] is also sensitive to color forgery, but this method do not possess the rotation invariant property for arbitrary angle. However, it is experimentally observed that the proposed method is robust against arbitrary rotation. The hash length of proposed method is little bit large but it is still below 1 kb size which is usually considered to be maximum size of the hash [28]. It is shown that TPR when FPR ≈ 0 of proposed method is 0.9982, which is comparatively better than the other methods. Similarly, when TPR ≈ 1, the FPR is observed to be lowest amongst all other method considered during this study. It is expected that the method that provides high TPR and low FPR may be considered as better hashing technique. From this perspective it may be inferred that the proposed method may be considered as a better hashing technique as compared to the other existing methods. Figure 5 shows the ROC performance of compared methods, inferred the efficacy of the proposed method.

Table 5 Performance comparison with different existing techniques

Parameters	RT-DFT based [6]	Invariant moment based [27]	RE based [16]	Ring and Vector distance based [18]	Proposed method
Sensitive to image color changes	No	Yes	No	No	Yes
Robust against arbitrary rotation	No	No	Yes	Yes	Yes
Hash length (digits)	15	42	64	40	64
TPR at optimal threshold	0.8863	0.9323	0.9612	0.9769	0.9901
FPR at optimal threshold	0.1321	0.0312	0.0414	0.0031	0.0002
Average time (sec)	3.14393	0.7192	0.1265	0.2839	0.3245

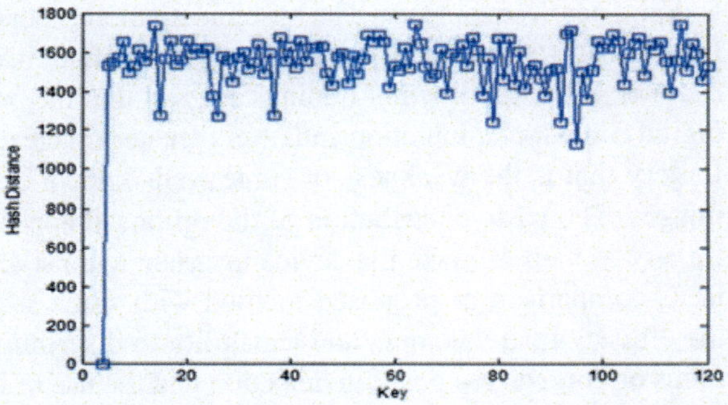

Fig. 4 Key dependance of hash

The proposed hash generation technique has been implemented in Windows 10, 8 GB RAM desktop system and MATLAB 2018a. The other methods that has been used for comparative analysis has also been implemented on the same platform to evaluate the computational complexity. The proposed method computationally efficient as compared to some of the exiting techniques. However, it is computationally little bit demanding as compared to baseline ring partition based models.

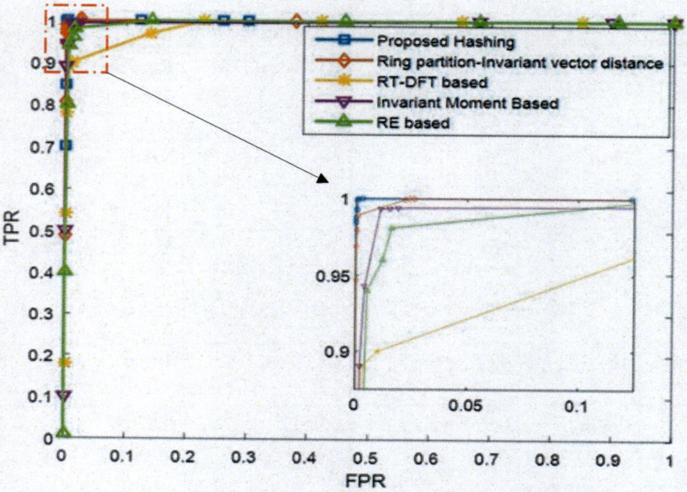

Fig. 5 Performance comparison of the proposed method with some existing methods

4 Conclusions and Future Works

This paper introduces a perceptual hashing of the image using modified ring partition and invariant vector space, The baseline method is depend on only intensity channel. In the proposed method, the chromatic information is added along with intensity channel. The experimental findings suggest that the proposed solution is resilient to digital content manipulation and offers greater image diversity. It can detect the color forgery that is the weakness of some well- known existing hashing algorithms for images. The basic contribution of the proposed approach is that it can detect color forgery as well as make the device invariant against arbitrary angle of rotation. The ROC comparison of proposed method with some state-of-the-art methods depicts the efficacy for robustness and capability to discriminate. The future work may be focus on forged area position detection and the use of local features to enhance hash efficiency to reduce the possibility of authentication error.

Acknowledgements. The writers would like to thank all the members of speech and Image' Processing' Laboratory, Silchar National' Institute of 'Technology, India, for providing assistance and essential facilities to do this work.

References

1. Du L, Ho AT, Cong R (2020) Perceptual hashing for image authentication: A survey. Signal Process Image Commun 1(81):115713

2. Paul M, Karsh RK, Talukdar FA (2019) Image hashing based on shape context and speeded up robust features (SURF). In: 2019 International Conference on Automation, Computational and Technology Management (ICACTM), pp 464–468. IEEE
3. Saikia A, Karsh RK, Lashkar RH (2017) Image authentication under geometric attacks via concentric square partition based image hashing. In: TENCON 2017–2017 IEEE Region 10 Conference, 5 November 2017, pp 2214–2219. IEEE
4. Monga V, Banerjee A, Evans BL (2006) A clustering based approach to perceptual image hashing. IEEE Trans Inf For Secur 1(1):68–79
5. Swaminathan A, Mao Y, Wu M (2006) Robust and secure image hashing. IEEE Trans Inf For Secur 1(2):215–230
6. Lei Y, Wang Y, Huang J (2011) Robust image hash in radon transform domain for authentication. Signal Processing Image Commun 26(6):280–288
7. Tang Z, Wang S, Zhang X, Wei W, Zhao Y (2011) Lexicographical framework for image hashing with implementation based on DCT and NMF. Multimedia Tools Appl 52(2):325–345
8. Monga V, Mihcak MK (2007) Robust and secure image hashing via non-negative matrix factorizations. IEEE Trans Inf For Secur 2(3):376–390
9. Khelifi F, Jiang J (2010) Perceptual image hashing based on virtual watermark detection. IEEE Trans Image Process 19(4):981–994
10. Fouad K, Jianmin J (2010) Analysis of the security of perceptual image hashing based on non-negative matrix factorization. IEEE Signal Process Letter 17(1):43–46
11. Lu W, Varna AL, Wu M (2010) Forensic hash for multimedia information. In: Proceedings SPIE, Media Forensics and Security II, San Jose, CA. https://doi.org/10.1117/12.838745
12. Lu W, Wu M (2010) Multimedia forensic hash based on visual words. In: Proceedings IEEE Int'l Conference Image Processing, pp 989–992
13. Zhao Y, Wang S, Zhang X, Yao H (2013) Robust hashing for image authentication using zernike moments and local features. IEEE Trans Inf For Secur 8(1):55–63
14. Li YN, Wang P, Su YT (2015) Robust image hashing based on selective quaternion invariance. IEEE Signal Process Lett 22(2):2396–2400
15. Qin C, Chen X, Ye D, Wang J, Sun X (2016) A novel image hashing scheme with perceptual robustness using block truncation coding. Inf Sci 361:84–99
16. Tang Z, Zhang X, Huang L, Dai Y (2013) Robust image hashing using ring-based entropies. Signal Process 93(7):2061–2069
17. Tang Z, Zhang X, Zhang (2014) Robust perceptual hashing based on Ring partition and NMF. IEEE Trans Knowl Data Eng 26(3):711–724
18. Tang Z, Zhang X, Li X, Zhang S (2016) Robust image hashing with ring partition and invariant vector distance. IEEE Trans Inf For Secur 11(1):200–214
19. Karsh RK, Saikia A, Laskar RH (2018) Image authentication based on robust image hashing with geometric correction. Multimedia Tools Appl 77(19):25409–25429
20. Karsh RK, Laskar RH (2015) Perceptual robust and secure image hashing using ring partition-PGNMF. In: TENCON 2015–2015 IEEE Region 10 Conference, 1Nov 2015, pp 1–6. IEEE
21. Karsh RK, Laskar RH (2017) Robust image hashing through DWT-SVD and spectral residual method. EURASIP J Image Video Process 2017(1):31
22. Karsh RK, Mishra H, Vashist T, Laskar RH (2018) Image authentication using image hashing based on ring partition with corner inclusion. In: 2018 International Conference on Communication and Signal Processing (ICCSP), 3 Apr 2018, pp 0557–0562. IEEE
23. Connolly C, Fliess T (1997) A study of efficiency and accuracy in the transformation from RGB to CIE Lab color space. IEEE Trans Image Process 6(7):1046–1048
24. USC-SIPI Image database (2007). http://sipi.usc.edu/database/
25. CASIA Tampered image detection evaluation database. http://forensics.idealtest.org/
26. Petitcolas FA (2000) Watermarking schemes evaluation. IEEE Signal Process Mag 17(5):58–64
27. Tang Z, Dai Y, Zhang X (2012) Perceptual hashing for color images using invariant moments. Appl Math Inf Sci 6(2S):643S–650S
28. Decker BD, Lapon J, Naessens V, Uhl A (2011) Communication and multimedia security. Springer, Ghent

A Single Element Dual Feed MIMO Cylindrical Dielectric Resonator Antenna with High Isolation for 5.6 GHz WLAN Application

Biplob Biswas and Anshul Gupta

Abstract In this communication, a single element cylindrical shape dielectric resonator antenna (DRA) excited by two different feeding techniques; co-planar wave guide (CPW) and aperture coupled feeding, is presented. The proposed antenna is working in the frequency range of (5.39 GHz to 5.85 GHz) for port-1 and (5.46 GHz to 5.85 GHz) for port-2 with fractional bandwidth of 8.15% and 6.87% respectively. By triggering orthogonal modes in the DRA, mutual coupling has been reduced, and the isolation (S_{12}) is improved exceeding −33 dB between both the ports. The multiple input multiple output (MIMO) diversity performance parameters (ECC, DG and CCL) are also calculated using MATLAB code; and it is contained within the agreeable limits. The simulation tool used in the designing of the proposed antenna is Ansys High Frequency Structure Simulator (HFSS) software. HFSS is 3D electromagnetic (EM) simulation software which uses Finite Element Method (FEM) for simulation.

Index Terms Cylindrical dielectric resonator antenna (CDRA) · MIMO antenna · Co-planar waveguide (CPW) · Slot coupled · isolation

1 Introduction

The demand of low loss, high gain, high efficiency, and miniaturization of the antenna system is pushing the antenna designers towards multiple input multiple output (MIMO) antenna systems. The issue of mutual coupling between different ports is one of the major problems in a MIMO antenna system; which arises due to the sharing of common ground plane. There are different types of mutual coupling reduction methods available in the literature, such as electromagnetic band-gap structure

B. Biswas (✉) · A. Gupta
National Institute of Technology Raipur, Raipur, Chhattisgarh, India

A. Gupta
e-mail: agupta.etc@nitrr.ac.in

V. Nath and J. K. Mandal (eds.), *Proceeding of Fifth International Conference on Microelectronics, Computing and Communication Systems*, Lecture Notes in Electrical Engineering 748,
https://doi.org/10.1007/978-981-16-0275-7_6

(EBG) [1], parasitic element [2], defected ground structure (DGS) [3], neutralization line [4], orthogonal mode generation inside the DRA [5], etc.

With the introduction of the dielectric resonator antenna (DRA), the losses involved in the patch antenna has greatly reduced; since a dielectric resonator has negligible metallic loss and very low surface current loss [6]. Also a dielectric resonator antenna possesses high radiation efficiency and greater impedance bandwidth; when compared to a traditional patch antenna [6]. There are various DRA based MIMO antennas in the literature; with multiband [7], wideband [8], and circular polarization [9] characteristics. A combination of more than one different feeding techniques can be found in the literature to excite the DR; such as a combination of co-planar wave guide (CPW) and probe feeding [10], CPW and microstrip line fed [11], CPW and aperture coupled feeding [12], slot coupled and probe feeding [13] etc.

In this paper a single element cylindrical DRA is proposed which is fed by two different feeding techniques; CPW at port-1 and slot coupled at port-2. The feeding structures are designed orthogonal to each other which are generating orthogonal modes $HEM_{11\delta}^{x}$ and $HEM_{11\delta}^{y}$ in the DRA. An optimized conformal strip (for port-1) and an optimized open circuited stub (for port-2) are used for the accomplishment of desired impedance bandwidth and return loss. The proposed antenna is intended for use with WLAN application (5.47 GHz to 5.85 GHz).

2 Antenna Design

The antenna design is analyzed using Ansys HFSS software. The substrate used in this design is made up of FR4 epoxy material whose dielectric permittivity is $\epsilon_r = 4.4$ and dimensions are $L_S \times W_S \times H_S$. The ground plane dimensions are $L_G \times W_G \times H_G$.A cylindrical shape dielectric resonator (DR) with dielectric constant $\epsilon_r = 9.8$ is used as a radiating element whose optimized height and diameter are H1 and D respectively. To excite the DRA co-planar waveguide fed conformal strip excitation is utilized at port-1. On the other hand at port-2, the aperture coupled feeding is used to excite the DR as illustrated in Fig. 1.

3 Results and Discussion

Optimization of $|S_{11}|$ for port-1 (CPW feed) is done by performing parametric analysis for different values of height of conformal strip. In case of a slot coupled feeding mechanism, by adjusting the slot dimensions along with open circuited stub length; good impedance matching, and desired bandwidth can be achieved [14, 15].Parametric analysis is carried out for different values of stub length to obtain optimized return loss for port-2 (slot coupled feed). The plot of isolation S_{12} along with optimized scattering parameters (S_{11} and S_{22}) is illustrated in Fig. 2.

Fig. 1 Illustrated figure of proposed MIMO antenna; (a) Isometric view; (b) Top view; (c) Side view [$L_S = L_G = 50$, $L_1 = 20$, $L_2 = 25.85$, $L_3 = 6.7$, $W_S = W_G = 50$, $W_1 = 2$, $W_2 = 3$, $W_3 = 2.6$, $H_S = 1.6$, $H_G = 0.035$, $H_1 = 9$, $H2 = 5.65$, $D = 6$, $g = 0.5$, $L_b = 2.8$]. All dimensions are in mm

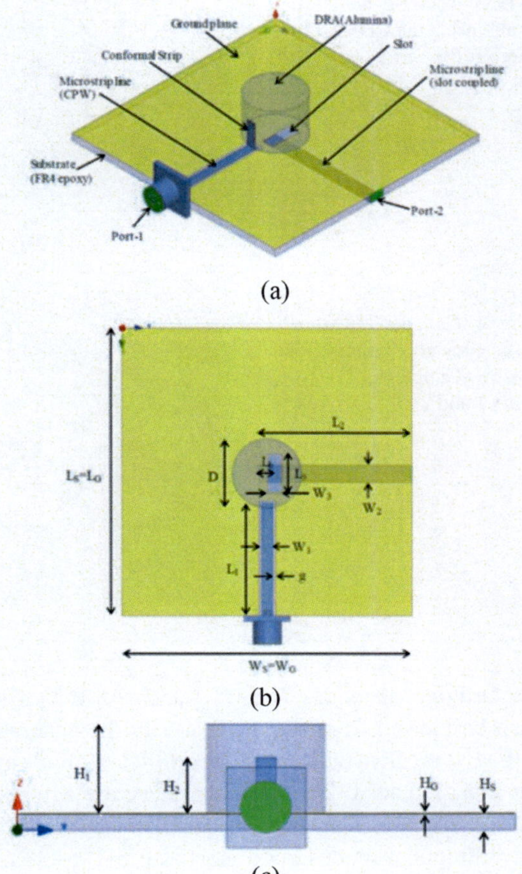

Fig. 2 Characteristics of simulated scattering parameter for proposed MIMO antenna

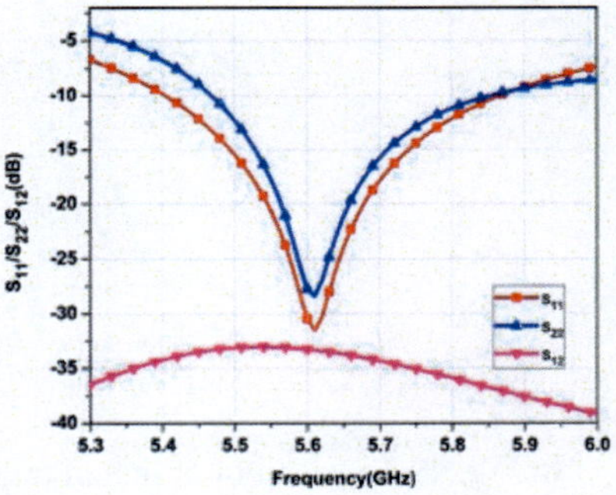

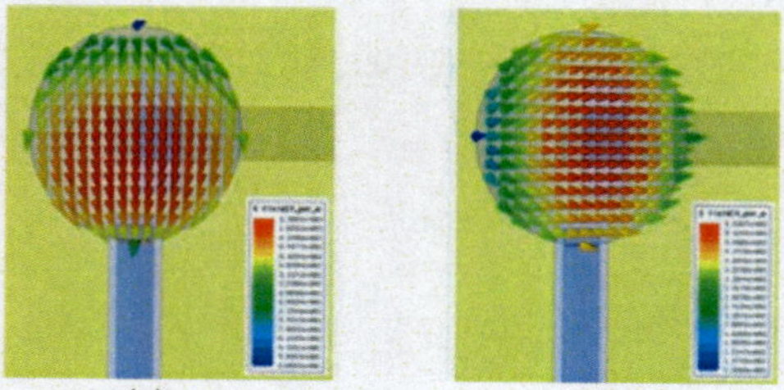

(a) (b)

Fig. 3 Vector E-field/near field inside the CDRA (a) port-1; (b) port-2

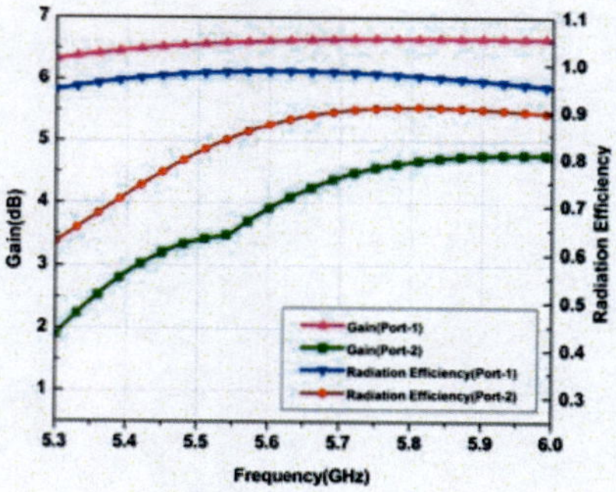

Fig. 4 Gain and radiation efficiency graph of proposed MIMO CDRA (for both port-1 and 2)

Orthogonal modes $HEM^{x}_{11\delta}$ and $HEM^{y}_{11\delta}$ are excited in the DRA [16] which can be viewed from Fig. 3. To verify the generated mode; resonant frequency for $HEM_{11\delta}$ is calculated using the empirical formula available in the literature [17] given by equation no. 1, 2 and 3. There is a close agreement found between the calculated (5.8 GHz) and simulated (5.61 GHz) resonant frequency. Due to the generation of orthogonal modes good isolation; $S_{12} <-33$ dB is attained within the working frequency range.

$$f_{r(HEM11\delta)} = \frac{6.321 \times c}{2\pi R\sqrt{\varepsilon_{reff} + 2}}\left[0.27 + 0.36\left(\frac{R}{2H_{eff}}\right) + 0.02\left(\frac{R}{2H_{eff}}\right)^{2}\right] \tag{1}$$

$$H_{eff} = H_{sub} + H_{alumina} \tag{2}$$

$$\varepsilon_{reff} = \frac{H_{eff}}{\frac{H_{sub}}{\varepsilon_{rsub}} + \frac{H_{alumina}}{\varepsilon_{ralumina}}} \tag{3}$$

In Fig. 4; the graph of gain and radiation efficiency is illustrated. It can be clearly viewed that the range of the gain varies from 3.28 dB to 4.73 dB for port-1 and 6.46 dB to 6.66 dB for port-2 within the operating frequency range. And the radiation efficiency reaches approximately 90% for both port-1 and port-2 inside the working frequency band.

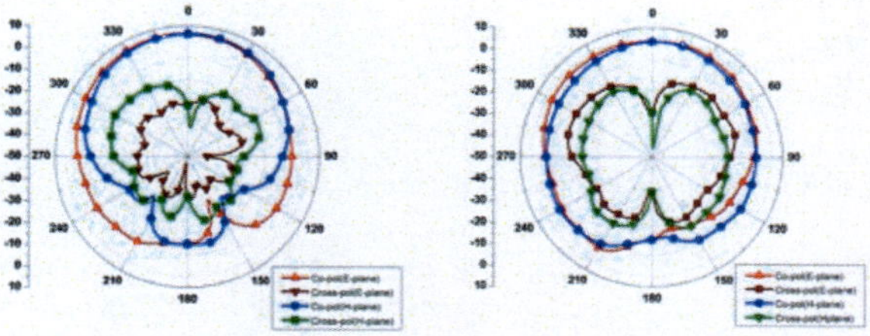

(a) (b)

Fig. 5 Simulated far field characteristics at 5.61 GHz frequency (a) for Port-1; (b) for Port-2

Fig. 6 Plot of ECC and DG versus frequency

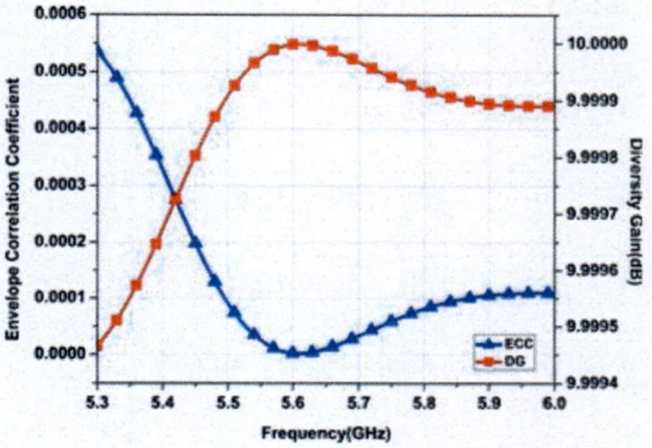

Simulated far field radiation pattern for E-plane and H-plane at 5.61 GHz resonant frequency for port-1 and port-2 are plotted in Fig. 5(a) and (b) respectively. From Fig. 5, it can be realized that; in the broad side direction i.e.; ($\theta = 0$, $\phi = 0$) the proposed MIMO antenna is showing maximum radiation which is due to the triggering of $HEM_{11\delta}^{x}$ and $HEM_{11\delta}^{y}$ modes in the DRA. Also the gap between the co-polarization and cross-polarization levels is greater than 30 dB; which is desirable for low loss MIMO antenna systems.

Diversity performance parameter characteristics of the proposed MIMO antenna such DG, ECC, and CCL [18, 19] is also calculated. Figure 6 shows the characteristics of ECC and DG versus frequency. From the plot, it is concluded that ECC is below 0.0003 and DG is nearly 10 dB inside the operating frequency band. Figure 7 shows the plot of CCL characteristics versus frequency which is below 0.5 bits/s/Hz inside the working frequency band. In Table 1 the comparison of the presented antenna has been done with different antennas available in the literature.

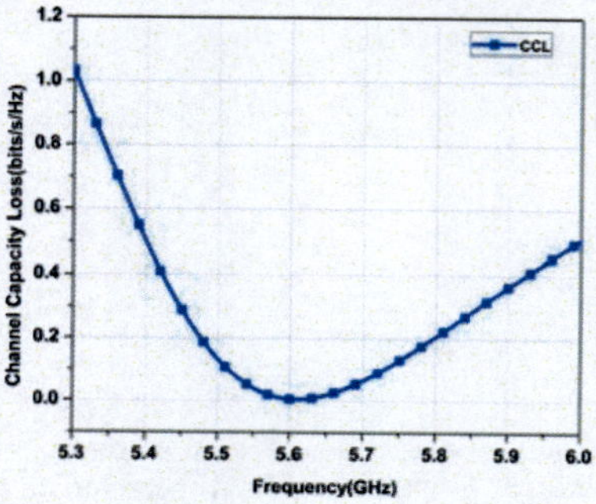

Fig. 7 Plot of channel capacity loss versus frequency

Table 1 Comparison table of presented antenna with different antennas available in the publications

Ref No	Size (mm^3)	$\lvert S_{12}\rvert$ (dB)	Gain (dB)	Fabrication complexity
[10]	70 × 70 × 11.6	20 dB	4.97 dB(P1) 4.51 dB(P2)	Easy
[11]	50 × 50 × 12.6	28 dB	5.5 dB(P1) 5.5 dB(P2)	Easy
[12]	50 × 50 × 10.1	32 dB	4.5 dB(P1) 5 dB(P2)	Complex
[13]	56.6 × 56.6 × 14.09	20 dB	8.1 dB(P1) 7.5 dB(P2) 7.4 dB(P3)	Easy
This Work	**50 × 50 × 10.6**	**33 dB**	**6.6 dB(P1)** **3.9 dB(P2)**	**Easy**

4 Conclusion

In this communication, two different feeding techniques have been studied for a two port, cylindrical shaped, single element MIMO DRA. The presented design consists of an orthogonal feeding structure which is found responsible for providing high isolation exceeding −33 dB inside the operating frequency band. This proposed antenna is found relevant for 5.6 GHz WLAN application .

References

1. Yang F, Rahmat-Samii Y (2003) Microstrip antennas integrated with electromagnetic band-gap (EBG) structures: a low mutual coupling design for array applications. IEEE Trans Ant Propag 51(10):2936–2946
2. Li Z, Du Z, Gong K (2009) A dual-slot diversity antenna with isolation enhancement using parasitic elements for mobile handsets. In: 2009 Asia Pacific Microwave Conference, pp 1821–1824. IEEE

3. Xiao S, Tang MC, Bai YY, Gao S, Wang BZ (2011) Mutual coupling suppression in microstrip array using defected ground structure. IET Microwaves Ant Propag 5(12):1488–1494
4. Zhang S, Pedersen GF (2015) Mutual coupling reduction for UWB MIMO antennas with a wideband neutralization line. IEEE Ant Wirel Propag Lett 15:166–169
5. Nasir J, Jamaluddin MH, Khalily M, Kamarudin MR, Ullah I, Selvaraju R (2015) A reduced size dual port MIMO DRA with high isolation for 4G applications. Int J RF Microwave Comput Aided Eng 25(6):495–501
6. Petosa A (2007) Dielectric resonator antenna handbook. Artech House Publishers, Norwood
7. Sharma A, Das G, Gangwar RK (2017) Design and analysis of tri-band dual-port dielectric resonator based hybrid antenna for WLAN/WiMAX applications. IET Microwaves Ant Propag 12(6):986–992
8. Kumari T, Das G, Sharma A, Gangwar RK (2019) Design approach for dual element hybrid MIMO antenna arrangement for wideband applications. Int J RF Microwave Comput Aided Eng 29(1):e21486
9. Das G, Sharma A, Gangwar RK (2018) Dielectric resonator based circularly polarized MIMO antenna with polarization diversity. Microwave Opt Technol Lett 60(3):685–693
10. Roslan SF, Kamarudin MR, Khalily M, Jamaluddin MH (2014) An MIMO rectangular dielectric resonator antenna for 4G applications. IEEE Ant Wirel Propag Lett 13:321–324
11. Das G, Sharma A, Gangwar RK (2017) A single cylindrical dielectric resonator based MIMO antenna system for WiMAX applications. In: 2017 Progress in Electromagnetics Research Symposium-Spring (PIERS), pp 3707–3712. IEEE
12. Das G, Sharma A, Gangwar RK (2017) Dual feed MIMO cylindrical dielectric resonator antenna with high isolation. Microwave Opt Technol Lett 59(7):1686–1692
13. Abdalrazik A, El-Hameed ASA, Abdel-Rahman AB (2017) A three-port MIMO dielectric resonator antenna using decoupled modes. IEEE Ant Wirel Propag Lett 16:3104–3107
14. Pozar DM (1985) Microstrip antenna aperture-coupled to a microstripline. Electron Lett 21(2):49–50
15. Shum SM, Luk KM (1994) Analysis of aperture coupled rectangular dielectric resonator antenna. Electron Lett 30(21):1726–1727
16. Kajfez D, Glisson AW, James J (1984) Computed modal field distributions for isolated dielectric resonators. IEEE Trans Microw Theory Tech 32(12):1609–1616
17. Mongia RK, Bhartia P (1994) Dielectric resonator antennas—a review and general design relations for resonant frequency and bandwidth. Int J Microwave Millimeter-Wave Comput-Aided Eng 4(3):230–247
18. Sharawi MS (2014) Printed MIMO antenna engineering. Artech House, Norwood
19. Choukiker YK, Sharma SK, Behera SK (2013) Hybrid fractal shape planar monopole antenna covering multiband wireless communications with MIMO implementation for handheld mobile devices. IEEE Trans Ant Propag 62(3):1483–1488

AlexNet-CNN Based Feature Extraction and Classification of Multiclass ASL Hand Gestures

Abul Abbas Barbhuiya, Ram Kumar Karsh, and Samiran Dutta

Abstract In computer vision, accurate recognition and classification of hand gestures play a vital part for developing gesture recognition systems for human–computer interaction (HCI). Error-free HCI can considerably improve systems that can recognize diverse class of hand gesture characters. In this manuscript, we propose a modified deep learning architecture based on AlexNet for recognizing American Sign Language (ASL) gestures. The objective of static gesture recognition is to characterize the data given by the hand gesture into predefined gesture classes. The ASL dataset taken in our methodology is composed of five different individuals. In contrast to most of the prior works, where data is split randomly for training and testing, we have trained the network on samples of four individuals and tested on the fifth individual. We have taken the original AlexNet architecture and modified the final fully connected layers to classify the ASL dataset. The modified network is trained using ASL dataset and is compared to the state-of-the-art AlexNet architecture to demonstrate the work's competence. The paper also demonstrates the training testing performance of the modified network with the ASL dataset.

Keywords Gesture recognition · ASL · Alexnet · Deep learning · Human–computer interaction

1 Introduction

Humans can easily recognize sign language, because of the combination of vision and synaptic connections formed together with the growth of the brain [1]. With the progression of technology, computers are widely persuading our day-to-day life [2] and further, they are used to mimic the human behavior and thinking process. To imitate human skills in computers some of the problems need to be attended like distinct objects of interest in images, more appropriate classification techniques. Recently [3], most of HCI depends on mechanical input devices like joystick, mouse,

A. A. Barbhuiya (✉) · R. K. Karsh · S. Dutta
Department of ECE, National Institute of Technology Silchar, Silchar, India

V. Nath and J. K. Mandal (eds.), *Proceeding of Fifth International Conference on Microelectronics, Computing and Communication Systems*, Lecture Notes in Electrical Engineering 748,
https://doi.org/10.1007/978-981-16-0275-7_7

keyboard, etc., but the ability to distinguish human gestures naturally has developed an increasing interest in computer vision-based approaches. Lately hand gestures become an essential part of HCI, which is a motivating force for research in the area of hand gestures analysis, modeling and recognition.

In sign language interpretation, medical application and smart environments hand gesture recognition have been widely used [4]. Three basic types of sensors [5] are used mainly to sense hand gestures: 1) Vision-based sensors, 2) Mount based Sensors, and 3) Multi-touch screen sensors. The vision-based sensors have no physical interfaces [6] with the user so they are easier compared to mounted sensors. At the same time, vision-based sensors deals with much larger working distances, unlike multitouch screen sensors.

Hand gesture recognition can be classify in static and dynamic gesture recognition [7]. This study provides a discriminative study among static gesture classes to recognize the true fingerspelled letters of ASL. 36 characters of ASL data have been considered for classification.

Generally, in a vision-based recognition system, the images are initially processed to train the networks. Deep learning-based networks enjoy more biologically encouraged architectures and learning algorithms, compared to conventional feed-forward net-works. Deep networks allow the representation of more complex problems and the representation [8] of highly nonlinear functions by training in a layer-wise fashion. So, the complex problems can be better modeled in deep learning architectures.

The remaining paper is arranged as follows. Section 2 describes the background study pertaining to recognition of ASL gestures. The methodology proposed is represented in the Sect. 3. Section 4 howss the results of experimental analysis. Lastly, the conclusions and further works are drawn in Sect. 5.

2 Background Studies

Static gesture recognition is a problem of pattern recognition, where the extraction of feature is a crucial pre-processing stage [9]. Extraction of features is the first step towards the application of standard pattern recognition algorithms. The features correspond to the most discriminative image information. A fair amount of research has been done in this area, some of the approaches and concepts are detailed below.

Zhou et al. [10] suggested an effective algorithm for extracting fingers directly from prominent edges of the hand. The fingers are molded as cylinder-shaped objects for their parallel edge features. The authors show how the hand posture is segmented. Considering the geometry of hand physiognomies, the hand posture is then defined depending on palm center location, wrist position and finger positions. The suggested method not only extracts extensional fingers with great precision, but also flexional fingers. In addition, the work also has no effect on the performance of algorithm when considering variations in the hand rotation and the finger angle. The authors performed experiments to validate that the system can straightly extract high-level hand features besides the estimation of hand poses in real-time.

Wang et al. [11] proposed a novel hand gesture recognition model using super pixels. The suggested procedure uses the distance metric of the novel superpixel earth mover, composed of a Kinect depth camera. Here Kinect's depth and skeleton evidence is effectively exploited to create markless hand extraction.

Hasan et al. [12] applied a shape exploration method for hand gesture recognition. Authors classify through a neural network-based method among six static hand gestures to execute tasks such as opening and closing objects, minimizing and maximizing a current window, etc. They attain an accuracy of 86.38% by employing a back propagation-based algorithm which is predominantly designed on several layers of neural network.

Zi Li et al. [13] suggested a novel feature learning methodology for recognizing human actions. The system is designed depending on principal component analysis (PCA) and sparse auto-encoder aimed at RGBD images. The features are initially learned accordingly from the four channeled RGBD images via the sparse auto-encoder using convolution neural networks. Secondly, features learned are concatenated from both the channels and then fed to a multi-layer PCA for acquiring the ultimate feature. Here the authors achieve recognition accuracy of 99.05%.

Ju et al. [14] presented a novel architecture that uses earthmover's distance and Lasso algorithms to extract the features of hand gesture from RGB-D images. This combined procedure uses the distance algorithm for finger earth movers to detect the image of the palm and extract features of the fingertips. A Lasso algorithm is also proposed to proficiently obtain feature of the fingertip from contour curve of a hand.

Zhu et al. [15] suggested convolutional LSTM Networks and 3-D convolution for multimodal gesture recognition system. The algorithm initially studies short-term spatio-temporal features of movements via the 3-D convolutional neural network. Further, the model acquires long-term spatio-temporal features through convolutional LSTM networks, depending on short-term spatio-temporal features extracted. Also, refinement among multimodal data is assessed and found out that it could be measured as a probable ability to avert overfitting once no pre-trained models exist.

Gupta et al. [16] applied the collective features of HOG and SIFT algorithms. The authors combined both the HOG and SIFT features in a single array for classification. A standard KNN classifier is used for classification. A total of 179 gestures are classified correctly out of 200 gestures.

Kumar et al. [17] presented a novel multimodal algorithm through sensor devices meant for isolated Sign Language Recognition. The authors used Leap Motion and Microsoft Kinect sensors for gesture acquisition. Together both the horizontal as well as vertical information of fingers are acquired by placing the Leap motion sensor and Kinect accordingly. Bi-directional LSTM Neural Network and HMM sequential classifiers are used separately for performance recognition. The authors have shown a comparative result using a dataset composed of 7.5 thousand gestures of Indian sign language (ISL).

Oyedotun et al. [18] work with gesture recognition database from Thomas Moeslund by applying deep learning algorithm. The authors used deep learning for the challenging task of recognition entire 24 hand gestures. The authors presented more complex neural networks with lesser error rates that are capable of learning complicated hand gestures and classifying them.

Nagarajan et al. [19] applied Edge-Oriented Histogram for feature extraction, where all the input hand gesture images are denoted by the count of edge histogram. The authors applied a multiclass SVM classifier and achieved an overall accuracy of 93.75%.

Feng et al. [20] proposed an integrated model based on three depth projection maps generated from the projected depth map of three orthogonal planes. The three depth projection maps are used to extract the bag of contour fragment for the extraction of structural data information and hand shapes from the depth maps. It is then concatenated as a concluding structure description of the initial depth data. The authors used a SVM with a linear kernel for classification.

Ren et al. [21] presented an integrated framework considering hand contour as a hand gesture feature and classifying the feature using a support vector machine. It is suggested that normalization based on multiscale weighted histogram of contour direction confirms good performance in recognition. Towards the improvement of efficiency, the most significant hand feature is counted for the histogram. The weight of the position of individual contour point and the direction are together considered using the direction-angle map for the histogram. According to the paper's experimental results, it gives a 97.1% accuracy of recognition on a personal computer via a frame rate of 30 fps.

Li et al. [22] suggested a novel approach to train CNN on soft consideration mechanism in an end to end way. This helps in routinely locating hands and arrange gestures through a unique network, instead of depending on traditional way of hand recognition and classification stage-wise. The algorithm computes the weight generated from the whole image to comment on the presence of hand in a defined section. Finally, a global sum is realized concentrating primarily on the corresponding weights to represent the complete image. The authors calculated the viability and effectiveness of the technique through widespread experiments on two public datasets.

3 Proposed Methodology

The proposed method of gesture recognition systems (GRS) can be presented in the Block diagram shown in Fig. 1. The blocks of the proposed GRS are discussed in the Subsequent sub-sections.

3.1 Data Pre-processing and Expansion

Augmentation or data expansion for smaller image datasets [23] is an important technique to avoid the problem of underfitting. Replication of data is carried out before training the network. The publicly available American Sign Language Dataset (ASL)

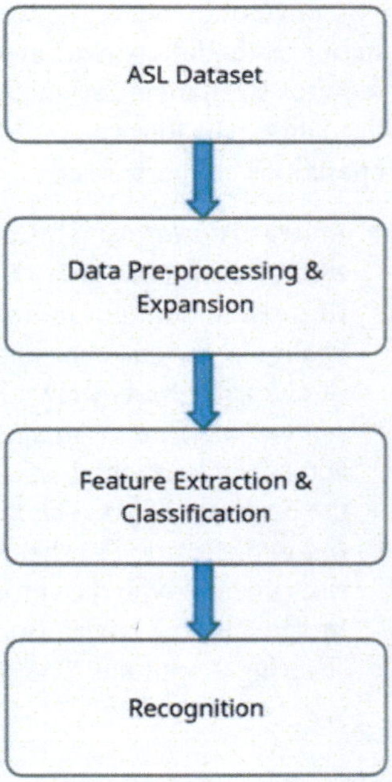

Fig. 1 A common hand gesture recognition model

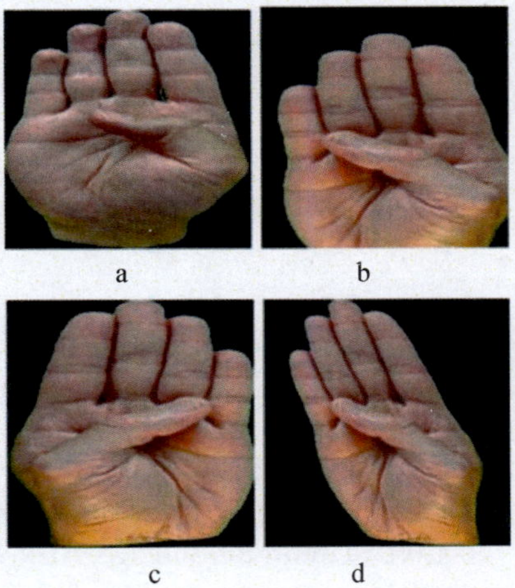

Fig. 2 Depicts the data expansion (a) Original image, (b) Translated image, (c) Flipped image and (d) Sheared image

[28] in our methodology contains 36 classes of static hand gesture sign characters. 26 classes are of the English alphabet (A–Z) and 10 classes are for Numerals (0–9). Gesture image of character 'b' is shown in Fig. 2(a). 70 images are present for each class in the dataset. So, the total number of images of hand gesture sign characters exist in the ASL dataset. The ASL dataset have sign images of 5 different individuals. The gesture images of 4 different individuals are separated out for training and gesture images of leftover individual are exclusively used for testing.

Convolution neural network requires a fixed dimension of images as input [24]. In our methodology, we have used AlexNet architecture as the base model. AlexNet requires sized input image. So, the images of the ASL dataset are resized to. Further, the data expansion is carried out on training image samples. The following expansion operations are performed:

- *Image Translation:* This performs a geometric transformation that maps the position of each pixel in an original image into a different position, where the size of input and output image is the same. Here Image translation is performed by changing various coordinates of the original image shown. The translated image of character 'b' is shown in Fig. 2(b).
- *Image Flip:* Flipped image is obtained by taking mirror-image of an original image across the horizontal axis or vertical axis. Here flipped image is obtained across the horizontal axis. The flipped image of character 'b' is shown in Fig. 2(c).
- *Image Shearing:* For shear image, pixels are shifted horizontally by a distance that rises linearly with the vertical distance from the horizontal line or vice-versa. Here Image shearing is performed for various horizontal shearing amplitude and vertical shearing amplitude. The sheared image of character 'b' is shown in Fig. 2(d).

3.2 Convolution Neural Network

CNN's or convolution neural networks are broadly used for classifying images, recognizing and detecting objects [25]. The CNN architectures can be summarized by three types of layers: convolution, pooling, and classification. For the task of image classification, an image portrayed by an uninformed color model is the input to a CNN. In our work, we have used CNN, AlexNet which was developed by [26].

- *AlexNet Convolution Neural Network (AlexNet CNN):* Because of the tradeoffs between speed and accuracy, AlexNet CNN is preferred as most studied CNN. Table 1 shows the complete AlexNet architecture. This model clas-sifies 1.2 million fine quality images into 1000 dissimilar classes in the contest of ImageNet Large Scale Visual Recognition Challenge 2010 [27]. The input to AlexNet is an RGB image having a size of. This means each image in training set and all the test images should have a size of. In the case of a grayscale input image, it should be converted into a RGB image. AlexNet's eight learned layers consists of initial five convolutional layers and the fully connected final three layers. Again final fully connected layer is connected to 1000 classes and remaining network is treated an extractor of features. The feature vector having dimension of 4096 for each image is generated using AlexNet, which has the activations of hidden layer just before the output layer.

- *Modified and Fine Tuned AlexNet:* The initial layers of AlexNet are frozen and retained as a feature extractor. Initially, for transfer learning, we substitute the last three layers by a set of fully-connected layers which can classify 36 classes to identify 36 characters of sign language. Secondly, in the AlexNet architecture after

Table 1 Alexnet architecture

Layer	Layer name	Layer type	Layer details
1	'Input 227 × 227 × 3'	Input image	Images of 227 × 227 × 3 with 'zero center' normalization
2	'Convolution1'	Convolution	96 filters of 11 × 11 × 3 kernel convolutions using stride, S = 4
3	'Relu1'	ReLU	ReLU
4	'Normalization1'	Cross channel normalization	Cross channel normalization
5	'Pooling1'	Max pooling	3 × 3 kernel max pooling using stride, S = 2
6	'Convolution2'	Convolution	256 filers of 5 × 5 × 48 kernel convolutions using stride, S = 1
7	'Relu2'	ReLU	ReLU
8	'Normalization2'	Cross channel normalization	Cross channel normalization
9	'Pooling2'	Max pooling	3 × 3 kernel max pooling using stride, S = 2
10	'Convolution3'	Convolution	384 filters of 3 × 3 × 256 kernel convolutions using stride, S = 1
11	'Relu3'	ReLU	ReLU
12	'Convolution4'	Convolution	384 filters of 3 × 3 × 192 kernel convolutions using stride, S = 1
13	'Relu4'	ReLU	ReLU
14	'Convolution5'	Convolution	256 filters of 3 × 3 × 192 kernel convolutions using stride, S = 1
15	'Relu5'	ReLU	ReLU
16	'Pooling5'	Max pooling	3 × 3 kernel max pooling using stride, S = 2
17	'Fullyconnected6'	Fully connected	4096 fully connected layer
18	'Relu6'	ReLU	ReLU
19	'Fullyconnected7'	Fully connected	4096 fully connected layer
20	'Relu7'	ReLU	ReLU
21	'Fullyconnected8'	Fully connected	1000 fully connected layer
22	'Probabilitu'	Softmax	Softmax classifier
23	'Classification layer'	Classification output	Cross entropy

Table 2 Modified and tuned alexnet

AlexNet [26]		Proposed modified and tuned AlexNet	
Layer name and type	Layer details	Layer type	Layer details
'fc7′, Fully connected	4096 fully connected layer	Fully connected	512 fully connected layer
'relu7′, relu	ReLU	ReLU	ReLU
'prob', Softmax	Softmax	Softmax	Softmax
'Classification layer', Classification output	Cross entropy	Classification output	Cross entropy

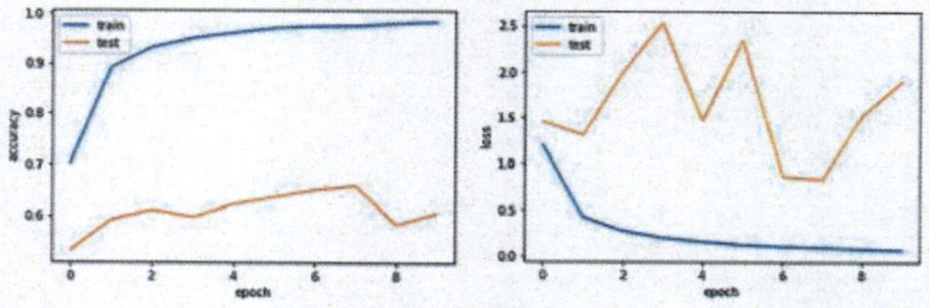

a b

Fig. 3 Training and testing performance of transfer learning on AlexNet, (a) Training and testing accuracy vs the number of epochs, (b) Loss, rate of training and testing vs number of epochs

the first 4096 fully connected layer we modified the second 4096 fully connected layer with 512 fully connected layer, succeeded by a ReLU layer shown in Table. 2. We added the Rectified Linear Unit layer to increase the non-linearity. ReLU is able to train the network many folds faster without producing any gradient vanishing effect. Finally, fully connected layer of 36 output neurons is added to facilitate classification of 36 ASL categories.

4 Experiments and Discussion

The tuned CNN architecture is trained on augmented ASL character images. The training is performed on online GPU available on Kaggle. A batch size of 16 and a total of 10 epochs are investigated for evaluating the training and testing performance of the architecture. Adaptive learning rate optimization is used as the training function. In leave one out method, the augmented ASL gesture images of four diverse individuals are used for training and, the gesture images of the fifth individual is used for testing only. This technique shows the effectiveness of the architecture, unlike the random 70–30 method. In the random 70–30 method, 70% of data is randomly chosen for training and rest 30% for testing.

Figure 3(a) demonstrations train and testing results of AlexNet network for classification of 36 ASL categories.

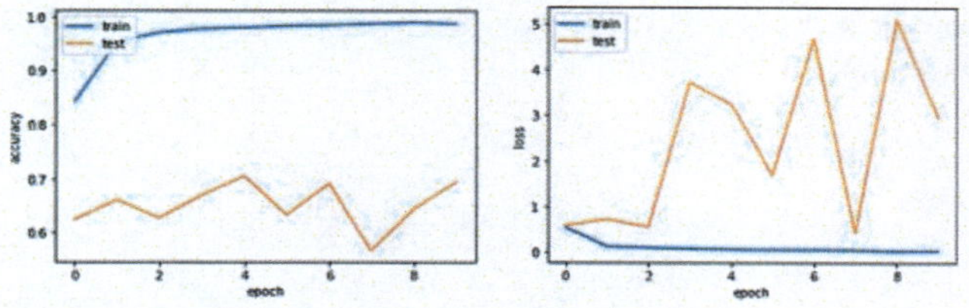

a b

Fig. 4 Training and testing performance of transfer learning on modified AlexNet (a) Testing and training accuracy vs epoch, (b) Loss rate vs Number of epochs

Table 3 Testing accuracy comparison for upto 10 epochs

Epoch	Accuracy	
	AlexNet [26] (%)	Proposed modified AlexNet (%)
1	53.06	62.50
2	59.17	66.11
3	61.11	62.71
4	59.72	66.94
5	62.22	70.56
6	63.61	63.33
7	65.00	69.17
8	65.83	56.67
9	58.33	64.44
10	60.56	69.44

Figure 3(b) shows loss rate of the network for both the training and testing for 10 epochs. The training loss is continuously decreasing and the testing loss is lowest for the 7th epoch. From Fig. 3(a) it can be seen that at epoch 7 maximum training accuracy is achieved.

Secondly, the results of transfer learning after modification of AlexNet in the latter layers are shown in Fig. 4. Here we achieved an increase of 5% in testing performance from the original AlexNet model. From Fig. 4(a) it can be seen that maximum testing accuracy is achieved at epoch 5.

The loss rate of the modified architecture for each epochs is shown in Fig. 4(b). The proposed network has a very trivial loss rate at epoch 5. So, at epoch 5 maximum testing accuracy is achieved.

From Table 3 the testing accuracy for each epoch of the proposed modified AlexNet architecture can be observed to be greater, except epoch 6 and 8. At epoch 5 the modified architecture shows an accuracy of 70.56%. At the same time the maximum training accuracy for the existing AlexNet architecture is 65.83%. Figure 5 shows a graphical representation of Table 3.

From Fig. 5 it can be seen that, the modified AlexNet outperforms the original AlexNet at each epochs.

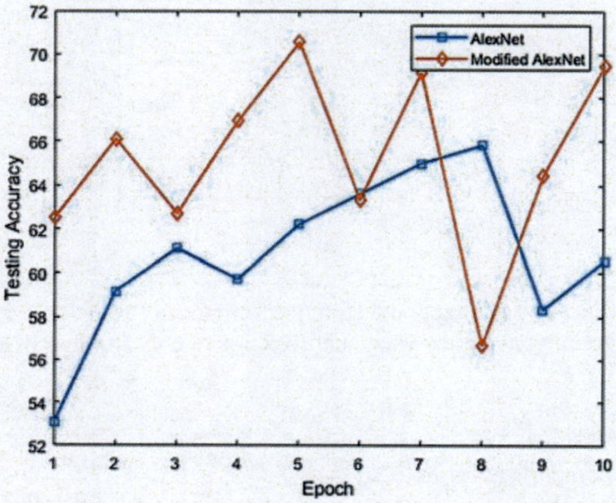

Fig. 5 Comparison of training accuracy along 10 epochs

Table 4 Number of correct predictions of asl characters out of 10 characters

Characters	Correct predictions (AlexNet [26])	Correct predictions (proposed and tuned AlexNet)
a	5	6
b	9	9
c	7	8
d	4	4
e	7	7
Eight	8	9
f	9	10
Five	4	4
Four	10	10
g	7	9
h	9	9
i	6	6
j	8	9
k	4	4
l	9	10
m	1	1
n	8	8
Nine	9	9

(continued)

Table 4 (continued)

Characters	Correct predictions (AlexNet [26])	Correct predictions (proposed and tuned AlexNet)
o	2	2
1	3	3
p	10	10
q	9	10
r	6	7
s	9	9
Seven	10	10
Six	5	5
t	8	9
Three	7	7
Two	6	7
u	10	10
v	6	7
w	3	3
x	7	8
y	2	2
z	1	1
Zero	9	10

Table 4 shows in each gesture classes, some of the gestures performs poorly in classification. The gestures images for characters ‘d’, ‘five’, ‘k’, ‘m’, ‘o’, ‘one’, ‘w’, ‘y’ and ‘z’ doesn’t classify well. The remaining characters performs better with the pro-posed modified ALexNet than the original AlexNet architecture. An overall accuracy of 71% is achieved with the proposed AlexNet architecture in testing. A 5% advance in accuracy from the existing AlexNet Framework is attained.

5 Conclusion

This work aims to recognize ASL gesture classes utilizing the modified AlexNet framework. We use leave one out method for train test split, where gesture images of four diverse individuals are used for training and fifth individual images are solitarily used for only testing. The training data has 540 ASL sign images of each gesture class and a total number of 360 images are used for testing. The performance for both the original and modified AlexNet architectures are evaluated on the leave one out data. It is observed that by modifying fully connected layers of the original AlexNet architecture we achieved an accuracy of 71%, an increase of 5% over the original architecture. Though evaluating the model in random train-test split method gives higher percentage accuracy, but leave one out method is better for assessment because it is tested on a formerly unseen individual's hand gesture. Thus a model having better performance in leave one out method will have a better performance in real-world scenario.

AlexNet has a better balance between speed and accuracy but for betterment recognition accuracy, more complex deep architectures may be implemented and performance evaluation should be done using leave one out method. Leave one out method of splitting data should be the testing criteria for a superior verdict on the model's competency.

References

1. National Research Council, How People Learn: Brain, Mind, Experience, and School: Expanded Edition, The National Academies Press, Washington, DC, USA (2000)
2. Hasan HS, Abdul Kareem S (2012) Human computer interaction for vision based hand gesture recognition: a survey. In: 2012 International Conference on Advanced Computer Science Applications and Technologies (ACSAT), pp 55–60. IEEE,
3. Manresa C, Varona J, Mas R, Perales F (2000) Real-time hand tracking and gesture recognition for human– computer interaction. Computer Vision Centre, University Autonomic, Barcelona
4. Hasan, HS, Abdul Kareem S (2012) Human computer interaction for vision based hand gesture recognition: a survey. In: 2012 International Conference on Advanced Computer Science Applications and Technologies (ACSAT), pp 55–60. IEEE
5. Pavlovic V, Sharma R, Huang TS (1997) Visual interpretation of hand gestures for human-computer interaction: A review. IEEE Trans Pattern Anal Mach Intell 1(7):677–95
6. Jadooki S, Mohamad D, Saba T, Almazyad AS, Rehman A (2017) Fused features mining for depth-based hand gesture recognition to classify blind human communication. Neural Comput Appl 28(11):3285–3294
7. Kılıboz NÇ, Güdükbay U (2015) A hand gesture recognition technique for human–computer interaction. J Vis Commun Image Represent 1(28):97–104
8. Kruger N, Janssen P, Kalkan S, Lappe M, Leonardis A, Piater J, Rodriguez-Sanchez AJ, Wiskott L (2013) Deep hierarchies in the primate visual cortex: What can we learn for computer vision? IEEE Trans Pattern Anal Mach Intell 35(8):1847–1871
9. Plouffe G, Cretu AM (2016) Static and dynamic hand gesture recognition in depth data using dynamic time warping. IEEE Trans Instrument Meas 65(2):305–316
10. Zhou Y, Jiang G, Lin Y (2016) A novel finger and hand pose estimation technique for real-time hand gesture recognition. Pattern Recogn 1(49):102–114
11. Wang C, Liu Z, Chan SC (2015) Superpixel-based hand gesture recognition with kinect depth camera. IEEE Trans Multimedia 17(1):29–39
12. Hasan H, Abdul-Kareem S (2014) Retracted article: static hand gesture recognition using neural networks. Artif Intell Rev. 41(2):147–81
13. Li SZ, Yu B, Wu W, Su SZ, Ji RR (2015) Feature learning based on SAE–PCA network for human gesture recognition in RGBD images. Neurocomputing. 5(151):565–573
14. Zhaojie J, Gao D, Cao J, Liu H (2016) A novel approach to extract hand gesture feature in depth images. Multimedia Tools Appl 75(19):11929–11943
15. Zhu G, Zhang L, Shen P, Song J (2017) Multimodal gesture recognition using 3-D convolution and convolutional LSTM. IEEE Access 17(5):4517–4524
16. Gupta B, Shukla P, Mittal A (2016) K-nearest correlated neighbor classification for Indian sign language gesture recognition using feature fusion. In: 2016 International Conference on Computer Communication and Informatics (ICCCI), pp 1–5. IEEE

17. Kumar P, Gauba H, Roy PP, Dogra DP (2017) Coupled HMM-based multi-sensor data fusion for sign language recognition. Pattern Recogn Lett 15(86):1–8
18. Oyedotun OK, Khashman A (2017) Deep learning in vision-based static hand gesture recognition. Neural Comput Appl 28(12):3941–3951
19. Nagarajan S, Subashini TS (2013) Static hand gesture recognition for sign language alphabets using edge oriented histogram and multi class SVM. Int J Comput Appl 82(4):28–35. https://doi.org/10.5120/14106-2145
20. Feng B, He F, Wang X, Yongjiang W, Wang H, Yi S, Liu W (2017) Depth-projection-map-based bag of contour fragments for robust hand gesture recognition. IEEE Trans Hum-Mach Syst 47(4):511–523
21. Ren Y, Xie X, Li G, Wang Z (2018) Hand gesture recognition with multiscale weighted histogram of contour direction normalization for wearable applications. IEEE Trans Circ Syst Video Technol 28(2):364–377
22. Li Y, Wang X, Liu W, Feng B (2018) Deep attention network for joint hand gesture localization and recognition using static RGB-D images. Inf Sci 1(441):66–78
23. Zheng S, Song Y, Leung T, Goodfellow I (2016) Improving the robustness of deep neural networks via stability training. In: Proceedings of the IEEE Conference on Computer Vision and Pattern Recognition, pp 4480–4488
24. Rocco I, Arandjelovic R, Sivic J (2017) Convolutional neural network architecture for geometric matching. In: Proceedings of the IEEE Conference on Computer Vision and Pattern Recognition, pp 6148–6157
25. Vargas AC, Paes A, Vasconcelos CN (2016) Um estudo sobre redes neurais convolucionais e sua aplicação em detecção de pedestres. In: Proceedings of the XXIX Conference on Graphics, Patterns and Images, vol 1, no 4
26. Krizhevsky A, Sutskever I, Hinton GE (2012) Imagenet classification with deep convolutional neural networks. In: Advances in Neural Information Processing Systems, pp 1097–1105
27. Russakovsky O, Deng J, Hao S, Krause J, Satheesh S, Ma S, Huang Z, Karpathy A, Khosla A, Bernstein M, Berg AC, Fei-Fei L (2015) Imagenet large scale visual recognition challenge. Int J Comput Vis 115(3):211–252
28. Barczak ALC, Reyes NH, Abastillas M, Piccio M, Susnjak T (2011) A new 2D static hand gesture colour image dataset for ASL gestures

Image Compression Based on DCT and Adaptive Grid Scanning

Venkatateja Jetti and Ram Kumar Karsh

Abstract A lot of attentions has been paid by the researchers for lossy image compression. Due to the high uses of multimedia, a high image compression ratio while keeping good image quality, is still a research issue. In this paper, we proposed an image compression using adaptive scanning, in which DWT(discrete wavelet transform) is performed to source image, DCT(discrete cosine transform) is performed to obtained DWT elements. Next, non zero quantized DCT (discrete cosine transform) coefficients are represented in less number of bits. First, the source image is divided into 16×16 grids and then DWT transformed, later the obtained elements are divided into 8×8 grids and then DCT is performed, quantized, differential encoded and an indexed vector is formed from the quantized matrix using adaptive scanning. Furthermore the indexed vector is divided into two indexed vectors namely $nonzero$ vector and *zerocount* vector. The nonzero coefficients are stored in $nonzero$ vector and the number of zeros preceding a nonzero coefficient is stored in *zerocount* vector. A new block is introduced to store the number of bits required to store highest value of nonzero coefficient in $nonzero$ vector so that all the nonzero elements can be represented in that many bits rather than a fixed number of bits. The experimental results show that the proposed method has achieved more compression ratio with slight reduction in image quality as compared to existing recent methods.

Keywords Lossy Color image compression · DWT · DCT · Adaptive grid scanning · Peak signal to noise ratio · Compression ratio

1 Introduction

In multimedia application, there is requirement huge amount of data. Using compression, actual amount of data can be represented into smaller quantity yields less storage and efficient transmission. Actual data can be compressed via removing redundant information.

V. Jetti (✉) · R. K. Karsh
National Institute of Technology Silchar, Assam 788010, India

V. Nath and J. K. Mandal (eds.), *Proceeding of Fifth International Conference on Microelectronics, Computing and Communication Systems*, Lecture Notes in Electrical Engineering 748,
https://doi.org/10.1007/978-981-16-0275-7_8

Image compression approaches may be typically categorized into two types such as lossless and lossy. In lossless, the actual image has to be reconstructed without loss in information. This approach is mainly used in medical diagnosis. In some applications, loss of some information can be tolerated. This type of compression is called lossy, which is typically applied in television applications. The lossy image compression is widely studied and is typically classified [1] as direct and indirect methods. In first one, image samples are directly controlled as discussed in Vector Quantization (VQ) [2] and Block Truncation coding (BTC) [3] based methods. In later one, generally the mathematical transformations are used to map energy of image into lesser number of coefficients. Some of the transform based methods are Discrete Wavelet Transform (DWT) [4–6], Discrete Cosine Transform (DCT) [7–10], and Principal component analysis (PCA) [11].

DCT based approaches are computationally efficient, however a small loss in signal to noise ratio is observed compared with DWT [7, 12]. It has been observed from the experimental results [4] that the important factors in the image encoding is entropy coder and quantizer, irrespective of difference between DCT and wavelet transform. This is main factor, the existing video and image coding standards favor DCT above DWT [12, 13]. For many image and video processing applications DCT is the basic building block. So DCT finds its application in many image and video processing standards like JPEG [14–16].

In this paper, lossy image compression based on hybrid transform i.e. both DWT-DCT with novel encoder is proposed. First, the source image is divided into 16 × 16 blocks. Next, DWT is performed, the elements are divided into 8 × 8 blocks. Later, DCT, quantization, and thresholding are performed successively. Finally, using the proposed encoder adaptive block scanning is perform to obtain the compressed image.

The remaining of the manuscript are arranged as follows. Section 2 reveals the steps of lossy image compression method. In Sect. 3, the proposed encoder is presented. Section 4 deals with experimental results and their observations. Finally, the conclusion and future works are drawn in Sect. 5.

2 Steps of Lossy Image Compression Methods:

The compression and decompression phases are shown in Fig. 1 and the steps are as follows:

Compression phase:

- Reading the image.
- Color conversion of image from RGB to YCbCr.
- Segregating pixels into small 16 × 16 matrices.
- Discrete Wavelet Transform of segregated 16 × 16 matrices.
- Segregating pixels into small 8 × 8 matrices.
- Discrete Cosine Transform of Segregated 8 × 8 matrices.

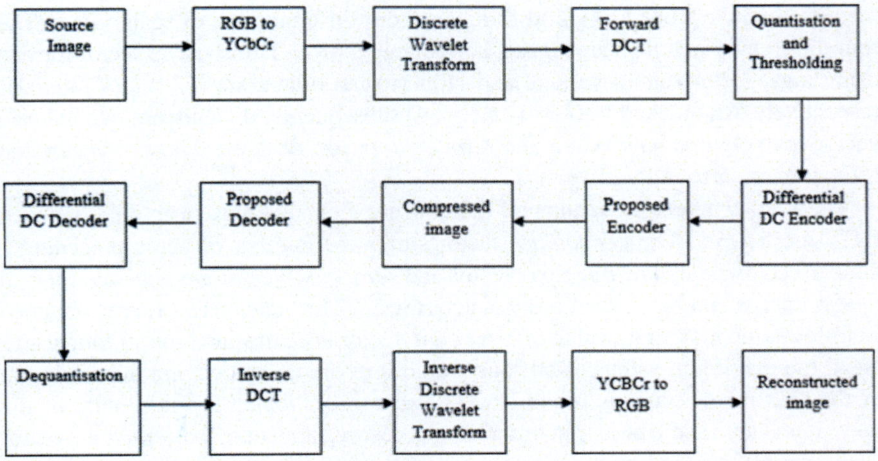

Fig. 1 Block diagram showing the steps of compression and decompression

- Thresholding and Quantization of obtained DCT matrices.
- Adaptive block scanning using following scans:
  1. Zigzag scan
  2. Horizontal scan
  3. Vertical scan
  4. Hilbert scan
- Determining the $zerocount_i$, $nonzero$ vectors.
- Using the vectors generate a bit stream (i.e., compressed image).

Decompression phase:

- Reconstructing the quantized matrix from the bit stream.
- Dequantizing the reconstructed quantized matrices.
- Inverse Discrete Cosine Transform of the obtained dequantized matrices.
- Inverse Discrete Wavelet Transform of the constructed DCT matrices.
- Color conversion from YCbCr to RGB.
- Reconstruct the source image.

3 Proposed Encoder

The proposed approach used for encoding is simple and efficient. In this method, after finding the DWT of the source image, perform DCT to the LL part and discard remaining. Later thresholding and quantization is done yields quantized DCT coefficients. Next the whole image is divided into 8×8 grids, then the differential encoding is done for every 8×8 grid. The differential encoding is discussed in the Sect. 3.1. Further, for every grid an indexed vector V_i ($i = 1, 2, \ldots, 4$) is generated from the differential encoded 8*8 grid following four different type of scan orders. The four different scan orders are zigzag, horizontal, vertical, Hilbert. The vector formed using zigzag, horizontal, vertical and Hilbert scan orders are V_1, V_2, V_3 and V_4 respectively. All indexed V_i $(i = 1, 2, \ldots, 4)$ subsequently divided into two indexed vectors namely *nonzero* vector and *zerocount* vector. Both the *nonzero* vector and *zerocount* are also indexed vectors.

The *nonzero* vector contains all the nonzero elements present in the quantized DCT coefficients. The *zerocount* vector stores the number of zeros preceding a nonzero coefficient. The *nonzero* vector and *zerocount* vector are generated for all the four scans and better scan order is determined. The better scan order is obtained as follows: the maximum value of *zerocount* vector is determined for all four scans. Next, the maximum values of all four scan orders are compared and the minimum of the maximum values is found out. The scan order which has minimum of the maximum value is the best scan order because every element in the *zerocount* vector can be represented in binary in least number of bits compared to other scans. Also a variable AS of 2 bits is used to store which scan order is used while encoding so that it can be successfully decoded in the decompression phase. The AS value is stored as 00,01,10,11 for zigzag scan, horizontal scan, vertical scan, hilbert scan respectively. Further for the best scan order, the maximum value of nonzero element in *nonzero* vector is found out and the minimum number of bits required to store that value is stored in third block in Table 1. As depicted in Table 1, for every 8 × 8 grid all the six blocks are constructed orderly. For every grid we get six blocks while encoded. The generation of bit stream is discussed as follows.

- The first block is termed as nonzero count and it stores the number of nonzero elements in the grid.
- The second block is termed AS block and it stores the information of which scan is stored. 00 for zigzag scan, 01 for horizontal scan, 10 for vertical scan, and 11 for Hilbert scan.
- The third block stores the least number of bits required to store the highest pixel value in each grid. If we can represent the highest pixel value in k bits, then we can represent every pixel in k bits. Let the value be *a*.
- The fourth block stores all the elements of best scan's *nonzero* vector in *a* bits.

Table 1 Construction of bit stream

Number of nonzero elements in each 8 × 8 grid	The selected scan (AS) represented in binary	Value of '*a*' represented in binary	Nonzero elements each represented in '*a*' bits	Value of '*b*' represented in binary	Zero count vectors elements each represented in '*b*' bits
P bits	2 bits	3 bits	m* *a* bits	3 bits	m**b* bits

'*a*': least number of bits required to represent highest pixel value in *nonzero* vector
'*b*': least number of bits required to represent highest value in *zerocount* vector
'm': number of nonzero elements in each 8 × 8 grid

- The fifth block stores the least number of bits required to store the highest value in $zerocount_i$ vector for selected scan. If we can represent the highest value in k bits, then we can represent every pixel in k bits. Let the value be b.
- The sixth block stores all the elements of best scan's *zerocount* vector in b bits.

3.1 Differential DC Coding

The DC coefficient is a measure of the average value of the all pixels within the block. The DC coefficient is the 1×1 element in the 8×8 grid. Generally, DC coefficient contains an important fraction of the total energy image and for this, the DC coefficient value will be higher in every grid and will be the largest pixel value in the grid. So to reduce number of bits required to represent the DC coefficient, its value will be modified to difference of original DC value and the DC value of first grid. So the DC value for every grid will be less after the differential encoding and number of bits required to represent elements will become less and better compression can be obtained. Differential decoding is the opposite of differential encoding, the DC coefficient in every grid is summed with the DC coefficient of first grid to obtain its original value.

3.2 Illustrative Example for Compression

In order to clarify the idea, for example, the 16×16 intensities matrix B depicted from arbitrary image of size 512×512.

$$B = \begin{bmatrix}
80 & 80 & 80 & 80 & 80 & 80 & 79 & 79 & 81 & 81 & 78 & 78 & 81 & 81 & 80 & 80 \\
80 & 80 & 80 & 80 & 80 & 80 & 79 & 79 & 81 & 81 & 78 & 78 & 81 & 81 & 80 & 80 \\
81 & 81 & 80 & 80 & 80 & 80 & 80 & 80 & 81 & 81 & 78 & 78 & 81 & 81 & 80 & 80 \\
81 & 81 & 80 & 80 & 80 & 80 & 80 & 80 & 81 & 81 & 78 & 78 & 81 & 81 & 80 & 80 \\
80 & 80 & 81 & 81 & 80 & 80 & 79 & 79 & 81 & 80 & 79 & 78 & 81 & 80 & 80 & 80 \\
80 & 80 & 81 & 81 & 80 & 80 & 79 & 79 & 80 & 81 & 79 & 78 & 81 & 80 & 80 & 80 \\
81 & 81 & 81 & 81 & 79 & 79 & 79 & 79 & 81 & 80 & 79 & 78 & 81 & 81 & 80 & 81 \\
81 & 81 & 81 & 81 & 79 & 79 & 79 & 79 & 80 & 81 & 79 & 78 & 80 & 80 & 81 & 80 \\
80 & 80 & 80 & 81 & 80 & 79 & 81 & 78 & 81 & 81 & 78 & 79 & 81 & 81 & 81 & 80 \\
80 & 80 & 80 & 81 & 80 & 78 & 81 & 78 & 80 & 80 & 79 & 78 & 80 & 80 & 80 & 81 \\
81 & 81 & 81 & 81 & 80 & 81 & 81 & 80 & 80 & 81 & 80 & 81 & 79 & 80 & 80 & 79 \\
81 & 81 & 81 & 81 & 81 & 80 & 79 & 77 & 79 & 77 & 79 & 80 & 80 & 79 & 79 & 80 \\
80 & 80 & 79 & 77 & 81 & 81 & 79 & 79 & 80 & 81 & 79 & 81 & 80 & 81 & 79 & 80 \\
80 & 81 & 79 & 77 & 81 & 80 & 79 & 78 & 79 & 80 & 80 & 79 & 81 & 80 & 81 & 79 \\
79 & 79 & 79 & 78 & 78 & 79 & 79 & 78 & 79 & 79 & 80 & 78 & 81 & 80 & 80 & 79 \\
79 & 81 & 78 & 79 & 79 & 78 & 78 & 79 & 80 & 81 & 80 & 77 & 79 & 81 & 79 & 79
\end{bmatrix}$$

After the DCT transform of B, the obtained matrix is B_{dwt}

$$B_{dwt} = \begin{bmatrix} 160 & 160 & 160 & 159 & 161 & 156 & 161 & 159 \\ 161 & 160 & 159 & 159 & 161 & 155 & 161 & 159 \\ 160 & 161 & 160 & 158 & 161 & 156 & 162 & 159 \\ 161 & 161 & 159 & 159 & 161 & 156 & 161 & 159 \\ 160 & 160 & 160 & 159 & 161 & 156 & 162 & 159 \\ 162 & 162 & 157 & 154 & 159 & 159 & 157 & 159 \\ 159 & 159 & 162 & 157 & 159 & 161 & 158 & 155 \\ 157 & 157 & 155 & 155 & 157 & 158 & 155 & 157 \end{bmatrix}$$

After the DCT transform of B_{dwt}, the obtained matrix is B_{dct}

$$B_{dct} = \begin{bmatrix} 1271.5 & 3.6 & 2.8 & 1.3 & 0 & -1.5 & -4.1 & 5.8 \\ 5.8 & -0.1 & 0.5 & -3.1 & 2.4 & 2.4 & -4.9 & 3.2 \\ -4.4 & 0.5 & -1.5 & 0.5 & -0.3 & 0.3 & 2.5 & -1.5 \\ 2.1 & 1.0 & 0.8 & 0.7 & -0.5 & -0.7 & 0.8 & 0.2 \\ -1.2 & -1.5 & -0.5 & -1.2 & 1.7 & 0.9 & -1.3 & -0.2 \\ 1.8 & 1.0 & -1.7 & 0.2 & -2.2 & 0.3 & 1.3 & 0.2 \\ -1.6 & -0.5 & 2.7 & 0.9 & 0.4 & -2.5 & -0.1 & -0.9 \\ 1.5 & -0.7 & -2.5 & -1.5 & -1.7 & 2.0 & 0 & -1.5 \end{bmatrix}$$

The matrix used for quantizing the obtained DCT matrix B_{dct} is

$$\text{Quantiser matrix} = \begin{bmatrix} 16 & 11 & 10 & 16 & 24 & 40 & 51 & 61 \\ 12 & 12 & 14 & 19 & 26 & 58 & 60 & 55 \\ 14 & 13 & 16 & 24 & 40 & 57 & 69 & 56 \\ 14 & 17 & 22 & 29 & 51 & 87 & 80 & 62 \\ 18 & 22 & 37 & 56 & 68 & 109 & 103 & 77 \\ 24 & 35 & 55 & 64 & 81 & 104 & 113 & 92 \\ 49 & 64 & 78 & 87 & 103 & 121 & 120 & 101 \\ 72 & 92 & 95 & 98 & 112 & 100 & 103 & 99 \end{bmatrix}$$

After quantization with a quantizer matrix the obtained matrix is:

$$Q = \begin{bmatrix} 63 & -1 & -1 & 0 & 0 & 0 & 0 & 0 \\ 0 & -1 & 1 & 0 & 0 & 0 & 0 & 0 \\ -1 & 1 & 0 & 0 & 0 & 0 & 0 & 0 \\ 0 & 0 & 0 & 0 & 0 & 0 & 0 & 0 \\ 0 & 0 & 0 & 0 & 0 & 0 & 0 & 0 \\ 0 & 0 & 0 & 0 & 0 & 0 & 0 & 0 \\ 0 & 0 & 0 & 0 & 0 & 0 & 0 & 0 \\ 0 & 0 & 0 & 0 & 0 & 0 & 0 & 0 \end{bmatrix}$$

Four scans are generated by the adaptive block scanning. The best scan is the scan that gives the lowest value of maximums of the index vectors $zerocount_i$ (the last zeros sequence is not considered). The *zerocount* vector contains length of the zero-run sequence preceding a nonzero DCT quantized coefficient. The different generated scans are:

1. **Zigzag scan:**
   63 -1 0 -1 -1 -1 0 1 1 0 0 0 0 0 0 0 0 0 0 0 0 0 0 0 0 0 0 0 0 0 0 0 0 0 0 0 0 0
   0 0 0 0 0 0 0 0 0 0 0 0 0 0 0 0 0 0 0 0 0 0 0 0 0 0
   nonzero: 64 -1 -1 -1 -1 1 1
   zerocount: 0 0 1 0 0 1 0
   AS: 00

2. **Horizontal scan:**
   63 -1 -1 0 0 0 0 0 0 -1 1 0 0 0 0 0 -1 1 0 0 0 0 0 0 0 0 0 0 0 0 0 0 0 0 0 0 0 0 0
   0 0 0 0 0 0 0 0 0 0 0 0 0 0 0 0 0 0 0 0 0 0 0 0 0
   nonzero: 64 -1 -1 -1 -1 1 1
   zerocount: 0 0 0 6 0 5 0
   AS: 01

3. **Vertical scan:**
   63 0 -1 0 0 0 0 0 -1 -1 1 0 0 0 0 0 -1 1 0 0 0 0 0 0 0 0 0 0 0 0 0 0 0 0 0 0 0 0 0
   0 0 0 0 0 0 0 0 0 0 0 0 0 0 0 0 0 0 0 0 0 0 0 0 0
   nonzero: 63 -1 -1 -1 -1 1 1
   zerocount: 0 1 5 0 0 5 0
   AS: 10

4. **Hilbert scan:**
   63 -1 -1 0 -1 0 0 1 0 0 0 0 0 1 -1 0 0 0 0 0 0 0 0 0 0 0 0 0 0 0 0 0 0 0 0 0 0 0 0
   0 0 0 0 0 0 0 0 0 0 0 0 0 0 0 0 0 0 0 0 0 0 0 0 0 0
   nonzero: 63 -1 -1 -1 -1 1 1
   zerocount: 0 0 0 1 2 5 0
   AS: 11

Here the best scan is the zigzag scan as the maximum of $zerocount_i$ vector is 1 and is minimum of the all 4 scan vectors. So the encoded bit stream for this 8 × 8 grid is.

000111 00 110 11111110111111101111111011111110111111110000011000001 001 0010010

In this way bit stream is calculated for all the grids in the image matrix.

3.3 Encoding and Decoding Algorithm

Encoding algorithm:

Algorithm 1: Encoding algorithm (The algorithm used to generate the image into a bitstream)

Input: F (Source image gone through quantisation, thresholding, differential encoding followed by the Discrete wavelet transformation and Discrete Cosine transformation, size $M \times N$

Output: Compressed image in the form of bitstream.

1: The image is divided into 16×16 grids and every 16×16 grid is Discrete Wavelet transformed and it forms a 8×8 grid.

2. After the Discrete Wavelet Transformed image is obtained, it is quantized, thresholded and differential encoded followed by Discrete cosine transform, the image is divided into 8×8 grid and each grid is read.

3: Perform the scan operations (zigzag, horizontal, vertical and hilbert) and determine the indexed vector V_i for all the four scan orders.

4: From the obtained vectors V_i determine the *nonzero* vector and *zerocount* vector for all the four scans.

5: The *nonzero* vector stores the non zero elements in respective scan order. The *zerocount* vector for each scan order stores number of zeros before a nonzero coefficient for respective scan order.

6: $a = 0$;

7: f=0;

8: ***for*** b =1 **to** length(V_i) **do**

9: **If** V (b) !=0 **then**

10: *nonzero* (a) = V (b);

11: $zerocount_i\,(a)$ = f;

12: f=0;

13: $a = a + 1$;

14: Else

15: *f=f+1;*

16: ***endif***

17: ***Endfor***

18: Finding AS the index of the best scan
AS = "00" if $zerocount_1$ is the min(max($zerocount_1$), · · · , max($zerocount_4$))
AS = "01" if $zerocount_2$ is the min(max($zerocount_1$), · · · , max($zerocount_4$))
AS = "10" if $zerocount_3$ is the min(max($zerocount_1$), · · · , max($zerocount_4$))
AS = "11" if $zerocount_4$ is the min(max($zerocount_1$), · · · , max($zerocount_4$))

19: Find the value of 'a' and 'b' where 'a' : least number of bits required to represent highest pixel value in *nonzero* vector
'b' : least number of bits required to represent highest value in *zerocount* vector and form the bitstream as shown in Table 1.

Decoding algorithm: To determine the DCT vector.

Algorithm 2: Decoding algorithm (The algorithm used to reconstruct the compressed image from obtained bitstream)

Input: Bitstream of a compressed image
Output: The DCT coefficients are reconstructed and is obtained as a matrix.

```
Get AS bit, nonzero, zerocount vectors, value of a, b from the bitstream.
Determine the scan order used from AS bits
Initialize a new block to store decoded quantised (DQ) values and initiate each value to zero
DQ=0;
for b =1 to length(nonzero ) do
a = a + zerocount(b);
DQ(a)=nonzero(b);
End for
Construct a Matrix from the reconstructed DQ vector doing the inverse of respective scan order, the differential decoding is performed.
Thus from the reconstructed Quantised matrix, the DCT matrix is obtained.
Inverse DCT is performed to the obtained DCT matrix and thus the Discrete wavelet Transformed matrix is obtained.
Inverse DWT is performed to the obtained DWT matrix , the reconstructed image is obtained.
```

4 Experimental Results and Discussion

To show the efficacy of the proposed compression method with the state-of-the-art methods, experiments have been conducted. In this experiment, six mostly used standard images are used. These images are peppers, Lena, Airplane, Zelda, Couple and Baboon. The images are of size 512×512 each. The five test images are compressed with the proposed method and results are compared with the previous works. The (Bits per pixel)Bpp and $PSNR$ (Peak signal to noise ratio) of the reconstructed images are taken for the comparison.

Let the source image be $f(x, y)$ and the reconstructed image is $f'(x, y)$.

Error between two images:

$$e(x, y) = f'(x, y) - f(x, y) \tag{1}$$

Total error between two images is:

$$\sum_{x=0}^{M-1} \sum_{y=0}^{N-1} [f'(x, y) - f(x, y)] \tag{2}$$

Rms error is given by:

$$e_{rms} = [1/MN \sum_{x=0}^{M-1} \sum_{y=0}^{N-1} [f'(x, y) - f(x, y)]^2]^{\frac{1}{2}} \quad (3)$$

Peak signal to noise ratio ($PSNR$) is given by:

$$PSNR = 20 \log_{10} \frac{255}{MSE} \quad (4)$$

where MSE is mean squared error e_{rms}.

Also to determine the compression ratio, we use a term known as bits per pixel (Bpp). Bpp is the number of bits required to store one pixel.

$$Bpp = \frac{Total, number\ of\ bits\ in\ Bitstream}{number\ of\ pixels\ in\ image} \quad (5)$$

The compression is performed on the six test images and the results are compared with the [23] method and JPEG method [24] as shown in the Table 2. For the test image peppers, the $PSNR$ was better in the JPEG [24] method which is 35.63 where as in the proposed method it was 26.73, but the Bpp was better in the proposed method which is 0.43 whereas the Bpp in the JPEG [24] method was 3.50 which is too high when compared with proposed method's Bpp. When we compare the average results of six test images the $PSNR$ was still better in JPEG [24] method with value of 35.97 whereas the $PSNR$ of proposed method is 27.44, but the Bpp was still better in proposed method with a value of 0.41 whereas the bpp in the JPEG [24] method is 3.81 which is still higher than the proposed method's bpp.

The results of proposed method are compared with state-of-the-art methods [23, 24]. In all the cases even though the $PSNR$ is slightly lesser in the proposed method, the Bpp is far better and it compensates the little loss in image quality. All the experiments were perfomed with the Matlab (R2018 a) on a Intel® core™ I5-5200U CPU @2.20 GHz with a RAM of 8 GB.

Table 2 Performance comparison between the proposed method, [23] method and JPEG[24] algorithms

Image	Proposed method		[23]		JPEG[24]	
	$PSNR$	Bpp	$PSNR$	Bpp	$PSNR$	Bpp
Peppers	26.73	0.43	31.60	0.90	35.63	3.50
Lena	28.84	0.38	31.55	0.73	34.39	3.76
Airplane	27.13	0.39	32.17	0.64	35.63	3.50
Zelda	32.73	0.30	32.24	0.72	37.94	3.28
Couple	27.57	0.39	33.65	0.58	40.31	3.60
Baboon	21.68	0.57	29.79	0.80	31.93	5.25
Average	27.44	0.41	31.83	0.72	35.97	3.81

5 Conclusion and Future Work

In this paper, an image compression techniques is proposed using adaptive scanning, in which the non-zero quantized DCT coefficients are represented in less number of bits. In the proposed method,the source image is divided into 16×16 grids and then DWT transformed, later the obtained elements are divided into 8×8 grids and then DCT is performed, quantized, differential encoded and an indexed vector is formed from the quantized matrix using adaptive scanning. Furthermore the indexed vector is divided into two indexed vectors namely $nonzero$ vector and *zerocount* vector. The nonzero coefficients are stored in $nonzero$ vector and the number of zeros preceding a nonzero coefficient is stored in *zerocount* vector. A new block is introduced to store the number of bits required to store highest value of nonzero coefficient in $nonzero$ vector so that all the nonzero elements can be represented in that many bits rather than a fixed number of bits. The experimental results show that the proposed method gives high compression ratio and better than the state-of-the-art methods with slight reduction in image quality.

In the future work, some other transformation approaches may be explored for further improvement in compression ratio.

References

1. Khalid S (2006) Introduction to Data Compression, 4th edn. Elsevier, San Francisco
2. Feng YS, Nasrabadi NM (1991) Dynamic address-vector quantisation of RGB colour images. IEE Proc. I Commun. Speech Vision 138(4):225–231
3. Kurita T, Otsu N (1993) A method of block truncation coding for color image compression. IEEE Trans Commun 41(9):1270–1274
4. Xiong Z, Ramchandran K, Orchard MT, Zhang YQ (1999) A comparative study of DCT and wavelet-based image coding. IEEE Trans Circuits Syst Video Technol 9(5):692–695
5. Pearlman W, Islam A, Nagaraj N, Said A (2004) Efficient, low complexity image coding with a set partitioning embedded block coder. IEEE Trans Circuits Syst Video Technol 14(11):1219–1235
6. Shoitan R, Nossair Z, Isamil I, Tobal A (2017) Hybrid wavelet measurement matrices for improving compressive imaging. Signal Image Video Process. 11(1):65–72
7. Wallace GK (1992) The JPEG still picture compression standard. IEEE Trans Consum Electron 38(1):18–34
8. Dagher I (2010) Highly-compacted DCT coefficients. Signal Image Video Process. 4(3):303–307
9. Ponomarenko N, Lukin V, Egiazarian K, Astola J (2005) DCT based high quality image compression. In: Scandinavian Conference on Image Analysis, pp 1177–1185. Springer, Berlin
10. Ponomarenko N, Lukin V, Egiazarian K, Astola J (2008) ADCTC: Advanced DCT-Based Image Coder. In: Proceedings of LNLA, Switzerland
11. Clausen C, Wechsler H (2000) Color image compression using PCA and back propagation learning. Pattern Recognit 33(9):1555–1560
12. Ohm JR, Sullivan GJ, Schwarz H, Tan TK, Wiegand T (2012) Comparison of the coding efficiency of video coding standards including high efficiency video coding (HEVC). IEEE Trans Circuits Syst Video Technol 22(12):1669–1684

13. Sullivan GJ, Ohm JR, Han WJ, Wiegand T (2012) Overview of the high efficiency video coding (HEVC) standard. IEEE Trans Circuits Syst Video Technol 22(12):1649–1668
14. Bauermann I, Steinbach E (2004) Further lossless compression of JPEG images. Picture Coding Symposium. In: Proceedings of PCS San Francisco, 15–17 December 2004
15. Ponomarenko N, Egiazarian K, Lukin V, Astola J (2005) Additional lossless compression of JPEG images. In : Proceedings of 4th Symposium on Image and Signal Processing and Analysis, Zagreb, Croatia, September, pp 117–120
16. Silveira TLTD, Oliveira RS, Bayer FM, Cintra RJ, Madanayake A (2017) Multiplierless 16-point DCT a pproximation for low-complexity image and video coding. Signal Image Video Process 11(2):227–233
17. Messaoudi A, Benchabane F, Srairi K () DCT-based color image compression algorithm using adaptive block scanning (2019)
18. Rahul K, Tiwari AK (2018) Saliency enabled compression in JPEG framework. IET Image Process 12(7):1142–1149. https://doi.org/10.1049/iet-ipr.2017.0554
19. Dhara BC, Chanda B (2007) Color image compression based on block truncation coding using pattern fitting principle. Pattern Recognit 40(9):2408–2417
20. Douak F, Benzid R, Benoudjit N (2011) Color image compression algorithm based on the DCT transform combined to an adaptive block scanning. AEU Int J Electron Commun 65(1):16–26
21. Boucetta A, Melkemi KE (2012) DWT based-approach for color image compression using genetic algorithm. In: Elmoataz A, Mammass D, Lezoray O, Nouboud F, Aboutajdine D (eds) Image and Signal Processing. ICISP 2012. Lecture Notes in Computer Science, vol 7340. Springer, Heidelberg
22. Messaoudi A, Srairi K (2016) Colour image compression algorithm based on the DCT transform using difference lookup table. Electron Lett 52(20):1685–1686
23. Nigam AK, Khare P, Srivastava VK (2020) Image compression using hybrid approach and adaptive scanning for color images. In: Dutta D, Kar H, Kumar C, Bhadauria V (eds) Advances in VLSI, Communication, and Signal Processing. Lecture Notes in Electrical Engineering, vol 587. Springer, Singapore
24. Rabbani M, Joshi R (2002) Eastman kodak company, rochester, NY 14650, USA. An overview of the JPEG2000 still image compression standard. Signal Process Image Commun 17:3–48

Low Power Bandgap Reference Using Chopper Amplifier

Sujata S. Kotabagi, Chetan Hanakanahalli, and Abirmoya Santra

Abstract This paper describes design and implementation of low power Bandgap Reference (BGR) using chopper stabilized amplifier. The design focuses on reduction of power, output variation due to device mismatch and noise. First order compensated BGR with chopper amplifier is designed in CMOS-65 nm technology. Simulated across 18 process corners with temperature variation from $-30\,^{\circ}\mathbf{C}$ to $125\,^{\circ}\mathbf{C}$ along with Monte Carlo analysis, achieves $1.0012 \pm 0.56\%$ **V** and dissipates power less than 3.12 μW. Due to the implementation of chopping technique, output variation due to device mismatch is decreased by 95% (137.6 mV to 6.51mv) as compared to BGR without chopping technique. 1/f Noise reduction of 88% ($116\,\boldsymbol{\mu V/\sqrt{Hz}}$ to $14.06\,\boldsymbol{\mu V/\sqrt{Hz}}$ at 1 Hz) was observed. Curvature trimming has reduced the variation of output across Process, Voltage and Temperature (PVT) by 71% (20.3 mV to 5.79 mV).

Keywords Band gap reference · Chopper amplifier · Low power · Curvature trimming · Device mismatch

1 Introduction

Bandgap reference [4] play an important role in analog and mixed signal circuits by providing constant reference voltage. With the development of technology, low power design has become critical in analog circuits. In order to achieve low power MOSFETs are operated in weak inversion (sub-threshold) region [7] where the drain current is in nano amperes range.

S. S. Kotabagi (✉)
Department of ECE, B V B College of Engineering and Technology, Hubli, India
e-mail: sujatask@kletech.ac.in

C. Hanakanahalli
B V B College of Engineering and Technology, Hubli, India

A. Santra
Sankalp Semiconductors Pvt. Ltd., Hubli, India

V. Nath and J. K. Mandal (eds.), *Proceeding of Fifth International Conference on Microelectronics, Computing and Communication Systems*, Lecture Notes in Electrical Engineering 748,
https://doi.org/10.1007/978-981-16-0275-7_9

In mixed signal circuits, especially high resolution ADC and DAC, the accuracy and stability of BGR is very important. Temperature stability can be achived by using higher order temperature compensation but the accuracy depends on fabrication technique. In sub-micron CMOS technology, device mismatch is a major problem which degrades the accuracy of the circuit especially those which include amplifier [1]. In OPAMP based BGR, beside noise (1/f and thermal), DC offset is added to the input of OPAMP due to the device mismatch. This results in the variation of output voltage which can be analyzed using Monte Carlo (MC) analysis.

There are three techniques to remove the input offset voltage of an OPAMP: Trimming, Auto-zeroing and chopping [1, 5, 6]. Trimming is done during the fabrication to eliminate offset. Auto-zeroing and chopping are dynamic offset reduction techniques. Auto-zeroing is a sampling technique in which offset is sampled at one clock phase and then subtracted it from the input in the next clock phase. Chopping is a modulation technique. In chopping, offset is modulated to higher frequency and is eliminated using low pass filter. Both these techniques remove 1/f noise as well as offset drift over temperature. Since, chopping is continuous in time, it does not produce noise folding. Auto-zeroing technique is discrete in time and requires switched capacitor circuits which produces noise folding [2] and also chopping is more simple compared to auto-zeroing. Therefore, chopping technique is chosen over auto-zeroing technique to remove the offset of OPAMP.

2 Conventional BGR

Figure 1 represents conventional voltage mode BGR whose output voltage is sum of voltage across resistors R_2, R_3 and voltage across transistor Q2.

Collector current of a transistor [3] can be rewritten as

$$I_C = I_o e^{\frac{V_{BE}-V_{Go}}{V_T}} \tag{1}$$

Fig. 1 Conventional voltage mode BGR

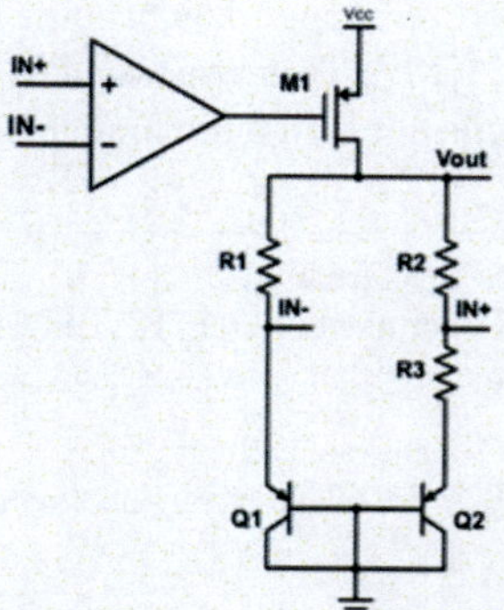

Where, I_o is reverse saturation current and V_{Go} is bandgap voltage of silicon.

$$V_{BE} = V_{Go} - \frac{KT}{q}\ln\left(\frac{I_o}{I_C}\right) \quad (2)$$

$$\frac{\partial V_{BE}}{\partial T} = -\frac{K}{q}\ln\left(\frac{I_o}{I_C}\right) \quad (3)$$

From the above equation, it can be inferred that V_{BE} is CTAT in nature.

In Fig. 1, $IN+$ and $IN-$ is connected to the input of OPAMP, therefore both the terminals are at same potentials and resistors R_1 and R_2 are equal in value. As a result, equal currents flow through both the branches.

The voltage across resistor R_3 is given as,

$$V_{R3} = V_{BE1} - V_{BE2} = \Delta V_{BE}$$

Where, V_{BE1} and V_{BE2} are voltages across transistors Q1 and Q2 respectively.

$$\Delta V_{BE} = \frac{KT}{q}\ln\left(\frac{I_{C1}I_{S2}}{I_{C2}I_{S1}}\right) \quad (4)$$

Here, $\Delta V_{BE} \propto T$

$$V_{out} = V_{BE2} + \left(1 + \frac{R_2}{R_3}\right)\Delta V_{BE} \quad (5)$$

In (6), V_{BE2} is CTAT voltage and ΔV_{BE} is PTAT voltage. By varying the ratio of R_2 to R_3 an output voltage of zero temperature coefficient is obtained.

3 Chopping Technique

In chopping technique, input signal is modulated by a square wave as shown in Fig. 2. The modulated signal V_m is given as

Fig. 2 Chopping technique

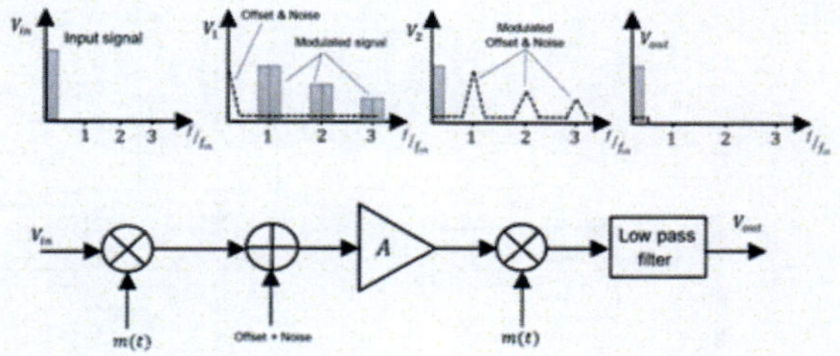

$$V_m = \sum\nolimits_{n=-\infty}^{+\infty} C_n V_{in}(f - nf_{chop}) \tag{6}$$

$$C_n = 2e^{\frac{-jn\pi}{2}} \frac{\sin\left(\frac{n\pi}{2}\right)}{\frac{n\pi}{2}} \text{ and } C_o = 0$$

Where, C_n is Fourier series coefficient of square wave. The square wave exhibits half wave symmetry; therefore, the modulated signal consists of only odd harmonics of frequency f_{chop}. To avoid signal aliasing,

$$f_{chop} > 2f_{sig} + f_c$$

Where, f_{sig} is input signal frequency and f_c is noise corner frequency.

Noise and offset get added to modulated signal in the amplifier. The output of amplifier is demodulated using square wave. In the process, signal is shifted to even harmonics and noise to odd harmonics. Amplified base band input signal is obtained at the output of low pass filter.

4 Proposed Bandgap Reference

The architecture shown in Fig. 3 consists of three blocks: (A) BGR core, (B) Chopper Amplifier, (C) Start-up circuit.

Chopper amplifier shown in Fig. 3 is a two-stage amplifier where first stage is in cascode configuration. Output of the amplifier taken at the drain of M_{10} and M_{11} is fedback through BGR core. It is a shunt-series feedback topology where current in the BGR core branch is sampled and fedback to the input of amplifier as a voltage.

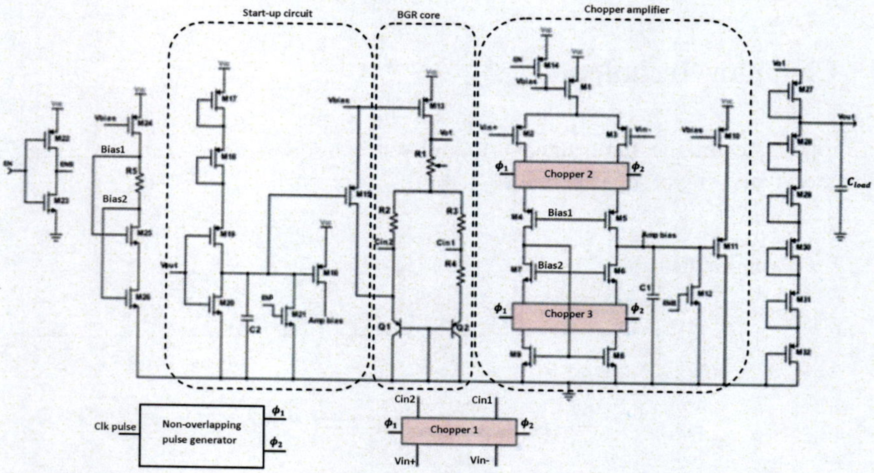

Fig. 3 BGR with chopper amplifier

Since the circuit is self-biased, it requires start-up circuit to bring the MOSFETs to proper operating point. The start-up circuit will be ON only during transient state. Once all the MOSFETs reach proper biasing point, the start-up circuit will automatically turn OFF.

4.1 BGR Core

In the BGR core block shown in Fig. 3, Q_1 and Q_2 are diode connected transistors and voltage across these transistors be V_{BE1} and V_{BE2} respectively. Terminals C_{in1} and C_{in2} are the input of chopper amplifier. The voltage and current through resistor R_4 are given as,

$$V_{R4} = V_{BE1} - V_{BE2}$$

$$I_{R4} = \frac{\Delta V_{BE}}{R_4}$$

In self-biased BGR circuit, the current of BGR core branch which is dependent on the value of resistor R_4 bias transistors M_1, M_{10} and M_{13}. Therefore, for low power design, R_4 value should be as high as possible. The output of BGR core V_{o1} is given as,

$$V_{o1} = V_{BE2} + \left(\frac{2R_1 + R_3 + R_4}{R_4}\right)\Delta V_{BE} \tag{7}$$

The linear temperature dependency of V_{BE2} and ΔV_{BE} are given by (8) and (9), which was found by performing DC analysis in simulation software.

$$V_{BE2} = K_1 - 1.9908 \times 10^{-3} \times T \tag{8}$$

$$\Delta V_{BE} = K_2 + 0.1838 \times 10^{-3} \times T \tag{9}$$

Where, K_1 and K_2 are constants.
The output can be independent of temperature by setting

$$\left(\frac{2R_1 + R_3 + R_4}{R_4}\right) = 10.83 \tag{10}$$

The voltage divider at the output provides desired voltage of 1 V.

Fig. 4 Trimming block

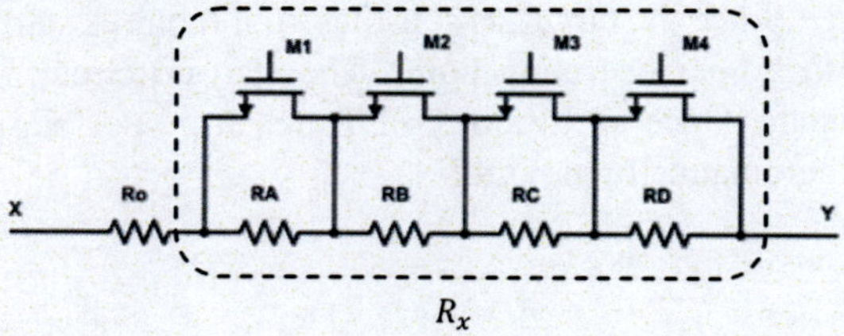

1) *Curvature trimming*

Across different process corners the output voltage is either PTAT or CTAT. Use of Curvature trimming technique reduces the temperature dependency of output voltage. To obtain the desired curvature the value of resistance R_1 of BGR core is varied.

$$R_1 = R_o + R_x$$

Where, R_o is fixed and R_x is variable resistance. In 4-bit trimming technique, R_x can take 16 different values depending on the control bits. In Fig. 4,

$$R_A = 8R;\ R_B = 4R;\ R_C = 2R;\ R_D = R$$

Where, R is the resolution of trimming. Maximum possible value of R_x in 4-bit trimming is $15 \times R$. In typical condition $R_x = 8R$ (control bits are "1000"). The value of R_x is either increased or decreased based on the nature of output curve i.e. CTAT or PTAT.

4.2 Chopper Amplifier

Input referred offset voltage V_{offset} of an amplifier occurs because of mismatch in the threshold voltage of differential pair (M_2 and M_3) and the active load (M_8 and M_9). It is given by

$$V_{offset} = \Delta V_{TH,diff} + \left(\frac{g_{m9}}{g_{m2}}\right)\Delta V_{TH,load} \tag{11}$$

Correspondingly, the overall variation in the output is given by

$$\Delta V_{out} = \left(\frac{2R_1 + R_3 + R_4}{R_4}\right)V_{offset} \tag{12}$$

Using (10),

$$\Delta V_{out} = 10.83 \times V_{offset} \tag{13}$$

Table 1 Mismatch analysis before chopping

Parameter	Mean(V)	Sigma(V)	Mean – 3*Sigma (V)	Mean + 3*Sigma(V)	Minimum (V)	Maximum (V)
Vout	1.005	0.02846	0.9196	1.09	0.945	1.0826
Input offset (OPAMP)	-0.281m	2.66m	-8.27m	7.7m	-7.18m	5.487m
Vth offset (diff pair)	-0.0455m	1.14m	-3.48m	3.388m	-3.344m	2.893m
Vth offset (Load pair)	-0.171m	2.272m	-8.339m	7.997m	-8.055m	6.239m

Fig. 5 Chopper block

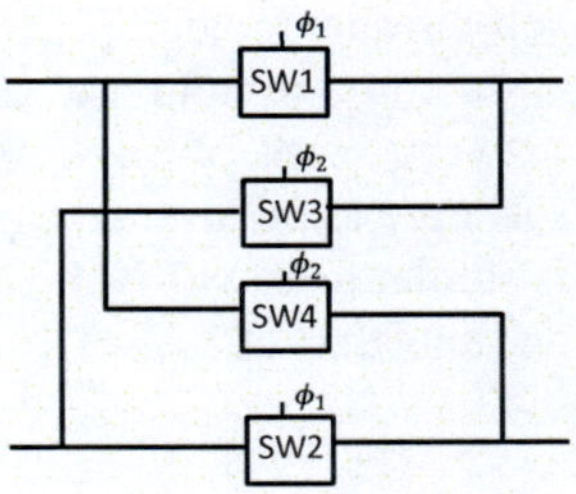

Fig. 6 Non-overlapping clocks

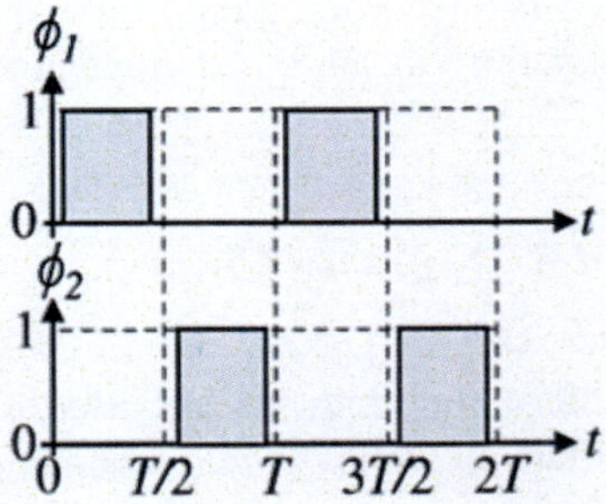

Equation (13) shows that input referred offset of an amplifier will cause more variation at the output than any other component in the circuit (refer Table 1). If input offset of an amplifier is eliminated then overall output offset will be reduced by 99%.

1) *Chopper circuit*

Figure 5 represents a chopper circuit which consists of four switches [7]. These switches are actuated by two non-overlapping clock pulses (Fig. 6) whose frequency is f_{chop}. Switches SW1 and SW2 will be ON during the first clock pulse $\phi 1$ while

SW3 and SW4 will be ON during second clock pulse $\phi 2$. Thus, chopper circuit acts as a ring modulator in which input signal is modulated to odd harmonics of f_{chop}.

In Fig. 3, "Chopper 1" modulates the input signal present at terminal Cin1 and Cin2. "Chopper 2" acts as a demodulator. It also acts as a modulator for the DC offset, $\Delta V_{TH,diff}$ caused due to mismatch in threshold voltage of differential pair. "Chopper 3" modulates the error current ($g_{m9} \cdot \Delta V_{TH,load}$) caused due to mismatch in threshold voltage of active load. The output of amplifier connected to the BGR core consists of input signal and the modulated DC offset. The BGR core along with load capacitor which acts as a low pass filter eliminates the offset.

2) *Sub-threshold operation*

Sub-threshold conduction is a dominant current leakage mechanism that occurs in MOSFET when $V_{GS} < V_{TH}$ and the current is called sub-threshold current. Magnitude of this current is very less (nanoamperes) compared to that of saturation region current (microamperes). Therefore, in low power design, MOSFETs are operated in sub-threshold region [7].

In sub-threshold, the surface potential V_S of substrate is $\phi_b < V_S < 2\phi_b$ i.e. between onset of week and strong inversion. ϕ_b is potential difference between substrate fermi level and intrinsic fermi level. The behavior of sub-threshold current is similar to that of BJT, with source and drain as emitter and collector respectively and substrate as base. The drain current I_D is of the form,

$$I_D \, \alpha \, e^{\frac{(V_{GS}-V_{TH})}{V_T}} \tag{14}$$

In this work, MOSFETs of chopper amplifier and output voltage divider are operated in sub-threshold region.

4.3 Start-up Circuit

Since, the circuit is self-biased, a start-up circuit is required to fix the operating point. Start-up circuit should automatically turn OFF when all the devices are biased. In Fig. 3, when enable signal is low M_{12}, M_{14} and M_{21} will be ON, therefore $V_{out} = 0V$. M_{19} and M_{20} combinedly act as a NOT gate therefore drain of M_{19} will be at higher voltage which will turn ON the M_{15} and M_{16}. As a result, gate of M_{11} and input of amplifier will be at higher potential. This will cause current to flow through M_{13}, M_{10} and M_1. As V_{out} increases, output of NOT gate decreases, which will turn OFF M_{15} and M_{16}. Resistance (r_{ds}) of transistors M_{17}, M_{18} and capacitor C_2 control transient behavior of the circuit.

5 Simulation Results

5.1 DC Analysis

In Fig. 7a, it is observed that variation in the output across PVT is 20.3 mV. Trimming has reduced it to 5.79 mV, which is 71% reduction (Figs 8 and 9).

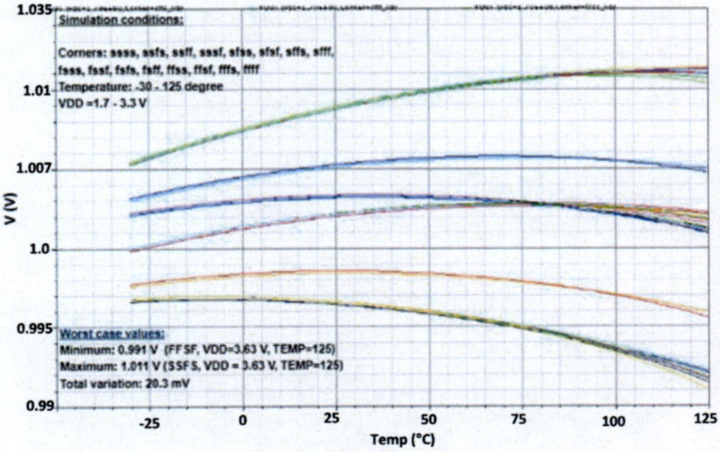

(a)

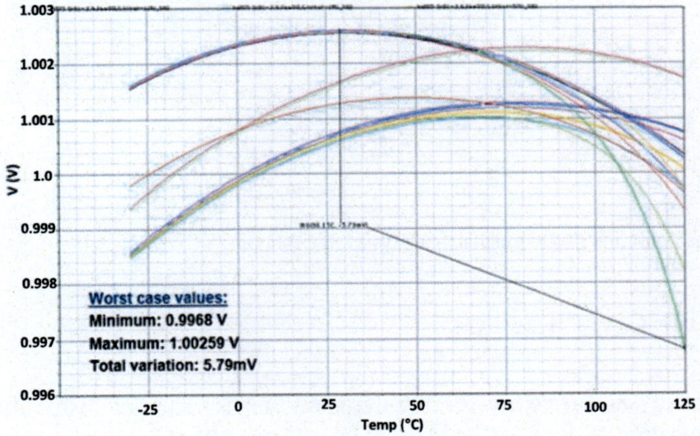

(b)

Fig. 7 Output voltage variation across PVT (a) Untrimmed (b) trimmed

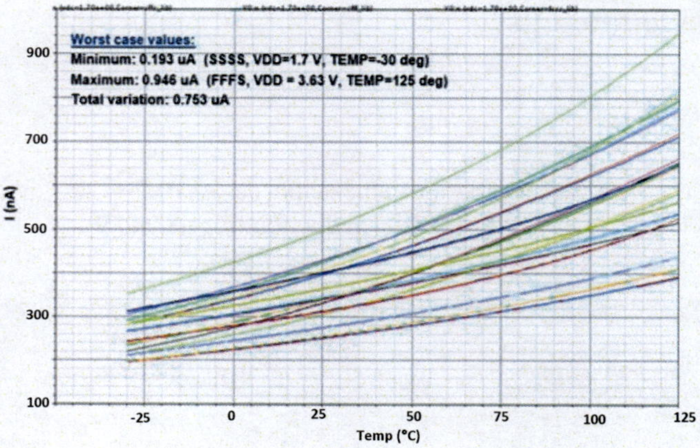

Fig. 8 Quiescent current across PVT

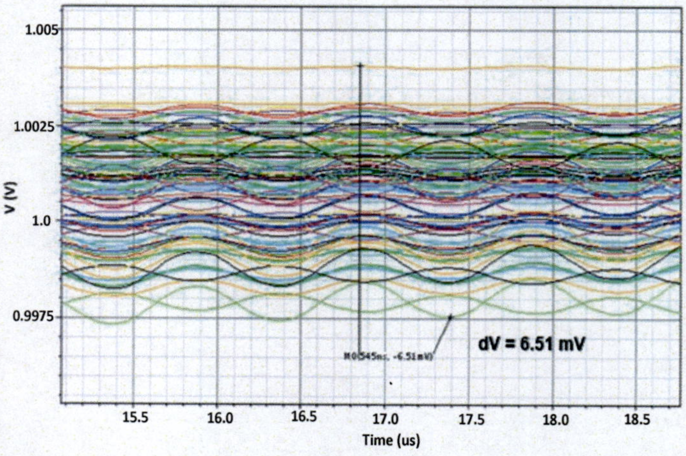

Fig. 9 Output variation due to mismatch analysis after chopping

5.2 *Monte Carlo Analysis*

Mismatch analysis carried out for 300 iterations shows Gaussian distribution. The standard deviation of output voltage is 28.46 mV and that of input offset is 2.6 mV. From (14), the variation caused at the output due to input offset is 10.83×2.6 mV $= 28.158$ mV, which is 99% of the output standard deviation. Maximum variation of the output is 137.6 mV.

Maximum deviation observed at the output after chopping is 6.51 mV, a reduction of 95.26%. Small fluctuations of frequency f_{chop} are present at the output. This is due to the non-ideal low pass filter. The fluctuations can be decreased by increasing the chopping frequency or load capacitance (Table 2).

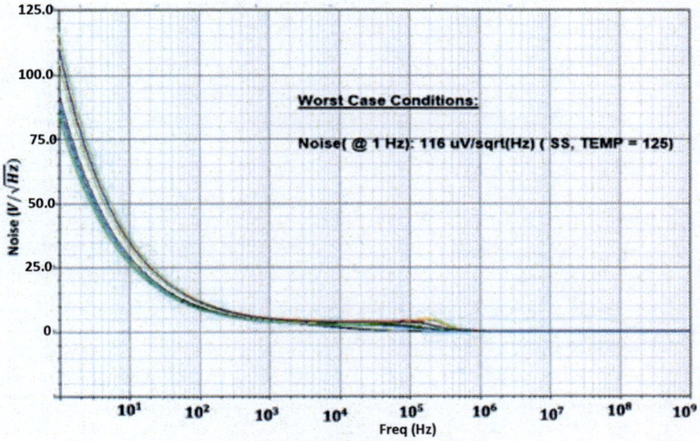

(a)

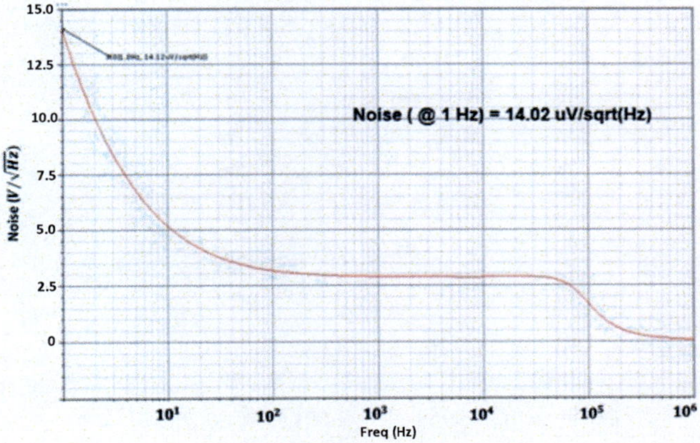

(b)

Fig. 10 Output noise (a) without chopper (b) with chopper

5.3 *Noise Analysis*

In Fig. 10, it is observed that due to chopping technique, 1/f noise is reduced by 88%.

5.4 *Transient Analysis*

The settling time and peak overshoot are found to be 47.06 μs and 0.387 V respectively as shown in Fig. 11.

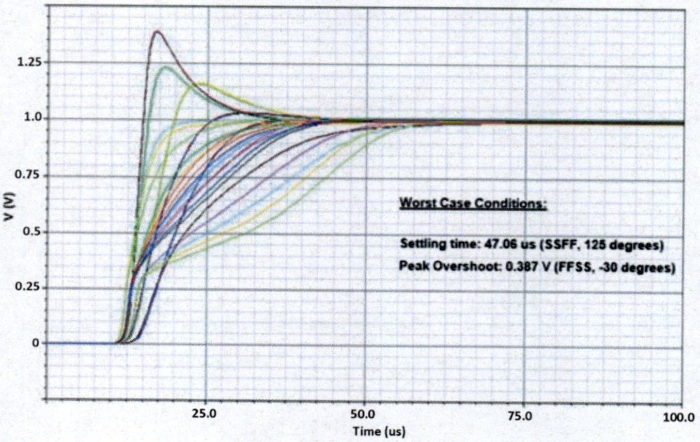

Fig. 11 Step response

Table 2 Simulated results

Parameters		Achieved values			Units	Comments
		Min	Typ	max		
Supply voltage		1.7	3.3	3.63	V	
Load capacitance			1		pF	
Temperature range		-30	27	125	°C	
Output voltage	PVT (untrimmed)	0.991	1.002	1.011	V	20.3 mV (PVT)
	PVT (trimmed)	0.9968	1.0012	1.00259	V	5.79 mV (PVT)
	MC (without chopper)	0.94521	1.0054	1.0827	V	Sig = 28.16 mV
	MC (with chopper)	0.9975		1.004	V	Sig = 1.085 mV
Quiescent current		0.193	0.322	0.946	μA	
AC analysis	Phase margin	57.5	69.13	70.69	deg	
	Gain margin	15.07	17.65	24.97	dB	
	DC loop gain	49.85	90.62	93.23	dB	
Settling time	98% of final value	10.4	26.07	47.06	μs	

6 Conclusion

This paper presents a low power bandgap reference using chopper amplifier, with a focus on chopping to remove offset and 1/f noise of OPAMP. Curvature trimming reduces the temperature dependency by 71%. The comparison of the simulated results of BGR with and without chopper shows an offset reduction of 95% and 1/f noise by 88%. The simulated BGR exhibits 20.844 ppm/°C over temperature range of −30 °C to 125 °C at 3.3 V. The proposed error minimization technique can be used in low power and precision CMOS bandgap reference designs.

Acknowledgements The authors would like to thank Sankalp Semiconductors Pvt Ltd., for providing this opportunity, especially the analog design team, without whose guidance and support the project would not have been completed. Our sincere thanks to Dr P. Subbanna Bhat, Professor Emeritus, B. V. B College of Engineering and Technology for his guidance and support throughout the project.

References

1. Gao YH, Fu DB, Chen GB, Ye R, Zhang L, Zhu C (2016) High precision bandgap reference with chopping offset reducing technique. In: 13th IEEE international conference on solid-state and integrated circuit technology (ICSICT)
2. Rong W, Huijsing JH, Makinwa KAA (2013) Precision instrumentation amplifiers and read-out integrated circuits. Springer, New York
3. Razavi B (1999) Design of analog CMOS integrated circuits. McGraw-Hill, New York
4. Xing H, Jin L, Chen D (2006) Characterization of a current mode bandgap circuit structure for high-precision reference applications. In: 2006 IEEE international symposium on circuits and systems
5. Fan Q, Makinwa KAA (2017) Capacitively coupled chopper amplifiers. Springer, Heidelberg
6. Chen C, Hei Y, Hu X (2012) A chopperstabilized operational amplifier for sensor signal detection. Microelectronics 42(1):17
7. Baker RJ (2019) CMOS circuit design, layout, and simulation. Wiley, Hoboken

Analytical Investigation of Transconductance and Differential Conductance of Ultrathin ID-DG MOSFET with Gradual Channel Approximation

Arpan Deyasi, Riya Chakraborty, Deepanwita Mondal, Nabanita Pramanik, and Swarnav Mukhopadhyay

Abstract Transconductance (g_m) and differential conductance (g_d) of ultra-thin double-gate symmetric MOSFET is analytically investigated considering independent-gate architecture. Ortiz-Conde's model is considered for the computation purpose, where symmetric potential is applied at both the gates. Centre potential is derived from gradual channel approximation, and modification is introduced at the drain current calculation. Following static and transfer characteristics, both type of conductances are calculated in presence of high-K dielectric. Results are compared with that obtained for SiO_2 material, and subthreshold swing is also derived from the result. Simulated findings speak for sharp change in differential conductance and transconductance for ultrathin dielectric thickness, and peak of transconductance is attained at moderate gate bias. Very low subthreshold swing is observed at lower gate voltage, and 10–100 times lower g_d is obtained along with approximately 2 times higher g_m for equivalent EOT when compared with Ortiz-Conde model. Findings are significant for ID-DG MOSFET based nano-device circuit design.

Keywords Transconductance · Differential conductance · Tied-gate architecture · Taur's model · High-K dielectric · Subthreshold swing · Gradual channel approximation

---

A. Deyasi (✉) · N. Pramanik
Department of Electronics and Communication Engineering, RCC Institute of Information Technology, Kolkata, India
e-mail: arpan.deyasi@rcciit.org.in

R. Chakraborty · D. Mondal
Department of Electronic Science, A.P.C College, Barasat, India

S. Mukhopadhyay
Department of Electronics and Telecommunication Engineering, Jadavpur University, Kolkata, India

V. Nath and J. K. Mandal (eds.), *Proceeding of Fifth International Conference on Microelectronics, Computing and Communication Systems*, Lecture Notes in Electrical Engineering 748,
https://doi.org/10.1007/978-981-16-0275-7_10

1 Introduction

The most critical problem in the existing VLSI technology is nothing but ever-increasing short-channel effect [1, 2], which forces the researchers to think about alternative novel architectures [3–5] with precise gate control. Not only structural modifications get prime importance, but also the role of material parameters received equal concern from both material and design engineers. High-K dielectrics are now-a-days replacing the conventional materials [6, 7], thus helping to attain ultrathin dimensions in nanometric range. New substrates are also considered [8] in order to reduce both leakage current and DIBL effect. Among the different architectures proposed, a few of them are now practically realizable owing to the progression of fabrication methodologies with precise growth control mechanism [9, 10]. In this context, double-gate MOSFET becomes one prime candidate for added gate control, and already proved for effective reduction of short-channel effect [11]. Though other designs/devices/architectures are proposed later on, but DG MOSFET provides lowest complexity issues in terms of fabrication, and therefore, becomes one of the choicest candidates for nano-circuit applications.

Several models are already proposed by theoretical researchers in recent years for analyzing electrical characteristics of DG MOSFET, where two different topologies are considered for various application requirements. Tied-gate architecture, among the two, is preferred for higher drain current [12], which leads to higher conductance, but also large subthreshold swing. Solution of DG MOSFET from simultaneous solution of Schrödinger and Poisson's equations are obtained for undoped [12] and lightly doped symmetric [13, 14] structures. Though researches are mostly carried out on independent-gate topology because of lower subthreshold current [3, 15]; but it generally follows constant potential model for surface potential calculation following Ortiz-Conde [16, 17]. Following Taur's approach [18–20], surface potential is modified and corresponding drain current. In the present paper, works is based on Taur's model, and from the static and transfer characteristics, differential conductance and transconductance are evaluated. This work is compared with that obtained from Ortiz-Conde model with similar architecture, electrical and dimensional parameters; and thus the effect of GCA is established in presence of high-K dielectric. Results are extremely significant as comparison reveals which model s appropriate under specific conditions, and role of dielectric plays a very critical role in this aspect.

2 Mathematical Formulation

Ortiz-Conde model is initially considered for long-channel DG MOSFET structure without consideration of quantum-mechanical effects. In order to obtain surface potential, one has to solve 1D Poisson's equation [21] following the procedure in published literature

$$\phi(z) = \phi_C - 2\phi_t \ln\left[\frac{t_{sub}}{2\beta}\sqrt{\frac{qn_i}{2\varepsilon_{sub}\phi_t}}\cos\left(\frac{2\beta z}{t_{sub}}\right)\right] \tag{1}$$

where the parameter is defined as [21], and the term 'β' as proposed by Taur, is given as

$$\beta = \frac{t_{sub}}{2}\sqrt{\frac{qn_i}{2\varepsilon_{sub}\phi_t}}\exp\left[\frac{\phi_0 - \phi_C}{2\phi_t}\right] \tag{2}$$

where substrate layer thickness is denoted by t_{sub}, and quasi-Fermi potential due to the channel carriers is represented by φ_C.

Drain-to-source current for that potential is given by [21]

$$I_{DS} = \mu_{neff}\frac{W}{L}\frac{4\varepsilon_{sub}}{t_{sub}}(2\phi_t)2[f(\beta_s) - f(\beta_d)] \tag{3}$$

where

$$f(\beta) = \beta\tan\beta - 0.5\beta^2 + \frac{\varepsilon_{sub}t_{ox}}{\varepsilon_{ox}t_{sub}}\beta^2\tan^2\beta \tag{4}$$

3 Results and Discussions

3.1 Results on Differential Conductance

Following Eq. (3), drain current is computed as a function of both V_{DS} and V_{GS}, and corresponding pinch-off voltage is determined. From that analysis, differential conductance is evaluated as a function of drain bias. In Fig. 1, result is obtained for different dielectrics, whereas in Fig. 2, effective oxide thickness is varied to investigate the effect on the conductance. For both the cases, drain bias is limited within 0.3 V, and monotonous fall of the profile is reported. It is found that higher dielectric constant reduces the conductance, whereas steep curve is seen for conductance profiles where SiO_2 is used. It is found from the graph that differential conductance decreases with increase of dielectric constant. Materials with a higher dielectric constant show a tendency to be associated with a lower band gap, which reduces the effectiveness of these materials in suppressing gate leakage current. For higher V_{DS}, nature of dielectric becomes almost immatematerial, as evident from Fig. 1.

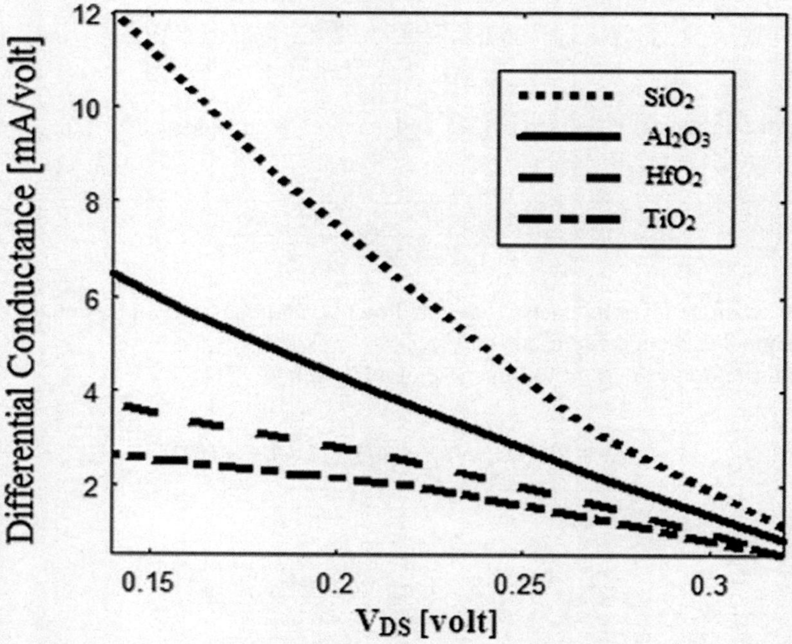

Fig. 1 Effect of high-k dielectric on differential conductance

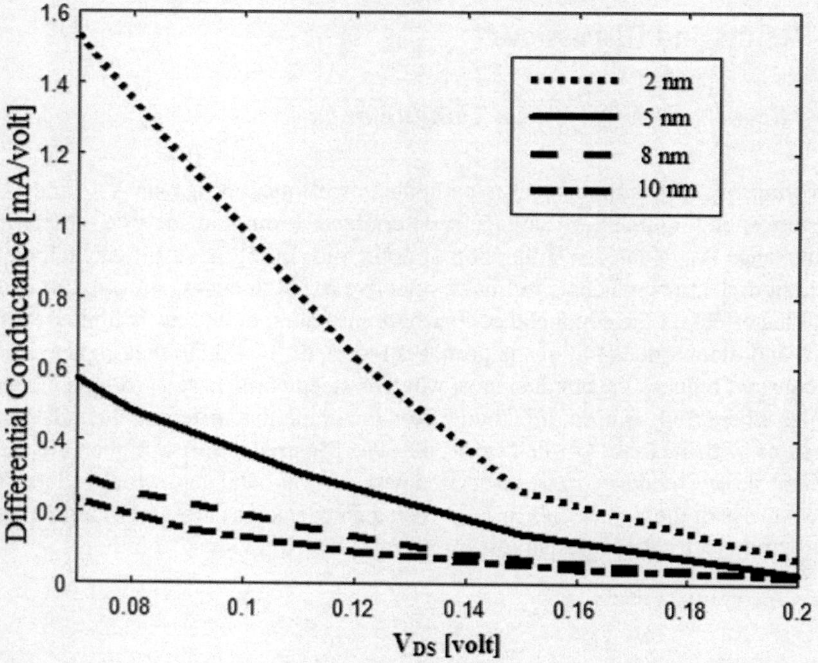

Fig. 2 Effect of dielectric thickness on differential conductance

Effect of dielectric thickness is investigated in Fig. 2. With increase of dielectric thickness, differential conductance decreases, It is also found that high gate tunnel current can flow between gate and substrate as the thickness of the gate oxide is reduced below a few nanometers. As lower thickness of dielectric ultimately enhances the conductance magnitude, so a sharp difference is observed for lower bias. For Fig. 2, simulation is carried out for HfO_2, and monotonous nonlinearly decreasing trend of differential conductance is observed for lower thickness. However, as effective oxide thickness increases, conductance profile becomes almost linear.

Effect of back-gate voltage is calculated and plotted in Fig. 3. It is found from the plot that differential conductance increases with increase of gate voltage. It leads to higher saturation current due to DIBL factor. However, significant point may be noted out in this context that for lower bias, conductance remains almost constant, whereas noticeable difference is observed once back-gate bias is increased.

A comparative study is carried out with the already published results as well as the data obtained following Ortiz-Conde model.

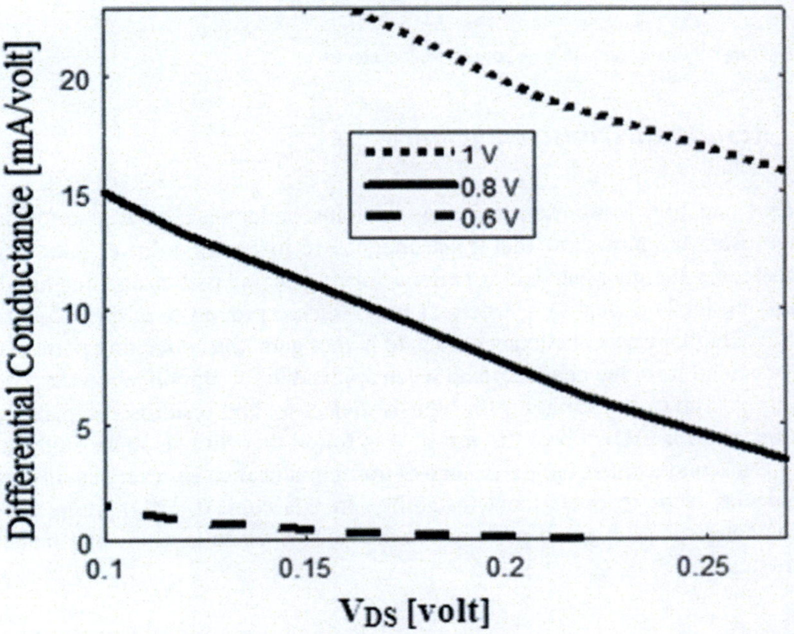

Fig. 3 Effect of back-gate voltage on differential conductance

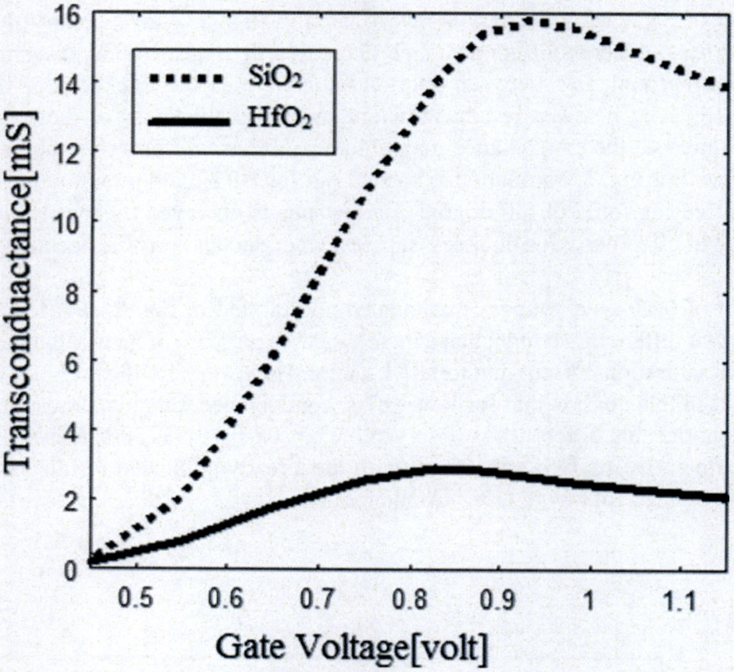

Fig. 4 Transconductance with V_{GS} for different dielectrics

3.2 Results on Transconductance

Figure 4 and Fig. 5 represent variation of transconductance with gate-to-source voltage. Both the plot show that transconductance increases at lower gate voltage and decreases at higher gate voltage after achieving peak. Position and magnitude of the peak critically depends on structural and electrical properties of dielectric material. Since higher transconductance leads to higher gain, thus operating point can be easily decided from the characteristic when applicable for amplification purpose.

Figure 4 shows the variation for high-K dielectric, and result is compared with that obtained for SiO_2. From the result, it is found that SiO_2 is more suitable for higher transconductance, but the change of transconductance with very small change of gate bias leads to operational instability. In this context, HfO_2 offers a near-flat characteristic with small peak which provide wider zone of biasing range for amplification.

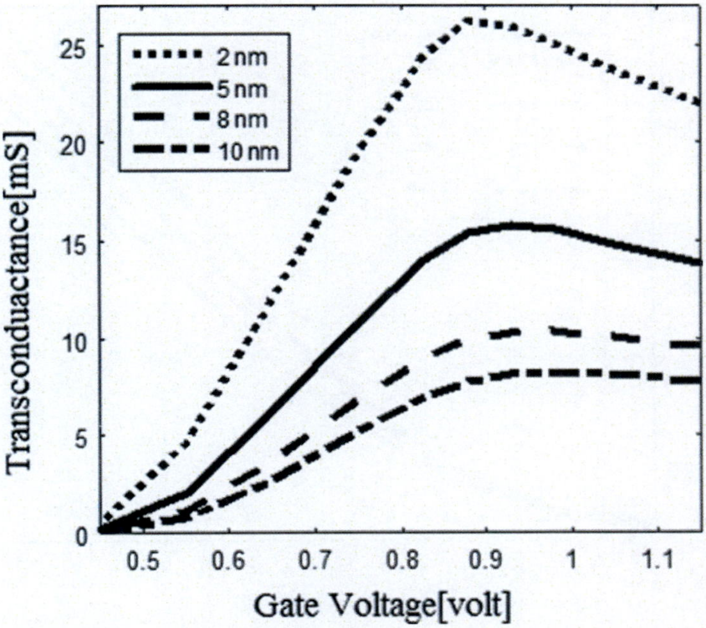

Fig. 5 Effect of dielectric thickness on transconductance

Effect of dielectric thickness is investigated in Fig. 5. With increase of dielectric thickness, it is found that transconductance decreases, as evident from Fig. 2. This is quite obvious, but another interesting fact that devices with thick oxide layer have less control of gate on channel barrier height. For higher width of the dielectric layer, transconductance remains almost constant after attaining the peak.

3.3 Results on Subthreshold Swing

Subthreshold swing is computed from the electrical characteristics, and plotted considering with various factors of dielectric materials. At lower gate bias, it becomes almost constant, whereas with increase with applied voltage, it increases non-linearly. As dielectric thickness is lowered, swing increases, as evident from Fig. 6. So for better performance higher layer width is required, which again reduces the differential conductance. Thus an optimization is required.

High-K factor is investigated in Fig. 7, whereas Fig. 8 represents drain bias effect. From Fig. 7, it is seen that with increase of gate bias, swing increases. However, the change is not linear, and increases rapidly for high-K material. Thus low operating voltage is required. But the swing is hardly affected by drain bias, precisely when gate voltage is moderate or comparatively large. In the present case, drain bias is varied in wider range of 0.2 to 0.8 V.

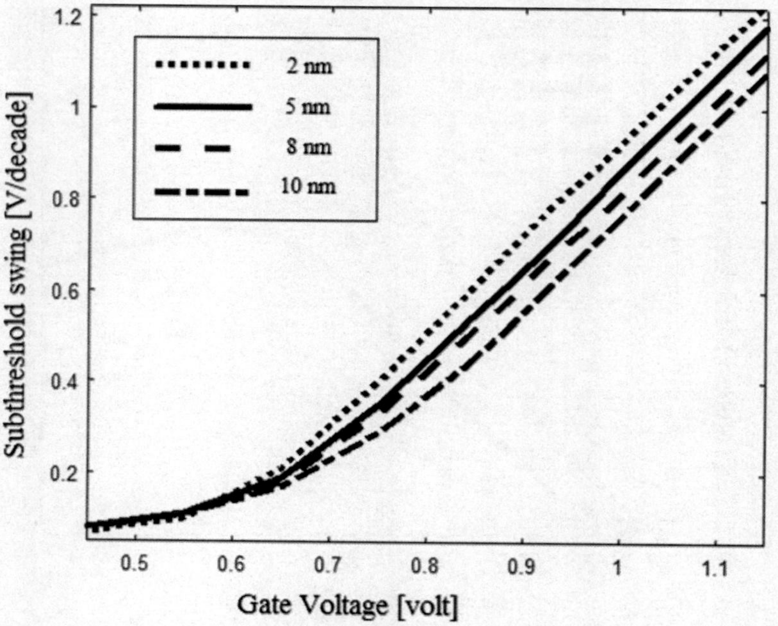

Fig. 6 Effect of dielectric thickness on subthreshold swing

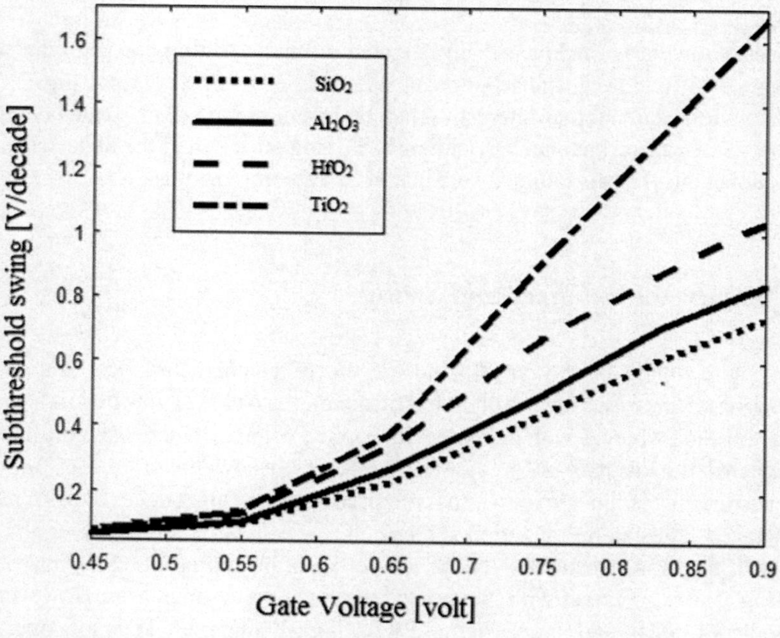

Fig. 7 Effect of high-k dielectric on subthreshold swing

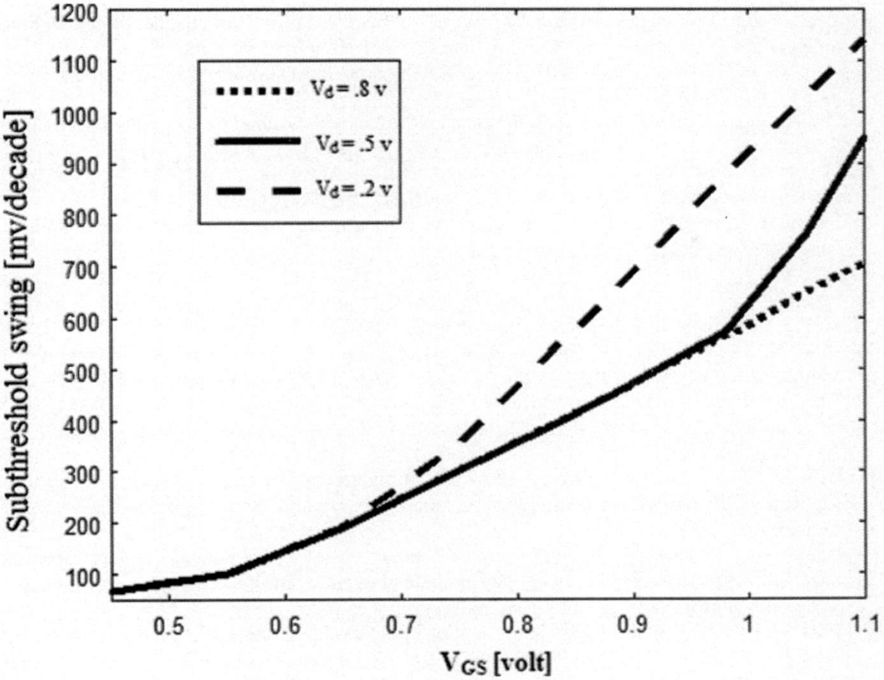

Fig. 8 Effect of drain voltage on subthreshold swing

4 Conclusion

Differential conductance and transconductance of double-gate symmetric MOSFET is analytically computed following Ortiz-Conde model,. Though result speaks in favor of Ortiz-Conde model in general due to higher g_d and g_m (10-100 times), but when comparison is performed with equal EOT, higher g_m is achieved (approximately 2 times). Subthreshold swing characteristics here speaks in favor of lower V_{GS} but higher V_{DS}, contrary to the result obtained from Ortiz-Conde model, where higher V_{GS} and lower V_{DS} is preferred for better amplification. Lowering dielectric thickness results for higher transconductance, and its peak becomes prominent as we use dielectrics with lower permittivity of layer width. Thus an optimization is required both in terms of applied bias as well as dielectric thickness and permittivity, which will ultimately provide required gain for amplification purpose.

References

1. Tsai MC, Wang CI, Chen YC, Chen YJ, Li KS, Chen MC, Chen MJ (2018) Suppression of short channel effects in FinFETs using crystalline ZrO_2 high-K/Al_2O_3 buffer layer gate stack for low power device applications. Semicond Sci Technol 33(3):

2. Zareiee M (2017) High Performance nano device with reduced short channel effects in high temperature applications. ECS J. Solid State Sci Technol 6(7):M75–M78
3. Yu F, Huang G, Lin W, Xu C (2018) An analytical drain current model for symmetric double-gate MOSFETs. AIP Adv 8:
4. Gola D, Singh B, Tiwari PK (2019) Subthreshold characteristic analysis and models for tri-gate SOI MOSFETs using substrate bias induced effects. IEEE Trans Nanotechnol 18:329–335
5. Deyasi A, Sarkar A (2018) Analytical computation of electrical parameters in GAAQWT and CNTFET using NEGF Method. Int J Electron 105(12):2144–2159
6. Swami Y, Rai S, (2018) Ultra-thin high-K dielectric profile based NBTI compact model for nanoscale bulk MOSFET. Silicon 1–11
7. Jing KH, Arshad MKM, Huda ARN, Ruslinda AR, Gopinath SCB, Nuzaihan MNM, Ayub RM, Fathil MFM, Othman N, Hashim U (2016) Gate dielectric scaling in MOSFETs device. In: AIP Conference Proceedings, vol 1733, p 020073
8. Duan X, Zhang J, Chen J, Zhang T, Zhu J, Lin Z, Hao Y (2019) High performance drain engineered InGaN heterostructure tunnel field effect transistor. Micromachines 10(1):75
9. Pott V, Pott V, Moselund KE, Moselund KE, Bouvet D, Michielis LD, Ionescu AM (2008) Fabrication and characterization of gate-all-around silicon nanowires on bulk silicon. IEEE Trans Nanotechnol 6(7):733–744
10. Park D (2008) Silicon nanowire CMOSFETs: fabrication, characteristics, and memory application. In: IEEE 9th International Conference on Solid-State and Integrated-Circuit Technology
11. Panigrahy S, Sahu PK (2013) Performance enhancement and reduction of short channel effects of nano-MOSFET by using graded channel engineering In: International Conference on Circuits, Power and Computing Technologies
12. Woo HJ, Jin KC, Kyu CY (2008) Universal potential model in tied and separated double-gate MOSFETs with consideration of symmetric and asymmetric structure. IEEE Trans Electron Devices 55(6):1472–1479
13. Palanichamy V, Balamurugan NB (2013) Analytical modeling of drain current, capacitance and transconductance in symmetric double-gate MOSFETs considering quantum effects. Int J Nanosci 12(1):
14. Bhartia M, Chatterjee AK (2016) Modeling the drain current and its equation parameters for lightly doped symmetrical double-gate MOSFETs. J Semiconductors 36(4):
15. Thakur PK, Mahapatra S (2011) Large-signal model for independent DG MOSFET. IEEE Trans Electron Devices 58(1):46–52
16. Ortiz-Conde A, García-Sánchez FJ, Muci J, Malobabic S, Liou JJ (2007) A review of core compact models for undoped double-gate SOI MOSFETs. IEEE Trans Electron Devices 54(1):131–140
17. Ortiz-Conde A, Garcia-Sanchez FJ, Muci J (2005) Rigorous analytic solution for the drain current of undoped symmetric dual-gate MOSFETs. Solid State Electron 49(4):640–647
18. Taur Y, Liang X, Wang W, Lu H (2004) A Continuous, Analytic Drain-Current Model for DG MOSFETs. IEEE Electron Device Lett 25(2):107–109
19. Taur Y (2001) Analytic solutions of charge and capacitance in symmetric and asymmetric double-gate MOSFETs. IEEE Trans Electron Devices 48(12):2861–2869
20. Lu H, Taur Y (2006) An analytic potential model for symmetric and asymmetric DG MOSFETs. IEEE Trans Electron Devices 53(5):1161–1168
21. Bhattacharyya AB (2009) Compact MOSFET Models for VLSI Design. 1st ed Wiley, Singapore
22. Pradhan KP, Mohapatra SK, Sahu PK, Behera DK (2014) Impact of high-K gate dielectric on analog and RF performance of nanoscale DG-MOSFET. Microelectron J 45:144–151
23. Deyasi A, Sarkar A, Roy K, Roychowdhury A (2019) Effect of high-k dielectric on differential conductance and transconductance of ID-DG MOSFET following ortiz-conde model. Microsyst Technol
24. Prasher R, Dass D, Vaid R (2013) Performance of a double gate nanoscale MOSFET (DG-MOSFET) based on novel channel materials. J Nano Electron Phys 5(1):01017

Throughput Radix-8 Based FFT Architecture

Ajay Singh Raghuvanshi, Ashutosh Kumar, and Shashank Gavel

Abstract Fast Fourier Transform (FFT) processor plays key role in Orthogonal Frequency Division Multiplexing (OFDM) system. Power consumption, speed, accuracy and area are the key aspects of FFT. FFT finds vast application in the field of Digital and Signal Processing (DSP). The FFT processor can be used in the areas such as Digital Audio Broadcasting, Worldwide interoperability for Microwave access (Wi-Max), Digital Video Broadcasting-Terrestrial (DVB-T). In this article, we present an optimized high speed and throughput 64 × 64 non-pipelined and pipelined 2D FFT architecture with a radix-8 block structure. The designed structure is a parallel architecture that takes less computation time and also requires less area. The proposed design is realized using the Vitex-7 FPGA tool as a supplementary hardware.

Keywords 2D FFT · Radix-8 · Virtex-7 · Slices · Pipeline · OFDM · High throughput · High speed

1 Introduction

Fast Fourier Transform (FFT) is a very important aspect of Digital signal processing (DSP), especially in the applications involving Orthogonal Frequency Division Multiplexing (OFDM) Systems such as wireless-LAN, ADSL, VDSL systems and Wi-MAX. Very Large Scale Integrated (VLSI) Circuits implementation of FFT demands an architecture that provides high speed, low power consumption and reduced chip area. For an N-point transformation, the Discrete Fourier Transform (DFT) has a complexity of $O(\mathrm{N}^2)$. Cooley and Turkey explained the concept of Fast Fourier Transform (FFT) which reduces the complexity to $O(N l\mathrm{og}_2 N)$. Fast Fourier

A. S. Raghuvanshi (✉) · A. Kumar · S. Gavel
Department of Electronics and Communication Engineering, National Institute of Technology, Raipur, Raipur, India
e-mail: asraghuvanshi.etc@nitrr.ac.in

S. Gavel
e-mail: sgavel.phd2016.etc@nitrr.ac.in

V. Nath and J. K. Mandal (eds.), *Proceeding of Fifth International Conference on Microelectronics, Computing and Communication Systems*, Lecture Notes in Electrical Engineering 748,
https://doi.org/10.1007/978-981-16-0275-7_11

Transform (FFT) is an efficient structure of DFT that provides better performance by taking less number of operations. In general, the complexity of DFT is reduced by utilizing FFT [1].

FFT has found major applications in the field of image and signal processing due to its less complex structure and processing. One of the applications is the reorganization of the image from raw data by using 2D FFT. This necessitates systematic implementations to practical applications. Since to support the size of an image, the processor requires more memory. So, an efficient architecture is needed in order to optimize the memory as well as to achieve better performance.

To implement FFT, many hardware and software solutions have been utilized [2, 3]. In this article, we present an efficient FPGA based non-pipelined and pipelined 2D FFT architecture. We have used the structured behaviour of FFT to create a parallel-pipelined model of architecture that is executed in FPGA. 2D FFT is evaluated from $2\times M$ 1D FFT. Thus, the performance of 2D FFT is directly influenced by 1D FFT [4].

To implement 64 × 64 FFT architecture, we have used radix-8 algorithm that is built on a novel radix-8 butterfly unit [5]. 64 × 64-point FFT is computed using a cascade of two radix-8^2 blocks. The reordering of outputs is not needed as it is already ordered in our design that improves memory optimization [6]. The main contributions of this study are summarized below:

- Non-pipelined and pipelined 2D FFT architecture built on radix-8 algorithm with maximum frequency using lesser number of slices and registers.
- A systematic output reordering technique implemented using six-bit up-counter with three bits needed for each stage of radix-8 butterfly unit.
- Twiddle factors are generated and used directly rather than storing it in memory which minimizes memory requirement.
- FPGA implementation of non-pipelined and pipelined 2D FFT architecture.
- Analyzation of our work with other works in terms of frequency, slices and registers.

The remaining paper is organized as shown. Section 2 provides brief idea of 1D and 2D FFT algorithms. A short study of available literature is also described. Section 3 presents the architecture. Section 4 presents VLSI execution and results. Section 5 presents the conclusion.

2 Related Study and Framework

2.1 Related Study

There are many architectures and algorithms available on FFT. Pipeline based architecture consumes more power and provides more area overhead [8, 9]. [10] presents radix-2 real-valued FFT through rigorous formula deduction. Memory-based designs provide an advantage in terms of power consumption but require more area. [11] present several memory access architectures.

For 2D FFT also different 2D architectures have been proposed. [13] presents a spiral software solution which provides high performance but also consumes more power. Hardware solutions are also available in [5]. A constructive architecture to calculate the 2D DFT for huge sized input is implemented in [14]. Hardware acceleration platform for reconstruction of the image is presented in [15]. Mixed radices.

When designing FFT, output reordering is a big issue. With reordering, we produce output in natural order i.e., in a reordered way. In parallel-pipelined FFT, the reordering of output is complex and is introduced in [17].

Even and odd outputs are saved using two buffers having size M. [17] presents one memory of size M alternately for even and odd outputs. $3 \times M$ memories are needed for data reordering for feedback pipelined architectures having a multi-path delay.

2.2 Framework

The evolution of computationally systematic algorithm is made possible if we adopt a divide-and-conquer approach. Cooley-Tukey is the usual algorithm to calculate FFT [1]. Due to this algorithm, its complexity reduces to $O(Mlog_2M)$ from $O(M^2)$. M-point sequence of $g(m)$ is given by-

$$G(r) = \sum_{\mathrm{m}=0}^{M-1} g(m) W_m^{mr}, \quad r = 0, 1, 2, ..., M-1 \tag{1}$$

where, W_M is the twiddle factor and is described as-

$$W_M = e^{-2\pi i/M} \tag{2}$$

M-point inverse DFT is calculated as

$$g(m) = \frac{1}{M} \sum_{r=0}^{M-1} G(r) W_M^{-mr}, \quad m = 0, 1, 2, ..., M-1 \tag{3}$$

1D FFT can be used to calculate the 2D FFT. Using row-column deposition algorithm, a 1D FFT can be used to calculate the 2D FFT as shown below -

$$G(r_1, i_2) = \sum_{i_1=0}^{M-1} g(i_1, i_2) W_M^{r_1 i_1}$$

where,

$$r_1 = 0, 1, 2, ..., M - 1 \tag{4}$$

$$Q(r_1, r_2) = \sum_{i_2=0}^{M-1} G(r_1, i_2) W_M^{r_2 i_2}$$

where,

$$r_2 = 0, 1, 2, ..., M - 1 \tag{5}$$

Parallel radix-8 butterfly unit is employed in this study.

3 Proposed 2D FFT Architecture

This portion explains the radix-8^2 algorithm that is used to perform the two-dimensional FFT. If input is $g(i_1, i_2)$, then 2D FFT for size 64×64 is expressed by-

$$q(r_1, r_2) = \sum_{i_2=0}^{63} \left[\sum_{i_1=0}^{63} g(i_1, i_2) W_{64}^{r_1 i_1} \right] W_{64}^{r_2 i_2} \tag{6}$$

where, $r_1, r_2 = 0, 1, 2, ..., 63$
Taking into account the inner block of summation

$$\sum_{i_1=0}^{63} g(i_1, i_2) W_{64}^{r_1 i_1}$$

The above expression is 64-point FFT which is performed using radix-8^2 algorithm. In a similar manner, we can also compute the outer summation. So, two radix-8^2 blocks are coupled to produce 2D FFT. 1D FFT is computed with the help of radix-8^2 architecture shown in Fig. 3. Radix-8^2 consists of two levels of radix-8 butterfly units. Now onwards, radix-8^2 is denoted as R 8^2 and radix-8 as R8.

R 8^2 architecture is inspired by the signal flow graph [1]. The signal flow graph follows decimation in frequency pattern. The first phase consists of eight radix-8 operations and the second phase employs a single radix-8 operation. $g(0), g(8), g(16), g(24), g(32), g(40), g(48), g(56)$ are inputs of first R8 unit producing $G(0), G(8), G(16), G(24), G(32), G(40), G(48), G(56)$ as outputs. A fully unrolled $R8^2$ architecture using a parallel R8 butterfly unit is our fundamental block. The Radix-8 block consists of eight inputs, producing single output at a time depending on a three-bit control signal. Depending upon the three-bit control signal i.e., control, outputs are produced in a specified order. In standard radix-8 operation, outputs are produced in a fixed sequence. The R8 butterfly component equation is shown in (7).

$$
\begin{aligned}
G(0) &= g(0) + g(1) + g(2) + g(3) \\
&\quad + g(4) \\
G(1) &= g(0) + w_8^1 g(1) + w_8^2 g(2) + w_8^3 g(3) \\
&\quad + w_8^5 g(5) + w_8^6 g(6) \\
G(2) &= g(0) + w_8^2 g(1) + w_8^4 g(2) + w_8^6 g(3) \\
&\quad + w_8^2 g(5) + w_8^4 g(6) \\
G(3) &= g(0) + w_8^3 g(1) + w_8^6 g(2) + w_8^9 g(3) \\
&\quad + w_8^7 g(5) + w_8^2 g(6) \\
G(4) &= g(0) + w_8^4 g(1) + g(2) + w_8^4 g(3) + g(4) \\
&\quad + w_8^4 g(5) + g(6) + w_8^4 g(7) \\
G(5) &= g(0) + w_8^5 g(1) + w_8^2 g(2) + w_8^7 g(3) \\
&\quad + w_8^4 g(4) \\
G(6) &= g(0) + w_8^6 g(1) + w_8^4 g(2) + w_8^2 g(3) \\
&\quad + w_8^6 g(5) + w_8^4 g(6) \\
G(7) &= g(0) + w_8^7 g(1) + w_8^6 g(2) + w_8^5 g(3) \\
&\quad + w_8^3 g(5) + w_8^2 g(6) \\
&\quad + w_8^1 g(7)
\end{aligned}
\qquad (7)
$$

The Fig. 1 and 2 shows the non-pipelined and pipelined architecture of radix-8 butterfly unit respectively.

Outputs are produced in an ordered way using six-bit control signal which is implemented using a six-bit up- counter. Each phase is controlled by three-bit of up-counter which is shown as control in Figs. 1 and 2.

The parallel unrolled architecture is shown in Fig. 3 and 4 for non-pipelined and pipelined architecture respectively. The first phase is having eight R8 blocks and the second phase is having a single R8 block. W1 represent the twiddle factors of the first stage. The twiddle factors are not stored in memory; rather it is generated and used directly. Wherever twiddle factor is having unity value, we do not need to use complex multipliers for that particular case. So, we will be using lesser number of multipliers. In each phase consisting of M twiddle factor values, $M/4$ will be having unity twiddle factor values, so no need for multipliers while doing complex

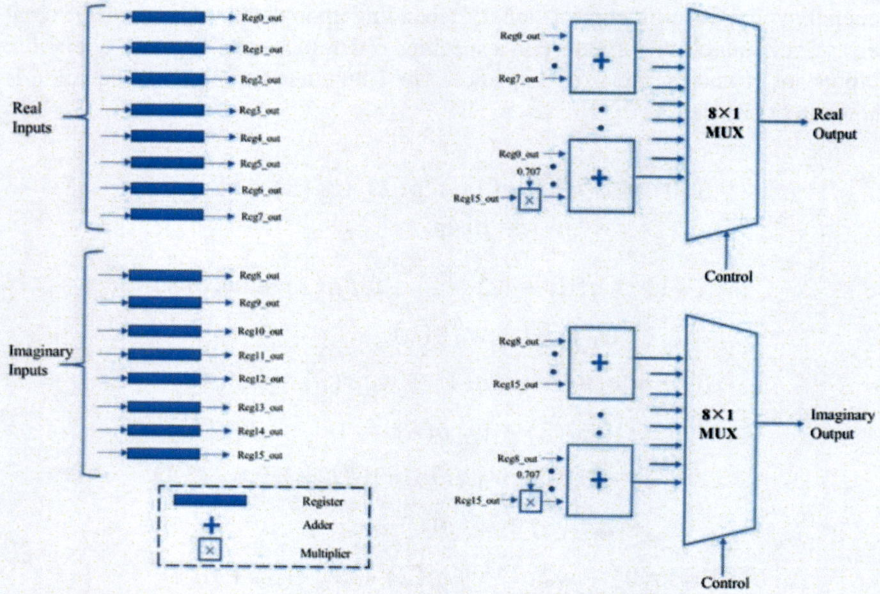

Fig. 1 Non-pipelined radix-8 butterfly architecture are used for 1D FFT implementation. [16] presents two-dimensional FFT where more datasets occupy off-chip in DRAM. The inner loop unrolling technique is employed to design two-dimensional FFT on FPGA. [4] presents FFT in which M^2 intermediate memories are needed for the first phase but in our work, M intermediate memories are needed. In 1D FFT itself, RAM requirement is less

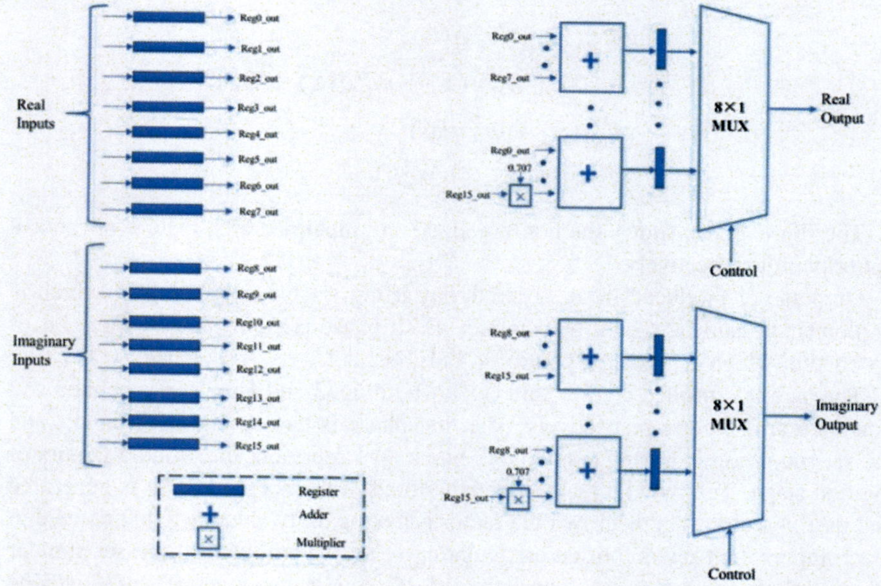

Fig. 2 Pipelined radix-8 butterfly architecture

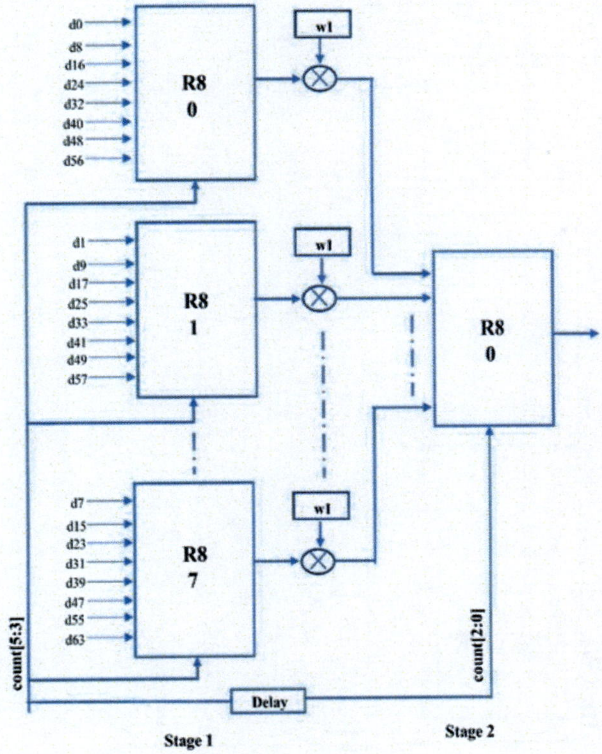

Fig. 3 Non-pipelined Radix-8^2 Architecture

multiplication. Suppose 64 × 64 input, which we have stored in memory is described by matrix X, where ‘p’ will denote the row number and ‘q’ will denote the column number. In cycle 1, the mode is set to be 0 and the first column of X is provided as input. Its FFT will be calculated and it represents the first element of the first row of matrix Y. In cycle 2, the mode is set to be 0 and the second column of X is provided as input. Its FFT will be calculated and it represents the second element of first row of matrix Y. In this way, all the elements of matrix Y are produced.

In the 65th cycle, when the entire first row of matrix Y is produced, it is provided to the second $R8^2$ block. Registers are used to store the intermediate results. In each cycle of the second block, the mode is changed from 0 to 63 resulting in the first row of the matrix Z.

A parallel unrolled architecture is used to implement the $R8^2$ block. The ordering of output is controlled by a six-bit up-counter. The implementation in this way is using less memory and improved latency. First $R8^2$ block is used to implement the 1D FFT and its output is provided to the next $R8^2$ stage to produce 2D FFT. The $R8^2$ blocks which are used for the first phase and second phase are alike. The 2D FFT architecture block diagram is shown in Fig. 5.

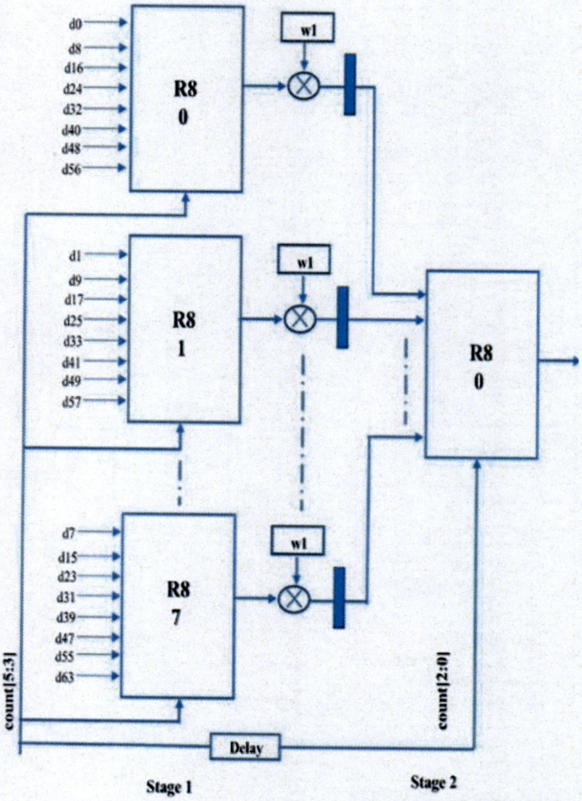

Fig. 4 Non-pipelined Radix-8^2 Architecture

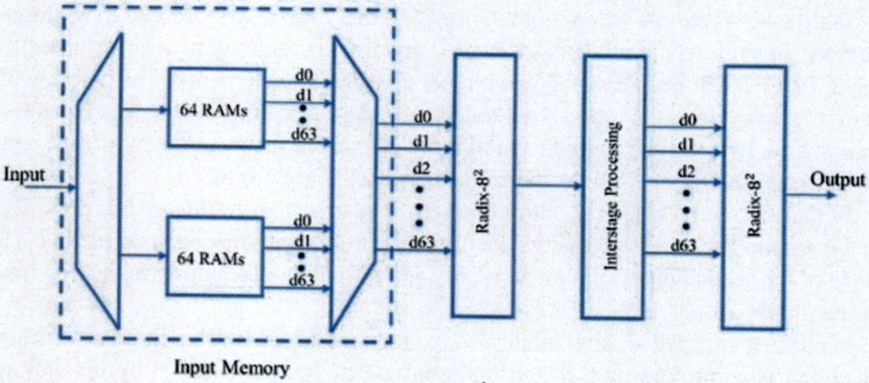

Fig. 5 2D FFT architecture

Fig. 6 Input memory

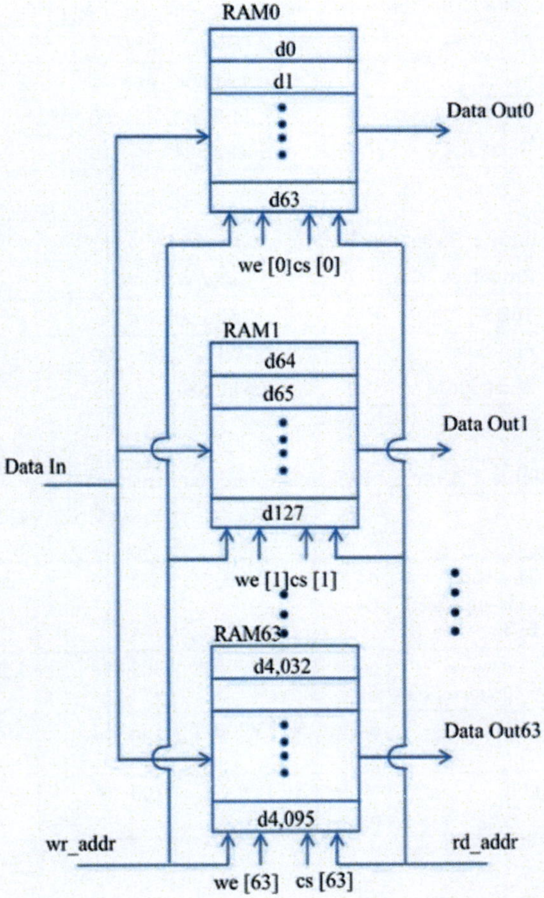

We have used two banks of 64 RAMs for the input memory. One memory is used to read-in the data while the other one is used to read-out the data. One bank of RAM is shown in Fig. 6. RAM0 receives first 64 inputs; next 64 inputs are received by RAM1 and so on. The RAMs are read in a parallel way. For all RAMs, the read address is the same.

4 Implementation Results and Analysis

Xilinx Virtex 7 (XC7V2000T) device is used for FPGA implementation. FPGA results are presented in Table 1 and Table 2 for 64 $\times$ 64 non-pipelined and pipelined FFT respectively. Table 3 shows a comparison of different 2D FFT architectures [7, 12].

Table 1 Resources used for non-pipelined 64 × 64 FFT

Resources	Used/Available	Percentage
Slices	4,511/305,400	1.477
Flip-flops	11,203/2,443, 200	0.459
Block RAM	64/2,584	2.47

Table 2 Resources used for pipelined 64 × 64 FFT

Resources	Used/Available	Percentage
Slices	7,506/305,400	2.458
Flip-flops	11,112/2,443, 200	0.455
Block RAM	64/2,584	2.47

Table 3 Different FFT architecture comparison

	Size	FPGA	Freq.(MHz)	Slices	Clk cycles	Throughput	**Cont. Flow**
This work (non-pipelined FFT)	64 × 64	XC7V2000T	205.125	4,511	4096	205.125	Yes
This work (pipelined FFT)	64 × 64	XC7V2000T	308.436	7,506	4096	322.436	Yes
[6]	64 × 64	XC7V2000T	156.254	5,692	4096	156.254	Yes
[17]	64 × 64	XC4VSX35	100	6,472	4516	90.6	No

From Table 3, it is evident that our architecture takes fewer cycles and area (slices) with improved frequency. The operating frequency is 205.125 MHz and 308.436 MHz for non-pipelined and pipelined architecture respectively. The memory is reduced considerably in our architecture. The FFT architecture uses a continuous flow of data.

4096 cycles are consumed by memory block. To read data from memory, one clock cycle is needed. Two $R8^2$ blocks need two cycles each and interstage processing need 65 cycles. So, total latency requirement is 4167 clock cycles but effectively for 64 × 64 FFT, 4096 clock cycles are needed.

5 Conclusion

In this article, we have designed an efficient 64 × 64-point 2D FFT architecture with the radix-8 block structure. The output of first radix-8^2 block is provided as input to the next radix-8^2 block. To obtain output in normal order, the six-bit counter is used. There is a significant improvement in terms of resources used and throughput when compared with existing architectures for FFT. 8-bits are used to represent the real part and 8-bits for the imaginary part. Verilog HDL is used to implement the 2D FFT architecture in RTL. As the twiddle factors are generated and used directly rather than storing in memory, it improves computation time and area. The designed architecture takes 4096 clock cycles to perform 64×64 2D FFT. FPGA implementation of 2D FFT is done with an operating frequency of 205.125 and 308.436 MHz for non-pipelined and pipelined architecture respectively.

References

1. Proakis JG (2001) Digital signal processing: principles algorithms and applications. Pearson Education India
2. Puschel M, Moura JM, Johnson JR, Padua D, Veloso MM, Singer BW, Xiong J, Franchetti F, Gacic A, Voronenko Y et al (2005) SPIRAL: code generation for DSP transforms. Proc IEEE 93(2):232–275
3. Frigo M, Johnson SG (1998) FFTW: an adaptive software architecture for the FFT. In: Proceedings of the 1998 IEEE international conference on acoustics, speech and signal processing, ICASSP 1998 (Cat. No. 98CH36181), vol 3, pp 1381–1384. IEEE
4. Kee H, Bhattacharyya SS, Petersen N, Kornerup J (2009) Resource efficient acceleration of 2-dimensional fast Fourier transform computations on FPGAs. In: 2009 third ACM/IEEE international conference on distributed smart cameras (ICDSC), pp 1–8. IEEE
5. Babionitakis K, Chouliaras VA, Manolopoulos K, Nakos K, Reisis D, Vlassopoulos N (2010) Fully systolic FFT architecture for Giga-sample applications. J Sig Process Syst 58(3):281–299
6. Kala S, Mathew J, Jose BR, Nalesh S (2018) Radix-43 based two dimensional FFT architecture with efficient data reordering scheme. IET Comput Digital Tech 13(2):78–86
7. Kala S, Nalesh S, Maity A, Nandy S, Narayan R (2013) High throughput, low latency, memory optimized 64k point FFT architecture using novel radix-4 butterfly unit. In: 2013 IEEE international symposium on circuits and systems (ISCAS 2013), pp 3034–3037. IEEE
8. He S, Torkelson M (1998) Design and implementation of a 1024- point pipeline FFT processor. In: Proceedings of the IEEE 1998 custom integrated circuits conference (Cat. No. 98CH36143), pp 131–134. IEEE
9. Oh J-Y, Lim M-S (2005) New radix-2 to the 4th power pipeline FFT processor. IEICE Trans Electron 88(8):1740–1746
10. Yin X-B, Yu F, Ma Z-G (2016) Resource-efficient pipelined architectures for radix-2 real-valued FFT with real datapaths. IEEE Trans Circuits Syst II Express Briefs 63(8):803–807
11. Xia K, Wu B, Zhou X, Xiong T (2016) A generalized conflict-free address scheme for arbitrary 2k-point memory-based FFT processors. In: 2016 IEEE international symposium on circuits and systems (ISCAS), pp 2126–2129. IEEE
12. Xing Q-J, Ma Z-G, Xu Y-K (2017) A novel conflict-free parallel memory access scheme for FFT processors. IEEE Trans Circuits Syst II Express Briefs 64(11):1347–1351
13. Chen R, Prasanna VK (2014) Energy optimizations for FPGA-based 2-D FFT architecture. In: 2014 IEEE high performance extreme computing conference (HPEC), pp 1–6. IEEE
14. Lenart T, Gustafsson M, O¨ wall V (2008) A hardware acceleration platform for digital holographic imaging. J Sig Process Syst 52(3), 297–311
15. Yu C-L, Kim J-S, Deng L, Kestur S, Narayanan V, Chakrabarti (2011) Fpga architecture for 2D discrete fourier transform based on 2D decomposition for large-sized data. J Signal Process Syst 64(1):109–122

16. Kristensen F, Nilsson P, Olsson A (2003) Flexible baseband transmitter for OFDM. In: Proceedings of the IASTED conference on circuits signals systems, pp 356–361
17. Rodríguez-Ramos J, Castelló EM, Conde CD, Valido MR, Marichal-Hernández J (2008) 2D-FFT implementation on FPGA for wavefront phase recovery from the CAFADIS camera. In: Adaptive optics systems, vol 7015, p 70153. International Society for Optics and Photonics

Transportation Network Analysis Using Fuzzy Logic Through GIS Techniques: A Study of Central Kolkata

Amrita Sarkar

Abstract Traffic congestion has become a very undesirable aspect of urban living throughout the world. From this paper it is inferred that the conventional way of determining congestion using volume and capacity is not matching with the actual traffic conditions. Directly and precisely measurable quantities such as Speed and Inter Vehicular Distance (IVD) were the two parameters considered for input in the Fuzzy model to quantify the congestion in par with reality. The network analysis such as the most direct path between two points and the optimum route between many points using GIS was carried out in Central Kolkata. This paper describes the process of quantifying congestion using Fuzzy tools and techniques and also discusses the process of network analysis. This paper combines the advantages of both Fuzzy logic and GIS to offer public, policy makers and traffic managers a new means to assess congestion as short and medium term measures.

Keywords Fuzzy logic · Traffic congestion · Road network analysis · GIS techniques

1 Background

Congestion is the travel time or delay in excess of that normally incurred under light or free-flow travel conditions [6]. Traffic congestion is a dynamic phenomenon. Congestion does not occur everywhere, all-at-once. Instead, congestion occurs in specific locations and propagates through network over time as congested conditions on a link spread to nearby links These incidents result in congestion patterns that propagate from a localized incident through many portions of the network, potentially resulting in serious flow disruption [17]. Unfortunately, urban traffic congestion will likely worsen over the future due to structural trends in both developed and developing regions of the world. In developing regions, these trends combine with rapid population growth and rural-to-urban migration to create transportation systems

A. Sarkar (✉)
Department of CSE, Sarala Birla University, Ranchi, India
e-mail: amritasarkar@bitmesra.ac.in

V. Nath and J. K. Mandal (eds.), *Proceeding of Fifth International Conference on Microelectronics, Computing and Communication Systems*, Lecture Notes in Electrical Engineering 748,
https://doi.org/10.1007/978-981-16-0275-7_12

that operate at near-standstill for substantial portions of the day in many urban areas. Consequently, in most urban areas, increases in transportation system capacity cannot match the corresponding increases in the volume of vehicles [7]. Roadway traffic congestion may be further partitioned into two types, recurring and non-recurring. Recurring congestion is characterized by the repetitive nature of the traffic occurring during predictable times of the day when traffic volumes exceed the design capacities of the roadways and when the geometric design of the roadway does not allow a normal traffic flow. Such congestion occurs during the morning and afternoon peak travel periods common in urbanized areas. The problem of recurring congestion is exacerbated by random events, such as accidents, vehicle disablements, unexpected weather conditions, roadway construction or maintenance activities or other special events. These unexpected events produce non-recurring traffic congestion [13]. The increasing level of traffic congestion in urban areas is creating substantial negative externalities to every aspect in urban life. The type and intensity of congestion depend on many quantifiable factors such as volume, speed, headway, ratio of slow moving and fast moving vehicles etc. [14].

Network analysis in geospatial information system (GIS) provides strong decision support for users in searching optimal route, finding the nearest facility and determining the service area. Searching optimal path is an important advanced analysis function in GIS. Delavar et al. applied GIS route finding modules, heuristic algorithms to carry out its search strategy and also addressed the problem of selecting route to a given destination on an actual map under a static environment [4]. GIS has provided an efficient means for organizing basic transportation related data in order to facilitate the input, analysis, and display of transport networks. This method was applied to present a data model to design an object-relational geodatabase for roads network [5]. Claramunt et al. developed a new framework for the integration, analysis and visualization of urban traffic data within GIS. The integration of urban traffic data within GIS requires a sequence of manipulations that include pre-processing functions, selection and derivation of traffic data, and visualization and animation tasks [3]. Transportation planners have used GIS and remote sensing in several aspects that include forecasting the location and type of growth within metropolitan areas as well as urban roads mapping. Also it is used for bus facilities including bus routes and bus stops [9]. Jha and Schonfeld presented a model for highway alignment optimization that integrates a GIS with genetic algorithms, examines the effects of various costs on alignment selection, and explores optimization in constrained spaces that realistically reflect the limits on road improvement projects [10]. Ahmed et al. developed an enhanced network analysis model that was applied with the enormous capabilities of Geographic Information System (GIS) to identify the best route from the location of an incident for any healthcare service providers in the Greater Cairo metropolitan area [2].

In the field of decision making to control the traffic, Fuzzy logic controller system is more appropriate and has given accurate result and better performance over conventional controller system. Th application of fuzzy logic technique also gives the benefits in the field of traffic control system [1]. Lu and Cao also applied a new method to evaluate congestion from traffic flow information based on fuzzy logic. Level of congestion was considered as a continuous variable from free flow to traffic jam. After a simulation, adaptive neuro-fuzzy inference system and trained a series of fuzzy logic rules are employed to estimate the congestion [12]. Hoogendoorn et al. presented an viewpoint of the theoretical and methodological principles of the fuzzy logic approach on the perspectives of using fuzzy logic techniques in traffic and transportation systems analysis and control. An outline was specified for the fuzzy logic applications in the transportation and traffic-engineering field, especially in the areas of estimation and prediction of traffic engineering parameters, modeling the behavior of road users and traffic control [8]. Kruse et al. introduced two aspects of a traffic management system for multi-lane highways with variable road signs. First, fuzzy logic was employed to take into account the uncertainties of traffic data, and to detect traffic congestion in isolated road sections. Second, a traffic control approach using a fuzzy model based on experience are explained [11]. Pongpaibool defined Road-traffic estimation system by image processing data using manually adjusted fuzzy logic and adaptive neuro-fuzzy techniques. The technique was developed to mimics human's expertise on describing three levels of traffic congestion within Bangkok Metropolitan Area [15].

In this study an approach is made to identify the suitability and type of technique that can be effective and also a model is built to quantify the congestion using fuzzy logic. In this paper the directly and precisely measurable quantities such as Speed and Inter Vehicular Distance (IVD) are the two parameters considered for inputs in the Fuzzy model. In this paper, GIS has been used to perform three main tasks: to map the road network configuration; to link or associate the different traffic conditions and to carry out network analysis using those conditions. This paper also combines the advantages of both Fuzzy logic and GIS to offer public, policy makers and traffic managers a new means to assess and to alleviate congestion.

Kolkata, is one of the oldest metropolises of India. The study area, constitute the central areas of Kolkata, is situated in old part of the city and at the same time very well linked to the other parts of the city. Being the Central Business District of the city, the area caters heavy load of commuters and traffic and the area is also characterized by highly anastomosing network of roads of varied width, which caters huge traffic volume throughout the day.

2 Methods

2.1 *Quantifying Congestion Using Fuzzy Techniques*

In this research paper the two-inputs: speed and inter vehicular distance (IVD); one-output: congestion and five-rules discussed below were used to quantify congestion. Figure 1 shows diagrammatically the basic structure of this problem.

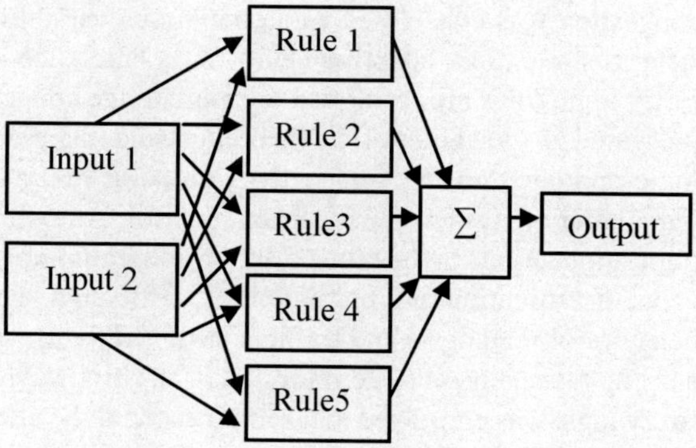

Fig. 1 Shows the diagram explaining the relationship among the input variables, fuzzy rules and output

The above flowchart was based on the principle that information flows from left to right. The parallel nature of the rules was one of the most important aspects of fuzzy logic systems. In fact in the Fuzzy Logic toolbox, there were five steps of the fuzzy inference process which is explained contextually below.

Step 1. Fuzzify Inputs
The first step was to take the inputs and determine the degree to which they belong to each of the appropriate fuzzy sets via membership functions. In the Fuzzy Logic toolbox, the input was always a crisp numerical and the output was a fuzzy degree of membership in the qualifying linguistic set. The 'if' part of the rules denote antecedent and 'then' part of the rules denote consequent.

Step 2. Apply Fuzzy Operator
Once the inputs were fuzzified, the degrees were known to which each part of the antecedent had satisfied for each rule. If the antecedent of a given rule had more than one part, the fuzzy operator AND, OR was applied to obtain one number that represented the result of the antecedent for that rule.

Step 3. Apply Implication Method
After applying fuzzy operator the implication method, the process of shaping the fuzzy set in a Fuzzy Interface System was implemented. The consequent, a fuzzy set represented by a membership function, was reshaped using a function associated with the antecedent. The input for the implication process was a single number given by the antecedent, and the output was a fuzzy set. Implication was implemented for each rule. Two built-in were used for implication methods. AND method or min (minimum), which truncated the output fuzzy set, and prod (product), which scaled the output fuzzy set.

Step 4. Aggregate All Outputs
The rules here were combined in order to make a decision. The fuzzy sets that represented the outputs of each rule were combined into a single fuzzy set by aggregation which occurred once for each output variable. The list of truncated output functions

returned by the implication process for each rule was the input of the aggregation process. The output of the aggregation process was one fuzzy set for each output variable. Three built-in methods were supported this method: max (maximum), probor (probabilistic or), and sum (simply the sum of each rule's output set).

Step 5. Defuzzify
The input for the defuzzification process was generally a fuzzy set and the output was a single number. Here since the aggregate of the fuzzy set encompassed a range of output values, hence it was defuzzified in order to resolve a single output value from the set. The defuzzification method applied was the centroid calculation, which returns the center of area under the curve.

2.2 *Process of Performing Network Analysis*

A transport network is used for transport network analysis to determine the flow of vehicles through it within the field of transport engineering, typically using graph theory.

The optimum routes delineated from both these conditions were compared. In the present research paper, the network analysis performed here primarily considers the rule of Network Analyst that is described below.

To perform network analysis, four different sets of parameters were used as follows.

1. Standard road network attributes like road length, average journey speed.
2. Congestion were quantified using standard road network attributes as well as road network attributes like IVD and times.
3. Congestion types were determined by using the fuzzified road parameters viz. speed and IVD with the help of fuzzy logic. The congestion type before being used for network analysis was modeled on the basis of rules.
4. Network modeling was used to include the rules relating to the arcs in association with solving transportation problems. Representation of one-way or both ways street, closed or opened road were modeled by the following guidelines to restrict travel to certain directions (Table 1).

In all, determination of the optimum routes was performed for twelve important road segments occurring within the study area. The analyst results obtained from the network analysis comprise the following discussions.

The optimum routes considering the highly congested roads and averting the highly congested roads were delineated respectively. Under this, the analysis considers two continuous time intervals pertaining to the peak hours, i.e. 8 am–2 pm and 2 pm–8 pm.

Table 1 Shows travel directions guidelines

To set these rules	Use these codes as values
Travel is permitted from the start to the end of the line only, which is the same as the digitized direction	FT, ft (i.e. FROM TO)
Travel is permitted from the end of the line to the start of the line only, which is the opposite the digitized direction	TF, tf (i.e. TO FROM)
Travel is permitted in neither direction; the line is closed for travel	N, n (NEITHER DIRECTION i.e. CLOSED road)
Travel is permitted in both directions	Any other value or no data (say 1 or 2 or any other value)

Majority of the road segments considered in the present study, the optimum routes delineated for these two time periods were found to be same. These optimum routes follow those roads are referred as 'unidirectional' in the present study which implies that the traffic flows on these roads are in the same directions. Contrary to this, there occur optimum routes that follow 'unidirectional' as well as 'bi-directional' roads, which imply that the direction of the traffic flow is reversed in the nature during the two different time periods The optimum routes delineated for different traffic conditions and for different time periods as and where applicable are shown in different sets of diagram in this paper.

The delineation optimum route based on distance and time factors were carried out by considering highly congested roads and averting this respectively. The total distance covered and total time taken to traverse from the origin to destination is also mentioned under both types of traffic conditions.

2.3 *Algorithm of Network Analysis*

Dijkstra's algorithm is the simplest pathfinding algorithm, even though to this day many other algorithms have been developed. Dijkstra's algorithm reduces the amount of computational time and power needed to find the optimal path. The algorithm strikes a balance by calculating a path, which is close to the optimal path, which is computationally manageable. The algorithm breaks the network into nodes (where lines join, start or end) and lines represent the paths between such nodes. In addition, each line has an associated cost representing the cost (price) of each line needing to reach a node. Dijkstra's algorithm finds the length of an *optimal* path between two vertices in a graph. Optimal can mean shortest or cheapest or fastest or optimal in some other sense: it depends on how you choose to label the edges of the graph [16]. The algorithm works as follows:

1. Choose the source vertex
2. Define a set S of vertices, and initialize it to empty set. As the algorithm progresses, the set S will store those vertices to which a shortest path has been found.
3. Label the source vertex with 0, and insert it into S.
4. Consider each vertex not in S connected by an edge from the newly inserted vertex. Label the vertex not in S with the label of the newly inserted vertex + the length of the edge.
   - But if the vertex not in S was already labelled, its new label will be min(label of newly inserted vertex + length of edge, old label)
5. Pick a vertex not in S with the smallest label, and add it to S.
6. Repeat from step 4, until the destination vertex is in S or there are no labelled vertices not in S.

If the destination is labelled, its label is the distance from source to destination. If it is not labelled, there is no path from the source to the destination.

Following the approach mentioned above, discussion for the few road segments covering important location is provided below.

3 Results and Discussion

3.1 Derivation of Optimum Routes

11, Woodmount Street - 26, Marx Engle's Bithi (During 8 am–8 pm)

Considering congestion on all roads, while identifying optimum route it may be inferred that under such condition, for the shortest path taken, the time taken was 10.21 min. which was more than the time taken when highly congested roads were avoided i.e. 9.94 min. For identification of optimum route, considering congestion but avoiding highly congested roads, distance traveled will be more but, as stated earlier, time consumed will be less. Also, all roads considered in this case are unidirectional; hence the optimum delineated was found to be same throughout the peak hour (8 am–8 pm) of the day. Therefore adopting this optimum route avoiding highly congested roads will facilitate a commuter to move within the study area in lesser time without considering the fuel cost (Fig. 2).

Fig. 2 **a** including highly congested roads **b** averting highly congested roads.

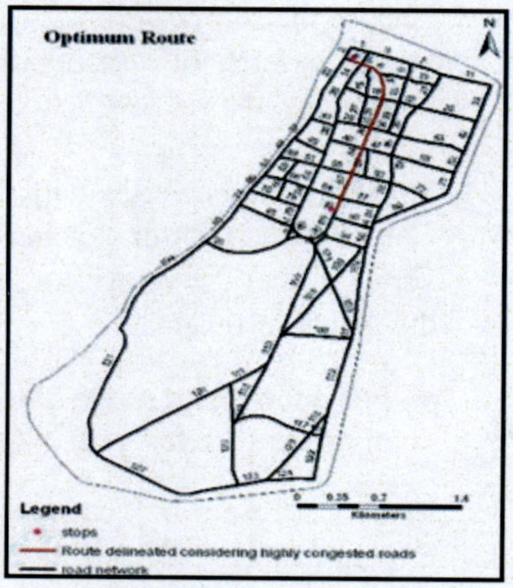

(a) including highly congested roads

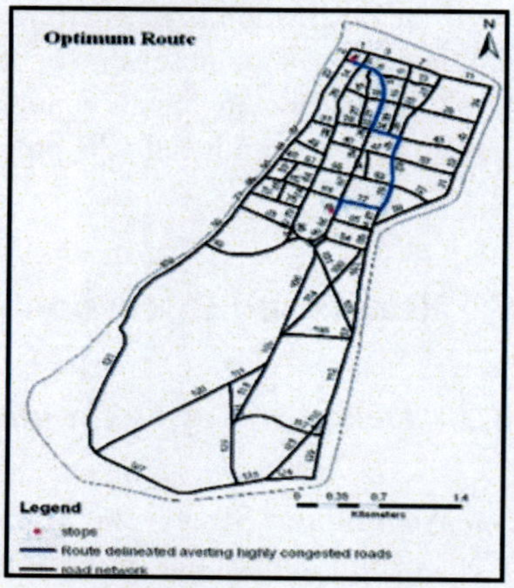

(b) averting highly congested roads

31, Woodmount Street - 8, Gurunanak Sarani (During 8 am–8 pm)

Considering congestion on all roads, while identifying optimum route it may be inferred that under such condition, for the shortest path taken, the time taken is 10.6 min. which is more than the time taken when highly congested roads are avoided i.e. 10.32 min. For identification of optimum route, considering congestion but avoiding highly congested roads, distance traveled will be more but, as stated earlier, time consumed will be less. Also, all roads considered in this case are unidirectional; hence the optimum route delineated is found to be same throughout the peak hour (8 am–8 pm) of the day. Therefore adopting this optimum route avoiding highly congested roads will facilitate a commuter to move within the study area in lesser time without considering the fuel cost (Fig. 3).

Fig. 3 **a** including highly congested roads **b** averting highly congested roads

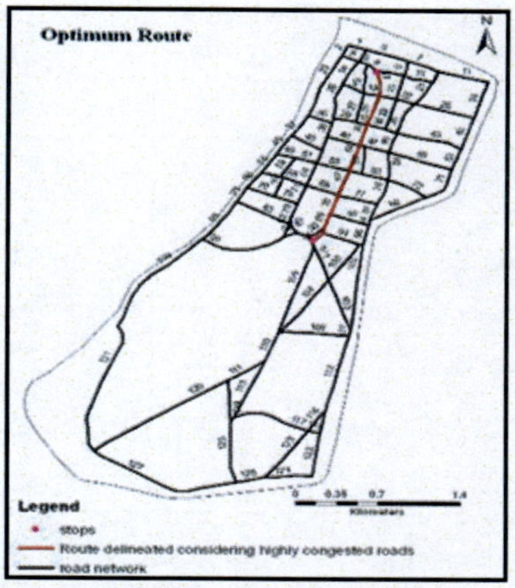

(a) including highly congested roads

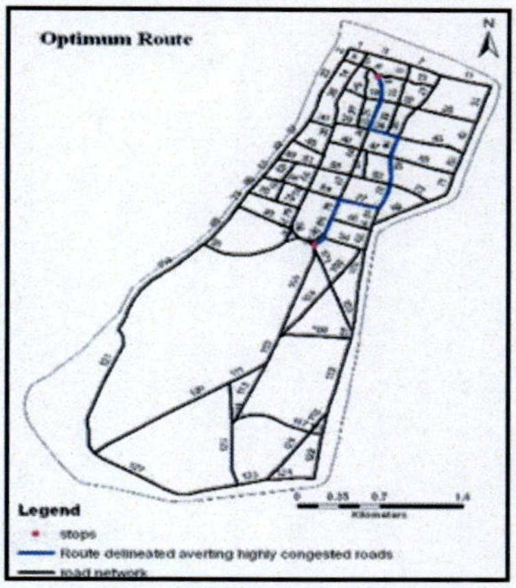

(b) averting highly congested roads

16, Bentinck Street - 84, Eden Garden Road (During 8 am–8 pm)
Considering congestion on all roads, while identifying optimum route it may be inferred that under such condition, for the shortest path taken, the time taken is 9.21 min. which is more than the time taken when highly congested roads are avoided i.e. 6.07 min. For identification of optimum route, considering congestion but avoiding highly congested roads, distance traveled will be more but, as stated earlier, time consumed will be less. Also, all roads considered in this case are unidirectional; hence the optimum route delineated is found to be same throughout the peak hour (8 am–8 pm) of the day. Therefore adopting this optimum route avoiding highly congested roads will facilitate a commuter to move within the study area in lesser time without considering the fuel cost (Fig. 4).

Fig. 4 **a** including highly congested roads **b** averting highly congested roads

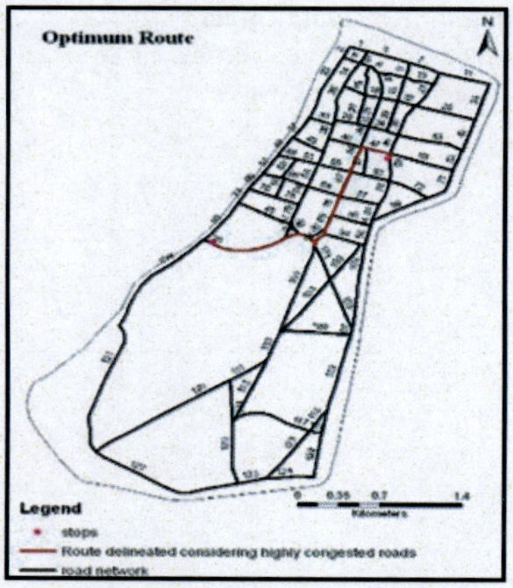

(a) including highly congested roads

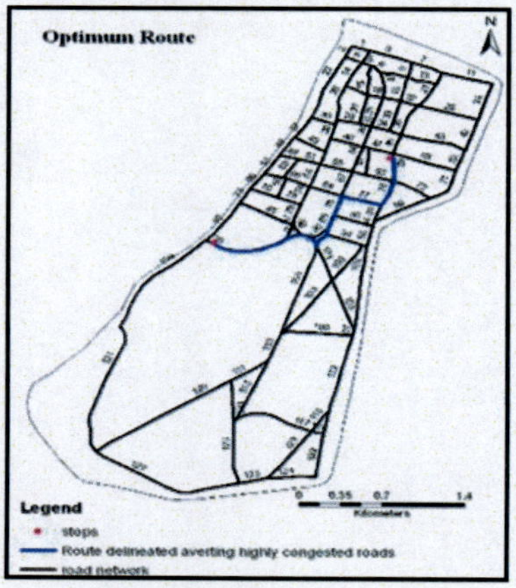

(b) averting highly congested roads

Example of Optimum Routes (8 am–2 pm) for a School Bus Between Several Pickup Points

This example illustrates how a school bus can move within the study area covering several pick-up points considering congestion including highly congested road and avoiding highly congested roads respectively.

Considering congestion on all roads, while identifying optimum route it may be inferred that under such condition, for the shortest path taken, the time taken (Fig. 5(a)) is more than the time taken when highly congested roads are avoided (Fig. 5(b)). For identification of optimum route, considering congestion but avoiding highly congested roads, distance traveled will be more but, as stated earlier, time

Fig. 5 **a** including highly congested roads **b** averting highly congested roads

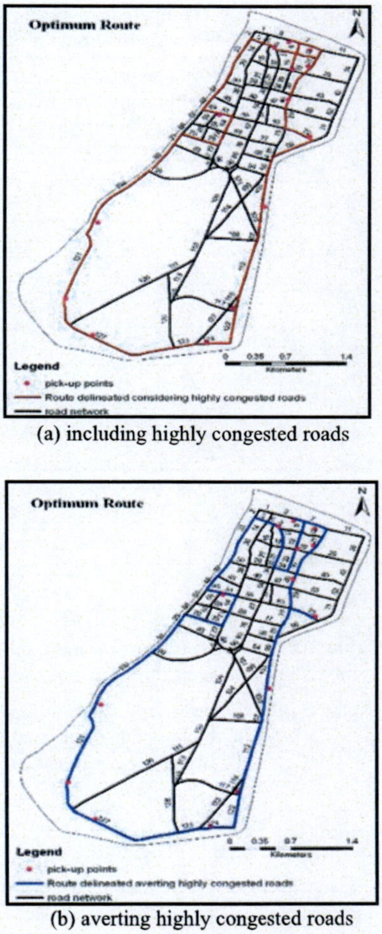

consumed will be less (Fig. 5(b)). Also, all roads considered in this case are unidirectional; hence the same route is followed throughout the day. Therefore adopting this optimum route avoiding highly congested roads will facilitate the school bus to move within the study area in 50.15 min time, which is lesser than normal time taken by 58.87 min. This typical example shows the net gain in time availed by a school bus, which might also be applicable for other commuters availing other modes of communication within this area.

Results of network analysis using twelve important road segments occurring within the study area are shown below in the Table 2a, 2b, 2c, 2d.

Table 2(a) Shows comparison between length of optimum route considering and avoiding highly congested roads during 8 am–2 pm

Road segment	Total length considering highly congested roads (8 am–2 pm) (km)	Total length avoiding highly congested roads (8 am–2 pm) (km)	Comparisons length
1	1.74	2.32	Increased
2	1.85	2.2	Increased
3	1.86	2.4	Increased
4	2.56	3.07	Increased
5	2.34	2.34	Same
6	2.64	2.8	Increased
7	3.22	3.36	Increased
8	3.4	3.54	Increased
9	1.84	2.49	Increased
10	3.98	4.72	Increased
11	5.5	5.7	Increased
12	4.7	4.9	Increased

Table 2(b) Shows comparison between drive time of optimum route considering and avoiding highly congested roads during 8 am–2 pm

Road segment	Drive time considering highly congested roads (8 am–2 pm) (min)	Drive time avoiding highly congested roads (8 am–2 pm) (min)	Comparisons drive- time
1	10.21	9.94	Decreased
2	10.62	7.22	Decreased
3	10.6	10.32	Decreased
4	13.06	10.31	Decreased
5	9.21	6.07	Decreased
6	9.38	8.56	Decreased
7	7.63	6.29	Decreased
8	8.4	7.07	Decreased
9	8.81	7.83	Decreased
10	11.65	6.68	Decreased
11	20.22	16.63	Decreased
12	18.69	15.11	Decreased

Table 2(c) shows comparison between length of optimum route considering and avoiding highly congested roads during 2 pm–8 pm

Road segment	Total length considering highly congested roads (2 pm–8 pm) (km)	Total length avoiding highly congested roads (2 pm–8 pm) (km)	Comparisons length
1	1.74	2.32	Increased
2	1.85	2.2	Increased
3	1.86	2.4	Increased
4	2.56	3.07	Increased
5	2.34	2.34	Same
6	2.64	2.8	Increased
7	3.22	3.36	Increased
8	3.4	3.54	Increased
9	1.84	2.49	Increased
10	3.98	3.98	Same
11	4.7	4.9	Increased
12	5.68	5.88	Increased

Table 2(d) Comparison between drive time of optimum route considering and avoiding highly congested roads during 2 pm–8 pm

Road segment	Drive time considering highly congested roads (2 pm–8 pm) (min)	Drive time avoiding highly congested roads (2 pm–8 pm) (min)	Comparisons drive-time
1	10.21	9.94	Decreased
2	10.62	7.22	Decreased
3	10.6	10.32	Decreased
4	13.06	10.31	Decreased
5	9.21	6.07	Decreased
6	9.38	8.56	Decreased
7	7.63	6.29	Decreased
8	8.4	7.07	Decreased
9	8.81	7.83	Decreased
10	11.65	6.68	Decreased
11	20.22	16.63	Decreased
12	18.69	15.11	decreased

4 Conclusion

The technique used in the analysis above may be inferred to be more accurate and perfect in calculation of congestion and ultimately finding the shortest route in transportation planning. While on one hand, it was possible to quantify congestion for each and every road through the application of fuzzy logic which lead to a definite result so as to state the condition of each road with respect to congestion, parallelly it was also possible to find the optimum route that might suit best to a commuter trying to reach a particular destination in shortest time avoiding congestion.

This analytical research which in depth dealt with how various factors like speed and inter-vehicular distances determine congestion, did not limit itself only in understanding the level of congestion in various roads of Central Kolkata which like magnets are pulling people from hinterlands due to sheer benefits of economics associated with this area. This understanding process was carried out using Fuzzy tools and techniques. As a further extension of this research, this congestion data was used for road network analysis using GIS tools and techniques. In fact the network analysis was carried out based on two different assessments; first one being related to identification of route which helped in calculating the travel distance and travel time on that particular road from one point to another considering congestion on all roads and second being the same but avoiding highly congested roads. This finding basically lead to the conclusion that for the second type of assessment mode, for most cases the distance traveled is more but the time consumed is less as compared to the first case considering traffic flow at the peak hour.

The methodology discussed above, worked out and applied for the Central Kolkata housing within the Central Business District of the city may find its applicability across the globe in areas having similar types of constraints and problems.

References

1. Acharya S, Dash KK, Chaini R (2020) Fuzzy logic: an advanced approach to traffic control. In: Applications of robotics in industry using advanced mechanisms. 10.1007/978-3-030-30271-9_17
2. Ahmed S, Ibrahim R, Hefny HA (2017) GIS-based network analysis for the roads network of the greater Cairo area. In: Proceedings of the international conference on applied research in computer science and engineering ICAR 2017, Lebanon. https://ceur-ws.org
3. Claramunt C, Jiang B, Bargiela A (2000) A new framework for the integration, analysis and visualisation of urban traffic data within geographic information systems. Transp Res Part C 8:167–184
4. Delavar MR, Samadzadegan F, Pahlavani P A GIS-assisted optimal urban route finding approach based on genetic algorithms
5. Dodge S, Alesheikh AA (2005) Transportation data model implementation for Iranian roads network. Map Middle East
6. Hanks JW, Lomax TJ (1992) Roadway congestion estimates and trends. Cooperative Research Program, Texas Transportation Institute, The Texas A&M University System, College Station. Texas Research Report Number 113, pp 1-4

7. Hanson S (1995) Getting there: urban transportation in context. In: Hanson S (ed) The geography of urban transportation, 2nd. The Guilford Press, New York, pp 305–341
8. Hoogendoorn S, Hoogendoorn-Lanser S, Schuurman H (2000) Fuzzy perspectives in traffic engineering
9. El-Shair IM (2003) GIS and remote sensing in urban transportation planning: a case study of Birkenhead, Auckland. In: Map India conference
10. Jha MK, Schonfeld P (2004) A highway alignment optimization model using geographic information systems. Transp Res Part A 38:455–548
11. Krause B, Altrock CV, Pozybill M (2002) Intelligent highway by fuzzy logic: congestion detection and traffic control on multi-lane roads with variable road signs. In: Proceedings of IEEE 5th international fuzzy systems, New Orleans, LA, USA https://doi.org/10.1109/FUZZY.1996.552649
12. Lu J, Li C (2004) Congestion evaluation from traffic flow information based on fuzzy logic. In: Proceedings of the 2003 IEEE international conference on intelligent transportation systems, Shanghai, China https://doi.org/10.1109/ITSC.2003.1251919
13. Mark AM, Kayin L (1994) An investigation of the costs of roadway traffic congestion: a preparatory step for IVHS benefit's evaluation
14. Narayanan R, Udayakumar R, Kumar K, Subbaraj L (2003) Quantification of congestion using fuzzy logic and network analysis using GIS
15. Pongpaibool P, Tangamchit P, Noodwong K (2008) Evaluation of road traffic congestion using fuzzy techniques. In: TENCON 2007 - 2007 IEEE region 10 conference, Taipei, Taiwan. https://doi.org/10.1109/TENCON.2007.4429119
16. Singal P, Chhillar RS (2014) Dijkstra shortest path algorithm using global position system. Int J Comput Appl. https://doi.org/10.5120/17690-8654
17. Wu YH, Miller HJ, Hung MC (2001) A GIS-based decision support system for analysis of route choice in congested urban road networks. J Geogr Syst 3:3–24

Integration of IoT Based PLC for Smart Relaying of a PV-Fed Induction Motor Driven Conveyor Belt

Sourav Chakraborty, P. Arvind, Suprava Poddar, Alok Kumar Acharya, and S. Deepak Kumar

Abstract In the era of industry 4.0, PLC plays a crucial role to engross the automation sector in the industrial unit. PLC can be considered as the hidden workforce that silently executes with the help of a Ladder Logic. PLC is not only applicable to the manufacturing network but is often used in the power industry too. A substantial portion of this work circumscribes the use of PLC for smart relaying. The relays are operated using the programmable logic control to increase the life span of a Squirrel Cage Induction Motor (SCIM), thereby protecting it from overheating. It also ensures that the conveyor unit maintains its tare weight within permissible limits so as to work in its fullest possible efficiency. This paper also focuses on renewable energy source (solar PV) to supply power to the three-phase induction motor in the manufacturing unit. With Industrial Automation, the data is being stored securely in cloud platform in large capacity to handle the Big Data being generated. Here, the concept of IoT has also been used, which is an autonomous system composed of sensor and intelligent device networks. The efficacy of PLC as a realistic tool has been shown through a simulation in Matlab/Simulink® environment, which serves to predict the real behaviour of the system. Thus, saving and optimizing the time, compared to costly trial and error experiments.

Keywords PLC · IoT · Relay · SCIM · Solar PV · MPPT

S. Chakraborty · P. Arvind (✉) · S. Poddar
Department of Electrical and Electronics Engineering, Birla Institute of Technology, Mesra, Ranchi, India

A. K. Acharya
School of Electrical Engineering, KIIT University, Bhubaneswar, India

S. D. Kumar
Department of Production Engineering, Birla Institute of Technology, Mesra, Ranchi, India
e-mail: dks10@iitbbs.ac.in

V. Nath and J. K. Mandal (eds.), *Proceeding of Fifth International Conference on Microelectronics, Computing and Communication Systems*, Lecture Notes in Electrical Engineering 748,
https://doi.org/10.1007/978-981-16-0275-7_13

1 Introduction

It has always been a blessing after Industrial Automation came into existence. With up-gradation in technology, there has been improvement in better quality of production. PLC can be said to be an industrial computer which is programmable and can be used for the sequential task to be performed in the industry hence controlling technological process. It enables simultaneous cooperation with many analog and digital signals. It is field programmable, which means that the user can write and change program in the field without rewriting/sending the unit back to the manufacturer. This paper also focuses on the renewable energy source (solar PV) to supply the 3-phase squirrel-cage induction motor in the manufacturing unit. In this paper, the motor being running with the help of green energy which means that PV power is being used to generate energy. In industry with mass production, much energy is being utilized as well as exhausted, so this is a small way in which carbon emission can be reduced, thus providing a better environment Depending on the solar insolation and atmospheric temperature, peak power can be derived from the Photovoltaic array at the optimum point of operation. A Boost Converter amplifies the low voltage from the PV panels and feeds it to the inverter [1]. This is accomplished by the incorporation of P&O method into the MPPT system [2, 3]. The latter algorithm seeks to regulate the duty cycle. The motor is being protected through relays which are operated by Ladder Logic in the PLC. With the help of Time Delay Relay, the motor's cooling Fan turns On and Off with a rise in the temperature, which is eventually caused due to losses [4]. In this paper lays emphasis on how the motor will continue to operate normally even though there is a temperature rise, provided PLC being programmed accordingly. Here the ladder programming will be done in a way to set out a limit of weights the conveyer unit will be carrying in terms of torque. Damage to the motor will result in additional expenses and also would lead to shut down of a process causing enormous loss to the company. There are many protection guidelines for a motor laid down by the IEEE Std 3004.8-2016 [5]. However, for this study, the focus is laid for overheating and overload protection.

The PLC programming for relaying is conditioned to work in two cases: (a) When the motor gets overheated due to losses along with ambient temperature, the motor' gets turned off, while once the motor is cooled down, it will turn on. This will result in increasing the lifespan.of the motor. However, if the motor heat is less than the alarming rate, the cooling fan gets turned on to cool the motor further (b). When the motor is operating in normal condition then, in this case, two conditions are considered: firstly when the conveyer unit is carrying the specified range of allowable weight, motor will be in ON state. Later, when the conveyor unit is carrying weight more than specified, then the motor will be in OFF State The tare weight is transduced in terms of torque measured by the torque cum speed sensor which is equipped with a rotary encoder. This sensor is connected to the PLC and by using counter logic [6, 7].

With industrial automation, another problem that arises is to store the vast data being generated and storing it securely. Data can be stored in the cloud by means of the Internet of Things (IoT), but cyber-security is the major challenge. The emerging development of "smart" technology nowadays has morphed conventional devices and environments into "intelligent" entities with the advent of cloud computing [8]. Hence in this work, the central idea is to exhibit the capability of PLC in protection of the motor along with coordination with IoT that can expand the possibilities of remote monitoring and management. yet preventing the system from being compromised. The remaining sections are discussed as follows:

Section 2 discusses the solar PV power generation and delivery. Section 3 elaborates the control methodology of the process. Section 4 gives a brief idea on IoT integration with PLC. The results and finally, the conclusions drawn are given in Sect. 5 and 6, respectively.

2 Solar PV MPPT

The increase in carbon footprint in the atmosphere led to the idea of green energy. The simulation is carried out using Kyocera KC 200GT PV array in MATLAB integrated with a DC-DC boost converter and a DC-AC inverter to run the induction motor. The perturb & observe (P&O) algorithm is used to detect the peak power and accordingly adjusts the duty cycle to trigger the switch (IGBT) of the converter, the output of which is fed to a three-phase VSI that converts dc voltage to three-phase ac voltage. The I-V relation is depicted in (1).

$$I = I_{\text{pv}} - I_0\left[\exp\left(\frac{V + IR_s}{V_{\text{thm}}\alpha}\right) - 1\right] - \frac{V + IR_s}{R_{\text{sh}}} \tag{1}$$

where I_{pv} represents the photo-current and I_0 represents the saturation current of the diode. The module's thermal voltage V_{thm} is given as (2).

$$V_{\text{thm}} = \frac{N_s KT}{q} \tag{2}$$

where N_s is the number of series-connected cells, K is the Boltzmann Constant [9].

The study is performed with an irradiance of 1000 W/m^2 and temperature of 25 °C. The I vs V and P vs V curves are shown in Fig. 1.

The erratic nature of solar photovoltaic can be overcome by using a battery. (Preferably nickel-metal hydride battery is used). The unit is having 14 Ah with 220 V having one series and parallel stack. The battery can be charged and discharged based on its SoC (State of Charge) and availability of irradiance [10]. If Q be the battery charge capacity and i_b is the charging current then:

$$\text{SoC} = \left[1 - \frac{\int_0^t i_b dt}{Q}\right] \times 100\% \tag{3}$$

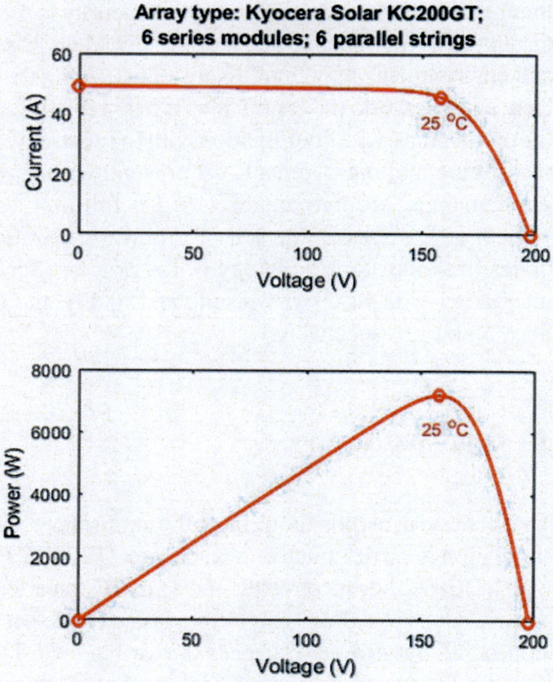

Fig. 1 Solar photovoltaic: I vs V & P vs V characteristics

3 Methodology

3.1 PLC: An Overview

Automation is nothing but the elimination of human interference in the production line. As discussed earlier, this can be done with the help of PLC (Programmable Logic Control). This helps in carrying out the sequential task in the industry efficiently as well as reducing the chance of mistakes, as a single mistake can result in a massive loss in the production unit. Nearly any assembly line, machine operation or process can be enhanced with this control system sort [11].

The ability to manipulate and replicate the process while collecting and communicating relevant data accounts for the PLC framework's primary advantage. The user specification can, therefore, be better suited by matching the form of input and output devices. The PLC architecture is shown in Fig. 2. The input modules accept and convert sensor signals to a logic e.g., Switches, Pushbuttons. The output module transforms the control instructions into a signal that the actuators will be triggered.

Several works related to PLC based motor protection have been done in the past. Some of which were of over-voltage, over-current, or over-speed-based protections [12–14]. However, the literature surveyed have either not incorporated IoT or have not combined a renewable energy source. Thus, in this work, comprehensive relaying

Fig. 2 PLC components

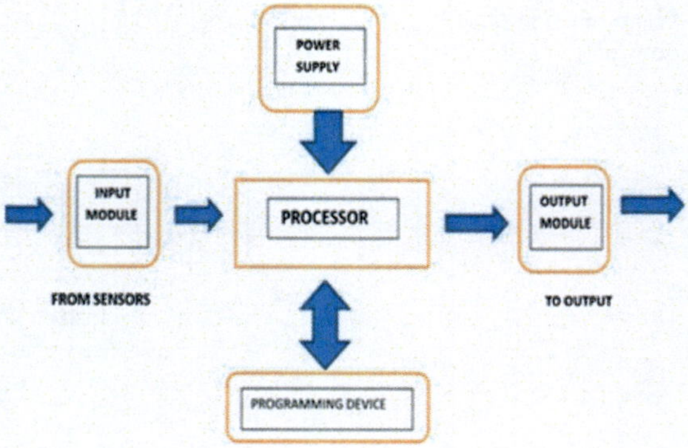

of a motor-driven conveyor belt involving IoT, solar a PLC has been described. The control methodology has been elaborated in the subsequent section.

3.2 *Working of PLC for the System Under Study*

The PLC has been programmed according to our required needs. The specification of the Squirrel Cage Induction Motor (SCIM) taken is of 10 HP, with a rated output of 7.5 kW, rated voltage 415 V, rated current of 13 A, rated speed of 2945 rpm and rated frequency of 50 Hz and rated torque of 2.5 kgf-m or 24 Nm.

Two conditions considered in this paper is as follows:

(1) When motor is sensing the torque if it is higher than the rated torque motor stops. If it is less than the rated torque, the next conditioned is being checked, i.e. if the temperature is less or greater than 63.5% of rated temperature.
(2) If the temperature is higher than the above-said temperature then the cooling fan is programmed to be turned on, and simultaneously the OFF-Delay Timer is toggled for a delay time of 90 s, the temperature is checked in tandem. If it is not below 80% of the rated temperature, the condition is repeated or else is jumped into the next condition where the torque is again checked. If the torque is greater than or equal to 80% of rated torque. The motor is stopped, instantaneous relay comes into picture and triggers the breaker to stop the motor. For real-time, the virtual sensor measures the temperature curve and forecasts the effect of temperature on running the motor at various loads. The control strategy can be understood through the flowchart shown in Fig. 3. The emulation of the control logic complies with IEC 61,131-3, where, the PLC's control program is transcribed into Matlab/Simulink® environment [15].

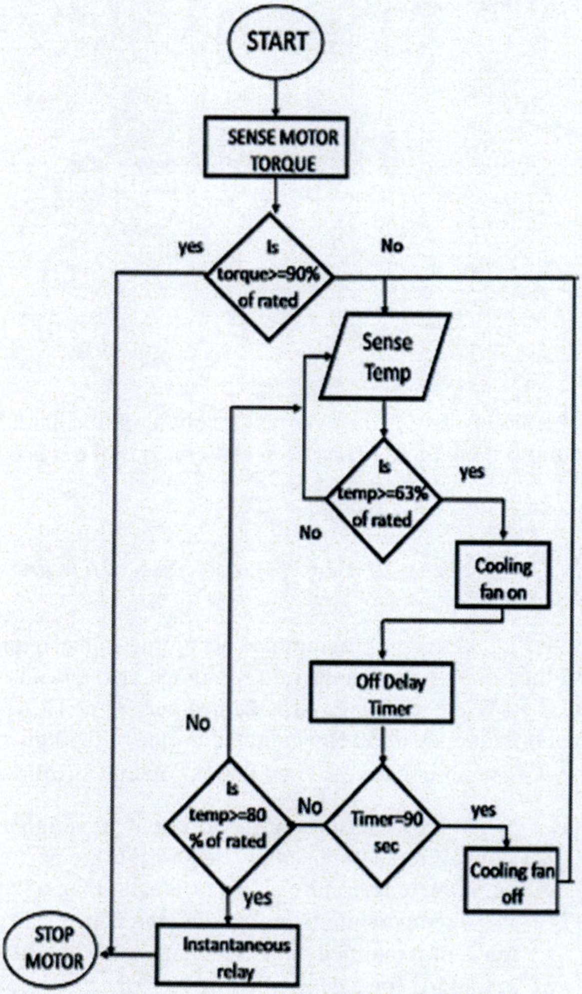

Fig. 3 PLC control flowchart

4 Integrating IoT with PLC

The Internet of Things (IoT) can be broadly defined as the cyber-physical existence of an identifiable entity that can potentially communicate with devices without the need of human interaction [16]. It encompasses a three-level structure that comprises of constrained devices, Smart devices/gateways, and IoT platforms along with wireless or wired interfaces. There are several issues in the use of wired mode. The first downside in the use of wired communication is the amount of wiring needed, which leads to an exorbitant cost of installing and repairing the cables. Moreover, these cables must always be correctly terminated. Once they are set down, it would be impractical to make any modifications in the future, as it is inconvenient to restructure and re-implement the whole framework [17].

A significant portion lies in the manner in which a PLC exposes its inputs and outputs (sensors and actuators) to external hardware and software. The PLC's core control loop is completely isolated from any new structure or add-ons that require networking as the external network communication is unsuitable for involvement in driving real-time controls owing to its non-deterministic nature. It should be kept in mind that the platform should be vendor-independent, allowing it to connect seamlessly.

On the one hand, the program running on the PLC may be updated to make different values available as outputs or to process new external inputs such as remote control requests. On the other hand, the PLC's existing outputs may be ingested using a standard industrial communication protocol by leveraging a gateway-based architecture [18].

A critical factor that governs the smooth operation of PLCs networked in cyberspace is cybersecurity. If there is a loophole, an insecure network may allow a threat to spread rapidly throughout the facility. Suspicious activities can be thwarted by extracting metadata from the network-flow through passive sensors, monitoring and learning normal behaviour through statistical and behavioural descriptions, recommending preventative actions by triggering an incident response based on the evidence of a breach [19].

Hence, in this work, Zigbee protocol is chosen as It is explicitly designed for a cost-effective Wireless personal Area (WPAN's) networks with low-data rate, and low energy usage thereby eliminating the redundancies of the wired mode of communication. Its mesh topology network and its 128-bit AES-based encryption system makes it best suited for industrial automation [20]. It is based upon the IEEE 802.15.4 platform, which allows transmission of data at different frequency bands [21]. An

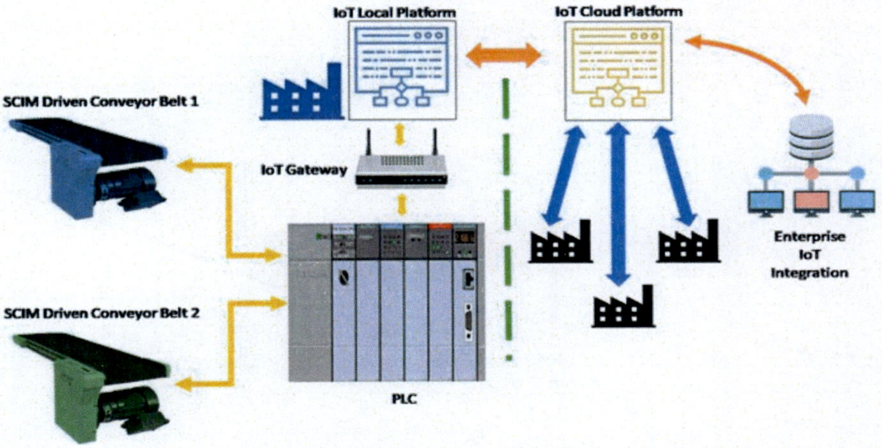

Fig. 4 Schematic diagram of IoT based PLC connected to the conveyor belt system

elaborative description of its functionality has been discussed in [22, 23]. Figure 4 shows a schematic diagram of the synchronization of PLC controlled system with IoT.

5 Results and Discussions

The dual-stage inverter topology finds its application in an industrial unit in which the frequency of 50 Hz is to be maintained throughout. The three-phase Line to Line voltage supply of 415 V is given to the motor. The RMS voltage waveform is exhibited in Fig. 5 (Fig. 6).

The charging and discharging status can be depicted by SoC of the battery unit. The motor will turn off or turn on based on the PLC command. The conveyor weight and the heating status is measured by the sensors and send the trip signal to the breaker. The THD (Total Harmonic Distortion) is also kept below 5%. The Fig. 7 below depicts the speed and torque graph. The motor is turned off at time 0.96 to 2.5 s, after that again the normal operation is restored, and hence the motor operates. Due to the inertia of the motor, the rotor will oscillate and hence during the restoration period; it will operate normally. The PLC with its ladder logic, the time taken to

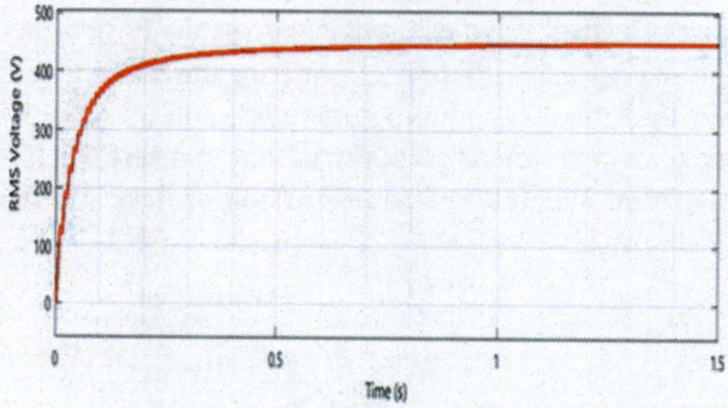

Fig. 5 Output RMS line-line voltage from inverter

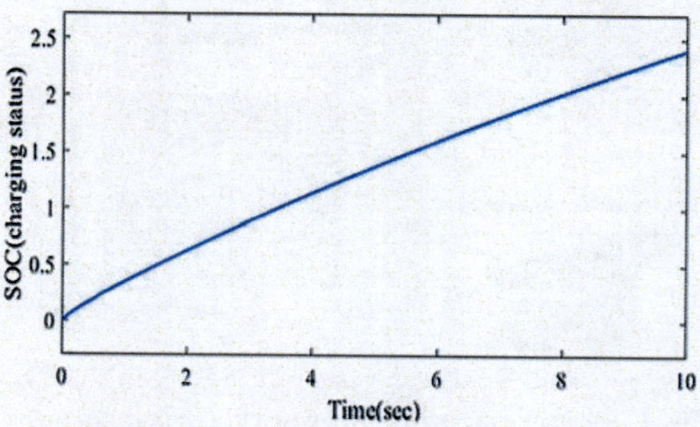

Fig. 6 Battery SOC

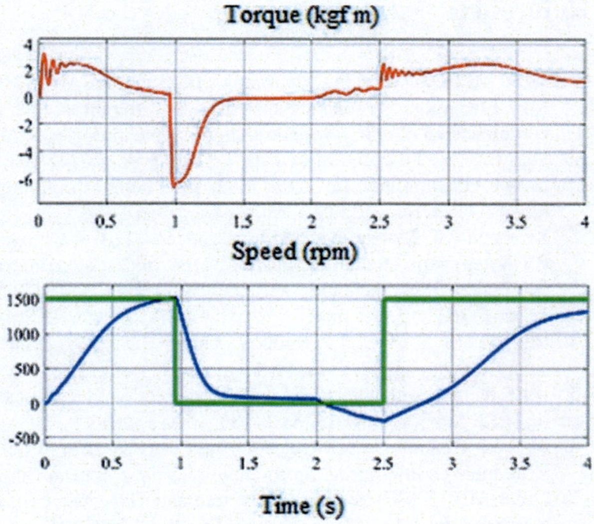

Fig. 7 Torque and speed curves

Table 1 Comparative analysis for the observed response time with and without PLC for different loading

S.no	Percentage loading (%)	Time of response (s)	
		With PLC	Without PLC
1	40	1.28	2.33
2	70	0.96	2.14
3	90	0.81	1.29

trigger the breaker is very less, and one can control the complete system just sitting in the control room.

A comparative analysis based on the simulation results have been summarised in Table 1.

6 Conclusion

Regulation of several processes can be a daunting job in the present scenario. There is a significant saving of time and money with the aid of smart control methods and the incorporation of cyber-physical systems. The above method gives an idea of how PLC's can easily manage complicated controls at the same time. With the incorporation of green energy, there is an appreciable decrease in the carbon footprint and thus makes it environment friendly. The operation of a large number of motors can be monitored/controlled using low bandwidth secured communication network. Therefore, the scope of IoT can be broadened to all spheres of power system operations.

References

1. Azad ML, Das S, Sadhu PK, Satpati B, Gupta A, Arvind P (2017) P & O algorithm based MPPT technique for solar PV system under different weather conditions. In: 2017 international conference on circuit, power & computing technologies (ICCPCT), pp 1–5
2. Selmi T, Abdul-Niby M, Devis L, Davis A (2014) P&O MPPT implementation using MATLAB/Simulink. In: 2014 ninth international conference on ecological vehicles and renewable energies (EVER), Monte-Carlo, pp 1–4
3. Krishnan AR, Mohammed SS, Manafudeen S (2019) Comparison of P&O MPPT based solar PV system with interleaved boost converter. In: 2019 2nd international conference on intelligent computing, instrumentation and control technologies (ICICICT), Kannur, Kerala, India, pp 1370–1376
4. Chakraborty S, Arvind P, Sinha S, Kumar SD (2019) Implementation of IoT based PLC in smart relaying applications. Int J Adv Innov Res 6(2):249–252
5. IEEE recommended practice for motor protection in industrial and commercial power systems. In: IEEE Std 3004.8-2016, pp 1–163, 15 May 2017
6. Bhavesh Count and Pack Objects from Conveyor using PLC Ladder Logic. https://instrumentationtools.com/plc-program-to-count-and-pack-objects-from-conveyor/
7. Lahari MNS (2017) Venkatesan DP automated conveyor belts for object counting in small-scale industries. Int J Adv ManagTechnol Eng Sci 7:287–294
8. Machidon OM, Stanca C, Ogrutan P, Gerigan C, Aciu L (2018) Power- system protection device with IoT-based support for integration in smart environments. PloS One 13(12):e0208168
9. Chakraborty S, Arvind P, Kumar D (2019) Integrated solar PV MPPT and V-*f* Control for stand-alone microgrid. In: 2019 international conference on electrical, electronics and computer engineering (UPCON), Aligarh, India, pp 1–6
10. Wahyuddin MI, Priambodo PS, Sudibyo H (2018) State of charge (SoC) analysis and modeling battery discharging parameters. In: 2018 4th international conference on science and technology (ICST), Yogyakarta, pp 1–5
11. bin Zainon M, bin Zulkifli MZ, Tamjis MR (2016) Design and construction of an automated system for parts assembly machine. Int J Appl Eng Res 11(8):5545–5548
12. Ioannides MG (2004) Design and implementation of PLC-based monitoring control system for induction motor. IEEE Trans Energy Convers 19(3):469–476
13. Colak I, Bayindir R, Bektas A, Sefa I, Bal G (2007) Protection of induction motor using PLC. In: 2007 international conference on power engineering, energy and electrical drives, Setubal, Portugal, pp 96–99
14. Kumar D, Basit A, Saleem A, Abbas EG (2019) PLC based monitoring & protection of 3-phase induction motors against various abnormal conditions. In: 2019 2nd international conference on computing, mathematics and engineering technologies (iCoMET), Sukkur, Pakistan, pp 1–6
15. Pereira A, Lima C, Martins JF (2011) The use of IEC 61131-3 to enhance PLC control and Matlab/Simulink process simulations. In: 2011 IEEE international symposium on industrial electronics, Gdansk, pp 1243–1247
16. Al-Fuqaha A, Guizani M, Mohammadi M, Aledhari M, Ayyash M (2015) Internet of Things: a survey on enabling technologies, protocols, and applications. IEEE Commun Surv Tutor 17(4):2347–2376
17. Bhattacharjee S, Gokalgandhi B, Singh A (2014) Industrial automation and sensing using ZigBee. Int J Comput Appl 105(10):37–40
18. What You Need to Know About PLC Programming in the Age of IoT, January 2019. https://blog.temboo.com/plc-programming-for-IoT-systems
19. Nugent E, August MR (2016) SCADA cybersecurity in the age of the Internet of Things: supervisory control and data acquisition (SCADA) systems' traditional role is changing as the Industrial Internet of Things (IIoT) continues to take a larger role. SCADA systems need to adjust. Control Eng 63(9):36
20. Wang Y, Chen C, Jiang Q (2019) Security algorithm of Internet of Things based on ZigBee protocol. Cluster Comput 22(6):14759–14766

21. Joshi P, Singh G, Raghuvanshi AS (2020) Comparative study of different routing protocols for IEEE 802.15. 4-enabled mobile sink wireless sensor network, pp 161–170
22. Kantilal VM, Lakhani AP (2014) Zigbee based wireless monitoring and controlling of automation system using PLC & SCADA. Int J Adv Res Electr Electron Instrum Eng 3(1):6768–6775
23. Long S, Miao F (2019) Research on ZigBee wireless communication technology and its application. In: 2019 IEEE 4th advanced information technology, electronic and automation control conference (IAEAC), Chengdu, China, pp 1830–1834

Stacking Based Ensemble Learning for Improved Software Defect Prediction

Sweta Mehta and K. Sridhar Patnaik

Abstract The process of developing a good quality software requires rigorous testing of the software modules. The effort and resources required for testing can be reduced by early prediction of the defects present in various modules of the software. This paper aims to do a comparative research on different classification algorithms taking into consideration the data imbalance and high dimension of the defect datasets. Artificial Neural Network (ANN), Decision Trees, K-nearest neighbour, SVM and Ensemble Learning are some of the algorithms in machine learning that have been used for classifying the modules in software as defect-prone and not defect-prone. For datasets from PROMISE repository, multiple software metrics have been evaluated with feature selection (FS) techniques such as Recursive Feature Elimination (RFE) and correlation based FS combined with Synthetic Minority Oversampling Technique for imbalanced datasets. In this research work, Stacking Ensemble technique gave best results for all the datasets with defect prediction accuracy more than 0.9 among the algorithms used for this experiment.

Keywords Defect prediction · Feature selection · Data imbalance · Machine learning algorithms · Stacking ensemble classifier

1 Introduction

Software systems have become an inseparable part of our lives. Any disruption in the working of software requiring high dependency can have serious consequences. To ensure error free working of the systems, software must undergo a thorough testing process. As long as the software is small, testing it is easy. But in recent years, size and complexity of software have increased immensely thus increasing

S. Mehta (✉) · K. S. Patnaik
Department of Computer Science and Engineering, Birla Institute of Technology, Mesra, Ranchi 835315, India
e-mail: mtit10008.18@bitmesra.ac.in

K. S. Patnaik
e-mail: kspatnaik@bitmesra.ac.in

V. Nath and J. K. Mandal (eds.), *Proceeding of Fifth International Conference on Microelectronics, Computing and Communication Systems*, Lecture Notes in Electrical Engineering 748,
https://doi.org/10.1007/978-981-16-0275-7_14

the complexity and time for software development and testing process [1, 2]. The software once developed, goes through multiple updates depending on change is user requirements followed by testing phase. To enhance the testing productivity, an early detection of defect prone software modules is very helpful. This helps to focus more on the testing on specific modules of the software which ensures defect free software. In case of detection of fault prone modules before the release of the software, developers can be cautious of not committing an error during the development or updating process of the software. This in turn helps to reduce the software delivery time, cost and effort. With the development of multiple machine learning algorithms many researchers have focused in the area of developing defect prediction models. These models require dataset which contains data related to the software metrics of each module in the software. Detection of defects at an early stage as defect prone and not defect prone ensures efficient usage of testing resources in the software development cycle [3, 4].

2 Literature Review

Defect prediction modelling has been a very important area of research in past studies. Researchers' have been making use of various Machine Learning Algorithms for classifying the software modules as defect prone or not defect prone. Logistic Regression, Support Vector Machines, Artificial Neural Network, Decision Trees, K-Nearest Neighbour, Random Forest etc. are some of the machine learning technique based models and statistical models based on which multiple research work has been conducted by many researchers. Burcu et al. [5] used seven classification algorithms for their performance evaluation in defect prediction in software and identify the best performing algorithm by using NASA datasets that are publically available on PROMISE repository [6]. The research indicated that performance of Random Forests (RF) and Bagging belonging to ensemble learner category produced best results. Two major issues that researchers have faced while using software defect datasets are data imbalance and higher dimensionality of data. Researchers have implemented multiple techniques of feature selection from the dataset. A new technique of feature selection using support vector machine was proposed by Ajit et al. [7]. Technique of feature selection based on filter was proposed by Bora [8]. Their study used datasets from multiple software to create model based on rank. Since software have very few defect prone modules, software defect datasets are highly skewed towards Not-Defect-Prone modules. In recent years, the issue of imbalanced class due to skewed dataset has been researched a lot. In [9], Benin et al. investigated the significance and impact that data resampling has on defect prediction models. Research work conducted by Wang et al. [10] using AUC (Area under Curve) and G-mean verified that using balanced dataset provides better results than imbalanced dataset with an exception to Naïve Bayes classifier.

3 Software Defect Metrics

Software has different kind of features. These features indicate the characteristic of the software. Software metrics is used to measure these characteristics. Software performance and quality can be determined using these metrics. Maintainability and reliability of the software can be estimated using this. They are mainly of four types: Traditional, Object Oriented, Hybrid and Miscellaneous Metrics [5, 19]. Traditional metrics mainly consists of metrics that measure complexity and functional size of the software. It includes size metrics e.g.: Lines of Code, metrics given by McCabe and Halstead. Properties of Object-Oriented (OO) software like cohesion, inheritance, coupling etc. of Object-Oriented classes are measured using Object-Oriented metrics. Metrics defined by Chidamber and Kemerer belong to this category. The most important OO metrics are Response for a Class, Coupling between Objects along with LOC as they improve efficiency of prediction when used with FS methods [11].

4 Methodology

This section consists of the machine learning algorithms used and the dataset description.

4.1 Machine Learning Algorithms Used

In this experiment of predicting software defects at an early stage of the developing the software, the research of Malhotra et. al. [11, 13] was taken into consideration for selecting the machine learning algorithms. Among multiple techniques mentioned in this research such as, Bayesian Learners, Ensemble Learners, SVM and Artificial Neural Networks, we have considered the techniques mentioned below along with Stacking and Extra Trees (ET) [17].

In order to increase the accuracy of prediction, usage of feature selection (FS) techniques such as Recursive Feature Elimination (RFE), correlation based FS, Data Balancing using Synthetic Minority Oversampling Technique (SMOTE) is done in this research work. SMOTE is one of the oversampling methods which help to handle the data imbalance problem. It randomly increases minority class examples by replicating them to balance the class distribution [14, 16]. Once we get the balanced dataset feature selection is done. Correlation based feature selection method generates correlation values between the attributes and rank them accordingly while RFE repeatedly constructs a model and picks either the topmost performing feature or the least performing feature, saves that feature and then repeats the same process with the leftover features. This process is repeated as long as features of the dataset are not over. Selecting the features repeatedly using lesser size of feature set for each iteration is the main aim of RFE.

This processed dataset is then used with machine learning algorithms for software defect prediction. Regularization technique such as dropout is also used for ANN to overcome overfitting conditions. Ensemble learning techniques have been used for the same datasets and comparative analysis was done based on their results.

1) *Artificial Neural Network:* ANN consists of a network of nodes to represent human brain network [12]. In this research work Multilayer Perceptron is used which has one input and output layer each with multiple hidden layers in between them. Dropout is also used in the ANN model. It is a regularization technique to overcome over-fitting conditions by adding a penalty to the loss function. Dropout is implemented per-layer in a neural network where the nodes are dropped randomly while training.
2) *Support Vector Machine:* The aim of this technique is to find decision planes using which decision boundaries can be designated. It places a boundary between the instances of the two specified classes. In this research, Linear Kernel SVM was used along with regularization parameter [15].
3) *Decision Trees*: It is a technique to create a model for classification by training it using decision rules generated using the training data.
4) *K-Nearest Neighbour*: It is an algorithm which classifies data based on measure of similarity such as distance functions.
5) *Logistic Regression*: It solves classification problems by using independent variables to get the probabilistic value as predicted output for a binary dependent variable.
6) *Ensemble Learning:* Ensemble method is an algorithm that combines the output of multiple models, which in this case are different classifier models, into one model in order to enhance predictive capability of the model.

- Random Forest: Random Forest makes use of numerous decision trees to create a model along with randomly selected training data as well as feature subset.
- Extra Trees: This technique is alike Random Forest as Extra Trees are its more randomized version. This randomization is achieved by selecting random splits for the decision trees instead of best splits as in case of Random Forest.
- Bagging: Bagging combines the output of several models to get better output of the model than a single one. In this research work, Decision Tree has been used as the base model.
- Max Voting: This ensemble learner uses output values of each data set from multiple models to predict the final output. The output of each model is interpreted as a 'vote'. The output that is present for most of the models is considered as the final output.
- Stacking: Stacked Generalization is an advanced ensemble technique in which a new model is trained to consolidate the predictions from two or more models that are already trained on the dataset. The models at the base level generally are

Fig. 1 Flowchart diagram of stacking ensemble

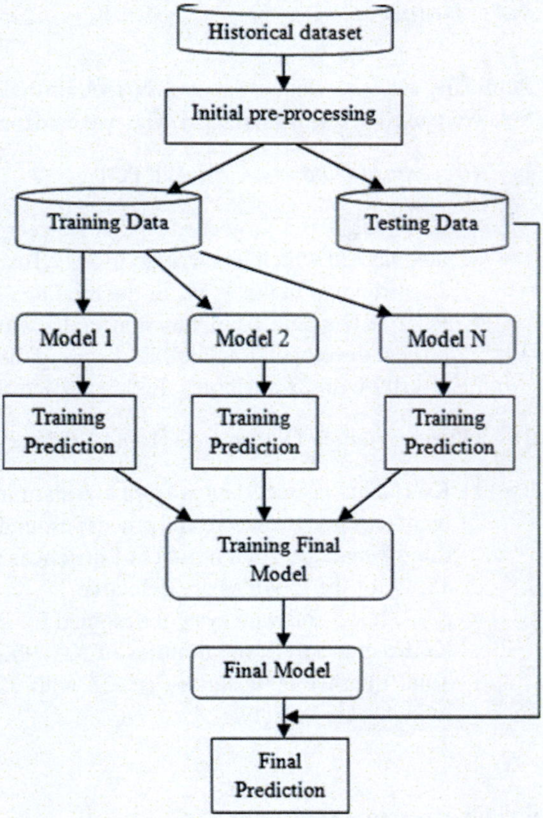

dissimilar learning algorithms. Thus, stacking ensemble models are of heterogeneous type. The test dataset is then used for prediction using this meta-classifier [18] (Fig. 1).

Stacking Algorithm

Input: InpData = { (a, b) | a $\in$ A, b $\in$ B}
Output: Stacking Ensemble Classifier Model, S

1. Step 1: Pre-process InpData
2. Step 2: Train classifiers at 1st level
3. For y $\leftarrow$ 1 to Y do
4. Based on InpData train a base classifier s_y
5. Step 3: Using InpData create a new set of data
6. For m $\leftarrow$ 1 to z do
7. Form a new input data set, InpData containing $\{a_m^{new}, b_m\}$
   $a_m^{new} = \{ s_n(a_m)$ for n=1 to Y$\}$
8. Step 4: Train classifier at 2nd level
9. Using $InpData^{new}$ train a new classifier s^{new}
10. Return $S(a) = s^{new}(s_1(a), s_2(a) \ldots s_Y(a)))$

4.2 Datasets

Publically available datasets such as **NASA Data Sets, PROMISE Repository Data Sets** were used in this research [2]. The two categories of dataset is mentioned below:

1) *Procedural Datasets*: CM1 and PC1.

- CM1: CM1 is a project of NASA spacecraft instrument. The programming language in which it is written in "C". Total number of instances is 498 with 22 attributes. In this 9.7% of the modules are defective.
- PC1: It has data from function codes written in "C" of a flight software project developed for satellites orbiting the earth. Total number of instances is 1109 with 22 attributes. In this 6.9% of the modules are defective.

2) *Object Oriented Datasets*: KC1 and KC2.

- KC1: This project data is from a system implemented for storage management required to receive and process ground data written using C++ programming language. Total number of instances is 2109 with 22 attributes. In this 15.4% of the modules are defective.
- KC2: It is a software system designed for data processing as an extension of KC1 using only some libraries of KC1 that belonged to 3rd party software. Total number of instances is 522 with 22 attributes. In this 6.3% of the modules are defective.

5 Evaluation Criteria

In order to do a comparative analysis we need to compare the predictive capability of the used machine learning models. Evaluation of the performance of techniques used for this research was done by using confusion matrix. Values of the confusion matrix are listed below (Tables 1 and 2):

Table 1 List of values of confusion matrix

Values	Meaning
True Positive (TP)	Output predicted by model and output in test data is true
True Negative (TN)	Output predicted by model and output in test data is false
False Positive (FP)	Output predicted by model is true while output in test data is false
False Negative (FN)	Output predicted by model is false while output in test data is true

Table 2 List of evaluation metrics

Evaluation metrics	Accuracy
Accuracy	$\frac{TN+TP}{TP+FP+TN+FN}$
Recall	$\frac{TP}{FN+TP}$
Precision	$\frac{TP}{FP+TP}$
F-measure	$\frac{2*Precision*Recall}{Precision+Recall}$

6 Experimental Results

The main purpose of this research is to assess the performance of multiple machine learning algorithms with various ensemble models using the performance measures mentioned above. The summary of results obtained in this research work is described in Table 4. It has been observed that the performance of a Multi Layer Perceptron model for procedural datasets (PC1 and CM1), accuracy of the model after carrying out feature selection based on correlation was better in comparison to other MLP models while MLP model with SMOTE and dropout implementation performed better for Object Oriented Datasets (KC1 and KC2). For SVM model, RFE gives best results for PC1.

For CM1, accuracy of the model is same even after reducing the dimension. For KC1, as the attribute have high correlation among them, therefore selecting certain features for building the model result in decrease in accuracy of the model. For KC2, accuracy of the model using correlation based dimension reduction technique is better among the other models. In case of Random Forest accuracy of the classifiers is better when used with correlation based feature selection for dimension reduction. RFE gives better results compared to correlation based feature selection for dimension reduction. RFE and smote performs better for procedural data set while it does not show much improvement for OO datasets. Performance of Extra Tree is similar to that of Random Forest for the considered software defect dataset. Bagging implementation of decision tree gives better result for correlation based feature selection approach. Bagged decision tree performs better than individual decision tree model. Using MaxVoting technique to the ensemble of Decision Trees, KNN and Randon Forest gives better result for correlation based feature selection approach.

The best accuracy in defect prediction among the mentioned models for the experimental dataset was achieved by the use of stacking based ensemble learning technique. In stacking ensemble, DT+KNN+LR and KNN+DT+RF clearly give better accuracy for majority of the datasets (PC1, CM1, and KC2). While for KC1, almost all the combination of classifier gives better result except KNN+RF+LR (Table 3).

Based on the work of Burcu et al. [5], the accuracies of best performing machine learning techniques are listed in Table 5. In their experiment the best accuracy of defect prediction was achieved by using Bagging for PC1 and CM1, Random Forest for KC1 and MLP for KC2 dataset. Based on the analysis of data in Table 4 and Table 5, it is concluded that stacking based ensemble technique used in this research gives

Table 3 Summary of results of stacking ensemble

Project	Metrics	KNN+RF+LR	KNN+DT+RF+LR	DT+LR+KNN	DT+LR+RF	DT+KNN+RF
PC1	Accuracy	0.9639	0.9639	**0.977**	0.9639	**0.977**
	Precision	0.96	0.96	0.98	0.96	0.98
	Recall	1.00	1.00	1.00	1.00	1.00
	F-measure	0.98	0.98	0.99	0.98	0.99
CM1	Accuracy	0.96	0.96	**0.96**	0.96	**0.96**
	Precision	0.96	0.96	0.96	0.96	0.96
	Recall	1.00	1.00	1.00	1.00	1.00
	F-measure	0.98	0.98	0.98	0.98	0.98
KC1	Accuracy	0.914	**0.971**	**0.969**	**0.971**	**0.969**
	Precision	0.91	0.97	0.97	0.97	0.97
	Recall	1.0	1.0	1.0	1.0	1.0
	F-measure	0.95	0.98	0.98	0.98	0.98
KC2	Accuracy	0.942	0.942	**0.96**	0.942	**0.96**
	Precision	0.93	0.93	0.95	0.93	0.95
	Recall	1.0	1.0	1.00	1.0	1.00
	F-measure	0.97	0.97	0.98	0.97	0.98

Table 4 Result summary of machine learning algorithms

Project	Metrics	MLP	LR	DT	KNN	SVM	RF	ET	Bagging	Max voting	Stacking
PC1	Accuracy	0.9234	0.9324	0.918	0.9324	0.9399	0.949	0.945	0.936	0.941	**0.963**
	Precision	0.93	0.94	0.95	0.94	0.94	0.96	0.96	0.95	0.94	**0.96**
	Recall	0.99	1.0	0.96	1.0	1.00	0.99	0.99	0.98	1.0	**1.0**
	F-measure	0.96	0.96	0.96	0.96	0.97	0.97	0.97	0.97	0.97	**0.98**
CM1	Accuracy	0.93	0.91	0.86	0.92	0.93	0.91	0.92	0.92	0.91	**0.977**
	Precision	0.95	0.94	0.93	0.93	0.93	0.94	0.94	0.94	0.94	**0.98**
	Recall	0.98	0.97	0.91	0.99	1	0.97	0.98	0.98	0.97	**1.00**
	F-measure	0.96	0.95	0.92	0.96	0.96	0.95	0.96	0.96	0.95	**0.99**
KC1	Accuracy	0.853	0.838	0.7938	0.831	0.845	0.841	0.855	0.822	0.848	**0.96**
	Precision	0.86	0.86	0.87	0.86	0.85	0.86	0.88	0.86	0.86	**0.96**
	Recall	0.98	0.97	0.89	0.95	0.99	0.97	0.95	0.94	0.98	**1.00**
	F-measure	0.92	0.91	0.88	0.90	0.91	0.91	0.92	0.90	0.92	**0.98**
KC2	Accuracy	0.866	0.828	0.819	0.838	0.8571	0.876	0.838	0.857	0.857	**0.969**
	Precision	0.88	0.86	0.87	0.87	0.85	0.88	0.85	0.87	0.85	**0.97**
	Recall	0.96	0.94	0.90	0.94	1.0	0.98	0.96	0.96	0.99	**1.0**
	F-measure	0.92	0.90	0.89	0.90	0.92	0.93	0.90	0.91	0.92	**0.98**

Table 5 Best performing techniques as per Burcu et al.'s experiment

Dataset	MLP	Bagging	Random Forest
PC1	0.936	**0.941**	0.937
CM1	0.876	**0.898**	0.888
KC1	0.859	0.86	**0.867**
KC2	**0.847**	0.837	0.833

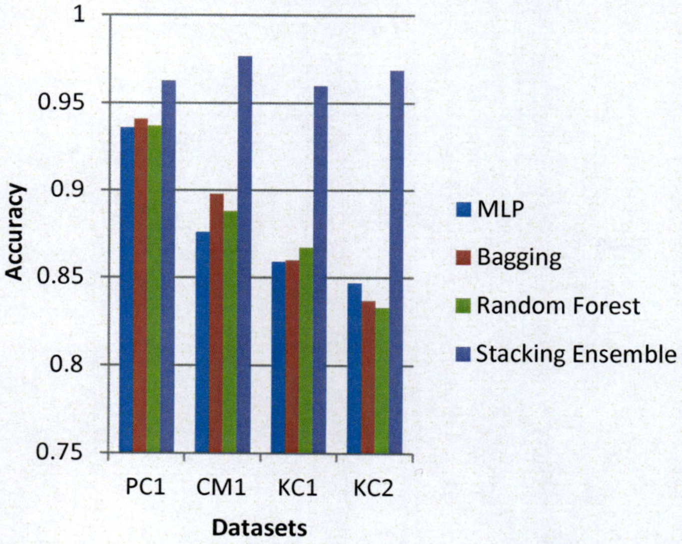

Fig. 2 Comparison of performance between best techniques in Burcu et al.'s experiment and stacking based ensemble

better results when compared to the techniques used in previous research papers. This comparison is depicted in Fig. 2 for all the datasets used in this research.

7 Conclusions

The requirement of defect prediction techniques of higher accuracy and efficiency for software systems will always be a field of interest for researchers. This research work helps to predict the defect before the actual testing and hence reduce the time and cost of the software projects. Using Recursive Feature Elimination (RFE), correlation based FS, Data Balancing using Synthetic Minority Oversampling Technique (SMOTE) improved the results of multiple Machine learning algorithms. Stacking ensemble learning method provided best results among the techniques used in this research. Further research could be planned to create many other hybrid models so as to develop models which would predict the defects of software systems with more accuracy and minimum errors.

References

1. Yang F (2005) On software engineering technology development. J Softw 16(01):1000–9825
2. Hauer F, Pretschner A, Schmitt M, Grötsch M (2017) Industrial evaluation of search-based test generation techniques for control systems. In: The 28th international symposium on software reliability engineering (ISSRE), Toulouse, France, 23–26 October 2017
3. Sun F-B, Zhang J (2017) Dealing with fractional failures and survivors in data analysis of reliability engineering. In: 2017 annual reliability and maintainability symposium (RAMS), Orlando, FL, USA, 23–26 January 2017
4. Costa EO, Pozo ATR, Vergilio SR (2010) A genetic programming approach for software reliability modeling. IEEE Trans Reliab 59:222–230
5. Burcu, Merve (2019) 4rd international conference on computer science and engineering (UBMK 2019). IEEE
6. Sayyad Shirabad J, Menzies TJ (2005) The PROMISE repository of software engineering databases. School of Information Technology and Engineering, University of Ottawa, Canada
7. Shenvi AA (2009) Defect prevention with orthogonal defect classification. In: Proceeding ISEC 2009 proceedings of the 2nd India software engineering conference
8. Caglayan B (2010) Usage of multiple prediction models based on defect categories. In: Proceeding PROMISE 2010 proceedings of the 6th international conference on predictive models in software engineering
9. Bennin KE, Keung J, Monden A, Phannachitta P, Mensah S (2017) The significant effects of data sampling approaches on software defect prioritization and classification. In: Proceedings of the 11th ACM/IEEE international symposium on empirical software engineering and measurement, pp 364–373. IEEE Press, November 2017
10. Wang S, Yao X (2013) Using class imbalance learning for software defect prediction. IEEE Trans Reliab 62(2):434–443
11. Malhotra R (2015) A systematic review of machine learning techniques for software fault prediction. Appl Soft Comput J 27:504–518
12. Reddivari S, Raman J (2019) Software quality prediction: an investigation based on machine learning. In: IEEE 20th international conference on information reuse and integration of data science
13. Wang F, Ai J, Zou Z (2019) 19th international conference on software quality, reliability and security (QRS). IEEE
14. Song Q, Guo Y, Shepperd M (2019) A comprehensive investigation of the role of imbalanced learning for software defect prediction. IEEE Trans Softw Eng 45(12):1253–1269
15. Arora I, Saha A (2018) Software defect prediction: a comparison between artificial neural network and support vector machine. In: Advanced computing and communication technologies, pp 51–61. Springer, Singapore
16. Sohan MF, Kabir MA, Jabiullah MI, Rahman SSMM (2019) Revisiting the class imbalance issue in software defect prediction. In: 2019 international conference on electrical, computer and communication engineering (ECCE), Cox'sBazar, Bangladesh, pp 1–6
17. Delphine Immaculate S, Farida Begam M, Floramary M (2019) Software bug prediction using supervised machine learning algorithms. In: 2019 international conference on data science and communication (IconDSC), Bangalore, India, pp 1–7
18. Awad MA, ElNainay MY, Abougabal MS (2017) Predicting bug severity using customized weighted majority voting algorithms. In: 2017 Japan-Africa conference on electronics, communications and computers (JAC-ECC), Alexandria, pp 170–175

19. Meiliana, Karim S, Warnars HLHS, Gaol FL, Abdurachman E, Soewito B (2017) Software metrics for fault prediction using machine learning approaches: a literature review with PROMISE repository dataset. In: 2017 IEEE international conference on cybernetics and computational intelligence (CyberneticsCom), Phuket, pp 19–23

Design and Implementation of Single Stage Resonant Inverter for Electronic Ballast

Y. Sukhi, Y. Jeyashree, and S. Anita

Abstract Even though the market demand for compact fluorescent lamp has increased from a long period, these lamps have the problem of power factor in household applications. This has led to the design of resonant inverter for electronic ballast with discontinuous conduction using single stage. Buck boost converter at the input stage improves power factor. The duty cycle of the MOSFET switches are varied to control the load power. Based on the analysis and mode of operations the design equations are derived. The design equations are used to calculate the value of the circuit parameters. Experimental setup is built for 40-W electronic ballast of a fluorescent lamp.

Keywords Single stage · Resonant inverter · Electronic ballast · Discontinuous conduction mode · Power factor correction

1 Introduction

Nowadays environmental friendliness along with energy saving are the two things followed everywhere. The cost for the protection of environment need not be equal to the amount of energy saving. Non-biodegradable wastes are reduced due to the usage of electronic ballasts instead of electromagnetic ballasts. Generally magnetic ballasts last for 30 years and it can be recycled only after this period of life. Fluorescent lamps are replaced with electronic ballast due to energy conservation [1]. The driver circuit used to drive the fluorescent lamp should be energy efficient [2]. Fluorescent lamps

Y. Sukhi (✉) · S. Anita
EEE Department, R.M.K. Engineering College, Kavaraipettai, Tamilnadu, India
e-mail: ysi.eee@rmkec.ac.in

S. Anita
e-mail: saa.eee@rmkec.ac.in

Y. Jeyashree
EEE Department, SRMIST, Kattankulathur, Tamilnadu, India
e-mail: jeyashry@srmist.edu.in

V. Nath and J. K. Mandal (eds.), *Proceeding of Fifth International Conference on Microelectronics, Computing and Communication Systems*, Lecture Notes in Electrical Engineering 748,
https://doi.org/10.1007/978-981-16-0275-7_15

are used due to its characteristic properties like low running cost and high efficiency [3]. Even though electromagnetic ballast costs less and has high reliability, it has low efficiency and is bulky, heavy. This is because of the operation of the circuit at line frequency [4, 5]. Therefore electronic ballasts are used as compact fluorescent lamps [6]. This lamp reduces the amount of electrical energy used for lighting. The converters used for compact fluorescent lamps are to be designed to produce minimum harmonic content. The harmonic content standards are given by the IEC standards [8]. The lamp size can be reduced by high frequency of operation [9]. This also increases the life of the lamp. Load resonant converters are used in compact fluorescent lamp because it is easy to produce the ignition voltage of required level using this this converter. [10–13]. This also provides a stable arc current. DC voltage is obtained using a peak detection rectifier [11]. Generally power factor correctors are used in the input side of the converter to provide better shape for the current flowing in the input. This in turn changes the input current to be in phase with the input voltage. High frequency operation also provides better THD content for the input current. Filer circuit is designed in the input side to eliminate high frequency components generated produced due to high frequency of operation in the input side [14]. The input power factor correction is influenced by the harmonics in the supply current. This may cause interference. More over the power factor correction has to be improved along with total harmonic distortion. There are some standards like International Electro-technical Commission IEC61000-3-2 to give information about limits of THD. In case of low power applications, IEC61000-3-2 Class C standards are to be followed. The important point to be noted is that the shape of the input current has to be varied to obtain better power factor. It may lead to low THD. The input side produces a pulsating power. The brightness of the lamp depends upon the input current. Therefore driver circuit should deliver a constant power to the output. Power imbalance is produced between the input and the output side [7]. In order to meet out the energy imbalance, energy storage elements like inductors, capacitors are used. Control circuits are used to limit the power flow to the load from the source. The cost of control circuit does not vary with the rating of the lamp. So it costs more for low power lamps and the cost is comparatively less for high power applications. Analog control circuit uses more number of integrated chips. But microcontroller based control circuits have the advantages of low cost, easy modification of the circuit, high reliability, less affected by environmental effects and more flexible design [12]. The design of compact fluorescent lamps operating at high frequency provides solution for lighting applications [15–19]. This paper aims to implement electronic ballast operating at high frequency having high input power factor. This is constructed using buck boost converter integrated with resonant inverter. This developed topology provides improved input power factor. The circuit parameters are designed to fulfill the IEC standards. The values of circuit elements are selected to have discontinuous conduction mode (DCM) of operation. The proposed compact fluorescent lamp is designed with minimum number of energy transfer elements. Due to this the cost of the circuit is reduced and also provides better efficiency. The following Sect. 2 explains about the topology used for the fluorescent lamp. The analysis of the proposed topology is done in Sect. 3 which leads to the design of the lamp. The design equations are derived in Sect. 4. Section 5 explains the experimental setup with proto type developed and the experimental output results. Section 6 gives the conclusion of the paper.

2 Proposed Converter Topology

The circuit diagram of conventional converter with buck boost converter stage at the input side and half bridge resonant converter stage at the output side for electronic ballast is shown in Fig. 1. This circuit has resonant elements to produce sinusoidal current using the half bridge resonant inverter. The first stage is in the input side and the second stage in the output side. Discontinuous mode of operation is used in the first stage with buck boost converter. Half bridge inverter with series inverter is used in the second stage. The DCM of operation provides almost in-phase relationship between the input current and voltage in the input side. The number of switches used by two stage converter is four but the single stage converter uses two switches only. MOSFET switches are used as switching device. Bidirectional current flow is allowed in the circuit due to the reverse connected diode. This circuit is applicable to low and medium power applications. Power factor correction is done using the buck boost converter in the input side. Discontinuous mode of operation is ensured in the input stage by selecting proper value of input inductance L_{in}. This inductor is used to store the input energy and transfer it to load. The resonant converter in the output side reduces switching losses. Capacitor is connected across the lamp. The switches in the proposed converter are reduced to two but the number of switches in two stage converter is four. MOSFET device with anti-parallel diode is used as switch. Anti-parallel connection of diode allows the current in reverse direction. This circuit can be used for small and intermediate power applications. The input side phase relationship between voltage and current of the converter can be enhanced by the integration of the input dc-dc converter and output series resonant inverter. The resonant inverter has resonating elements made of capacitor and inductor. These resonant circuit elements are in series with the tube and parallel capacitor is connected across the tube to balance the fluctuations in voltage. The frequency at which the circuit is operating is slightly above or below the resonant frequency due to development of high current at resonant frequency. During the initial stage ignition, the resistance of the lamp is high. The electronic ballast is selected to operate at a high frequency of 50 kHz.

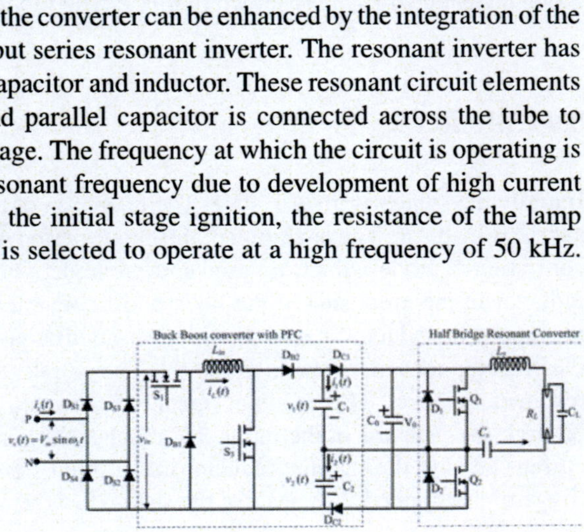

Fig. 1 Conventional converter

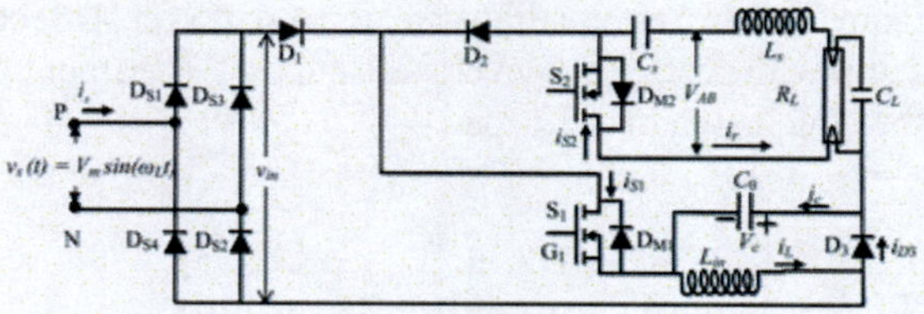

Fig. 2 Proposed single stage converter

Generally lighting loads except filament lamp are inductive in nature. The capacitor across the lamp is used to store energy and supplies a constant dc voltage.

3 Operation of the Circuit

The single stage converter is designed to change the duty cycle of dc-dc converter switch to have the buck operation and also the boost operation. A detailed discussion on the circuit structure of dc-dc converter is explained in this section. At present there are various techniques implemented to change voltage conversion ratio for dc-dc converters. The required output voltage is more and also less than the input voltage leads to boost operation and buck operation respectively. In case of buck operation the duty ratio is adjusted to values to lower than 0.5 and vice versa. The discontinuous conduction mode of operation is proposed in this converter. The operation of the control circuit is to change the duty cycle based on the load voltage to maintain the required value. Whenever the load voltage is more than the set voltage, the width of triggering pulse is reduced till it reaches the reference value. In case of lesser output voltage, the width of the gate pulse is increased till it reaches the reference value. The key waveforms for one cycle of the dc-dc buck boost converter for discontinuous mode under steady state is shown in Fig. 2. The modes are explained here.

3.1 *Mode 1*

Initially the switch is in OFF state. The operation of the circuit starts as soon as the switch is made on. The operation of this mode period is $t_0 < t < t_1$. The rectifier circuit consisting of diodes which are used convert ac-dc. The rectifier output is given to the inductor in the input side of the buck boost converter. Initially the current through the input side inductor is zero due to the discontinuous mode of operation. Therefore the inductor current increases from zero. The output capacitor C_0 discharges through resonant elements. This mode of operation comes to an end as the value of current through the inductor in the input side reaches the maximum value. As soon as the current flowing through the inductor in the input side reaches the maximum value, trigger pulse applied to the gate of the switch S_1 is removed and is made OFF. At this

Fig. 3 Mode 1 ($t_0 < t < t_1$)

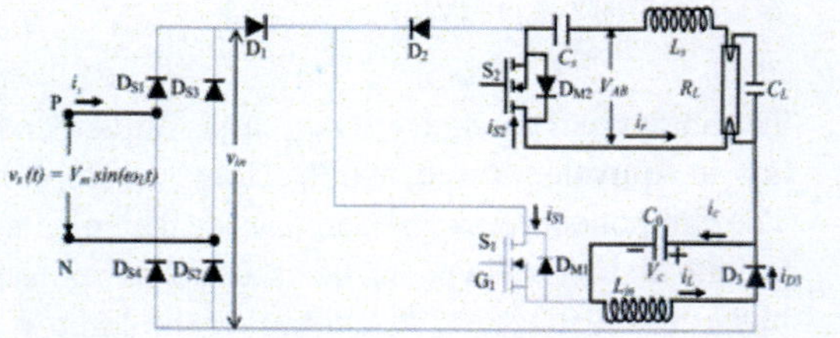

Fig. 4 Mode 2 ($t_1 < t < t_2$)

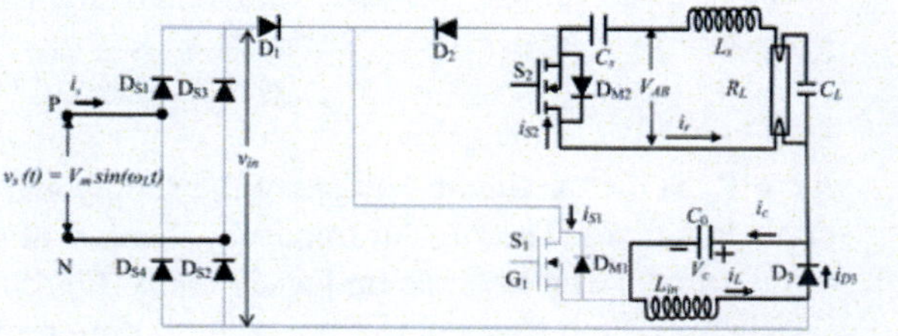

instant of operation, zero voltage switching (ZVS) happens due to the conduction of diode D_{M1} (Fig. 3).

3.2 *Mode 2*

The state of switch during the operation of this mode changes from on to off. The duration of time interval for the operation of this mode is $t_1 < t < t_2$. During the starting of this mode, the states of switch S_1 is off and the state of switch S_2 is on. In this mode the path of current flow is through switch S2, resonant inductor, the output capacitor and the diode D_3. At the end this mode, the current flowing through the resonant inductor becomes zero. Diode D_{M2} is conducting in this mode which ensures ZVS of the switch S_2 (Fig. 4).

3.3 *Mode 3*

During this mode, the state of switch S_2 is retained in the ON state. The time interval for this mode of operation is $t_2 < t < t_3$. The current flowing through the resonant circuit flows through the switch S_2. The diode D_3 does not conduct during mode 3. This mode comes to end only when next cycle starts by turning on switch S_1.

4 Circuit Analysis

The analysis of the converter circuit is implemented under DCM. In its analysis, lamp has an equivalent resistance R_L. Some assumptions are made for simplified analysis. The components are considered as ideal elements. By the use of large output capacitor C_0, the dc link voltage can be assumed as constant. Before the start of ignition takes place in the tube, the resistance of the lamp is very high. The expression for the working of the converter is obtained using the modes of operation.

The ac input voltage is sinusoidal in nature and its equation is

$$v_{in}(t) = V_p \sin \omega t \quad (1)$$

where V_p is the maximum voltage of the supply and ω is the angular frequency of supply in radians. The angular frequency in terms of supply frequency is given as $\omega = 2\pi f$ where f is the source frequency in Hz (Fig. 5).

Generally, high frequency of operation is preferred for the switches as compared to source frequency. Since switching is done at high frequency, the switching time interval is less and the voltage magnitude is assumed to be constant during this interval. When the switch is made ON in the first mode of operation, the supply current passes through inductor L_{in} and rises from zero linearly. The inductor current expression is written as

$$i_{Lin}(t) = \frac{V_p|\sin(2\pi f)|}{L_{in}} t \quad 0 \le t \le t_1 \quad (2)$$

Here the interval time t_1 is the on duration of switch S_1 and can be represented in terms of switching period T_{sw} and duty cycle as $t_1 = D \times T_{sw}$ and $D = T_{on}/T_{sw}$ where T_{on} is the ON time for S_1 and T_{sw} is the switching time period. When the first mode ends, the time during which the inductor current rises to the maximum value is given by

$$DT_{sw} = \frac{L_{in}}{V_m I_{Linm}} \quad (3)$$

The waveform of the current flowing through input inductor is shown in Fig. 6. During the mode 2, the state of switch S_1 changed from ON to OFF condition. As

Fig. 5 Mode 3 ($t_2 < t < t_3$)

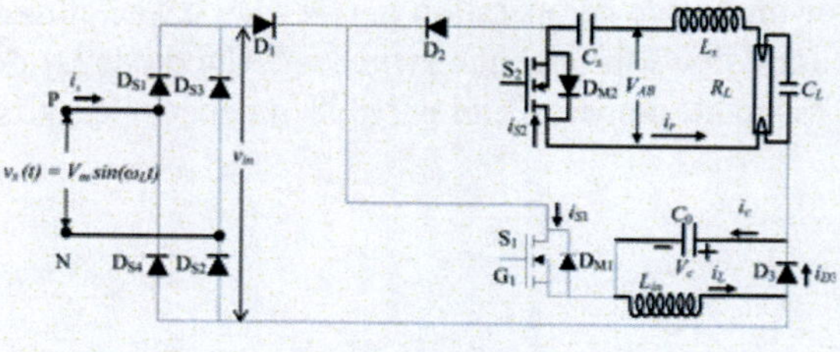

Fig. 6 Key waveforms

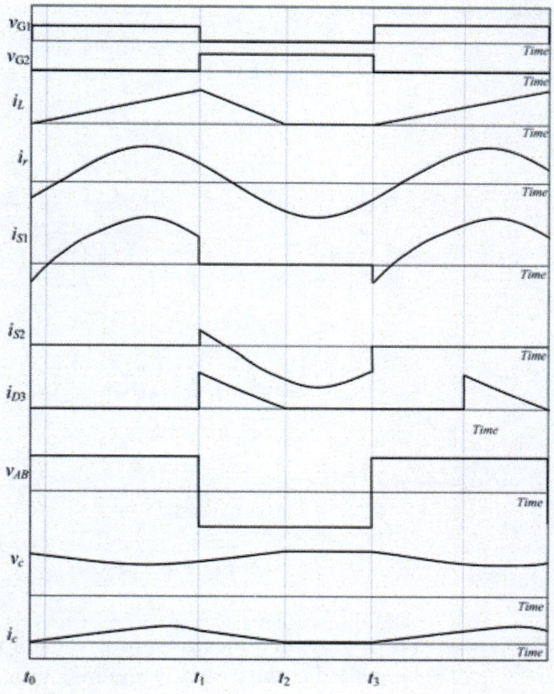

the second mode 2 ends, the input inductor current is zero. During the mode 2, energy transformation is between the inductor and capacitor. The voltage across the capacitor and current flowing through the input inductor are $V_C(t) = DV_p sin\omega t/(1\text{-}D)$ and $i_L(t) = 0$ respectively. Since discontinuous mode of operation is taking place, the switch S_2 is changed to OFF condition as the inductor current value is zero. The maximum input inductor current is given by

$$I_{L\max} = \frac{V_c}{(t_2 - t_1) \times L_{in}} \tag{4}$$

The discharging time interval is less than the off time of the switch S_1. The average of inductor current is the input current which is given by

$$I_L = \frac{1}{T_S}\int_0^{T_S} i_L(t)dt = \frac{V_m T_S D^2}{2L_{in}} \sin(\omega_s t) \tag{5}$$

It can be noted from the input voltage Eq. (1) and the inductor current Eq. (5), the input current and voltage are in phase (Fig. 7).

The power supplied by the source is given by the following expression

Fig. 7 Inductor current $i_L(t)$

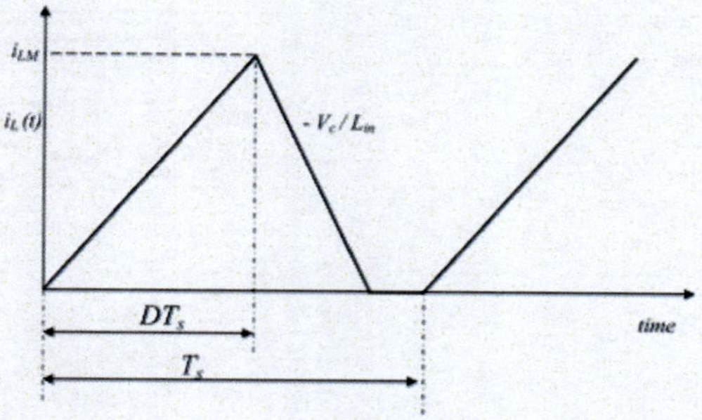

$$P_{in} = \frac{1}{2\pi}\int_0^{2\pi} v_{in}(t) \times i_{in}(t) d(\omega_s t) = \frac{V_m^2 D^2}{4L_{in} f_s} \tag{6}$$

Now, the power output to the lamp is be calculated using its efficiency (η) as below

$$P_0 = \eta \times P_{in} = \frac{\eta V_m^2 D^2}{4L_{in} f_S} \tag{7}$$

The output power supplied to the lamp is related to duty cycle, D. The capacitor voltage can be related to duty ratio D and maximum value of source voltage as

$$V_c \geq V_P D/(1 - D) \tag{8}$$

The expression for quantity of output power supplied to the lamp is

$$P_{lamp} = \eta V_c^2 (1 - D)^2 / 4L_{in} f \tag{9}$$

It can be noted from the above expression that the power applied to the load can be varied by changing the duty cycle. The theoretical calculation shows that the source voltage in phase with input current leads to UPF. Figure 8(a) shows the instantaneous variation of input current waveform and the input voltage waveform. Figure 8(b) shows the current and voltage waveform in the output side. The expression for input current is

$$i(t) = I_p \sin \omega t \tag{10}$$

The expression for instantaneous power supplied is calculated using the source voltage and source current as given below

$$P(t) = v(t) \times i(t) \tag{11}$$

Thus the input power is given by

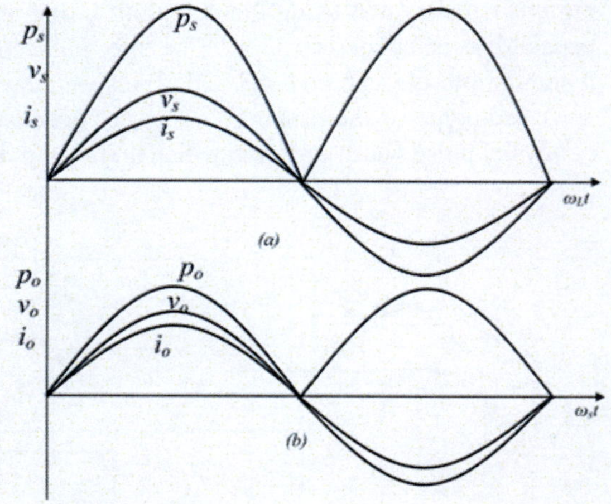

Fig. 8 **a** Representation of source voltage, source current and power supplied **b** Representation of load voltage, load current and load power

$$P(t) = V_p \times I_p \times \sin^2 \omega t \tag{12}$$

The average value of input power can be obtained from using the equation given below

$$P(t) = \int_0^{2\pi} (V_p I_p (1 - \cos 2\omega t)/2) d\omega t \tag{13}$$

On solving the above equation, the input power can be expressed as

$$P = P_{lamp}/\eta = V_p I_p/2 \tag{14}$$

Under ideal condition, there is power loss is zero in the converter circuit. The power obtained from the converter is equal to the input power supplied. The circuit elements are designed to operate in DCM mode with ZVS. This reduces the switching losses occurring in the circuit. This results in high frequency of operation. Since the operation of the circuit at resonant frequency leads to large value of current in the circuit, the circuit is not designed for resonant frequency.

5 Experimental Results

The design of the converter is done for a lamp of rating 40 W, 230 V. The converter circuit is operated at a high frequency of 50 kHz while the supply frequency is 50 Hz. A lamp having a resistance 591 Ω is considered as the load. The efficiency of the converter circuit is assumed as 95%. The input inductor value is calculated from Eq. (8) as 3.13 mH. The output capacitor value is considered as 220 μF. Input current is calculated using the equation. Using empirical formula, series resonant capacitor is calculated as 33 nF. The filter inductance and capacitance is calculated using thumb rule as 5 mH and 220 nF respectively.

A prototype of the proposed converter is constructed using microcontroller for control of pulse width variation which is shown in Fig. 9. The source voltage applied

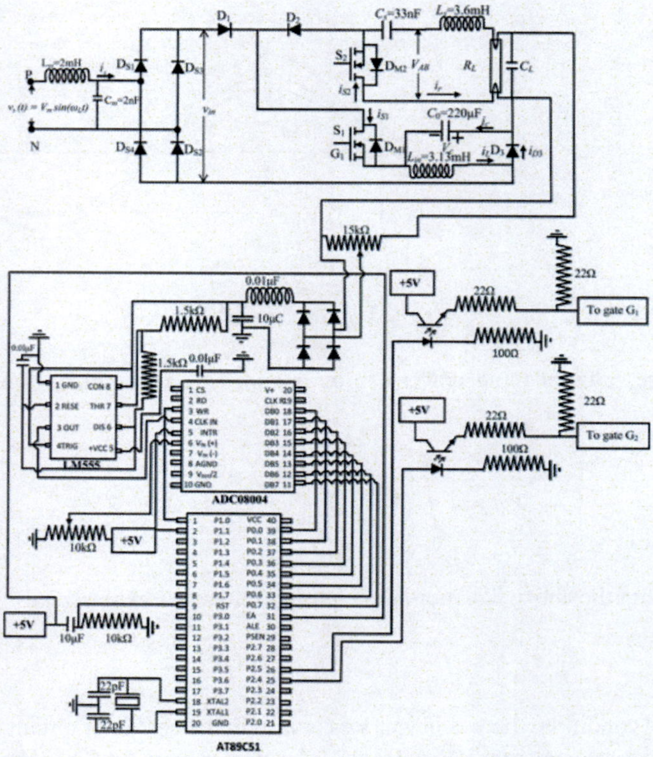

Fig. 9 Converter with its control circuit

Fig. 10 Source current and Source voltage

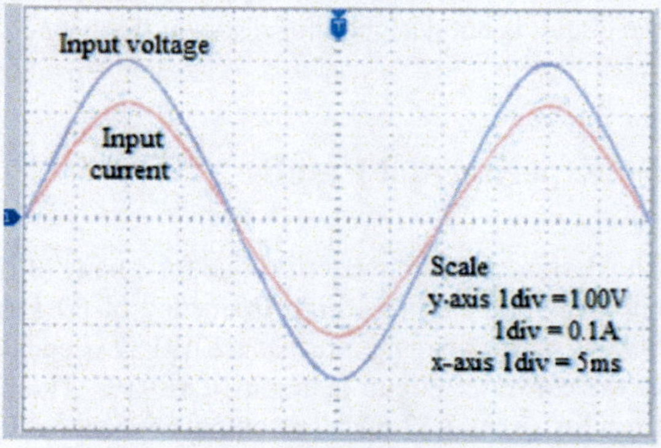

Fig. 11 Rectified voltage and inductor current

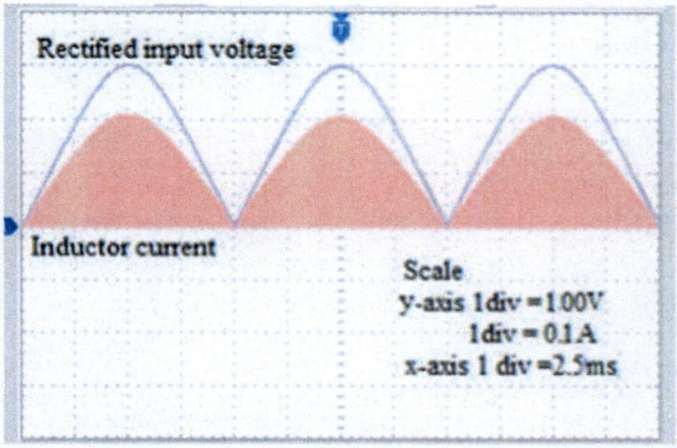

Fig. 12 Resonant circuit voltage and current

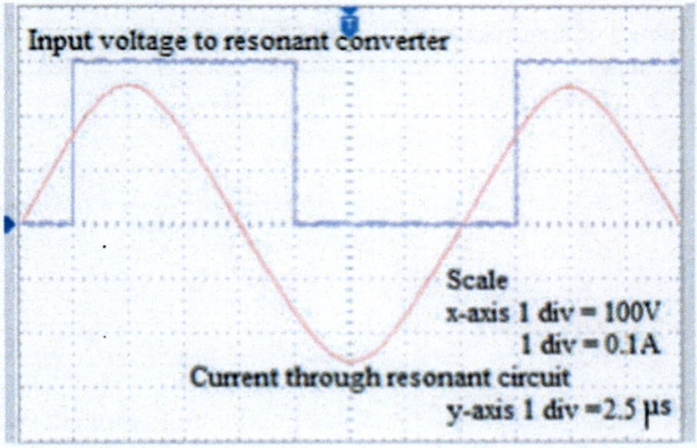

to the converter through low pass filter. Because of the discontinuous conduction the in-phase relation between input current and voltage are shown in Fig. 10. Figure 11 shows the voltage obtained from the rectifier and current flowing through the input inductor is shown in Fig. 11.

The switching losses are reduced by the resonance producing elements. The sinusoidal current produced by resonance circuit and the voltage applied waveform are shown in Fig. 12. The current through the resonant elements also flows through the fluorescent lamp. The voltage developed across the tube of the lamp and the current that flows through the lamp are shown in Fig. 13.

The converter efficiency for the lamp is found to vary between 87.06% and 90.71%. Comparison of efficiency of some of single stage converters are done to validity of the proposed new driver circuit (Table 1).

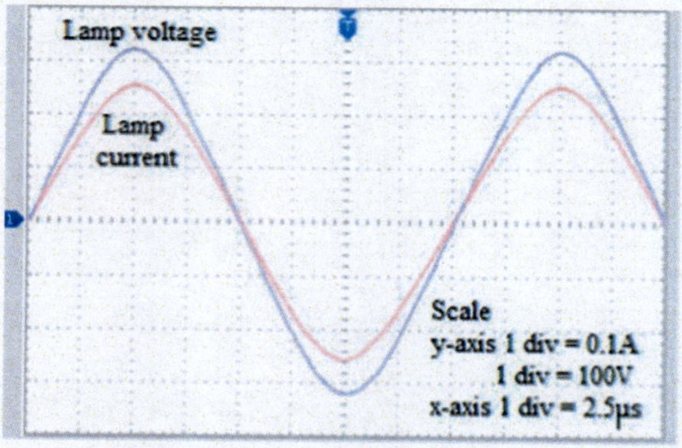

Fig. 13 Lamp voltage and current

Table 1 Comparison of efficiency of converter

S. no.	Reference paper no.	% Efficiency	No. of switches
1	[13]	84.25%	2
2	[14]	87.31%	2
3	[12]	90.01%	2
4	Proposed	90.71%	2

6 Conclusion

Novel electronic ballast is designed using resonant inverter. DCM operation of the converter provides better power factor in the input side. The integration of buck boost converter with resonant inverter satisfies the specification of the standards. The size of components is reduced due to the high frequency operation. In order to provide energy transfer from input side to output side, minimum number of energy transfer elements is used. This results in the reduced cost of the circuit. Input voltage fluctuations do not affect the operation of the lamp.

References

1. Lam JCW, Hui JCY, Jain PK (2012) A dimmable high power factor single-switch electronic ballast for compact fluorescent lamps with incandescent phase-cut dimmers. IEEE Trans Industr Electron 59(4):1879–1888
2. Fuerback AL, da S Postiglione C, Nascimento CB, Martins DC, Perin AJ (2012) Near-unity power factor electronic ballast based on integration techniques to drive high-intensity discharge metal halide (HID–MH) lamps. IEEE Trans Industr Electron 59(4):1760–1769
3. Jeyashree Y, Sukhi Y (2015) LCC resonant converter with power factor correction for power supply units. J Chin Inst Eng 38(7):843–854
4. Lin R-L, Shih H-M (2008) Piezoelectric transformer based current source charge pump power factor correction electronic ballast. IEEE Trans Power Electron 23(3):1391–1400

5. Yesuraj S, Yesuraj J, Muthaiah P, Balaraman S (2017) Standalone PV-fed LED street lighting using resonant converter. Electr Power Compon Syst 45:548–559
6. Chiu H-J, Song T-H, Cheng S-J, Li C-H, Lo Y-K (2008) Design and implementation of a single-stage high-frequency HID lamp electronic ballast. IEEE Trans Industr Electron 55(2):674–683
7. da Silva MF, Fraytag J, Schlittler ME, Marchesan TB, Dalla Costa MA, Alonso JM, do Prado RN (2013) Analysis and design of a single-stage high-power-factor dimmable electronic ballast for electrodeless fluorescent lamp. IEEE Trans Industr Electron 60(8):3081–3091
8. Ekkaravarodome C, Chunkag V, Jirasereeamornkul K, Kazimierczuk MK (2014) Class-D zero-current-switching rectifier as power-factor corrector for lighting applications. IEEE Trans Power Electron 29(9):4938–4948
9. Cheng C-A, Cheng H-L, Chen-Wei Ku, Chung T-Y (2014) Design and implementation of a single-stage acoustic-resonance-free HID lamp ballast with PFC. IEEE Trans Power Electron 29(4):1966–1976
10. Hsieh J-C, Lin J-L (2011) Novel single-stage self-oscillating dimmable electronic ballast with high power factor correction. IEEE Trans Industr Electron 58(1):250–262
11. Yesuraj S, Ramadas G, Yesuraj J (2016) High-power-factor single-switch AC to DC converter for LED lighting. IEEJ Trans Electr Electron Eng 12(6):1–11
12. Ekkaravarodome C, Nathakaranakule A, Boonyaroonate I (2010) Single-stage electronic ballast using class-DE low-dυ/dt current-source-driven rectifier for power-factor correction. IEEE Trans Industr Electron 57(10):3405–3414
13. Pan S, Lam JCW, Jain P (2019) Single-stage single-switch high power factor driving circuit for lighting applications. IET Power Electron 12(15):4005–4015
14. Tahan M, Tingshu Hu (2017) Multiple string LED driver with flexible and high-performance PWM dimming control. IEEE Trans Power Electron 32(12):9293–9306
15. Lin R-L, Shih H-M (2008) Piezoelectric transformer based current-source charge-pump power-factor-correction electronic ballast. IEEE Trans Power Electron 23(3):1391–1400
16. Shrivastava A, Singh B (2013) Power factor-corrected DCM-based electronic ballast. Electr Eng 95:403–411
17. Lee ES, Choi BH, Nguyen DT, Jang GC, Rim CT (2016) Versatile LED drivers for various electronic ballasts by variable switched capacitor. IEEE Trans Power Electron 31(2):1489–150
18. Zhang X, Dong K, Wang Y, Xu D, Jiang J (2018) Research on the parameter optimization of electronic ballast for UV-lamps considering its lifetime and UVC irradiance. IEEE Access 6:11931–11939
19. Ramadas G, Nadesan MK, Yesuraj S, Yesuraj J (2016) High power factor electronic ballast using resonant converter for compact fluorescent lamp. Int J Circuit Theory Appl 45:95–109

A New Hybrid Feature Selection-Classification Method to Identify Churned Customers

Chandra Das, Arghadip Chakraborty, Sohel Raja Molla, Disha Sinha, Shankhadeep Giri, and Shilpi Bose

Abstract Customer churn is a common problem within customer service oriented industries and the companies have been always trying to identify the possible churned customers for taking necessary steps to prevent churn. In this paper, data mining technique based a hybrid feature selection-classification model is proposed that can be used to predict which customers are probable to churn. The proposed model works in two phases. In first phase, decision tree, K-prototype clustering and apriori algorithm based a new feature selection technique have been used to form reduced dataset via selection of relevant features. In second phase, an ensemble classifier is formed via combining KNN, Naïve Bayes, SVM, Decision tree, and Logistic regression to predict churned customers. The proposed model is applied on a telecom service customer churn dataset. From the experimental results, it can be said that the proposed model gives better results than other existing methods.

Keywords Customer churn · Decision tree · K-prototype clustering · Apriori algorithm · KNN · Naïve Bayes · SVM · Logistic Regression

1 Introduction

To survive in today's competitive business world, different companies under different industries have been adopting different novel and innovative business strategies. For any industry, customers or clients are the most valuable resource as they are main reason of success and sustainability for that industry. Actually, to get new customers than to retain existing customers is much more difficult task. So, companies have been giving much effort not only to catch the attention of new customers but also to keep their on-hand customers. Churners are those customers for a company who stop doing interaction or business with that company and start interaction with a new company. Churn rate measure counts the number of individual users or customers quit from a collective group or a company or any service during a specific time

C. Das · A. Chakraborty · S. R. Molla · D. Sinha · S. Giri · S. Bose (✉)
Department of CSE, Netaji Subhash Engineering College, Kolkata, India

V. Nath and J. K. Mandal (eds.), *Proceeding of Fifth International Conference on Microelectronics, Computing and Communication Systems*, Lecture Notes in Electrical Engineering 748,
https://doi.org/10.1007/978-981-16-0275-7_16

period. Among several important factors, it is an important factor that determines the steady-state level of customers for a business or company or the particular industry. Customer churn directly affects the revenue and success of a company or industry. So, companies have been developing new methods to understand customers' demands from the company and also the reasons for dissatisfaction with the service. If they are successful to identify customers who are moving out of service, they can take proper actions for the maximum benefit of the company. So, prediction of churned customers is very important for success and survival of any company [1–4]. So, the companies are going mile to attract customers and retain them.

A Customer Relationship Management (CRM) support system for a company analyzes customer behavior to identify most profitable customers and several measures to retain them in order to improve customer acquisition, retention, and profitability for that company [5, 6]. In CRM support system, customers' transactions and interactions with the company are recorded and then stored in CRM databases and then computational techniques are used to perform different types of analysis. One of the vital analysis parts of CRM support system is customer churn prediction.

Data Mining [7] is an area of computer science which deals with several computational techniques to extract implicit and valuable information from large data repositories. Data mining techniques [7] have been used extensively to identify churned customers in different marketing areas such as electronic banking services, mobile telecommunications, television viewership, the wireless carrier, and newspaper subscription [1, 8–11] etc. Although data mining technique based different prediction methods [8–11] are already developed by researchers to predict churned customers, still researchers have been trying to develop new methods to improve prediction accuracy. In this regard, here data mining technique based a new predictive model is proposed to predict customer churn.

2 Proposed Work

In this study, a new hybrid feature selection-classification model (NEFS-EC) is proposed and developed to identify churned or probable churned customers from a dataset. The proposed hybrid model performs two tasks (1) feature selection (2) classification and it works in two phases. In the first phase, relevant features are selected by using a new ensemble feature selection method (NEFS) based on majority voting technique. In this method, Decision tree classifier [7, 12, 13], *K*-prototype clustering [14], and Aprioi association rule mining technique [7, 15] are applied individually and for each method a subset of feature is selected. Finally, from these subset of features using majority voting technique most relevant features are selected and reduced dataset is formed. Then in the second phase an ensemble classifier (EC) is formed based on majority voting technique by combining predictive accuracy K-nearest Neighbors, Logistic Regression, Support Vector Machine, Decision Tree, Random Forest and Naive Bayes Classifier [7, 16] and applied on the reduced dataset. The

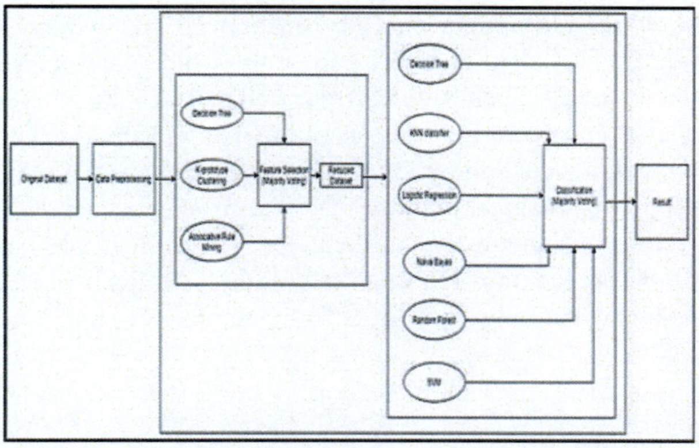

Fig. 1 Architecture of the proposed model

proposed model used for this study is illustrated diagrammatically in Fig. 1. Every phase of this model is discussed next.

2.1 *Feature Selection Phase*

In feature selection phase, significant features are selected using NEFS technique. This ensemble technique is formed using Decision tree, K-Prototype method and Aprioi technique. The relevant feature selection phase works in fourth phases.

In the first phase, two well known decision tree algorithms CART [12, 13] and ID3 [12, 13] are applied on the dataset several times using LOOCV method for different heights of the decision tree to find optimal tree. The optimal tree is selected based on the highest classification accuracy. Then a number of most significant features according to their class relevance are chosen from every optimal decision tree and after that using majority voting technique a number of most significant features are chosen from those already selected features.

In the second phase K-prototype clustering technique is applied on the dataset. In this phase, the dataset is first divided into two parts on the basis of class attribute. One part consists of churned customers and other part consists of non-churned customers. Then K-prototype clustering technique is applied separately on each part of the dataset and two set of clusters are formed: one set is for the churn-customers and another set is for the non-churn customers. The optimal number of clusters is selected using Elbow method [17] (discussed in the result section). Now the two clustering results (cluster centroids) are merged considering the cluster centroids as objects and original features in the whole dataset as features, and the reduced dataset is formed. Then the correlation matrix is calculated for this reduced dataset. Then features are sorted in descending order according to their correlation value in the correlation matrix and after that a number of most relevant features are selected.

In the third phase, aprioi algorithm, one of the well known algorithms of association rule mining, is applied on every part of the whole dataset (whole dataset is fragmented as in phase 2). Then features are sorted in descending order according to their class association value for every part of the whole dataset and after that using majority voting technique a number of most relevant features are selected.

In the fourth phase, the most relevant features are selected using majority voting technique from those features which are selected in phase1, phase2, and phase3.

In the fourth phase, the most relevant features are selected using majority voting technique from those features which are selected in phase1, phase2, and phase3.

2.2 *Classification Phase*

In this phase an ensemble classifier is formed via applying majority voting technique on classification accuracy of several classification models like K-nearest applying majority voting technique on classification accuracy of several classification models like K-nearest Neighbors, Logistic Regression, Support Vector Machine, Decision Tree, Random Forest and Naive Bayes Classifiers.

The algorithm of the proposed model is discussed below.

Algorithm: NEFS-EC

Input: The churn customer data matrix $D_{M\times(N+1)}$, where M represents the number of customers and N represents the attributes or features of customers. The last attribute is the decision attribute representing whether the customers is churned or not. The dataset is divided into training set $T^1_{k\times(N+1)}$ and testing set $T^2_{l\times(N+1),}$, where k and l are number of customers in each set.

Output: Identified relevant feature set and the classification accuracy of testing dataset.

Phase 1: NEFS

Step 1: The raw data $D_{M\times(N+1)}$ is first preprocessed to get complete non redundant cleaned dataset.

Step 2: The complete dataset is divided into training set $T^1_{k\times(N+1)}$ and testing set $T^2_{l\times(N+1),}$, where k and l are number of customers in each set.

Step 3: Identification of relevant features from $T^1_{k\times(N+1)}$:

a. Phase1: Selection using decision tree
   i. Apply CART algorithm to get optimal decision tree.
   ii. Apply ID3 algorithm to get another optimal decision tree.
   iii. Optimum no. of features are selected using majority voting from the feature set obtained in step a and step b.

b. Phase2: Selection using *K*-Prototype clustering
   i. First the dataset is divided into two parts, one for churned class and another for non churn class.
   ii. Apply K-Prototype clustering for each part. The no. of clusters is decided using Elbow method.
   iii. The reduced dataset is formed using cluster centroids as the objects with all attributes or features.
   iv. The relevant features are selected whose correlation with the class level is high.
c. Phase3: Selection using Association rule mining
   i. Frequent itemset is formed using Apriori algorithm.
   ii. Apply association rule mining to find the association rules for the itemset.
   iii. The optimum feature set is selected based on the higher confidence in the generated association rules.
d. Phase4: Ensemble of results obtained in Phase1, Phase2 and Phase3 using majority voting to get optimal feature set.

Phase 2: EC (Ensemble classifier formation).

Step 4: Apply K-nearest Neighbors, Logistic Regression, Support Vector Machine, Decision Tree, Random Forest and Naive Bayes classifiers on training dataset $T^1_{k\times(N+1)}$ individually and then using majority voting technique an ensemble classifier is formed.

Testing

Step 5: Now this hybrid method is applied on test dataset to predict unknown churn customers.
Step 6: End.

3 Result

In this work, a new hybrid feature selection-classification method is planned and developed to predict churned customers from a telecom service customer churn dataset [18].

3.1 Dataset Overview

The dataset or data matrix used for this study is the raw data of a telecom company (source: IBM sample datasets), with a purpose to predict the churned customers. The data matrix consists of 7043 rows and 21 columns. In this data matrix, each row represents a customer, and each column represents a customer's features except the last column. The last column is the decision attribute. The features of the dataset are described below:

- Customer ID: Customer ID
- Gender: Sex of the customer
- Senior Citizen: Binary attribute representing aged customer or a senior citizen or not
- Partner: Binary attribute representing the customer is married or not
- Dependents: Binary attribute representing the customer has dependents or not
- Tenure: An attribute representing the time period for which the customer has taken service from the company
- Phone Service: Binary attribute representing whether the customer has a phone service or not
- Multiple Lines: An attribute representing whether the customer has multiple lines or not
- Internet Service: Category of Customer's internet service
- Online Security: An attribute representing whether the customer has online security support or not
- Online Backup: An attribute representing the customer has online backup support or not
- Device Protection: An attribute representing the customer has device protection service or not
- Tech Support: An attribute representing the customer has tech support service or not
- Streaming TV: An attribute representing the customer has streaming TV service or not
- Streaming Movies: An attribute representing the customer has streaming movies service or not
- Contract: The term period of contract of the customer with the company
- Paperless Billing: An attribute representing the customer has paperless billing service or not
- Payment Method: An attribute representing the customer's paying method
- Monthly Charges: An attribute representing the monthly charge of the customer for getting the service
- Total Charges: An attribute representing the total amount charged to the customer
- Churn: An attribute representing whether the customer will churn or not

The 'Churn' column is the decision attribute for this study as aim of this study is to predict whether a customer will churn or not.

3.2 Data Preprocessing

Data preprocessing is all about handling the missing data, cleaning the absurd and inconsistent data and transforming the data so that the dataset becomes ready to send it as input data for the required machine learning techniques. In the next step, training and testing dataset is prepared by dividing the dataset into two portions after random sampling. Here 6338 rows of data is taken for training the machine learning method and 705 rows (10% of total) of data is taken for testing and validating the method.

3.3 Feature Selection

In this study, the important features are selected using NEFS technique. This ensemble technique is formed using decision tree, K-Prototype clustering and Apriori technique.

The proposed ensemble technique works in four phases. In the first phase decision tree is used to select relevant features.

In the second phase K-prototype clustering based method is used to select relevant features. In the third phase, Aprioi algorithm is used to select relevant features. Then using majority voting technique most relevant features are selected in the fourth phase. The ensemble feature selection process is shown phase by phase below on the used dataset.

1) *Decision Tree Classifier based Feature selection:*

In this study, two decision tree algorithms CART and ID3 are run with different heights to find the optimal decision trees. Optimality is measured in terms of classification accuracy. In CART, gini index and in ID3 entropy measures is used for taking splitting decision. The top 8 attributes as splitting attribute obtained from these two decision trees along with their importance is shown in Table 1.

Table 1 Attributes and their splitting value

Attributes	ID3	CART
Contract	0.57892	0.54418
Tenure	0.12428	0.149562
Online Security	0.10114	0.133837
Monthly Charges	0.08495	0.054220
Internet Service	0.06650	0.085264
Streaming Movies	0.01453	0.007104
Senior Citizen	0.01341	0.008986
Phone Service	0.007080	0.007084

Table 2 Attributes and their corresponding correlation value

Attributes	Correlation Value
Phone Service	0.707107
Internet Service	0.333333
Partner	0.333333
Monthly Charges	0.333333
Tenure	0.198356
Contract	0.333333
Gender	−0.333333
Senior Citizen	0.333333
Dependents	0.000000
Multiple Lines	−0.333333
Online Security	−0.447214
Online Backup	0.000000
Device Protection	−0.447214
Tech Support	0.000000
Streaming TV	0.000000
Streaming Movies	0.000000
Paperless Billing	0.000000
Payment Method	−0.218218

2) *K-Prototype Clustering based Feature selection:*

In this phase, K-prototype clustering based reduced dataset is formed and then the correlation matrix is calculated for this reduced dataset. After that a number of most significant features are selected according to their correlation value in the correlation matrix. The optimal value of K is 3 in this study using Elbow method.

From the correlation matrix, the top 8 important features are obtained for this dataset. The attributes and the corresponding correlation value is shown in Table 2.

3) *Apriori Algorithm based feature selection:*

In phase 3, Apriori algorithm [7] finds all the rules existing in the dataset depending on different support count values (20–40) and for different confidence (70, 80, 90%) values. Significant association rules are found for support value 30 and for confidence value 90%.

The most important features obtained by implementing Association Rule Mining on the datasets divided on the basis of Non-Churn and Churn Customers are shown in Table 3 and and Table 4. respectively. Most important attributes obtained are tenure, Monthly Charges, Internet Service and Phone Service.

In phase 4, using majority voting technique based on the results of Decision Tree classifier, K-Prototype Clustering and Apriori algorithm (association rule mining technique) the following relevant features are obtained - tenure, Monthly Charges, Internet Service and Phone Service.

Table 3 Most important attributes when Churn = No/0

Confidence	Itemset in association rule → Non Churn
1.000000	(tenure)
1.000000	(Monthly Charges)
1.000000	(Monthly Charges, tenure)
0.909042	(Phone Service)
0.909042	(tenure, Phone Service)
0.909042	(Internet Service)
0.909042	(Monthly Charges, Phone Service, Internet Service)
0.909042	(Monthly Charges, tenure, Phone Service)

Table 4 Most important attributes when Churn = Yes/1

Confidence	Itemset in association rule → Churn
1.0000	(Monthly Charges)
0.9978	(tenure)
0.9978	(Monthly Charges, tenure)
0.9010	(Phone Service)
0.9010	(Internet Service)
0.9010	(Monthly Charges, Phone Service, Internet Service)

Table 5 Classification accuracy report of multiple classifiers with all features

Model	Accuracy score % (ACC)
Logistic Regression	80.57%
K-nearest Neighbors	81.84%
Support Vector Machine	81.28%
Naive Bayes	76.31%
Decision Tree	74.61%
Random Forest	80.28%

To measure the importance of the selected relevant features, the classification accuracies of different classifiers named K-nearest Neighbors, Logistic Regression, Support Vector Machine, Decision Tree, Random Forest and Naive Bayes Classifiers by running them individually taking all features of the dataset are shown in Table 5 and also the classification accuracies obtained from the above mentioned classifiers by running them individually taking selected features of the dataset by the proposed method NEFS are shown in Table 6.

From the results of Table 5 and Table 6, it can be said that after selection of relevant features accuracy of every classifier is improved except logistic regression.

Table 6 Classification accuracy report of multiple classifiers considering the selected features

Model	Accuracy score % (ACC)
Logistic Regression	78.72%
K-nearest Neighbors	82.41%
Support Vector Machine	82.13%
Naive Bayes	80.71%
Decision Tree	81.99%
Random Forest	82.41%

Table 7 Accuracy report of proposed method and Ensemble Classifier

Model	Accuracy score % (ACC)
Proposed NEFS-EC model	82.41% (considering selected features)
Ensemble Classifier (EC)	80.28% (considering all features)

3.4 Ensemble Classifier

In this phase an ensemble classifier (EC) is formed via applying majority voting technique on classification accuracy of several classification models like K-nearest Neighbors, Logistic Regression, Support Vector Machine, Decision Tree, Random Forest and Naive Bayes Classifiers and EC is applied on the reduced dataset. The final accuracy of the proposed method is 82.41% while the classification accuracy of the ensemble classifier (EC) is 80.28% considering all features as shown in Table 7. From these results it can be said that the proposed method gives highest prediction accuracy than all existing classifiers.

In order to check the predictive accuracy of the model, it has been compared with previous prediction works for predicting the churned customers in the telecommunication industry. The previous works which mentioned here are: Ning Lu et al.'s predictive model [19], Abbas Keramati et al.'s predictive model [20]. The results are shown in Table 8. From the results, it can be said that the proposed NEFS-EC model gives better results in all cases.

Table 8 Comparative analysis of the proposed method with existing method

Researchers	Accuracy score % (ACC)
Proposed NEFS-EC model	82.41%
Ning Lu et al.'s predictive model [19]	*80.7%*
Abbas Keramati et al.'s predictive model [20]	*82%* (SVM*)*
Amjad Hudaib et al.'s predictive model (1) [21] Amjad Hudaib et al.'s predictive model (2) [21] Amjad Hudaib et al.'s predictive model (3) [21]	*79.1%* *80.8%* *81.70%*

3.5 Conclusion

Customer churn analysis has become a major concern in almost every industry that offers products and services. In this study data mining technique based a hybrid feature selection-classification model is developed to predict probable churned customers. The first phase that means feature selection phase in the proposed model is designed using Decision-tree classifier, K-prototype Clustering algorithm and Apriori association rule mining technique. Then based on the selected relevant features an ensemble classifier is developed in the second stage. From the results it can be said that the proposed model is superior to well-known existing works.

References

1. Tsai C-F, Chen M-Y (2009) Variable selection by association rules for customer churn prediction of multimedia on demand. Expert Syst Appl
2. Idris A, Khan A (2016) Churn prediction system for telecom using filter-wrapper and ensemble classification. Comput J Adv Access 60:410–430
3. Swetha P, Dayananda RB (2020) Customer churn prediction and upselling using MRF (modified random forest) technique. IJITEE 9
4. Sabbeh SF (2018) Machine-learning techniques for customer retention: a comparative study. Int J Adv Comput Sci Appl 9(2):273–281
5. Kumar DA, Ravi V (2008) Predicting credit card customer churn in banks using data mining. Int J Data Anal Tech Strat 1(1):4–28
6. Oyeniyi AO, Adeyemo AB (2015) Customer churn analysis in banking sector using data mining techniques. Afr J Comput ICT 8(3):165–174
7. Han J, Kamber M, Pei J (2012) Data mining: concepts and techniques. Morgan Kaufmann Publishers, Elsevier. https://doi.org/10.1016/C2009-0-61819-5
8. Ahn J-H, Han S-P, Lee Y-S (2006) Customer churn analysis: churn determinants and mediation effects of partial defection in the Korean mobile telecommunications service industry. Telecommun Policy 30(10–11):552–568
9. Chu B-H, Tsai M-S, Ho C-S (2007) Toward a hybrid data mining model for customer retention. Knowl-Based Syst 20(8):703–718

10 Coussement K, Van den Poel D (2008) Churn prediction in subscription services: an application of support vector machines while comparing two parameter-selection techniques. Expert Syst Appl 34(1):313–327
11. Kim S, Choi D, Lee E, Rhee W (2017) Churn prediction of mobile and online casual games using play log data. PLOS One 12(7):e0180735
12. Sharma H et al. (2013) A survey on decision tree algorithms of classification in data mining. Int J Sci Res (IJSR) 5(4):2094–2097
13. Kotsiantis SB (2013) Decision trees: a recent overview. Artif Intell Rev 39:261–283
14. Ji J et al (2012) A fuzzy k-prototype clustering algorithm for mixed numeric and categorical data. Knowl-Based Syst 30:129–135
15. Adebayo OS et al (2019) Improved malware detection model with apriori association rule and particle swarm optimization. Secur Commun Netw 2019:1–13. https://doi.org/10.1155/2019/2850932
16 Amancio DR et al (2014) A systematic comparison of supervised classifiers. PLOS ONE 9:e94137
17. Syakur MA et al (2018) Integration K-means clustering method and elbow method for identification of the best customer profile cluster. IOP Conf Ser Mater Sci Eng 336:012017
18. https://www.kaggle.com/palashfendarkar/wa-fnusec-telcocustomerchurn#WA_Fn-UseC_-Telco-Customer-Churn.csv
19. Ning Lu, Lin H, Jie Lu, Zhang G (2014) A customer churn prediction model in telecom industry using boosting. IEEE Trans Industr Inf 10(2):1659–1665
20. Keramati A, Jafari-Marandi R, Aliannejadi M, Ahmadian I, Mozaffari M, Abbasi U (2014) Improved churn prediction in telecommunication industry using data mining techniques. Appl Soft Comput 24:994–1012
21. Hudaib A, Dannoun R, Harfoushi O, Obiedat R, Faris H (2015) Hybrid data mining models for predicting customer churn. Int J Commun Netw Syst Sci 8(05):91

A Comparative Study of MOSFET (Single and Double Gate), Silicon Nanowire FET, and CNTFET by Varying the Oxide Thickness

Mayank Kumar Jethwa, Hardiki Mukesh Devre, Yash Agrawal, and Rutu Parekh

Abstract In this work, performance analysis of electrical properties of the single gate MOSFET (SG-MOSFET), double-gate MOSFET (DG-MOSFET), silicon nanowire FET (SNWFET) and carbon nanotube FET (CNTFET) devices is done using FETTOY simulator at room temperature by varying the oxide thickness for 0.3, 0.7 and 1.2 nm. The electrical parameters include average velocity, quantum capacitance, ratio of transconductance/drain current and drain current. The simulation for CNTFET and SNWFET were carried out using CNT and SNW as channel materials, silicon dioxide as the gate dielectric, and silicon substrate as a base material. Quantum capacitance has an impact in calculating the gate capacitance of a FET device at nanometer scale. The simulations performed show that CNTFET has 35% better average velocity when compared to SNWFET for 0.3 nm oxide thickness. The average velocity of DG-MOSFET and SG-MOSFET is zero. It was also observed that with the increase in gate oxide thickness, average velocity decreases. CNTFET and DG-MOSFET have 12.6 and 52% better drain current than SNWFET and SG-MOSFET respectively. The drain current decreases with the increase in gate oxide thickness. Also, it can be observed that the quantum capacitance for CNTFET (for 0.3 nm oxide thickness at 1 V) is less than SNWFET by 6.38%. The transconductance/drain current ratio is better for DG-MOSFET which is around 1.2% more when compared to SG-MOSFET.

Keywords SG (Single gate) MOSFET · DG (Double gate)-MOSFET · SNWFET (Silicon Nanowire FET) · CNTFET (Carbon nanotube FET) · Inversion layer capacitance · Quantum capacitance

---

M. K. Jethwa (✉) · H. M. Devre · Y. Agrawal · R. Parekh
DA-IICT, Gandhinagar, Gujarat, India

Y. Agrawal
e-mail: yash_agrawal@daiict.ac.in

V. Nath and J. K. Mandal (eds.), *Proceeding of Fifth International Conference on Microelectronics, Computing and Communication Systems*, Lecture Notes in Electrical Engineering 748,
https://doi.org/10.1007/978-981-16-0275-7_17

1 Introduction

The reduction in the size of the device dimension is the essence of circuit miniaturization. According to Moore's law, every 18 months, in an Integrated circuit, the number of transistors doubles [1]. The main reason for circuit miniaturization is the portability, added functionality and low cost. As the device dimensions are getting smaller, there are some problems that are faced which limit further scaling down of these devices. Problems like high power dissipation, short channel effect, high leakage current, reliability issues etc. occur due to circuit miniaturization. There are many ways proposed to conquer these scaling obstacles. Some of the ways can be done by means of expanding the present systems and technologies which can lead to extending the scalability of devices, on the other hand some solutions include the replacement of the exciting silicon MOSFETs (Metal Oxide Semiconductor Field-effect Transistor) with other materials [2]. Technologies that extend the traditional MOSFET functionalities include replacing device channels with N-type III-V material or using Si nanowire FET (SNWFET), graphene nanowire or carbon nanotube FET (CNTFET) in addition to Double gate MOSFET (DG-MOSFET).

In the process of transistor size reduction, single gate metal oxide semiconductor field-effect transistors (SG-MOSFET) is expected to exhibit a problem of short channel effects which will lead to less scaling capabilities [3]. In the deep nanometer regime, DG-MOSFET has greater advantages over SG-MOSFETs. This is due to better control on the channel of DG-MOSFET which leads to reduced leakage current and short channel effects (SCEs) [4]. In the trend of miniaturization other potential candidates are silicon nanowire FET (SNWFET) and carbon nanotube FET (CNTFET). In such devices the channel is replaced by nanotubes, whose diameter are few nanometers which leads to quantum confinement behaviour (1-D conduction, this leads to a reduction of short channel effects) [5]. This FETs have more advantages like charge carriers that have high mobility and the potential to minimize the subthreshold slope which inturn leads to minimizing the short channel effects [5]. In the case of CNTFET, a major deciding role is of the quantum capacitance which makes it more favourable for use than other MOSFETs [6].

A study done by Prasher et al. investigated the performance of a double gate MOSFET (DG-MOSFET) on the basis of novel channel materials [3]. A comparative study of electrical properties of carbon nanotube (CNT) and silicon nanowire (SNW) MOSFET devices for performance parameters like quantum capacitance, drain current, average velocity and carrier injection velocity has been presented in [7]. In another paper by the same group, comparison of SG-MOSFET and DG-MOSFET is done for parameters like drain current, quantum capacitance, mobile electrons and gm/Id [8]. Sinha et al. studied the effect of variation of gate oxide thickness on gate capacitance. This experiment was done for MOSFET, nanowire FET and CNTFET devices. The analysis was done for quantum capacitance [2]. Chaudhary et al. presented the future semiconductor devices applications for carbon nanotube and nanowires [9]. An analysis of I_{on}/I_{off} ratio of silicon nanowire transistor with planar MOSFET is done by Gupta et al. [10]. Comparison of nanotransistors which are based on CNT and GNR materials is discussed in [11]. In addition to this the effect of variation in gate oxide thickness and dielectric constant has been presented by them. An analytical study was done on quantum capacitance effects in ultra-thin-body III-V transistors in [12]. In addition, the effect of variation gate dielectric on gate capacitance, drain current, and trans-conductance was done for MOSFET, nanowire FET and CNTFET devices in [13]. The exclusive advantage of CNTFET over traditional MOSFET devices because of reduced Leakage Power have been shown in [14]. An investigation on how different device parameters affect the CNTFET at the nanometer scale is carried out in [15]. The detailed study of the electrostatics of triangular cross sectioned NW transistors is done by Vashaee et al. [16]. A 2D simulation study on analysis of symmetric DG-MOSFET with gate and channel engineering is explored in [17].

In this paper, the comparison of SG-MOSFET, DG-MOSFET, SNWFET and CNTFET is done. By simulation, we have investigated and compared the performance parameters like drain current, average velocity, quantum capacitance and the ratio of transconductance/drain current.

2 Working of MOSFETs

2.1 CNTFET

The schematic of the CNTFET is shown in Fig. 1. A CNTFET is a FET device which uses a single carbon nanotube or sometimes it uses a collection of carbon nanotubes as the channel material. As shown in the figure, the semiconducting carbon nanotube is placed in between source and drain. The extraordinary electrical properties of carbon nanotubes like high electron mobility, drain current as well as better gate capacitance versus gate voltage makes CNTFET unique amongst other devices. The reason for these unique properties like high electron mobility and high electrical conductivity is due to the electronic structure of graphene (rolled graphene sheets are known as carbon nanotube). Remarkable progress is being achieved in the structure of CNTFET by sorting carbon nanotubes with chemical techniques which helps in achieving a distribution of nanotubes. This helps in controlled bandgap energy and hence improves the purity of single chiral semiconducting nanotubes to ~99% [5]. In addition to this, a CNTFET device with an oxide thickness of 0.3 nm at a constant

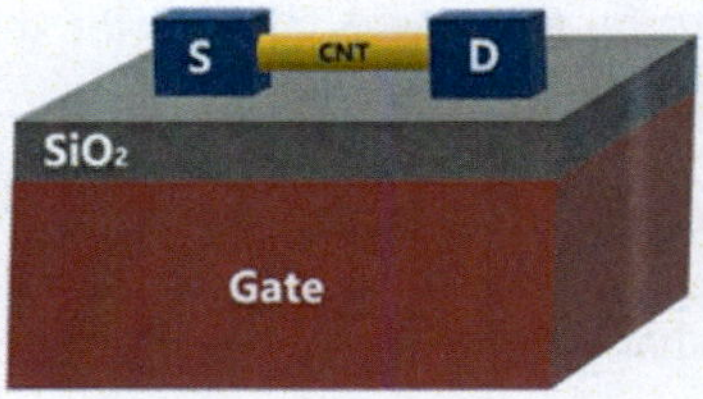

Fig. 1 Schematic of carbon nanotube FET

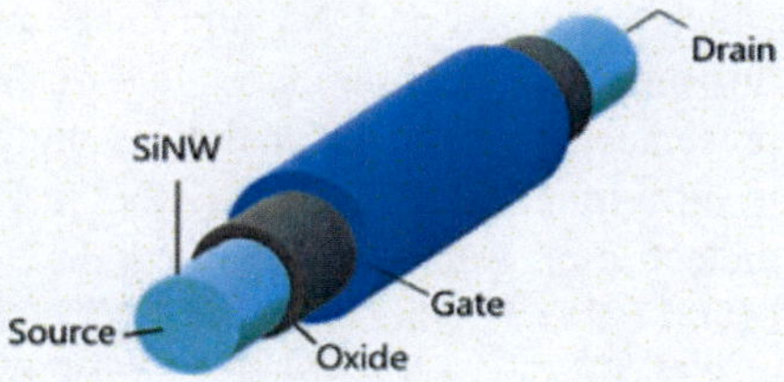

Fig. 2 Schematic of silicon nanowire FET

drain voltage of 1 V is shown to have a higher average velocity of an electron than other three MOSFET devices. For a detailed study of CNTFET, please refer to [5, 18].

2.2 Silicon Nanowire FET

The schematic of silicon nanowire FET is shown in Fig. 2. SNWFET structures are different from planar MOSFET. The conventional planar MOSFET channel is replaced with a semiconducting nanowire. Semiconductor nanowire is prepared with high yield along with uniform electronic properties and hence they are used for large-scale integrated systems. For a semiconductor transistor, during fabrication process the doping of transistor is done. Silicon is widely used in all electronic devices today. Other alternatives to silicon are group III-V semiconductors (these are compounds made out of elements from III and V groups from the periodic table). Some of the examples include GaAs and InAs. The problem with widespread use of these materials is that they are more expensive than Si. Most of the single-walled nanotubes (SWNT) that are used have a diameter approximately equal to 1 nm and the length of tube is almost millions of times longer [5]. For a detailed study on the Electronic and transport properties of silicon nanowires, please refer [5, 9].

2.3 Single Gate MOSFET and Double Gate MOSFET

The structure of SG-MOSFET is shown in Fig. 3a. Traditionally, in a metal oxide semiconductor structure, a layer of silicon dioxide (SiO2) is grown on the top of a silicon substrate as shown in the figure. The silicon dioxide is a dielectric material and hence the structure of silicon dioxide is equivalent to a planar capacitor. Tox mentioned in the figure is the thickness of gate oxide. TSi is the thickness of the silicon substrate (body).

A double gate metal oxide semiconductor (DG-MOSFET) is the same as an ordinary metal oxide semiconductor but with two gates. Schematic of DG-MOSFET is shown in Fig. 3b. In the DG-MOSFET, the current is controlled by both the gates as shown in the figure. The DG-MOSFETs are more resistant to short channel effects

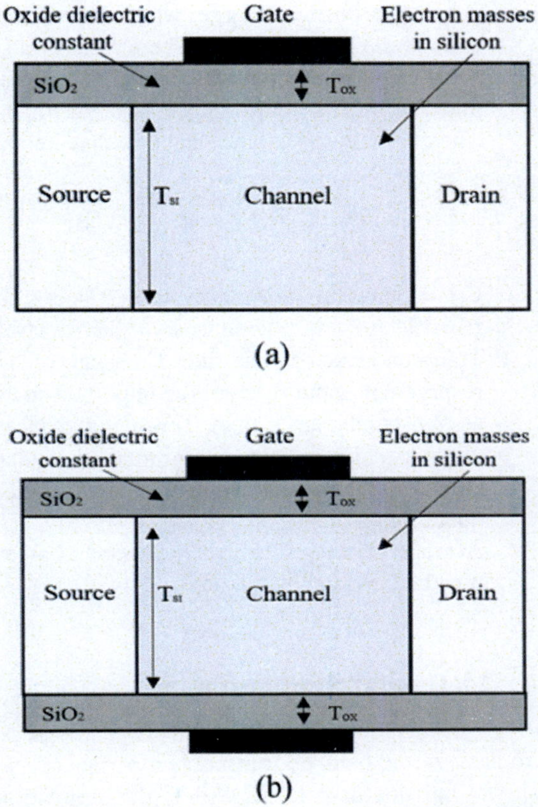

Fig. 3 Schematic of a MOSFET. **a** SG-MOSFET **b** DG-MOSFET

(SCEs) and have high conductivity because of two gates on either side which gives better control on the channel when compared to SG-MOSFET. And due to this property, we can scale double-gated FETs to shorter dimensions for the same channel thickness when compared to SG-MOSFETs [8].

3 Electrical Properties of FET Devices

The following are some of the physical and electrical properties of MOSFET devices:

1. Quantum Capacitance: Quantum capacitance comes into actions when one or both the plates of capacitors are semiconductors (as in the case of MOSFET). So, in MOSFET the Galvani's potential ($Q = CV$) is not the only capacitance that contributes to whole capacitance. As the voltage is applied to the capacitor plates, electrons are filled up at the negative plate of the MOSFET. Hence, in the band structure, the negative plate occupies higher-energy states. The opposite phenomenon happens at the positive plate which loses electrons and thereby

acquires a positive charge, which leaves behind electrons with lower energy state of the band structure. There comes into action one other effect that is the band-filling/band-emptying effect. These effects affect the original capacitance and a second new capacitance is added which is the quantum capacitance [19]. The quantum capacitance can be calculated as [7],

$$C_Q = e^2 \int_{-\infty}^{+\infty} f(E)\left(\frac{\partial f(E - E_F)}{\partial E}\right) dE \tag{1}$$

C_Q = Quantum Capacitance e = Electron charge.
E = Electrostatic potential E_F = Fermi level.

2. Transconductance: The rate of change of output current from a device with respect to its input voltage is an important characteristic property and it is termed as transconductance (g_m). Transconductance has great importance because it plays a major role in determining the switching speed of an active device. Circuits created from high transconductance devices have a high speed of operation.
3. Average velocity: The average speed of an electron is defined as the distance travelled divided by the time.

4 Method of Simulation

To analyze the ballistic transport properties of carbon nanotube, silicon nanowire, single and double-gate MOSFET, the simulation and modelling were achieved through FETTOY [14]. FETTOY tool is a numerical simulator that calculates the various parameters for the four MOSFET devices. For conventional MOSFET, FETTOY assumes nanowire and nanotube MOSFETs as cylindrical geometry.

Modelling of the device is done by choosing the appropriate device type (carbon nanotube, silicon nanowire, single gate, and double gate MOSFET). Setting the oxide thickness from sets of values 0.3, 0.7, 1.2 nm, gate voltage 0–1 V and drain voltage of 1 V with other parameters fixed which are mentioned in Table 1. Then the simulation is performed by the tool to get the results of the device chosen. The values of average velocity, quantum capacitance, g_m/I_d and drain current were obtained and their results were analyzed for all four devices (Fig. 4).

5 Simulation Results

The simulation results are shown below for single gate MOSFET, double gate MOSFET, CNTFET and SNWFET devices for all the performance parameters (drain

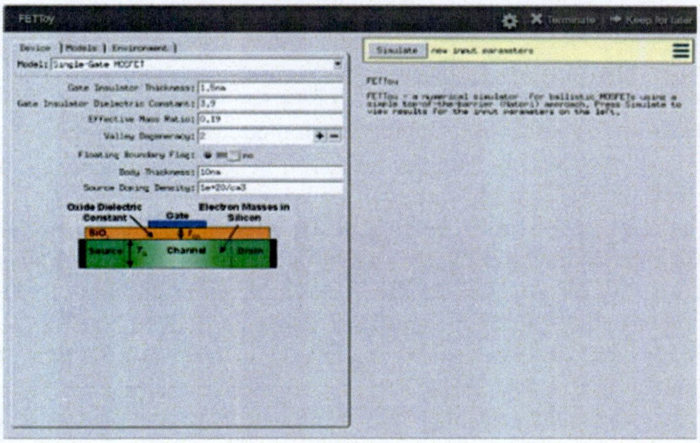

Fig. 4 FETTOY software interface

Table 1 Input parameters used for simulation in FETTOY tool [20]

FETTOY input parameters	Values
Oxide thickness	0.3, 0.7, 1.2 (nm)
Insulator dielectric constant	3.9
Number of bias points (gate)	13
Number of bias points (drain)	13
Temperature	300 (K)
Initial gate voltage	0 (V)
Final gate voltage	1 (V)
Drain control parameter	0.035
Initial drain voltage	0 (V)
Final drain voltage	1 (V)
Threshold voltage	0.32
Series resistance	0 (ohms)
Doping density	1e+26 (/m^3)
Gate control parameter	0.88
Si body thickness	1e−08 (m)
Transport effective mass	0.19
Diameter	1.0 nm
Valley degeneracy	2

current, average velocity, quantum capacitance and g_m/I_d ratio). The results are simulated at different gate voltage, at constant drain voltage of 1 V with diameter 1.0 nm and oxide thickness of 0.3, 0.7, 1.2 nm.

5.1 Drain Current Vs Gate Voltage

For all oxide thickness, all the four MOSFETs show half parabolic-like curve. It is observed from the simulation and the graph plotted that drain current of all four MOSFET devices increases with the reduction in oxide thickness. It can also be observed that, the CNTFET has more drain current when compared to SNWFET. The drain current for CNTFET at 1 V for gate oxide thickness of 0.3 nm is 59.8 μA and for SNWFET the drain current is 53.1 μA. Hence CNTFET has 12.6% better drain current than SNWFET. And DG-MOSFET has more drain current (18,700 μA/μm) when compared to SG-MOSFET (which has drain current of 12,300 μA/μm). Hence DG-MOSFET has 52% better drain current at 1 V than SG-MOSFET. The drain current simulation was done in [8] shows that the drain current for SG-MOSFET at 0.83 V for oxide thickness of 0.3 nm is around 8110 μA/μm which is almost the same as it can be observed from the Fig. 5. The drain current for the same parameters for DG-MOSFET, CNTFET and SNWFET is 12300 μA/μm, 43.2 and 37 μA respectively which is similar to the simulation results in [8] (Fig. 6).

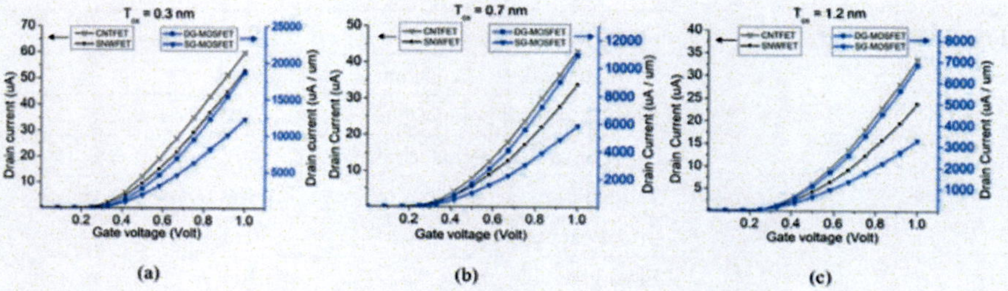

Fig. 5 Drain current vs gate voltage characteristics for different oxide lengths **a** 0.3 nm **b** 0.7 nm **c** 1.2 nm

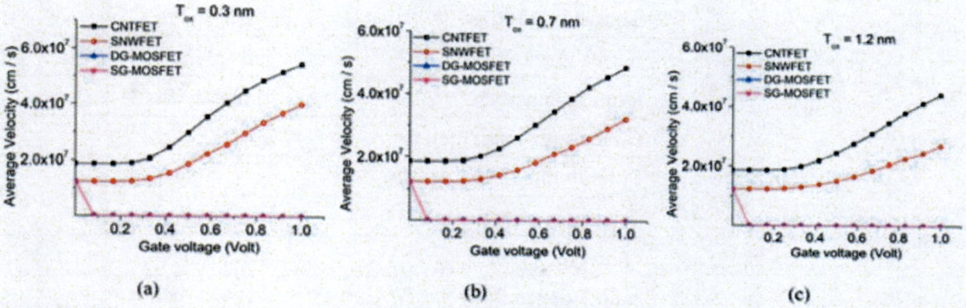

Fig. 6 Average velocity vs gate voltage **a** 0.3 nm **b** 0.7 nm **c** 1.2 nm

5.2 Average Velocity Vs Gate Voltage

It is observed from the graph and the simulation that the average velocity for DG-MOSFET and SG-MOSFET coincides for all oxide thickness. It can also be noticed that for SG-MOSFET and DG-MOSFET, there is a drastic drop in the average velocity from 0 V to approximately 0.1 V after which there is no rise in average velocity. For a certain change in gate voltage for SG-MOSFET and DG-MOSFET, the average velocity approaches to zero. It means gate voltage does not affect average velocity in the case of SG-MOSFET and DG-MOSFET. It has been observed from the graphs, that as the thickness of oxide is downsized from 1.2 to 0.3 nm, the average velocity of mobile electrons for CNTFET and SNWFET increases. Till gate voltage of 0.167 V, the average velocity for CNTFET and SNWFET remains constant after which there is a rise in average velocity. We can conclude that CNTFET at an oxide thickness of 0.3 nm with a constant drain voltage of 1 V has a higher average velocity of the electron which is 5.43E + 07 cm/s (among the three oxide thickness chosen i.e., 0.3, 0.7 and 1.2 nm) as the average velocity decreases with increase in oxide thickness. Whereas SNWFET has an average velocity 4E + 07 of at 1 V for 0.3 nm oxide thickness. Hence CNTFET has 35% more average velocity than SNWFET (Fig. 7).

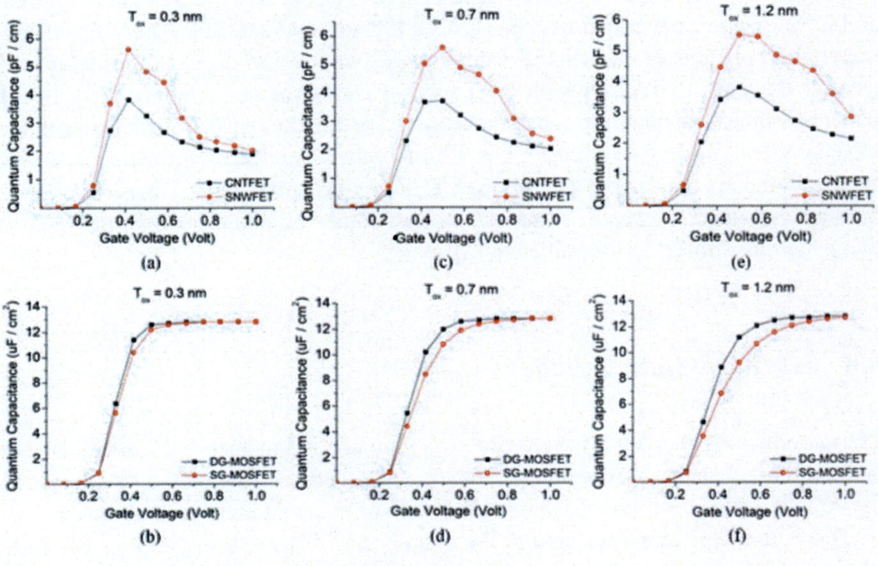

Fig. 7 Quantum capacitance vs gate voltage **a, b** 0.3 nm, **c, d** 0.7 nm, **e, f** 1.2 nm.

5.3 *Quantum Capacitance Vs Gate Voltage*

The quantum capacitance characteristics are different for all four FETs. For CNTFET and SNWFET, there is an abrupt rise in quantum capacitance from 0.25 till 0.416 V (the value of quantum capacitance is 3.82 pF) after which there is a drastic fall in QC. Not a similar trend is shown by DG-MOSFET and SG-MOSFET, they show an abrupt rise in QC from 0.25 to 0.5 V and then it becomes steady (which used to fall in the case of CNTFET and SNWFET). Due to this property, CNTFET and SNWFET have a greater advantage over DG-MOSFET and SG-MOSFET. It can also be observed that for DG-MOSFET and SG-MOSFET, as the thickness of gate oxide decreases, the QC tends to increase. Whereas in contrast, in the case of CNTFET and SNWFET, as the thickness of oxide decreases from 1.2 to 0.3 nm, the value of QC also tends to decrease. This takes place when a gate voltage of 0.5 V and above is applied. SNWFET shows similar characteristics as that of CNTFET. Also, it is observed that the quantum capacitance for CNTFET (for 0.3 nm oxide thickness at 1 V, the QC is 1.88 pF/cm) is less than SNWFET (for 0.3 nm oxide thickness at 1 V, the QC is 2.0 pF/cm) by 6.38%. As a gate voltage of 0.5 V and above is applied, the QC values for CNTFET and SNWFET shows a dropping trend. As the quantum capacitance is low for oxide thickness of 0.3 nm, it (0.3 nm oxide thickness) is ideal for all MOSFETs. So, from the above results, we observe that the SNWFET and CNTFET are beneficial over SG-MOSFET and DG-MOSFET because of low quantum capacitance. Higher the value of QC, higher the propagation delay and leakage current in circuits which leads to degradation of performance [19]. In the paper by Sinha et. al quantum capacitance for oxide thickness of 1.2 nm at gate voltage of 0.83 V for single gate MOSFET is 12.4 $\mu F/cm^2$. The Quantum capacitance for same parameters for Double gate MOSFET, SNWFET and CNTFET is 12.7 $\mu F/cm^2$, 4.37 pF/cm and 2.47 pF/cm respectively. The values obtained using simulation in this paper are similar to the values in [2].

5.4 *G_m/I_d Vs Gate Voltage*

Transconductance (g_m) is also referred to as mutual conductance. Transconductance is the electrical characteristic that relates the current through the output of a device to the voltage across the input of a device. The ratio of transconductance to drain current, (g_m/I_d) is used for transistor sizing. We can observe that, as the thickness of oxide goes from 1.2 to 0.3 nm, the ratio increases for SG-MOSFET and DG-MOSFET but decreases for CNTFET and SNWFET. As the thickness of oxide decreases, the gate oxide capacitance increases which causes the ratio to increase. The curve for all the four MOSFETs shows a similar curve structure, initially highest (is nearly around 34 V^{-1}). The g_m/I_d ratio starts decreasing as the operating point moves toward strong inversion. Since this ratio of DG-MOSFET is maximum in all the devices, it justifies it is better in downsizing of transistor (Fig. 8).

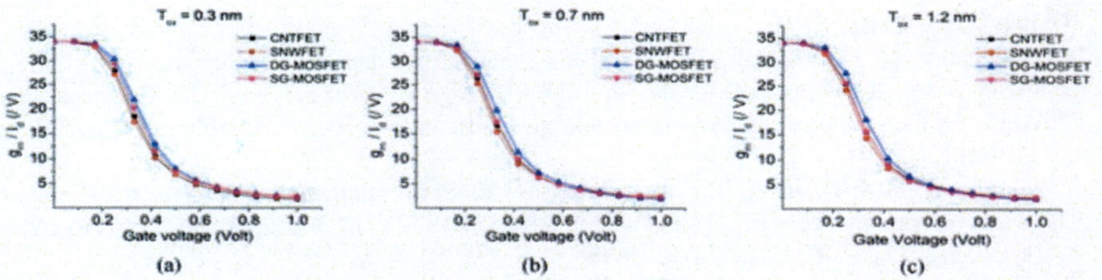

Fig. 8 G_m/I_d vs gate voltage **a** 0.3 nm **b** 0.7 nm **c** 1.2 nm

6 Conclusion

In this work, we have studied, simulated and investigated about the electrical characteristics like quantum capacitance, average velocity, g_m/I_d ratio, drain current at different gate voltages for four different MOSFETs that are, SG-MOSFET, DG-MOSFET, SNWFET and CNTFET. All the simulation was done using the FETTOY simulation tool at nanohub.org. The simulation results of this FETs were analyzed for all the different parameter at different oxide thicknesses of 0.3, 0.7 and 1.2 nm. We can conclude that, CNTFET has an average velocity 35% better when compared to SNWFET for 0.3 nm oxide thickness. The average velocity of SG-MOSFET and DG-MOSFET is zero. In the nanometer regime, the quantum capacitance for CNTFET (for 0.3 nm oxide thickness at 1 V) is less than SNWFET by 6.38%**.** SNWFET and CNTFET devices show a greater advantage when compared to the SG-MOSFET and DG-MOSFET due to their lesser quantum capacitance and due to this property, they have improved (lower) threshold voltage, reduced leakage, and propagation delay. In contrast, in the case of SG-MOSFET and DG-MOSFET, the value of quantum capacitance keeps on rising; this, in turn, leads to higher propagation delay and hence degradation of performance which is undesirable. It was observed CNTFET and DG-MOSFET have 12.6 and 52% better drain current than SNWFET and SG-MOSFET respectively. The results also show that the average velocity of CNTFET is highest when compared to SNWFET. DG-MOSFET and SG-MOSFET have zero average velocity. The g_m/I_d ratio is higher for DG-MOSFET and lowest for CNTFET.

References

1. Moore GE (1965) Cramming more components onto integrated circuits. Electronics, 19 April 1965. Accessed 01 July 2016
2. Sinha SK, Chaudhury S (2013) Impact of oxide thickness on gate capacitance-a comprehensive analysis on MOSFET, nanowire FET and CNTFET devices. IEEE Trans Nanotechnol 12(6):50–62
3. Prasher R, Dass D, Vaid R (2013) Performance of a double gate nanoscale MOSFET (DG-MOSFET) based on novel channel materials. J Nano Electron Phys 5:01017
4. Balhara A, Punia D (2014) Design and analysis of double gate MOSFET devices using high-k dielectric. Int J Electr Eng 7:53–60

5. ITRS-2011 ITRS, ERD
6. Shailendra SR, Ramakrishnan VN (2017) Analysis of quantum capacitance on different dielectrics and its dependence on threshold voltage of CNTFET. In: 2017 international conference on Nextgen electronic technologies: silicon to software (ICNETS2), Chennai, pp 213–217
7. Galadanci GSM, Tijjani A, Babaji G, Gana SM (2018) Comparative study of electrical properties of carbon nano tube (CNT) and silicon nanowire (SNW) MOSFET devices. Bayero J Pure Appl Sci 11(1):55–63. ISSN 2006 – 6996
8. Galadanci GSM, Tijjani A, Gana SM (2018) Performance analysis of electrical characteristics of single gate and double gate nano-MOSFET devices. Am J Eng Res (AJER) 7(6):248–259
9. Chaudhury S, Sinha SK (2019) Carbon nanotube and nanowires for future semiconductor devices applications. https://doi.org/10.1016/B978-0-12-813353-8.00014-2
10. Gupta BK, Mishra M (2018) Comparative study of Ion/Ioff ratio of silicon nanowire transistor (SNWT) with planner MOSFET. Bull Pure Appl Sci Phys. 37d(1). https://doi.org/10.5958/2320-3218.2018.00001.5
11. Kumari A, Rani S, Singh B (2019) Parameterized comparison of nanotransistors based on CNT and GNR materials: effect of variation in gate oxide thickness and dielectric constant. J Electron Mater 48(5):3078–3085
12. Tiwari M, Agarwal N, Saxena AK (2019) Impact of gate dielectric on drain current, gate capacitance and trans-conductance for MOSFET, nanowire FET and CNTFET devices. In: 2019 international conference on cutting-edge technologies in engineering (ICon-CuTE), Uttar Pradesh, India, pp 93–96
13. Yadav C, Agarwal A, Chauhan YS (2016) Analysis of quantum capacitance effect in ultra-thin-body III-V transistor. In: 2016 29th international conference on VLSI design and 2016 15th international conference on embedded systems (VLSID), Kolkata, pp 571–572
14. Sinha SK, Chaudhury S (2014) Advantage of CNTFET characteristics over MOSFET to reduce leakage power. In: 2014 2nd international conference on devices, circuits and systems (ICDCS), pp 1–5
15. Sinha SK, Chaudhury S (2015) Effect of device parameters on carbon nanotube field effect transistor in nanometer regime. J Nano Res 36:64–75
16. Vashaee D, Shakouri A, Goldberger J, Kuykendall T, Pauzauskie P, Yang P (2006) Electrostatics of nanowire transistors with triangular cross sections. J Appl Phys 99:054310–054310. https://doi.org/10.1063/1.2168229
17. Pradhan KP, Mohapatra SK, Agarwal PK, Sahu PK, Behera DK, Mishra J (2013) Symmetric DG-MOSFET with gate and channel engineering: a 2-D simulation study. Microelectron Solid State Electron 2:1–9
18. Cai L, Wang C (2015) Carbon nanotube flexible and stretchable electronics. Nanoscale Res Lett 10:320
19. https://en.wikipedia.org/w/index.php?title=Quantum_capacitance&oldid=940543792. Accessed 15 Apr 2020
20. Nanohub.org

Road Littering Preventive System Using Object Detection and Tracking

Jyotiranjan Biswal and Anil Kumar Shukla

Abstract This paper proposes a prototype system, for efficient and accurate detection of illegal dumping of trash preventing road littering. The dumped garbage not only make the view of city unpleasant but also creates a lot of health issues, here Trash is a rather complicated object to detect because of its nonuniform shape, size, texture, color and grade. By using Mask CNN with python implemented, it provides both segmentation and object detection and thus can be made more accurate and reliable. By using polygon-based method for allocating the trash in the, images captured or surveillance videos.

1 Introduction

Object detection is a technique that uses computer vision and processing of images for creation of certain sets of data set that can be classified as semantic objects [1]. The image or video are processed over the protoc under RCNN python implementation for faster and accurate identification of the object every time. Some examples are Face detection, Face Recognition, Lane detection, Pavement detection, etc.

Computer Vision or generally termed as CV works on the high-level understanding of the image to the computer. Computer vision is one of the most popular part of artificial-intelligence and machine learning. So a Computer vision basically uses the camera and various methods for gaining the capability of human vision. It also helps in decreasing overall cost, Labor replacing human. It may also become reliable as it does not fell fatigue as it works on computer hardware. One of the significantly important factors behind the growth of CV (computer vision) is the amount of data we generate nowadays that is then used to train and make computer vision better.

J. Biswal (✉) · A. K. Shukla
ASET, Amity University Uttar Pradesh, Noida, India

A. K. Shukla
e-mail: akshukla2@amity.edu

V. Nath and J. K. Mandal (eds.), *Proceeding of Fifth International Conference on Microelectronics, Computing and Communication Systems*, Lecture Notes in Electrical Engineering 748,
https://doi.org/10.1007/978-981-16-0275-7_18

This paper focuses on the detection of the trash dumped by the people. There is a major issue of illegal dumping of trash by the people. However manual patrol for the garbage identification is done to keep the road or highways clean though it is time consuming and increases labor, cost. This a computer vision-based model based over the RCNN and Python, will be used to keep observing the highway division and tracking the objects (trash) thrown out on the road or highways.

Trash is rather very complicated object to detect as it does not have any definite shape, color, texture, design. As a trash may vary time to time. For example, water bottle on table may not be trash whereas water bottle fallen on the road is trash [2].

Bircanoglu et al. [14] developed a CNN model namely Recycle-Net for garbage classification. This achieved around 81% in terms of accuracy for Trash-net, it decreased the time complexity by lowering the numberof parameters from seven million to three million.

In the year 2017, A deep learning-based model for recognizing various types of mostly dumped garbage in the city was proposed. To save memory space Nvidia TensorRT with jetpack 2.3.1 is used as it significantly reduced the model size for edge computing. They used various approaches for training the model and conducted experiments on reference models like AlexNet and GoogleNet [11].

In the year 2018, a proposed system that uses improved Yolo V2 to do waste detection and recognition, as yolo Increases the overall speed of the model but it sacrifices somewhat the accuracy of the model compared to RCNN [13].

Excluding methods shown above various other methods developed for trash classification [15], several famous CNN models such as Res-Next [16], ImageNet [17], VGG [18], Res-Net [19], and Dense-Net [20] for image classification also can be used to classify trash as a based model.

The present paper is arranged in the following: Sect. 2 presents proposed designing; Sect. 3 presents Result & Discussion of the dataset and Sect. 4 presents the Conclusion of the paper.

2 Proposed Model for Detection

The Regional Conventional Neural Network or RCNN proposes certain number of boxes and checks for, if it contains any object. These boxes are called regions. The regions are identified using the varying textures, colors, Scales, and enclosure; The "Selective search" checks for these patterns to form a region over a image [3].

The ConvNet [4] or Convolutional Network then analyses the segmented regions and the wrapped image regions are categorized for further classification over the SVMs [5] i.e. Support Vector Machine.

Python implementation-MaskRCNN [6] gives the benefit of both object detection and segmentation [2]. Thus, this requires more manual annotation of the images: as MaskR-CNN uses polygons as its input. Thus, annotation of images was done with the VGG Image Annotator tool [7].

Fig. 1 Model framework (based on FPN)

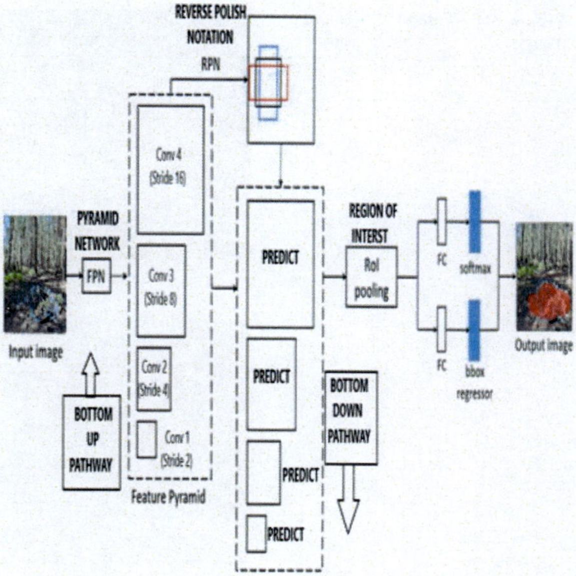

It's based on Feature Pyramid Network [8] (FPN) and a ResNet101 backbone (Fig. 1).

Detecting objects in diverse scales is difficult for small objects [9]. Thus, using a pyramid of the similar image at varied proportion to detect objects. Though, processing different proportion images is very time consuming and the memory used is too high to train simultaneously. Therefore, using it in inference to increase accuracy as much as possible, when speed is not a concern. Alternatively, a pyramid of features and using them for object detection is created.

In Fig. 2(a) a video with human and trash in the foreground is used as confidence map. The heatmap is then generated from color reversal method of the main image taken from the video. The dumping can be seen as human getting separated from the carrying object. In Fig. 2 (d) the human is masked and then checked for object getting separated. Figure 2(c) shows the dumping of trash is detected and is bounded in a box [10].

An initial box surrounds the object. If a human is holding the object then,region box of human and object will move together in the video by comparing the foreground and heatmap. And if the region containing the object in human moves away from the human then it is detected as trash, this all follows CNN [11] based model, for detecting trash and human. So, using the preexisting model for the detection of human hand, a model for tracking trash was created which check for human hand and object being carried by the human. This basically works as this model checks the initial position of the human and object and if both of them get separated from initial position then the object is allocated as trash.

Fig. 2 Color reverse masked image

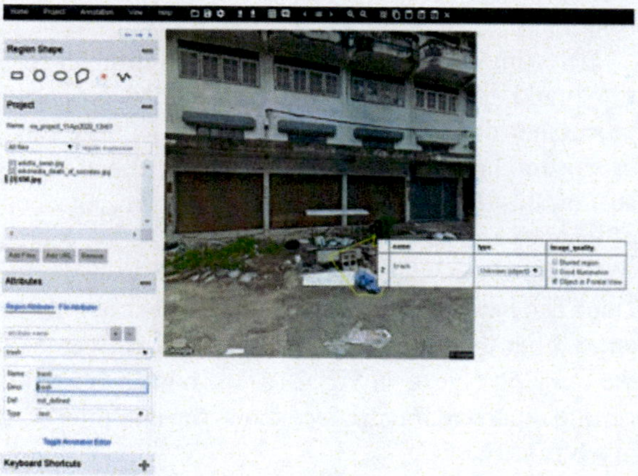

Fig. 3 VIA region shaping and interface

3 Result and Discussion

VIA (VGG image annotation) is a manual image annotation tool [12]. This tool does not need to be downloaded on the computer as it web based. The manual annotation may be time consuming but it increases the overall reliability, efficiency and accuracy of the work (Fig. 3).

In Computer Vision, such a one images are essential for testing & training of a lot of computer vision algorithms. Publicly accessible image datasets such as & contain such manual annotations and also have portrayed a significantly important role in the advancement of CV analysis.

During the approach found out that the there are certain reasons which may lead to insufficient accuracy i.e., if resolution of camera is low or if the camera lens in not clean [13].

In Fig. 4, the trash in foreground is detected and bound box follows the region accurately and also masks the trash within the bounding box (Fig. 5).

The Fig. 6, shows that human and the trash being carried by the human, has been detected accurately by the proposed model using Raspberry-pi System and pi-cam for CV. The trash has been marked by the blue box region and human is marked by cyan color box region with the prediction time count (Table 1).

Fig. 4 Output with bounding box and trash masked in foreground

Fig. 5 Only human in the foreground with no trash in hand

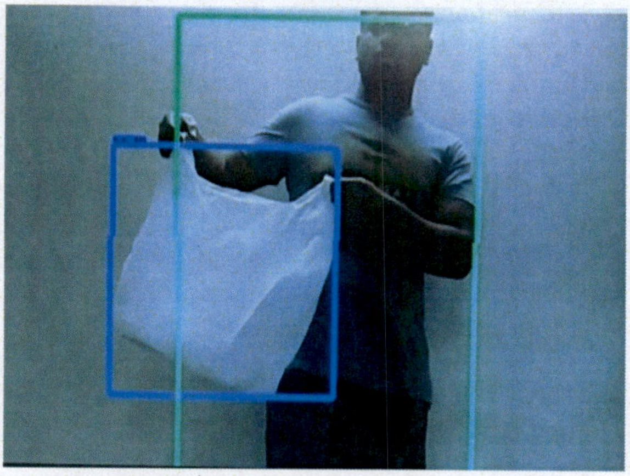

Fig. 6 Raspberry Pi and Picam based detection

Table 1 Computation time

Module	Time(seconds)
Foreground Region	0.048
Trash Tracking	0.00012
Human Tracking	0.0132

Table 2 Comparative study

	[5]	[14]	[15] Yolo-V2	Proposed Model
Detection Accuracy	63%	81%	89.1%	89.9%
GPU (ms)	-	15.9	-	21
Object Tracing Precision	0.24	0.69	0.71	0.77
Recall	0.35	0.34	0.3	0.38

The Table 2 shows the various comparative study.

To conclude, the model runs around 5 frames per second with single gpu system. And around 0.6–0.9 frames per second when using raspberry pi. With accuracy around 50% to 90% when using raspberry pi and pi-cam.

4 Conclusion

In todays world everything is moving towards automated system. As it overall increase's reliability, accuracy and efficiency. Computer vision basically means replicating human eye, and also it is one of the most important part of deep learning and machine learning systems. So, data collection and analysis became most important for running the advance system smoothly. Now a days, many technicians are actively involved in the development of computer vision systems.

In this paper the dumping of trash can be detected automatically using computer vision and deep learning methodology. The suggested model is trained using over 900 trash dumping images with accuracy from 60 to 100%.

References

1. Girshick R, Donahue J, Darrell T, Malik J (2014) Rich feature hierarchies for accurate object detection and semantic segmentation. In: Proceedings of IEEE conference on computer vision and pattern recognition (CVPR)
2. Kerge K, Cowger W, Haamer K, Ehala K, Kivistik K, Tammiste T, Vares M (2019) Wade AI, June 2019. https://ai.letsdoitworld.org
3. Wu Q, Liu Y, Li Q, Jin S, Li F (2017) The application of deep learning in computer vision. In: 2017 Chinese automation congress (CAC), Jinan, pp 6522–6527
4. Ruiz V, Sánchez Á, Vélez JF, Raducanu B (2019) Automatic image based waste classification. In: Proceedings of the IWINAC, vol 2, pp 422–431
5. Yang M, Thung G (2016) Classification of trash for recyclability status. Machine Learning, Stanford, CA, USA, Project Report CS229
6. He K, Gkioxari G, Dollár P, Girshick R (2017) Mask R-CNN. In: 2017 IEEE international conference on computer vision (ICCV), Venice, pp 2980–2988
7. Ren S et al (2015) Faster R-CNN: towards real-time object detection with region proposal networks. In: Advances in neural information processing systems
8. Lin T-Y, Dollár P, Girshick R, He K, Hariharan B, Belongie S (2017) Feature pyramid networks for object detection. In: CVPR
9. Ding S, Zhao K (2018) Daily objects detection based on deep neural network. In: IOP conference series: materials science and engineering, pp 3–4
10. Yun K, Kwon Y, Oh S, Moon J, Park J (2019) Vision-based garbage dumping action detection for real-world surveillance platform. Electron Telecommun Res Inst (ETRI) J 41:494–505
11. Dabholkar A, Muthiyan B, Srinivasan S, Ravi S, Jeon H, Gao J (2017) Smart illegal dumping detection. In: IEEE third international conference on big data computing service and applications, pp 255–260
12. He K, Zhang X, Ren S (2016) Deep residual learning for image recognition. In: Proceedings of the IEEE conference on computer vision and pattern recognition, pp 770–778
13. Reddy KR, Priya KH, Neelima N (2015) Object detection and tracking – a survey. In: 2015 international conference on computational intelligence and communication networks (CICN), Jabalpur, pp 418–421
14. Bircanoglu C, Atay M, Beser F, Genc O, Kizrak MA (2018) RecycleNet: intelligent waste sorting using deep neural networks. In: Proceedings of the INISTA, pp 1–7
15. Liu Y, Ge Z, Lv G, Wang S (2018) Research on automatic garbage detection system based on deep learning and narrowband Internet of Things. In: IOPs conference, pp 1–6
16. Xie S, Girshick RB, Dollár P, Tu Z, He K (2017) Aggregated residual transformations for deep neural networks. In: Proceedings of the CVPR, pp 5987–5995

17. Krizhevsky A, Sutskever I, Hinton GS (2017) ImageNet classification with deep convolutional neural networks. Commun ACM 60(6):84–90
18. Simonyan K, Zisserman A (2014) Very deep convolutional networks for large-scale image recognition. arXiv:1409.1556. https://arxiv.org/abs/1409.1556
19. He K, Zhang X, Ren S, Sun J (2016) Deep residual learning for image recognition. In: Proceedings of the CVPR, pp 770–778
20. Huang G, Liu Z, van der Maaten L, Weinberger KQ (2017) Densely connected convolutional networks. In: Proceedings of the IEEE conference on computer vision and pattern recognition, July 2017, pp 2261–2269

Fixed-Point Divider Using Newton Raphson Division Algorithm

Pawan Kumar Pandey, Dilip Singh, and Rajeevan Chandel

Abstract A fixed point divider is needed for determining the result of division up to a fixed number of points in its fractional part. The divider does so with a good accuracy so that the result can be used for further applications, where the accuracy as well as speed of different modules should be high. This paper presents a 32-bit fixed point divider design based on Newton Raphson division algorithm. The design comprises of two units viz., one is reciprocal unit and the other is the multiplication unit. For the reciprocal unit the divider uses Newton Raphson method and for multiplication unit it uses the multiplication operator. The Verilog HDL coding of the proposed divider design is simulated using Xilinx Vivado tool and synthesized using Cadence EDA tool.

Keywords Verilog · Newton–Raphson method · Slack · FPGA · Divider

1 Introduction

Division algorithms mainly fall into two categories namely, slow division and fast division methods. The slow division methods produce one digit of the final quotient per iteration, which includes algorithms like restoring and non-restoring algorithms [1, 2]. These are based on add and shift operations, consequently, these are time consuming methods. Fast division method starts with a close approximation of final result, so the computation time of final result is much less. Division using Newton Raphson technique comes in the category of fast division algorithms as it uses an initial approximation [3, 4]. Fixed point arithmetic [2, 5, 6] is commonly used for computation because floating point calculation increases the size and complexity of modules [2].

Dividers play an important role in digital signal processing algorithms [1, 7, 8], like the algorithms used to compute the linear prediction coefficients in linear prediction coder. Hence, the study on various types of dividers is essential. Reference [9] gives a

P. K. Pandey (✉) · D. Singh · R. Chandel
Electronic and Communication Engineering Department, National Institute of Technology Hamirpur, Hamirpur, Himachal Pradesh, India

V. Nath and J. K. Mandal (eds.), *Proceeding of Fifth International Conference on Microelectronics, Computing and Communication Systems*, Lecture Notes in Electrical Engineering 748,
https://doi.org/10.1007/978-981-16-0275-7_19

review on various dividers of which the Vedic divider [10] comes out to be faster and area centric. In [5] an array structure for high speed division algorithm is reported in which pipelining method is used to increase the speed of divider. But the pipelining along with increasing the speed, also increase area and cost of the divider. Reference [5] also includes an Iterative Restoring divider which comes out to acquire small area and cost, but the clock cycles used for producing the results are high. Reference [11] shows floating point division using Taylor series expansion algorithm which is basically a third order Newton Raphson method with a reduction in lookup tables causing a reduction is area. In [6] a 32-bit 231 MHz arithmetic unit is given in which division is carried using logarithmic manipulation. It includes 8- region piecewise linear approximation for logarithmic at arithmetic for reducing the number of clock cycles needed for obtaining the final results. But the ASIC and FPGA implementation are not carried out. Reference [12] shows introduction and FPGA implementation of division using Goldschmidt algorithm which also comes under the fast division algorithms. The FPGA implementation of VLSI divider circuits is also important [2, 5, 7, 12] as it helps in the practical verification of the design. Other types of divider reported in the literature are high-redux; serial; integer and floating dividers [13–15]. The division using Newton Raphson method is also represented in [3, 4] using different techniques but the simulation and FPGA implementation are not carried out. The present paper is an attempt in this direction.

The present paper presents a 32-bit fixed point divider using Newton Raphson division algorithm in which the first bit from MSB is the sign bit, the next three bits are for integer part and the remaining 28 bits are for the fractional part. The large number of bits for the fractional part results in a good accuracy. This also includes multipliers in its operation for that the data representation in Q-format [16] is used.

A 32-bit fixed-point divider is also implemented in the present work using shift and subtract method. Shift and subtract method use the conventional method of division. In this for the integer part of the result it uses simple subtraction method where the divisor is repeatedly subtracted from dividend until the result is less then divisor. The number of times divisor is subtracted then become the integer part. For the fractional part it uses shifting and subtraction method, but the drawback of this method is that it produces one bit of result per clock cycle for fractional places. Hence, it takes large number of clock cycles for producing the result and is not suitable for delay centric designs.

This paper is organised as follows. The present section introduces the topic and provide a brief background on different division algorithms. In Sect. 2 Newton Raphson method, its use as division algorithm along with the flow chart are presented. Section 3 provides the simulation and synthesis results obtained using various EDA tools, their discussion and conclusion is drawn in Sect. 4, followed by references.

2 Newton Raphson Method

Newton Raphson method is used to find the root of an algebraic function with the help of some initial approximation. Let the function be,

$$y = f(X) \tag{1}$$

Let the initial approximation of the root of this function is X_0. Then according to the Newton Raphson method, the next approximation of the root say X_1 is given as,

$$X_1 = X_0 - (f(X_0)/f'(X_0) \tag{2}$$

where, f'(X) is the first derivative of the function $f(X)$.

With the help of the second approximation the next approximation is computed for the same i.e.,

$$X_2 = X_1 - (f(X_1)/f'(X_1)) \tag{3}$$

This computes the next approximation of the root. Thus, a generic relationship is given for calculating the next approximations as,

$$X_{n+1} = X_n - \left(f(X_n)/f'(X_n)\right) \tag{4}$$

where, X_{n+1} is the next approximation while X_n is just the previous approximation.

The approximations are computed till the next approximation is nearly the same as the previous approximation.

2.1 Newton Raphson Division Algorithm

Suppose two numbers say *a* and *b* are to be divided and let *Y* be the result after the division, which can be written as,

$$Y = a/b \tag{5}$$

Alternately *Y* can be written as,

$$Y = a \times (1/b) \tag{6}$$

Now the reciprocal of *b* i.e. *1/b* is computed with the help of Newton Raphson method. Then it is multiplied by a to get the output *Y*.

The function $f(X)$ for calculating the reciprocal of b is,

$$f(X) = (1/X) - b \tag{7}$$

Its root is $1/b$.

So,

$$f'(X) = -(1/X^2) \tag{8}$$

Let the initial approximation for the root be X_0, then we can calculate the next approximation of the function using Newton Raphson method, which comes out to be,

$$X_1 = X_0(2 - b \times X_0) \tag{9}$$

2.2 *Computation of the Initial Approximation for Newton Raphson Division Algorithm*

Newton Raphson method needs an initial approximation, based on that the subsequent approximations are calculated. For division algorithm the initial approximation is calculated by making a linear approximation for the reciprocal function.

In Fig. 1 the black line shows the actual reciprocal function in which x-axis ranges from 0.5 to 1, and y-axis is the reciprocal of x-axis points, thus y-axis ranges from 1 to 2.

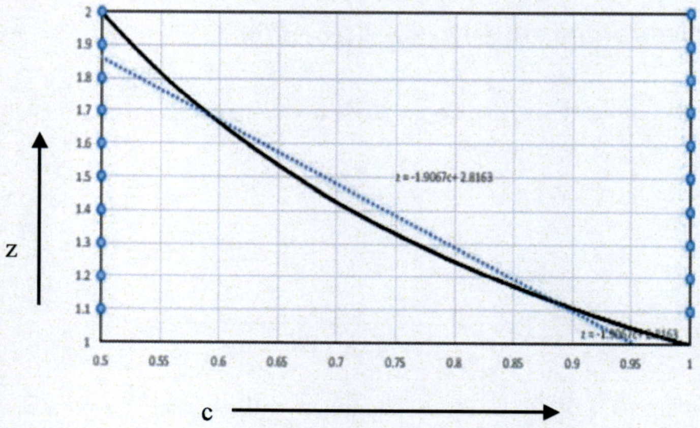

Fig. 1 Linear approximation for the reciprocal function

The dotted blue line shows the best linear approximation using MS office Excel, for the reciprocal function which comes out to be,

$$z = -1.9067c + 2.8163 \quad (10)$$

2.3 Flow-Chart for Newton Raphson Division Algorithm

Figure 2 illustrates the flow-chart for the Newton Raphson division algorithm.

As seen from the flow-chart in Fig. 2, firstly the inputs *a* and *b* are read. One of the inputs let *b* is normalized in the range of 0.5 to 1. In order to normalise it we have to divide or multiply *b* by suitable powers of 2. Divide or multiply by 2 in Verilog HDL [16] is equivalent to left or right shift by 1 respectively. After normalising the input, the Newton Raphson division algorithm is applied to find the best approximate result for 1/*b*. Again, we must de-normalise the answer by dividing or multiplying the answer by the same powers of 2 by which it is normalised previously. To get the final answer we multiply *a* to de-normalised value i.e. 1/*b*.

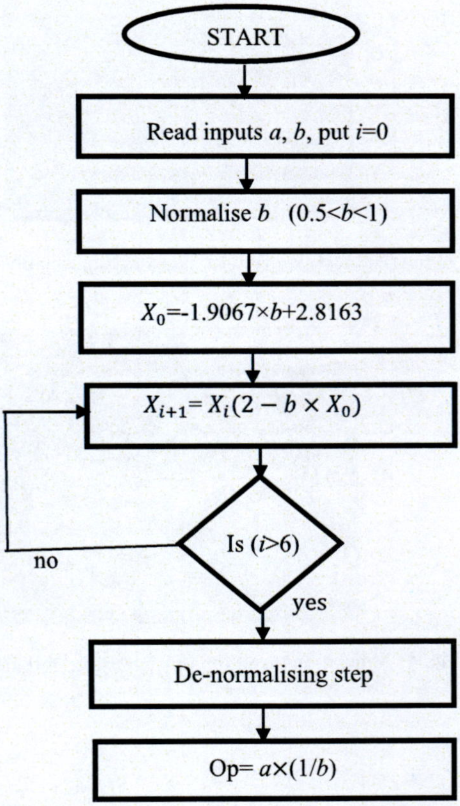

Fig. 2 Flow-chart for the Newton Raphson division algorithm

3 Results and Discussion

The simulation of Newton Raphson division algorithm is performed using Xilinx Vivado 2015.2 tool [17]. For comparative analysis of the designed Newton Raphson division algorithm, the shift and subtract division algorithm is also simulated in the present work.

From Fig. 3, it can be seen that the output of division using Newton Raphson method is attained after every 6 clock pulses. However, for Fig. 4 output is coming after every 27 clock pulses as it produces one bit at a time for its fractional part, thus it can be inferred that the Newton Raphson division algorithm has higher latency than that of shift and subtract method. The divider using Logarithmic arithmetic [6] can produce the result only in 2 clock cycles with an error of 0.2%. However, the percentage error in cases of both of the designed dividers is nearly 0. Consequently, for better latency and high accuracy the division using Newton Raphson method is the best.

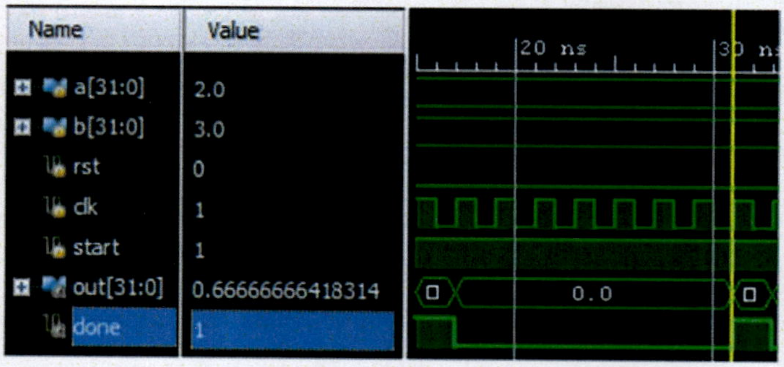

Fig. 3 Simulation result for Newton Raphson division algorithm

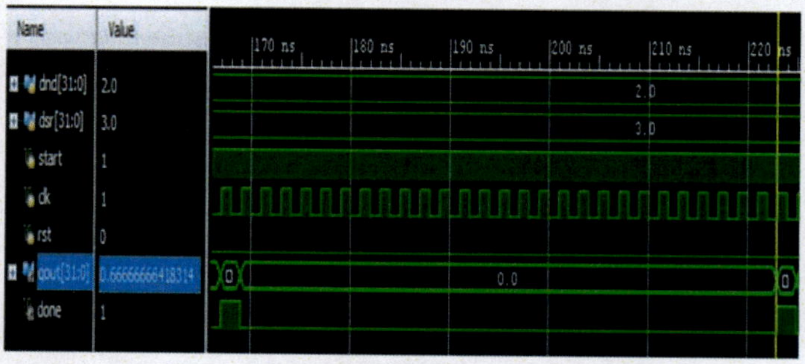

Fig. 4 Simulation result for division using shift and subtract method

3.1 Analysis of Synthesis Results

The synthesis of the two types of dividers is carried out using two different types of EDA tools. For FPGA synthesis Xilinx Vivado 2015.2 tool [17] is used by selecting Zybo board on it. For ASIC synthesis Genus tool from Cadence [18] is used. The coding is accomplished in Verilog HDL [19] (Table 1).

Figure 5 shows the total on-chip power consumed (P_t) by the Newton Raphson divider and is the sum of static power and dynamic power. Here it is,

$$P_t = 0.102W + 0.007W = 0.109W \tag{11}$$

The total on-chip power consumed by the shift and subtract divider is (Fig. 6).

$$P_t = 0.102W + 0.003W = 0.105W \tag{12}$$

The clock frequencies for these two types of dividers is calculated on the basis of slack and for both dividers the clock frequency comes out to be 25 MHz As seen from the power results the power consumed by the divider using Newton Raphson method is nearly 3.7% more than that of shift and subtract method. This is because the divider using Newton Raphson method uses 4 multipliers which consume more power than any other RTL component.

Table 1 Description of RTL components used in dividers using Xilinx Vivado 2015.2

RTL components	Divider using Newton Raphson's method	Divider using shift and subtract method
Adders	4	5
XOR gates	1	1
Multipliers	4	0
Multiplexers	9	11
Registers	8	9

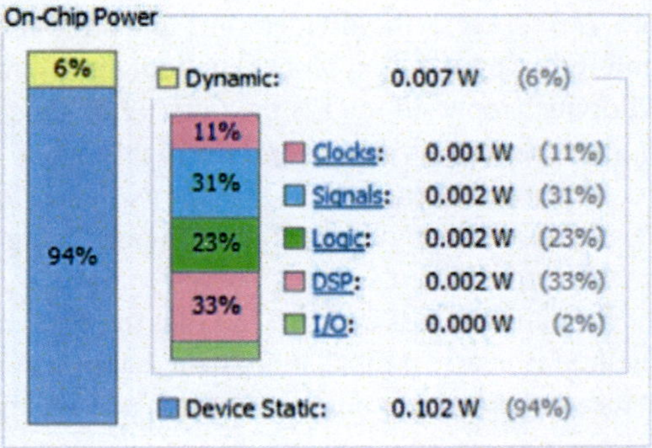

Fig. 5 Power results for divider designed using Newton Raphson method

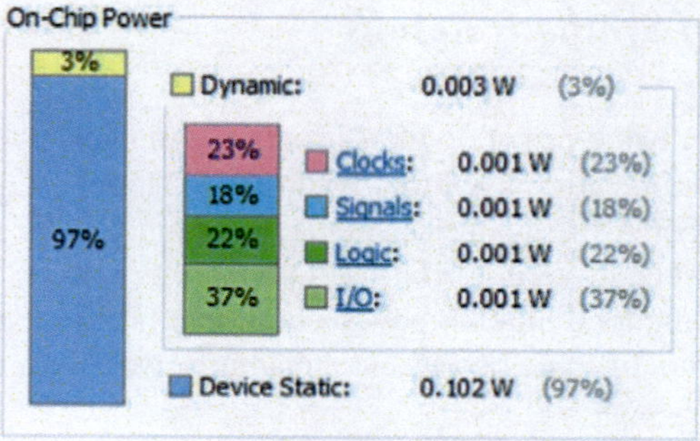

Fig. 6 Power results for divider designed using shift and subtract method

Table 2 FPGA results for the designed 32-bit fixed point dividers

Performance metrics	Goldschmidt algorithm [12]	Shift and subtract method (This work)	Newton Raphson method (This work)
Slice LUTs (look up tables)	9281	348	9281
Flip flops	3025	192	3025
IOBs (input output blocks)	161	100	161

Table 3 ASIC area requirement for various dividers

Performance metric	Taylor series method [11]	Shift and subtract method (This work)	Newton Raphson method (This work)
Area (mm^2)	0.037	0.003215	0.016113

The FPGA implementation of the designed 32-bit fixed point dividers is presented in Table 2. These are also compared with the divider designed using Goldschmidt Algorithm reported in [12].

From Table 2 it can be seen that the number of Look up tables (which is basically a block of SRAM that is indexed by LUT's inputs) used in Newton Raphson divider are 753 whereas the divider shift and subtract method uses 348 also the flip flops and IOBs(input output blocks) in both the dividers are comparable. However, in case of divider reported in [12] the number of slice LUTs, flip flops and IOBs are much higher than the dividers presented in this paper. Tables 3 and 4 show the ASIC results for dividers designed in the present work and for the comparative analysis the ASIC results for Vedic Divider [10] and divider using Taylor series expansion method [11] are also considered.

From Table 3, it can be seen that the area of the divider using Newton Raphson method is nearly 5 times more than that of add and subtract method. This is because it uses 4 number of multipliers which need a larger area. However, the area of divider

Table 4 ASIC dynamic power dissipation in different dividers

Performance metric	Vedic divider [10]	Taylor series method [11]	Shift and subtract method (This work)	Newton Raphson method (This work)
Power (mW)	0.024	>0.812	0.0142	0.812

using Taylor series expansion method is nearly twice than that of the Newton Raphson divider. In [10] the ASIC area of Vedic divider results are not given.

From Table 4 it is analysed that the power dissipation in case of shift and subtract divider is least of the four divider designs given in Table 4 because it does not require any multipliers for its operation and is therefore power centric design. The dynamic power results for division using Taylor series in not given in [11] but it also uses four number of multipliers for its operation [11], so its power consumption is assumed to be equal to or more than division using Newton Raphson method because its area (Table 4) is also more than that of Newton Raphson method. The Vedic divider [10] comes out be better in case of dynamic power consumption than that of division using Newton Raphson method.

4 Conclusion

The paper presents the design and analysis of 32-bit divider using Newton Raphson algorithm. The comparative analysis of this divider is performed by designing and analysing another divider based on shift and subtract method along with other dividers given in references. From the results it can be seen that the Newton Raphson divider comes out to be a faster division algorithm. But since it uses 4 number of multipliers, this causes increment in the area and power. The Vedic divider is consuming less dynamic power than divider using Newton Raphson division algorithm bur its accuracy and latency is not proposed in [10]. For reducing the power and area of divider using Newton Raphson method the normal multipliers can be replaced by Booth multipliers [20] and by proper pipelining the area and speed requirement can be improved. However, it is analysed that for power and area centric designs, divider designed using shift and subtract method proves to be better candidate. Thus, the choice of divider circuits depends upon the applications and the usage.

Acknowledgements The technical support through SMDP-C2SD project awarded to NIT Hamirpur by MeitY (Ministry of Electronics and Information Technology), Government of India, New Delhi, is duly acknowledged.

References

1. Meyer-Baese U (2004) Digital signal processing with field programming gate arrays. Springer, Heidelberg
2. Sorokin N (2006) Implementation of high-speed fixed-point dividers on FPGA. J Comput Sci Technol 6(1):8–11
3. Rodriguez-Garcia A, Pizano-Escalante L, Parra-Michel R, Longoria-Gandara O, Cortez J (2013) Fast fixed-point divider based on Newton-Raphson method and piecewise polynomial approximation. In: International conference on reconfigurable computing and FPGAs, ReConFig 2013, vol 2, no 2
4. Louvet N, Muller JM, Panhaleux A (2010) Newton-Raphson algorithms for floating-point division using an FMA. In: International conference on application-specific systems, architectures and processors, pp 200–207
5. Narendra K, Shabirahmed BJ, Swaroop Kumar K, Asha GH (2015) FPGA implementation of fixed-point integer divider using iterative array structure. Int J Eng Tech Res (IJDTR) 3(4):170–179
6. Kim H, Nam BG, Sohn JH, Yoo HJ (2005) A 231MHz, 2.18mW 32-bit logarithmic arithmetic unit for fixed-point 3D graphics system. In: IEEE Asian solid-state circuits conference ASSCC 2005, pp 305–308
7. Malik P (2015) High throughput floating-point dividers implemented in FPGA. In: IEEE 18th international symposium on design and diagnostics of electronic circuits & systems (DDECS), pp 291–294
8. Rodellar V, Sacristan MA, Alvarez A, Diaz A, Peinado V, Gómez P (2001) A divider-multiplier high level synthesis library element for DSP applications. In: IEEE international conference on electronics, circuits and systems, vol 1, pp 545–548
9. Vemula R, Chari KM (2018) A review on various divider circuit designs in VLSI. In: Conference on signal processing and communication engineering systems SPACES 2018, January 2018, pp 206–209
10. Saha P, Banerjee A, Bhattacharyya P, Dandapat A (2011) Vedic divider: novel architecture (ASIC) for high speed VLSI applications. In: International symposium on electronic system design ISED 2011, pp 67–71
11. Kwon TJ, Draper J (2009) Floating-point division and square root using a Taylor-series expansion algorithm. Microelectron J 40(11):1601–1605
12. Singh N, Sasamal TN (2016) Design and synthesis of Goldschmidt algorithm based floating point divider on FPGA. In: International conference on communication and signal processing ICCSP 2016, pp 1286–1289
13. Aoki T, Nakazawa K, Higuchi T (2000) High-radix parallel VLSI dividers without using quotient digit selection tables. In: International symposium on multiple-valued logic, pp 345–352
14. Mishra R (2017) An efficient VLSI architecture for a serial divider. In: 2nd international conference on devices for integrated circuit, DevIC 2017, pp 482–486
15. Saadat H, Javaid H, Parameswaran S (2019) Approximate integer and floating-point dividers with near-zero error bias. In: Design automation conference, pp 1–6
16. Hussain MZ, Parvin KN (2016) Q-format data representation and its arithmetic. Int J Electron Commun Technol (IJECT) 7(2):57–62
17. Xilinx Inc. (2015) Vivado Design Suite User Guide. Ug903, vol 4, pp 1–173
18. Xcelium C (2019) Tempus, Voltus Cadence RTL-to-GDSII Flow Course Version 2.0 Developed By University Support Team Cadence Design Systems, Bangalore
19. Palnitkar S (2003) Verilog HDL, a guide to digital design and synthesis. IEEE 1364-2001 Compliant
20. Chen YH, Li CY, Chang TY (2011) Area-effective and power-efficient fixed-width Booth multipliers using generalized probabilistic estimation bias . IEEE J Emerg Sel Top Circuits Syst 1(3):277–288

Real-Time Globally Accessible Accident Notifications and Tracking for the Motor-Vehicles

Vismay Patel, Pruthvish Rajput, Abhi Zanzarukiya, and Rutu Parekh

Abstract To provide spontaneous emergency services in the cases of vehicular accidents, the accident notifications have to be fast (real-time), accurate, guaranteed and easily accessible world-wide. The earlier and existing solutions for accident notification services are on commercial levels which are limited to particular voice call assistance only. The proposed approach focuses on globally accessible accident notifications. The aim here is to provide globally accessible real-time accident notifications that are to be delivered in form of a 'Marker' on Google Map. A 'WebApp' or 'Webpage' is suggested that displays google map with accident location marker on it. The Marker should contain the information like date & time of accident, vehicle and driver details and accident severity. The proposed 'WebApp' can be accessed by any hospital (emergency services) in the world. Meanwhile an emergency voice call and an SMS is also placed at registered emergency number. They can provide emergency medical services for the accidents in their region. The 'Marker' on Google Map can be deleted by medical service centre once the victim is provided the required help.

Keywords Accident detection · Accident notification · Microcontroller · Firebase · Motor-vehicle · Vehicle tracking

1 Introduction

In today's world, the inventions of powerful automobile engines have increased the speed of transportation. As a result the situation has become more vulnerable to the fatal accidents. According to the World Health Organization (WHO) database [1], there were 1.25 million road traffic deaths globally in 2013. In order to reduce these fatalities, the Emergency Medical Service (EMS) has to be quick enough. For that the accident notifications must be accessible by hospital in the world. There are several accident emergency service providers like Onstar, Verizon Vehicle, Automatic, Mojio, Zubie, etc. But, the emergency services provided by each of them are

V. Patel · P. Rajput · A. Zanzarukiya (✉) · R. Parekh
DA-IICT, Gandhinagar, India

V. Nath and J. K. Mandal (eds.), *Proceeding of Fifth International Conference on Microelectronics, Computing and Communication Systems*, Lecture Notes in Electrical Engineering 748,
https://doi.org/10.1007/978-981-16-0275-7_20

based on emergency voice calls and they are limited to their voice call assistance only. Still there is a need to link hospitals world-wide for emergency services. Considering this aim, we propose such a robust system that consists of separate accident detection and a separate notification system.

There has been similar work carried out by various groups. The papers published in [2–5] utilized Android smartphone sensors for car crash detection. The papers by Ali et al. [2] and Fernandes et al. [4] utilizes some of the automotive concepts to decide the accident detection threshold. But they have not considered the possibility of accidents below the threshold. In the paper by Imran et al. [3], an accident severity classification is suggested. But the severity criterion is established based on a lab experiment by colliding two metal pieces at different intensities. That is absolutely unreliable because it does not include any automotive criteria. For accurate crash detection, the accelerometer has to be rigidly mounted to the vehicle's body and hence using smartphone sensors is not always reliable. The smartphone resting on a smooth surface in a car may get into the motion if car gets into a serious frontal crash. As smartphone gets in motion, it will not detect the deceleration experienced by the car. Ahirrao et al. [6], suggested a separate collision detection circuit installed in the car and a separate notification system using Android App. The circuitry utilizes LDR and photodiode as the sensors for collision detection but it is logically an invalid approach. The work presented in [7] explains an accident detection and notification system using an accelerometer and Raspberry Pi respectively. The constraint is, this method utilizes Raspberry Pi for notifications which needs separate hardware and a SIM.

[8–11] uses image based techniques for detecting accidents. Ravi et al. [8], uses blinking of eyes to check whether driver is feeling drowsy or not. Haar-Cascased technique is used for processing of captured image. Some research work proposes to leverage the installed CCTV camera on highway and detect accidents from segmented images from live video [9]. They use fusion of Convolutional Neural Network (CNN) layers and Long Short Term Memory (LSTM) layers to carry out the detection of accident. The work presented in [10], uses three step procedure to detect the accident of vehicle. They use Gaussian Mixture Model (GMM) to capture vehicle from video frame. After that, that vehicle is tracked using mean shift algorithm and finally three decisive parameters are obtained to make final decision. However, in all these vision based approaches, unless the environmental conditions like light, weather and visibility are favorable, the detection cannot be performed efficiently. There are several works that uses sensor like vibration sensor to detect the accident [8, 12–14]. However, there are chances that high and sudden thrust is produced due to bump or bad road conditions in which case, the system might falsely detect it as accident.

The unique thing about this work is, it has a module which can be customized to detect accident with severity for any type of motor-vehicle. The notification system delivers accident notifications in form of google map markers on a webpage using real-time database that can be accessed by any hospital in the world. The previous efforts suggested notifications transmission through email and Http protocol that might be slow or unreliable. The details of an accident will be shown in the info window of the marker. Such an approach covers all the hospitals world-wide and

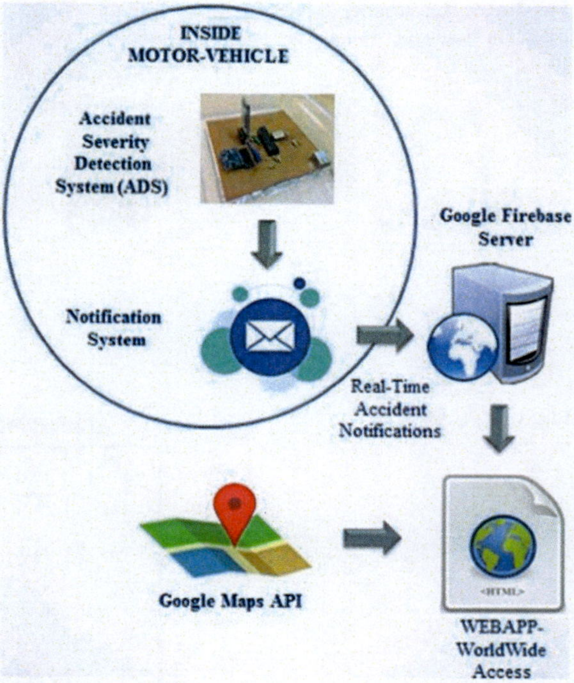

Fig. 1 Overview of the system

makes sure that none of the victims remains helpless. The graphical overview of the complete system is shown in the Fig. 1. The motor-vehicle is equipped with Accident Severity Detection module (ADS) module and a notification system connected to Google Firebase Database.

2 Methodology

The proposed approach incorporates a separate Accident Severity Detection module (ADS) and a separate notification system to be installed in the motor-vehicle.

2.1 Accident Severity Detection System (ADS)

The ADS device design is described in Fig. 2. The proposed ADS Hardware consists of ATMega328P microcontroller, 3-axis accelerometer MPU-6050 mounted on GY521 Breakout Board and a Bluetooth Device HC-05. These components are chosen as they satisfy the requirements of ADS Hardware.

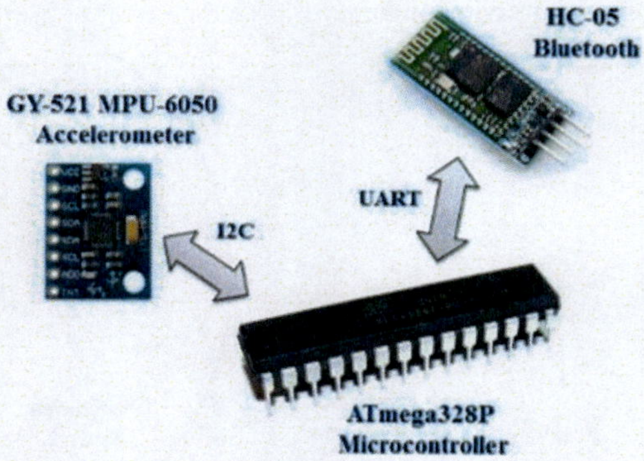

Fig. 2 Accident severity detection system hardware

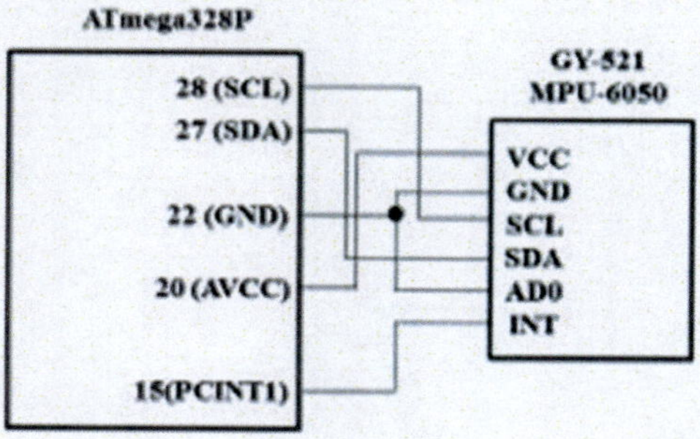

Fig. 3 MPU-6050 interfacing

Accelerometer MPU-6050 is interfaced with microcontroller ATmega328P via I2C protocol as shown in Fig. 3, and a bluetooth device HC-05 is interfaced with ATmega328P via Universal Asynchronous Receiver Transmitter (UART) serial connection as shown in Fig. 4.

Accelerometer MPU-6050 supports maximum range of ± 16 g that is sufficient for severity estimation. It supports output sample rate of 200 Hz with each sample of size 16 bits. Therefore, to store the acceleration samples along x (driving) axis and y (side) axis for 200 ms we need at least 160 Bytes of memory. Microcontroller ATmega328P has 2 KB of SRAM that is sufficient to store the acceleration samples. Moreover, ATmega328P could be simply programmed using Arduino loader and utility during research and development phase. Therefore, ATmega328P is selected in our implementation. Bluetooth device HC-05 is chosen as wireless device because it provides reliable transmission at 2.4 GHz and it is compatible with proposed notification system (explained later) for short distance communication (approx. 10 m). Above described individual components used for ADS Hardware can be powered ON using a 5 V power supply that is easily available. This ADS device estimates the impact accident severity. The severity estimation is carried out based on the classification Table 1.

Table 1 is derived based on the references [15] and [16]. Here, P is maximum acceleration or deceleration magnitude, $g = 9.8\ m/s^2$ and DeltaV is change in velocity

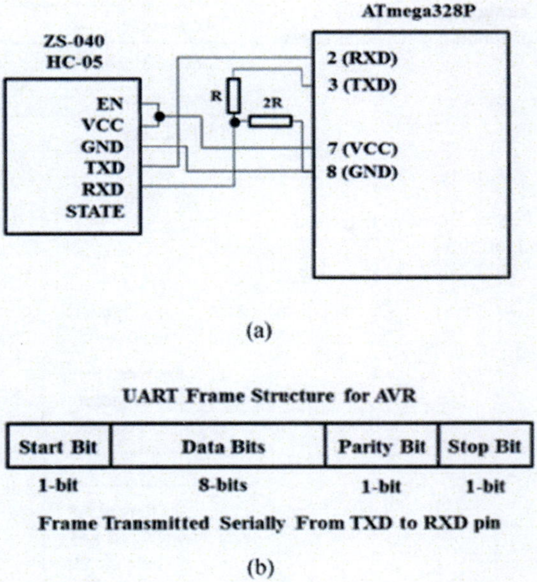

Fig. 4 Microcontroller and Bluetooth module connection **a** HC-05 interfacing. **b** UART Frame

Table 1 Accident severity classification

Condition	Severity
(1.5 g < P < 5 g) and (DeltaV < 12.8 km/h)	Moderate
(5 g < P < 13 g) or (12.8 km/h < DeltaV < 22.5 km/h)	High
(P > 13 g) or (DeltaV > 22.5 km/h)	Extreme

of the vehicle during the impact. As it is based on airbag deployment criteria, it is valid for frontal impacts and side impacts as well. The acceleration or deceleration can be directly measured by using an accelerometer. Velocity is basically an integration of acceleration signal. But in order to measure the change in velocity during crash, a systematic method is required. One such automotive research paper [17] gives an indirect idea about the estimation of change in velocity during crash.

According to data given at [15], the fastest car among all the common cars driven on the road, Bugatti Veyron has maximum acceleration of 1.5 g. Therefore, it is considered as the threshold for accident detection in the proposed algorithm. Here |Acc| is the magnitude of overall acceleration along front and side directions of the vehicle. The tilt angle for different two-wheelers may be different when they fall down. To determine the tilt angle threshold, an experiment was conducted to find tilt angles practically for different types of two-wheelers. We have observed that before the tire dimensions come into the picture, the side body and the side accessories of a two wheeler play a major role. Therefore, we have performed the fall down of different types of two wheelers.

Table 2 Tilt angle of different two-wheeler cases

Two wheeler	Tilt (roll) angle
Honda Shine	49
Honda Activa	48
TVS Scooty	50

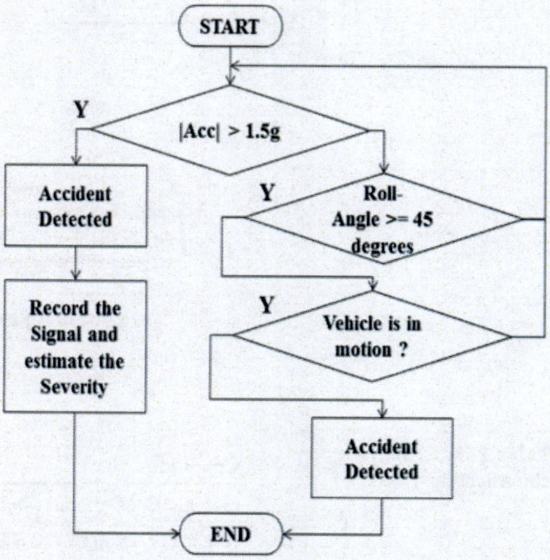

Fig. 5 Accident detection algorithm

Table 2 shows the practical values of tilt angles for different two-wheelers when they fall down on the ground. Here we are considering tilt angle as roll angle that corresponds to rotation w.r.t driving direction (axis). These values were measured using a standard inclinometer. From the analysis of roll angles obtained through experiment, it can be inferred that for most of the two-wheelers the value of roll angle exceeds 47° when they fall down. Considering some margin, the roll angle threshold for two wheelers is taken as 45° when they fall down. It has been reflected in Fig. 5. That is if any vehicles roll angle exceeds 45° while it is in motion, it signifies the roll-over accident. The proposed algorithm is applicable to all types of motor-cars and other heavy vehicles. Now, the block diagram shown in Fig. 6 explains the method to estimate the severity.

The processing on recorded acceleration signal to retrieve P and DeltaV values was referenced from [17] This paper actually describes the novel approach to determine ending time of crash (crash pulse) based on acceleration signal processing. According to this method, from the beginning of crash (1.5 g point) the acceleration/deceleration signal samples should be recorded and stored in memory for approximately 200 ms. This recorded signal should be filtered using 30 Hz Digital Butterworth Low Pass Filter of order 2. Equation 1 represents the frequency response of such filter [18]. It effectively removes unwanted high frequency components.

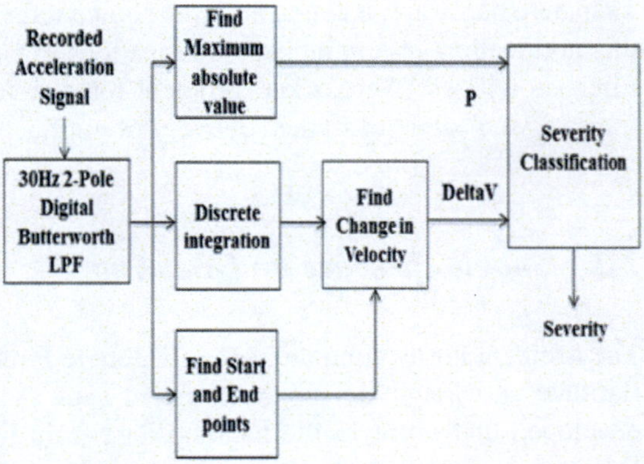

Fig. 6 Accident severity estimation

$$H(z) = \frac{(a_0 + a_1 z^{-1} + a_2 z^{-2})}{(1 - b_1 z^{-1} + b_2 z^{-2})} \tag{1}$$

Where,

$$a_0 = \frac{\omega_a^2}{\left(1 + \sqrt{2}\omega_a + \omega_a^2\right)} \quad b_1 = \frac{2(1 - \omega_a^2)}{\left(1 + \sqrt{2}\omega_a + \omega_a^2\right)}$$

$$a_1 = 2a_0 \qquad b_2 = \frac{\left(1 - \sqrt{2}\omega_a + \omega_a^2\right)}{\left(1 + \sqrt{2}\omega_a + \omega_a^2\right)}$$

$$a_2 = a_0 \qquad \omega_a = \tan(\pi \cdot 37.5 \cdot T)$$

$T\ is\ Sampling\ Interval$

In the output filtered signal, the maximum acceleration magnitude is peak acceleration ‘P’. According to the novel approach presented in [17], the ending of crash or end point is considered as the time where acceleration magnitude becomes 5% of ‘P’ after ‘P’ point. The acceleration signal is integrated to get velocity waveform and the velocity difference is calculated between start and end points of the crash that gives, DeltaV. In the block diagram shown in Fig. 6, the severity classification is carried out based on the Table 1.

2.2 *Notification System*

The accident severity detected by ADS is delivered to Google Firebase Database using a separate notification system. The notification system requires GPS, GSM/3G/4G network module to deliver the notifications. The notification system can be implemented on a vehicles’ driver’s smartphone (Using Android/iOS Apps) or on a separate hardware. It depends on the choice/requirement of the user. But ultimately this notification system delivers notifications to Google Firebase Database. Google Firebase utilizes ‘WebSocket’ protocol for data transmission that gives “real-time transmission” and guaranteed delivery of notifications.

2.3 Google Firebase and WebApp

The accident notification delivered to Google Firebase Server contains the location (latitude and longitude) of the accident spot. A Firebase WebApp (Webpage) is developed that retrieves the location data from Firebase Database and displays a ‘Marker’ on Google Map along with the information like date-time of the accident, accident severity, vehicle and driver’s details. The proposed WebApp/Webpage can be accessed by any hospital in the world. The hospitals can provide the emergency services for the accidents in their region. For further enhancement in the proposed WebApp, the ‘Marker’ on Google map can be deleted by the emergency service providers once the victims are provided the required medical help.

3 Accidents Tests, Setup and Results

This section includes the real world testing of the proposed system.

3.1 Testing of Accident Severity Detection Algorithm

After developing accident severity detection algorithm, it was tested using standard crash test data (accelerometer signal data) available on NHSTA website. Two different crash tests were tested, one for frontal crash and the other for the side crash as detailed in Table 3. Based on the results it can be concluded that the developed algorithm can successfully classify the severities for corresponding impact directions for different test cases. In Fig. 9(a) for test case 2, there is slow speed crash and not significant damage. The algorithm also outputs it as moderate severity shown in Fig. 9(b). Similarly in Fig. 8(a) the damage done to the vehicle can be compared with the output of the algorithm.

3.2 Practical Test Setup for Accident Detection

After verifying the validity of accident severity detection algorithm, the practical test setup was carried out to verify the overall performance of the complete system

Table 3 Test cases

Attributes	Test case 1 [11]	Test case 2 [12]
NHSTA test no	1847	2344
Test config	Vehicle into barrier	Impactor into vehicle
Target vehicle	Ford tempo	Chevrolet caprice
Impact direction	Front	Left
Impact speed (km/h)	15.5	8.40
Accelerometer location	Vehicle CG	Vehicle CG
Accelerometer axis	(x, y)	(x, y, z)
Accelerometer sampling rate	8 kHz	12.5 kHz

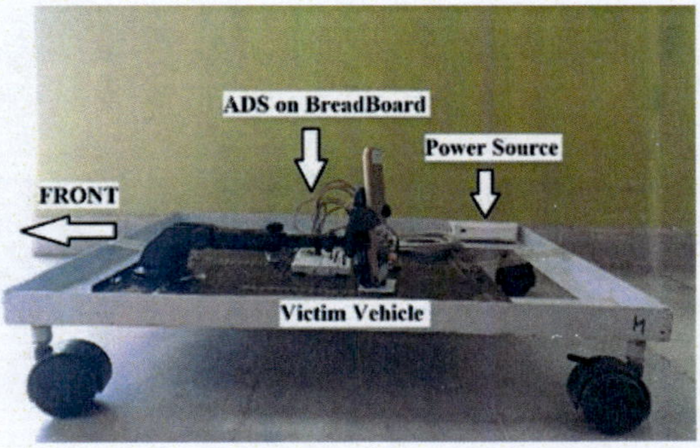

Fig. 7 Vehicle model for accident test

including notification system and Google Firebase Database WebApp. Figure 7 shows the accident severity detection system installed on a four-wheeled iron frame. Detailed design of ADS is presented in Sect. 2.1. Here, Android phone is used as a notification system. Utilizing GPS, data and cellular network features of android smart phone by using various Android and Google APIs, the Android App was developed. ADS and Android App communicates via Bluetooth protocol.

As shown in Fig. 7, the accident test was carried out on the scaled models. The stationary victim vehicle was hit by a moving heavy impactor vehicle (approx. 20 km/h) with high intensity. The relative mass(m) of both the models were chosen such that this accident would generate the same motion in a real car when it is hit by a real heavy truck. ADS was implemented on a bread board as shown in Fig. 7. The Android Smartphone was used to deliver the notifications SMS, Voice call and also data to Google Firebase Database. The notifications were delivered as shown in the Fig. 10. The notifications delivered to Firebase WebApp (Google Map) are as shown in Fig. 11(a). The corresponding entries in Google Firebase Database is also shown in Fig. 11(b). As shown in Fig. 11(a), it displays the markers on google map

Fig. 8 ADS test case 1. **a** Vehicle after crash [19] **b** Output of the algorithm

(a)

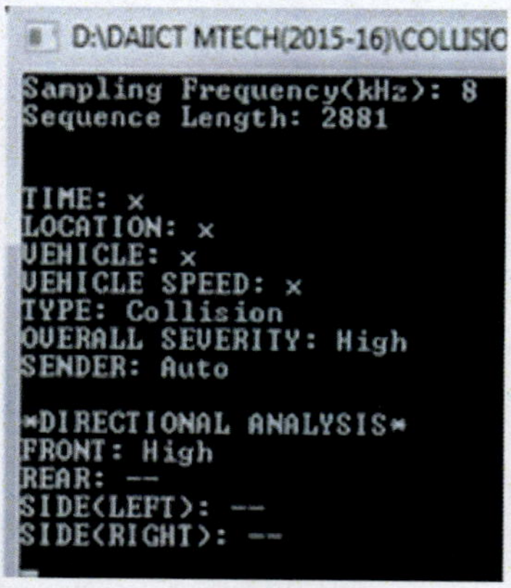

(b)

at the accident locations. The marker is automatically added on the google map as soon as the notification is delivered to the Firebase Database. Figure 11(b) shows the corresponding data in Firebase Database. Now by left clicking on the marker, an info window opens that displays the accident related important information like date-time, vehicle, driver and accident severity details. By right clicking on the marker, it displays a 'Delete' button. It can be used to delete the marker after the emergency medical help is provided to the victim. Aforementioned, the Google Firebase uses the 'WebSocket' protocol for data transmission. Therefore, the accident notification could be received in real time. In absence of the internet connectivity, an SMS would also be sent over some cellular network that might take several seconds.

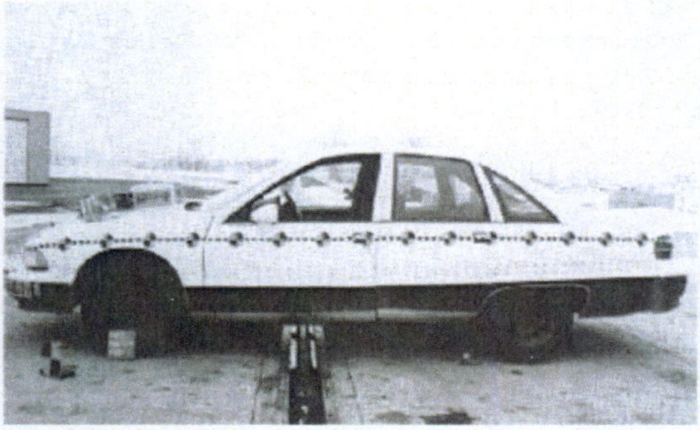

(a)

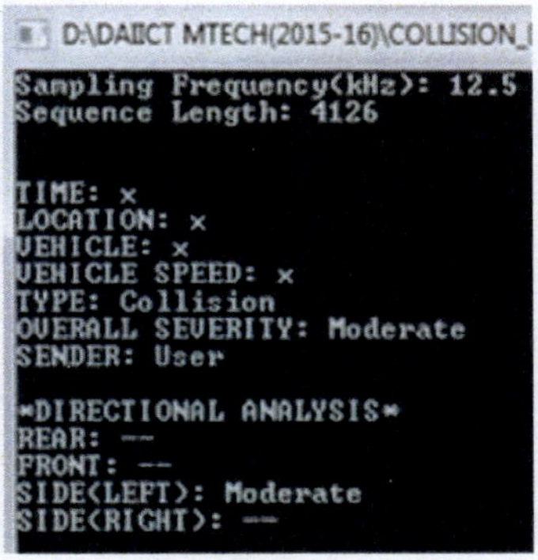

(b)

Fig. 9 ADS test case 2. **a** Vehicle after crash [20] **b** Output of the algorithm

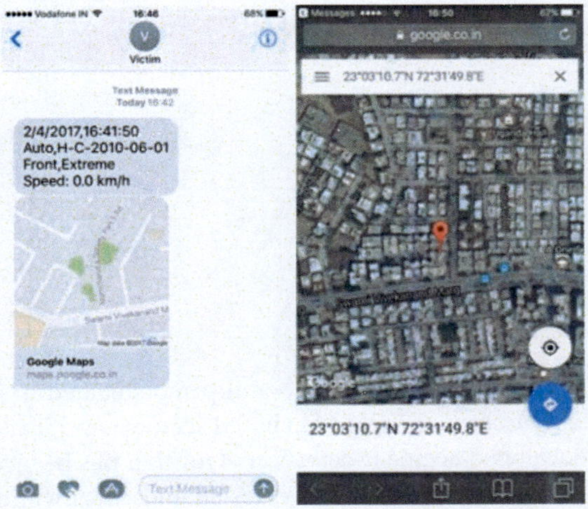

Fig. 10 Accident notification at registered mobile

Fig. 11 Accident notification at emergency service. **a** Accident location **b** Firebase Database

(a)

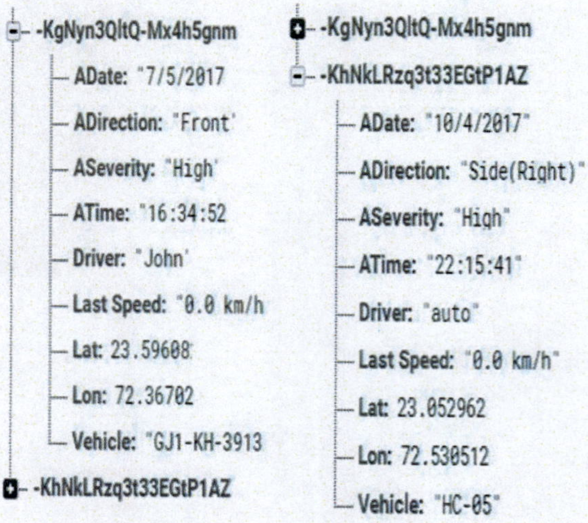

(b)

4 Conclusion

In this paper we have successfully implemented an accurate and low cost ADS system and Firebase-Google Map based accident notification system for four-wheelers. The developed accident detection algorithm has been verified using standard crash test data from NHSTA and also tested under real world environment. As a result, emergency service is availed on time at the correct location without delay from the nearest emergency service provider. Meanwhile a voice call is placed and an SMS is delivered on the registered mobile number. The SMS will contain accident severity and accident location information. This would provide global coverage for emergency medical services even at affordable economic costs to the user. It can be extended to the heavy vehicles also.

References

1. Traffic deaths numbers: https://www.who.int/gho/road_safety/mortality/traffic_deaths_number/en/
2. Ali HM, Alwan S (2017) Car accident detection and notification system using smartphone. LAP LAMBERT Academic Publishing, Saarbrucken
3. Aloul F, Zualkernan I, Salma R, Ali H, Merri M (2015) iBump: smartphone application to detect car accidents. Comput Electr Eng 43:66–75
4. Fernandes B, Gomes V, Ferreira J, Oliveira A (2015) Mobile application for automatic accident detection and multimodal alert. In: 2015 IEEE 81st vehicular technology conference (VTC spring), pp 1–5, May 2015
5. Dar BK, Shah MA, Islam SU, Maple C, Mussadiq S, Khan S (2019) Delay-aware accident detection and response system using fog computing. IEEE Access 7:70975–70985
6. Ahirrao SM, Dhanrale P, Mahant L, Kotwal H (2015) Accident detection and notification system using Android. Int J Recent Innov Trends Comput Commun 3(3):1084–1086
7. Reddy V, Sree L, Kumar V (2014) Design and development of accelerometer based system for driver safety. Int J Sci Eng Technol Res (IJSETR) 3:3463–3468
8. Ravi, Phanigna T, Lenina Y, Ramcharan P, Teja P (2020) Automatic accident detection and rescue system using image processing and IOT. J Anal Comput XIII(III)
9. Ghosh S, Sunny SJ, Roney R (2019) Accident detection using convolutional neural networks. In: International conference on data science and communication (IconDSC), 1 March 2019, pp 1–6. IEEE
10. Hui Z, Yaohua X, Lu M, Jiansheng F (2014) Vision-based real-time traffic accident detection. In: Proceeding of the 11th world congress on intelligent control and automation, pp 1035–1038. IEEE
11. Yu Y, Xu M, Gu J (2019) Vision-based traffic accident detection using sparse spatio-temporal features and weighted extreme learning machine. IET Intell Transp Syst 13:1417–1428
12. Taj FW, Masum AKM, Reza ST, Chy MKA, Mahbub I (2018) Automatic accident detection and human rescue system: assistance through communication technologies. In: 2018 international conference on innovations in science, engineering and technology (ICISET), pp 496–500. IEEE
13. Balaji S, Gumber A, Santhakumar R, Kumar MR, Tiwari A, Debnath H (2020) An IOT-based vehicle tracking system with user validation and accident alert protocol. In: Soft computing for problem solving, pp 717–729. Springer, Singapore
14. Tushara DB, Vardhini PH (2016) Wireless vehicle alert and collision prevention system design using Atmel microcontroller. International Conference on Electrical, Electronics, and Optimization Techniques (ICEEOT), 3 March 2016, pp 2784–2787. IEEE
15. Order of magnitude: https://en.wikipedia.org/wiki/Order_of_magnitude
16. Kendall J, Solomon K (2014) Airbag deployment criteria: Institute of Risk & Safety Analyses. The Forensic Examiner
17. Locey CM, Espana J, Toh A, Belwadi A, Arbogast K, Maltese M (2012) Homogenization of vehicle fleet frontal crash pulses from 2000–2010. In: Annals of advances in automotive medicine/annual scientific conference, vol 56, p 299. Association for the Advancement of Automotive Medicine
18. Alem N, Perry M (1995) Design of digital low-pass filters for time-domain recursive filtering of impact acceleration signals. United States Army Aeromedical Research Laboratory

19. NHTSA test case 1: https://www-nrd.nhtsa.dot.gov/database/VSR/veh/TestDetail.aspx?LJC=1847&existreport=Y&r_tstno=1846&database=v&tstno=1847
20. NHTSA test case 2: https://www-nrd.nhtsa.dot.gov/database/VSR/veh/TestDetail.aspx?LJC=2344&existreport=Y&r_tstno=2344&database=v&tstno=2344

Avenues to Improve Channel Estimation Using Optimized CP in STBC Coded MIMO-OFDM Systems—A Global Optimization Approach

Shovon Nandi, Narendra Nath Pathak, and Arnab Nandi

Abstract A new semi-blind channel estimation with optimized Cyclic Prefix (CP) assisted Space Time Block Coded Multi-Input Multi-Output Orthogonal Frequency Division Multiplexing (STBC-MIMO OFDM) system is proposed. The main hurdle of high complexity and low convergence in earlier systems are avoided by our proposed scheme in flat fading environment. In our work, the hyper parameters are optimized with proposed Lévy Krill-herd (LKH) algorithm and it is clear that the channel estimation performance is varied with this parameter values and by this global optimization technique the incorrect selection of hyper parameters (local optima) are eliminated. The selection process of this algorithm can be simplified with the number of bounds used. The improvement performance is shown by using BER vs SNR plot of Forward-backward (FB) Kalman helical approach and different pilot carrier insertions. Also a comparative plot is shown among FB Kalman, Krill-herd (KH) and finally LKH approach by using Matlab software.

Keywords MIMO-OFDM · Kalman helical process · Lévy Krill-herd algorithm · Cyclic prefix · Alamouti-STBC · BER

1 Introduction

In current focused world, we make an endeavor to deliver a most extreme growth or advantage from the confined measure of usable asset. The high speed data communication process is grateful to the advancement of optimization process. Several

S. Nandi (✉)
Department of Electronics and Communication Engineering, Bengal Institute of Technology, Kolkata, W.B, India
e-mail: shovon.nandi@bitcollege.in

N. N. Pathak
Department of Electronics and Communication Engineering, Dr. B C Roy Engg. College, Durgapur, W.B, India

A. Nandi
Department of Electronics and Communication Engineering, National Institute of Technology, Silchar, Assam, India

V. Nath and J. K. Mandal (eds.), *Proceeding of Fifth International Conference on Microelectronics, Computing and Communication Systems*, Lecture Notes in Electrical Engineering 748,
https://doi.org/10.1007/978-981-16-0275-7_21

optimization techniques are established from the inspiration of naturally happened phenomena of different bodies. The burden of low convergence and high complexity in data transmission has been mitigated by the proposed modified global approach [1]. The population diversity of the solutions, improvement of the whole search region covering up and to increase the convergence rate of the multi objective heuristic algorithm is possible and shown here. The MIMO combining technology with the OFDM (i.e. MIMO-OFDM) is generally employed in wireless communications systems [2] and it is a well-known method for high-data-rate wireless transmission, specially in 4G, 5G or beyond standards [3]. The utilization of MIMO-OFDM upgrades the channel limit and improves the communication unwavering quality [4, 5]. The bandwidth efficiency of OFDM along with diversity technique has led to the use of this technology in many advanced wireless communication systems [6, 7]. The OFDM eliminates the Inter-Symbol Interference and Inter Carrier Interference. Additionally, on account of the utilization of a cyclic prefix and an orthogonal transform [8, 9].

In OFDM framework, channel estimation stays a present worry, since the general execution relies emphatically upon it. Especially for enormous MIMO systems the Channel State Information (CSI) turns out to be more testing [10, 11]. Insufficient CP in OFDM transmission prompts ISI and ICI, which influences both channel estimation and data detection. Whenever overlooked, it can prompt a mammoth detection error in MIMO-OFDM. To ease this issue, various channel estimation and data detection techniques for inadequate CP frameworks have been proposed in writing [12, 13]. Employing CP in multi-carrier systems not only protects the signal from ISI but also allows circular interpretations of the channel which improves the estimation and equalization process.

To the extent data identification occurred, a time domain (TD) finite impulse response filter can regulate insufficient CP MIMO-OFDM circulation [14]. The FD (Frequency-domain) per tone equalization (PTEQ) was proposed and the precoding procedures are utilized to evacuate the distortion because of deficient CP [15]. ISI and ICI induced by insufficient CP length can complicate channel estimation and cause large estimation error. To lighten this issue, a lot of channel estimation strategies can be suggested in writings [16].

Different techniques have been proposed in writing for estimating channel impulse response in OFDM frameworks. These join pilot based techniques, blind and semi-blind algorithms [17–19]. Semi-blind strategies offer a trade-off dealing with pilot formed and blind estimation plan. A preliminary channel estimation utilizing pilots has to be acquired by the strategies, stated earlier and make use of a grouping of from the prior prerequisites to improve it further [20]. Regularly, channel estimation inside MIMO-OFDM systems relies upon the PACE—Pilot-aided channel estimation process. The channel estimation in a canister without pilot images is acquired by right initiation [22].

The surrounding of this paper is sorted out as follows. The background study or literature survey portion is given in Sect. 2. Section 3 depicts the system model, where we introduce the MIMO Alamouti-STBC architecture and we present the proposed EM based helical FB Kalman Filter approach. Performance metrics are given in next section. The obtained simulated results of various graphs have been produced in Sect. 5 to prove the supremacy of the proposed approach over the existing mechanisms. Finally, conclusions and few glimpses on future scope is drawn in Sect. 6.

2 Related Work

Pham et al. [13] have proposed the joint channel estimation and identification for orthogonal frequency-division multiplexing systems utilizing lacking cyclic prefix and pilot sub-carriers. The bandwidth usage is being minimum for the shorter CP in OFDM architecture. Alternatively, their methods can enable range extension for OFDM transmission with a certain CP length. The cost was increased by ISI and ICI that corrupts the pilot subcarriers and leads to a large channel estimation error. They then formulated a PSB-MMSE—Pilot sub-carrier based minimum mean square error scheme to estimate the TD channel response for SIMO—Single input multiple output systems within the sight of ISI and ICI. From that, an iterative reception process of channel estimation and data detection is constructed.

Tae-Jun Lee and Young-Chai Ko [23] have proposed the moderating plan to lessen the impacts of phase noise in MIMO-OFDM framework using autonomous oscillator in every RF chain. Their suggesting plans comprising two phases—channel estimation dais and information decoding dais. In the main stage, they proposed the channel estimation algorithm dependent on MAP—Maximum a posteriori estimator and a strategy for selecting those sequence of training used for channel estimation with numerical investigation of recommended plot. In the subsequent moment, MAP estimators were utilized to mutually appraise phase noise at transmitter as well as receiver and distinguish data symbols. For examination of MSE—Mean Square Error exhibitions, they infer Bayesian Cramer-Rao lower bound (BCRLB) at every phase for estimation issue of different parameter.

Chih-Yu Chen, and Wen-Rong Wu [24] proposed another joint Angle-of-departure (AoD), Angle-of-arrival (AoA), and channel estimation process for pilot-based MIMO-OFDM frameworks. Initial, a compressive detecting system was utilized to evaluate the channel impulse response, abusing the sparsity property of remote channels. At that point, AoA and AoD were mutually evaluated for each recognized way by the Maximum Likelihood (ML) strategy. The Cramer-Rao Lower Bound (CRLB) was moreover construed and a transmit beam forming plan was proposed in like way. In the situation of accessible prior data, a most extreme posteriori estimation was proposed. The BCRLB issue was additionally inferred and a transmit beam framing plan was additionally proposed. Incidentally, just two training OFDM images were required for the estimation.

A novel algorithm is likewise use the various estimation vectors at the getting radio wires. In any case, they alter the known Multiple Sparse Bayesian Learning (MSBL) calculations for pilot-based a-bundle meager divert estimation in MIMO-OFDM structures. Thusly, they summarize the MSBL calculation to get a novel J-MSBL figuring for joint a-bunch sparse channel estimation and data disclosure (Prasad and Murthy [25]).

Amir Aminjavaheri et al. [26] has investigated the channel distortions without CP blur away and effects on the performances of the standard Maximum Ratio Combining (MRC) receiver. Their examination uncovers that in this recipient, there reliably remains some leftover obstruction inciting immersion of signal-to-interference-pulse noise ratio. To determine that issue, they proposed utilizing the Time Reversal (TR) strategy. In addition, so as to additionally lessen the multiuser interference, they proposed a zero-forcing equalization out to be sent after the TR joining.

Some major crisis may arise during channel estimation and we found that these exert influence on efficiency of MIMO. Due to huge lacking CP in OFDM, the huge sum of detection error and channel estimation hitch come about along with the performance delay [27]. The heart of the work i.e. proposed system model is discussed below.

3 System Model

In association with CP, the channel estimation and data detection problem occurs and in this paper, the center is essentially around the optimization of these techniques. Figure 1 shows the framework model of STBC MIMO OFDM, where we can configure the transmitting and receiving antenna as per needed. The codes of ST blocks are utilized for MIMO frameworks for empowering the sending of different stream of duplicate information over various reception antennas for abusing different collected variants to enhance the authenticity of communication [16]. Also Fig. 2 shows Expectation maximization (EM) based helical Forward-Backward (FB) Kalman Filter approach.

The CP is modified using enhanced Kalman filter technique. The Kalman filter is enhanced further by using the proposed Krill-herd and Lévy flight krill-herd (LKH) algorithm techniques [1] which improves the quality of the cyclic prefix assisted data. This is a *bio-inspired metaheuristic approach* (BIMA) which mimics the herding behavior of ocean krill individual. The best fitness parameters are used to further

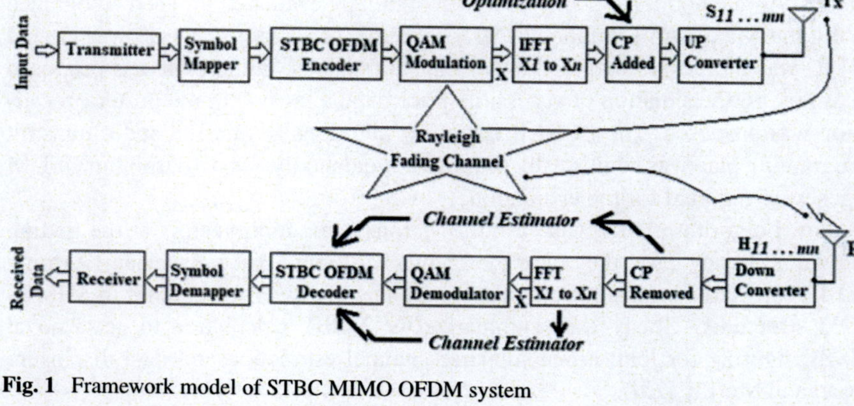

Fig. 1 Framework model of STBC MIMO OFDM system

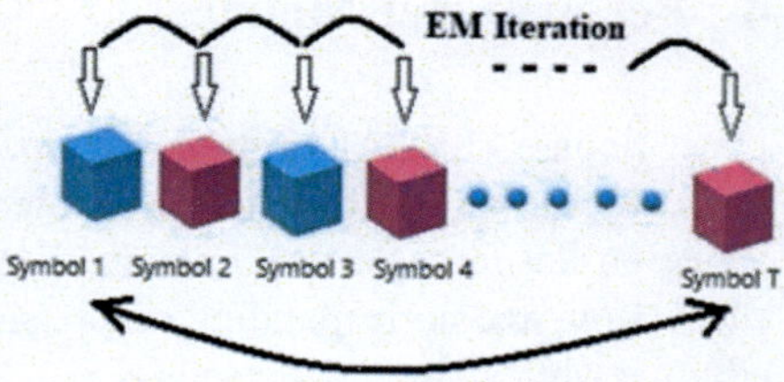

Fig. 2 EM based helical FB Kalman Filter approach

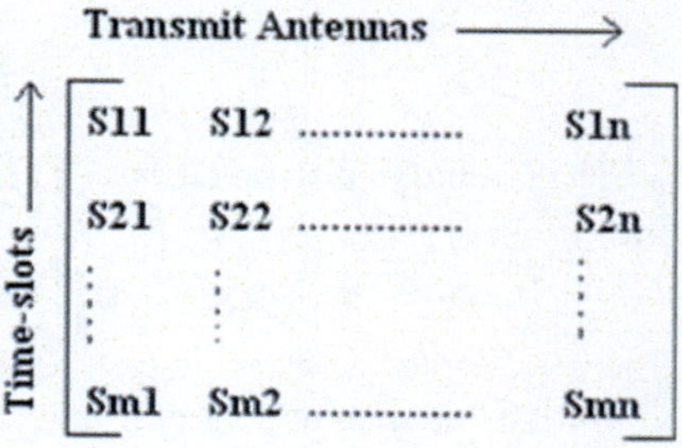

Fig. 3 Representation of Matrix form of STBC

modify the KH approach. The spatial and temporal form of data are used jointly to construct space time block code [17]. Space-time coding consolidates each of the duplicates of the received signal in an ideal manner to remove however much data from every one of them as could reasonably be expected. Alongside this MRC receiver, it gives better diversity and boosts the SNR.

The propelled mobile communication framework consequently utilizes this plan to demonstrate the best execution. Other assorted variety plot can also be hired through this scheme [13]. STBC is ordinarily addressed by a grid as appeared in Fig. 3, where columns are proportional to the amount of transmit radio antenna and its rows are identical to the amount of the schedule vacancies required to transmit facts [21]. The code structure is symmetrical. The recipient is experienced about the straightforward and ideal deciphering plan.

The space time codes are supplanted by the Alamouti STBC code that compensates for the channel hazards, for example, fading and warm clamor. The STBC Alamouti contrive was made subject to space time coding framework. The scientist S M Alamouti presented a two reception apparatus STBC in 1998. It doesn't require any extended form of bandwidth at the sending end. In perspective on expanding the volume of block information it summed up to receiving wires and plan to join several antenna in the framework [28].

Using the channel least square (LS) method the channel covariance in matrix form and different way outs are obtained for the sub-carriers during the pilot insertion. Then the maximum likelihood task is used to optimize the time-scale. The proposed algorithm utilizes both the time and frequency information on behalf of correlation. By using this technique, we experienced a better outcome in data detection and iterative joint estimation in the MIMO OFDM systems containing space time block code.

4 Performance Metrics

Least Square (LS) Algorithm: We first check the free noise MIMO channel by this calculation. Lower bound is availed by immaculate approximation. Expected Rayleigh flat-fading MIMO channel [29] is described with S and H, the training succession, and corresponding output signal as Y [30]. Then, Y = SH + N, where additive white Gaussian noise is N.

The algorithm simplified $\hat{H}$ so that

$$S\hat{H} \approx Y \tag{1}$$

The Euclidian distance $(S\hat{H} - Y)$ is minimized by using the ensuing stages,

$$\left\| S\hat{H} - Y \right\|^2 = (S\hat{H} - Y)^H (S\hat{H} - Y) \tag{2}$$

Derive this with respect to H and calibrate the equation as zero, it reduces to,

$$\hat{H} = (S^H.S)^{-1}.S^H.Y \tag{3}$$

This solution is used to LS channel estimation algorithm.

Channel Estimation Algorithm MMSE—Minimum Mean Square Error: Mean square error crop up the average mistake linking the evaluated wave and the genuine info wave. Clearly, the smaller the MSE is, the better efficiency of the proposed work [31].

$$\hat{h}_{n,MMSE} = [S^H R_{nn}^{-1} S + R_{hh}^{-1}]^{-1} S^H R_{nn}^{-1} Y_n \tag{4}$$

Where $R_{hh} = E\{h_n h_n^H\}$ and considering the presence of additive white noise, the MMSE Channel estimates as

$$\hat{h}_{n,MMSE} = [S^H S + \frac{\sigma_n^2}{\sigma_h^2}]^{-1} S^H Y_n \tag{5}$$

So the Least Square channel is estimated by putting the term $\frac{\sigma_n^2}{\sigma_h^2}$ to zero.

The channel is considered as frequency particular and time variant. The channel response is represented by $h_{r_x}^{t_x}$ where p distinct paths are created between transmit antenna t_x and receiving antenna r_x. Sequence vector $t_x = 1.....Tx$ and $r_x = 1......Rx$. The consisting ST (Space Time) blocks, channel decay factor and the Doppler frequency are also assumed, then the acknowledgement form of the transmit and receive antenna is presented by,

$$h_{t+1} = (I_{TxRx} \otimes F)h_t + (I_{TxRx} \otimes G)u_t \text{ } (\hat{h} \text{ is an estimation of } h) \tag{6}$$

Where power detain is G and Doppler detain is F. The detection of transmitted data will be more visualized by the joint estimation detection scheme under the CP based EM environment.

Two possible extreme cases are analyzed (a) in first case the information is totally known at the recipient (b) secondly when the load is totally undisclosed at the receiver. These two cases are set up inferable from its semi blind nature.

(a) The information is totally known at the recipient: Consider an input time series is $T + 1$, also the ST (space time) symbols $\bar{I}_0^T (= \bar{I}_0, \bar{I}_1, \bar{I}_2, \ldots.., \bar{I}_T)$ and $\tilde{j}_0^T$ The limit of posteriori estimate of h_0^T is controlled by boosting the log probability function,

$$£ = \ln p(\tilde{j}_0^T | \bar{I}_0^T, h_0^T) + \ln p(h_0^T) \tag{7}$$

Which minimizes to,

$$£ = -\sum_{t=1}^{T} \|\tilde{j}_t - (I_{Rx} \otimes \bar{I}_t)\, h_t\|^2_{\frac{1}{\sigma_n^2}} - \|h_0\|^2_{\pi_0 - 1} - \sum_{t=1}^{T} \|h_t - (I_{TxRx} \otimes F)\, h_{t-1}\|^2_{(GR_u G^H)^{-1}} \tag{8}$$

Value of h_t is determined by incorporating the prototype with FB Kalman viewpoint as mentioned by [16] and [18]. Beginning from starting condition $h_{0|-1} = 0$ and $p_{0|-1} = \Pi_0$.

Two approaches are considered in Kalman filter as mentioned below,

Forward run approach:

For $i = 1, \ldots\ldots, T$, Determine,

$$R_{e,t} = \sigma_n^2 I_{TxRxN} + (I_{Rx} \otimes \bar{X}_t) P_{t|t-1} (I_{Rx} \otimes \bar{X}_t^H) \tag{9}$$

$$K_t = P_{t|t-1} (I_{Rx} \otimes \bar{X}_t^H) R_{e,t}^{-1} \tag{10}$$

$$\hat{h}_{t|t} = (I_{TxRx(P+1)} - K_t (I_{Rx} \otimes \bar{X}_t)) \hat{h}_{t|t-1} + K_t \tilde{j}_t \tag{11}$$

$$\hat{h}_{t+1|t} = (I_{TxRx} \otimes F) \hat{h}_{t|t} \tag{12}$$

$$P_{t+1|t} = (I_{TxRx} \otimes F)(P_{t|t-1} - K_t R_{e,t} K_t^H)(I_{TxRx} \otimes F^H) . GR_u G^H \tag{13}$$

Say, $N = 2(N_{CP} + N_{Data})$ where N_{CP} and N_{Data} are the CP length and information part of the OFDM symbol [32].

Backward run approach:

Beginning from $\lambda_{T+1|T} = 0$ and for $t = T, T - 1, \ldots.., 0$; Determine,

$$\lambda_{t|T} = (I_{P+N} - (I_{R_x} \otimes \bar{I}_t^{H}) K_t^{H})(I \otimes F^{H}) \lambda_{t+1|T} + (I \otimes \bar{I}_t) \\ .R_{e,t}^{-1}(\tilde{j}_t - (I \otimes \bar{I}_t)\hat{h}_{t|t-1}) \quad (14)$$

$$\hat{h}_{t|T} = \hat{h}_{t|t-1} + P_{t|t-1}\lambda_{t|T} \quad (15)$$

So, $\hat{h}_{t|T}$ is the final estimation.

5 Simulated Result

The performance investigation of the optimized CP cooperated MIMO-OFDM framework with the transmit diversity of Alamouti STBC is uncovered in this work. We reproduce a several plots using Matlab to make a comparative study to finding out the best fitness value. The Rayleigh fading channel, experiences receive diversity of MRC procedure which furthermore shown in this paper. The modulation scheme is taken as the 16-QAM. Figure 4 shows the bit error rate performance when two different pilot sub-carriers (8 and 16) are used for the same algorithm. It is proved that the higher valued pilot sub-carriers perform well. Figure 5 compares the BER vs SNR simulation of EM based FB Kalman helical coded and uncoded data format when 16 pilot subcarriers are inserted [33].

Obviously the coded helical structure outperforms the uncoded one. Figure 6 compares the BER performance of proposed optimization scheme (Lévy-flight Krill-herd) along with the other three cases of without optimization, EM based FB Kalman and Krill-herd global optimization schemes. It is easily established that the nature inspired Lévy-flight Krill-herd optimization has the ability to improve the performance of BER and accuracy over FB Kalman algorithm and also Krill-herd algorithm [34].

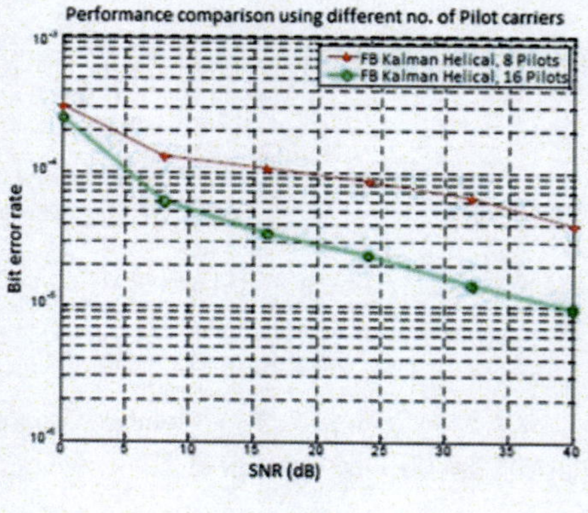

Fig. 4 Performance comparison between 8 and 16 Pilots FB Kalman helical approach

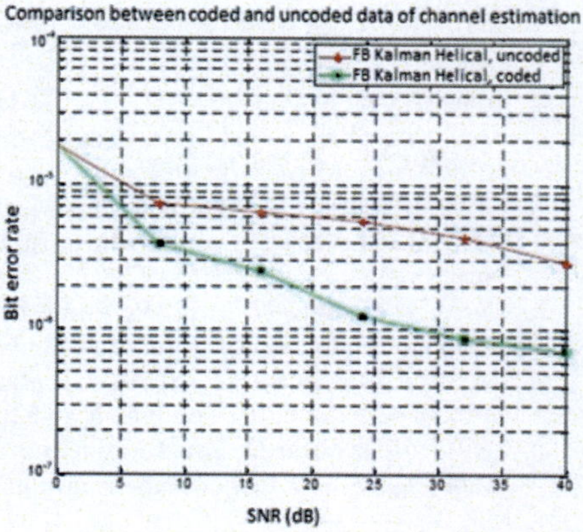

Fig. 5 Comparison between EM based FB Kalman Helical coded and uncoded simulation

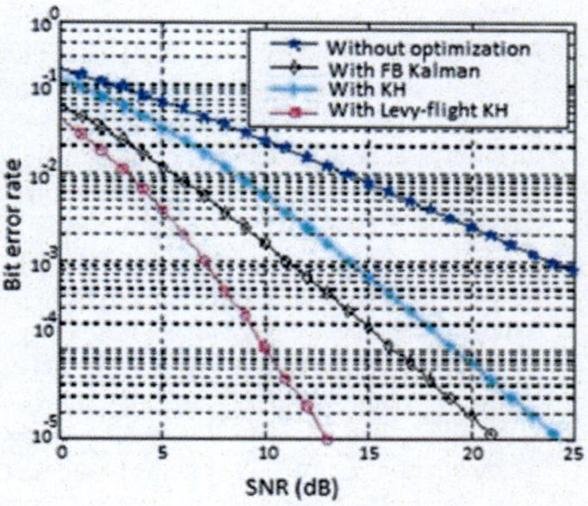

Fig. 6 BER vs SNR plot for (i) Without Optimization (ii) With FB Kalman (16 Pilots) (iii) With Krill-herd (iv) With Lévy-flight Krill-herd algorithm

6 Conclusion and Future Work

Continuous research since few decades is still a huge problem and in practical impact on the high complexity and low convergence speed in MIMO-OFDM network.

In this paper, the novelty of proposed Lévy-flight Krill-herd algorithm proved an optimized way-out over expectation maximization based forward-backward Kalman technique and Krill-herd approach. The higher order pilot inserted coded EM based FB Kalman helical model provide better SNR performance. Also the proposed optimization process (Lévy-flight Krill-herd algorithm) applied in the receiving end with the existing system deliver better performance using Alamouti space time block coded semi-blind nature of fading environment. Theoretical concept, system model and simulation results are given to support the statement.

In future we will investigate several bioinspired algorithms to find-out the more improved fitness parameters which in turn produce a better system to mitigate a strong demand of efficient spectral bandwidth in wireless communication.

References

1. Wang G, Guo L, Gandomi AH, Cao L, Alavi AH, Duan H, Li J (2013) Lévy-Flight Krill Herd Algorithm. Hindawi Publishing Corporation, Mathematical Problems in Engineering, Volume 2013, Article ID 682073, p 14
2. Ladaycia AH, Anissa Mokraoui, KA-M, Belouchrani A (2017) Performance bounds analysis for semi-blind channel estimation in MIMO-OFDM communications systems. IEEE Trans Wirel Commun **16**(9): 5925–5938
3. Shoba B, Jayanthi K (2012) Performance improvement of MIMO OFDM systems through channel estimation. Int J Wirel Mob Networks 4(5):49
4. Larsson Erik G, Edfors O, Tufvesson F, Marzetta TL (2014) Massive MIMO for next generation wireless systems. IEEE Commun Mag **52**(2): 186–195
5. Fredrik R, Persson D, Lau BK, Larsson EG, Marzetta TL, Edfors O, Tufvesson F (2013) Scaling up MIMO: Opportunities and challenges with very large arrays. IEEE Signal Process Mag **30**(1): 40–60
6. Roopa J, Kurian NC (2015) ISI Reduction in MIMO-OFDM with insufficient cyclic prefix-a survey ISI 3, no. 8 (2015)
7. Payaswini P, Manjaiah DH Analysis of effect of cyclic prefix on data rates in OFDM modulation techniques. Int J Adv Comput Math Sci (2012)
8. Nandi S, Nandi A, Pathak NN, Sarkar M (2017) Performance analysis of cyclic prefix OFDM using adaptive modulation techniques. IJEECS **6**(8), 214–220. ISSN 2348–117
9. Nandi S, Sarkar M, Nandi A, Pathak NN (2017) Performance analysis of CO-OFDM system in a CR network. In: Computer, Communication and Electrical Technology, Guha, Chakraborty & Dutta (Eds) ©Taylor & Francis Group, ISBN: 978–1-138-03157-9
10. Amar Al-Jzari, Iviva Kostanic (2015) Cyclic prefix length determination for orthogonal frequency division multiplexing system over different wireless channel models based on the maximum excess delay spread. Am J Eng Appl Sci 8(1):82
11. Peter Fertl, Matz Gerald (2010) Channel estimation in wireless OFDM systems with irregular pilot distribution. IEEE Trans Signal Process 58(6):3180–3194
12. Tongliang Fan, Haowei Wu, Huang Hongcheng (2011) Channel estimation and interference cancellation for OFDM systems based on total least squares solution. JCM 6(8):640–647
13. Tri Pham, Le-Ngoc Tho, Woodward Graeme K, Martin Philippa A (2017) Channel estimation and data detection for insufficient cyclic prefix MIMO-OFDM. IEEE Trans Veh Technol 66(6):4756–4768
14. Ksekkaya Y, Bariş, Toker VC (2017)Joint transceiver FIR filter design for multiuser MIMO channel shortening equalization and full equalization using channel duality. Turkish J Electr Eng Comput Sci **25**(5): 4077–4090
15. Donatella Darsena, Gelli Giacinto, Paura Luigi, Verde Francesco (2012) Blind channel shortening for space-time-frequency block coded MIMO-OFDM systems. IEEE Trans Wireless Commun 11(3):1022–1033
16. Amo D, Prieto C, Julia Fernández-Getino García M (2013) Iterative joint estimation procedure for channel and frequency offset in multi-antenna OFDM systems with an insufficient cyclic prefix. IEEE Trans Vehicular Technol **62**(8): 3653–3662
17. Swamy MNS (2018) Comb type pilot arrangement based channel estimation for spatial multiplexing MIMO-OFDM Systems. IRJET **5**(02) (2018)
18. Obaid UM, Alsusa E (2015) a virtually blind spectrum efficient channel estimation technique for MIMO-OFDM systems. University of Engineering and Technology Taxila, Techn. J. **20**(3): 91
19. Dey A, Nandi S, Sarkar M (2018) Security measures in IOT based 5G networks. In: 3rd International Conference on Inventive Computation Technologies (ICICT), Coimbatore, India, pp 561–566, https://doi.org/10.1109/ICICT43934.2018.9034365
20. Quadeer AA, Sohail MS (2010) Enhanced channel estimation using cyclic prefix in MIMO STBC OFDM systems. In: IEEE international symposium signal processing and information technology (ISSPIT), 2010. IEEE, pp 277–282

21. Nandi S, Nandi A, Pathak NN (2017) Performance analysis of Alamouti STBC MIMO OFDM for different transceiver system. In: IEEE conference on ICISS 2017, pp 883 – 887, https://doi.org/10.1109/iss1.2017.8389305
22. Dragan O, Na C, LupasScheiterer R, Szabo A (2008) EM-based semi-blind channel estimation method for MIMO-OFDM communication systems. Neurocomputing **71**(10–12): 2388-2398
23. Lee Tae-Jun, Ko Young-Chai (2017) Channel estimation and data detection in the presence of phase noise in MIMO-OFDM systems with independent oscillators. IEEE Access 5:9647–9662
24. Chen, CYu, Wu W-R (2018) Joint AoD, AoA, and channel estimation for MIMO-OFDM systems. IEEE Trans Vehicular Technol (2018)
25. Ranjitha Prasad, Murthy Chandra R, Rao Bhaskar D (2015) Joint channel estimation and data detection in MIMO-OFDM systems: a sparse Bayesian learning approach. IEEE Trans Signal Process 63(20):5369–5382
26. Amir A, Farhang A, Reyhani AR, Doyle LE, Farhang-Boroujeny B, OFDM without CP in massive MIMO. IEEE Trans Wirel Commun **16**(11): 7619–7633 (2017)
27. Banerjee JS, Nandi S, Chakrabory A (2019) A text book on mastering digital electronics: principle, devices and applications. Aryan Publishing House, ISBN: 9788193894217
28. Alamouti SM (1998) A simple transmit diversity technique for wireless communication. IEEE J Sel Areas Commun **16**(8)
29. Nandi A, Kundu S (2011) Energy level performance of error control schemes in WSN over Rayleigh fading channel. In: IEEE symposium on industrial electronics and applications, pp: 194–199
30. Nandi S, Pathak NN, Nandi A (2020) A novel adaptive optimized fast blind channel estimation for cyclic prefix assisted space-time block coded MIMO-OFDM systems. Wireless Pers Commun 115:1317–1333. https://doi.org/10.1007/s11277-020-07629-z
31. Nandi S, Pathak NN, Nandi A (2019) Efficacy of channel estimation and efficient use of spectrum using optimised cyclic prefix (CP) in MIMO-OFDM. IJEAT **9**(2) ISSN: 2249–8958. https://doi.org/10.35940/ijeat.b4093.129219
32. Sohail MS, Al-Naffouri TY (2010) An EM based frequency domain channel estimation algorithm for multi-access OFDM systems. Sig Process 90(5):1562–1572. https://doi.org/10.1016/j.sigpro.2009.10.026
33. Al-Naffouri TY, Quadeer AA (2008) A Forward-Backward Kalman Filter-based STBC MIMO OFDM receiver. EURASIP J Adv Signal Process **2008**, Article ID 158037. https://doi.org/10.1155/2008/158037
34. Kundu K, Pathak NN (2019) Circular antenna array optimization using flower pollination algorithm. In: Advances in computer, communication and control. Springer, pp 407–414

A Comparative Study on Fault Detection Using Twin Support Vector Machines for Wireless Sensor Networks

Jyotsana Singh, Ajay Singh Raghuvanshi, Namrita Shukla, and Shashank Gavel

Abstract Wireless Sensor Networks (WSNs) are composed of distributed small sensors that are deployed in various environmental conditions to collect data such as temperature, moisture, light speed, etc. These sensors are subjected to various damages, noises and low battery life as a result of which sensors account for abnormal data readings. These anomalies must be detected for the reliable and efficient working of a WSN. This study presents a comparative study for fault detection using Twin Support Vector Machines (TSVM) by utilizing different kernel functions. TSVM uses two non-parallel hyperplanes to classify data and is insensitive to class size imbalance. The choice of kernels significantly influences the performance of TSVM. The proposed work is estimated using different standard WSN datasets by inserting various types of fault. The performance of the work is measured using standard performance indexes.

Keywords Wireless Sensor Networks · Twin Support Vector Machine · Kernel functions · Anomaly detection · Machine learning

1 Introduction

Wireless Sensors Networks (WSNs) are emerging technology as well as a leading research field. It is used in various fields such as defence, agriculture and transportation for monitoring and data acquisition. Data is collected by physical devices called sensors. WSNs Sensor Nodes collect the data from the surrounding area where they are deployed. Sensors are devices that sense a measurable entity that describes the

J. Singh (✉) · A. S. Raghuvanshi · N. Shukla · S. Gavel
Department of Electronics and Communication, National Institute of Technology Raipur, Raipur 492010, India

A. S. Raghuvanshi
e-mail: asraghuvanshi.etc@nitrr.ac.in

S. Gavel
e-mail: sgavel.phd2016.etc@nitrr.ac.in

V. Nath and J. K. Mandal (eds.), *Proceeding of Fifth International Conference on Microelectronics, Computing and Communication Systems*, Lecture Notes in Electrical Engineering 748,
https://doi.org/10.1007/978-981-16-0275-7_22

physical conditions of the surrounding area such as heat, light, temperature, and pressure, etc., where WSN is thus described as a network of small information gathering devices which can deliver the information collected from wirelessly supervised field [1]. The collected information is further redirected to other networks via numerous network nodes.These nodes comprises of a gateway which helps in establishing a connection between various networks like a wireless Ethernet. It act as an interface between the digital and physical world due to its capability to collect, process and communicate data smartly and also due to its low cost and its large domain of applications. Sensors have disadvantages including processing, storing capacity and their constraint in energy. Sensor, as an electronic device is prone to break down. In addition, most sensors are deployed in a hazardous environment. Failures that can occur or WSNs are prone to, are classified in three types that are communication failures, software failures and hardware failures. Software failure is due to problems in sensor programs and hardware failure is caused due to hardware units: sensor unity, location unity, power unity. Failure of the battery also causes error this will lead to error in the whole network by malfunctioning of the sensor [2, 3].

In the past decade, the machine learning techniques have been widely implemented for various tasks, such as classification, density estimation and regression. These techniques have been tested and shown significant performance in major application areas i.e. detection, bioinformatics, speech recognition, computer vision, and advertising networks. The origination of algorithm is from basic field of mathematics, statistics, and computer science. However, they have been extensively used as an application in sensor networks.

In the past few years, many machine learning models, techniques and tools have been used for detecting faults in WSNs such as Artificial Neural Networks (ANN) and Support Vector Machines (SVM) [4, 5]. Another technique called Twin Support Vector Machines has been in the main focus of research to make fault detection more efficient. As the name suggests it uses to nonparallel planes instead of single like in SVM and reduces the computational complexity by four times.

In this article, we present a comparative study for fault detection by utilizing Twin Support Vector Machines for Wireless Sensor Networks. We have utilized different kernel functions to attain the maximum performance in terms of classification and detection of fault. We have utilized standard WSNs datasets for testing and evaluating the utilized scheme. The set of standard performance indexes have been utilized for evaluation.

Rest part of the research paper is as follows: Sect. 2 covers the related work in the field, Sect. 3 covers the brief introduction to the TSVM. Section 4 describes the proposed fault detection by utilizing TSVM, Sect. 5 covers the results and discussion followed by conclusion.

2 Related Work

SVM has proved itself as a classifier tool showing efficient performance in the area of ML (Machine Learning) [5, 6]. The foundation and design of SVM are based on the theory of structural risk minimization (SRM) and statistical learning [7, 8]. Different versions of the SVM has been utilized in the past years, for Eg. Lagrangian Support Vector Machine (LSVM) [9] followed by a Smooth Support Vector Machine (SSVM) for binary classification [10], Least Squares Support Vector Machine (LS-SVM) [11] and Proximal Support Vector Machine (PSVM)[12]. By considering the parallel hyperplane based classifiers like SVM, author in [13] proposed a Generalized Eigenvalue Proximal SVM (GEPSVM) that has a nonparallel hyperplane classifier. This generates two numbers of hyperplane instead of generating one. TSVM is one such classifier that has been motivated by GEPSVM that utilizes two hyperplanes and has speed almost four times than that of SVM [14, 15]. TSVM has four times faster computation speed than SVM and is insensitive to class imbalance. The author has shown that TSVM shows better accuracies than ANNs and SVM in [16]. In our research work, performance TSVM has been analyzed by equipping it with different kernel functions.

3 Twin Support Vector Machine

TSVM is insensitive to the imbalance present in the size of classes and moreover it has less computational complexity than SVM. Based on the optimization theory given by Karush-Kuhn-Tucker theorem [17, 18], the nonlinear decision function is defined:

$$f(x) = \text{sign}(\sum_{i=1}^{N} \alpha_i y_i K(x_i, x_j) + b) \tag{1}$$

where $K(x_i, x_j)$ is the kernel representation of function.

The nonlinear TSVM seeks two nonparallel hyperplanes:

$$w_1'(x) + b_1 = 0 \tag{2}$$

$$w_2'(x) + b_2 = 0 \tag{3}$$

In order to find the solution of required hyperplanes, the solutions to the original problem must be obtained.

Minimize:

$$U(w_1, \zeta_2) = \frac{1}{2}\|Aw_1 + e_1 b_1\|^2 + c_1 e_2' \zeta_2 \tag{4}$$

subjected to

$$-(Bw_1 + e_2b_1) + \zeta_2 \geq e_2, \tag{5}$$

$$\zeta_2 \geq 0,$$
$$c_1 > 0$$

and
Minimize:

$$U'(w_2,\zeta_1) = \frac{1}{2}\|Bw_2 + e_2b_2\|^2 + c_2e_1'\zeta_1 \tag{6}$$

subjected to

$$(Aw_2 + e_1b_2) + \zeta_1 \geq e_1, \tag{7}$$

$$\zeta_1 \geq 0,$$

$$c_2 > 0$$

Here the constants c_1 and c_2 are referred as the punishment parameters whereas e_1 and e_2 are the two vectors of suitable dimensions consisting of ones.

4 Proposed Methodology

In the proposed work, we present a comparative study for fault detection using TSVM for different kernel function for WSNs. Figure 1 shows the detailed flow for

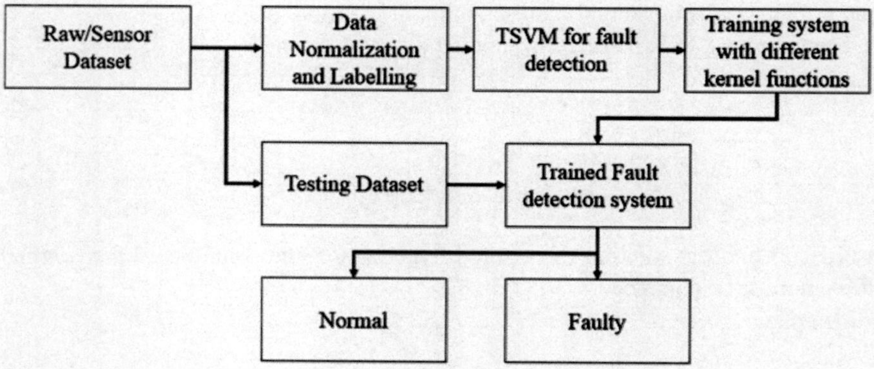

Fig. 1 Flow diagram of proposed work

the proposed work. Here the dataset is distributed in two subsets i.e. training and testing dataset. The training dataset is a type of labelled datasets that is utilized to train the fault detection system. The labelled dataset is then normalized in order to scale the data of different dimensions into a single dimension.

Kernel functions or Kernel trick enables a linear classifier to classify a non-linear dataset. A kernel function maps a non-linear data points into high-dimensional space in which they are separable. The details of different kernel functions utilized are added below:

- **Gaussian RBF:** Radial Basis Function is one of the most common kernel function. It is mathematically represented as:

$$K(x_i, x_j) = \exp(-\gamma \|x_i - x_j\|^2)$$

- **Linear Kernel:** It is mathematically represented as

$$K(x_i, x_j) = x_i.x_j + c$$

- **Polynomial Kernel Function:** This function is mathematically represented as:

$$K(x_i, x_j) = (x_i.x_j + c)^d$$

  where d denotes degree of polynomial.
- **Sigmoid Kernel:** It is mathematically represented as:

$$K(x_i, x_j) = \tanh(\alpha x_i^T.x_j + c)$$

5 Results and Discussions

5.1 Experimental Setup

In our experiment, seven datasets collected from Wireless Sensor Networks scenarios are utilized to analyze and evaluate the performance of our proposed algorithm. Each dataset has been divided into two parts: learning set and test set. The learning set consist of 70% dataset and the rest 30% dataset is used for the test set. In datasets, normal data is labeled with '0' and anomalous data with '1'. TSVM algorithm equipped with different kernel functions has been applied to each of the seven datasets. The seven different datasets has been used from IBRL lab datasets and WSN lab dataset. Manual fault has been inserted to IBRL datasets that generally occurs in the network i.e. utilized in [1].

5.2 Performance Evaluation

It is a crucial aspect to decide which model is best among various other models. This evaluation is done by comparing algorithms based on various performance parameters such as Accuracy, F1 - score and ROC curves. Performance measurements and evaluation for various kernel functions- RBF, linear, Polynomial, Sigmoid Kernel in Twin SVM classifier have calculated for each of 7 datasets.

Accuracy:- Accuracy is the measurement that provides the correct percentage of classification. It can be obtained simply using the following formulae -

$$Accuracy = \frac{correctly\ classified\ items}{all\ classified\ items}$$

F1-score:- F1-score is defined as the function that calculates the harmonic mean of precision value and recall value. This score is used for comparison among various classifier models. This measure recognize '1' as its best value and '0' as its worst value. It is obtained by following formulae-

$$F1\_score = \frac{2 \times Precision \times Recall}{Precision + Recall}$$

Confusion-Matrix – Confusion metric is the metric that provides the relation between the sets of predicted as well as the actual variables. Then confusion matrix is shown in the form of Table 1.

The four parameters of a confusion matrix are as follows:-

- **True Positives value (TP)**: It is defined as the count of positive cases whose prediction has been done correctly.
- **False Positives value (FP)**: It is defined as the count of negative cases whose prediction has been done incorrectly.
- **True Negatives value (TN)**: It is defined as the count of negative cases whose prediction has been done correctly.
- **False Negatives value (FN)**: It is described as the count of positive cases whose prediction has been done incorrectly.

AUC-ROC Curve - AUC - ROC which stands for 'Receiver Operating Curve' is a probability curve that determines performance in classification. AUC is obtained by calculating the area defined under the curve of ROC. AUC is known for representing degree of separability. The value of AUC ranges from 0.5 to 1. A classifier having

Table 1 Confusion matrix

	Predicted +ve	Predicted −ve
Actual +ve	TP	FN
Actual −ve	FP	TN

AUC values below 0.7 is bad, between 0.7 to 0.8 is considered average, between 0.8 to 0.9 as excellent and above 0.9 as an outstanding classifier.

The ROC curve is a relative plot representing the curve between True Positive Rate (TPR) and the False Positive Rate (FPR).

5.3 *Simulation Results*

The simulation results has been evaluated using different kernel functions. Figure 2(a) and Fig. 2(b) shows the accuracies and F1_score of various kernels functions applied to Twin SVM for two singlehop datasets. Figure 3(a) and Fig. 3(b) shows the accuracies and F1_score of various kernels functions applied to Twin SVM for two Multihop datasets and Fig. 4(a) and Fig. 4(b) shows accuracy and F1_score for remaining three

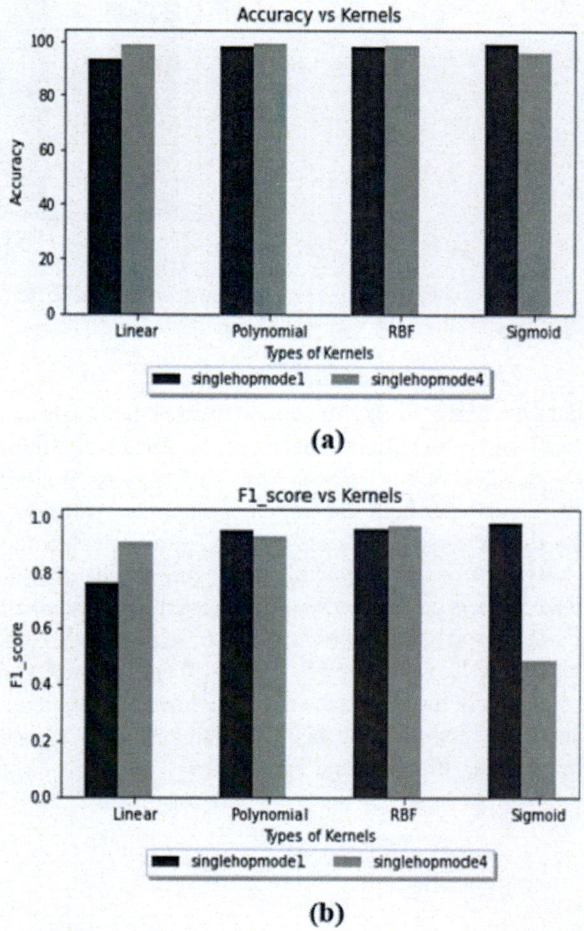

Fig. 2 **a** Detection Accuracy for Singlehop WSN dataset **b** F1_score for Singlehop WSN dataset

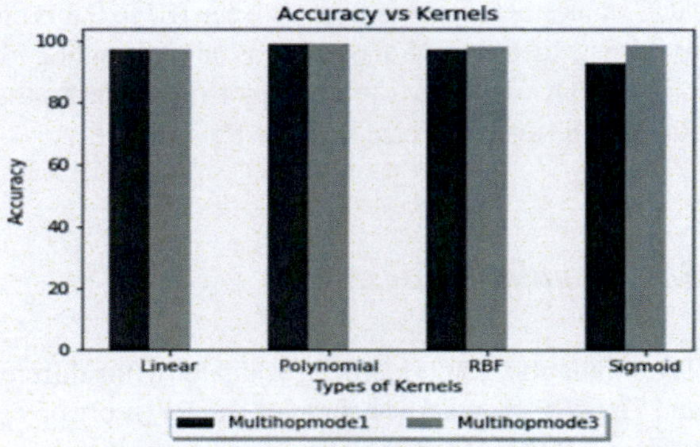

(a)

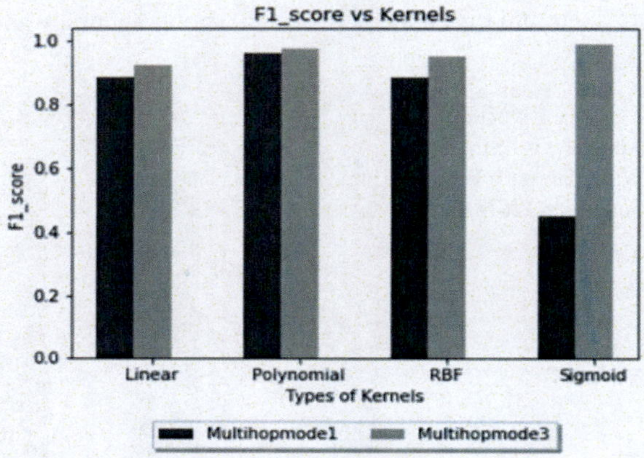

(b)

Fig. 3 **a** Detection Accuracy for Multihop WSN dataset **b** F1_score for Multihop WSN dataset

datasets respectively. To support the results obtained using the proposed work the ROC curve for different datasets has also been plotted. Figure 5 and Fig. 6 shows the ROC curve plot for both Multihop datasets. Similarly Fig. 7 and Fig. 8 shows the ROC curve for both the singlehop datasets. And Fig. 9, 10 and 11 shows the ROC for different mode datasets utilized from IBRL laboratory.

The various above mentioned Figures reveal that on an average Polynomial and RBF kernels perform exceptionally well with minute differences in their accuracies, F1-score and ROC curves followed by Linear kernel function. Sigmoid kernel notices highest AUC values 0.965, 0.965, 1.0 and 1.0 in Fig. 6, Fig. 7, Fig. 10 and Fig. 11 respectively for some datasets while lowest AUC values of about 0.5 in Fig. 5, Fig. 8 and Fig. 9 for other datasets. The Highest accuracy values are quite ideal as shown in Fig. 2(a), Fig. 3(a) and Fig. 4(a).

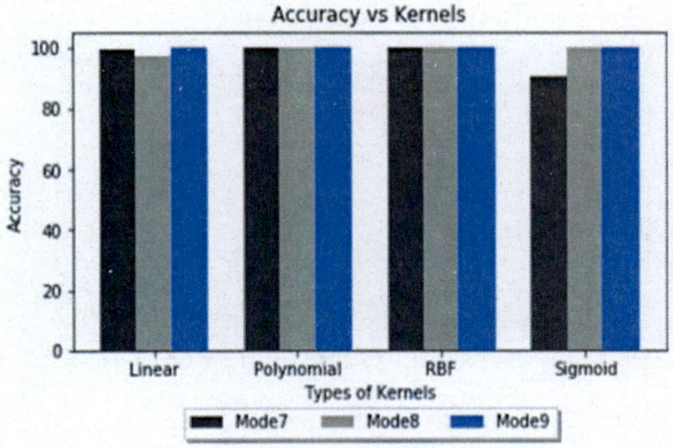

(a)

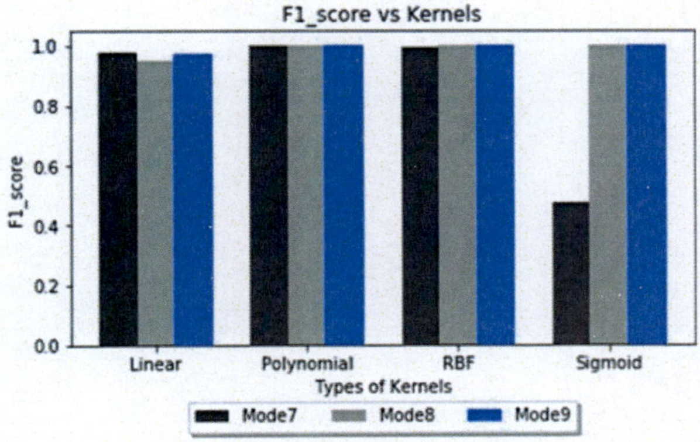

(b)

Fig. 4 **a** Detection Accuracy for various IBRL datasets **b** F1_score for various IBRL datasets

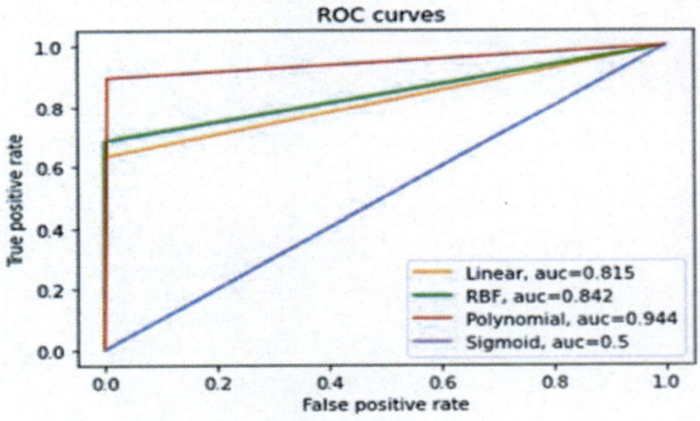

Fig. 5 ROC curve analysis: Multihopmode1 dataset

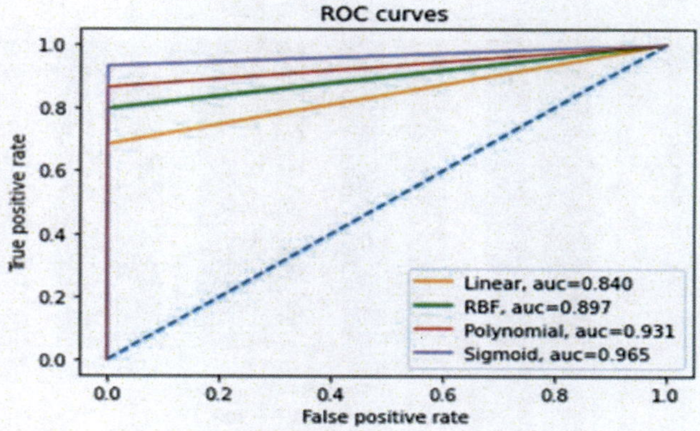

Fig. 6 ROC curve analysis: Multihopmode3 dataset

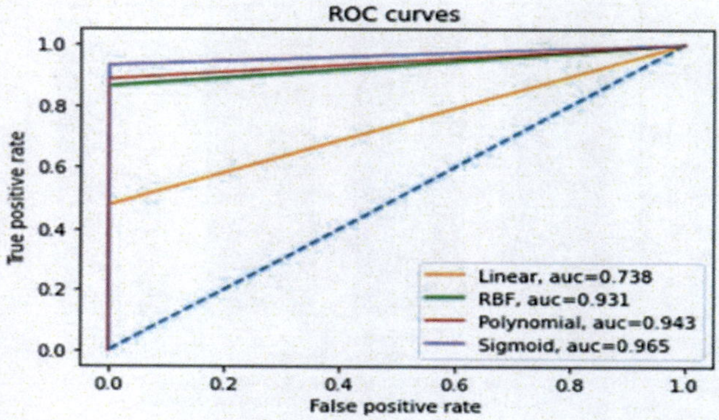

Fig. 7 ROC curve analysis: Singlehopmode1 dataset

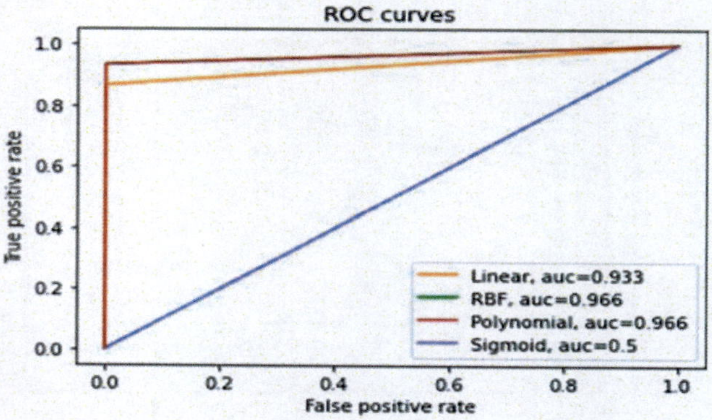

Fig. 8 ROC curve analysis: Singlehopmode4 dataset

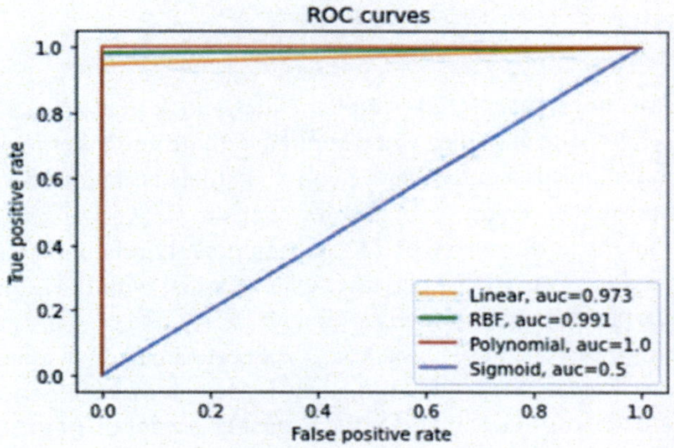

Fig. 9 ROC curve analysis: mode7 dataset

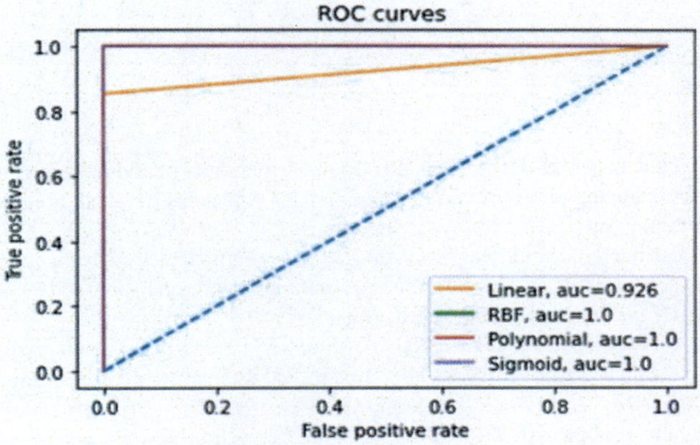

Fig. 10 ROC curve analysis: mode8 dataset

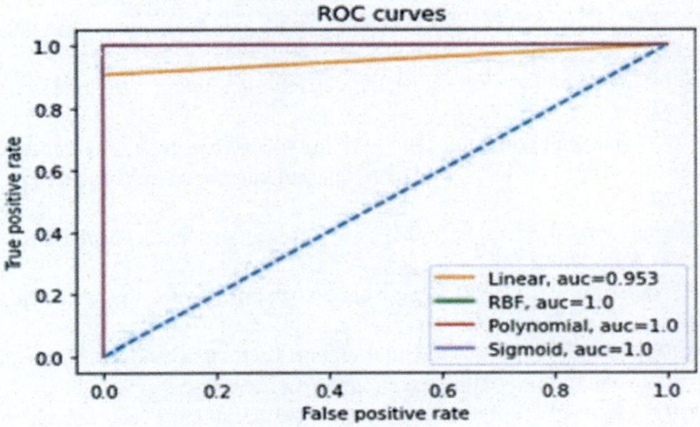

Fig. 11 ROC curve analysis: mode 9 dataset

6 Conclusion

In this work we have successfully implemented and tested the TSVM for fault detection in WSNs. The classifier when applied with different kernel functions like Linear, Polynomial, Gaussian RBF and Sigmoid kernels is suitable for the detection of anomalies present in datasets of Wireless Sensor Networks. The experimental results show that the performance of TSVM significantly rely upon the selection of kernel function used. The TSVM classifier when applied with Polynomial and RBF kernel functions perform exceptionally well with all datasets used in our experiments whereas with Sigmoid kernel function gives best performance for some datasets and worst for others. Linear kernel function always gives an average performance. The overall experimental result accounts for the High Accuracies of Twin Support Vector Machines. In the Future, We will perform a similar experiment on multiclass datasets and assess the effect of varying the kernel function on the model.

References

1. Gavel S, Raghuvanshi AS, Tiwari S (2020) Comparative study of anomaly detection in wireless sensor networks using different kernel functions. In: Advances in VLSI, communication, and signal processing, pp 81–89. Springer, Singapore
2. Biswas P, Charitha R, Gavel S, Raghuvanshi AS (2019) Fault detection using hybrid of KF-ELM for wireless sensor networks. In: 2019 3rd international conference on trends in electronics and informatics (ICOEI), pp 746–750. IEEE, April 2019
3. Liu Q, Cui X, Abbod MF, Huang SJ, Han YY, Shieh JS (2011) Brain death prediction based on ensembled artificial neural networks in neurosurgical intensive care unit. J Taiwan Inst Chem Eng 42(1):97–107
4. Yang J, Singh H, Hines EL et al (2012) Channel selection and classification of electroencephalogram signals: an artificial neural network and genetic algorithm-based approach. Artif Intell Med 55(2):117–126
5. Cortes C, Vapnik V (1995) Support-vector networks. Mach Learn 20(3):273–297
6. Burges CJ (1998) A tutorial on support vector machines for pattern recognition. Data Min Knowl Disc 2(2):121–167
7. Vapnik VN (1999) An overview of statistical learning theory. IEEE Trans Neural Netw 10(5):988–999
8. Vapnik V (2000) The nature of statistical learning theory. Springer, Heidelberg
9. Mangasarian OL, Musicant DR (2001) Lagrangian support vector machines. J Mach Learn Res 1:161–177
10. Lee Y-J, Mangasarian OL (2001) SSVM: a smooth support vector machine for classification. Comput Optim Appl 20(1):5–22
11. Suykens JA, Vandewalle J (1999) Least squares support vector machine classifiers. Neural Process Lett 9(3):293–300
12. Mangasarian OL, Wild EW (2001) Proximal support vector machine classifiers. In: Proceedings KDD 2001: knowledge discovery and data mining. Citeseer
13. Mangasarian OL, Wild EW (2006) Multisurface proximal support vector machine classification via generalized eigenvalues. IEEE Trans Pattern Anal Mach Intell 28(1):69–74
14. Jayadeva, Khemchandani R, Chandra S (2007) Twin support vector machines for pattern classification. IEEE Trans Pattern Anal Mach Intell 29(5):905–910

15. Khemchandani R (2008) Mathematical programming applications in machine learning. PhD thesis, Indian Institute of Technology Delhi New Delhi-110016, India
16. Ding M, Yang D, Li X (2020) Fault diagnosis for wireless sensor by twin support vector machine. Burges C (1998) A tutorial on support vector machines for pattern recognition. Data Min Knowl Disc 2:1–43
17. Ding S, Zhang N, Zhang X, Wu F (2017) Twin support vector machine: theory, algorithm and applications. Neural Comput Appl 28(11):3119–3130
18. Khemchandani R, Goyal K, Chandra S (2016) TWSVR: regression via twin support vector machine. Neural Netw 74:14–21

Effect of Different Channel Material on the Performance Parameters for FinFET Device

Himanshu Kumar, Mayank Kumar Jethwa, Ashwin Porwal, Rasika Dhavse, Hardiki Mukesh Devre, and Rutu Parekh

Abstract This paper studies and investigates the effect of channel material on various performance parameters for FinFET. Channel material like Silicon (Si), Gallium Arsenide (GaAs), Gallium Antimonide (GaSb), Gallium Nitride (GaN), Germanium (Ge) and Indium Arsenide (InAs) were used. Effect of different channel materials on parameters like temperature, drain induced barrier lowering (DIBL), subthreshold swing (SS), threshold voltage (V_t) and transconductance (g_m) were simulated and studied. From the simulation performed, it is observed that amongst all the channel materials used for the experiment GaAs channel is highly dependent on temperature. It was also determined that the best channel material for DIBL is GaN and GaAs DIBL is around 223 mV/V at gate length (L_g) = 46 nm. Further, it was found that GaN has better SS of 108 mV/dec at L_g = 46 nm, which is less than other materials. Simulations performed show that the threshold voltage for GaN is highest and for GaSb, it is lowest. In addition, it was concluded that transconductance for GaSb is better than GaN. It was hence confirmed that GaN and GaAs, when used as channel materials, are the most efficient channel materials which are effective for most of the performance parameters. Whereas Si, when used as channel material, is most stable for all the parameters and GaSb is the worst channel material for performance parameters.

Keywords MOSFET · FinFET · MuGFETS · DIBL · SS · SCE · Threshold voltage · Transconductance

H. Kumar · M. K. Jethwa (✉) · A. Porwal · H. M. Devre · R. Parekh
DAIICT, Gandhinagar, Gujarat, India

R. Parekh
e-mail: rutu_parekh@daiict.ac.in

R. Dhavse
SVNIT, Surat, Gujarat, India
e-mail: rsk@eced.svnit.ac.in

V. Nath and J. K. Mandal (eds.), *Proceeding of Fifth International Conference on Microelectronics, Computing and Communication Systems*, Lecture Notes in Electrical Engineering 748,
https://doi.org/10.1007/978-981-16-0275-7_23

1 Introduction

As the Metal Oxide Semiconductor Field Effect Transistor (MOSFET) reaches its downscaling restriction, a few new FET structures are being prospected widely. As the device dimensions are scaled towards the nanometer level, traditional single gate MOSFETs experience specific short channel effects (SCEs), which weaken the drive current and transitions to off-state. To address such scaling down issues, alternate multiple-gate field effect transistors (MuGFETS) structures and materials are being explored and are under continuous study. MuGFETS show better performance on SCE which can replace the existing single-gate planar device structure. A feasible implementation of the multi-gate MOSFET is a FinFET. It is one of the potential alternative to eliminate the SCEs while downscaling the FETs to follow the projections of ITRS. FinFET is considered to be the most efficient thin-body device due to the self-aligned gate electrodes [1]. One of the main differences between a FinFET and a MOSFET is that multiple small unit fins make up the FinFET device. FinFET circuits can achieve lower functional voltage supply as opposed to CMOS circuits and lower effective energy consumption. In addition, FinFET's subthreshold region has greater tolerance to soft error. But, further downsizing of the FinFET structure appears to be significantly more troubling despite various rational limitations, such as parasitic resistance and capacitance, DIBL, subthreshold swing (SS), transconductance and threshold voltage roll-off and gate leakage by hot carrier tunneling. Further improvements can be done to improve the speed and performance of the devices while reducing its size in order to meet the Moore's law and ITRS roadmap [2]. This is possible by replacing the channel materials. Compound semiconductors like Gallium Arsenide (GaAs), Gallium Antimonide (GaSb), Gallium Nitride (GaN), Germanium (Ge) and Indium Arsenide (InAs) are prominent candidates for future applications because of their high mobility and high on-currents at low supply voltage which leads to high performance of devices. In this paper, we have simulated and investigated performance of a FinFET device with such different channel materials.

There is much research done on FinFETs. Saini et al. studied the physical scaling limits of a FinFET structure [3]. It was found that the effective fin length to fin thickness (Leff/Tfin) ratio limits the scaling capabilities of the FinFET structure. An examination of TG n-FinFET parameters by changing channel doping concentration and gate length was done in the paper [4]. An investigation on change in parameters on electrical characteristics of FinFET with high-k dielectric was carried out in [5]. A study on the threshold voltage variability originated by statistical parameter fluctuations in nanoscale bulk and SOI-FINFETs is discussed in [6]. FinFET with scaled fin-dimensions and oxide thickness has been analyzed at 10 nm [7]. There was a similar study in the paper by Bhat et al. on the study of SCE in n-FinFET structure for Si, GaSb, GaAs and GaN channel materials [8]. Study of FinFET as a temperature nano sensor based on semiconductor channel was reported in [1].

InGaSb FinFET's have been fabricated with fin widths and height down to 30 and 100 nm respectively with a high transconductance (g_m) of 122 uS/um [9]. The study on the effects of decreasing channel dimensions on electrical properties like

I_{ON}/I_{OFF} ratio, SS, etc., have been implemented in [10]. Investigation of various kinds of FinFETs like SiGe, Ge and GaN have been studied for optimized performance [11–13]. In addition, FinFET performance in terms of SCEs such as DIBL, SS, and threshold voltage (V_t) have been optimized in [14]. Another work reports on designs of tri-gate and double gate FinFETs using a numerical simulator (Silvaco TCAD) [15]. In our paper, we have taken Si, Ge, GaAs and InAs as channel type semiconductor for investigation of temperature sensitivity. Furthermore, we include two more channel which are made up of GaN and GaSb for analysis of DIBL, SS, V_t and g_m.

2 FinfET Design and Simulation Framework

A FinFET is classified as a multi-Gate type MOSFET. In FinFET, the body is formed by a thin Silicon film wrapped over the conducting channel, as shown in Fig. 1(a). A gate-length Lg of a FinFET is said to be the distance between source junctions and Drain junctions. The channel width (W_{ch}) of FinFET is said to be width of channel between the source and drain as depicted in Fig. 1(b). First of all, we simulate using MuGFET tool [16] for its I-V characteristics using different materials at different

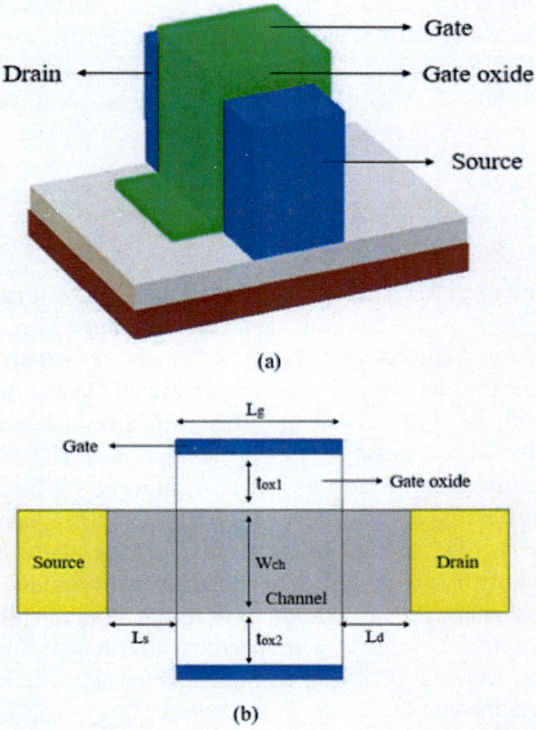

Fig. 1 FinFET structure **(a)** 3-D **(b)** 2-D planer

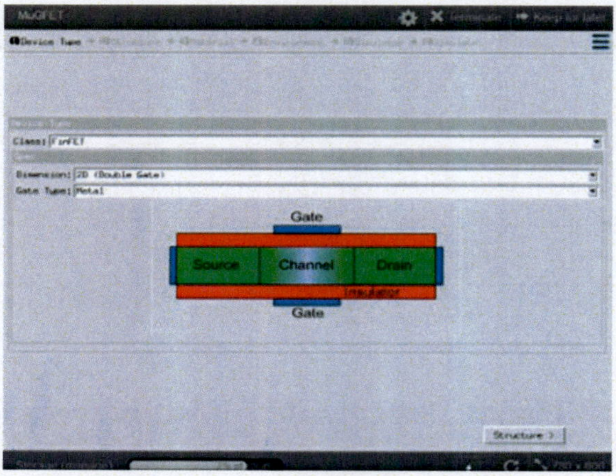

Fig. 2 MuGFETS tool interface

temperature. In order to investigate with temperature parameter, we keep channel concentration (P-type) = 10^{16} cm^{-3}, channel width = 30 nm, source and drain concentration (N-type) = 10^{19} cm^{3}, source length = drain length = 50 nm, channel length = 45 nm and oxide thickness = 2.5 nm for all channels.

The MuGFET tool as shown in Fig. 2, simulates for the performance parameters of FinFET. Researchers using this simulation tool can select one of two, PADRE or PROPHET simulators. The physical and electrical parameters of the channel applied in MuGFET tool are given in Table 1. Firstly, we investigate the different types of FinFET channels for thermal behaviour. Furthermore, we take other two parameters of FinFET, which are gate length and gate width to study SCE parameters.

3 Results and Discussion

Firstly, I-V characteristic is simulated for investigation with temperature parameter. We keep changing the temperature in range from −25 to 125 °C. Figure 3 shows the I-V characteristic for all six channel materials. Temperature characteristics for the FinFET with Si, GaAs, GaN, GaSb, Ge and InAs as a FinFET channel at drain voltage V_d = 1 V with different temperature estimates (−25, 50 and 125 °C) is shown. It can be seen from the figure that, at 125 °C ($V_g = V_d$ = 1 V), the drain current (I_d) for Si is low, at 1.1 A/mm and for InAs, it is high, around 2.6 A/mm. It is because of effective electron mass is less in InAs (0.023 m_0). When electric field is applied through gate voltage, due to less effective electron mass, electrons will easily flow from source to drain. So, drain current in InAs is high compared to other channels. Figure 4 shows I_d vs temperature qualities at $V_d = V_g$ = 1 V for a wide variety of channels, with specific temperature estimates (−25, 50 and 125 °C). It can be seen that FET with channel material as Si, GaAs, Ge has lower drain current as compared to InAs and Si the lowest.

Table 1 Various properties of channels at 300 K [17–19]

S. No.	Electrical properties	Si	GaAs	Ge	InAs	GaSb	GaN
1.	Energy-Band Gap (eV)	1.12	1.454	0.66	0.35	0.726	3.2
2.	Dielectric Constant	11.7	12.9	16.2	15.15	15.7	8.9
3.	Electron Affinity (V)	4.05	4.07	4	4.9	4.06	4.1
4.	Electron effective mass	$0.2 \times m_0$	$0.041 \times m_0$	$1.6 \times m_0$	$0.023 \times m_0$	$0.063 \times m_0$	$0.20 \times m_0$
5.	Density of state effective mass of Electrons	$1.18 \times m_0$	$0.57 \times m_0$	$0.22 \times m_0$	$0.29 \times m_0$	$0.6 \times m_0$	$0.57 \times m_0$
6.	Density of state effective mass of Holes	$0.81 \times m_0$	$0.8 \times m_0$	$0.34 \times m_0$	$0.41 \times m_0$	$1.5 \times m_0$	$0.8 \times m_0$
7.	Light hole effective mass	$0.16 \times m_0$	$0.076 \times m_0$	$0.044 \times m_0$	$0.026 \times m_0$	$0.05 \times m_0$	$0.3 \times m_0$
8.	Heavy hole effective mass	$0.49 \times m_0$	$0.050 \times m_0$	$0.33 \times m_0$	$0.41 \times m_0$	$0.4 \times m_0$	$1.4 \times m_0$
9.	Electron Mobility (cm^2/V. s)	1450	8500	3900	40000	3000	1000
10.	Hole Mobility (cm^2/V. s)	500	400	1900	500	1000	200
11.	Saturation Velocity of Electron (cm/s)	1.0×10^7	0.72×10^7	0.70×10^7	8×10^7	1.34×10^7	3×10^7
12.	Saturation Velocity of Holes (cm/s)	0.704×10^7	0.9×10^7	0.63×10^7	8×10^7	1.1×10^7	1.0×10^7

Note: $m_0 = 0.91 \times 10^{-30\,kg}$ (rest mass of electron)

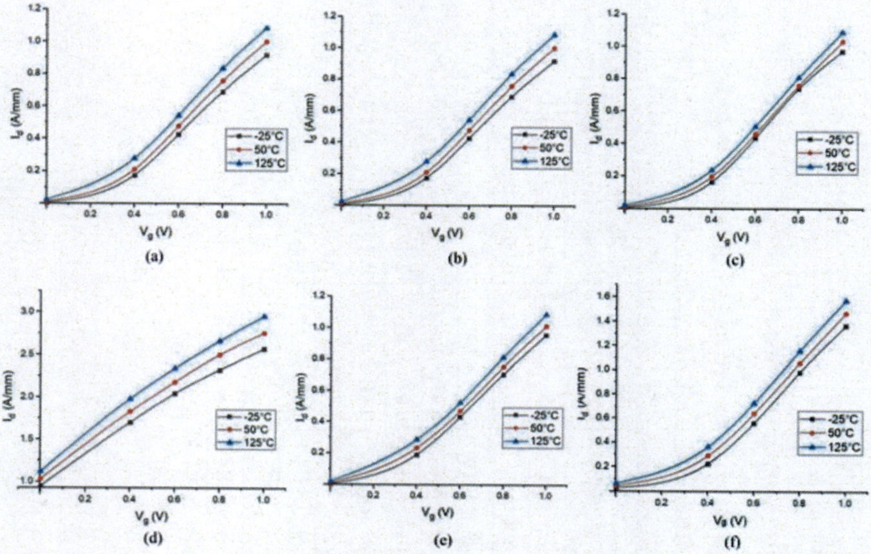

Fig. 3 I-V Characteristics of FinFET for channels, **(a)** GaAs **(b)** Si **(c)** GaN **(d)** InAs **(e)** GaSb **(f)** Ge

Fig. 4 Drain current (I_d) vs temperature with $V_d = V_g = 1$ V

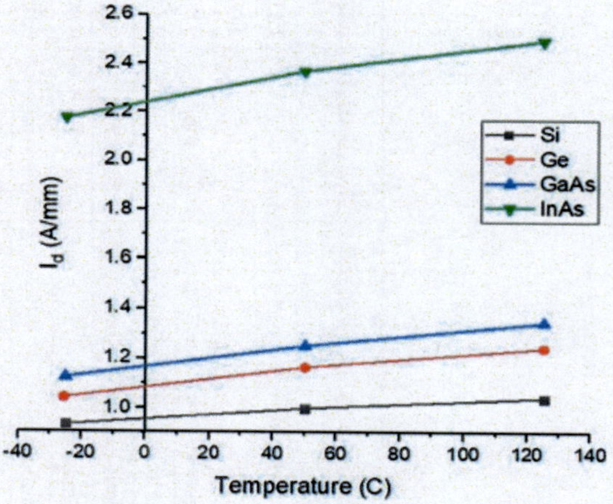

Figure 5 depicts current change (ΔI) with temperature for FinFET with different types of channel material. It can be seen that, ΔI decreases with increase in temperature for all different channels, but with InAs it falls faster. Current change between 0 to 50 °C for InAs is around 0.058 A/mm. and further current change between 50 °C to 125 °C is around 0.0544 A/mm. As we take a large range of temperature to investigate the best channel, we have to normalize the current change at specific temperature to test sensitivity of the channel. We normalized the current change at 25 °C for FinFET with various types of semiconductor channels. The graph in Fig. 6

Fig. 5 Current change (ΔI) vs temperature with $V_g = V_d = 1$ V

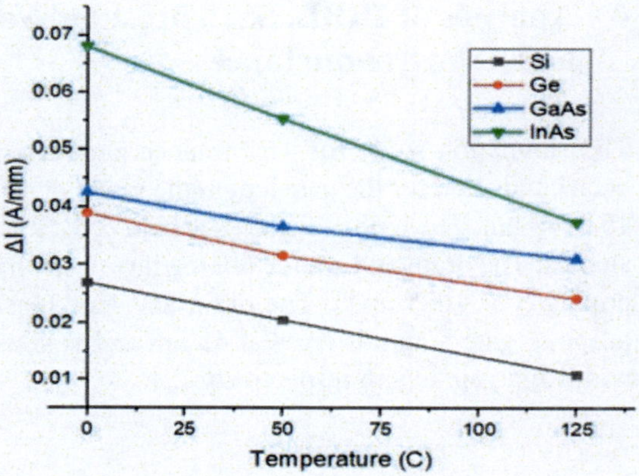

Fig. 6 Normalized current change (ΔI) vs temperature with $V_d = V_g = 1$ V

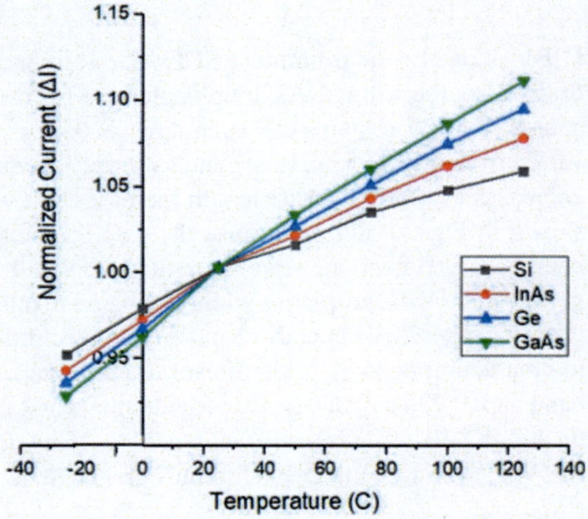

shows that GaAs channel is more sensitive to temperature and has the highest increment of 10.8% (ΔI) at 125 °C of its value at 25 °C. It is therefore, possible to use GaAs FinFET as temperature nano-sensor. Whereas, Si, InAs, Ge has increment of 6, 7.9, 9.5% respectively.

It can hence be inferred that Si channel is a better stable FINFET, because it has the minimum (ΔI) with higher temperature.

4 Analysis of DIBL, SS, Threshold Voltage and Transconductance

The simulation study for performance parameters like DIBL, SS, V_t and g_m, has been conducted for the gate length the range of 40 to 55 nm and channel width from 15 to 40 nm. The oxide thickness is held constant at 2 nm throughout the simulation studies. The drain and source doping have been fixed at 1×10^{20} cm^{-3} and channel doping 5×10^{16} cm^{-3}. The drain and gate bias are fixed at 1 V. For simulation purpose, gate length is fixed at 45 nm and gate width to 30 nm while varying gate width and gate length respectively.

4.1 *DIBL Vs Gate Length (L_g) and Channel Width (W_{ch})*

DIBL is one of the prominent SCEs due to reduced channel length and increased drain to source voltage. As drain depletion area moves closer to the depletion of the source, FinFET remains ON even if $V_d = 0$. For simulation purpose, gate length is varied from 40 to 55 nm range and gate width in the range of 20 nm to 35 nm. DIBL decreases as FinFET's gate length increases and with decrease in channel width as shown in Figs. 7 and 8, because the gate has better control on the channel. It has been observed from the simulation that, GaN and GaAs channel FinFET structures gives better DIBL properties when compared to other type of materials. Due to wider energy gap (3.2 eV) in GaN channel type, electrons cannot jump from valence band to conduction band with significant drain voltage. whereas GaSb have less energy band gap (0.726 eV) and hence provides the worst DIBL characteristics. As shown in the figure DIBL for GaN and GaSb at $L_g = 46$ nm is 223 and 268 mV/V respectively. For $W_{ch} = 25$ nm, the DIBL for GaN and GaSb is 125 and 175 mV/V respectively.

Fig. 7 DIBL vs Gate length (L_g)

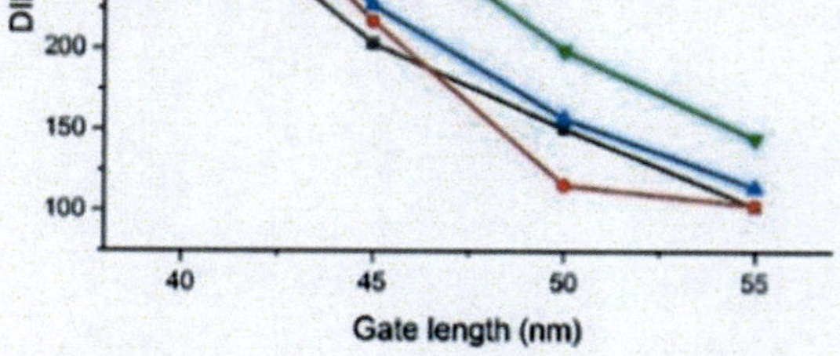

Fig. 8 DIBL vs Gate width (W_{ch})

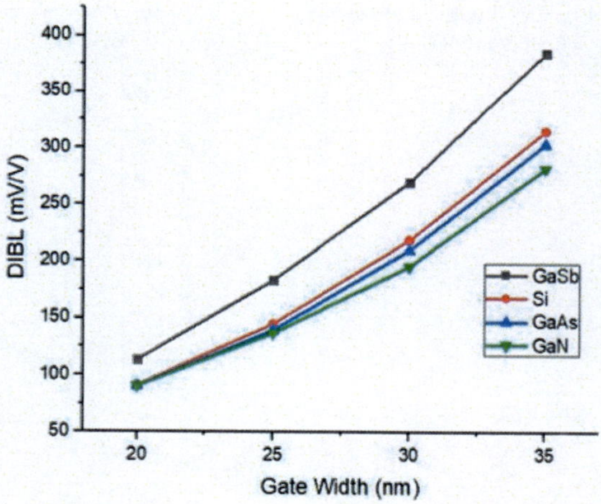

4.2 *Subthreshold Swing Vs Length (L_g) and Channel Width (W_{ch})*

For a transistor, the subthreshold swing is defined as the ratio of the change in gate voltage so as to generate ten times increase in the output current. The lower its value better the device performance. SS reflects the sharpness of a FET's ON–OFF switch. SS gives an overview of the leakage currents associated with characteristics of the device. This increases with reduction in channel length and decreases as channel width decreases. As the length of the channel decreases, gate loses control over the channel due to SCE, and therefore increases SS. From Fig. 9, it can be observed that Silicon and Gallium Arsenide channel FinFET display nearly same SS characteristics, whereas the GaN-channel provides better SS characteristics in comparison to other material. It is because of high saturation velocity (3×10^7 cm/s)

Fig. 9 Subthreshold swing vs Gate length (L_g)

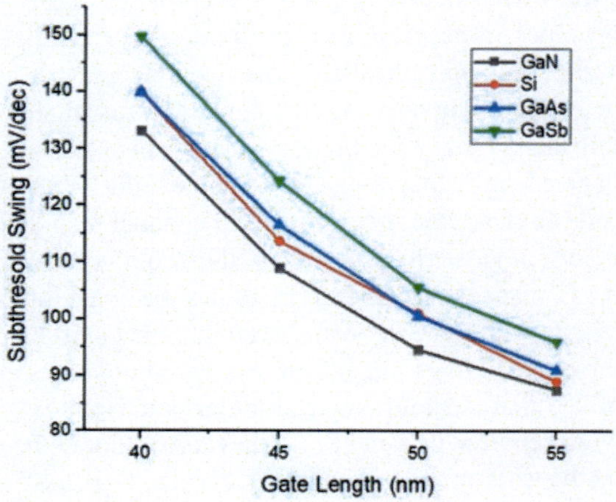

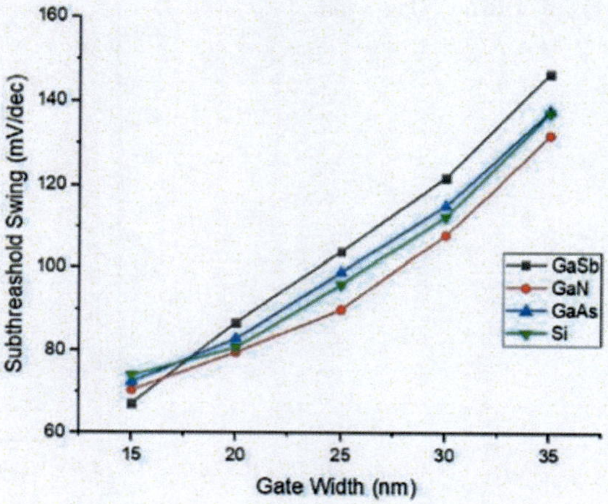

Fig. 10 Subthreshold swing vs Gate width (W_{ch})

of GaN compared to other channel type. The worst SS characteristics are obtained for GaSb channel material structures. It can be observed that SS for GaN and GaSb at $L_g = 55$ nm is 88 mV/dec and 97 mV/dec respectively. Furthermore, in Fig. 10, SS for GaN and GaSb at $W_{ch} = 25$ nm is 85 mV/V and 102 mV/dec respectively. GaSb channel n displays the worst SS characteristics in comparison with other materials.

4.3 *Threshold Voltage Vs Length (L_g) and Channel Width (W_{ch})*

The threshold voltage is defined as the minimum gate-to - source voltage required by the source and drain terminals to create a channel. In general, the threshold voltage proportional to gate length of the device structure which is shown in Fig. 11. It rolls off with the reduction of gate length. This is because, by lowering the length of the gate, when the distance between drain and source is reduced, the channel potential becomes more prominent in draining electric field encroachment, resulting in an earlier gate bias threshold. The two-side gates (e.g. - Double Gate FinFET) offers a heavy coupling over the channel region for smaller channel width and this holds the threshold voltage at a higher value. It can be observed from Fig. 11 that with decrease in gate length, Si, GaAs, GaN-channel FinFET show almost similar threshold voltage roll off characteristics with GaSb channel displaying the worst. As shown in figure below, at gate length of 40 nm, threshold voltage for GaN and GaSb is 0.258 V and 0.112 respectively. Figure 11 shows the behaviour of threshold voltage with respect to gate width. It can be observed that the GaSb-channel again shows a worst case of threshold voltage roll-off for W_{ch} variations. As shown in figure below, at gate width of 25 nm, threshold voltage for GaN and GaSb is 0.48 and 0.375 V respectively. In a work reported by Bhat et al., they have studied the effect of L_g and W_{ch} dependence on the various short channel effects viz., DIBL, SS and threshold voltage roll-offs for

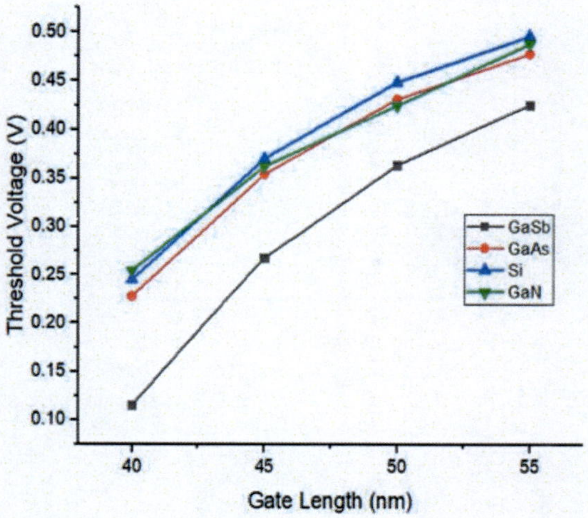

Fig. 11 Threshold Voltage vs Gate length (L_g)

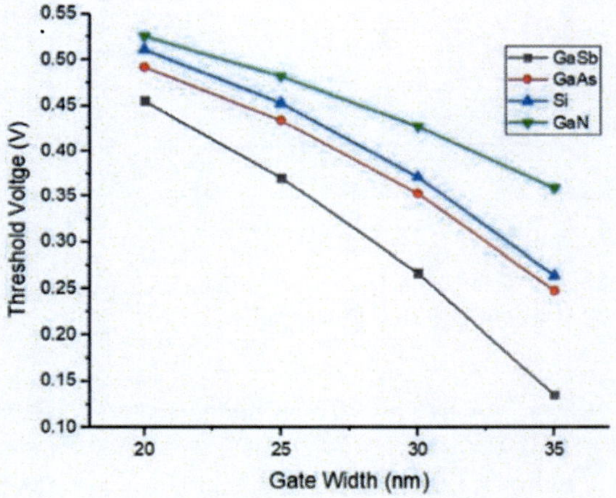

Fig. 12 Threshold Voltage vs Gate width (W_{ch})

various channel material (Si, GaAs, GaSb and GaN) [8] and they are in accordance with our results (Fig. 12).

4.4 Transconductance Vs Gate Length (L_g)

Transconductance tells how much strong the device can convert a small amount of voltage into desirable current. It is the ratio of the change in output current to the change in input voltage of a FET. And it can be described using the Shichman–Hodges model as [20]:

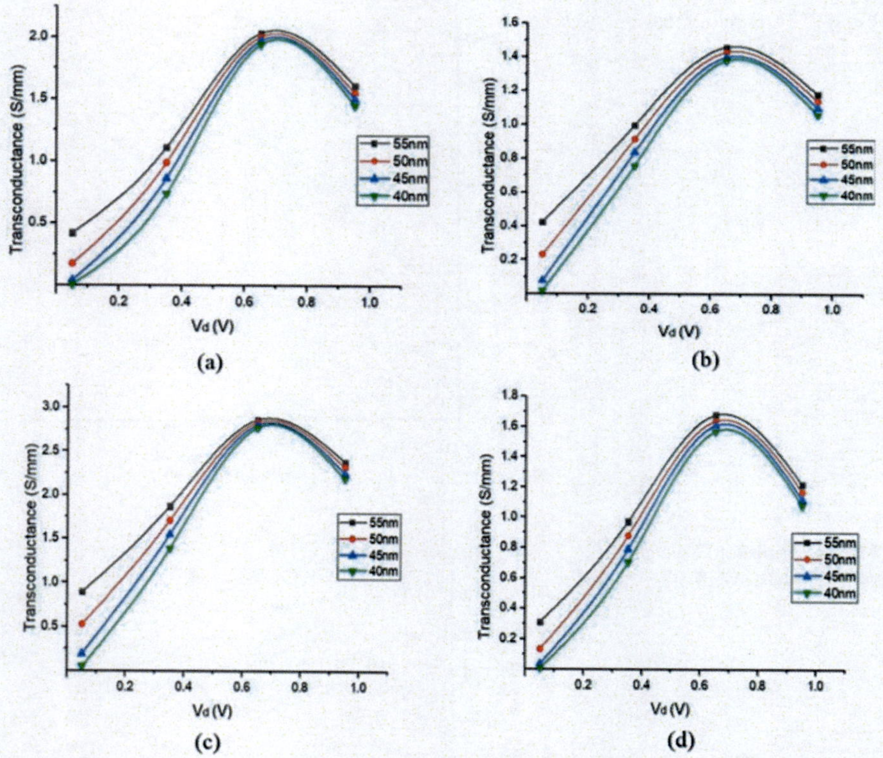

Fig. 13 Transconductance vs V_d with different gate lengths (L_g). (a) Si channel (b) GaAs channel (c) GaSb channel (d) GaN channel

$$\text{Transconductance } (g_m) = 2 \times I_d / V_{eff} \quad (1)$$

In Eq. 1, I_d is drain current, and V_{eff} is the effective voltage, which is a difference between gate source voltage and the threshold voltage ($V_{eff} = V_{GS} - V_{TH}$). As we seen in Fig. 13, transconductance vs V_d have been plotted for Vg equal to 1 V. We can see that GaSb gives the best transconductance among all the channels and its value is 2.79 S/mm at V_d equal to 0.7 V. Transconductance is directly proportional to V_d in linear region as we see in above graphs. When current goes linear to saturation region, the downfall in graphs shows that, in saturation region, transconductance is not depend upon only drain V_d.

5 Conclusion

In this paper we have successfully investigated the effect of various parameters like temperature, DIBL, SS, threshold voltage and transconductance on different FinFETs channel materials. The materials are Silicon (Si), Gallium arsenide (GaAs), Gallium antimonide (GaSb), Gallium nitride (GaN), Germanium (Ge) and indium arsenide (InAs). The outcomes show that both GaN and GaAs show better DIBL qualities. Investigation of SS attributes shows that GaN device provides stronger SS qualities contrasted with other three channel materials. Threshold voltage roll-off characteristics has indicated that Si, GaAs and GaN-channel structures give nearly identical V_t characteristics for variation in channel length, while GaN appeared to delivered significantly better V_t qualities for channel width. It is important that GaSb has demonstrated the most pessimistic scenario for all short channel effect characteristics. For transconductance we see that GaSb gives the good result compared to all other channels.

References

1. Hashim Y (2017) Investigation of FinFET as a temperature nano-sensor based on channel semiconductor type. IOP Conf. Series: Mater Sci Eng **226:** 012123
2. http://www.itrs2.net/ww.itrs2.net
3. Saini G, Rana AK (2011) Physical scaling limits of FinFET structure: a simulation study. Int J VLSI Des Commun Syst (VLSICS) 2(1):26–35
4. Boukortt S, Hadri N, Patanè B (2017) Investigation on TG n-FinFET parameters by varying channel doping concentration and gate length. Silicon 9:885–893
5. Gopinadh D, George A (October 2015) Variation in Parameters on Electrical Characteristics of FinFET with High-k dielectric. IJAREEIE **4**(10)
6. Rathore RS, Rana AK, Sharma R (2017) Threshold voltage variability induced by statistical parameters fluctuations in nanoscale bulk and SOI FinFETs. In: 2017 4th international conference on signal processing, computing and control (ISPCC), Solan, pp 377–380
7. Dargar SK, Srivastava VM (2019) Performance analysis of 10 nm FinFET with scaled fin-dimension and oxide thickness. In: 2019 international conference on automation, computational and technology management (ICACTM), London, United Kingdom, 2019, pp 1–5
8. Bhat, Mustafa T, Beigh M, Rafiq M (2015) Study of short channel effects in n-FinFET structure for Si, GaAs, GaSb and GaN channel materials. J Nano- Electron Phys **7**: 03010
9. Lu W, Kim JK, Klem JF, Hawkins SD, del Alamo JA (2015) An InGaSb p-channel FinFET. In: 2015 IEEE international electron devices meeting (IEDM), Washington, DC, 2015, pp: 31.6.1–31.6.4
10. Ahmed M, et al. (2019) Effects of downscaling channel dimensions on electrical characteristics of InAs-FinFET transistor. Int J Electr Comput Eng (2088-8708), 9
11. Guo D, et al. (2016) FINFET technology featuring high mobility SiGe channel for 10 nm and beyond. In: 2016 IEEE symposium on VLSI technology, Honolulu, HI, 2016, pp: 1–2
12. Duriez B, et al. (2013) Scaled p-channel Ge FinFET with optimized gate stack and record performance integrated on 300 mm Si wafers. In: IEEE international electron devices meeting, Washington, DC, pp: 20.1.1–20.1.4
13. Das R, Baishya S (2018) Analysis of GaN nanoscale FinFET for low power circuit applications. Micro Nano Lett. **13**(4): 568–571
14. Das RR et al (2018) Temperature-dependent short-channel parameters of FinFETs. J Comput Electron 17(3):1001–1012
15. Zaki ZA, Tanjila N, Saha JK (2018) Short channel effects characterization of 3-D FinFET for High-k gate dielectrics. In: IEEE ICISET, pp 36–40
16. http://nanohub.org/resources/NANOFINFET
17. http://www.ioffe.rssi.ru/SVA/NSM/Semicond/Ge/electric.html

18. http://www.ioffe.rssi.ru/SVA/NSM/Semicond/GaAs/electric.html
19. http://www.ioffe.rssi.ru/SVA/NSM/Semicond/InAs/electric.html
20. https://en.wikipedia.org/wiki/Transconductance

Single Inverter Control to Resolve Power Quality Issues in Fuel Cell Grid Integration

Jeevisha Sehgal, Divya Asija, Pranjal Singh, and Tejasvi Bhatnagar

Abstract Electricity has become a necessity in everyone's life these days and an essential commodity which is very integral to all our work. Our lives today would be unimaginable without continuous and reliable supply of electricity. Fossil fuels, coal, etc. are going to deplete soon. People are realizing this and have developed techniques to use renewable energy sources as much as possible with current technology. Due to ever increasing power demand, new emerging technologies and depleting fuels, we have finally adopted renewable energy in our lives and want to make it a part of our daily lives, as it never runs out. Integration of renewable energy sources with the grid helps the availability of a reliable and clean energy supply. This paper deals with simulation of fuel cell and battery power conditioning system with grid inverter to show output while integrating renewable sources with the grid using the software MATLAB/Simulink R2019b.

Keywords Grid integration · Fuel cell · Battery · Boost converter · Buck-Boost converter · Inverter

1 Introduction

Grid Integration methods involving connectivity of renewable energy sources with the grid is one of the obvious solutions that is seen to reduce CO_2 emissions and other greenhouse gases which harm the environment. But there are also many challenges that are faced and need to be resolved to make the grid integration as efficient as possible. Most of the challenges that are faced in power arrangement are due to the interconnection between different components like transformer, transmission lines, etc. The least disturbance can affect or hamper the stability of the whole system. These disturbances occur due to the loading effect (of transmission lines). To reduce this effect to minimal, the trend is to use Distributed Generation (DG) by the distribution network service providers. The bidirectional flow of power due to

J. Sehgal (✉) · D. Asija · P. Singh · T. Bhatnagar
Department of Electrical and Electronics Engineering, Amity University, Uttar Pradesh, India

V. Nath and J. K. Mandal (eds.), *Proceeding of Fifth International Conference on Microelectronics, Computing and Communication Systems*, Lecture Notes in Electrical Engineering 748,
https://doi.org/10.1007/978-981-16-0275-7_24

distributed generation results in the reduced capacity of transmission line and also the feeder. But while integrating the DGs at the load end we face several power quality issues which need to be resolved for improving the overall reliability and efficiency of the power system. Thus in the proposed work we have tried to accommodate single inverter with controllers to connect the renewable DG with the grid. We can call DG as individual generation, which is on much smaller scale. Although generation capacity is much lower, it has many advantages for people to consider its installation. It's incredible flexibility in size and operation, its reliability and is expandable and has better power characteristics, no transmission costs and also it uses RES, makes it a good choice and source for power [1].

2 Grid Integration

Grid Integration is the technique of combining or integrating renewable energy sources with the existing power grid to obtain clean energy and reduce carbon emissions.

India is a developing country which has target of utilizing huge renewable resources in order to compete with its ever growing demand. It aims to accomplish its total load demand by utilizing solar, wind and biomass type of fuels which are available in abundance. Thus with more renewable resources coming in to play for power contribution with in the network via electric grids, challenges for integration with the grid has become more cumbersome due to variable supply from solar and wind sources. Grid operations and grid integration strategies need to be strong so that these challenges are managed accordingly. India is taking steps for creation and implementation of frameworks which support a higher share of renewable energy in the grid on a larger scale than currently exists and is more cost-effective.

3 Issues/Challenges of Grid Integration

The renewable energy sources integration to the grid is considered quite challenging as they are intermittent in nature. These issues can be regarding power quality, grid operation-related, grid stability, harmonic distortions, voltage fluctuations, grid interface, load balancing, etc. There are also some other issues with variable Renewable Energy like nature of the energy, issues with utility scale and technological issues [1, 2].

Most of such issues are related to solar PV and wind energy systems. These issues can be categorized as either technical or non-technical:

- *Technical issues*
- Issues related to Power Quality
- System stability

- Harmonics
- Frequency and voltage fluctuations
- Power Fluctuation issues
- Fluctuations for a small period (small time)
- Fluctuations for a long period (long time)
- Islanding.
- Storage/Battery requirements
- Optimal placement of Energy Sources (Renewable)
- Protection issues
- Load balancing
- *Non- Technical issues*
- Technical skilled workers are less
- Technologies related to the renewable energy sources are not encouraged for installing new power plant as a reserve
- Transmission line are not easily available for accommodation of RES.
- *Issues precise to fuel cell integration*

Fuel cells are slowly gaining attention as a renewable energy source. They convert chemical energy into electrical energy using a fuel, usually hydrogen, without any kind of pollution and is hence a clean, reliable and efficient source of energy. As it is clear that they have many advantages, yet there are still certain challenges that are faced during its integration with the grid [2, 7]. These challenges are:

- High cost
- Low durability
- Problems regarding energy management
- Problems related to system control and hybridization
- High uncertainty and non-linearity

These are not yet as commercialized as other renewable energy sources, but, much attention is being paid to overcome the disadvantages faced.

In this project, our aim is to improve power quality. Power quality is the behavior of a system in an environment (electromagnetic). The lesser the disturbance caused in the system due to the electromagnetism, the better the power quality of the system.

4 Prospective Solutions to Common Issues

- *For solar and wind energy integration issues*
- More usage of new and emerging power electronics technologies
- Water pumping system which uses solar energy to be implemented in place of standard storage system.
- Forecasting tools to be used for variable energy sources

- If the power is generated from distributed generation, that is, distributing the RE sources in small units to a larger area instead as a large concentrated unit, we can control the intermittence of power generation
- Smart grid
- Use of Maximum Power Point Tracker for reducing power fluctuations [2]
- *For fuel cell integration issues*
- Improving the material and the assembling
- Improving the performance of auxiliary system components
- Suitable modeling
- Advanced monitoring
- Appropriate control system designing [7].

5 Topology Utilized for Grid Integration

The main aim is to evaluate the concept of single source energy systems and hybrid systems connected to the grid and the ways in which it is ensured that the energy source is utilized in the best way possible. The backup needed, in case of failure of one supply source, is provided by the fuel cell and also the battery along with their respective power conditioning systems. The system becomes safer [10]. The integration is done with the help of following components: -

- A DC-DC (uni-directional) converter for obtaining regulating high voltage DC. This stops the negative current going back. It causes the reversal of cell and results in damaging of the fuel cell. The boost converter switching causes ripple current which needs to be low.
- An inverter along with a filter so that the DC voltage can be accommodated to the desired AC voltage.
- The bi-directional DC-DC converter is used for the charging and discharging of the batteries (so that the energy supply security is increased and load dynamics are improved) [4, 10].

6 Operation Modes of Grid

- **Utility supplying power directly**

  Residential consumers get direct supply from the utility through static switch

- **Pre-charge operation mode**

In the Power Conditioning System, DC capacitors of the inverter can be charged in advance through the AC bus utility. After charging, the inverter can be instantly switched on. As it starts to run, the inverter keeps the DC capacitors are allowed to charge to such a level that it becomes higher than no-load mark of the cell. Although, the utility will still be supplying the consumers during this type of operation.

- **Normal operation mode**

The DC energy from the fuel cell is converted into AC by PCS and supplies the utility and the consumers.

- **Islanding or failure operation mode**

During normal operation, if the utility is short of tolerance, PCS turns to failure operation mode. It converts DC from fuel cell and battery and takes care of the loads like emergency alarms, auxiliary equipment, etc. which are called critical loads. This is also called failure operation [16, 19].

7 System Description

In the system, the aim is to use a single inverter, used for DC-AC conversion to interface the grid-connected consumer as well as independent stand-alone consumer (Fig. 1). By obtaining the inverter control, the operation is more efficient and the power quality requirements are accomplished.

The system is different from others because of the control methods used. We attempt to experiment with new and efficient inverter control methods [3, 9, 17, 19].

The system can operate in two modes: the normal mode and the backup mode (for emergency purposes).

- *Island Operation Mode*

The power supply for loads such as auxiliary system for fuel cell, alarms, etc. is essential for operation in this mode. First stage includes the power supplied by battery, then the fuel cell. A Simulink model is designed which is effective and simple. This is an energetic model which takes into consideration, at critical load,

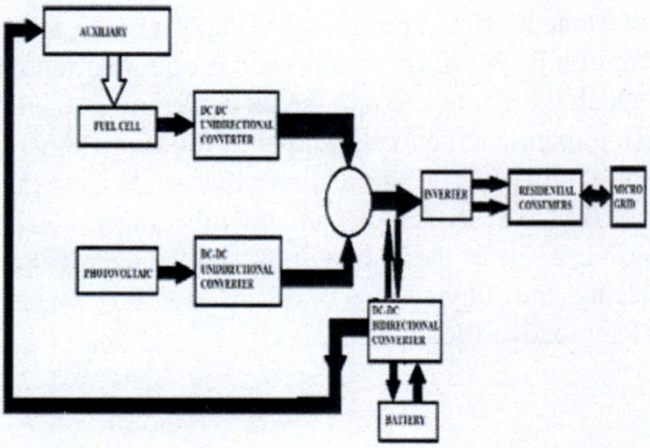

Fig. 1 Topology of the system [17, 19]

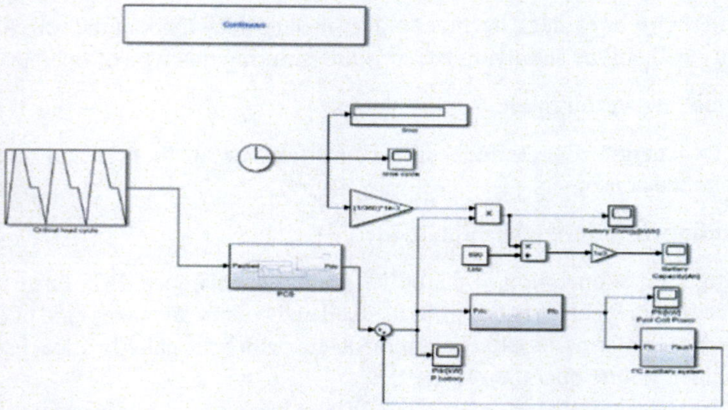

Fig. 2 Grid operation in island operation mode

Fig. 3 Power conditioning system losses [19]

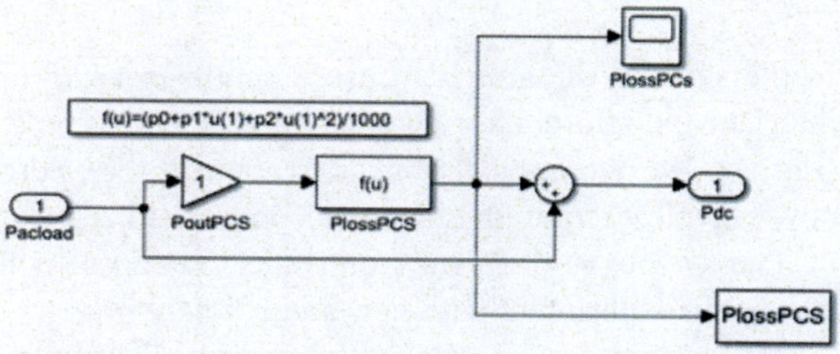

the inverter power losses. The load cycle of the consumer is represented using the 'repeating sequence' block. The required size of the battery pack can be figured out, if critical load power cycle is known. The output of the power inverter is same as the critical load power [10, 16] (Fig. 2).

The model depicting the losses in the Power Conditioning System is done using a commonly used quadratic function as shown in (Fig. 3). The critical loads' power supply must be assured by the power system. Stage 1 guarantees the power supplied from the battery connected, then the Solid Oxide Fuel Cell. So, we know the value of Pload_critical, the critical load power, and accordingly a model is designed using Simulink. We have considered an effective model which is very simple in design which takes into account the power losses of the inverter during critical load period. To implement the load cycle of the consumers, a repeating sequence block is used in the Simulink. The acquisition time which is made available for the load cycle is set as 10 s for every sample, and the time interval length is set as 2 days. If we know the power cycle of the critical load, we can decide the size of the battery pack required. Hence, the output power of the inverter, Po_inv, is the equal to the critical load power (Pac_load = Pload_critical) [10, 17, 19].

The Power Conditioning System losses ($P_{lossPCS} = P_{inv\_loss}$) by using a quadratic function are modelled, which is mostly used. The least-squares method is used for extracting power loss function which requires three parameters [17–19].

$$P_{inv\_loss} = p0 + p1\, P_{load\_critical} + p2\, P^{2^{load\_critical}} \tag{1}$$

*p*0—> load independent losses [W] are taken
*p*1—> the voltage drops in semiconductors
*p*2—> the magnetic losses [1/W] are included, which are called ohmic losses of the load.

We implement the model of the Power Conditioning System in Simulink, referring to the following:

$$f(u) = (p2 * u^2) + (p1 * u) + p0 \tag{2}$$

Where, the coefficient values are mentioned below:

$p2 = 0.01/P_r$, $p1 = 0.005$, $p0 = 0.0035{*}P_r$

(P_r being the rated power)

Power Conditioning Unit or System is used to monitor the battery voltage, solar output and load continuously and can also be used to charge battery bank.

$$P_{DC} = Po\_inv + P^{tot} P_{inv\_loss} \tag{3}$$

The input power and the output power of the inverter can be related to its efficiency:

$$\eta_{inv} = Po\_inv/P_{DC} \tag{4}$$

(Figs. 4, 5, 6, 7 and 8)

- *Normal Operation Mode*
- The Fuel Cell Power Conditioning System

It consists of stacks of fuel cell along with a DC power converter [6]. Fuel cell is a device in which DC power is directly produced from chemical energy. It is preferred as it has the advantage of high power density and does not emit any greenhouse gases.

The connection of fuel cell stacks is made in either series or parallel according to application. The key challenge faced in its design of converter is the current ripple reduction in fuel cell, also, the difficulty in maintaining a constant DC bus voltage. The first issue can be solved by the connection of an internal current-control loop

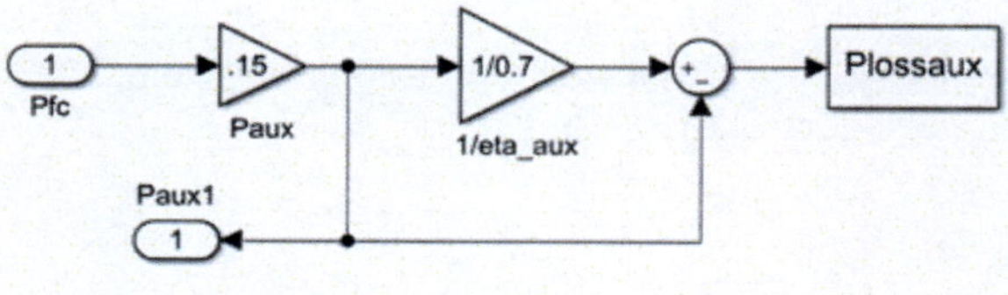

Fig. 4 Finding power loss in auxiliary circuits [19]

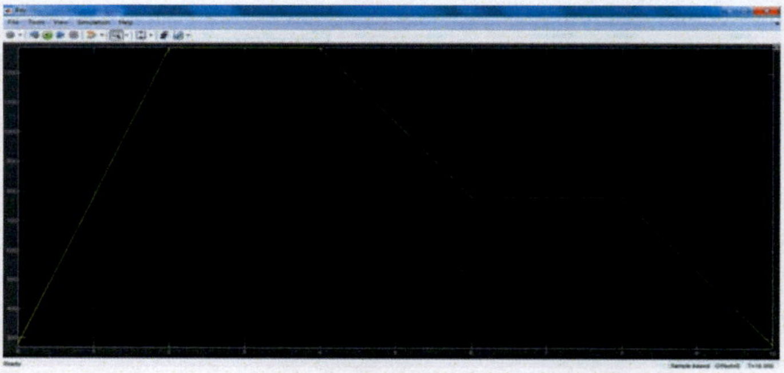

Fig. 5 Pdc, DC power waveform

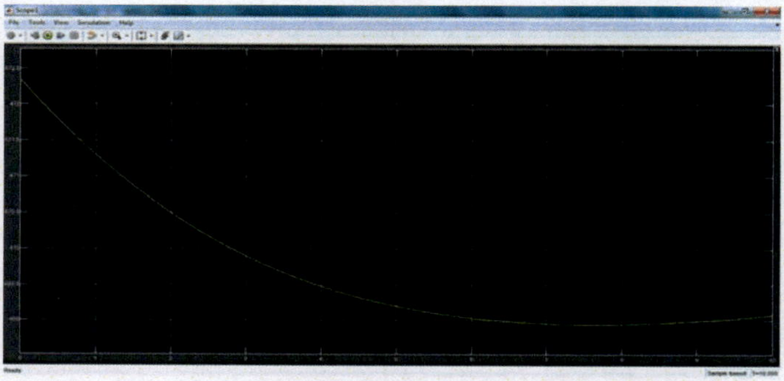

Fig. 6 Vfc, fuel cell voltage waveform

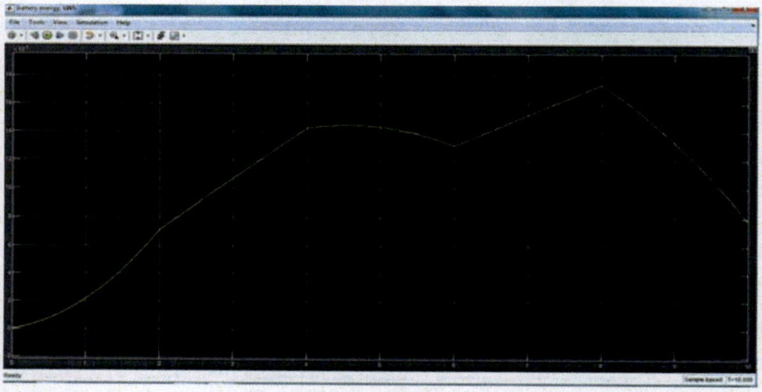

Fig. 7 Batter energy waveform [in kWh]

in the DC-DC power converter's control system. The other issue is solved using DC voltage control [6, 9, 10, 13].

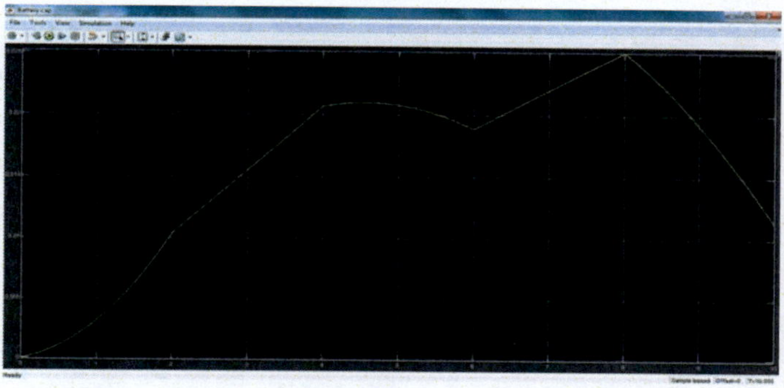

Fig. 8 Battery capacity

VARIABLE	*MEANING*	*VALUE*
T	Absolute Temperature	1273 K
F	Faraday's Constant	96.487 e6C/Kmol
R	Universal gas constant	8314J/(K mol K)
Eo	Ideal standard cell potential	1.18V
Umax	Maximum fuel utilization	0.9
Umin	Minimum fuel utilization	0.8
Uopt	Optimal fuel utilization	0.85
KH2	Valve molar constant for hydrogen	8.43 e-4 K mol/(s atm)
KO2	Valve molar constant for oxygen	2.52 e-4 K mol/(s atm)
KH2O	Valve molar constant for water	2.81 e-4 K mol/(s atm)
Tau_H2	Response time for hydrogen flow	26.1s
Tau_O2	Response time for oxygen flow	29.1s
Tau_H2O	Response time for water flow	78.3s
Ro	Ohmic loss per cell	3.2813e-004Ω
Te	Electric response time	0.8s
Tau_f	Fuel processor response time	5s
rHO	Ratio of hydrogen to oxygen	1.145
I	Initial current	100A
r	Load resistance	10 Ω

- Battery Power Conditioning System

It consists of a DC-DC power converter along with the battery pack, which produces load power (critical) and produces variable DC power. So, each battery is connected to other in either series or parallel combination as desirable [14].

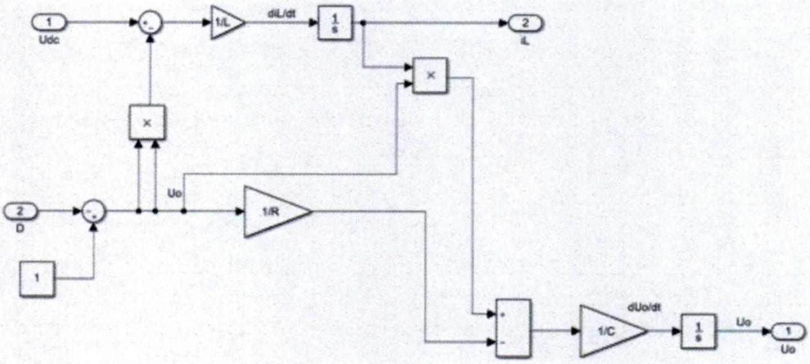

Fig. 9 Implementation of boost converter in simulink [19]

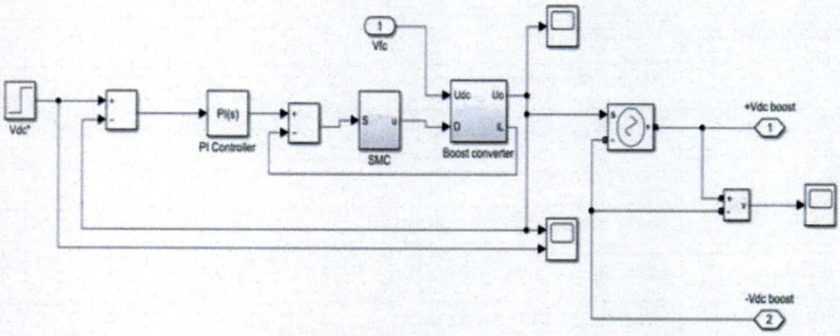

Fig. 10 Boost converter control [5, 19]

The battery model in MATLAB/Simulink is used. The charging and discharging of the battery according to the flow power as available is considered the main issue for its converter. The solution to this problem is similar to that of the fuel cell connection, which is, addition of an internal current loop (Fig. 9).

A boost converter is that converter which steps up the DC voltage while reducing current. It's a DC-DC converter. It contains at least one storage element and atleast two semiconductors. It can be either capacitor or inductor or both. Filters made of these elements reduce voltage ripple [4, 5] (Figs. 10, 11 and 12).

It's a type of DC-DC converter which can have the voltage magnitude of its output either greater or lesser than that of input.

Current controller: It assures the fast reference tracking at the same time while it delivers the duty cycles (represented by D), that is, if the inductor current reference is imposed. If an anti-parallel diode for every active power device introduced, we get a bi-directional buck-boost converter [4, 10, 12, 15].

Connection of the buck-boost converter with the battery stack (Ud = Udc) is a must for the power inverter system. The power demand of the consumers is more than the power which is generated and obtained from the fuel cell generator when it starts operating. One of the reasons we use the buck-boost converter is that it is used in recharging of the batteries from the other sources which are available. Because of

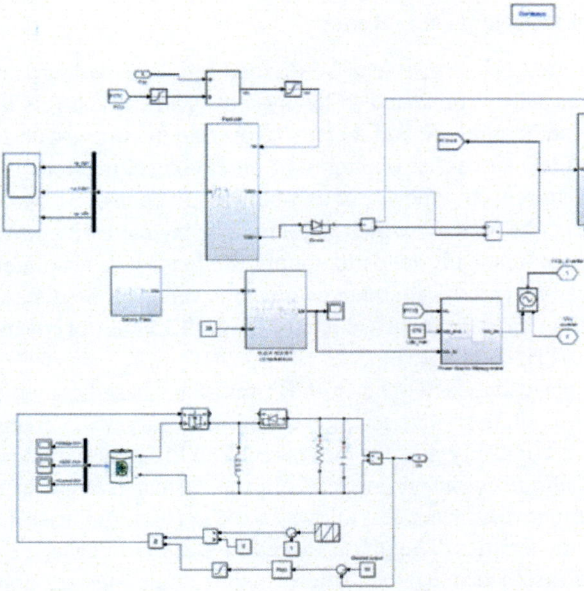

Fig. 11 Implementation of SOFC with buck boost converter in simulink [6]

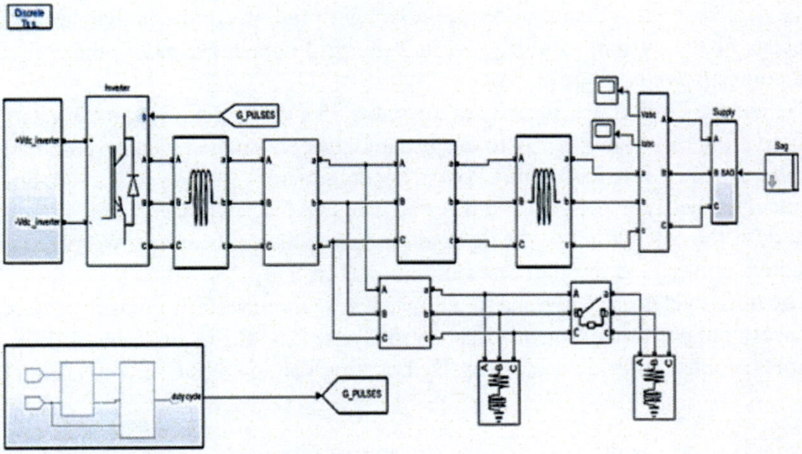

Fig. 12 Grid power inverter For RES with DC link load current estimator

the current controller for the buck-boost converter, the current of inductor will follow the current taken as reference. It is found that the delay of 001s is obtained in the output current. U_d, the input voltage, is about 390 V DC and then sent directly from the battery stack, and U_o, the output voltage, becomes 690 V DC [3, 10, 12, 19].

- modelling and control of the grid power inverter

The methods to control the grid inverter can be divided into two- voltage and current control. The inverter, when connected to the system (grid), it controls the inverter output's frequency and amplitude and it starts to operate in current-control mode. This mode of control leads to either obtaining of other control methods like the control of active power and reactive power or can be voltage control. If due to improper network parameters, the network which is going to be injected with power is unavailable, the inverter automatically tends to supply the load. It is absolutely unaffected by network blackouts and the alternative voltage is supplied. Such cases result in the inverter controlling the voltage. phase-locked loop (PLL) control ensures that the frequency remains 50 Hz [5, 10, 15].

The converter used here is an supposed to be an IGBT transistor-based converter (full-bridge) which tends to operate in inverter mode. This is because, the energy transfers either to the load and/or utility grid via the source. The PLL starts tracking the grid voltage due to synchronization purposes where the system is allowed to operate in grid-connection mode. But, when it is allowed to be in island/failure mod, the VSI cannot track its characteristics. The phase-locked loop starts to change the frequency, sending it to the installed integrator, where angle calculations are done while switching between the frequencies obtained from filter and frequency from the other reference which is fixed. The VSI must have one external frequency reference that should be provided, when it is in the island mode of operation. The PLL, is the key catalyst when disturbances are no longer present, for the re-synchronization purposes of the system to utility. During the grid connected mode, the frequency used comes from the filter [3, 5].

We use the Grid Power Inverter of the value 35 kVA for RES Integration and also for delivering of power directly to the grid and the consumers (3-phase parallel R-L-C load connected for simulation). Also a boost inductor (3-phase series R-L branch) is connected between the grid and inverter. The Feedback Signals Acquisition block is used for the calculation of the dq components of the grid current. We get to know the active power load through this simulation (Fig. 13).

The duty-cycles are delivered to the inverter by the inverter's control system. For the inverter to perform efficiently, its control system should be made carefully so that complete control of grid is achieved. By boosting voltage of DC-link to such a level

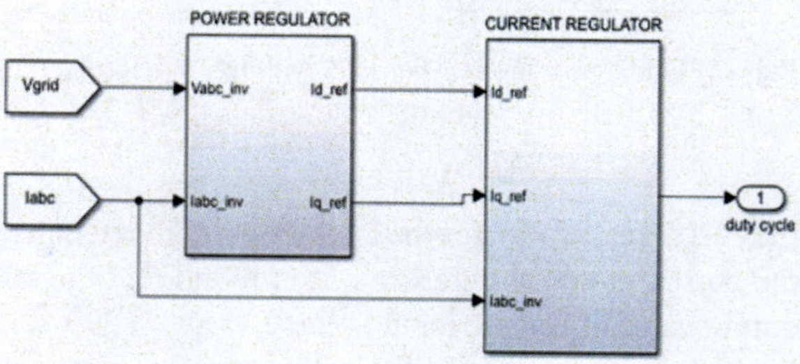

Fig. 13 Converter control system

so that it is more than grid's line voltage. To prevent this voltage from varying, it is necessary to control the inverter's power flow.

The structure or the type of control used here is vector control. The addition of load current feed-forward component is done for increasing the dynamic response.

For balancing the power of the system, the current is supplied to utility by the PWM inverter. The power flow is also controlled in PCS. SRF (the synchronous rotating frame) is required to control the active and reactive power.

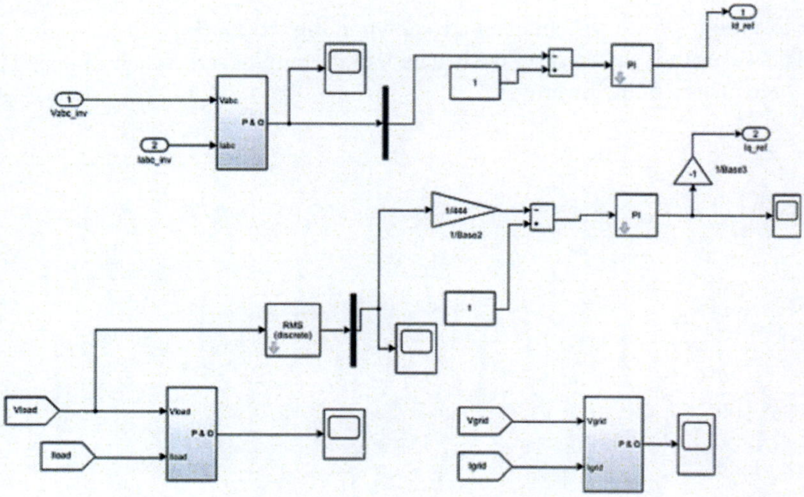

Fig. 14 Power regulator

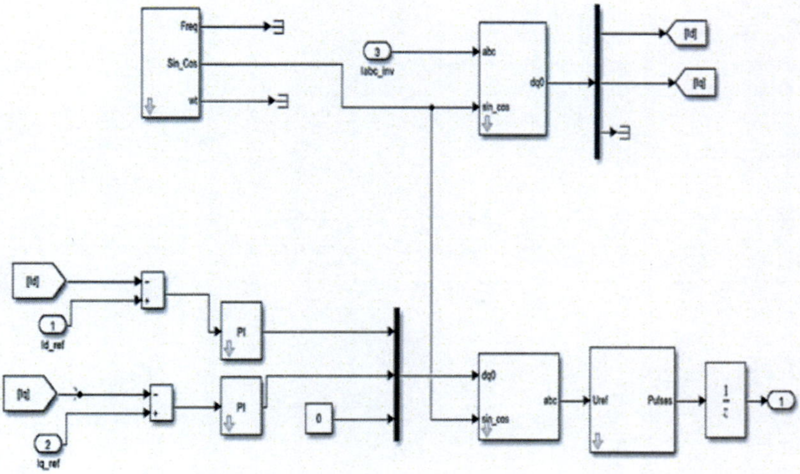

Fig. 15 Current regulator

We use the PI controller [5] because of its good performance when working with the inverter, even though it is sensitive to the parameters (Fig. 14).

The phase and the gain margins are the terms which helps measuring the stability (relatively) (Fig. 15).

8 Results

The resulting waveforms obtained are shown in Fig. 16(a), (b).

The waveforms shown show the behavior of the current and voltage obtained from the three phases of the inverter.

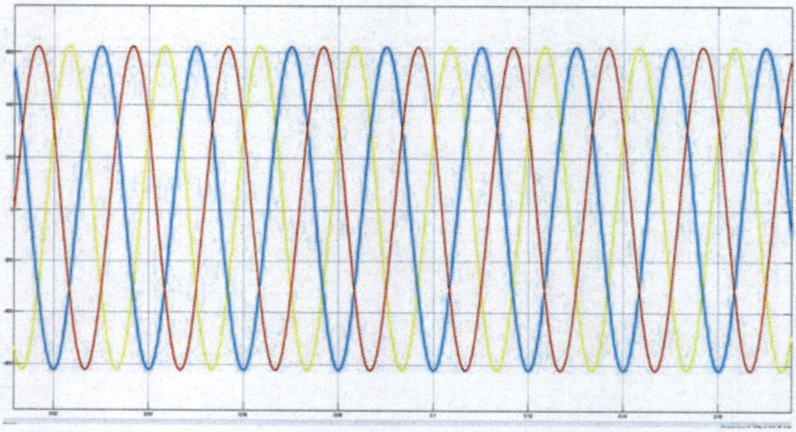

(a)

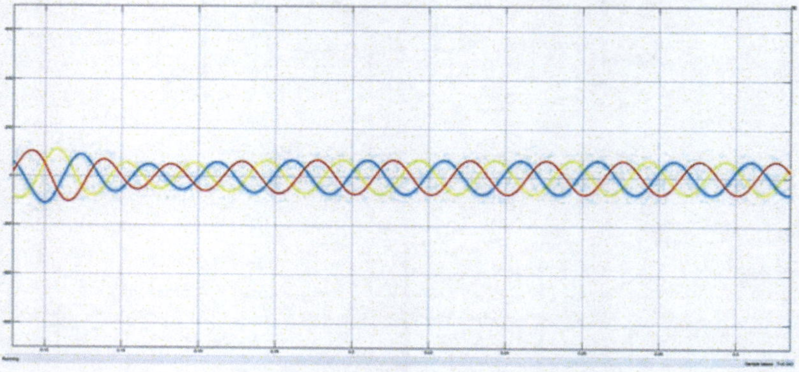

(b)

Fig. 16 **a** Grid voltage waveform **b** Grid current waveform

The waveforms obtained depend on the input values set for fuel cell, reference values, and values for inverter control.

The obtained values of output current and voltage is satisfactory, but, current waveform shows slight deviation in the beginning.

9 Conclusion

The project contributes to the renewable energy sources integration with the grid and other research work so that the projects can be made to produce better results.

Appropriate models were created for fuel cell and battery power conditioning systems, boost and buck-boost converter, and were implemented along with the inverter and connected to grid. Also, adequate controllers were designed for all the integrated systems for efficient operation of the components connected to the grid. This not only increases the stability of the system but also the power quality improves. Along with this, all connections were made to the inverter before connections to the grid were made to simulate grid integration. We obtained satisfactory results/waveforms, although current waveform can be improved.

References

1. Alsaif AK (2017) Challenges and benefits of integrating the renewable energy technologies into the AC power system grid Am J Eng Res (AJER) 6(4):95–100. e-ISSN: 2320-0847 p-ISSN: 2320-0936
2. Kumar V, Pandey AS, Sinha SK (2016) Grid integration and power quality issues of wind and solar energy system: a review. In: International Conference on Emerging Trends in Electrical, Electronics and Sustainable Energy Systems (ICETEESES–16)
3. Fouad MA, Badr MM, Ibrahim MM (2017) Modeling of micro-grid system components using matlab/simulink. Glob Sci J 5(5)
4. Bandyopadhyay S, Panda GK, Saha PK (2015) Non-linear control techniques for DC-DC buck converter. Int J Adv Res Electr Electron Instrum Eng 4(5)
5. Venumadhav Y (2018) Performance analysis of PI controller based grid connected renewable energy system. Int Res J Eng Technol (IRJET) 05(09)
6. Patel HB (2017) Mathematical modeling and simulation of solid oxide fuel cell power system for distributed generation J Emerg Technol Innov Res (JETIR) 4(5)
7. Li Z, Zheng Z, Xu L, Lu X (2019) A review of the applications of fuel cells in microgrids: opportunities and challenges. Li et al. BMC Energy 1, 1–23
8. Gaiceanu M (2017) Real-time implementation of the advanced control of the three-phase induction machine based on power inverters. https://doi.org/10.5772/intechopen.68217
9. AbdelHady R (2017) Modeling and simulation of a micro grid-connected solar PV system. Natl Water Res Center. https://doi.org/10.1016/j.wsj.2017.04.0011110-4929
10. Rajesh K, Kulkarni AD, Ananthapadmanabha T (2015) Modeling and Simulation of Solar PV and DFIG Based Wind Hybrid System Procedia Technol 21:667–675
11. Sakthivel K, Jayalakshmi V, Rajakumari G (2019) Modeling and simulation of a GRID-TIED solar PV system. Int J Recent Technol Eng (IJRTE) 7(6S2). ISSN: 2277-3878
12. Nithish M, Srikar M, Manjula A (2016) Performance enhancement of actively controlled hybrid DC microgrid incorporating pulsed load. Int J Eng Sci Comput 6(12)

13. Jayalakshmi V, Kumar S (2018) Simulation for smart grid integration of solar/wind energy conversion system. Int J Pure Appl Math 119(12) 2018
14. Corcau JI, Dinca L, Grigorie TL, Tudosie AL (2017) Fuzzy energy management for hybrid fuel cell/battery systems for more electric aircraft. In: AIP Conference Proceedings, 1836:020056
15. Anjali T, Lakshmi VSG (2017) Grid synchronization of a microgrid with renewable energy sources and storage assisted with controller area network. Int J Res Appl Sci Eng Technol (IJRASET) 5:9653. ISSN: 2321-9653
16. Soni KC, Belim FF (2015) MicroGrid during grid-connected mode and islanded mode - a review. In: National Conference on Recent Research in Engineering and Technology (NCRRET-2015), International journal of advance engineering and rresearch development (IJAERD) e-IS
17. Gaiceanu M, Buhosu R, Ghenea I, Vidan C (2017) Complete regenerative distributed drive system. In: 2017 5th International Symposium on Electrical and Electronics Engineering (ISEEE), pp 1–6. Galati, Romania. https://doi.org/10.1109/ISEEE.2017.8170687
18. Gaiceanu M, Epure S, Dache CR, Buhosu R, Ghenea I, Vidan C (2017) Laboratory power inverter platform for variable speed drive. In: 2017 5th International Symposium on Electrical and Electronics Engineering (ISEEE), pp 1–6. Galati. https://doi.org/10.1109/ISEEE.2017.8170689
19. Katsikis V MATLAB: A Fundamental Tool for Scientific Computing and Engineering Applications-Volume 3, eBook (PDF) ISBN: 978-953-51-5708-3. https://doi.org/10.5772/3339

A Survey on Data Security Challenges and Their Solutions in Cloud Computing

Ruchi Billore and Manish Pandey

Abstract As technology increases rapidly, the demand for cloud computing is also growing day by day due to its features like on-demand services, cost efficiency, and rapid elasticity. The work which was traditionally done on-premise, now performed remotely across the internet or off-premise. And with increasing popularization of cloud computing the security becomes a major concern. Because user stores their confidential data on the cloud so they require high-security policies from cloud service providers to keep user's data safe and confidential. In this paper, we are going to explain some major threats associated with data in cloud computing at all stages of the data life cycle and also discuss some solutions to emerge from the threats and these solutions are discussed based on structural properties of the cloud storage system.

Keywords Cloud computing · Cloud service provider · Cloud consumer · Data life cycle

1 Introduction

Cloud computing is a computing which provides on-demand, self-managed and efficient use of virtual resources. In recent years, cloud computing and storage have developed into well-liked topics. These two changing our way of living and have increased production efficiency in some areas. To use cloud base services, one requires only a decent internet connection. At present, if we store data locally, we've to accommodate numerous overheads along with the risks. Because of the defined storage resources and need for convenient usage, we select cloud servers to accumulate every kind of information, which is simple to access for companies and organizations too. For everybody, the cloud server gives an open and favorable storage platform, but it also announces risks of data security. Cloud computing is becoming famous for its three distinguishing qualities. Firstly, it can scale up or

R. Billore (✉) · M. Pandey
Computer Science Engineering, Maulana Azad National Institute of Technology, Bhopal, India

V. Nath and J. K. Mandal (eds.), *Proceeding of Fifth International Conference on Microelectronics, Computing and Communication Systems*, Lecture Notes in Electrical Engineering 748,
https://doi.org/10.1007/978-981-16-0275-7_25

scale down quickly as per the user requirement. Secondly, users have to pay only for what they use and require no capital expenditure to start using cloud services. And finally, it is self-service for which there is no need for IT experts to install or use any resources. Along with these, cloud computing also provides fast access to resources and applications, it gives you the latest applications for use without wasting your time and expenditure on installations.

As every coin has two sides, which means along with these many pros of clouds, there are some major cons associated with data security, as cloud consumer even stores confidential data on the cloud such as documents, organizations policies, etc. Due to easy and remote access control, i.e., we can use our data from anywhere in the world where the only requirement is a good internet connection, this feature attracts users the most and makes cloud computing popular for storing data and sharing information through it, due to which security becomes major concern among cloud service providers and we are going to explain some of the data security-related concerns of cloud computing and further techniques used to resolve security risks and make cloud computing more secure and most trusted third-party resource to store data.

Data of consumers are stored on a shared cloud system rather than the devices of cloud service providers, so consumers don't have the knowledge about how the data is stored on the cloud or how their data is encrypted and stored on the cloud to maintain integrity and confidentiality of data so that only authorized person can access it and if an attacker somehow manages to get the data, still the information cannot be decrypted easily. The NIST defines cloud computing as: "Cloud computing is a model for enabling ubiquitous, convenient and on-demand network access to a shared pool of configurable computing resources (e.g., networks, servers, storage, applications, and services) that can be rapidly provisioned and released with minimal management effort or service provider interaction. This cloud model is composed of five essential characteristics, three service models, and four deployment models" [1] explained in Fig. 1.

A. *Deployment models:*

- **Public Cloud:** In a public cloud, the cloud service provider owns the cloud infrastructure and makes it available to every user. By tapping on the public cloud, consumers can get new capabilities on demand just by paying the subscription fee for the resources they wish to use.
- **Private Cloud:** In a private cloud, a single organization owned the cloud infrastructure. It may be handled by the organization itself or by a third-party, where the storage network and hardware assume the top levels of security. Only the clients who owned this private cloud are able to access the data stored in the data centre.
- **Community Cloud:** In community Cloud, cloud infrastructure is shared among different organization (ex. security requirements, policy, etc.). It would be handled by either organization or by a third party and may be located locally or across the internet.

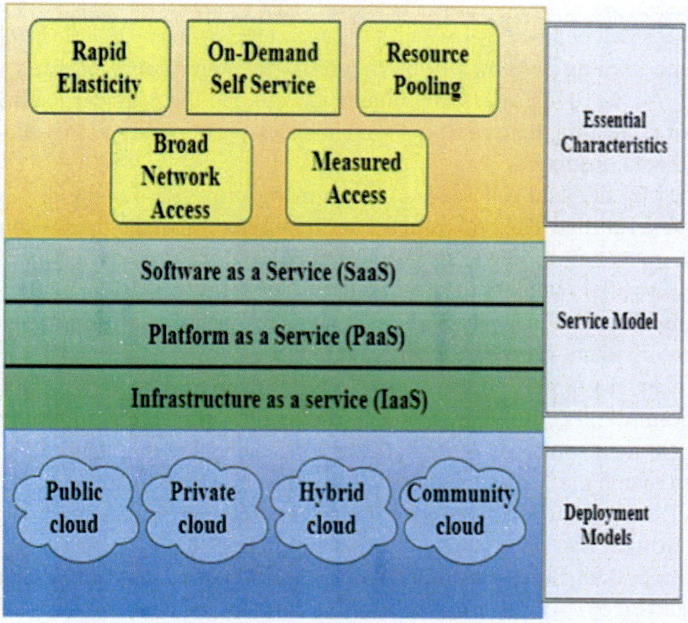

Fig. 1 Cloud model

- **Hybrid Cloud:** In a hybrid cloud, the cloud infrastructure is a combination of public and private cloud in a single environment and by standardized or proprietary technology. Clouds are bound together that permits application and data portability.

B. *Service models:*

- **Software as a Service (SaaS):** This service is totally managed and hosted by the cloud service providers in which a cloud consumer is directly going to use the applications. Consumers can use the services via a web browser or mobile application.
- **Platform as a Service (PaaS):** PaaS is concerned with abstraction and providing integrated development environments or platforms such as databases, application platforms (ex. to run Python, PHP, or other code), file storage and collaboration, or even proprietary application processing. The main inequality lies in the fact that with PaaS, there isn't any load of handling the underlying networks, servers, etc.
- **Infrastructure as a Service (IaaS):** IaaS is responsible for providing the infrastructure where the consumer can deploy and runs arbitrary software such as operating systems and applications. IaaS provides virtual resources to the consumers on a pay per usage basis so that consumers do not require to buy and maintain their own physical servers, they use resources in a convenient and as per their requirement.

C. *Essential characteristics:*

- Resource pooling is known to be the most elementary characteristic where the cloud service provider basically hide resources and the provider itself is responsible for collecting them in a pool which is then partitioned so as to allocate them to different consumers.
- By using on-demand self-service, Consumer's provision for the resources from the pool is fulfilled They themselves are required to manage their resources. The administrator has no role to play in this task.
- Broad network is actually the availability of the resources over a network without the requirement of any physical access. An important point to note here is that the network is not necessarily part of the service
- Rapid elasticity simply allows consumers to expand or contract the resources, i.e., provisioning and de-provisioning of resources they use from the pool of resources. This task, as it sounds is quite delicate but is accomplished automatically. This helps to match the level of resource consumption with the demands.
- The usage of resources is a measured service. Which is used to monitor and report that consumer can only use what they are allotted. And utility computing term comes from here. Since the cloud computing resources can now be consumed like water and electricity, so similarly user have to pay for what they use.

The remaining paper flow: Sect. 2 introduces the different types of data security threats, risks, and challenges present in cloud computing, Sect. 3 contains the solution to those problems discussed in Sect. 2, and then finally, Sect. 4 concludes the whole paper.

2 Related Works

Many researchers have published their paper on cloud computing by taking different aspects like: observed attacks, identified vulnerabilities, and some suggested remedies.

1. "Morsy et al. [18] explained cloud security issues in virtualization and service-oriented technologies along with security dimensions related to isolation and multitenancy. Based on the analysis they recommended for an integrated and adaptive configuration-based security model. However, their study does not have a detailed analysis of the security requirements, related threats, associated vulnerabilities, and corresponding countermeasures, and their mappings."
2. Zhang et al. [19] defined cloud architecture into four layers to provide the services of cloud. And highlighted some security measures as the challenges for the cloud like: Automated security provisioning, virtual machine migration, hardware server consolidation, energy management, software framework, data security, storage technologies.

3. Pearson [20] to address the cloud related trust and privacy issues, the classical techniques are no longer flexible. Also defines the relation between the threats, vulnerabilities and privacy issues.
4. Modi et al. [25] discussed security issues in the cloud architecture deliberated as a layered architecture. However, by taking the data storage and data life cycle management techniques into account they could have given more privacy-related issues and their solutions.
5. Dahbur et al. [24] provide guidance to address the threats, vulnerabilities, risks that cloud environment brings itself. However, they did not provide the insight view of these issues.
6. Cong Wang et al. [21] discuss about data security issues and proposed a scheme provides data manipulation and security.
7. Subashini and Kavitha [23], they give study of security issues on the basis of delivery models of cloud. Explains different threats possible on each model (SaaS, PaaS, and IaaS).
8. Raj Kumar [22] explained about the security threats using transmission of information in cloud computing. Encryption and decryption techniques for providing to the cloud computing.

3 Security Threats in Cloud Computing

3.1 Threats

In cloud computing, trust is a major issue between the cloud service provider and the cloud consumer and also raises the number of security issues. The cloud security alliance (CSA) did a survey aimed to raise awareness of threats, risks, and vulnerabilities in the cloud and find the top 11 threats listed in the below Table 1:

3.2 Security Risks of Data

From the generation of data to the destruction of it, the data goes from many stages called data life cycle [2, 3]. The data life cycle can be divided into 7 stages and these seven stages are categorized into 3 layers [4] illustrated in Fig. 2, Many data security issues in the cloud is involved in all stage of the data life cycle some of them are discussed:

[1] **Data creation:** The data creation phase involves two activities, creating new data sets or manipulating older data. Only authorized users can make changes to the existing data. In cloud computing, cloud consumers and cloud service providers communicate with the help of interfaces and APIs [5]. Cloud computing posses certain data protection risks for cloud entities. Between the

Table 1 List of threats

Threat no.	Name of threat
1.	"Data Breaches"
2.	"Misconfiguration and Inadequate Change Control"
3.	"Lack of Cloud Security Architecture and Strategy"
4.	"Insufficient Identity, Credential, Access and Key Management"
5.	"Account Hijacking"
6.	"Insider Threat"
7.	"Insecure Interfaces and APIs"
8.	"Weak Control Plane"
9.	"Metastructure and Applistructure Failures"
10.	"Limited Cloud Usage Visibility"
11.	"Abuse and Nefarious Use of Cloud Service"

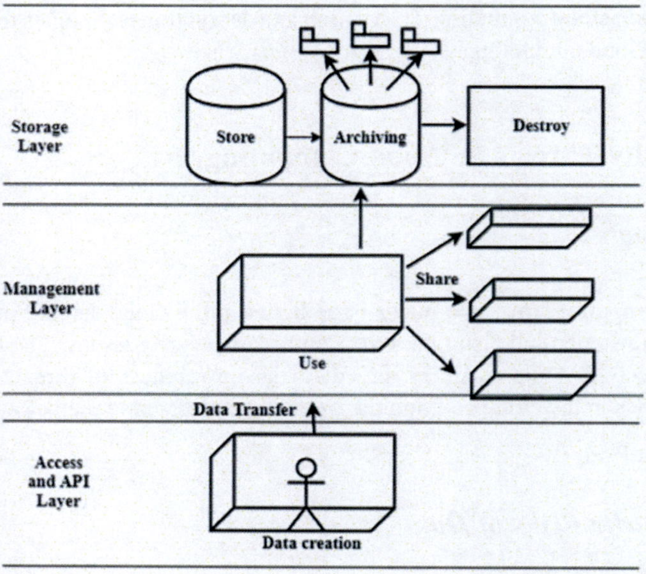

Fig. 2 Data life cycle

cloud entities, different kinds of service-level-agreement (SLA) are involved which lead to certain kinds of data leaks. It often happens, that checking the data handling practices of the provider becomes difficult for the cloud user [5].

[2] **Data transfer:** By default, data transmission is done through transport level security (TLS), which transfers data in a secured channel [6]. But there may be some chances of issues and threats possible. Tracing the path of data, auditing. It is tedious to trace the path because of the non-linear behaviour of the cloud

[7]. While data is transferred an eavesdropper can take confidential data and may inject some malicious information with original data. Therefore, along with the confidentiality, integrity of data is also required in the cloud.

[3] **Data use:** To gain better data access performance, the management layer caries out the task of collaboration between various storage devices through the cluster computing, distributed file system, and grid computing technology [4]. When a user accesses the data present in the cloud, to protect data from unauthorized user access is a major issue. Because once the attacker gets access privileges the privacy of data is broken and the attacker can change the encryption keys to prevent authorized users from the use services. Because of the metered feature of cloud computing many organizations adopting cloud computing without knowing about the threats [5].

[4] **Data share:** Increasing the usage range of data and information is made accessible between users, customers, and partners [8]. When data owner authorizes data access to other partners there may be chances that malicious insiders can take advantage and misuse the information and cannot maintain original security measures and usage restriction.

[5] **Data storage:** In the cloud, the user is not aware where their data is stored in the cloud due to the shared resources property of cloud i.e. data location is unknown, which leads questions of security, legal and requirement of regulatory compliance [6]. To store the massive amount of data, the data erasure technique is used by the service provider to satisfy users demand for storage. Due to this data integrity, transparency and availability issues arise in the cloud [4]. Multitenancy is also the main characteristic provided by cloud. Therefore, multiple consumers can store data on the same platform which makes data more vulnerable to data breach and loss of data [9].

[6] **Archiving:** The process of identifying and moving older inactive data to an independent storage device for long-term possession is called data archival. Archived data is important and can be used in the future so it focuses on the storage system. If storage media is portable then media is out of control and there is a risk of data leakage. If off-site archiving is not provided by the cloud service provider, the availability of the data will be threatened [2].

[7] **Destroy:** When data is no longer in use it is permanently destroyed using physical or digital means [8]. The data deleted may be still present and can be restored and inadvertently disclose sensitive information, due to the physical characteristic of storage medium [2].

4 Solutions for Security Issues

4.1 *Access and Application Programing Interface (API) Layer*

Protect the system from unauthorized user access. Before the user starts using cloud services, it is necessary to identify them whether the user is authorized or not. Roy I and Ramadan gave a fourth privacy protection system named "Airavat," which can prevent system from privacy leakage. This system uses decentralized information flow control (DIFC) and differential privacy protection protocols [10]. Khalid, develop a protocol for authentication and authorization. This protocol gave information about anonymous users and by using this, we can protect systems from unauthorized user access [6]. Lalitha proposed a work in which if the malicious user tries to access the system then an alert goes to admin with IP address through which admin can scrutinize the activities of the malicious user and can also block that IP address to prevent the system from unauthorized access [16]. Arora, build a secure cloud ecosystem, in which 2-step verification is applied, the first user enters the username and password after this they get one-time-password (OTP) to prove their identity, if the attacker gets password then also, he cannot enter into the system. For data transmission, SSL and TSL 1.2 are used and at the end uses a hybrid cryptographic system for encryption of data [9]. Puthalin proposed a security verification scheme using Dynamic prime numbers. Which reduces communication overhead [6].

4.2 *Management Layer*

To protect data from illegal user access, the basic requirement is the encryption of data. Which provides confidentiality to the user data. But the problem with encryption is the management of keys. User is not able to manage large amount keys therefore cloud service provider needs to maintain encryption keys. The Organization for the Advancement of Structured Information Standards (OASIS) Key Management Interoperability Protocol (KMIP) working to overcome key management problems [12]. With personal privacy information protection sharing of data is a challenge. Randike Gajanayake proposed a framework for privacy protection depends on Information Accountability (IA) components. IA agents monitor the activity of users. When the user tries to de illegal activity, IA agents detect it and apply some methods that force the user to accept his mistake [17]. Dong, defines a security policy for effective, scalable and privacy-preserving data sharing. The policy is made with the combination of two techniques i.e. CP-ABE and IBE [6].

4.3 Storage Layer

In cloud computing, it is necessary to use a secure storage system. R. Nivedhaa and J. Jean Justus, explain a secure erasure coding technique in which users data is divided and then encrypted using advanced encryption standard algorithm and proxy re-encryption [14]. Mowbray proposed a client-based security management tool. In which the user can configure it according to their requirement and can store and use their confidential information in the cloud in a secured manner [13]. Wei proposed a first protocol that makes storage and computation secure [15].

5 Conclusion

Over time, the use of cloud computing and inventing new features and techniques in it are increasing rapidly due to its prominent characteristics. At the same time, data security risks, threats, and challenges are also increasing and need to be solved as soon as possible. Because cloud computing is now used in health care, banks, private and public organizations, and many other fields. Where the user stores their confidential data, which cannot be compromised at any cost and for that it requires the highest security measures for data. In this paper, analysis of some data security threats and their solutions over the stages of the data life cycle and structural properties of the storage system are covered.

References

1. Mell P, Grance T. The NIST Definition of Cloud Computing, Version15, 10-7-09, http://www.wheresmyserver.co.nz/storage/media/faq-files/cloud-def-v15.pdf
2. Chen D, Zhao H (2012) Data security and privacy protection issues in cloud computing. In: 2012 International Conference on Computer Science and Electronics Engineering, Hangzhou, pp 647–651
3. The-Egregious-11-Cloud-Computing-Top-Threats-in-2019-Feb 2020: https://cloudsecurityalliance.org/press-releases/2019/08/09/csa-releases-new-research-top-threats-to-cloud-computing-egregious-eleven/
4. Zhe D, Qinghong W, Naizheng S, Yuhan Z (2017) Study on data security policy based on cloud storage. In: 2017 IEEE 3rd international conference on big data security on cloud (bigdatasecurity), ieee international conference on high performance and smart computing (hpsc), and ieee international conference on intelligent data and security (ids), Beijing, 2017, pp 145–149
5. Hendre A, Joshi KP (2015) A semantic approach to cloud security and compliance. In: 2015 IEEE 8th International Conference on Cloud Computing, New York, NY, pp 1081–1084
6. Subramanian N, Jeyaraj A (2018) Recent security challenges in cloud computing. Comput Electr Eng **7**(2018), 28–42. http://dx.doi.org/10.1016/j.compeleceng.2018.06.006
7. Bhadauria R, Sanyal S (June 2012) Survey on security issues in cloud computing and associated mitigation techniques. Int J Comput Appl (0975-888) June 2012
8. https://assets.cloudsecurityalliance.org/legacy/wp-uploads/2011/09/Domain-5.docx

9. Arora A, Khanna A, Rastogi A, Agarwal A (2017) Cloud security ecosystem for data security and privacy. In: 2017 7th international conference on cloud computing, data science & engineering - confluence, Noida, pp 288–292
10. Cloud computing security. http://en.wikipedia.org/wiki/Cloud_ computing_security
11. Aich A, Sen A, Ranjan Dash S. A Survey on Cloud Environment Security Risk and Remedy. Computer Science and Engineering, KIIT University Bhubaneswar, India{acrana.dgp09,alosen10}
12. OASIS Key Management Interoperability Protocol (KMIP) TC. http://www.oasis-open.org/committees/tc_home.php?wg_abbrev=kmip
13. Bowers KD, Juels A, Oprea A (2009) Proofs of retrievability: Theory and implementation. In: Sion R, ed. Proc. of the 2009 ACM Workshop on Cloud Computing Security, CCSW 2009, Co-Located with the 16th ACM Computer and Communications Security Conf., CCS 2009. Association for Computing Machinery, New York, pp 43–54 https://doi.org/10.1145/1655008.1655015
14. Nivedhaa R, Justus J.J (2018) A secure erasure cloud storage system using advanced encryption standard algorithm and proxy re-encryption. In: 2018 international conference on communication and signal processing (ICCSP), Chennai
15. Wei L, Zhu H, Cao Z, Dong X, Jia W, Chen Y, Vasilakos AV. Security and privacy for storage and computation in cloud computing. Inf Sci 2014 **258**:371–86. www.elsevier.com/locate/ins
16. Lalitha VP, Sagar MY, Sharanappa S, Hanji S, Swarup R (2017) Data security in cloud. In: 2017 international conference on energy, communication, data analytics and soft computing (ICECDS), Chennai, 2017, pp 3604–3608
17. Gajanayake R, Iannella R, Sahama T (2011) Sharing with care an information accountability perspective. IEEE Internet Comput **15**:31–38
18. Morsy MA, Grundy J, Müller I (2010) An analysis of the cloud computing security problem. In: Proceedings of APSEC 2010 cloud workshop, pp 1–6
19. Zhang Q, Cheng L, Boutaba R (2010) Cloud computing: state-of-the-art and research challenges. J Internet Serv Appl **1**(1), 7–18. http://dx.doi.org/10.1007/s13174-010-0007-6
20. Pearson S (2013) Privacy, security and trust in cloud computing. In: Pearson S, Yee G (eds) Privacy and Security for Cloud Computing, Springer, London, UK, pp 3–42. ISBN: 978-1-4471-4189-1, http://dx.doi.org/10.1007/978-1-4471-4189-1_1
21. Wang C, Wang Q, Ren K (2015) Ensuring data storage security in cloud computing. US National Science Foundation, pp 1–4
22. Kumar Raj (2015) Research on cloud computing security threats using data transmission. Int. J. Adv. Res. Comput. Sci. Softw. Eng. India 5(1):
23. Subashini S, Kavitha V.: A survey on security issues in service delivery models of cloud computing. J Netw Comput Appl **34**(1):1–11. ISSN: 1084-8045. http://dx.doi.org/10.1016/j.jnca.2010.07.006
24. Dahbur K, Mohammad B, Tarakji AB (2011) A survey of risks, threats and vulnerabilities in cloud computing. In: Proceedings of the 2011 international conference on intelligent semantic web-services and applications, ISWSA 2011. ACM, New York, pp. 1–6, ISBN: 978-1-4503-0474-0. http://dx.doi.org/10.1145/1980822.1980834
25. Modi C, Patel D, Borisaniya B, Patel A, Rajarajan M (2013) A survey on security issues and solutions at different layers of cloud computing. J Supercomput **63**(2), 561–592, ISSN: 1573-0484. https://doi.org/10.1007/s11227-012-0831-5

Analysis of Hybrid Vehicle Battery Models for Real-Time Parameter Estimation

Kausik Bhaumik and Divya Asija

Abstract A hybrid electric vehicle (HEV) consisting of primary source and usually a high voltage battery pack requires a stable battery management system (BMS). Also in electric vehicles (EVs) high power batteries are used. So to ensure safe and reliable operation BMS is necessary. This paper includes popular battery models for BMS & compares them accordingly along with some estimation techniques for SOC (state of charge) & SOH (state of health) are comprehensively surveyed. By comparing simulation results & measurements we choose the best model available and discuss their applicability in real-time control of Electric/hybrid vehicles. In this contribution the influence of time varying quantities such as charge-discharge cycle and capacity fade and its effect on EV batteries are also surveyed followed by a given severity index is used to quantify the extent of damage caused by the time varying quantities.

Keywords HEVs · BMS · Battery modelling · Battery state estimation

1 Introduction

As in 21st century we are very much concerned about the availability of fossil fuel and environmental needs, electric vehicles (EVs) and hybrid electric vehicles (HEVs) have been widely regarded as the most emerging needs in automobile industries. The history of hybrid cars may be much longer but in the last decade there is a rapid development in hybrid cars. In last ten years we customers, have begun interest in HEVs by either looking for a way to save fuel or in attempt to help reduce the negative effects caused on the environment. Batteries used for EVs & HEVs require particular care for smooth and reliable operation of the EVs as well as to prevent some improper operations which may lead to significant safety issues. In [1], key technologies in the field of battery modelling and state estimation are reviewed. On the other hand [2] suggests the BMS monitoring of the system to operate it correctly

K. Bhaumik (✉) · D. Asija
Amity University, Uttar Pradesh, India

V. Nath and J. K. Mandal (eds.), *Proceeding of Fifth International Conference on Microelectronics, Computing and Communication Systems*, Lecture Notes in Electrical Engineering 748,
https://doi.org/10.1007/978-981-16-0275-7_26

& perform the safety steps if any hazardous situation occurs. According to [3], accurate battery model helps the battery utilization more efficiently. In [4], various SOC and SOH estimation techniques and their advantages and limitations are suggested. We can summaries existing SOC estimation methods into three classifications such as adaptive, hybrid methods and data-driven [5, 6] which can be divided into two methods namely the direct measurement and the model-based methods. Various estimation techniques like coulomb counting method (CCM), open circuit voltage (OCV) method, and impedance method are includes under direct measurement methods. The coulomb counting method is widely used in battery management system in electric vehicle. To reduce fuel/energy consumption, greenhouse gases (GHG) or to improve the battery lifetime, advanced power management strategies have been studied [7, 8]. So it is necessary to develop an intelligent control module that is suitable for monitoring and controlling the SOC &SOH of batteries commonly used in EVs. In [9], an improvement of battery life is proposed though optimized partial charging. The driver's driving pattern determines the power requirement for EVs which will determine the battery voltage and current. As normally the driving patterns are very irregular so accurate state estimation is necessary along with others factors like depth of demand (DOD) and state of charge (SOC) for lifetime estimation. The main issue with battery packs used in current electric vehicles is their short life & health which is effected by charge-discharge cycles and temperature. In [10], an inter-cycle battery effects are considered while calculating aging. It is therefore important to model the ell level dynamics associating cell degradation with severity associated with driving cycles. On the other hand [11] suggested accounting of battery health degradation is necessary in EHV charging optimization. In this contribution, a novel state estimation method is proposed to accurately determine the battery SOC and SOH based on a runtime equivalent circuit model. The idea here is to update the actual capacity of the battery by calculating the loss in capacity due to severities associated with different drive patterns. The updated capacity is reflected in the SOC defined by Coulomb Counting method. In the following sections, the basic structure of BMS is explained along with the equivalent circuit modelling of the runtime battery model. With the help of mathematical equations, the proposed model is described along with a definition of its states. Finally the simulation results are presented.

2 Battery Management System

2.1 General Idea About HEV/EV Batteries

Depending upon the ability of recharging capability, batteries can be categorised in two parts i.e. i) primary battery which can't be reused or recharged after one use and ii) secondary battery, which can be charged discharged with electricity [1]. In [12] some of the current EV batteries are listed. In EVs or HEVs application secondary batteries are used as it has long life span, low internal resistance and better power density.

B. Tasks and importance of BMS

Every chemical equipment has some specific requirement & constraint which have to be taken under consideration to avoid accelerated aging or safety issues. Although there is no final definition of BMS [13] but as a general the BMS monitors the system to operate it correctly & perform the safety steps if any hazardous situation occurs [2]. The BMS is illustrated in Fig. 1. In this paper, some key models including some emerging models are surveyed and their real-time applicability are determined. Furthermore, the BMS also provides information about electric and thermal model from which relevant parameters like SOC and SOH are determined. In additions, BMS also operates the cooling system (air cooling or liquid cooling) which is essential as high temperature leads to accelerated aging. Besides all those necessities BMS does the most important responsibility by ensuring the safety of the battery i.e. the voltage, current and the temperatures are within the admissible limits. If any hazardous situation occurs then the BMS will take responsibility to determine the cause of the situation and eliminate the cases accordingly. The battery management structure is shown in the Fig. 1.

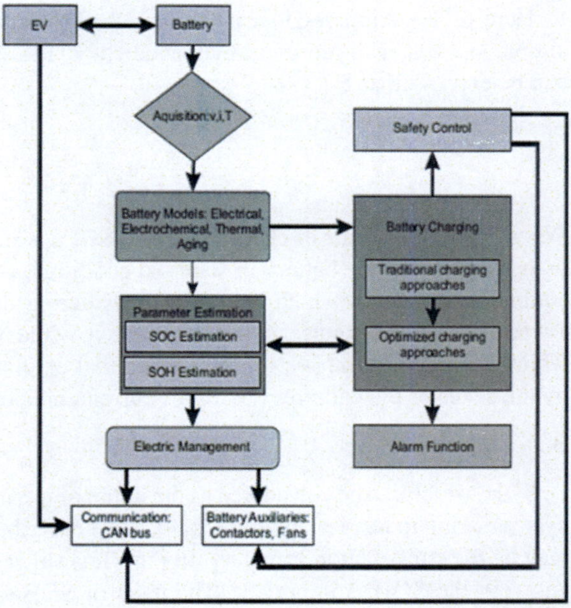

Fig. 1 Battery management system [1, 2]

3 Battery States and Their Estimation

There are several states of HEV/EV batteries. Among them two states are most important such as SOC and SOH estimation because of these states are directly responsible for performance and aging of the battery.

A. *SOC estimation*

Proper SOC modelling is necessary to utilize the battery more efficiently as well as complex. The most usual ways to accommodate the complexity of a SOC estimation is to estimate SOC by modelling. The percentage of available active charges remaining in the battery can be called as SOC, mathematically which is shown in Eq. (1)

$$\text{Soc(t)} = \frac{Qa(t)}{Qr} \cdot 100\% \tag{1}$$

Where Qa(t) is defined as the available active charge left in the battery at time point t, Qr is defined as the remaining or total charge capacity of the battery at current status. 100% stands for the battery is fully charged and 0% means the battery is fully discharged. So SOC can tell us the availability of energy or it gives us the knowledge about the usable energy left in the battery.

Here we use Ampere-Hour integral method to determine SOC because it is most simple and easier to represent mathematically. The Ampere-Hour integral method can be expressed as Eq. (2)-

$$\text{Soc(t)} = \text{Soc}_{0}\text{-} \frac{1}{Cn} \int \eta.\text{I.dt} \tag{2}$$

Where, SOC (t) is state of charge at time t, SOC_0 is state of charge at time t = 0, Cn is the capacity of the battery in standard condition, changing with service life, η is coulombic efficiency which is 1 while the battery is discharging and is smaller than 1 while charging, I is current which is negative while battery is charging and positive while discharging. In [14], it is suggested that coulombic efficiency are influenced by the states of the batteries (i.e. SOC, current, temperature etc.).

B. *SOH estimation*

SOH, which can be considered as the figure of merit of the condition of the battery by comparing to its ideal conditions [4]. SOH describes the condition of a battery or may be the stress factor of battery. In [15], it is suggested that for optimal charging knowing the SOH is important. The units of SOH are percent points (0% means the battery with no energy storage capability, while 100% means battery's condition matches with ideal condition or battery with full nominal capacity). SOH can be calculated as Eq. (3)

$$\text{SocH} = \frac{Qr}{Qn} \cdot 100\% \tag{3}$$

Where Q_r is the remaining capacity defined as available active charge left in the battery at a certain time, Q_n is the nominal capacity of the battery.

The aging of the batteries degrades its state of health.

According to [16] the parameters of the EHV/HV batteries defined correctly are rigid and calculations are relatively larger because of the complexity of detailed aging process of lithium-ion batteries, thus models are be used in the BMS in vehicle is usually impractical. The losses of the batteries degrades its state of health as [17] Eq. (4)

$$Q_{loss} = Ae^{\frac{-Ea}{RT}} \cdot (A_h)^Z \tag{4}$$

Where 'A_h' represents ampere-hour and its unit is Ampere-hour, 'Ea' represents active energy, 'R' is gas constant, 'T' represents time in hour and 'z' is exponent of time which can take 1/2 under simple conditions.

In [18] a framework for predicting battery life has been proposed, which uses models that relate to parameters such as battery life (EOL), such as Ah-throughput, number of cycles since production, or time since manufacturing. In this method Onori set a predetermined value of parameter and the battery is considered to have reached end of life when the battery exceeds that pre-determined value. In [19] battery health degradation are performed by accelerating aging for several months in laboratory. Battery degradation and lifetime can be estimated as a function of certain parameters like voltage, current, temperature, SOC, and SOH. These parameters are estimated based on various modelling techniques. Out of these, weighted Ah-throughput models relate battery EoL to Ah-throughput as the actual amount of current being drawn/supplied to the battery. The severity of current in/out of the battery depends primarily on the C-rate, temperature, and DoD. Based on the nominal/standard operating conditions (known C-rate, temperature, DoD), the actual operating conditions can be considered to be deviated from the standard by a severity factor σ as Eq. (5)-

$$\sigma(DoD, Tbatt) = \frac{Ah\_throughput_{nominal}}{Ah\_throughput_{actual}} \tag{5}$$

Where the $Ah\_throughput_{actual}$ is given by Eq. (6)

$$Ah\_throughput_{actual} = \int_0^{EoL} |I(t)|dt \tag{6}$$

with I(t) denoting the battery current.

We can consider the battery reaches its end of life when the capacity loss is 20% or more with respect to the original nominal capacity.

It is used to calculate $Ah\_throughput_{effective}$ as Eq. (7)

$$Ah\_throughput_{effective} = \sum \text{WE.nE.AhE} \tag{7}$$

where E denotes an event and w_E, the weight or severity associated with the event, n_E, the number of events, and Ah_E, the actual Ah-throughput associated with that event.

If the $Ah\_throughput_{effective}$ obtained from the Eq. 7 is greater than total $Ah\_throughput_{nominal}$ then the battery is considered to reach its end of life or the battery is considered to fail.

Here we refer severity factor as the weighting factors which is responsible for reduction battery life. This severity factors are highly nonlinear process.

The damage variables of the batteries can be defind with age factor and severity factor. The severity factor function depends on how severely the the changes of cycles are occuring i.e. temperature, SOC/DOD (depth of discharge).

4 Battery Models

To express the physical correlations between the battery input parameters (the current or power demand) and output parameters (SOC, SOH) developing a battery model is necessary. Till now many models have been developed which can primarily be categorized as Electro-chemical model, mathematical model and electric circuit model. Here we use Electrical models or equivalent circuit model because of its high accuracy and its error lies between around 1– 5%.

A. *Equivalent Circuit Model*

Electric models which are combination of voltage source, capacitor and resistance are very much useful for correlation with other electric circuit or electric systems [20] and those electric models are easier to handle to design a simulink model.

B. *Proposed Model*

A comprehensive, accurate and real run-time battery model is already proposed in [20] which is shown in Fig. (2). This model is a combination of other electrical models like thevenin model, impedance model and runtime based electric models which eases to capture the dynamical electrical characteristics. In this model on the left a capacitor and current control current source are there similar to the run-time based electric model. The RC circuits help to understand the transients of the model

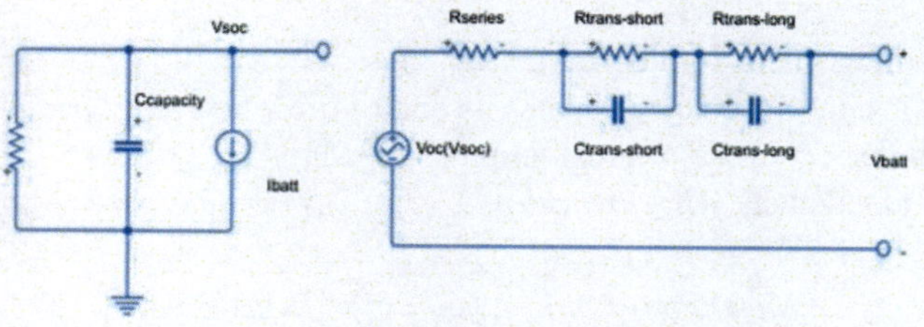

Fig. 2 Equivalent circuit model

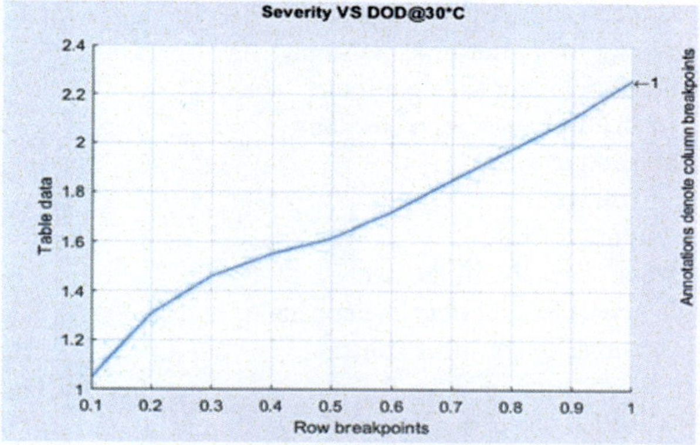

Fig. 3 Severity vs DoD

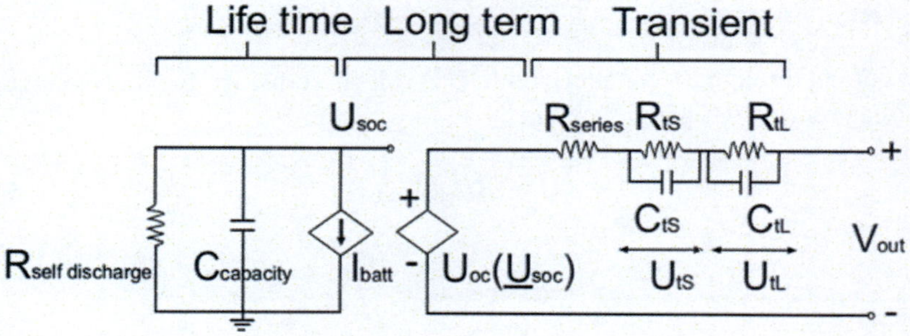

Fig. 4 Representation of RC networks

and a voltage source is used on the open-circuit voltage bridged from the SOC. Open circuit voltage is changed with the variation of SOC. In our paper we keep updating the capacity of the battery by considering battery losses over time.

With this model we can easily study the dynamic electric characteristics of the battery like usable capacity, open-circuit voltage response of RC transient circuit.

In this model we determine SOC with coulomb counting method. In [18] Onori concluded that high temperature and high DoD which leads to higher severity factor will cause accelerating battery aging as shown by the below Fig. (3). So it is necessary to include SOC, DoD and temperature distribution in account of lifetime estimation. By considering this severity function we determine Ah-effective which differs from Ampere-hour.

5 Mathematical Model

Mathematical model is shown as Fig. (4).

Rts = Resistance in shorter time constant.
Cts = Capacitance in shorter time constant.
R_{tL} = Resistance in longer time constant.
C_{tL} = Capacitance in longer time constant.
I_{batt} = Current source.
V_{out} = Output voltage.
U_{oc} = Open circuit voltage.
$C_{capacity}$ = Overall capacity of the battery.
U_{ts} = Voltage across RC shorter time constant.
U_{tL} = Voltage across longer time constant

If we denote capacitance in every RC network as parallel resistance R_2. Here C is denoted as R_2. Resistance is denoted as R1.

Then from the Fig. 4, equivalent resistance of a RC network, $R_{equivalent} = R_1 \| R_2$.

$$\Rightarrow R_{equivalent} = \frac{R1(-R2)}{R1 + (-R2)} \tag{8}$$

If I is current through RC parallel circuit, U is voltage across RC parallel circuit then,

Voltage across RC parallel circuit

$$U = (R_{equivalent}).I \tag{9}$$

$$\Rightarrow R_{equivalent} = \frac{U}{I} \tag{10}$$

Again, for capacitor C,

$$U = \frac{1}{c} \int I \Rightarrow \dot{U} = \frac{1}{c} \tag{11}$$

Now, capacitive resistance

$$R_2 = \frac{U}{I} = \frac{U}{\dot{U} * C} \tag{12}$$

Now from (8), $R_{equivalent} = \frac{R1(-\frac{U}{\dot{U}*C})}{R1+(-\frac{U}{\dot{U}*C})}$

$$\Rightarrow \frac{U}{I} - (\frac{R1 * U}{(R1 * \dot{U} * C) - U})$$

By solving in terms of $\dot{U}$,

$$\dot{U} = -(\frac{1}{c}) - \{\frac{U}{(R1 * C)}\} \tag{13}$$

Now, Voltage across series resistance, $U_s = -(R_s * I_{batt})$ From Eq. 13, voltage across RC short transient,

$$\dot{U}_{ts} = -\{\frac{Uts}{(Rts * Cts)}\} - (\frac{Ibatt}{Cts}) \tag{14}$$

Similarly, voltage across RC long transient,

$$\dot{U}_{tl} = -\{\frac{Utl}{(Rtl * Ctl)}\} - (\frac{Ibatt}{Ctl}) \tag{15}$$

Again, open circuit voltage, $U_{oc} = -(\frac{1}{C_{capacity}} \int Ibatt)$

$$\Rightarrow \dot{U}_{oc} = -(\frac{Ibatt}{C_{capactiy}}) \tag{16}$$

U_{soc} can also be express in terms of function of open circuit voltage such as (14)-

$$U_{soc} = f(U_{oc})$$

If we express Fig. 4 in terms of equivalent ordinary differential equation (ODE), then from Eq. 14, 15 and 16 will becomes

$$\dot{X} = \begin{bmatrix} 0 & 0 & 0 \\ 0 & -(Rts * Cts)^{-1} & 0 \\ 0 & 0 & -(Rtl * Ctl)^{-1} \end{bmatrix} X + \begin{bmatrix} -(C_{capactiy})^{-1} \\ -(Cts)^{-1} \\ -(Cts)^{-1} \end{bmatrix} \text{Ibatt}$$

Here, $X = \begin{bmatrix} Uoc \\ Uts \\ Utl \end{bmatrix}$

So the output voltage

$$V_{out} = g(x1) + x2 + x3 + R_s I_{batt} \tag{17}$$

6 Results

A Li-ion battery with nominal capacity 60 Ah, number of cells (33 in series) is chosen. A European drive cycle is used to represent the drive pattern of the driver. The corresponding charging-discharging of the battery leads to a state of charge curve with respect to time which is shown as Fig. (5). Here 1 means 100%.

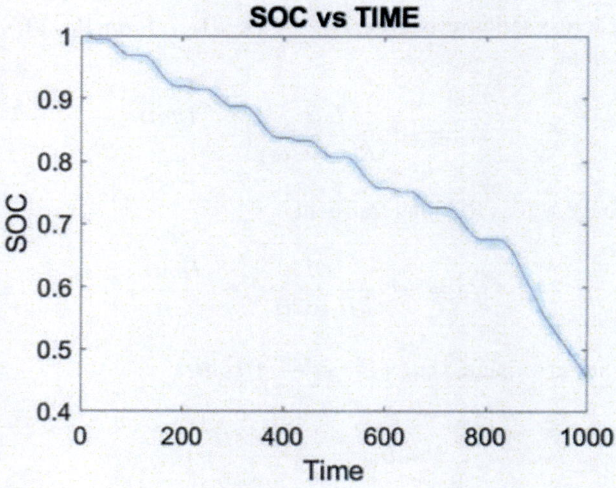

Fig. 5 SOC vs time

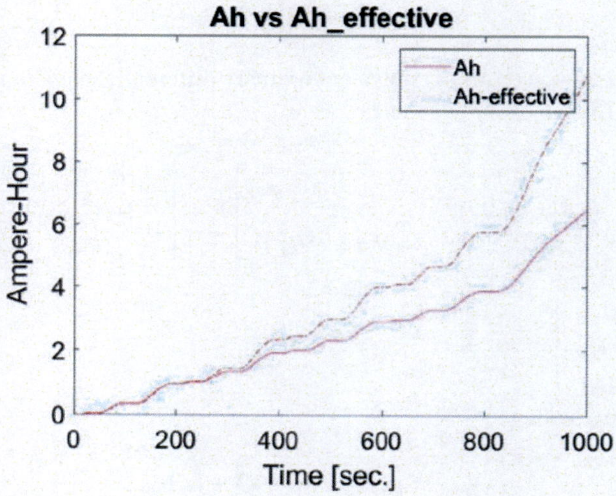

Fig. 6 Ah vs Ah-effective

The loss in capacity is calculated from the Ah-throughput and the updated capacity is also shown in the following figures. After considering severity function the updated Ah which is Ah-effective will differ from the Ah curve. In Ah-effective we consider the effect of severity. The differences between Ah and Ah-effective is shown in following Fig. (6).

The output curve between losses and capacity are shown in the figure below (7). From the figure it is clear that the overall losses is increasing from initial time which causes the overall decrease in capacity. The decrease in capacity is very marginal.

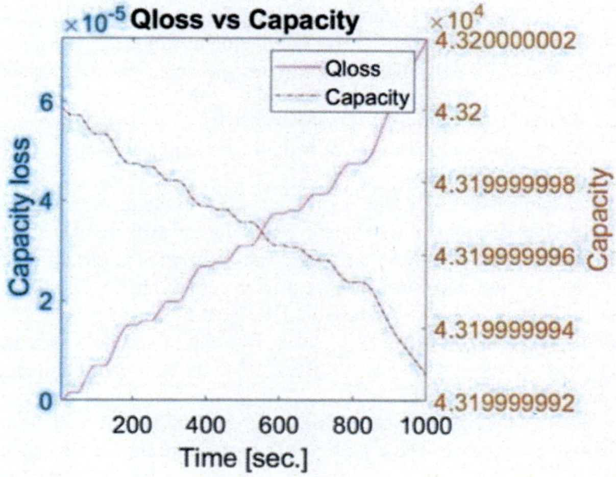

Fig. 7 Capacity vs loss

7 Conclusion

In this paper we investigate the influence of time varying quantities such as charge-discharge cycle and capacity fade and its effect on EV batteries. Capacity fade over time, due to varying charge-discharge cycle also leads to a change in SOC levels which has to be taken into account in order to have an accurate battery management system. In this research work both capacity fade as well as runtime parameters are used along with nominal capacity value considering the losses through Ah-throughput model to obtain a real-time value of battery SOC. We also define capacity loss as a function of Ah-throughtput and we also consider severity factor function.

References

1. Liu K, Li K, Qiao P, Cheng Z (2019) A brief review on key technologies in the battery management system of electric vehicles. Front Mech Eng 47–64 (2019)
2. Conte FV (2016) Battery and battery management for hybrid electric vehicles: a review e & i Elektrotechnik und Informationstechnik 424–431 (2006)
3. Li S, Ke B (2011) Study of battery modeling using mathematical and circuit oriented approaches. In: IEEE power and energy society general meeting, Detroit, MI, USA, pp 1–8
4. Lu L, Han X, Li J, Hua J, Ouyang M (2013) A review on the key issues for lithium-ion battery management in electric vehicles. J Power Sources 272–288
5. Watrin N, Blunier B, Miraoui A (2012) Review of adaptive systems for lithium batteries state-of-charge and state-of-health estimation. In: IEEE transportation electrification conference and Expo (ITEC), Dearborn, pp 1–6 (2012)

6. Prajapati V, Hess H, William E.J, Gupta V, Huff M, Manic M, Rufus F, Thakker A, Grover J (2010) A literature review of state of-charge estimation techniques applicable to lithium poly-carbon monoflouride (LI/CFx) battery. In: India international conference on power electronics (IICPE2010), pp 1–8
7. Shahverdi M, Mazzola MS, Grice Q, Doude M (2016) Pareto Front of energy storage size and series HEV fuel economy using bandwidth-based control strategy. IEEE Trans Transp Electrification 2(1): 36–51
8. Shahverdi M, Mazzola MS, Grice Q, Doude M (2017) Bandwidth-based control strategy for a series HEV with light energy storage system IEEE Trans Veh Technol 66(2):1040–1052
9. Hoke A, Brissette A, Maksimović D, Kelly D, Pratt A, Boundy D (2013) Maximizing lithium ion vehicle battery life through optimized partial charging. In: IEEE PES innovative smart grid technologies conference (ISGT), Washington, DC, pp 1–5
10. Petricca M, Shin D, Bocca A, Macii A, Macii E, Poncino M (2014) Automated generation of battery aging models from datasheets. In: IEEE 32nd international conference on computer design (ICCD), Seoul, pp 483–488
11. Hoke A, Brissette A, Smith K, Pratt A, Maksimovic D (2014) Accounting for lithium-ion battery degradation in electric vehicle charging optimization. IEEE J Emerg Select Top Power Electron 2(3): 691–700
12. Takami N, Inagaki H, Tatebayashi Y, Saruwatari H, Honda K, Egusa S High-power and long-life lithium-ion batteries using lithium titanium oxide anode for automotive and stationary power applications. J Power Sources 244:469–475
13. Andrea D (2010) Battery management systems for large lithium-ion battery packs first ed., Artech House, September 30
14. Li Z, Lu L, Minggao O (2010) Comparison of methods for improving SOC estimation accuracy through an ampere-hour integeration approach. J Tsinghua Univ (Sci. Technol.) 1293–1296+1301
15. Jiang T, Putrus G, Gao Z, Conti M, McDonald S (2013) Optimal charging strategy for EVs with batteries at different states of health. In: 4th International conference on power engineering, energy and electrical drives, Istanbul pp 326–329
16. Safari M, Morcrette M, Teyssot A, Delacourt C (2010) Life prediction methods for lithium-ion batteries derived from a fatigue approach: Li. capacity-loss prediction of batteries subjected to complex current profiles. J Electrochem Soc 157(6):A892–A898
17. Wang J, Liu P, Hicks-Garner J, Sherman E, Soukiazian S, Verbrugge M, Tataria H, Musser J, Finamore P (2011) Cycle-life model for graphite-lifepo4 cells. J Power Sources 196(8):3942–3948
18. Marano V, Onori S, Guezennec Y, Rizzoni G, Madella N (2009) Lithium-ion batteries life estimation for plug-in hybrid electric vehicles. In: IEEE vehicle power and propulsion conference, Dearborn, pp 536–543
19. Stroe D, Swierczynski M, Kær SK, Laserna EM, Zabala ES (2017) Accelerated aging of Lithium-ion batteries based on electric vehicle mission profile. In: IEEE energy conversion congress and exposition (ECCE), pp 5631–5637
20. Chen M, Rincon-Mora GA (2006) Accurate electrical battery model capable of predicting runtime and I-V performance. IEEE Trans Energy Conv 21(2):504–511

AES-Based Android Video Encryption and Decryption App

Anubhav Dinkar and B. Sahana

Abstract Encryption is a major requirement in today's world as it boosts the security of the data that everyone shares all over the Internet. Videos are the means of communication in several industries and can be private. Encrypting a video file to a text file using the AES encryption algorithm increases the security provided while sharing such content over the Internet. Currently, apps to encrypt text files and images exist for the Android OS, and this paper proposes the next step in that direction - an Android app for video encryption and decryption. The results obtained via the method proposed in this paper are nearly 10 times better than the existing results.

Keywords Android OS · Advanced encryption standard · Direct encryption standard · Symmetric key encryption · Security · Video encryption

1 Introduction

Encryption algorithms are of two broad types: Symmetric Key algorithms and asymmetric key algorithms. Some examples of symmetric-key ciphers are the Direct Encryption Standard and the Advanced Encryption Standard algorithms. In these cases, the encryption station and the decryption station both need to use a shared, private key, which is utilized in the encryption and decryption processes. Some examples of asymmetric key algorithms are the RSA and DSA algorithms.

The application presents two simple options for the user- selecting a video to be encrypted and selecting a text file to be decrypted. When the user selects an *mp4 (MPEG-4)* video from the device's internal storage, the byte stream - and eventually the 'String' is extracted from it. This string is encrypted using the AES algorithm [3] and this encrypted string is written onto a file in the phone's internal storage. The AES algorithm flow is shown here in Fig. 1.

A. Dinkar (✉) · B. Sahana (✉)
Department of Electronics and Communication Engineering, Rashtreeya Vidyalaya College of Engineering, Bengaluru, Karnataka, India
e-mail: sahanab@rvce.edu.in

V. Nath and J. K. Mandal (eds.), *Proceeding of Fifth International Conference on Microelectronics, Computing and Communication Systems*, Lecture Notes in Electrical Engineering 748,
https://doi.org/10.1007/978-981-16-0275-7_27

Fig. 1 Steps in the AES Algorithm

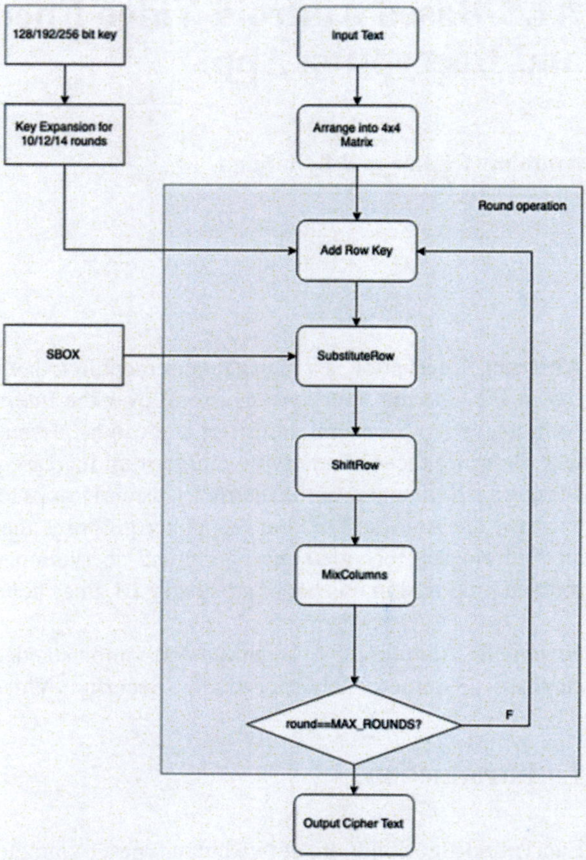

The decryption operation is exactly the opposite of the above procedure- the text from the selected *txt* file is first decrypted using AES, then converted into a byte stream. This byte stream is converted into a video file and stored in the phone's internal storage.

The Android app uses shared-key AES, meaning the key used to encrypt the video must be shared with someone who wants to decrypt the video and use/watch it.

The reason AES was chosen was that the load it causes on the CPU of the Android device is lower than other algorithms [6]. For an app of simple usage, it is of utmost importance that the CPU usage is minimal.

The key consideration in any application that involves encryption is how susceptible it is to attacks, which may be a brute force attack [13]. Cryptanalysis of the existing 128-bit symmetric cipher may be broken in 2^{64} operations [17].

2 Literature Survey

S. Tayade et al. [2] have used the AES algorithm to perform file encryption in the Android OS. This was limited to text and image encryption.

Seth et al. [5] made a detailed analysis of three algorithms - RSA, DES, and AES in terms of several parameters, like - computation time and memory usage. It was found that RSA takes the highest time for encryption and has the highest memory usage as well. This paper concluded that AES takes the least memory usage and the time difference between the DES and AES algorithms are very minor.

Rachmat [6] has compared the AES (Rijndael) algorithm with the Twofish and Serpent algorithms. This performance analysis is based on the performance parameters time, memory usage, and CPU Usage. It was found that while Rijndael took a marginally longer time for encryption and decryption and its memory usage was nearly identical to that of the other algorithms, its CPU usage was considerably lower.

P. Deshmukh et al. [11] made use of a modified AES algorithm for MPEG video encryption. There are many MPEG formats, one of them being MPEG-4, which is the format of the video files that can be used in this paper.

Rege et al. [16] have found that the AES algorithm, is highly efficient in the block encryption mode.

Further, Riaz et al. [19] have found that the 128-bit AES algorithm is invulnerable even to brute force attacks by supercomputers.

Thus, an Android app for *mp4* video encryption and decryption is proposed in this paper, using AES ECB with PKCS5 padding for the same.

3 Methodology

The application was coded in Android Studio, using Java and eXtended Markup Language (XML).

XML was used to create the layouts and widgets. Java was used to add functionality to the layouts and widget and to specify the behaviour of these components when the user interacts with them.

Encryption Algorithm

The encryption algorithm used in this app, as mentioned before, is the Advanced Encryption Standard electronic codebook encryption (AES ECB). AES ECB performs a 128-bit block encrypt. The AES encryption procedure is shown in Fig. 1.

The electronic code block (ECB) is a purely block cipher technique. The plaintext is broken into groups of N bits. Thus, the resultant ciphertext is also made of N bit blocks. N depends on the algorithm used. (Forouzan [7], p. 946).

In this case, N is equal to 128 bits, as the AES algorithm implemented in this app makes use of a 128-bit key. (Fig. 11).

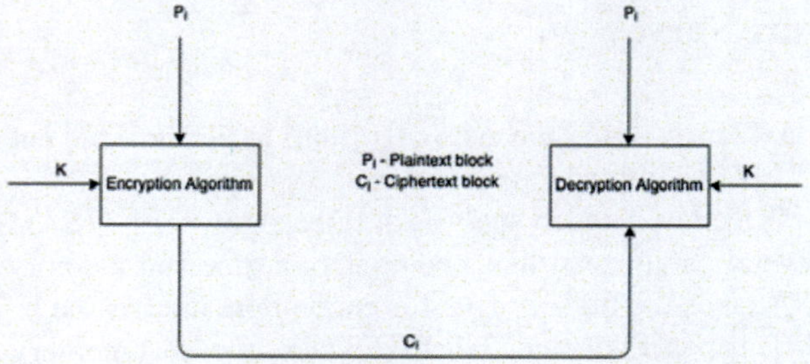

Fig. 2 ECB Block mode used in the AES Algorithm

Fig. 3 Algorithm to hash user input key

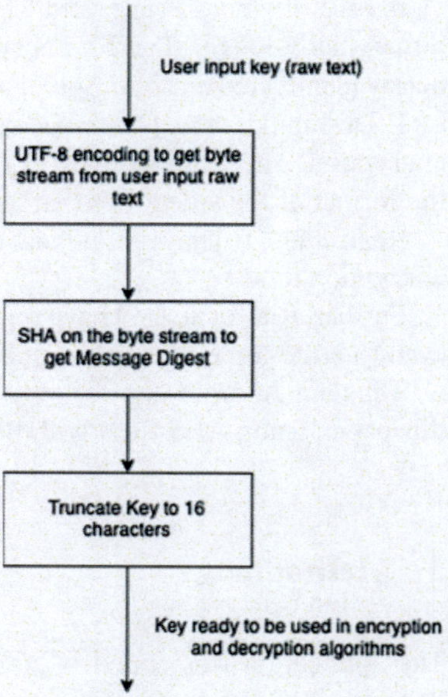

Also, PKCS5 padding is required to make sure that the size of the data blocks is divisible by 128 (mod 128 is equal to 0). The combination of AES ECB and PKC5 algorithms is used due to its high-security performance [18].

The algorithm for the same is shown below.

If len_bytes (input_text) % 8 == 7, P = M + 0 × 01

If len_bytes (input_text % 8 == 6, P = M + 0 × 0202

If len_bytes (input_text % 8 == 5, P = M + 0 × 030303

...

If len_bytes (input_text) % 8 == 0, P = M + 0 × 0808080808080808

Above, P is the padded message, while M is the message.

As mentioned later, the key is taken from the user of this app. To add security to this aspect as well, hashing is performed on this key input by the user. The 8-bit Unicode Transformation Format (UTF-8) is used to get the byte stream from the raw

Fig. 4 Block diagram of the video encryption process

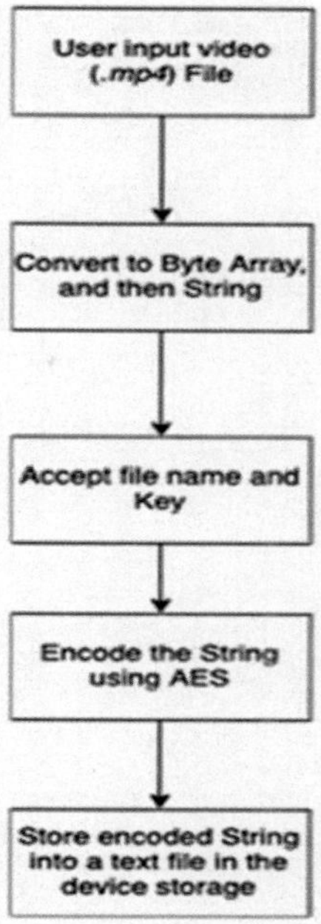

Fig. 5 Block diagram of the video decryption process

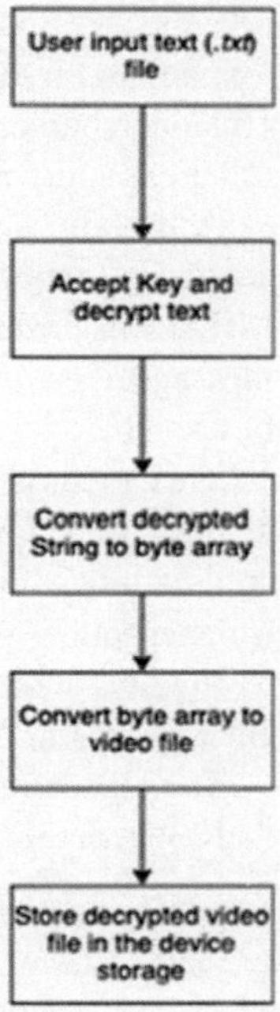

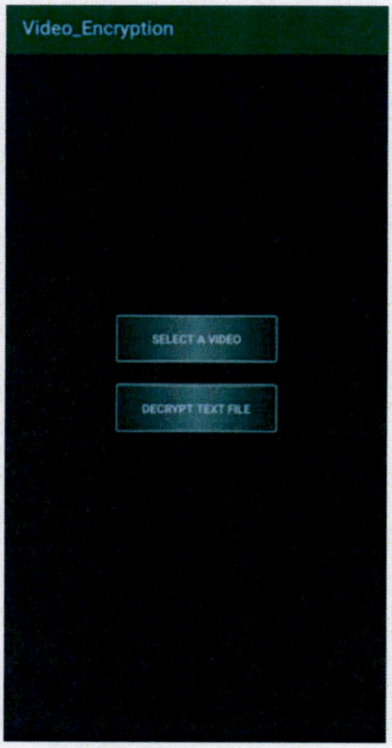

Fig. 6 The main screen of the app

text entered by the user. This transformation is capable of encoding over a million Unicode characters and is used in over 60% of the Web as of 2012, more than any other encoding mechanism [9].

A hashing algorithm is that which takes in a message input and returns a hex sequence (a message digest) as output. After encoding, the Secure Hash Algorithm (SHA-1) is performed on the generated byte stream. This algorithm accepts an input message, the output being a 160-bit message digest. The input is divided into 512-bit blocks [10].

This method is considered stronger than other hashing algorithms like MD5 against brute force attacks. SHA-1 is also quite simple and compact. On mobile applications, though the SHA1 encryption is shown to be stronger, the required time for encryption is higher in SHA1 than in MD5 [15].

Finally, the key is truncated to 16 bytes and is the final *secretKey* used for the encryption and decryption algorithms.

The algorithms for the above procedure are shown below:

Key generation:

- get Bytes from user input string (UTF-8)
- Get*MessageDigest* using SHA-1
- Truncate the key length to 16 bytes
- Construct secret key from this byte array using *SecretKeySpec* method in Java

Fig. 7 Video Encryption screen

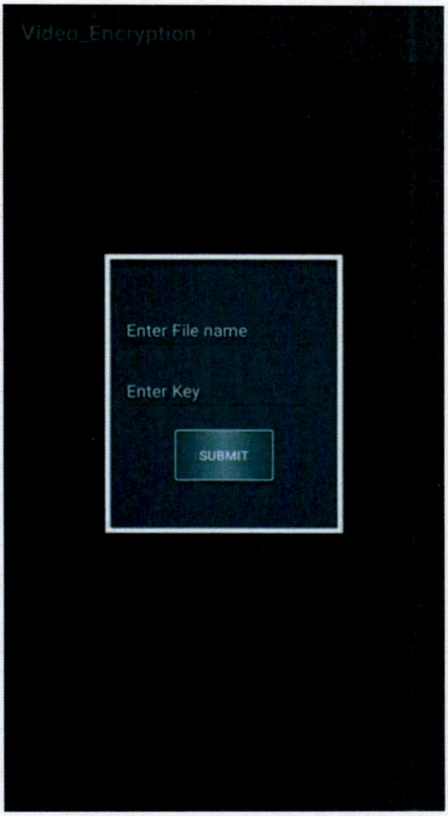

Encryption:

- Create the cipher (encryption algorithm) - AES ECB, with PKCS5 padding
- Initialize the cipher in *ENCRPT_MODE* along with the key as generated above
- Use the *Base64Encoder* in Java to encode the string that needs to be encoded

Decryption:

- Create the cipher (decryption algorithm) - AES ECB, with PKCS5 padding
- Initialize the cipher in *DECRYPT_MODE* along with the key as generated above
- Use the *Base64Encoder* in Java to decode the string that needs to be decoded

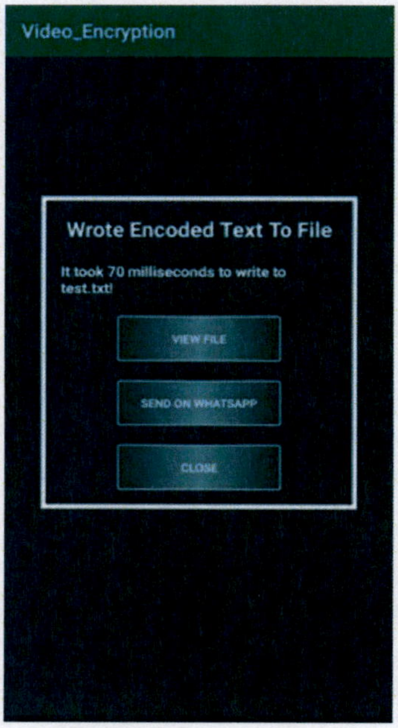

Fig. 8 Result screen after Encryption

Fig. 9 The popup displaying the contents of the text file created after encryption

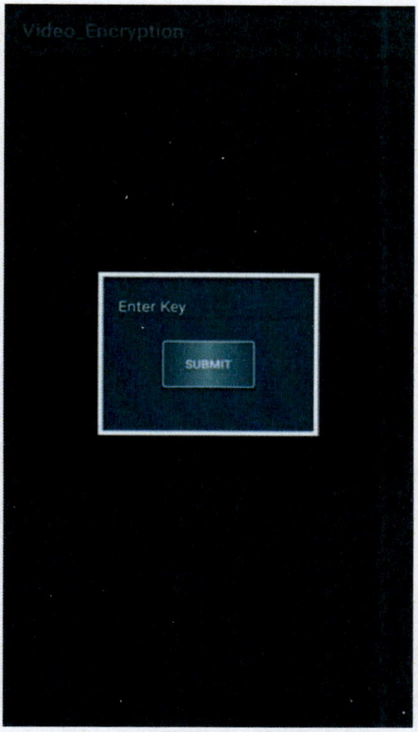

Fig. 10 The popup that requests the user to enter the key to allow decryption to take place

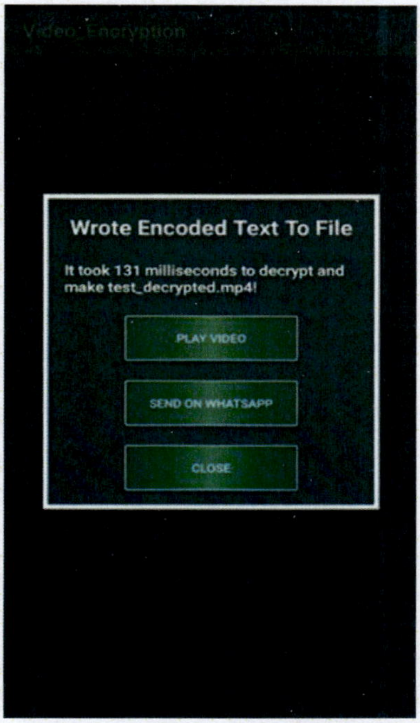

Fig. 11 The resulting popup presented to the user once the decryption process is successful. The user can play the video or send it on WhatsApp

4 Results

Working of the application

The Application is built using the following files:

1. *MainActivity.java* - This is the main Java file in which the entire code for the working of the app is written. Functionalities for all layouts are coded here.
2. *AES.java* - This java file is where the encryption and decryption methods are written. An object of the class defined in this script is called from *MainActivity.java* to encrypt/decrypt the text.
3. *activity_main.xml* - This is the XML file that is used to set the layout of the main screen.
4. *save_encrypted.xml* - This is an XML file for a popup window which is where the user inputs the file name which is used while saving the text and video files
5. *encoded.xml* - This XML file is for the popup which displays the encrypted String.
6. *decrypt.xml* - This is a popup window XML which accepts the user input for the 'key' value before decryption is performed.
7. *desc.xml* - This XML file is for the resulting popup after the encryption or decryption process is complete

The flow of the application is described in detail below (Fig. 5):

- When the application is opened for the first time, the user is requested for read and write permissions
- The main screen of the app is presented in Fig. 6. As mentioned earlier, the user is presented with two options when the application is opened.
- If the user chooses "Select a Video" in the main screen, the internal file manager is launched via an *Intent* (Marsicano et al. [1], p. 485) from the *MainActivity.java* file. Once a user selects a valid *mp4* video, the steps as per Fig. 3 are performed.
- Figure 7 shows the popup presented to the user upon successful conversion of the video to a byte stream, which *inflates* the *save_encrypted.xml* popup layout, by using a *LayoutInflater* (I. F. Darwin [4], p. 247).
- The user enters a file name (say *file)* and a secret key (which must be shared with whoever intends to decrypt the file), the app stores *file_encoded.txt* under *VideoEncryptorFiles* in the host phone's internal storage.
- Finally, the resulting popup is shown to the user (Fig. 8). This popup contains the total time taken to complete the encryption and storage process. Further, the user can choose to view the file, send the file on WhatsApp, or simply close the popup. Sending the file via WhatsApp itself does not present any security issues as it enables an end-to-end encryption by making use of a combination of the AES-256 and SHA256 algorithms [20, 21].
- If the user chooses to view the file within the app, a popup for the encoded text is created. The encoded text is then displayed, as shown in Fig. 9.
- In case the user taps "Decrypt Text File" in the app's main screen (Fig. 6), an *Intent* is used to launch a file manager from the *MainActivity.java* file.

Fig. 12 The video playing within the app

- Figure 10 shows the popup presented to the user upon selecting the text file to be decrypted. After this, the user is presented with a popup which requests the user to input a key-value.
- The flow in which the video is decrypted now is shown in the flowchart in Fig. 5
- If the user inputs the key which was used to encrypt the video (and thus, create this text file), then the text file is decrypted and a video file *file_decrypted.mp4* is created and stored inside. The resulting popup is also displayed (Fig. 11)
- The user can, as with the text file, view the video file (using a *VideoView* in *Android;* Fig. 12), or send it to any contact on WhatsApp, or simply close this popup. (I. F. Darwin [4] p. 407). This video is played according to the resolution and dimensions of the original video chosen by the user.

Currently, the video size limit for the encryption to be performed is 21 MB.

Further, the encrypted text which is stored in the phone's internal storage is extremely large - over 7 million characters for a video of size 384 KB. This makes loading the text inside a *TextView* within a popup inside the application slightly slower.

As a result of this, the storage requirements of the text file are slightly higher- a 9.82 MB video file is encrypted and stored within a 19.38 MB text file (contains over 19 million characters). Thus, we can see a trade-off between security and storage. The decrypted video, however, is of the same size as the original video.

The average time is taken to encrypt the sample video and decrypt the text files obtained via encryption are as shown below:

Table 1 Average time required to perform encryption and decryption

Video size	Time to encrypt (ms)	Time to decrypt (ms)
384 KB	83	108
1.57 MB	116	567
9.84 MB	529	1185
14.81 MB	811	1450

Table 2 Memory usage of the encryption and decryption algorithms

Original video size	Encrypted text file size	Decrypted video size
384 KB	777 KB	384 KB
1.57 MB	2.96 MB	1.57 MB
9.84 MB	19.55 MB	9.84 MB
14.81 MB	29.9 MB	14.81 MB

Table 3 A comparative study with existing methods

The method proposed by Deshmukh et al. [11]			The method proposed in this paper		
Video size (MB)	Time to encrypt (ms)	Time to decrypt (ms)	Video size (MB)	Time to encrypt (ms)	Time to decrypt (ms)
1.11	917	2011	1.06	219	169
4.45	2476	4823	4.72	332	601
1.26	1122	2540	1.34	275	183

It is also necessary to note that the times taken by the process proposed in this paper, as mentioned above, involves even writing the encrypted string (which can exceed over a million characters in length) to a text file.

5 Conclusion

This Android Application is a step further in the direction of secure file sharing in mobile devices. Video encryption for mobile devices is very important as a lot of communication in today's world is video-based.

Compared to the results obtained in the existing papers in this field, as shown in Table 3, it is clear that the method proposed in this paper is much faster. From the results in [11], the time taken to encrypt a *MPEG* file of size 1.26 MB is about 1122 ms. For an *mp4* file of size 1.57 MB, the encryption process in this paper takes about 116 ms.

Further development would be to provide for a more memory-efficient algorithm, across all mobile operating systems, to encrypt large video files and real-time video encryption in video conferencing, video chats, et cetera.

References

1. Marsicano K, Gardner B, Phillips B, Stewart C (2019) Android Programming: The Big Nerd Ranch Guide. Big Nerd Ranch (2019)
2. Tayde S, Siledar S (2015) File encryption, decryption using AES algorithm in android phone. Int J Adv Res Comput Sci Softw Eng 5:550–554
3. Daemen J, Rijmen V (November 2001) Announcing the Advanced Encryption Standard (AES). Federal Information Processing Standards Publication 197
4. Darwin IF (2011) Android Cookbook. O'Reilly Media Inc.
5. Mehrotra SS, Rajan M (2011) Comparative analysis of Encryption algorithm for data communication. Int J Comput Sci Technol **2**(2)
6. Rachmat N (2019) Performance analysis of 256-bit AES encryption algorithm on Android phone. J Phys
7. Forouzan BA (2007) Data Communication and Networking. McGraw Hill Higher Education
8. Pike, Rob; Thompson, Ken (1993). "*Hello, World or Καλημέρα κόσμε or こんにちは 世界*" (PDF). Proceedings of the Winter 1993 USENIX Conference
9. Davis M. (2012) Unicode over 60 percent of the web. Official Google Blog
10. Stallings W (1999) Cryptography and Network Security: Practices and Principles. Prentice-Hall, 2nd Edition
11. Deshmukh P, Kolhe V (2015) Modified AES Based Algorithm for MPEG Video Encryption. IEEE
12. Dumbere DM, Janwe NJ (2014) Video encryption using AES algorithm. In: Second international conference on current trends in engineering and technology, ICCTET 2014
13. Zeghid M, Machhout M, Khriji L, Baganne A, Tourki R (2007) A modified AES based algorithm for image encryption. World Acad Sci Eng Technol 1:745–750.
14. Jayasinghe D, Ragel R, Ambrose JA, Ignjatovic A, Parameswaran S (2014) Advanced modes in AES: are they safe from power analysis based side-channel attacks? IEEE
15. Setiadi R, Najib, Rachmawanto E, Sari C, Sarker, Rijati (2019) A comparative study MD5 and SHA1 algorithms to encrypt REST API authentication on mobile-based application. In: international conference on information and communications technology
16. Rege K, Goenka N, Bhutada P, Mane S (2013) Bluetooth Communication using Hybrid Encryption Algorithm based on AES and RSA", *International Journal of Computer Applications,* 2013
17. Ren W, Miao Z (2010) A hybrid encryption algorithm based on DES and RSA in Bluetooth communication. In: Second International Conference on Modelling, Simulation and Visualization Methods
18. Shrivas M, Amoako A, Boateng S Yeboah T (2015) Migration model for unsecure database driven software system to secure system using cryptography
19. Riaz M, Ikram A (2018) Development of a secure SMS application using advanced encryption standard (AES) on android platform. I.J, Mathematical Sciences and Computing
20. Fauziah N, Rachmawanto E, Setiadi DRIM, Sari A (2018) Design and implementation of AES and SHA-256 cryptography for securing multimedia file over android chat application, pp 146–151. https://doi.org/10.1109/ISRITI.2018.8864485
21. Whatsapp Inc., Whatsapp Security Overview (2017). https://www.whatsapp.com/security/whatsapp-security-whitepaper.pdf

Area Optimized Hardware Architecture of Piccolo-80 Lightweight Block Cipher

Shubham Mishra, Zeesha Mishra, and Bibhudendra Acharya

Abstract Lightweight cryptography is a field of cryptography where implementation for resource constrained devices are in big demand. Piccolo is an ultra-lightweight block cipher having block size of 64-bit and key size of 80-bit and 128-bit. Compact implementations becomes more important when they can provide sufficient security without compromise in performance. In this paper an efficient architecture for Piccolo-80 cipher have been proposed. This cipher requires extreamely low area in hardware implementation, and hence it is useful in RFID applications. Piccolo with its proposed architecture has been implemented on various platforms of FPGA.The proposed work shows 112% improvement in terms of area and 17.65% better results in terms of efficiency. Different devices of Virtex and Spartan have been used to get the results.

Keywords Ultra-lightweight · Piccolo cipher · Area efficient · FPGA · Throughput

1 Introduction

A huge formation of area constrained devices as RFID tags and sensor nodes in the field of IoT [1] and automation has become a subject undergoing intense study. These devices are used in regular basis so they need a level of security. Lightweight cryptography algorithms provide four basic necessity for transmission of data; those are integrity, privacy, authenticity and confidentiality. There are different kind of lightweight ciphers namely stream ciphers and block ciphers. In fact, there

S. Mishra (✉) · Z. Mishra · B. Acharya
Department of Electronics and Communication, National Institute of Technology, Raipur 492010, CG, India

Z. Mishra
e-mail: zmishra.phd2016.etc@nitrr.ac.in

B. Acharya
e-mail: bacharya.etc@nitrr.ac.in

V. Nath and J. K. Mandal (eds.), *Proceeding of Fifth International Conference on Microelectronics, Computing and Communication Systems*, Lecture Notes in Electrical Engineering 748,
https://doi.org/10.1007/978-981-16-0275-7_28

have been several blockciphers designed for a lightweight hardware implementation such as SIMON and SPECK [2], XTEA [3], Piccolo [4], LiCi [5], LEA [6], KATAN/KTANTAN [7] and QTL [8], PRESENT [9].

There are two methods using which these block ciphers can be categorized; Substitution Permutation Networks (SPNs) and Feistel-type Networks. SPN networks gives better diffusion property to the algorithms as compared to Feistel networks. Feistel networks provide better round operations and hence improves security [10]. Also, speed can be increased by inserting pipelined registers into different positions in the architecture [11].

Substitution Permutation Networks are basic structure of AES [12]. There are many more lightweight block ciphers which works on the same structure have been published. One of them is PRESENT cipher, mainly focused on the SPN networks, since it consists much more less number of gate equivalents as compared to other stream ciphers [13]. Stream ciphers like Grain and Trivium requires very less number of Gate Equivalents (GEs) as compared to many block ciphers, they are compact in nature. The other one is, Feistel-type structures mainly used in DES encryption algorithm as its basic building block and consist of Feistel networks and Generalized Feistel networks. This paper work has mainly focused on the efficient area utilization with optimum security. Piccolo cipher is extreamly low in area uses at the same time provides sufficient security for many applications like IoT and RFID tags. This work have been implemented with minimal number of slices as compared to previously proposed works on same FPGA platforms. These outcomes can be seen from graphs presented in result section.

A. *Organization of the Paper*

This paper composed of six sections. Section 2 provides the basic information about the Piccolo cipher. Proposed architecture of Piccolo cipher have been explained in Sect. 3. Theory and related methods have been defined in Sect. 4. Results and comparison are presented in Sect. 5. At last conclusion of the paper is outlined in Section 6.

2 The Piccolo Cipher

Piccolo [4] is a lightweight block cipher developed for highly resource constrained devices. Piccolo works on the 64-bit of block size and uses two key length for encryption process, they are 80-bit and 128-bit. Piccolo uses an alternative of generalized Feistel network for its iterative cycles. These iteration rounds depend on the key size and varies accordingly. There are 25 rounds for 80-bit key process and 31 rounds for 128-bit key. Top and bottom rounds of these iterations uses some additional whitening keys (16-bit), to create more confusion and diffusion to the algorithm. Piccolo uses some F-functions which consist of four S-boxes followed by Mix-column transformation and again four S-boxes. Hardware implementation of Piccolo-80 uses only 683 GEs and Piccolo-128 uses only 758 GEs, because of which Piccolo is also known as ultra-lightweight block cipher. Piccolo block cipher uses very compact hardware implementation and good security for RFID and resource constrained IoT devices. Piccolo algorithm [4] proposed by K. Shibutani et al. has been thoroughly explained further in this section in two parts as Key scheduling and Encryption process.

A. Key Scheduling Process

KEY-INPUT: 80-bit Key= [{Key_0 || Key_1 || Key_2 || Key_3 || Key_4}]

RESULT: Whitening keys are $wk_{[0\text{-}3]}$

Round keys $rk_{[0\text{-}2R]}$

- $wk_0 = Key_0^m \mid Key_1^R,$
- $wk_1 = Key_1^m \mid Key_0^R$
- $wk_2 = Key_4^m \mid Key_3^R,$
- $wk_4 = Key_3^m \mid Key_4^R$
- for round=0 to R-1
- $const_{2round}^{80} = \{Z_{round+1} | Z_0 | Z_{round+1} | \ 1'b0\} \oplus \{0F1E_{16}\}$
- $rk_{2round} = const_{2round}^{80} \oplus \begin{cases} (round) \bmod 5 = 0 \ ? \ Key_2: \\ (round) \bmod 5 = 1? \ Key_0: \\ (round) \bmod 5 = 4 \ ? Key_4 \end{cases}$
- $const_{2round+1}^{80} = \{1'b0 \mid M_{round+1} | M_0 | \ M_{round+1} | \ \} \oplus \ \{2D3C_{16}\}$
- $rk_{2round+1} = const_{2round+1}^{80} \oplus \begin{cases} (round) \bmod 5 = 2 \ ? \ Key_3: \\ (round) \bmod 5 = 3? \ Key_1: \\ (round) \bmod 5 = 4 \ ? Key_4 \end{cases}$
- stop

Representation of different symbols have some meaning which is described by Table 1.

Table 1 Representation of symbols

Representation	Definition
X	Plaintext
Z	Ciphertext
R	Total number of round
const	Constants
rk	Round key
$\oplus$	XOR Operation

B. Encryption Process

Plaintext: 64-bit Xi= {Xi_0 || Xi_1|| Xi_2 || Xi_3},

Round keys $rk_{[0\text{-}2R]}$

Ciphertext: 64-bit Z= {Z_0 || Z_1 || Z_2 || Z_3}

- $T_{0,0} = Xi_0 \oplus wk_0,$
- $T = Xi_1$
- $T_{0,2} = Xi_2 \oplus wk_1,$
- $T_{0,4} = Xi_3$
- for round=1 to R
- $T_{round,1} = [T_{round-1,1} \oplus F(T_{round-1,0}) \oplus rk_{2round}]$
- $T_{round,3} = [T_{round-1,3} \oplus F(T_{round-1,2}) \oplus rk_{2round+1}]$
- If (round ==R) ;$T_{round} = T_{round}$
- otherwise$T_{round} = RP(T_{round})$
- $Z_0 = T_{R,0} \oplus wk_2, Z_1 = T_{R,1}$
- $Z_2 = T_{R,2} \oplus wk_3, Z_3 = T_{R,3}$
- stop

3 Proposed Architecture of Piccolo Cipher

This proposed work have designed an architecture of Piccolo ultra-lightweight block cipher with as low as possible area requirement for hardware implementation. After design realization of algorithm, requirement have been full-filled, which can be seen by the results comparison table in result section. In this work there are two kind of modules namely Key scheduling process and Encryption process. Both modules have been explained in detail in this section.

A. *Key Scheduling Process*

Piccolo is lightweight block cipher consist of r rounds and takes input of 64 bits; $X \in \{0, 1\}$. Encryption process requires some constant and key values for performing these round operations. There are two kind of keys namely whitening keys, wki $(0 < = wki < 4)$ and round keys, rki $(0 < = i < 2r)$. Both round keys and whitening keys are of 16 bits. Encryption process made of mainly two modules, F-function and permutation operations for number of rounds. F-function module is non-linear in nature, because of this Piccolo is a non-linear block cipher. F-function has four S-boxes followed by diffusion matrix M and four output S-boxes. Round permutation part is also known as Round Permutation (RP) in algorithm.

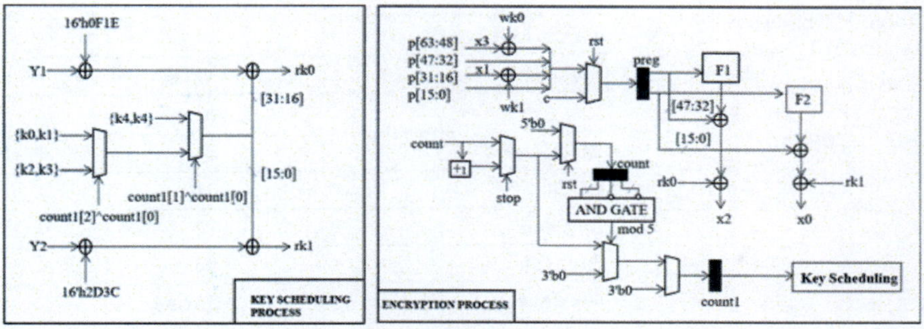

Fig. 1 Architecture of PICCOLO ultra-lightweight block cipher

B. *Data Encryption Process*

Key scheduling part of the proposed work consists generation of round key values and whitening keys. These values of the round keys are the output of some iterations. Piccolo-80 requires two round keys at each iteration hence total number of round keys are 2r, where r is the number of iterations. The key scheduling functions of Piccolo-80 is defined as KSr-80. To describe the scheduling operation more conveniently this algorithm use 16-bit constant named as $\text{const}^{80}_{\text{round}}$, which can be seen in Fig. 1 as 16′h0F1E and 16′h2D3C. Key scheduling process needs to be complex so that it can achieve both confusion and diffusion property for keys. This key scheduling process operates again and again until last round of encryption process produces ciphertext.

4 Theory and Related Methods

The proposed architecture have been implemented on Virtex-7 family of FPGA and results of different parameters can be seen in Table 2. Parameters related to ciphers like block size, key size, area specific specifications (Slices and LUTs), speed related entities (maximum frequency and throughput) are showing optimum values.

Area specified by slices in FPGA is having promising results at different platform with proposed architecture of Piccolo. In Virtex-7, number of slices required to design this architecture on FPGA is 93 only. Because of this much small number of components, it is very useful for RFID tags and resource constrained devices. Frequency of this design is also comparable to many previous works.

The family Virtex-7 and its device xc7vx330t-3 have been used in this work. Here -3 is the speed of family Virtex-7. Number of slices achieved in this implementation is very less and acceptable for resource constrained devices. As it requires much less slices, and hence Piccolo cipher is also known as ultra-lightweight cipher.

Table 2 Hardware synthesized results of Piccolo Cipher

Virtex-7 (xc7vx330t-3)	
Variable	Value
Key size	64
Block size	64
LUTs	210
Registers	72
Area (Slices)	**93**
Max.Frequency @ 100	**531.83**
Latency	26
Throughput	**1309.12**
Throughput/area	14.07
Static power	0.178
Dynamic power	0.040
Total power	0.217

Throughput of an algorithm is defined by total number of output bits per cycle in a critical delay path. It is closely related to maximum frequency, block size and clock cycle of the proposed architecture, which is in the unit of Mbps (Mega bit per seconds). Throughput must be as much as possible.

$$\text{Throughput} = \frac{\text{Number of output bit per cycle}}{\text{Critical path delay}}$$

Efficiency is also known as throughput per area which is in the unit of Mbps/ slice (Mega bit per seconds/slice). This relation between throughput and area is a factor of comparison between different algorithms.

$$\text{Efficiency} = \frac{\text{Throughput}}{\text{Area in slice}}$$

To define a parameter which is comprise of information having both area and speed is known as Efficiency.

5 Results and Comparison

Proposed architecture of this work has been implemented on various platforms and also been compared with similar properties of other lightweight cryptographic algorithms as shown in Table 3. Majority of the previous work have been used in this comparison table are presented in Virtex-5 family with different devices and packages and some of them at Spartan-3 devices. Proposed work results also have been obtained on same platforms for equal environment comparison. As this can be seen that LEA [14], HIGHT [15] and Hummingbird [16] are providing very poor efficiency as compared to proposed results on the same platform in Work-1 and Work-4. Efficiency defines, how fast an algorithm produces output bits hence it is an important factor to judge the area and speed both together. When comparing at the basis of area utilization; KLEIN [17] and Camalia [18] are having good results on Virtex-5 and Spartan-3 devices, respectively. On the same platform, proposed architecture is giving percentage improvement of 25% and 112% with respect to Virtex-5 and Spartan-3, respectively in design Work-2 and Work-5. Due to its moderate clock cycles and less number of slices, it provides better results in term of throughput at all platforms. XTEA [3] is a lightweight block cipher having simple algorithm, which is giving 111 Mbps throughput in Virtex-5 family. Comparing this results with same family, this proposed work is providing 791 Mbps throughput. Area in terms of number of LUTs are also a consideration for resource constrained devices and hence proposed work is having less LUT numbers as compared to LEA [19].

Figure 2 shows comparison of throughput values for varioud devices. From the Fig. 2, it can be seen that as technology of devices moves from Spartan-3 to Virtex-7, result shows better responses. In the same context Fig. 3 shows the comparison in number of slices on different families of FPGA.

Table 3 Comparison of synthesized hardware utilization of different algorithms

SR. no.	Algorithm	Device	Slices	LUTs	Latency	Speed (MHz)	Throughput (Mbps)	Efficiency
1	LEA [14]	xc5vlx330t	–	1130	32	126.23	505	0.071
2	LEA [19]	xc5vlx50t	–	432	29	340	1500	7.65
3	XTEA [3]	xc5vlx85	–	–	192	332	112	2.13
4	HIGHT [15]	–	–	–	160	162	64.8	0.72
5	Hummingbird [16]	xc3s200	273	473	20	40.17	160.4	0.59
6	KLEIN [17]	xc5vlx50t	110	327	28	407.84	932.19	8.47
7	Camalia [18]	xc3s50	318	–	–	18.41	2.69	0.06
8	Piccolo [20]	xc5vlx50t	131	343	–	523.01	1195.45	9.04
9	**Work-1**	**xc5vlx330t**	**85**	**240**	**26**	**304.92**	**750.57**	**8.83**
10	**Work-2**	**xc5vlx50t**	**88**	**243**	**26**	**321.35**	**791.02**	**9.00**
11	**Work-3**	**xc5vlx85**	**99**	**242**	**26**	**321.34**	**791**	**8.00**
12	**Work-4**	**xc3s200**	**154**	**293**	**26**	**126.95**	**312.5**	**2.02**
13	**Work-5**	**xc3s50**	**150**	**290**	**26**	**126.95**	**312.5**	**2.10**

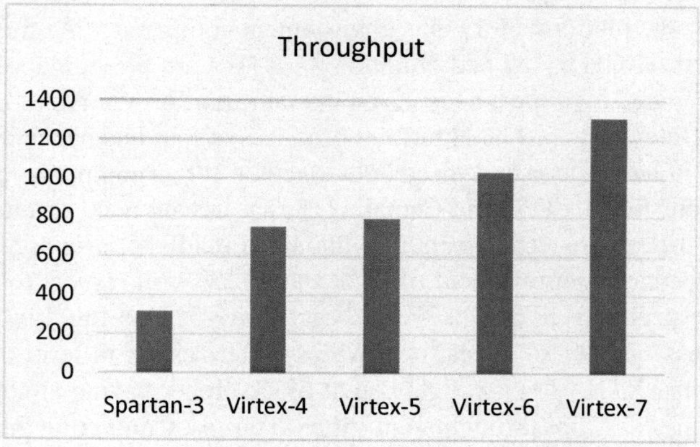

Fig. 2 Throughput comparison on various devices

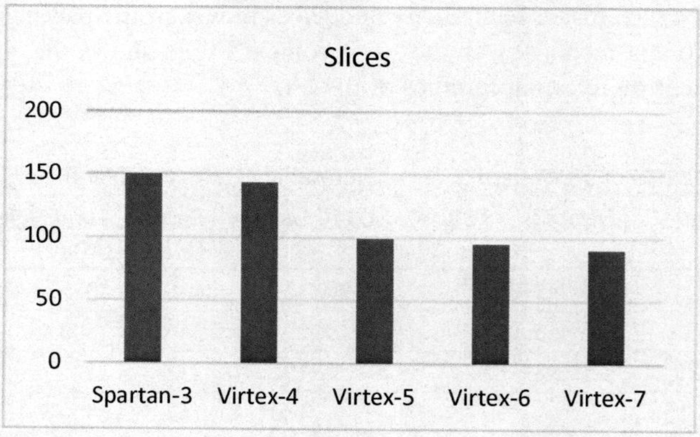

Fig. 3 Slices comparison on various devices

6 Conclusion

This work has been mainly concentrated on less area requirement and high efficiency in hardware implementation of Piccolo-80 cipher. To achive this, the paper proposed an efficient architecture for Piccolo-80 cipher. This paper used Verilog-HDL and synthesized by Xilinx tool. FPGA implementation results of various lightweight block ciphers are compared with proposed architecture. From the results, it is evident that proposed architecture requires very less area for hardware implementation, and this work gives promising outcomes. Hence, the proposed architecture of Piccolo lightweight block cipher is suitable for RFID and resource constrained IoT devices.

Acknowledgements This work has been carried out under Information Security Education Awareness (ISEA) project phase – II & SMDP-C2SD project funded by Ministry of Electronics and Information Technology (MeitY), Govt. of India in the Department of Electronics and Communication Engineering at National Institute of Technology Raipur, India. Authors are thankful to the Ministry for the facilities provided under this project.

References

1. Xu T, Wendt JB, Potkonjak M (2015) Security of IoT systems: design challenges and opportunities. In: IEEE/ACM Int. Conf. Comput. Des. Dig. Tech. Pap. ICCAD, vol. 2015-Janua, January, pp 417–423
2. Beaulieu R, Treatman-Clark S, Shors D, Weeks B, Smith J, Wingers L (2015) The SIMON and SPECK lightweight block cIPhers. In: Proc. - Des. Autom. Conf., vol. 2015-July 2015
3. Kaps JP (2008) Chai-tea, cryptographic hardware implementations of xtea. In: International Conference on Cryptology in India, pp. 363–375. Springer, Heidelberg
4. Shibutani K, Isobe T, Hiwatari H, Mitsuda A, Akishita T, Shirai T (2011) Piccolo: an ultra-lightweight blockcipher. In: International Workshop on Cryptographic Hardware and Embedded Systems. Springer, Heidelberg, pp 342–357
5. Patil J, Bansod G, Kant KS (2017) LiCi: a new ultra-lightweight block cipher. In: 2017 Int. Conf. Emerg. Trends Innov. ICT, ICEI 2017, pp 40–45
6. Hong D, Lee JK, Kim DC, Kwon D, Ryu KH, Lee D G (2014) LEA: 128-bit block cipher for fast encryption on common processors. Lect. Notes Comput. Sci. (including Subser. Lect. Notes Artif. Intell. Lect. Notes Bioinformatics), vol. 8267. LNCS, pp 3–27
7. De Cannière C, Dunkelman O, Knežević M (2009) KATAN and KTANTAN - A family of small and efficient hardware-oriented block ciphers.In: Lect. Notes Comput. Sci. (including Subser. Lect. Notes Artif. Intell. Lect. Notes Bioinformatics), vol 5747 LNCS, pp 272–288
8. Li L, Liu B, Wang H (2016) QTL: A new ultra-lightweight block cipher. Microprocess. Microsyst. 45:45–55
9. Bogdanov A, Knudsen LR, Leander G, Paar C, Poschmann A, Robshaw MJ, Seurin Y, Vikkelsoe C (2007) PRESENT: an ultra-lightweight block cipher. In: International workshop on cryptographic hardware and embedded systems, pp. 450–466. Springer, Heidelberg
10. Singh P, Acharya B, Chaurasiya RK (2019) A comparative survey on lightweight block ciphers for resource constrained applications. Int J High Perform Syst Archit 8(4):250–270
11. Singh P, Acharya B, Chaurasiya RK (2019) Efficient VLSI architectures of LILLIPUT block cipher for resource-constrained RFID devices. In: 2019 IEEE international conference on electronics, computing and communication technologies (CONECCT), pp 1–6. IEEE
12. NIST Fips Pub (2001) 197: Advanced Encryption Standard (AES). Federal Information Processing Standards Publication 197(441):0311
13. Lara-Nino CA, Diaz-Perez A, Morales-Sandoval M (2017) "Lightweight hardware architectures for the present cipher in FPGA. *IEEE Trans.* Circuits Syst I Regul Pap 64(9):2544–2555
14. Lee D, Kim DC, Kwon D, Kim H (2014) Efficient hardware implementation of the lightweight block encryption algorithm LEA. Sensors (Switzerland) 14(1):975–994
15. Yalla P, Kaps JP (2009) Lightweight cryptography for FPGAs. In: ReConFig'09 - 2009 Int. Conf. ReConFigurable Comput. FPGAs, pp 225–230
16. Fan X, Gong G, Lauffenburger K, Hicks T (2010) FPGA implementations of the Hummingbird cryptographic algorithm. In: Proc. 2010 IEEE Int. Symp. Hardware-Oriented Secur. Trust. HOST 2010, pp 48–51
17. Singh P, Acharya B, Chaurasiya RK (2019) High throughput architecture for KLEIN block cipher in FPGA. In: IEMECON 2019 - 9th Annu. Inf. Technol. Electromechanical Eng. Microelectron. Conf, pp 64–69

18. Yalla P, Kaps LP (2009) Compact FPGA implementation of Camellia. In: FPL 09 19th Int. Conf. F. Program. Log. Appl, pp 658–661
19. Mishra Z, Mishra S, Acharya B (2019) LEA 192: High Speed Architecture of Lightweight Block Cipher. Int. J. Innov. Technol. Explor. Eng. 9(2S):359–363
20. Ramu G, Mishra Z, Singh P, Acharya B (2020) Performance optimized architecture of Piccolo block cipher for low resource IoT applications. Int J High Perform Syst Archit 9(1):49–57

High Throughput Pipelined Architecture for AES Cipher

Piyush Modi, Pulkit Singh, Bibhudendra Acharya, and Shrish Verma

Abstract There has been a significant rise in internet and wireless device users, raising the need for security to protect consumer data shared over open networks. Field Programmable Gate Arrays (FPGAs) are particularly desirable alternative for hardware implementation of cryptographic algorithm. This paper discusses hardware implementation of AES algorithm on a Field Programmable Gate Array (FPGA) platform with focus on high throughput and low area constraint. In the proposed design, an efficient pipelined architecture for AES encryption/decryption is realized using unrolled and external pipelining techniques. The proposed design is implemented on different families Virtex-5 and Spartan-6 of FPGA platform. This design obtained maximum throughput of 37.57 Gbps and 32.44 Gbps on Virtex-5 and Spartan-6, respectively. Along with throughput, it also got maximum frequency and high throughput per slice with significant number of slices.

Keywords Advanced Encryption Standard (AES) · Field Programmable Gate Array (FPGA) · Encryption · Security · Throughput

1 Introduction

Cryptography plays a critical role in today's age of information technology and communication systems. The main concerns for any cryptographic system are security including data integrity, authenticity, and confidentiality. The purpose of any

P. Modi (✉) · P. Singh · B. Acharya · S. Verma
Department of Electronics and Communication, National Institute of Technology Raipur, Raipur, India

P. Singh
e-mail: pksingh.phd2015.elx@nitrr.ac.in

B. Acharya
e-mail: bacharya.etc@nitrr.ac.in

S. Verma
e-mail: shrishverma@nitrr.ac.in

V. Nath and J. K. Mandal (eds.), *Proceeding of Fifth International Conference on Microelectronics, Computing and Communication Systems*, Lecture Notes in Electrical Engineering 748,
https://doi.org/10.1007/978-981-16-0275-7_29

cryptographic algorithm is to ensure security against any threat. Security threats can be active or passive infringing on security services. Cryptographic algorithms implemented on hardware are effective solution in case where speed of communication network exceeds from Giga bits per second (Gbps). In addition, FPGA hardware implementation provides physically secure implementation.

Cryptography can be categorized as symmetrtic cryptography and asymmetric cryptography. DES, Triple-DES, AES, RC4 are examples of symmetric key or private key cryptography, and whereas RSA, Elliptic Curve cryptography are examples of asymmetric or public key cryptography. Private key cryptosystem uses only one key as opposed to public key cryptographic system which uses two keys (private and public key) one for encryption and other for decryption. Public key algorithms are difficult to implement due to the use of multiple keys and in contrast to the public key algorithm, private key algorithms are fastest.

There is always requirement of optimizing cryptographic algorithms for area, performance, and power consumption in hardware. However, it is hard to achieve different features like cost, throughput, and power/energy consumption at the same time [1, 2]. Conventional S-box uses the lookup table implementation of substitution box. It is easy to achieve high throughput rates using concept of pipelining and sub-pipelining. Also, it can easily change encryption and decryption modes using same architecture.

In 2001, National Institute of Standards and Technology (NIST) established Advanced Encryption Standard (AES), also known as Rijndael [3]. The aim of AES was to provide encryption of electronic data to secure U.S. government documents. Rijndael was developed to be immune to every specific attack and have characteristics such as speed and coding compactness on a broad range of platforms and flexibility in architecture. Every cryptographic algorithm has two fundamental principles: substittion and permutation. In the process of substitution, each byte in the plaintext is translated into a different byte and in permutation elements in the plaintext are re-arranged.

1.1 Contributions

In this paper, outer-round pipelined architecture of AES encryption/decryption core are proposed.

- Proposed a hardware structure with high throughput and having a maximum frequency with high throughput per slice with less critical delay, makes it suitable for high speed application.
- Also, there is significant reduction in number of slices as compared to other designs.

1.2 *Structure of the Paper*

The remaining of the paper is structured as follows. Sect. 2 briefly defines the literature survey performed for this work. In Sect. 3, the algorithm overview of AES is mentioned. In Sect. 4, proposed architecture for AES is elaborated. Sect. 5 consists of design methodology for the implementation. Section 6 consists of results and discussions. Section 7 provides concluding remark.

2 Related Work

Advanced Encryption Standard is the most favoured conventional cryptographic algorithm. Encryption and decryption processes with combined datapath along with reuse of all manipulating components are given in [4]. In this paper, a high-speed architecture of AES with logic optimization is proposed. In [5] a parallel fully-pipelined architecture on a single chip is presented. It suggested outer pipelined and sub-pipelined AES architecture and makes use of resource sharing. Also, it gave integration of fully-pipelined parallel units, with reduction in clock rates. It has advantage of less power and maintains decent throughput also. In [6], area and delay efficient multipliers and multiplicative inverse are designed making use of loop unrolling pipelining architecture. Also in this paper, authors explored both outer as well as fully-pipelined approach and high throughput implementation of AES in ECB and CTR modes are achieved. In order to make a high throughput AES system pipelining and loop unrolling techniques are used by many researchers. High throughput cryptographic design can also be achieved using parallel processing [7]. S-box is the crucial part of any cryptographic algorithm as it is the part which takes most of area in any cryptographic design and many researchers try to modify it in terms of its size and power to get an area and power efficient design. S-box used in AES is 8-bit S-box, all 8-bits are in the form of 16*16 table and it is used with big applications [8, 9]. Lightweight block ciphers such as PRESENT [10], mCrypton [11], KLEIN [12] and LBLOCK [13] are the examples of 4-bit S-box ciphers.

3 AES Algorithm Overview

AES is a symmetric block cipher, uses SPN based structure [3, 8]. It has 128-bits block size with three key variants. These key versions have different key sizes with distinct round count. The round function of AES cipher consists of four distinct phases, one of the phase is based on permutation and other three on substitution. There are 10 rounds in AES-128, all rounds except the last round consist of all four stages. The last round has three stages only. It comprises of two parts, namely Datapath and Key scheduling.

3.1 Datapath

Four different operations are performed in AES algorithm as:

- Substitute bytes/Inverse Substitute bytes-
  S-box contributes to non-linearity of AES and influences reliability and confidentiality of AES as well. S-box in the AES is classified as SubBytes (Substitution byte transformation) and InvSubBytes (Inverse Substitution byte transformation). All 256 bit input is divided into 16 states in the form of 4*4 matrix. Each state contains 8-bit or 1 byte. Every byte of the state is converted into different bytes. The left four bits of the byte is used as the value of the row, and the rest of the four bits of the byte is represented as the value of the column in the 16*16 table S-box. For a certain row and column size, the S-box provides a unique 8-bit output. To minimize the similarity between input bits and the output bits at the byte level, the substitution lookup tables are built based on bit scrambling.
  Each byte of the data in the state passed to the SubBytes stage which is a nonlinear operation. S-box consists of permutation of all input values. The S-box is created using defined transformation of values in Galois Field GF(2^8) by manipulating the product inverse of each byte in GF(2^8) with the fixed polynomial x^8 + x^4 + x^3 + x + 1.
- ShiftRows/Inverse ShiftRows-
  Inv-/ShiftRows is a cyclical left shift of each row of the state in encryption or to the right in case of decryption. The upper row is called row 0, and the lower row is called row 3. For example, the first row (row 0) remains as it is, while the successive row moves left side by one byte. Row 1 moves one byte left, row 2 moves 2 bytes left. Similarily, row 3 moves 3 bytes left.
- Mixcoloumns/Inverse Mixcoloumns-
  Inv-/MixColumns operation provide diffusion to cipher. Inv-/MixColumns treats each column independently as a four-term polynomial over GF(2^8) and converts each column into a different one by multiplying it with a fixed polynomial a(x) = {03} x^3 + {01} x^2 + {01} x + {02} modulo $x^4 + 1$. The first row of polynomial matrix consists of {02, 03, 01, 01} and next three rows of polynomial matrix are one bit right shift from the upper row i.e. {01, 02, 03, 01}, {01, 01, 02, 03} and {03, 01, 01, 02}. Inverse MixColumns is the reverse operation of forward mixcoloumn operation.
- AddRoundKey-
  The operation AddRoundKey is logical XOR between output of previous step with a round key created by key expansion. XOR action is an opposite of its own.

3.2 Key Scheduling

In AES-128, the input key used is 128 bit in length [8]. The input key in the form of 4-words is extended into an group of forty-four words. These forty-four words serve as 4-word round key for initial AddRoundkey and 10 rounds. Four different words are used as a round key for each round. Following are the operations performed in key expansion mechanism.

RotWord:- RotWord operation executes a one-byte circular left shift on a word. It ensures that the input word [x0, x1, x2, x3] is translated to [x1, x2, x3, x0].

SubWord:- SubWord operation executes a byte replacement using the S-box for each byte of its input word. It is the same operation used in datapath. It provides non-idealities in key schedule and based on Galois Field.

XOR:- The outcome of RotWord and SubWord stage is XORed with a round constant. The round constant is a term whose value relies on the round number as per Table 1.

4 Proposed Work of AES Cipher

In this paper, conventional pipelining approach is used to increase speed of execution. An outer-round pipelining is done by inserting registers before each round as shown in Fig. 1. Loop unrolling processes all 128 bit simultaneously, this boost speed at the expense of rise in area. Loop-unrolling technique eliminates all the loops involved in the AES algorithm which results in a critical path transition. For AES-128 encryption, data passes from round 1 to round 9 and then to round 10, final round does not contain mixcolumn step. Pipelined registers are inserted in between rounds to minimize the critical path delay. In the next round the data computed in each particular round is used successively as the input. With the help of resource sharing, it also managed to keep resources upto a certain extent. Shiftrows operation is eleminated completely by selecting elements diagonally from output of S-box. In this paper, encryption as well as decryption operations are assembled into one module. One multiplexer selects which process in between encryption and decryption has to be executed. Substitution

Table 1 Round constant values used in key scheduling

Round Number [I]	1	2	3	4	5	6	7	8	9	10
Round Constant RC[I]	01	02	04	08	10	20	40	80	1B	36

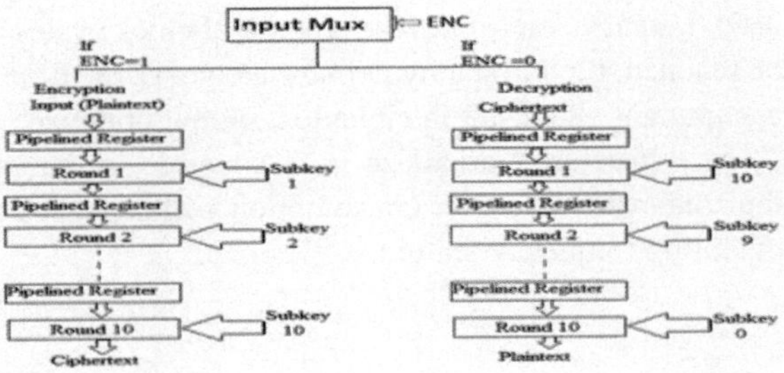

Fig. 1 Outer pipelining in AES Cipher

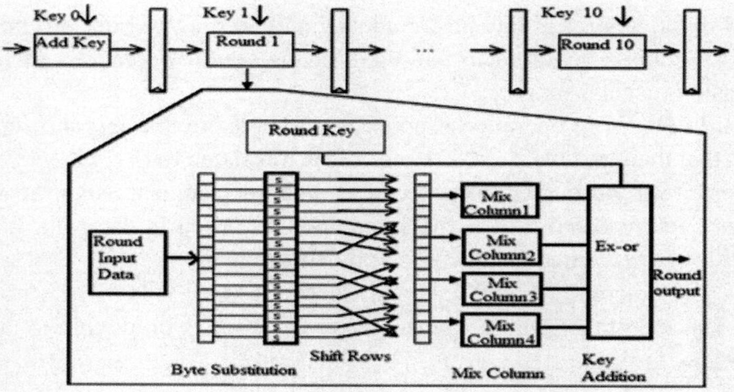

Fig. 2 Proposed pipelined architecture for AES Encryption

box is made using traditional lookup table approach. Figure 2 shows description of single round for AES cipher in detail.

5 Methodology

5.1 Environment

The proposed Pipelined AES architecture is synthesized by ISE design suite in Xilinx and simulation is performed by ISim simulator. Spartan-6 and Virtex-5 families of FPGA are used for comparison. There are different amount of LUTs, flipflops and slices, and different maximum frequency and throughput values in each unit.

5.2 Meaurement Parameters

The architecture realized on FPGA is evaluated based on different parameters including area and speed. This measures performance using frequency and latency. To describe some structure field, parameters like Slices, LUTs and Flip-flops are used. Latency determines the amount of clock cycles necessary for the first output to be reached. Cryptography defines latency as the time between the plaintext is given as input and the result in ciphertext output obtained.

Any structure's area size is responsible for energy or power consumption. As the area expands, power consumption will increase and vice versa. The output is an operating frequency feature.

6 Results and Discussion

For synthesis, timing analysis and placement and routing Xilinx ISE tool is used. The proposed pipelined architecture of AES cipher is implemented on the Xilinx Spartan-6 and Virtex-5 FPGA families. This architecture achieves higher throughput as compared to all other designs for a perticular device as shown in Table 2. It got throughput of 37.57 Gbps for Virtex-5 and 32.44 Gbps for spartan-6. The proposed design obtains higher throughput per slice (in *Mbps/slices*) with values 3.12 and 2.81 for Virtex-5 and Spartan-6 families, repectively. It shows maximum frequencies of 293.56 MHz and 253.47 MHz and low critical delay of 3.4 ns and 3.9 ns for Virtex-5 and Spartan-6, repectively. Despite of speed or performance parameters, the proposed design got a respectable low number of slices as compared to other devices.

Table 2 FPGA implementation results for various ciphers

Designs	Devices	Slices	BRAM	FMax (MHz)	Throughput (Gbps)	Critical delay (ns)	Throughput/Slice (Mbps/Slice)
[14]	XC2V4000	16938	0	184.16	23.57	5.4	1.39
[15]	Spartan-3 XC3S2000	17425	0	196.10	25.10	5.1	1.44
[16]	XCV1000E	11022	0	168.30	21.55	5.9	1.95
[17]	Spartan-3 XC3S4000	20720	0	240.90	30.83	4.1	1.48
[2]	XC2VP30	11720	0	206.84	26.64	4.8	2.26
[18]	Virtex-2 Pro XC2VP30	12556	100	373.00	47.74	2.6	3.80
[19]	Virtex-2 Pro XC2VP30	12450	0	168.30	21.54	5.9	1.73
[20]	Virtex-2 XC2V3000	139357	0	222.20	28.40	4.5	0.20
[21]	Virtex-E 3200E	15112	0	145.00	18.56	6.9	1.23
[17]	Spartan-3 XC3S4000	20720	0	240.90	30.83	4.1	1.48
Proposed design	Virtex-5 XC5VLX85T	**10377**	44	**293.56**	**37.57**	3.4	**3.12**
Proposed design	Spartan-6 XC6SLX100T	**11508**	37	**253.47**	**32.44**	3.9	**2.81**

7 Conclusion

In this work, a high-speed design for the AES cipher is realized on FPGA. As compared to related works, it can be noted that the proposed pipelined architecture achieved high performance in terms of speed or throughput, throughput per slice and maximum frequency for each device. Despite having high-speed performance for each device, it can be seen that this design also got respectable less area requirement valued at 10377 slices for virtex-5 family and 11508 for Spartan-6 family as compared to other related works.

Acknowledgment This work has been carried out under Information Security Education Awareness (ISEA) project phase – II & SMDP-C2SD project funded by Ministry of Electronics and Information Technology (MeitY), Govt. of India in the Department of Electronics and Communication Engineering at National Institute of Technology Raipur, India. Authors are thankful to the Ministry for the facilities provided under this project.

References

1. Singh P, Acharya B, Chaurasiya RK (2019) A comparative survey on lightweight block ciphers for resource constrained applications. Int J High Perform Syst Archit 8(4):250–270
2. Iyer NC, Anandmohan PV, Poornaiah DV, Kulkarni VD (2006). High throughput, low cost, fully pipelined architecture for AES crypto chip. In: 2006 Annual IEEE India Conference, pp. 1–6. IEEE
3. Pub NF (2001) 197: advanced encryption standard (AES). Federal Inf Proces Stan Publ 197(441):0311
4. Satoh A, Morioka S, Takano K, Munetoh S (2001) A compact rijndael hardware architecture with S-box optimization. In: International Conference on the Theory and Application of Cryptology and Information Security, pp. 239–254. Springer, Berlin
5. Alam M, Badawy W, Jullien G (2002) A novel pipelined threads architecture for AES encryption algorithm. In: Proceedings IEEE International Conference on Application-Specific Systems, Architectures, and Processors, pp. 296–302. IEEE
6. Soltani A, Sharifian S (2015) An ultra-high throughput and fully pipelined implementation of AES algorithm on FPGA. Microprocess Microsyst 39(7):480–493
7. Singh P, Acharya B, Chaurasiya RK (2019) High throughput architecture for KLEIN block cipher in FPGA. In: 2019 9th Annual Information Technology, Electromechanical Engineering and Microelectronics Conference (IEMECON), pp. 64–69. IEEE
8. Daemen J, Rijmen V (2013) The design of Rijndael: AES-the advanced encryption standard. Springer
9. Nyberg K (1991). Perfect nonlinear S-boxes. In: Workshop on the Theory and Application of Cryptographic Techniques, pp. 378–386. Springer, Berlin
10. Bogdanov A, Knudsen LR, Leander G, Paar C, Poschmann A, Robshaw MJ, Vikkelsoe C (2007) PRESENT: An ultra-lightweight block cipher. In: International Workshop on Cryptographic Hardware and Embedded Systems, pp. 450–466. Springer, Berlin
11. Lim CH, Korkishko T (2005) mCrypton–a lightweight block cipher for security of low-cost RFID tags and sensors. In: International Workshop on Information Security Applications, pp. 243–258. Springer, Berlin
12. Gong Z, Nikova S, Law YW (2011). KLEIN: a new family of lightweight block ciphers. In: International Workshop on Radio Frequency Identification: Security and Privacy Issues, pp. 1–18. Springer, Berlin

13. Wu W, Zhang L (2011). LBlock: a lightweight block cipher. In: International Conference on Applied Cryptography and Network Security, pp. 327–344. Springer, Berlin
14. Zambreno J, Nguyen D, Choudhary A (2004) Exploring area/delay tradeoffs in an AES FPGA implementation. In: International Conference on Field Programmable Logic and Applications, pp. 575–585. Springer, Berlin
15. Good, T., Benaissa M (2005). AES on FPGA from the fastest to the smallest. In: International workshop on cryptographic hardware and embedded systems, pp. 427–440. Springer, Berlin
16. Zhang X, Parhi KK (2004) High-speed VLSI architectures for the AES algorithm. IEEE Trans Very Large Scale Integr (VLSI) Syst 12(9): 957–967
17. Good T, Benaissa M (2007) Pipelined AES on FPGA with support for feedback modes (in a multi-channel environment). IET Inf Secur 1(1):1–10
18. Iyer N, Anandmohan PV, Poornaiah DV, Kulkarni VD (2011). Efficient hardware architectures for AES on FPGA. In: International Conference on Computational Intelligence and Information Technology, pp. 249–257. Springer, Berlin
19. Hodjat A, Verbauwhede I (2004) A 21.54 Gbits/s fully pipelined AES processor on FPGA. In: 12th Annual IEEE Symposium on Field-Programmable Custom Computing Machines, pp. 308–309. IEEE
20. Fan CP, Hwang JK (2008) FPGA implementations of high throughput sequential and fully pipelined AES algorithm. Int J Electr Eng 15(6):447–455
21. Standaert FX, Rouvroy G, Quisquater JJ, Legat JD (2003). Efficient implementation of Rijndael encryption in reconfigurable hardware: Improvements and design tradeoffs. In: International Workshop on Cryptographic Hardware and Embedded Systems, pp. 334–350. Springer, Berlin

Study and Design of Smart Embedded System for Smart City Using Internet of Things

Vidushi Goel, Ashish Kumar, Konthi Jyoti Prakash, G. S. Namith, Sanjay Kumar, Renuka, Ramashankar Yadav, Aamir Junaid Ahmad, Dharmveer Kumar Yadav, and Vijay Nath

Abstract In this paper a comprehensive view on the concept of smart cities is being provided along with the role of IoT technologies and the various electronic devices in making cities greener, safer and more efficient. A smart city is an urban development plan in which different electronic data collection sensors can be used to supply information which is required to manage assets and resources efficiently. IoT technologies include Radio-frequency Identification (RFID), IR's, Lasers, scanners etc.

Keywords RFID · Lasers · Scanners · IOT

1 Introduction

In this research paper, we have introduced the definition, components and standards of IoT (Internet of Things) and discussed possible models that can be implemented in smart city.

A smart city is an urban development plan in which different electronic data collection sensors can be used to supply information which is required to manage assets and resources efficiently [1].

V. Goel (✉) · K. J. Prakash · G. S. Namith · S. Kumar · Renuka · A. J. Ahmad · D. K. Yadav · V. Nath
Department of Electronics and Communication Engineering, Birla Institute of Technology Mesra, Ranchi 835215, Jharkhand, India

V. Nath
e-mail: vijaynath@bitmesra.ac.in

A. Kumar
Qualcomm India Pvt. Ltd., Madhapur Village, Serilingampally Mandal Ranga Reddy District, Hyderabad, India

R. Yadav
Neena Thapa Inter College Gorakhpur Cantt Gorakhpur, Gorakhpur, Uttar Pradesh, India

V. Nath and J. K. Mandal (eds.), *Proceeding of Fifth International Conference on Microelectronics, Computing and Communication Systems*, Lecture Notes in Electrical Engineering 748,
https://doi.org/10.1007/978-981-16-0275-7_30

A smart city is that which is well planned, provides cost effective a technologically sound services and environmental efficiency for the improvement of standard of living. The objective is to promote cities that provide advanced infrastructure, a tidy and sustainable environment and that applies ‘smart ‘solutions for existing problems.

The internet of things is the network of physical devices which are embedded with electronic sensors, transducers, actuators and are connected through internet through which these objects can collect and exchange information. This way objects can be sensed and controlled remotely across existing network infrastructure efficiently. The IoT prototype is composed to smart and self managing objects connected through a worldwide network infrastructure. IoT technologies include Radio-frequency Identification (RFID), IR’s, Lasers, scanners etc.

RFID is a technology that is used to record the presence of an object using radio signals. Electromagnetic/electrostatic coupling is being used in radio frequency region of spectrum to identify or detect any entity [2].

Scanner is an electromagnetic device that captures images from physical items and converts them into digital formats.

Laser is a device that emits light through the phenomenon of optical amplification which is based on the principle of simulated emission of electromagnetic radiation. Lasers provide many advantages. Air quality can be monitored through them.

IoT can be implemented in various projects concerning with sanitation, environmental pollution, smart traffic services, and security services etc.

Wireless Sensor Networks can provide different suitable data and are implemented in health care sector, government and environmental services etc. Internet is the reason behind the implementation of IoT facilities. The internet is the global system of interconnected computer networks to link devices worldwide. Thus for internet of things we install sensors for data collection and connected to internet through the use of certain protocols for exchanging and communicating allowing intelligent and smart recognition, tracking and also management. Sensors can be developed at various locations for collecting and analyzing data for proper usage. Smart cities thus focus on improving lives of citizens. They are advantageous as they free up resources by intelligently delivering essential services, thus investment can be invested in other services. Figure 1: Factors covered under smart cities are covered in this paper.

2 Methodology

The applications of IoT smart cities have been explained below:-

2.1 Waste Management

Management of waste is of great concern all over the world today for making cities cleaner and tidier. We can schedule the waste collection timing by keeping sensor

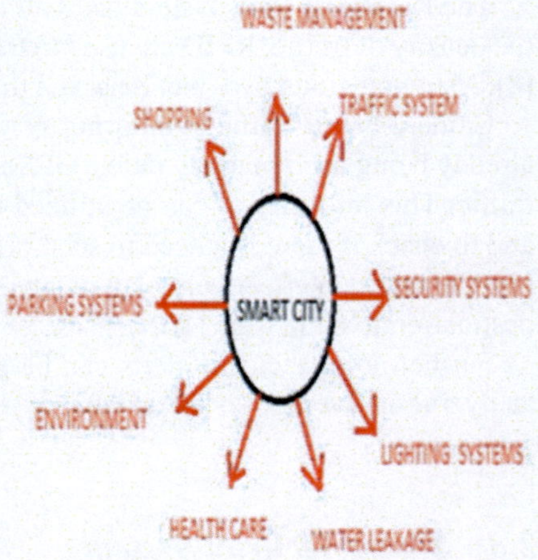

Fig. 1 Factors covered under smart cities

based bins which would check for the status if they are filled or empty. Smart dustbins provide ease to access them from anywhere and anytime by the concerned people. This way authority can send collection vehicle only when the bin is full also the collector truck route can be optimized, thereby minimizing the overall cost of waste collection. This system can be implemented at frequently visited public places like bus stops, railway stations etc [3].

Overflowing of the bins will be reduced this way and the cost and time requirement for same will be reduced too. We can make waste collection process efficient and pave way for recycling through the use of technologies like RFID's, IoT sensors etc. Cities become cleaner and environment quality is also improved by reduction in smell. Moreover availability of toxic substances in bin can also be detected.

Ultrasonic sensor can be used for level detection by the use of sound transmitter and receiver [4]. It creates an ultrasonic pulse known as ping and listens for the reflection of pulse. Arduino mega 2560 used in the proposed models consists of temperature and humidity sensors.

A better system may contain a sensor which would give warning if we dump biodegradable waste with non biodegradable. Warning should be sent to collectors somewhat before the bin is full [5]. In meanwhile till the collector arrives, while the bin is already full, an extra small flab should open up for those who want to dump.

2.2 Smart Traffic Services

Smart traffic management system would be able to bring real time change in traffic patterns, provide drivers with information and alerts about the traffic in various areas and manage traffic [6].

Also by proper implementation of sensors and scanners in roads, citizens will be able to determine the exact arrival time to a destination [7].

The Raspberry Pi in system takes all control. IR sensors are employed to identify the density of traffic. RFID are used to track the entry of ambulance and lost vehicles [8]. Many precious lives can be saved through the process.

Camera-based traffic monitoring systems, which take still pictures of traffic, are already being used in many cities. GPS installed on vehicles can be used to monitor traffic. This information can be utilized by authorities to control and manage traffic and to check if there is a need to send officers somewhere where there is congestion [9]. Citizens can effectively plan in advance the route they would take to reach their destination to be in time [10].

Furthermore, accident alerts can be generated; vehicle analysis can be done by analyzing its average speed on various road types and travel path [11].

2.3 Smart Security System

Security is the most important factor in any smart city. The entire city must be monitored. Hence it is the need of the hour to develop such IoT applications for providing safety and security, particularly in preventing the home crimes etc.

An IoT based wireless home security system sends alert message to the owner by using internet if there is any trespass alarm is raised in case of requirement. Home automation system can also be deployed on same lines. The system runs using a launchpad connected to Wi-Fi. Internet connectivity is not required for accessing the alerts and status of the IoT system and they can be tracked from anywhere. If person entering his house is not an intruder, the user can then open the door for him.

PIR motion sensors can be installed at the entrances of buildings for achieving this. These sensors are used for detecting the motion of human beings. An input trigger is sent to the micro-controller whenever sensor detects their presence.

In future, more specialized systems can be designed for bank security and also finger print sensors which would change the way we see and seek security.

2.4 Smart Health Care Systems

The field of health care has drastically changed and progressed by the advent of IoT as it has provided ease to patients and doctors. A system of connected apps and devices is being used by doctors to monitor, track and record patient's medical information. These devices include smart meters, wearable health bands, fitness shoes; RFID based smart watches and smart video cameras. Nowadays apps in smart phones which monitor heart beat, blood pressure etc. have made it easier to keep a track of individual's medical record. Sensors and microcontrollers can accurately measure the level of oxygen and glucose the body. In the pharmacy field there is application of sensors by using them into medicines and pills that are connected to a network and can send alerts about whether the patient has taken the required dose of prescribed medicines.

These way patients don't need to visit to doctor that frequently for little problems. This proposal can help in saving the time of both patients and doctors.

Sensors and microcontrollers can be employed to sense the condition of the patient's body and the data generated can be sent to the respective hospital's website.

Also in the emergency cases, wherein the patient's condition is critical, a smart ambulance can be arranged and sent to the house of the patient. Also integrated hardware using Arduino can be employed to avoid obstacles and accidents and also google maps can be used by the ambulance drivers to detect the location of patient instantly.

2.5 Smart Environment

A smart cannot include unhealthy people in it. Hence, the environmental pollution must be monitored and the necessary details of it must be provided to the citizens. Sensors can be manufactured in such a way that the amount of CO2 can be judged and also the amount of pollution let by the cars can be estimated precisely. Detectors can also be put in places where the quality of water can be judged, like in ponds, lakes, rivers and seas. It is important that we can also detect forest fires through the system which can check for combustion gases and can send alert stating fire conditions. Also the amount of noise present in the surroundings can be detected by the electronic devices in dB and alert the people to avoid noise pollution [12].

The quality of air can be monitored by urban IoT in various places like parks, fitness trails etc. Also many apps can be created in smart phones wherein the environment condition and the amount of pollution can be estimated and can alert people. Healthy paths for outdoor activities should be chosen by people by accessing different apps. Hence, it is quite necessary to deploy services in cities which determine the quality of air and also the purity of drinking water and that the collected data can be made available to all the citizens of the city publicly.

2.6 Smart Lighting Systems

This system should be able to design and execute the advanced development in embedded systems for energy saving of the street lights. The design should be able to eliminate the manual operation of the lights in the city. Also it must be able to save energy. We can use IR transmitter and IR receiver couple for achieving this purpose. Upon sensing the movement of the vehicle, the data will be transmitted to the microcontroller [13]. The light gets switched off as soon as the vehicle moves away. The sensor can be accessed anywhere and anytime through internet and also this must be done properly to ensure no errors creeping in. Smart embedded street lighting system can be used for making this project with an advantage of reducing street lighting costs as much as 50%−85%.

Light emitted from the bulbs can also be adjusted according to their usage and occupancy. Wireless based systems can also be installed to track and control the actual energy consumption of street lights so that measures can be taken to reduce power consumption.

Also, security can be ensured by street lighting systems. A panic button can be inserted on street lights which can be pressed in times of crimes or emergency cases and this can get redirected to the concerned Police station of that area.

So, street lighting systems have got a great scope in the future which can also eliminate many crimes in the cities.

For supporting 20-20-20 directive, the lighting system optimization is to be taken under consideration [14]. The street lamp intensity can be optimized according to the factors like time of the day, the weather condition, and the presence of people. For proper functioning, Smart city infrastructure must include this service of enhancing the facilities of smart lighting systems. Moreover, top of the street light controllers can be provided with a fault detection system.

2.7 Smart Parking System

Many problems are faced to park the vehicles in slots. Hence, in smart parking systems IoT sensors can be used to direct the user in finding nearest parking area and also guiding them in finding nearby parking slots. The time consumed in parking vehicles must be reduced by this system It should also reduce the fuel consumption of the vehicles [15].

A parking sensor can be set up which may have a camera to detect parking spaces which are empty and should be able to send this data to the server. The owners of the vehicles can then use this stored data. Even before start of their journey people can check the availability of parking slots. Effectively traffic congestion can be effectively searched through it. Embedded systems like arduino can also be used here. Even video sensors can be eventually developed which help the vehicles and viewing the back and the sides.

Verification system of parking can also be developed to permit the residents and the disabled to use the slots being reserved for them, thus in this way a better service can be offered to citizens who can legitimately use those slots. Violations can be quickly and effectively spotted through the use of this tool.

2.8 Conservation of Water Using IoT

Conservation of water is a very important factor to be dealt with in a smart city. A system should be developed in such a way that the right amount of water is present in the reservoirs and the overhead tanks by minimizing the cost of operation to pump water to houses.

Level sensors can be employed in reservoirs and overhead tanks; these can determine the level of overhead water and also be able to communicate the reading of the water level to the central server or the house owner. The volume of water in these tanks can be calculated by using predictive analysis method. In this way the water requirement is also estimated. Also by using appropriate information, water required for city can be measured by subtracting existing levels consumed by the people from the water demand in the city. Water leakage can also be avoided by setting up the sensors in washrooms, links of kitchen etc. Water conservation can be done appropriately through the system.

In the proposed infrastructure, intelligently we can utilize IoT sensors to generate predictive information to analyze the status and working conditions remotely. The information can then be communicated later, for example; we can open, close, or turn off a water gate remotely to adjust the flow of water through a water supply system [16]. Equipments with moving parts in the water infrastructure can also be sensed and easily be monitored for malfunction, thus it is an easy way to provide the maintenance which is otherwise difficult. The same utility can also be employed for fixing water failure thus sending alerts for the owner to fix it. These IoT based water treatment plants can tell if their filters are clean and functioning properly or not. The IoT can measure in the transported water the leakages faster and even the presence of many harmful chemicals in the consumable water supply [17].

2.9 Smart Shopping

Starting with an example of a grocery store where all items can be connected with each other to form a smart shopping system. An inexpensive RFID tag can be employed to each product which, when placed into a smart shopping cart, can be automatically read by a cart in which an RFID reader is already fixed. This will enable automatic billing being conducted from the shopping cart itself, resulting in saving customers from waiting in a long queue at billing counters. Additionally, addition of smart shelving into this system, equipped with RFID readers, can be used to monitor stock, by every time when an item is being added to cart, updating a central server. Another advantage of this kind of system is that inventory can be managed much easily and efficiently, as all items can be automatically read by an RFID reader instead of manually scanned by a worker at the checkout [18].

Information and communication technologies can together enable new forms of mobile marketing in respect to situated marketing communication, dynamic pricing models and dynamic product differentiation models.

3 Challenges/Limitations of These IoT Systems

3.1 Security and Privacy of IoT Systems

The IoT includes anything and everything right from wearable fitness bands and smart home appliances to factory control devices, smart solutions of real life problems in real time. But these devices cannot assure you the security you wish to have. It is not secure to collect data in a large common platform, as the system can be prone to several cyber attacks. Also the data leakage can be a big problem to the users, where their bank details, statements etc. can be hacked and even misused by others. Hence, development of properly secure IoT systems which may consists of unique ID's or password is the need of the hour which may consists of unique ID's or password.

3.2 Production of System in Large Scale and Proper Implementation

For proper implementation of applications in smart cities, we need to cover the entire region and we must be able to provide with the electronic devices and sensors, microcontrollers effectively for all proposed models like for smart security systems, the sensors and the alarm kits are be provided to all the houses it the city. Another issue is availability of Wi-Fi connection with all the users for getting the alert message. At least, he should have with him a smart phone for this. This may pose a big problem while implementing such a model of smart city as all people may not have proper facilities for these systems to run.

Citizens must be educated about the working of these instruments so that they can use them effectively by themselves. This may also be a big issue as everyone may not know enough about the usage.

4 Conclusion

Hence, the above research paper focuses on the various characteristics of IoT systems implemented in smart city model. It also says that how IoT can be used to make the city a much better place to live in by interconnecting various devices, vehicles and infrastructure everywhere around the city. Many of the useful applications of the sensors have been highlighted and their benefits have been listed. The challenges which arise while implementing the IoT in this model have been briefly explained. Thus, the technology covers many fields like smart vehicular systems, health care, waste management systems etc. becoming a more general class of cyber-physical system. Also, providing internet of secure things is very important.

Acknowledgements The authors wish to acknowledge especially to Dr. Vijay Nath, professor, department of Electronics and Communication Engineering, Birla Institute of Technology, Mesra, Ranchi, for his guidance, help and sharing fruitful ideas. The authors are deeply grateful to him for rendering all necessary information. The authors also wish him for his guidance, useful ideas, providing necessary suggestions and for his patience, continuous help and encouragement.

References

1. Wikipedia
2. Smart city using IoT, IEEE Xplore
3. How does an IoT system actually work, leverege
4. Research gate.net
5. www.IJRASET.com
6. Smart City Implementation Models Based on IoT Technology (2016) IEEE Paper
7. IoT Internet of Things 2014-research papers, engpaper.net
8. Evans D (2011) The Internet of Things: How the Next Evolution of the Internet Is Changing Everything
9. Bellavista P, Cardone G, Corradi A, Foschini L (2013) Convergence of MANET and WSN in IoT urban scenarios. IEEE Sens J 13(10):3558–3567
10. Hammi B, Khatoun R, Zeadally S, Fayad A, Khoukhi L (2018) IoT technologies for smart cities. IET Networks 7(1):1–13
11. Laya A, Bratu VI, Markendahl J (2013) Who is investing in machine-to-machine communications?. In: Proceedings 24th European registration ITS Conference, p. 20
12. Gubbi J, Buyya R, Marusic S, Palaniswami M (2013) Internet of Things (IoT): A vision architectural elements and future directions. Future Gener. Comput. Syst. 29:1645–1660
13. Ratti C (2011) The Social Nexus. Sci Am 305(3):42–48
14. Marsal-Llacuna M-L, Colomer-Llinàs J, Joaquim M-F (2015) Lessons in urban monitoring taken from sustainable and livable cities to better address the Smart Cities initiative. Technol Forecast Soc Chang 90:611–622
15. Nam T, Pardo TA (2011) Conceptualizing smart city with dimensions of technology, people, and institutions. In: Proceedings of the 12th Annual International Gigital Government Research Conference: Digital government innovation in challenging times, College Park, MD, ACM
16. Lombardi P, Giordano S, Farouh H, Yousef W (2012) Modelling the smart city performance. Innov: Eur J Soc Sci Res 25(2): 137–149
17. Harrison C, Eckman B, Hamilton R, Hartswick P, Kalagnanam J, Paraszczak J, Williams P (2010) Foundations for smarter cities. IBM J Res Dev 54(4): 1–16
18. Hall P (2000) Creative cities and economic development. Urban Stud 37(4): 639–649

Linear Precoding Techniques for MIMO Systems

P. N. Jayanthi, S. Rachana, and M. Ramya

Abstract Precoding is a method used in Multiple Input and Multiple Output t antennas (MIMO) technology to achieve parallel data at the receiver with reduced Bit Error Rate (BER) and with increased channel capacity. There two different precoding techniques available, namely Linear Precoding Technique (LPT) and Non-linear Precoding Techniques (N-LPT). LPT include different methods like Singular Value Decomposition (SVD), Block Diagonalization (BD) and N_LPT like Dirty Paper Coding (DPC). N-LPT have high computational complexity. Thus analysis is done for the existing LPT's which make use of Channel State Information (CSI) both at transmitter and receiver end to reduce computational complexity. In this paper, SVD and BD linear precoding techniques are simulated and compared their BER for different the modulation methods similar to BPSK, QPSK and 8-PSK using MATLAB for 4 × 4 MIMO. Two different equalization specifically Zero Forcing (ZF) and Minimum Mean Square Error (MMSE) are considered for BER calculation. BER is reduced by 5% with SVD precoding using QPSK modulation than without precoding. Comparing BER of SVD and BD techniques for a 4 × 4 MIMO using QPSK modulation, 90% better results have been achieved in BD. A 2 × 2 MIMO with SVD precoding system was implemented using Universal Software Radio Peripheral (USRP-2920) and the results are plotted.

Keywords MIMO · Precoding · ZF · MMSE · SVD · BD · USRP

P. N. Jayanthi (✉) · S. Rachana · M. Ramya
R V College of Engineering, Bangalore, Karnataka, India
e-mail: jayanthipn@rvce.edu

S. Rachana
e-mail: rachanas.ec16@rvce.edu.in

M. Ramya
e-mail: ramyam.ec16@rvce.edu.in

V. Nath and J. K. Mandal (eds.), *Proceeding of Fifth International Conference on Microelectronics, Computing and Communication Systems*, Lecture Notes in Electrical Engineering 748,
https://doi.org/10.1007/978-981-16-0275-7_31

1 Introduction

New upcoming web-based applications need communication services with higher and higher data rates to support interactions. The fundamental physical resources namely, the available bandwidth and radiated power are limited. In order to enhance the data rate for the digital communication and for reliable communications channel partitioning (MIMO) are employed. With the increase in the number of users of mobile technology, achieving higher bit rate and quality of service has become a primary importance. MIMO is a technology which utilizes space division multiplexing by employing multiple antennas at input and multiple antennas at output antennas [1–3]. The use of multiple antennas at both the transmitter and receiver are the main focus of the communication system to improve the data transmission rate, system coverage, link reliability, and throughput. But data received at the receiver consists of interferences due to different antennas and users. At the receiver to cancel these interferences different techniques are adopted but the complexity of the receiver increases. Hence there is a need for the precoding algorithms at the transmitter [4].

Precoding is a process used at the transmitter end to weight the information stream depending on the CSI [5]. Precoding helps in shifting the hardware complexity to the base station from the transmitting mobile station. In MIMO systems, to achieve better system performance, the information of complete CSI is required. To process the CSI at the receiver is quite difficult and the complexity increases. This gives an idea of processing user signals at transmitter so that, the user who is intended should only be able to receive the required signal by adopting the method called precoding. The available precoding algorithms are complex in achieving minimum interference and better system performance. This is the motivation for the research to design less complex, efficient precoding techniques to achieve better performance.

2 Precoding

Linear precoding can be done using Singular Value Decomposition (SVD) and Block Diagonalization (BD) [6, 7]. Dirty Paper Coding is a non-linear precoding technique. Non-linear precoding methods are complex in nature also involve a lot of computation time. This motivates us to analyze the available linear precoding methods which reduce computational complexity. In this paper, we consider two types of Linear Precoding that is SVD and BD [8].

2.1 Singular Value Decomposition

SVD precoding algorithm decomposes the impulse response of the channel H, to three matrices, presented as

$$\boldsymbol{H} = \boldsymbol{U}\boldsymbol{D}\boldsymbol{V}^{H} \tag{1}$$

where H is the impulse response of channel, D is the diagonal matrix with all the diagonal entries as the Eigen values of the channel matrix. U and V^H are unitary matrices [9].

The precoding matrix is multiplied with transmitter symbol matrix before transmission as shown in Eq. 2

$$\boldsymbol{c} = \boldsymbol{V}\boldsymbol{s} \tag{2}$$

The below Eq. 3, is the signal vector at the receiver **r and** is denoted by

$$\boldsymbol{r} = \boldsymbol{H}\,\boldsymbol{V}\boldsymbol{s} + \boldsymbol{n} \tag{3}$$

Matrix **n** adds AWGN the received vector. To separate the signal that is received into Eigen modes, the product of **r** and the matrix **U is** obtained as shown in Eq. 4, and simplified because of the unitary possessions of **U** and **V** [10].

$$\begin{aligned} \boldsymbol{y} = \boldsymbol{U}^{H}\boldsymbol{r} &= \boldsymbol{U}^{H}\boldsymbol{H} \cdot \boldsymbol{V}\boldsymbol{s} + \boldsymbol{U}^{H}\boldsymbol{n} \\ &= \boldsymbol{U}^{H}\boldsymbol{U}\boldsymbol{D}\boldsymbol{V}^{H} \cdot \boldsymbol{V}\boldsymbol{s} + \boldsymbol{U}^{H}\boldsymbol{n} \\ \boldsymbol{y} &= \boldsymbol{D}\boldsymbol{s} + \boldsymbol{U}^{H}\boldsymbol{n} \end{aligned} \tag{4}$$

The equalized data **y** summarize that every transmitted signal is prejudiced by its singular value, which is shown in Eq. 5.

$$\boldsymbol{yi} = \boldsymbol{d}_i \cdot \boldsymbol{s}_i + \boldsymbol{n}_i \tag{5}$$

With $i = 1,\ldots, Nt$. Equation 6 shows the SNR value.

$$\boldsymbol{SNR}_i = \boldsymbol{d}_i^2 \cdot \frac{|\boldsymbol{si}|^2}{2\sigma_n^2} \tag{6}$$

2.2 Block Diagonalization

Block Diagonalization is the application of Singular Value Decomposition to multiple antennas at the user side. The symbols are decoded at the receiver side after it passes through channel on combining the different symbols at the transmitter side.

On considering two antennas per user, the two channel matrices are given in Eqs. 7 and 8

$$\begin{aligned} \boldsymbol{H}_1 &= \boldsymbol{U}_1\boldsymbol{D}_1\boldsymbol{V}_1^H \\ \boldsymbol{H}_2 &= \boldsymbol{U}_2\boldsymbol{D}_2\boldsymbol{V}_2^H \end{aligned} \tag{7}$$

U and V^H are unitary matrices. The precoding matrices which are diagonal in nature are given by Eqs. 9 and 10

$$\begin{aligned} \boldsymbol{W}_1 &= \boldsymbol{V}_2(:, 3:4) \\ \boldsymbol{W}_2 &= \boldsymbol{V}_1(:, 3:4) \end{aligned} \tag{8}$$

$\boldsymbol{W_1}$ and $\boldsymbol{W_2}$ are precoding matrices formed using $\boldsymbol{V_2}$ and $\boldsymbol{V_1}$

$$\boldsymbol{C} = \boldsymbol{C}(:, 1:2)\boldsymbol{W}_1 + \boldsymbol{C}(:, 3:4)\boldsymbol{W}_2 \tag{9}$$

$$\begin{aligned} \boldsymbol{R}_1 &= \boldsymbol{H}_1\boldsymbol{C} + \boldsymbol{N}_1 \\ \boldsymbol{R}_2 &= \boldsymbol{H}_2\boldsymbol{C} + \boldsymbol{N}_2 \end{aligned} \tag{10}$$

$$\begin{aligned} \boldsymbol{EQ}_1 &= (\boldsymbol{H}_1\boldsymbol{W}_1)((\boldsymbol{H}_1\boldsymbol{W}_1)(\boldsymbol{H}_1\boldsymbol{W}_1)^H)^{-1} \\ \boldsymbol{EQ}_2 &= (\boldsymbol{H}_2\boldsymbol{W}_2)((\boldsymbol{H}_2\boldsymbol{W}_2)(\boldsymbol{H}_2\boldsymbol{W}_2)^H)^{-1} \end{aligned} \tag{11}$$

$$\boldsymbol{Y} = [\boldsymbol{EQ}_1\boldsymbol{R}_1 \ \boldsymbol{EQ}_2\boldsymbol{R}_2] \tag{12}$$

The transmitted symbol matrix is given by the matrix C. $\mathbf{R_1}$ and $\mathbf{R_2}$ are the received symbol matrices. On applying the equalization matrices obtained are $\mathbf{EQ_1}$ and **EQ2**. The final output matrix **Y** is obtained by combining the two equalization matrices.

3 Equalization

3.1 ZF Equalization

ZF Equalizer is a linear equalization technique where the converse of the channel matrix is multiplied with the acknowledged symbol matrix and is used in the MIMO systems. In an ideal channel zero forcing equalization reduces the ISI to zero [11].

$$\boldsymbol{Y} = \boldsymbol{HX} + \boldsymbol{NR}_2 \tag{13}$$

$$\boldsymbol{H}^{-1}\boldsymbol{Y} = \boldsymbol{H}^{-1}\boldsymbol{HX} + \boldsymbol{H}^{-1}\boldsymbol{NR}_2$$

$$\boldsymbol{H}^{-1}\boldsymbol{Y} = \boldsymbol{X} + \boldsymbol{H}^{-1}\boldsymbol{NR}_2 \tag{14}$$

If $\boldsymbol{N}$ matrix is considered to be zero, then the transmitted symbol is received without distortion.

3.2 MMSE Equalization

The Minimum Mean Square Error (MMSE) algorithm minimizes the error between the transmitted and the received symbols. It takes into consideration the noise in the channel and the equalization matrix is given by,

$$\boldsymbol{W} = \boldsymbol{H}^{\boldsymbol{H}}\left(\boldsymbol{H}\boldsymbol{H}^{\boldsymbol{H}} + \frac{\boldsymbol{\sigma}}{\boldsymbol{ea}} \cdot \boldsymbol{I}\right)^{-1} \tag{15}$$

The $\frac{\boldsymbol{\sigma}}{\boldsymbol{ea}} \cdot \boldsymbol{I}$ term accounts for the AWGN in the channel as a function of power.

4 Simulation Results and Discussions

In this section, the performance of various LPT can be observed through simulations in MATLAB and graph is plotted for BER vs SNR. The parameters used for simulation are listed in the Table 1.

Figure 1 illustrates the performance comparison of the system with and without precoding. With simple SVD precoding we can see that, the BER achieved is less compared to the system without precoding. Hence it is better to adopt precoding in the system to reduce the BER. The Table 2 shows the comparison of BER with and without precoding for 4 × 4 MIMO with QPSK at SNR equal to 5 dB.

BER for SVD and BD LPTs can be observed through simulation in MATLAB and graph is plotted for BER vs Eb/No in dB. Different Modulation schemes like BPSK, QPSK and 8-PSK with $N_T = N_R = 4$ are considered.

Figures 2, 3 and 4 shows the performance of SVD precoding with ZF and MMSE equalizers for three different modulation techniques, BPSK, QPSK and 8-PSK. It can be observed in the three graphs that the performance is better with MMSE equalizer

Table 1 Simulation parameters

BS Antennas	4
MS Antennas	4
Data bits transmitted	Variable (Number of bits used for testing = 10,000 bits)
Modulation techniques	QPSK, QAM
Channel	AWGN with Rayleigh fading
MIMO precoding	SVD and BD
Equalization techniques	Linear ZF and MMSE

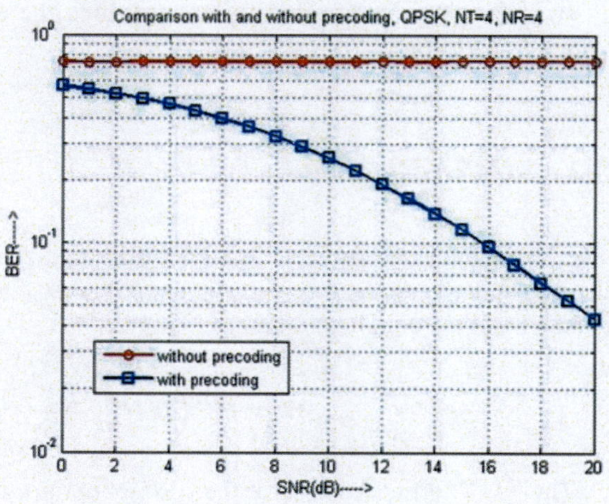

Fig. 1 Evaluation plot of with and without precoding

Table 2 Comparisons of BER with and without precoding.

4X4 QPSK @ SNR 5 dB	BER without precoding	BER with SVD precoding
ZF	0.0635	0.052
MMSE	0.0414	0.039

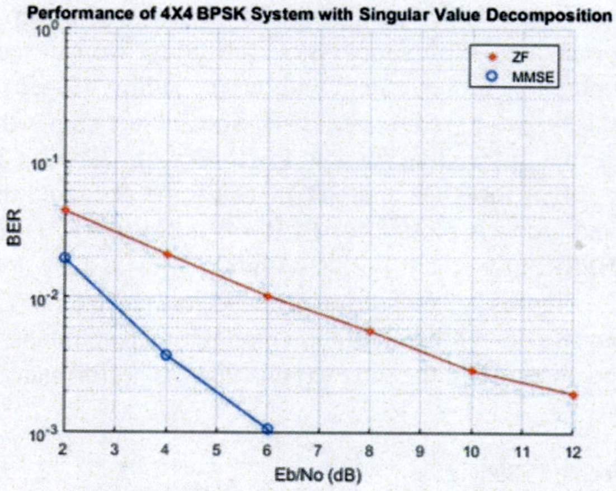

Fig. 2 Comparison plot for SVD using ZF and MMSE with BPSK

than with ZF equalizer. This is because; MMSE equalization takes AWGN in the channel into consideration unlike ZF.

It can also be observed that the performance decreases as the order of the modulation increases, as the data rate increase, error also increases. So, we can state that BPSK with MMSE equalizer gives better results.

Figures 5, 6 and 7 shows the performance of BD precoding with ZF and MMSE equalizers for three different modulation techniques, BPSK, QPSK and 8-PSK. As

Fig. 3 Comparison plot for SVD using ZF and MMSE with QPSK

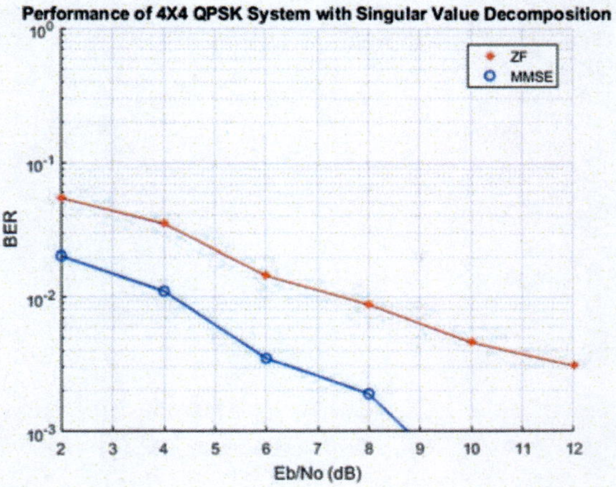

Fig. 4 Comparison for SVD using ZF and MMSE with 8-PSK

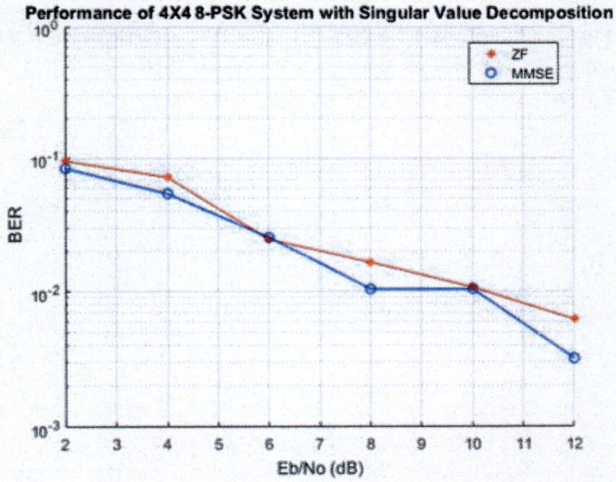

Fig. 5 Comparison for BD using ZF and MMSE with BPSK

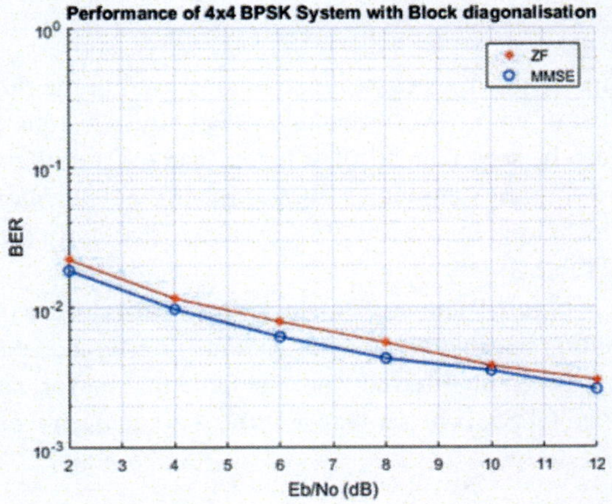

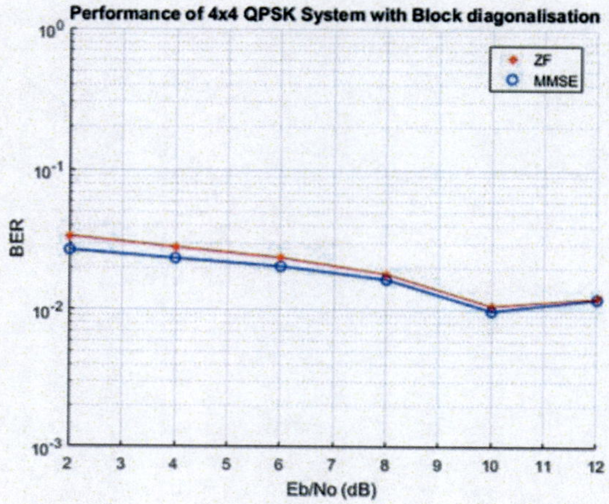

Fig. 6 Comparison for BD using ZF and MMSE with QPSK

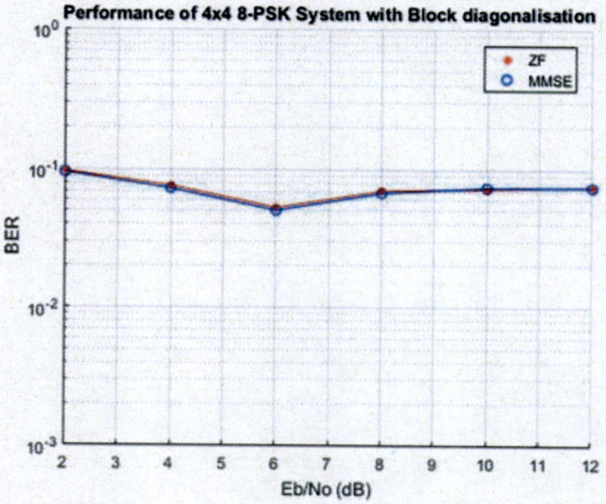

Fig. 7 Comparison for BD using ZF and MMSE with 8-PSK

in the previous case, it can be observed in the three graphs that the performance is better with MMSE equalizer than with ZF equalizer as mentioned before. Also, it can be seen that BPSK is better than other modulation techniques.

Figure 8 compares ZF and MMSE equalization without applying any precoding technique for 4 × 4 QPSK system. Clearly, we can see that MMSE is better than ZF (Table 3).

Figure 9 shows that Block Diagonalization is better than Single Value Decomposition because as the inter antenna interference per user is also reduced.

It can be inferred from the table that performance is better in BPSK than QPSK and QPSK is better than 8-PSK. Also, in each modulation, MMSE is better than ZF.

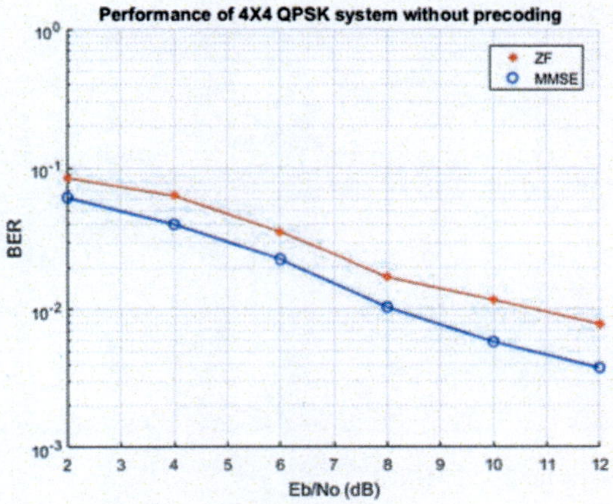

Fig. 8 Comparison of ZF and MMSE without precoding

Table 3 Comparisons of % number of correct received bits

Modulation techniques	ZF-SVD	MMSE-SVD	ZF-BD	MMSE-BD
BPSK	84.21	86.71	71.4	94.9
QPSK	61.3	68.3	52.3	84.9
8-PSK	1.3	2.2	3.5	25.4

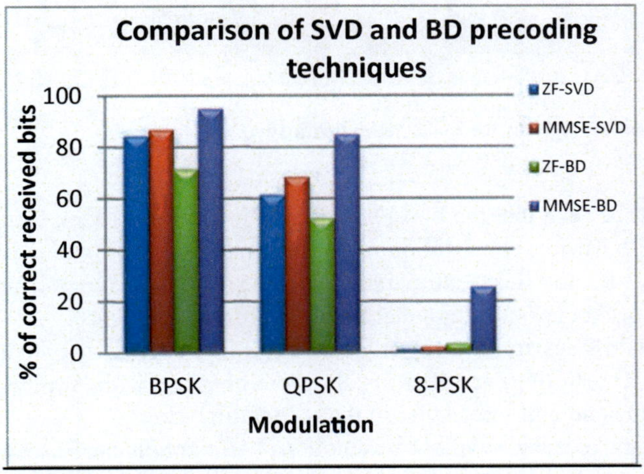

Fig. 9 No. of correct bits received Comparison of SVD and BD for the different modulation techniques

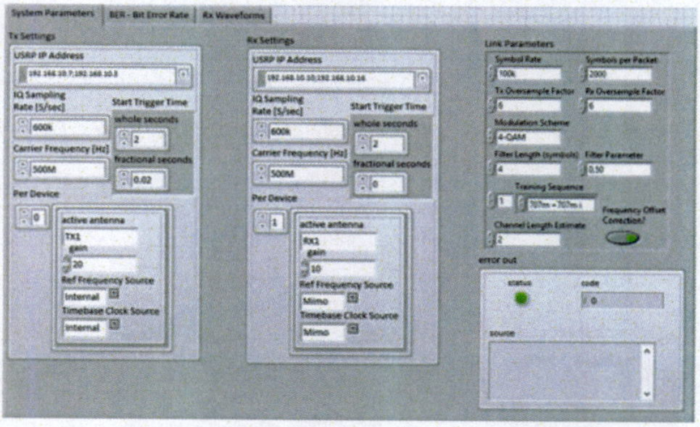

Fig. 10 Front panel view for 2 × 2 MIMO

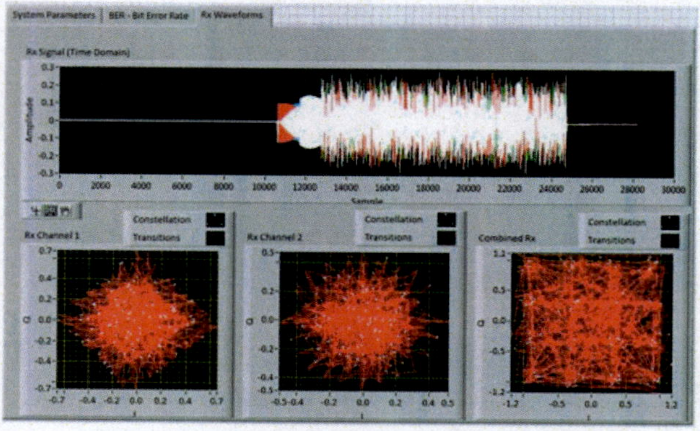

Fig. 11 Received waveforms for 2 × 2 MIMO with 16-QAM

And it can be seen that the best performance is given by Block Diagonalization Precoding with MMSE equalization for BPSK system.

In the Fig. 10, the link state parameters are set, 100 K symbol rate and 2000 symbols per packet are transmitted at 600 MHz carrier frequency.

A 2 × 2 MIMO system was implemented using USRP-2920. In two different time frames with 10 ms difference BER results were obtained without precoding in the first time frame and with precoding in the second time frame.

In Fig. 11, the received symbol constellation of 16 symbols can be seen for receiver 1, receiver 2 and a combination of receiver 1 and receiver 2. The time domain signal is also represented.

In Fig. 12, The received symbols can be observed for ICD mode and MIMO mode. The constellation diagram with SVD precoding is clearer than without, implying the lesser BER with SVD precoding.

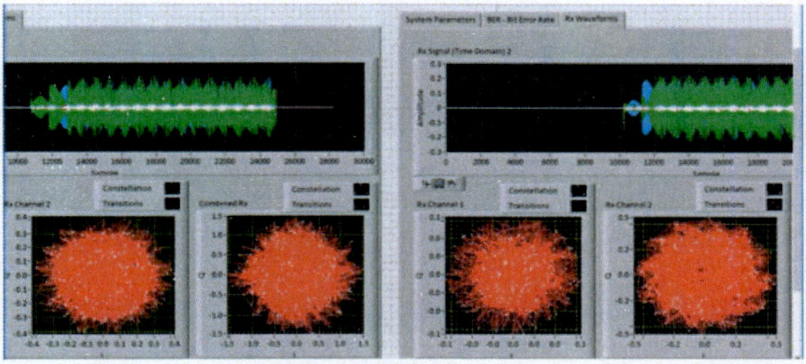

Fig. 12 Received waveform for 2 × 2 MIMO with SVD Precoding

5 Conclusion

BER is sufficiently reduced on the application of the precoding algorithms for MIMO. The MMSE Equalization technique gives better results as compared to ZF Equalization. And also observed that, as the order of modulation is increased the BER also increases. Block-Diagonalization with MMSE Equalization scheme gives the best performance in terms of BER. We have achieved 90.9% better results in Block-Diagonalization method with MMSE equalization for BPSK Modulation as compared to SVD precoding for ZF Equalization for 8-PSK system. A 2 × 2 MIMO with SVD precoding system was implemented using USRP-29200.

6 Future Scope

In this paper, how BER reduces with the use different precoding algorithms are verifies. To further reduce the BER, precoding with OFDM can be adopted. Also, better modulation schemes like QAM can be used for better performance. To accommodate multiple users, BD algorithm complexity will increase. So, there is a scope to improve the BD algorithm such that multiple users can be accommodated with minimum complexity.

References

1. Wang H, Li L, Song L, Gao X (2011) A linear precoding scheme for downlink multiuser MIMO precoding systems. IEEE Commun Lett 15(6):653–655
2. Wu CH, Liu CY, Tsai PY (2017) Design of an SVD engine for 8 X 8 MIMO precoding systems. IEEE.
3. Hwang YT, Chen KT, Wu CK (2015) A high throughput unified SVD/QRD precoder design for MIMO OFDM systems. IEEE

4. Lyu TK (2016) Capacity of multi-user MIMO systems with MMSE and ZF precoding. In: IEEE Infocom 2016 Poster Presentation
5. Tlili T, Rebhi N, Amiri H (2017) Massive MIMO approach for interference reduction using precoding. In: International Conference on Smart, Monitored and Controlled Cities (SM2C), Kerkennah, Tunisia, 17–19 February 2017
6. Busche H, Vanaev A, Rohling H (20009) SVD-based MIMO precoding and equalization schemes for realistic channel knowledge: design criteria and performance evaluation. Wirel Pers Commun 48(3):347–359
7. Gümüs M, Duman TM (2018) Multi-envelope precoding for massive MIMO systems. IEEE Wirel Commun Lett 7(5):720–733
8. Crâmariu VF, Arvinte MO, Enescu AA, Ciochină S (2016) Block-diagonalization applied in multi-user massive MIMO. IEEE
9. Crâșmariu VF, Arvinte MO, Enescu AA, Ciochină S (2016) Performance analysis of the singular value decomposition with block-diagonalization precoding in multi-user massive MIMO systems. IEEE
10. Guenther D, Leupers R, Ascheid G (2016) A scalable, multimode SVD precoding ASIC based on the cyclic Jacobi method. IEEE Trans Circ Syst I: Regular Papers 63(8):1283–1294
11. Grossmann M (2009) SVD-based precoding for single carrier MIMO transmission with frequency domain MMSE turbo equalization. IEEE Signal Process Lett 16(5):418–421

An Insight into the Existing Reversible Arithmetic and Logic Unit Designs

S. Girija and B. G. Sangeetha

Abstract International Technology Roadmap for Semiconductors-ITRS2.0 predicts an end to traditional scaling and shrinking of chips by 2028. The future depends on the alternative technology to fill the gap and perform better than the existing technology. There are numerous technologies emerging, one among them being the reversible logic is fast gaining the importance due to the quantum technology for minimal dissipation of energy whose operation is reversible in nature. Arithmetic and logic operations are the core of any processing system and its importance is found in all the modern devices. Reversible logic assures to overcome the limitations of the current technologies and several designs have been proposed to build optimized, efficient reversible arithmetic and logic unit. This paper makes an effort to compare the existing ALU designs based on the different design approaches.

Keywords Quantum · ALU · Reversible · Ancilla

1 Introduction

Power dissipation has become a critical factor in the design of recent integrated circuits as the density and capacity of the circuits exponentially increasing. The bit loss in the irreversible gates at higher speeds, circuit dissipate KTln2 joules of heat energy for every bit of loss where Boltzmann's constant is K and temperature of the circuit in Kelvin is T [1]. Moore's Law made the prediction in 1965, the number of transistors per square inch on integrated chip would be doubled every year since it was invented [2]. C.H. Bennet in 1973 proved that any circuit built using reversible logic can result in no heat dissipation due to the information loss [3].

S. Girija (✉)
Department of ECE, Dr. Ambedkar Institute of Technology, Bengaluru, India
e-mail: girija.ec@drait.edu.in

B. G. Sangeetha
Department of ECE, RNSIT, Bengaluru, India

V. Nath and J. K. Mandal (eds.), *Proceeding of Fifth International Conference on Microelectronics, Computing and Communication Systems*, Lecture Notes in Electrical Engineering 748,
https://doi.org/10.1007/978-981-16-0275-7_32

Operations carried out in the computational systems are typically arithmetic and logical operation for the manipulation of data bits. ALU is the brain of all the processing devices in the modern world, whose power consumption is the concern for the design to achieve and there is high demand to provide zero power dissipation. Reversible logic is a technology when one to one mapping among the input and the output signals exists and achieves zero power dissipation. The technology has the applications in the futuristic areas like quantum computing as the operations are fundamentally reversible, cryptography, nanotechnology, DNA computing and Image processing.

1.1 Basics of the Reversible Logic

Reversible Gate (RG): When equal number of input and output signals exists in a gate it is called as reversible gate and by nature it is bijective. The gate is represented as N * N, where N denotes number of input and output lines.

Reversible Circuit (RC): The circuits which are built with the existing reversible gates and/or novel or existing reversible gates in cascade are called reversible circuits.

Reversible circuits are optimized based on the several design parameters or the cost metrics.

Ancilla Input (AI): The reversible circuit design needs some of its input signals to be retained at constant value 0 or 1 to accomplish the functionality required [4]. One of the cost metric in reversible logic is to maintain constant input value as small as possible or ideally zero.

Garbage Output (GO): The unwanted output signals obtained across the gate along with the required output signals [5]. The power dissipation or heat generation in the reversiblecircuit is due to the garbage output and hence they must be maintained as least as possible, ideally zero.

Quantum Cost (QC): It is calculated by the number of all the 1 * 1 and 2 * 2 primitive reversible gates used to circuit is called as quantum cost. Any other reversible gates or circuits can also be realized using 1 × 1 NOT gate, 2 * 2 CNOT (Controlled NOT/Feynman gate), V (square root of NOT gate) and V^+ (Hermitian) gate [5].

Gate Count (GC): Total number of all the reversible gates expended to realize the functionality of the reversible circuit.

Logical Calculation: The irreversible gates namely two input XOR(α), two input AND(β) and unitary NOT(γ) required to recognize the logical calculation is also known as the hardware complexity.

Reversibility: The number of output pattern maps uniquely with the input pattern, the output of the reversible function is the combination of the inputs.

Universality: Any reversible gate is said to be universal, if the reversible gate achieves the logical operations of irreversible gate NOT, OR/ NOR, AND/NAND operations.

The drawbacks of the reversible gate in contrary to the conventional gates are

- Fan-out not supported
- Feedback and loops are not allowed.

1.2 Reversible Logic Gates

A Not gate is a 1 * 1 reversible gate, the quantum cost is zero is shown in the Fig. 1.

Feynman Gate is a 2 * 2 reversible gate having quantum cost of 1 as show in Fig. 2. It can perform inversion operation under the controlled signal A. The two inputs, the first input A is control input, B is known as the target input [6].

Toffoli Gate is used as a universal gate since any irreversible gates can be realized using the Toffoli gate and is as denoted in Fig. 3. It is a 3 * 3 reversible gate with 3 inputs and 3 outputs having quantum cost of 5 [7].

Fredkin Gate is a variation of SWAP gate with one or more control points is shown in the Fig. 4. One of the universal reversible gate as any of the basic irreversible gates NOT, AND and OR can be derived from this gate and whose quantum cost is 5 [8].

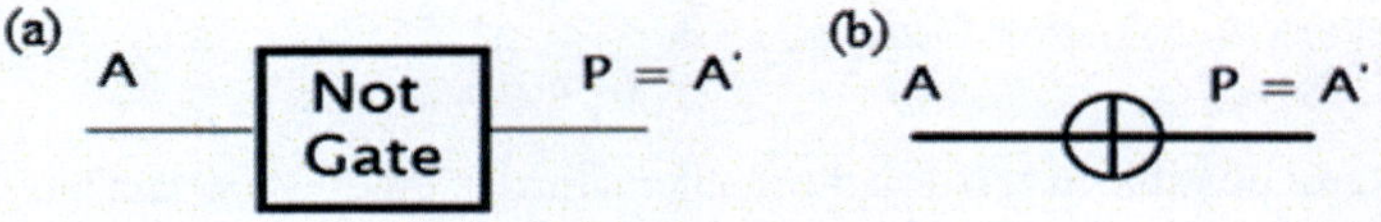

Fig. 1 NOT Gate (a) Block Diagram (b) Schematic

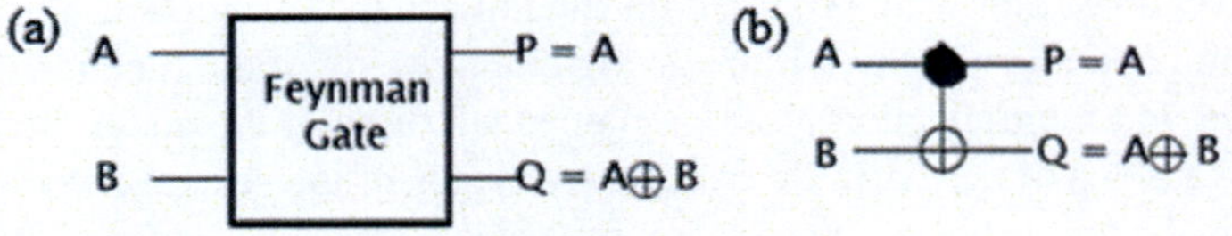

Fig. 2 Feynman gate **(a)** block diagram **(b)** schematic

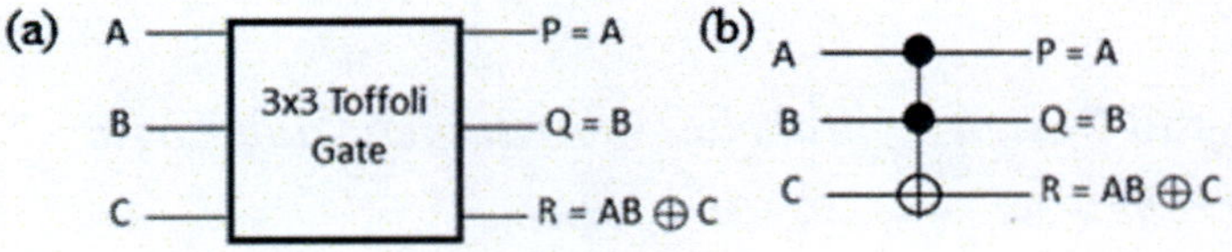

Fig. 3 Toffoli gate **(a)** block diagram **(b)** schematic

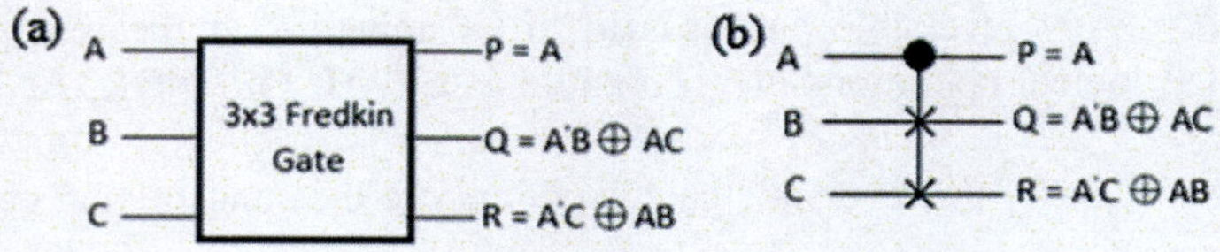

Fig. 4 Fredkin gate **(a)** block diagram **(b)** schematic

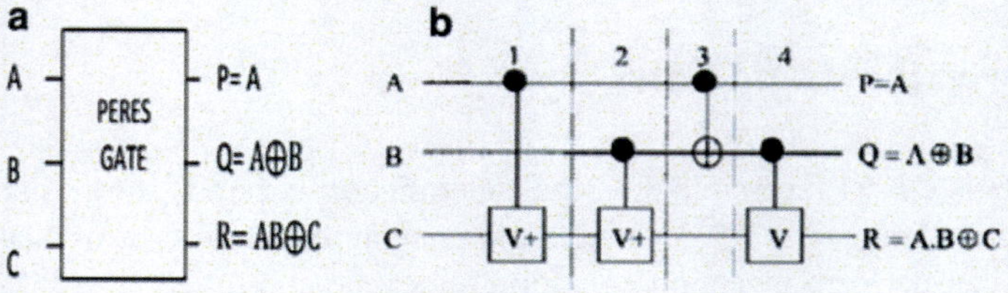

Fig. 5 Peres gate **(a)** block diagram **(b)** schematic

Peres Gate is a 3 * 3 gate but unlike the Fredkin and Toffoli gate it is not a universal gate as shown in Fig. 5. It is widely used in reversible designs owing to its quantum cost of 4 which is less in comparison other universal gate [9].

1.3 Reversible Adder Gates

ALU makes it possible to perform the mathematical operations on data bits, it is most important part of the Central Processing Unit. The operations which are crucial to the ALU is the addition apart from the subtraction, multiplication, division, logical operation and the shift operations which can be built as sub modules. In the literature, several reversible gates were proposed for the full adder realization, as it is the most fundamental operation in any arithmetic logic unit. Many of the functionalities like subtraction and multiplication also comprehended through the adder circuits. The logical operations of the ALU are realized by means of the several reversible gates available in the reversible gate repository. The gates which are used for the realization of full adder are TSG, HNG, MRG, MKG, DKG, DPG, PFAG [10–13]. The Table 1 lists the reversible full adder gates with their quantum cost.

1.4 Digital Design Approaches to Arithmetic and Logic Circuits

ALU operations are carried out simultaneously based on the control signals and data signals and the expected result is obtained at the output. After analysis of the circuits

Table 1 Reversible full adder gates

Gate	IO	Expression	QC
TSG	4*4	$P=A, Q=B, R=(A\oplus B)\oplus C$ $S=(A\oplus B)C\oplus AB$	10
HNG	4*4	$P=A, Q=B, R=(A\oplus B)\oplus C$ $S=(A\oplus B)\oplus C(AB\oplus C)$	4
MRG	4*4	$P=A, Q=A\oplus B, R\oplus(A\oplus B)\ \ C$ $S=(AB\oplus D)\oplus(A\oplus B)\oplus C)$	6
DKG	4*4	$P=A, Q=A'C'\oplus AD'$ $R=(A\oplus B)(C\oplus D)\oplus CD$ $S=(B\oplus C)\oplus D)$	6
DPG	4*4	$P=A, Q=A\oplus B, R=A\oplus B\oplus C$ $S=(A\oplus B)C\oplus AB\oplus D$	6
MKG	4*4	$P=A, Q=C, R=(A'D'\oplus B')\oplus D$ $S=(A'C'\oplus B')D\oplus(AB\oplus C)$	10
PFAG	4*4	$P=A, Q=A\oplus B, R=A\oplus B\oplus C$ $S=(A\oplus B)C\oplus AB$	6

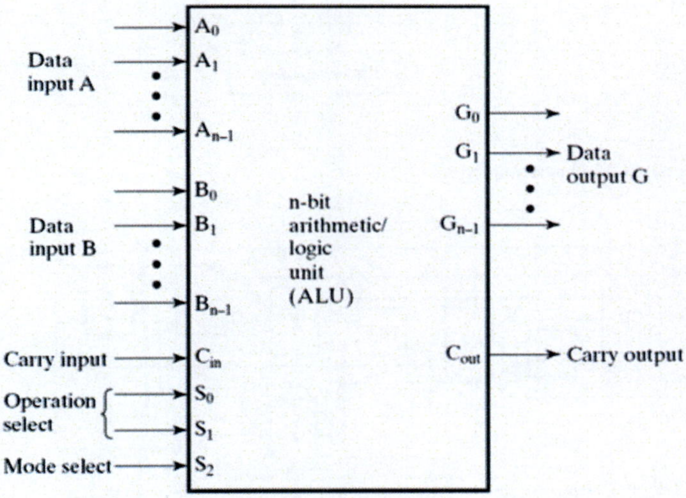

Fig. 6 Block diagram of ALU

of ALU in the literature it is concluded that, they can be categorized based on the architecture - dedicated design, control unit-based design with reversible adder and multiplexing the exclusively designed arithmetic and logical unit. Figure 6 depicts the general block diagram of the n-bit ALU.

The classifications of the designs of the ALU that are found in the literature are as follows.

1.5 Dedicated Design Single Output

Research work in paper [14–22] focuses on the implementation of the ALU as a dedicated unit to perform the arithmetic operation-addition with and without carry input, subtraction with and without borrow input, transferring of data and logical operations–AND, NAND, OR, NOR, EXOR and EXNOR operations. The general dedicated structure is as shown in the Fig. 7.

The Fig. 8 shows the construction of ALU using the dedicated design using basic reversible gates Feynman, Fredkin and Peres gate.

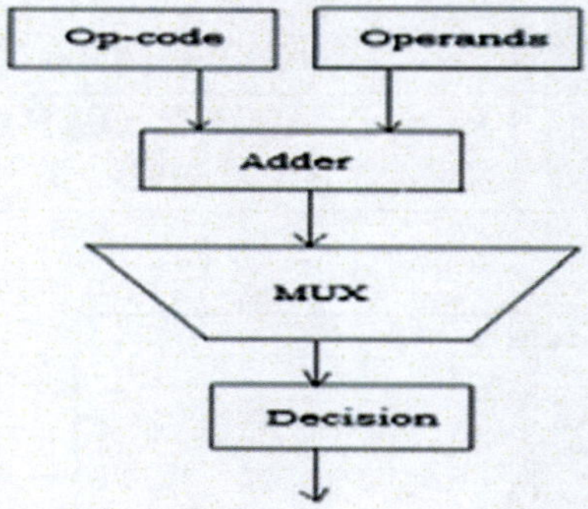

Fig. 7 Reversible ALU as a single unit

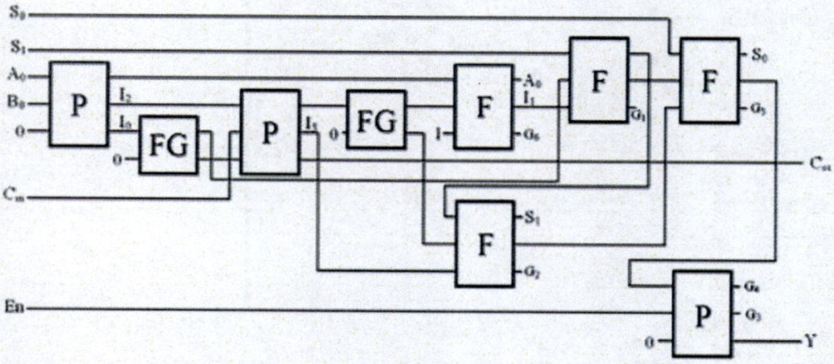

Fig. 8 Dedicated design RALU [14]

2 ALU Based on Control Unit

Reversible ALU based on the control structure is relatively slow as the logic depth is relatively higher and is complicated to design than the parallel ALU which has smaller logic depth. To achieve the more complex, controlled system is necessary to build the designs with higher level of control in reversible computing. The Reversible ALU based on the control unit is as shown in the Fig. 9.

The control unit implemented using the 7 3 × 3 Toffoli gate, 1 4 × 4 Toffoli gate and 7 Not gates costing 8 garbage outputs and 8 ancilla inputs [23]. In the paper [15] the control unit of reversible ALU is achieved by 3 Feynman gate, 3 Peres gate and Fredkin gate is used to realize the control unit is as shown in the Fig. 10.

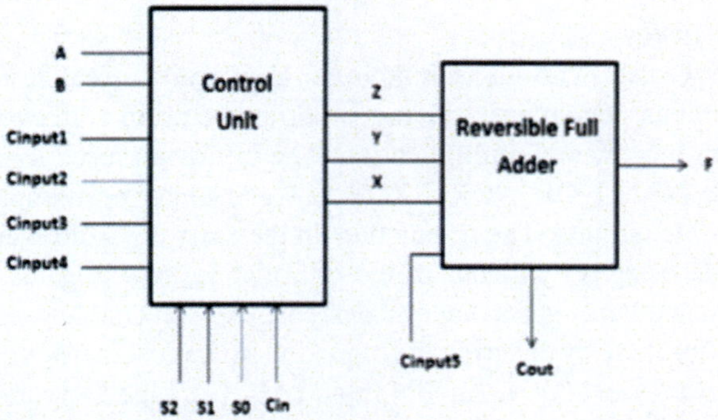

Fig. 9 Reversible ALU based on control unit [15]

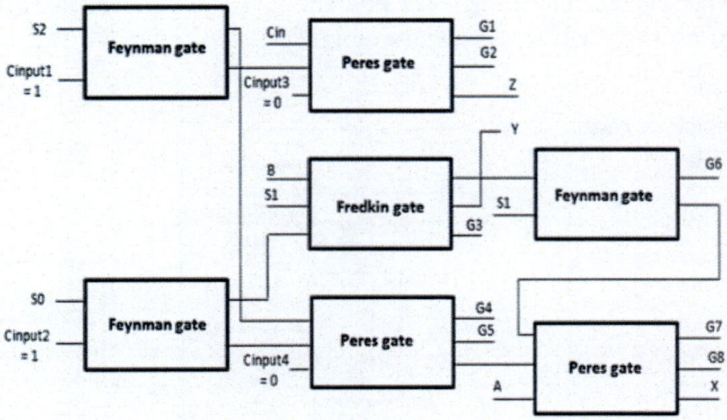

Fig. 10 Reversible ALU based on control unit [15]

3 Multiplexer Based ALU Design

In the multiplexer-based approach the arithmetic unit and logic units are designed individually to achieve all the required operations. The ALU is then realized by multiplexing the designed arithmetic unit and logical unit. The outputs of the arithmetic and logical unit are fed as the input to multiplexer and based on the selection inputs, required operation is implemented.

The Fig. 11 shows the arithmetic unit and logical unit integration. Input to the multiplexer is the output of the arithmetic or logical unit and the other input is the selection (S_2) to produce the output. If the input $S_2 = 0$ then arithmetic operations are carried out, if $S_2 = 1$ logical operation are implemented. The advantage of this approach it is possible to realize the greater number of operations.

II Literature Survey

Comparison between the various arithmetic and logical unit to realize arithmetic operation like increment, decrement, copying, addition, addition with carry, subtraction, subtraction with borrow and multiplication. The frequently used logical operations like NOT, AND, NAND, OR, NOR, EXOR, EX-NOR have been implemented using different reversible circuits. The researchers in the early days of reversible logic era carried out ALU designs by means of the primitive reversible gates like Feynman, Toffoli, Fredkin and Peres gates, later on designed new reversible gates to realize the various operations there by ensuring the reduction in the cost metrics of the reversible logic like Quantum Cost (QC), Ancilla Input (AI), Garbage Output (GO) and Gate Count (GC) and total number of operations performed by the design. Here we are comparing the ALU design based on the categories.

The implementation of ALU using two approaches, Design 1 is control structure based ALU is implemented using Peres, Fredkin and Feynman reversible gates which amounts to 9 reversible gates costing the design quantum cost of 34. The irreversible

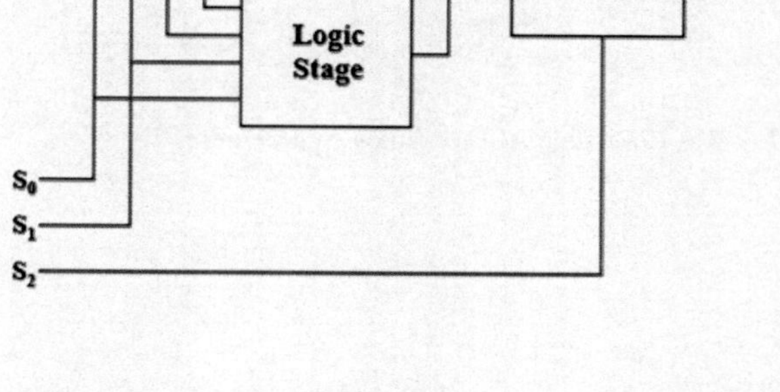

Fig. 11 Multiplexer based reversible ALU

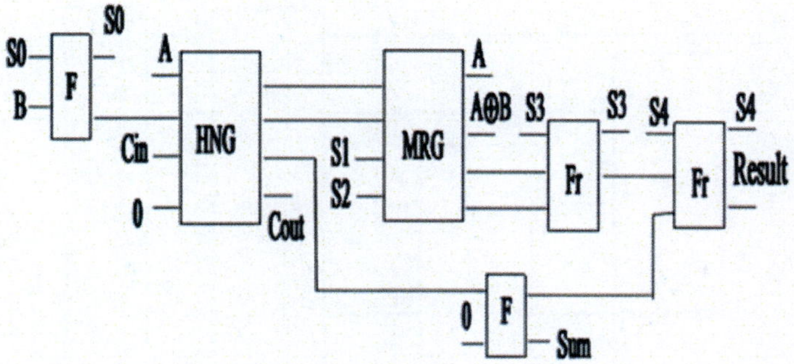

Fig. 12 Proposed reversible ALU [16]

hardware required to carryout the logical operation requires 8 NOT gates, 19 AND gates, 16 XOR gates, with ancilla input of 5 and garbage out of 6. Both the designs in the paper have low functionality and high quantum cost [14] (Fig. 12).

A novel reversible method that produces lesser number of garabage outputs is proposed in [16] which implements six operation with two designs. The Design 1 incorporates ALU using Feynman, Fredkin, HNG and MRG gates realising the operations - ADD, SUB, AND, NOR, OR, XOR and XNOR using a quantum cost of 24 is as shown in the Fig. 11. Design 2 uses Feynman, Fredkin, HNG and PAOG gates with operations ADD, SUB, OR, AND, NOR and NAND having a quantum cost of 24, it is low fuctionality design and having less number of constant inputs. The designs proposed by the author is not suitable for the practical usage due its fewer operations.

According to the paper [17] the design can be implemented using NCV gates which unitary realizing the 8 operations. The authors have demonstrated two designs 1-bit ALU using the elementary quantum gates such as CNOT, V, V^+. The design achieves optimization of cost-metrics and architectural complexity. Lowest quantum cost 24 for 12 operations is observed in the literature but no architecture is discussed and the claim to the quantum cost is through the quantum implementation [18].

The authors proposed that the ALU can be synthesized by optimized design using 4 * 4 CSA (Carry Save Adder) as shown in Fig. 13 with appreciable garbage and quantum cost, the number of operations performed for less than existing designs [19].

In the paper [20] researchers proposed two approaches by which they were able to achieve eleven arithmetic and eight logical operations. In the first approach the fault tolerant arithmetic and logical unit was implemented using new reversible gates KMD1, KMD2, KMD3 and KMD4 as shown in Fig. 14, second approach was designed by the proposed and the existing reversible gates.

In the recent paper [21] author achieved 73 operations by making use of the basic reversible FTRA and RMUX1 gate. Some of the operations are redundant in nature. The fault tolerant single design for 1-bit ALU was designed to achieve 35 operations

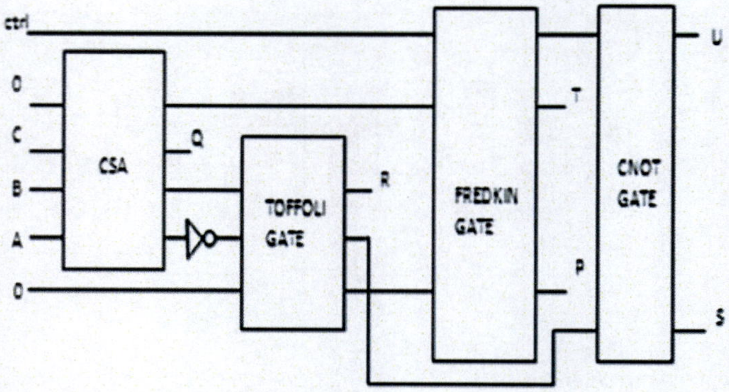

Fig. 13 Proposed reversible ALU [19]

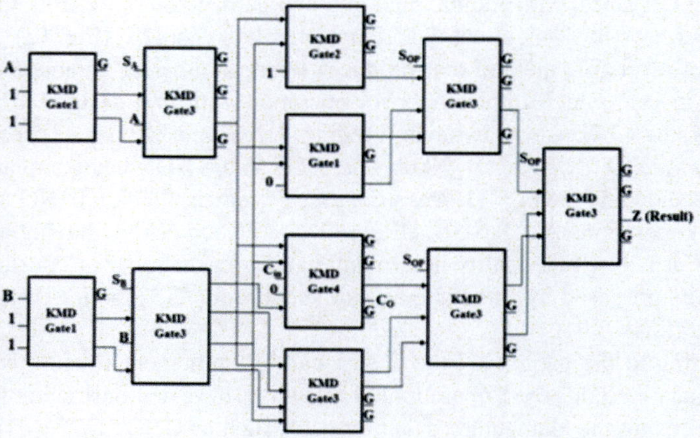

Fig. 14 Reversible ALU using KMD gates [20]

with the FTRA, RMUX and WG gate minimizing the Ancilla input and quantum cost of 33 [22] is shown below in the Fig. 15.

The Table 2 summarizes the literature review on dedicated designs.

The first ever high functionality ALU was proposed by the author Zhiin et al. [23], the multi-function ALU was based on the control unit depending on the selction inputs and data input manipulates the data to give controlled signals, these signal further applied to a full adder or function control to achieve the required operations. The author successfully tested 32 operations some of the operations being duplicated and with huge quantum cost.The design proposed by the authors are shown in the Fig. 16.

The authors have designed the control structure using the basic reversible gate in comparison with the existing structure in the literature the ancilla input, garbage output and gate count is improved [24]. The Fault tolerant reversible ALU [25] that can perform 16 arithmetic and 16 logical operations is a high functionality design

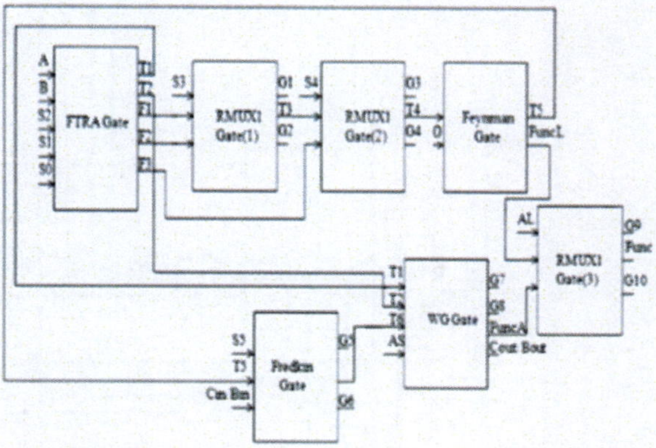

Fig. 15 Proposed reversible ALU [22]

Table 2 Comparison of dedicated design ALU

Paper	QC	AI	GO	GC	Operation
Design 1 Design 2 [14]	34 30	5 4	6 5	9 7	4 4
Design 1 Design 2 [16]	24 24	2 2	5 5	6 6	6 6
Design 1 [17]	9	0	1	19	8
Design 1 Design 2 [18]	21 24	2 4	2 8	21 24	10 12
Design [19]	24	1	2	8	8
Design 1 Design 2 [20]	118 99	7 8	21 22	11 11	18 18
Design [21]	45	6	14	6	73
Design [22]	33	1	10	7	35

adopted control-based structure with fault tolerance. However, the design is affluent in the number of optimizing parameters.

The control unit was derived by cascading the basic reversible gates such as Feynman, Fredkin and Peres gate as shown above in the Fig. 10. The functionalities of the ALU are realized through full adder. The PFAG gate is 4 * 4 gate with a quantum cost of 8. The ALU designed with a full adder using PFAG has gate count of 10, garbage output 10 with quantum cost of 28. The other design using the HNG gate which is 4 * 4 reversible gate whose quantum cost is 6. ALU designed had a gate count of 8, garbage output of 10 and quantum cost 26 [15].

Three ALU designs were proposed in the paper [27], two designs were dedicated designs performing 8 Boolean operations, control unit-based design with significant improvement in number of operations and functionality.

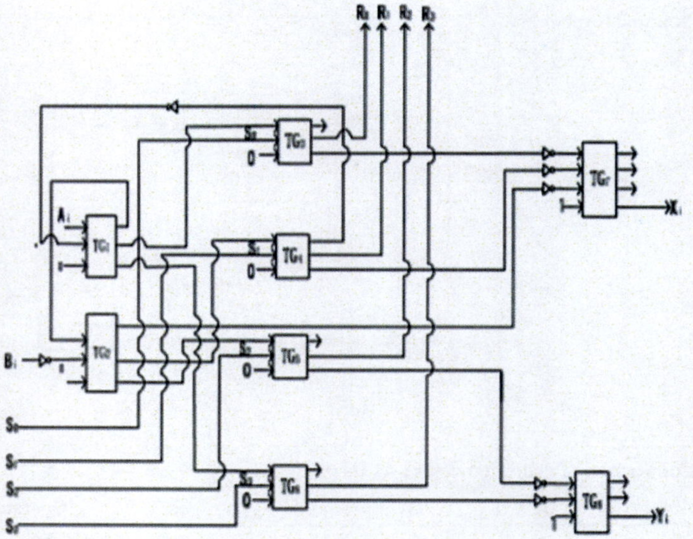

Fig. 16 Function generator [23]

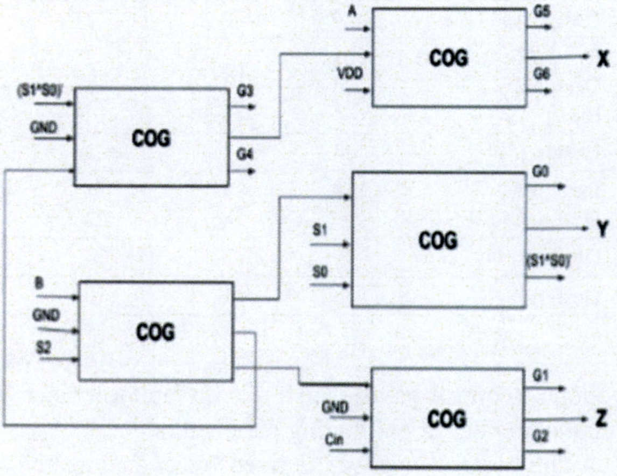

Fig. 17 Proposed control unit [29]

In the paper [28] proposed a design with high performance using UPPG gate with better hardware complexity, gate count and quantum cost. The design achieves 32 arithmetic and logical operations though some of them being redundant operations.

The transistor implementation of the ALU by designing the control structure using only the COG reversible gate is designed which has better optimizing parameters compared to the earlier designs of the same in the literature [29] (Fig. 17).

In the research work [30], the novel VSMT gate was used in the making of the control unit which reduces the garbage output and gate count considerably but the number of operations is limited to 8 arithmetic and logical operations (Fig. 18).

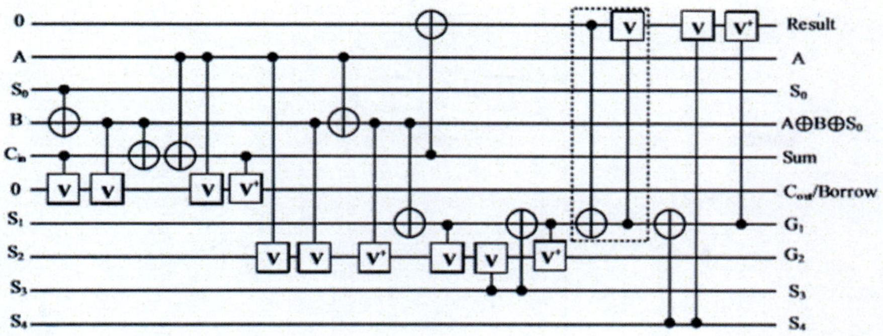

Fig. 18 Proposed RALU [18]

Table 3 Comparison of control unit based ALU design

Paper	QC	AI	GO	GC	Operations
Design [23]	70	12	12	24	32
Design [24]	26	3	08	07	12
Design [25]	29	4	8	10	12
Design [26]	–	23	27	13	32
Design 1 Design 2 [15]	28 26	5 5	10 10	08 08	12 12
Design [27]	25	4	5	06	12
Design [28]	77	25	25	16	32
Design [29]	26	5	9	6	12
Design [30]	–	4	7	4	8
Design [18]	24	2	4	6	12

According to the paper [18] control unit was designed in QCA technology owing to 11 operations with a quantum cost of 24 with fault tolerance but quantum cost of the ALU is not discussed. The design achieved through primitive gates such as CNOT, V, V^+ is shown above in the Fig. 16.

Summary of the control unit based ALU is as shown in the Table 3.

Various architectural designs were adopted to build the efficient, powerful and optimized reversible ALU by the authors in the literature either by V-shaped design or control unit-based design. The drawbacks of these styles were the limited functionality and unsatisfactory optimizing parameters. The other architecture which lately gaining popularity and widely adopted by the researcher is multiplexer based ALU design.

The idea of the multiplexer based ALU design was seen in the paper [31] though they were successful designing only the logical unit and multiplexer. The successful implementation of multiplexer based ALU was achieved by the author [32] through

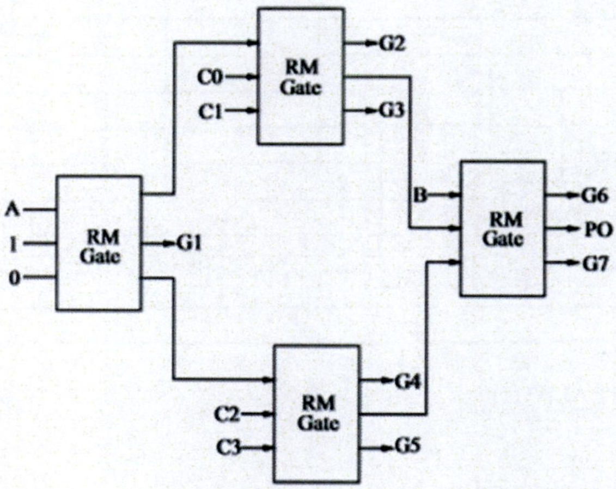

Fig. 19 Reversible logical unit [32]

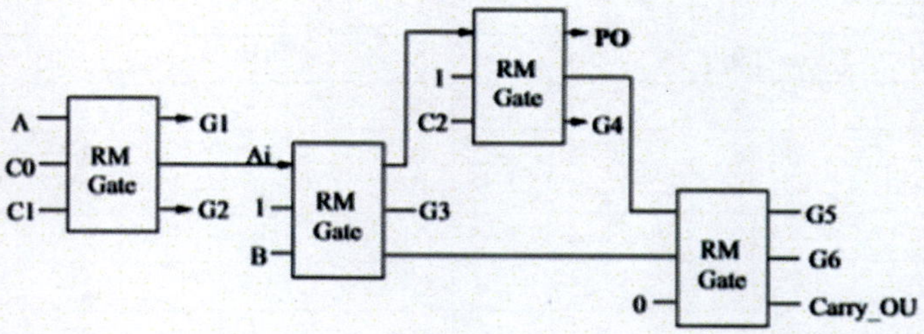

Fig. 20 Reversible arithmetic unit [32]

the QCA technology using RM gate, more functionality with large garbage output and ancilla input.

The Fig. 19 shows the design of logical unit using the RM gate (Figs. 20 and 21).

In the paper [33] ALU structure was built using RUG gate, In the research work aspects of the optimization of the reversible logic was successful not optimized, rather the design was area efficient in comparison with the other designs (Figs. 22 and 23).

The author [35] designed the ALU which shows the prodigious improvements in the cost metrics of reversible logic and the simulation constraints. The design was carried with proposed RG1, RG2, RG3, RG4 gates implemented with minimum number of cells in comparison with the existing reversible gates.

Arithmetic circuit was proposed in [34] with GN gate the quantum cost and garbage output is high and is able achieve only 8 operations. Authors in [36] has improvised on the quantum cost and garbage output at the increased cost of the ancilla input. He approached to the design using CNOT, COG, Toffoli and Peres gate. Available designs of the integrated ALU is as shown in the Table 4.

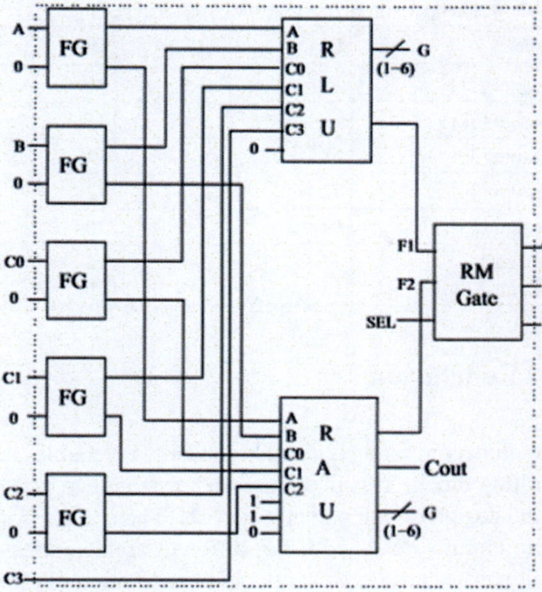

Fig. 21 Reversible RALU [32, 37]

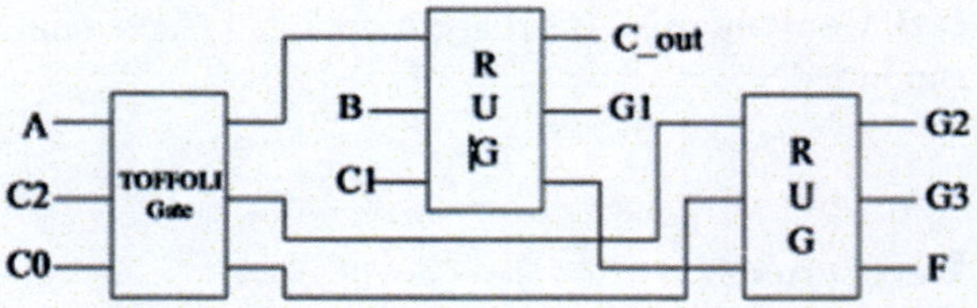

Fig. 22 Reversible arithmetic unit [33]

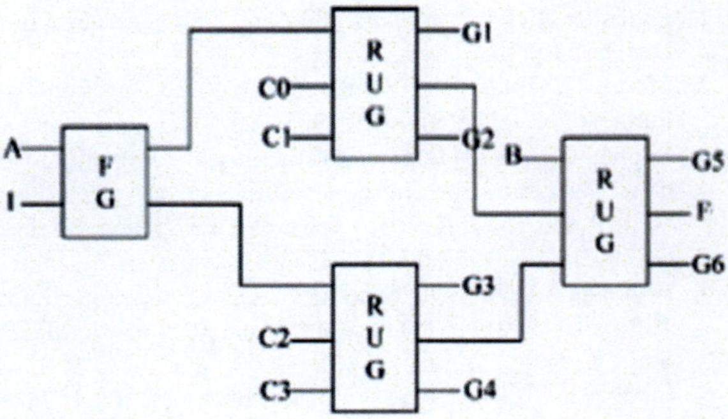

Fig. 23 Reversible logical unit [33]

Table 4 Comparison of multiplexer based ALU design

Paper	QC	AI	GO	GC	Operations
Design [31]	–	7	10	07	4
Design [32]	–	9	15	14	17
Design [33]	68	6	11	8	18
Design [34]	72	12	12	12	08
Design [35]	94	3	20	14	18
Design 1 Design 2 [36]	17 13	2 2	5 5	21 21	24 24

4 Conclusion

Quantum computing is characteristically reversible and reversible circuits is the basic building block. This paper presents extensive report on the architectures employed in the design of components of ALU. The comparison of the various arithmetic and logic circuits designed using different architectures available in the literature and design criteria adopted for the implementation of the ALU are presented. The ALU circuits are compared on the basis of the cost metrics of the reversible logic–Quantum cost, Constant input, Garbage output and Gate count and the number of operations supported.

References

1. Landauer R (1961) Irreversibility and heat generation in the computing process. IBM J Res Dev 3:183–191
2. Moore G (1965) The Future of Integrated Electronics. Fairchild Semiconductor Internal Publication, vol 38(8), April 1965
3. Bennett CH (1973) Logical reversibility of computation. IBM J Res Dev 17(Issue 6):525–532
4. https://spectrum.ieee.org/semiconductors/devices/transistors-could-stop-shrinking-in-2021
5. Wille R, Drechsler R (2010) Towards a design flow for reversible logic. Springer 2010. https://doi.org/10.1007/978-90-481-9579-4
6. Feynman R (1985) Quantum mechanical computers. Optic News 11(2):11–20
7. Toffoli T (1980) Reversible computing, Tech memo MIT/LCS/TM- 151. MIT Lab for Comp, Sci
8. Fredkin E, Toffoli T (1982) Conservative logic. Int J Theory Phys 21:219–253
9. Peres A (1985) Reversible logic and quantum computers. Phys Rev A 32(6):3266–3276
10. Roy VG, Indurkar PR, Khatri DM (2014) A survey on design approaches towards quantum ALU using reversible logic structures. IJSER 5:1372–1376
11. Thapliyal H, Srinivas M (2005) Novel reversible 'TSG' gate and its application for designing components of primitive reversible/quantum ALU. In: Tenth Asia Pacific Computer Systems Architecture Conference, 2005
12. Thapliyal H, Vinod AP Designing efficient online testable reversible adders with new reversible gate. In: Proceedings of IEEE International Symposium Circuits and Systems 2007, pp 1085–1088

13. Kumar P, Gour P, Soni BB (2014) An extensive survey: designing of ALU using reversible logic gates. Int J Emerg Technol Adv Eng 4(6):457–461
14. Syamala Y, Tilak AVN (2011) Reversible Arithmetic and Logic Unit. IEEE Publication, 978-1-4244-8679
15. Gopala L, Mohd NS, Chowdhary MK, Gopala AA, Singh AK (2014) Design and Synthesis of Reversible Arithmetic and Logic Unit (ALU). IEEE Publication, pp 289–293
16. Matthew M, Nagarajan R (2011) Design of reversible ALU on novel programmable reversible logic gate structures. IEEE Publication pp 126–131
17. Pal S, Vudadha C, Phaneendra S, Veeramachaneni S, Mandalika S (2014) A new design of an n-bit reversible arithmetic logic unit. International Symposium on Electronic System Design, IEEE, pp 224–225
18. Bolhassani A, Haghparast M (2017) Optimized designs of reversible arithmetic logic unit. Turkish J Electr Eng Comput Sci 25(2):1137–1146
19. Singh R, Upadhyay S, Jagannath K, Hariprasad S (2014) Efficient design of arithmetic logic unit using reversible logic gates. Int J Adv Res Comput Eng Technol (IJARCET) 3(4):1474–1477
20. Kamaraj A, Marichamy P (2019) Design of integrated reversible fault-tolerant arithmetic and logic unit. Microprocess Microsyst 69:16–23
21. Thakral S, Bansal D (2019) Fault tolerant arithmetic logic unit. In: 2nd International Conference on Emerging Current Trends in Computing and Expert Technology (COMET-2K19). Springer
22. Thakral S, Bansal D (2019) High Functionality reversible arithmetic and logic unit. Int J Electr Comput Eng (IJECE) 10:2329–2335
23. Guan Z, Li W, Ding W, Ni, YHL (2011) An arithmetic and logic unit design based on reversible logic. IEEE Publication, 2011
24. Aradhya HVR, Kumar BVP, Muralidharan KN (2011) Design of control unit for low power ALU using reversible logic. Int J Sci Eng Res 2:1–7
25. Dixit A, Kapse V (2012) Arithmetic and logic unit (ALU) design using reversible control unit. Int J Eng Innov Technol 1:55–60
26. Rakshith TR, Saligram R (2013) Parity preserving logic based fault tolerant reversible ALU. In: IEEE Conference on Information and Communication Technologies, IEEE, pp 485–490, 2013
27. Moallem P, Ehsanpour M, Bolhasani A, Montazeri M (2014) Optimized reversible arithmetic logic units. J Electr 31(5):394–405
28. Sharma N, Sachdeva R, Saraswat U, Yadav R, Kaur G (2015) Power efficient arithmetic logic unit design using reversible logic. Int J Comput Appl 128(6):36–41
29. Deeptha A, Muthanna D, Dhrithi M, Pratiksha M, Kariyappa BS (2016) Design and optimization of 8-bit ALU using reversible logic. In: IEEE, ICRTEICT, 20–21 May 2016
30. Shukla V, Singh OP, Misra GR, Tiwari RK (2017) Reversible realization of N-bit arithmetic circuit for low power loss ALU application. In: International Conference on Smart Computing and Communication, pp. 847–854
31. Rakshith S, Hegde SS, Kulkarni SA, Bhagyalaksh HR, Venkatesha MK (2013) Design of parity preserving logic based fault tolerant reversible arithmetic logic unit. Int J VLSI Des Commun Syst 4:53–68
32. Sena B, Duttaa M, Goswamia M, Sikdar BK (2014) Modular design of testable reversible ALU by QCA multiplexer with increase in programmability. Microelectr J 45(11):1522–1532
33. Sasamal T, Singh A, Mohan A (2016) Efficient design of reversible ALU in quantum-dot cellular automata. Optik 127(15), 6172–6182
34. Kalita G, Saikia N (2016) Designing reversible arithmetic, logic circuit to implement micro-operation in quantum computation. J Phys 1–5. Conference Series 759

35. Kamaraj A, Marichamy P (2017) Design and implementation of arithmetic and logic unit using novel reversible gates in quantum cellular automata. In: IEEE, ICACCS, 6–7 Jan 2017
36. Khatter P, Pandey N, Gupta K (2018) An arithmetic and logical unit using reversible gates. In: International Conference on Computing, Power and Communication Technologies, pp 476–480, 2018
37. Oskouei SM, Ghaffari A (2019) Designing a new reversible ALU by QCA for reducing occupation area. J Supercomput 1–27

Performance Analysis of Load Balancing Algorithms in Amazon Cloud

Ragini Kumari and Vijay Kumar Jha

Abstract Cloud computing has vast impact in the field of parallel and distributed computing systems today. It enables a massive number of users to access virtualized hardware and software infrastructures with the help of the internet. Consequently, it produces huge traffic into cloud servers. The abrupt growth of this load leads to serious challenges for load balancer of cloud to work out efficiently. The function of load balancer is to distribute the heavy load on servers in cloud environment to obtain maximum utilization of resources by achieving maximum throughput and minimum response time. Various researches have been done to implement better load balancing algorithms to minimize load on servers, but all existing research work in this direction have been developed as an analytical approach in which performance analysis of load balancing algorithms have been performed or compared by creating servers in their systems. The future analysis of this research has been left to implement the load balancing algorithms in real environments. This paper demonstrates the performance analysis of two most popular load balancing algorithms i.e. Round Robin and Least Connections in Amazon Web Service cloud in real-time with the help of performance measuring tool Apache JMeter.

Keywords AWS (Amazon Web Service) · Apache Jmeter · Load balancing · Least connection · Round robin

1 Introduction

Cloud computing [1] is internet-based computing in the field of information technology. It consists of three components such as cloud users, service providers and

R. Kumari (✉) · V. K. Jha
Department of Computer Science and Engineering, Birla Institute of Technology, Mesra, Ranchi, Jharkhand, India
e-mail: phdcs10053.18@bitmesra.ac.in

V. K. Jha
e-mail: vkjha@bitmesra.ac.in

V. Nath and J. K. Mandal (eds.), *Proceeding of Fifth International Conference on Microelectronics, Computing and Communication Systems*, Lecture Notes in Electrical Engineering 748,
https://doi.org/10.1007/978-981-16-0275-7_33

internet. The cloud users store, maintain and access resources such as hardware, software, network, servers, applications and services with the help of internet from cloud service providers instead of using their personal systems. According to NIST [2], cloud is model for enabling ubiquitous, on demand network access to a shared pool of computing resources that can be rapidly provisioned within minimal management effort. The rapid development of network technology makes network application quite popular in public. The widely used network applications nowadays are online shopping, online tickets booking systems and other services. It enables huge number of users to approach towards this emerging technology. This leads to sometimes instant and sudden increase of users request on web servers, causing system to slow down or even crash the website. The websites such as Google, Amazon and Microsoft have millions of user hits per day [3]. For handle huge number of requests to respond smoothly and quickly the sites use a technology called load balancing. There are many cloud service providers which offer cloud load balancing technologies such as Amazon Web Services (AWS), Google and Microsoft Azure. AWS offers Amazon Elastic Load Balancing, which distributes traffic among EC2 instances. Google Cloud Platform offers Google Compute Engine, which distributes network traffic between VM instances. Microsoft Azure's Traffic Manager distributes traffic for its cloud services across multiple data centers. These providers provide their services pay-as-per-use model. This feature increases immense number of users. Ultimately load on servers increase and performance gets degraded. The various researches of load balancing algorithms have been developed in this direction to minimize load on server for better performance analysis. Mustafa [15] has implemented and compare load balancing algorithms such as round robin and least connection on load balancer to distribute network traffic efficiently on 8 servers by using VMware in personal system. The performance can be evaluated using OPNET software and they have found that round robin load balancing algorithm on 8 servers are less efficient than least connection load balancing algorithm on these servers. The proposed work shows live performance evaluation of load balancing algorithms such as round robin and least connection on AWS cloud. The paper consists of four sections. The first section explains related work. The second section describes load balancing in cloud. The third section demonstrates proposed work and the fourth section explains conclusion and future work.

2 Literature Survey

The tremendous developments of cloud services and applications approach large number of users to shift their work over cloud platform with the help of internet to save time, space and cost. As a result, the load on cloud servers increases. This affects the performance of servers to work efficiently. Behal and Kumar [4] have taken a theoretical framework combining load balancing and service broker policy–named it active tracking load balancing algorithm and further comparing the same with round robin load balancing algorithm using cloud analyst tool. Authors have shown that their proposed algorithm has better integrate performance in terms of total cost of different data centers, average response time and data center service time. Radojevic and Zagar [17] have given framework for load balancing models and algorithms for better working efficiency of servers. Authors have mentioned regions where load balancing policies have been applied also describing that load balancing algorithms have been used on load balancers either on application layer, network layer or processor load balancing mode. Models for implementing various load balancing algorithms in virtualized computer environment have been given therein. Singh et al. [9] compare the load balancing algorithms in data centers using cloud analyst tool for performance analysis. They found that the least connection gives effective results as compared to round robin. However, there is no significant difference in response time. Similarly using cloud analyst tool, Duggal and Dave [19] have given comparative results in terms of various performance metrics such as cost, response time and data center processing time and concluded that round robin algorithm has lower response as well as data center processing time than ESCE & Throttled while cost being same for all these load balancing algorithms. Mishra et al. [18] have shown a comparative view of existing research on load balancing algorithms as well as presented some results using cloud sim tool while mentioning their desire to do the same evaluation in real world cloud environment as future scope of their work. Alternatively, authors use qualitative approach such as stability, process migration etc. showing the working environment of load balancing algorithms in which workload is assigned either in compile time i.e. static or run time i.e. dynamic [10]. Various algorithms have been analyzed such as round robin, central queuing or randomized in terms of response time and space using cloud simulator. It has been concluded that the stability of round robin is better than other and in distributed environment dynamic algorithms perform better than static algorithms. Chen et al. [13] have compared two types of cloud load balancing architectures namely CLB enabled web servers and CLB enabled physical servers. Also, they have used two algorithms namely least connection and weighted round robin. They have analyzed the above-mentioned algorithms in both the servers and found that the throughput of web server better than physical server. The future work for researchers is analysis of live performance of load balancing algorithms in cloud environment for comparison of various performance metrics so as to select the best algorithms to minimize huge load on servers of cloud. The authors have surveyed the load balancing in cloud computing system [14]. Definition, classification and example of cloud vendors have been mentioned as well as describing Amazon load balancing in which incoming traffic is distributed in a number of Amazon EC2 instances. Panwar and Mallick [16] has shown comparative studies of two most popular load balancing algorithms such as round robin and least connection. For performance analysis they have used cloud analyst tools. They will perform their work by configuration of various components of this tool such as four datacenters in four different regions and each data centers having twenty virtual machines. The simulation work has been done by using algorithms mentioned therein for 24 h. They have found that least connection performs better than round robin in terms of cost and time. The above-mentioned papers deploy analytical approach for comparison of different load balancing algorithms.

The performance analysis of load balancing algorithms has been measured using simulation tools. This tool does not produce accurate statistical data for analysis of performance metrics of load balancing algorithms in cloud.

3 Load Balancing in Cloud

Load balancing is process of using server to distribute traffic to multiple servers so as to share the workload among web servers. For better understanding of load balancing, let us consider an example when a user searches a site such as Gmail. To start user types domain name such as www.gmail.com. The user web browser sends domain name to special servers on the web called domain name server DNS which in turn return sites internet protocol address IP, the browser than uses IP address to contact the server. When the site uses load balancing, IP address returned by DNS related to load balancing server. The load balancer receives the browser requests, it then send request to one of the servers. If demand on the sites increases additional servers can be added to which load balancer can distribute requests. Elastic Load Balancing distributes incoming application traffic across multiple Amazon Elastic Compute Cloud (Amazon EC2) instances [5]. There are two logical components in the Elastic Load Balancing service architecture: load balancers and a controller service. The load balancers are used to watch traffic and distribute clients/users request over the Internet. The controller service monitors the load balancers, adds and removes capacity as needed, and verifies that load balancers are behaving properly. There are two types of elastic load balancers Application Load Balancers and Classic Load Balancers. In Classic Load Balancer, user registers instances with the load balancer where as in Application Load Balancer user registers the instances as targets in a target group, and route traffic to a target group. With a Classic Load Balancer, the load balancer receives the user requests and chooses servers to distribute request using the round robin routing algorithm for TCP listeners and the least connection requests routing algorithm for HTTP listeners. The main aims of elastic load balancing are:

- To increase the performance significantly.
- To have a backup plan in case the system failure.
- To keep the system stable.

4 Load Balancing Strategies

The load balancing strategy is use to maximize resource utilization on cloud servers for efficient working of servers if load on server increases. The various load balancing algorithms have been developed and analyzed on load balancers to distribute requests among servers. In this paper two most popular load balancing algorithms on load balancer have been used to distribute users request on web servers such as round robin and least connection. The performance evaluation of these algorithms has been done in AWS cloud environment.

5 Round Robin Load Balancing Algorithm

Round robin algorithm is simple scheduling algorithm in which users request to load balancer in cloud environment [7]. The load balancer sends back IP address of each cloud servers in round fashion, the first request to first server IP address, second to second server IP address and so on. Let, there are 'n' servers s_0, s_1, s_2, ……, s_{n-1} and 'm' is IP of last request to server node. The algorithm is given below:

Round Robin (m, n).

1. Let 'j' be the initial variable such that j = m;
2. Do:
3. j = (j + 1) mod n;
4. if (s_j is free)
5. return s_m;
6. else { }
7. while (j! = m) //if agree again go to step 2
8. return NULL;

6 Least Connection Load Balancing Algorithm

Least connection algorithm is scheduling algorithm in which the user connection to each server to which the user connects [8]. The load balance architecture is designed based on the quantity of user connections. The distributor server is connected to the HTTP request, and each new user sends an HTTP request that will be assigned to the minimum number of connections with the server. For example, there are three servers in which first server have 3 active transactions, second server has 15 active transactions and third server has 0 active transactions. The next incoming user request to web server goes to third server as this server has least active transaction among all servers. Let there are 'n' number of servers such as s_0, s_1, s_2, ……,s_{n-1}. The algorithm is given below:

Least connection (m, n).

1. Traverse all servers of AWS cloud;
2. Determine whether the 'm' server works properly if not return to step1 and move to next server if yes continue to step 3
3. Again traverse the servers from m + 1 server;
4. Then find server with least connections;
5. Find the corresponding server, return s_m;
6. Return NULL if not find the corresponding server.

For better performance analysis of above-mentioned algorithms in cloud environment, average response time should be minimum and throughput should be maximum and the results will be recorded by simulation tool. The most popular simulation tool such as cloud analyst has been used for performance analysis of load balancing algorithms in cloud environment. Reference [11, 12] explains cloud simulation and modeling of cloud computing infrastructures and services using cloud simulation tool. Authors used this tool for exploring the results of experimental works repeatedly for better evaluation in terms of time, costs and space. Authors have also observed that the relationship between time and space is linear i.e. as time increases exponentially number of space increases in this simulation environment. We used Apache Jmeter for performance evaluation of load balancing algorithms in AWS cloud.

7 Apache Jmeter

The Apache Jmeter [6] is open source software to measure heavy load on servers to test its strength or to analyze overall performance in form of tables, graphs, etc. It was originally designed for testing Web Applications. It is developed by Stefano Mazzocchi. It is a Java desktop application with a graphical interface that uses the Swing graphical API. Apache JMeter is a framework for Java, so the very first requirement is to have JDK 1.6 or above installed in systems. It can run on any platform that accepts a Java virtual machine, for example Windows, Linux, Mac, etc. For proposed work Apache Jmeter version 3.1 have been used. The performance is used to check efficiency of the systems. This has to be improved at reasonable cost, reducing response time and maximizing throughput. The performance metrics calculated by Apache Jmeter tool:

Throughput: This metric is used to estimate the total number of tasks whose execution has been completed successfully. High throughput is necessary for overall system performance.

$$\text{Throughput} = \frac{1}{\text{TotalTime}}$$

$$=> \text{Totaltime} = (\text{Averagebits}) \times \left(\frac{1}{\text{KB/sec}}\right)$$

Average Response Time: It can be measured as the time interval between sending a request and receiving its response. It should be minimized for better performance.

$$\text{Average}, \mu = \frac{\sum_{i=1}^{n} x_i}{n}$$

Standard Deviation: It gives how many error cases have been found while samples execution and deviating from the average value of the receiving response time from the server. For better performance analysis standard deviation should minimum. The standard deviation (σ) measures the mean distance of the values to their average (μ).

It gives good idea of the dispersion or variability of the measures to their mean value. The standard deviation (σ) is calculated as:

$$\sigma = \sqrt{\frac{1}{n}\sum_{i=1}^{n}(x-\mu)^2}$$

8 Problem Description

The proposed work compares two most popular load balancing algorithms available for cloud computing i.e. Round Robin and Least Connections. This comparison uses Apache JMeter as performance testing tool that will record results of performance test. To conduct this test, a data center has been created in Mumbai using AWS cloud services. Within this data center 20 single core servers and 1 elastic load balancer have been created. This elastic load balancer distributes traffic to these 20 servers. Web application has been hosted on these servers. This web application has been accessible at www.awsagile.in. Apache JMeter has been used to sends TCP and HTTP requests respectively to this elastic load balancer via web application and load balancer. Based on data collected by this performance measuring tool, conclusions have been drawn on performance of load balancing algorithms in cloud computing. For above mentioned works to analyze performance of load balancing algorithms in cloud computing following steps would be taken in this section:

- Selection of cloud providers
- To create and configure data center in cloud
- To create servers in data center
- To create and configure load balancer in data center
- To create and configure load balancing algorithm for HTTP request
- To create and configure load balancing algorithm for TCP request
- Apache Jmeter for set up for sending HTTP request to 20 Servers
- JMeter Result for 20 Servers for HTTP Request
- Apache Jmeter has been set up for sending TCP request to 20 Servers
- JMeter Result for 20 Servers for TCP Request
- Comparison of Performance analysis of Round Robin and Least Connection load balancing algorithm

Table 1 Datacenter configuration – AWS

Datacenter configuration – AWS	
Server type (AWS specific)	t2.micro
# of servers	20
Geographical location	Mumbai
Monthly cost of servers	$7.8
Monthly cost of load balancer	$22.39
Total monthly cost of datacenter	$30.19

9 Selection of Cloud

There are various service providers in cloud environment such as Amazon Web Services (AWS), Google cloud, and Microsoft azure. AWS is selected for experiment purpose. The AWS Cloud has 42 Availability Zones within 16 geographic Regions around the world. They have distributed resources as data warehousing to deployment tools, directories to content delivery and 50 services are available in a second.

10 To Create and Configure Data Centre in AWS Cloud

Architecture diagram of Data Centre in AWS cloud is shown in Table 1. It shows how a typical user will access servers in cloud through Elastic Load Balancer via internet. It also shows that data center is physically located in Mumbai and has 20 servers and 1 load balancer. All of these resources are accessible through internet. The data center consists of 20 virtual machines.

11 To Create Servers in Data Center

Once data center is created, thereafter 20 servers are created in this data center for testing purpose.

12 To Create and Configure Load Balancer in Data Centre

Load balancer in AWS cloud is also known as Elastic Load Balancer because it manages load automatically. This load balancer is available in Software as a Service (SaaS) model from AWS cloud. Once this load balancer is created and configured, all 20 servers in the data center are attached to this load balancer. This elastic load balancer will use different routing algorithms to manage load among all these 20

Table 2 Performance testing configuration – AWS – least connection algorithm

Performance testing configuration – AWS – least connection requests algorithm	
Application type	Web application
Programming language	PHP, HTML5, JavaScript
Concurrent users	1000
Total number of requests	1000
Type of requests	HTTP
Test duration	180 s
Number of times test run	1
Performance testing tool used	Apache JMeter

servers. Following figure also shows that 20 servers are running and these 20 servers are attached to this load balancer. Elastic load balancer in AWS cloud allows to configure various parameter such as ports, listeners etc. The load balancer provisioned to handle requests for specific application sends requests to the Amazon EC2 instances that are running in application. In following figure, HTTP and TCP protocol for elastic load balancer are configuring.

13 To Create and Configure Load Balancing Algorithm for HTTP Request

Apache JMeter will send HTTP requests to access servers in cloud through Elastic Load Balancer via internet and load balancer distributes these HTTP requests using least connections algorithm across servers. The configuration used for the same, is shown in Table 2.

14 To Create and Configure Load Balancing Algorithm for TCP Request

Apache JMeter will send TCP requests to access servers in cloud through Elastic Load Balancer via internet and load balancer distributes these TCP requests using round robin algorithm across servers. The configuration used for the same, is shown in Table 3.

Table 3 Performance testing configuration AWS round robin algorithm

Performance testing configuration AWS round robin algorithm	
Application type	AWS elastic load balancer
Concurrent users	1000
Total number of requests	1000
Type of requests	TCP
Test duration	180 s
Number of times test run	1
Performance testing tool used	Apache JMeter

Table 4 Jmeter results for least connections algorithm

Label	# Samples	Average	Min	Max	Std. Dev	Throughput	Received KB/sec	Sent KB/sec	Avg. bytes
HTTP request	1000	810	128	9540	1087.97	5.47157/sec	20.05	0.98	3752

15 Apache Jmeter for Sending HTTP Request to 20 Servers

Apache Jmeter has been used for sending HTTP requests to web application www.awsagile.in, ramping up 1000 users in 180 s i.e. 1000 HTTP requests sent to elastic load balancer in 180 s.

16 JMeter Result for 20 Servers for HTTP Request

Table 4 shows results for 1000 HTTP requests being sent to load balancer through Apache JMeter tool in respect of Least Connections algorithm.

17 Apache Jmeter for Sending TCP Request

Apache Jmeter has been used for sending TCP requests to web application www.awsagile.in, ramping up 1000 users in 180 s i.e. 1000 TCP requests sent to elastic load balancer in 180 s.

Table 5 Jmeter results for round robin algorithm

Label	# Samples	Average	Min	Max	Std. Dev	Throughput	Received KB/sec	Sent KB/sec	Avg. bytes
TCP request	1000	44,615	125	69,502	26,220.13	4.12213/sec	0.12	0	30.3

18 JMeter Result for 20 Servers for TCP Request

Table 5 shows results for 1000 TCP requests being sent to load balancer through Apache JMeter tool in respect of Round Robin algorithm.

19 Results

The following performance metrics have been compared through performance measuring tool.

- Standard deviation
- Throughput
- Average response time

The first performance metric is standard deviation, which is deviation from average i.e. smaller the value better the performance. Comparing results from both tables, it has been found that the standard deviation of least connection is smaller than round robin algorithm. The second metric is throughput. It represents the ability of server to handle heavy load. If the throughput is large then the performance is better. Comparing results from both tables, it has been found that least connection algorithm is better than round robin load balancing algorithm. The third performance metric is average response time, which should be minimum for better performance of servers. Again, it has been found that average response time which is measured in milliseconds is larger for Round Robin algorithm than Least Connections algorithm in AWS cloud. So, 'least connection' is better than 'round robin' load balancing algorithm.

20 Conclusion

The analysis of performance metrics like throughput, standard deviation, average response time and cost using Apache Jmeter show that the round robin algorithm is inferior to the least connection algorithm for load balancing in cloud environment if the volume of traffic is huge. Hence, it is concluded out of these two most popular load balancing algorithms, the least connection algorithm is more suited for users. It is recommended to include multiple cloud solutions such as Microsoft Azure and

Google Cloud along with AWS for future work. Various other algorithms should also be compared to improve the performance metrics of load balancing in cloud environment.

References

1. Erl T, Mahmood Z, Puttini R (2014) Cloud Computing Concepts, Technology & Architecture published by Pearson, pp 58–78, 90–95, 97–112 (2014)
2. Mell P, Grance T (2011) The NIST Definition of Cloud Computing. NIST Special Publication 800–145, September 2011
3. Jamsa K (2013) Cloud computing SaaS, PaaS, IaaS, Virtualization, Business Models, Mobile, Security and more. Jones & Bartlett Learning, 46–48, 178, 278–280
4. Behal V, Kumar A (2014) Comparative study of load balancing algorithms in cloud environment using cloud analyst. Int J Comput Appl 36–40
5. https://docs.aws.amazon.com/elasticloadbalancing/latest/userguide/elb-ug.pdf, 2020
6. https://jmeter.apache.org/, 2013
7. Zongyu X, Xingxuan W (2014) A modified round-robin load-balancing algorithm for cluster-based web servers. In: Proceedings of the 33rd Chinese Control Conference, pp 3580–3584
8. Irkal S, Sunag V (2017) Load balancing algorithms in cloud computing. Int J Innov Res Comput Commun Eng 5:1–6
9. Singh A., Bedi RK, Gupta SK (2014) Comparative analysis of load balancing algorithms in cloud computing. Int J Adv Technol Eng Res 18–21
10. Ray S, De Sarkar A (2012) Execution analysis of load balancing algorithms in cloud computing environment. Int J Cloud Comput. Serv Archit 2:1–13
11. Wickremasinghe B (2010) Cloud Analyst: A Cloud Sim-based Visual Modeller for Analysing Cloud Computing Environments and Applications, 2010
12. Calheiros RN, Ranjan R, De Rose CAF, Buyya R (2011) CloudSim: A Novel Framework for Modeling and Simulation of Cloud Computing Infrastructures and Services Grid Computing and Distributed Systems (GRIDS) Laboratory Department of Computer Science and Software Engineering The University of Melbourne, Australia, 2011
13. Chen SL, Chen YY, Kuo SH (2011) CLB: A novel load balancing architecture and algorithm for cloud services. Comput Electr Eng 154–160
14. Khiyaita A, Bakkali HE, Zbakh M, Kettani DE (2012) Load balancing in cloud computing: state of art. IEEE - National Days of Netw Secur Syst 106–109
15. Mustafa MEG (2017) Load balancing algorithms round robin, least connection and least loaded efficiency. GESJ: Comput Sci Telecommun 25–29
16. Panwar R, Mallick B (2015) A comparative study of load balancing algorithms in cloud computing. Int J Comput Appl 117(24):33–37
17. Radojevic B, Zagar M (2011) Analysis of issues with load balancing algorithms in hosted (Cloud) environments. In: IEEE–34th International Convention on Information and Communication Technology, Electronics and Microelectronics (MIPRO), pp 416–420, 2011
18. Mishra SK, Sahoo B, Parida PP (2020) Load balancing in cloud computing: a big picture. J King Saud Univ Comput Inf Sci 149–158
19. Duggal AK, Dave M (2020) A comparative study of load balancing algorithms in a cloud environment. In: Advances in Computing and Intelligent Systems, Springer Nature Singapore Pte Ltd, 115–126

Unacknowledged Mode LAPDm Protocol Development at MS Side of GSM Network

Saumya Borwankar, Rishi Pandya, and Rachna Sharma

Abstract In GSM, the data link layer protocol is LAPDm (Link access procedure for Dm channel). In this paper, an implementation of the Layer 2 protocol – LAPDm in unacknowledged mode at the mobile station (MS) of Global System for Mobile communications (GSM) network has been implemented using layered architecture. In the layered architecture related system's tasks are grouped into different layers and describe the desired functionality of each layer. Development of each layer takes place independently and communication between layers can be implemented with the help of inter process communication.

Keywords LAPDm · GSM · BTS · IPC

1 Introduction

In signalling protocol architecture in GSM as shown in Fig. 1 layer 2 is a modified version of the link access procedure for D channel (LAPD) protocol used in the integrated service digital network (ISDN), called LAPDm.

Link Access Procedure for Dm channel (LAPDm) [1] is a data link layer protocol and is predominantly used for air interface. For designating all signalling channels obligatory to GSM we use the term Dm channel. The main function of the LAPDm protocol of MS is to establish/suspend/resume/terminate the connection with the base transceiver station (BTS) over air interface and to perform unacknowledged/multiframe information transfer operation.

S. Borwankar (✉) · R. Pandya · R. Sharma
Electronics and Communication Department, Institute of Technology, Nirma University, Ahmedabad, India
e-mail: 17bec095@nirmauni.ac.in

R. Pandya
e-mail: 17bec086@nirmauni.ac.in

R. Sharma
e-mail: rachna.sharma@nirmauni.ac.in

V. Nath and J. K. Mandal (eds.), *Proceeding of Fifth International Conference on Microelectronics, Computing and Communication Systems*, Lecture Notes in Electrical Engineering 748,
https://doi.org/10.1007/978-981-16-0275-7_34

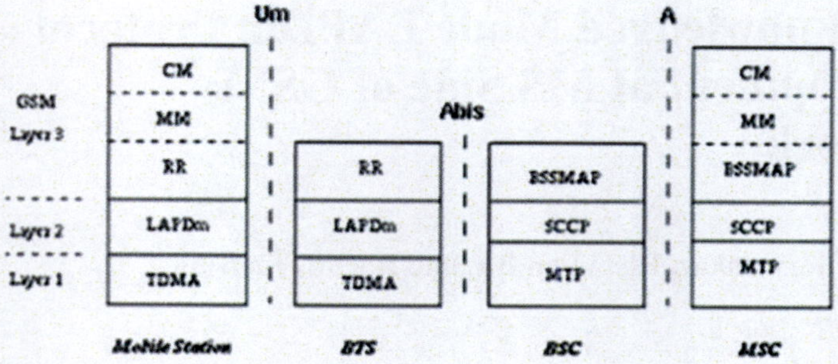

Fig. 1 GSM protocol architecture

With the help of primitives, the LAPDm protocol provides services to the upper layer by utilising services from the lower layer. As the focus of the work presented is on unacknowledged mode LAPDm protocol development at MS side of GSM network, in order to demonstrate the functionality of LAPDm protocol the stubs of upper layer and that of lower are create to pass the test primitives. The LAPDm is implemented as a process in C and the Inter-process communication (IPC) concept is used in order to realize communication of LAPDm with upper and lower layer of GSM protocol stack. Stub testing is proposed here to testify the functionality of each layer.

The Sect. 2 explains in details the LAPDm protocol. The LAPDm implementation model is discussed in Sect. 3. Section 4 explains the software requirement specification (SRS). In Sect. 5, we have explained the high-level design (HLD). The Sect. 6 describes low-level design (LLD). The final two sections deal with description of testing model and conclusion of the work carried out.

2 LAPDM Protocol

DLL (Data link layer) provides services to the layer 3 (call control (CC), radio resource management (RR), mobility management (MM) in case of GSM) and receives services from layer 1 (Physical layer) as shown in Fig. 1.

Peer to Peer protocols are used to provide cooperation between DLL entities at the MS side and at the network side. Before information exchange begins between entities as described in layer 3 in Fig. 2, it is imperative to establish a data link connection.

Through service primitives, the DLL is requested for providing services by Layer 3. The physical layer and DLL interaction also works with the same principle [3]. The apercu of logical information exchange and control is delineated by primitive by epitomizing the functioning between DLL and vicinal layers. As shown in Fig. 3, we can categorise the primitives exchanged between DLL and adjacent layers broadly into 4 types:

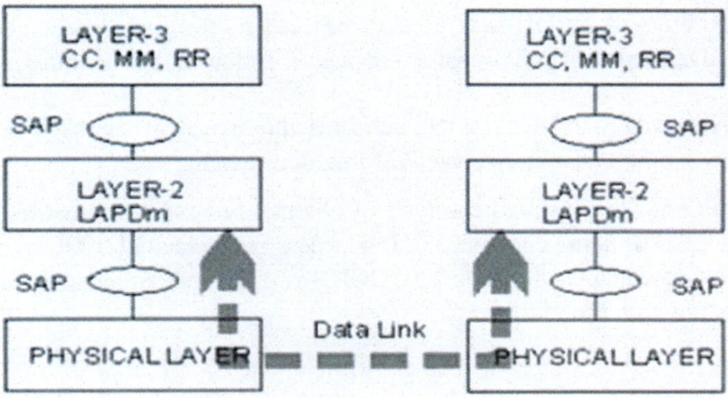

Fig. 2 Peer to peer data link establishment

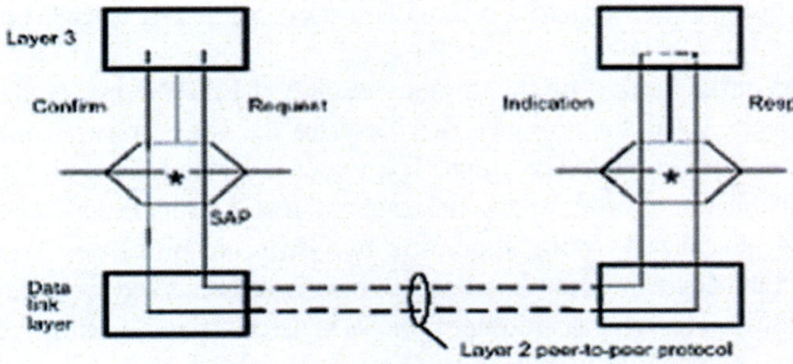

Fig. 3 Primitive action sequence

For transmitting layer 3 messages, any one of the following two operating modes may be taken into practice by LAPDm protocol:

1. Unacknowledged Operation
2. Multiple Frame Operation

The salient features of the unacknowledged information transfer service [1, 2] are:

a) Provision of a data link connection between layer 3 entities for unacknowledged information transfer of layer 3 message units.
b) Recognition and discerning of endpoints of data link connection in order to allow a layer 3 entity to intuit another layer 3 entity.
c) Transferring frames in congruence with priority given to the message.
d) No verification of arrival of signal within the DLL.

DL-UNIT DATA- -REQUEST and DL_UNITDATA-INDICATION are primitives auxiliary with the unacknowledged information transfers service. The functions of each primitives are:

- DL-UNIT DATA-REQUEST: used to beseech that proper procedure of unacknowledged information transfer service is followed for sending a message unit.
- DL-UNIT DATA-INDICATION: indicates the arrival of message unit that is received through unacknowledged information transfer.

Message unit, priority and the type of channel being used (Broadcast control channel (BCCH), paging channel (PCH) + access grant channel (AGCH) or specific type of dedicated control channel (DCCH)) are the fundamental specifications corresponding with the primitives.

3 LAPDM Implementation Model

The generic implementation model of LAPDm protocol at MS of GSM network is shown in Fig. 4.

As the focus of the work is on the implementation of LAPDm protocol, the upper layer to whom the LAPDm provides services and the lower layer from which it takes services are implemented as stubs. The upper layer and the lower layer stubs will provide primitives in order to demonstrate and test the functionality of LAPDm layer. LAPDm is realized by implementing two different processes. One process called as LAPDm down process handles the primitives received from layer 3 and pass these primitives to layer 1. Another process is used to handle primitive received from the lower layer which is end up with the indication to layer 3 after required processing. In order to share variables between two different processes of LAPDm, shared variables concept of IPC is used. The communication the between layers is done with the help of message queues. Various steps are involved in implementation of LAPDm protocol are:

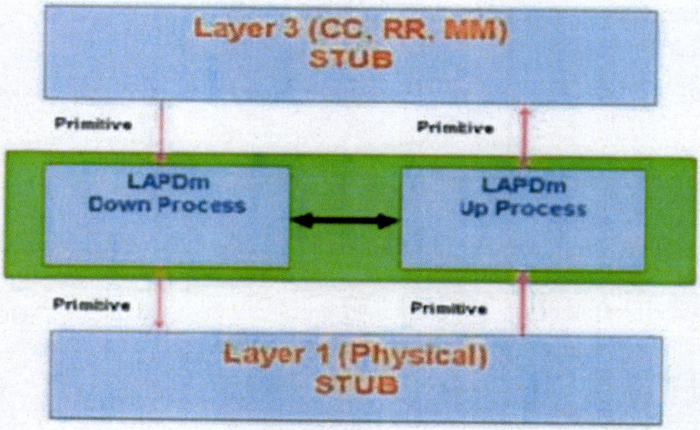

Fig. 4 Generic LAPDm implementation model

1) SRS *(software requirement specification)*
2) *HLD (high-level design)*
3) *LLD (low level design)*
4) *Testing*

4 Software Requirement Specification

The LAPDm is required to provide information transfer service to layer 3. The various functions should be performed by the LAPDm in order to provide this service [4–6].

In SRS document, LAPDm functionality in unacknowledged mode is spitted into individual software requirements. Each requirement is assigned a requirement ID and a small description.

Req. Id JIITU_GSM_LAPDm_001
Summary Transmission of unacknowledged information of layer 3 with normal L2 header.
Description LAPDm will receive the information from Layer 3 in unacknowledged information (UI) frame using DL_UNIT DATA_REQUEST primitive and will attach normal L2 header to it. It will then transmit the frame to the peer entity.
Req. Id JIITU_GSM_LAPDm_002
Summary Transmission of unacknowledged information of layer 3 with short L2 header type 1.
Description LAPDm will receive the information from Layer 3 in UI frame using DL_UNIT DATA_REQUEST primitive and will attach short L2 header type 1 to it. It will then transmit the frame to the peer entity.
Req. Id JIITU_GSM_LAPDm_003
Summary Receipt of UI frame with invalid service access point identifier (SAPI) value
Description On receipt of UI frame with invalid SAPI value the UI frame will be discarded.
Req. Id JIITU_GSM_LAPDm_004
Summary Receipt of UI frame with length indicator set to 0
Description UI frames received with the length indicator set to 0 shall be ignored.

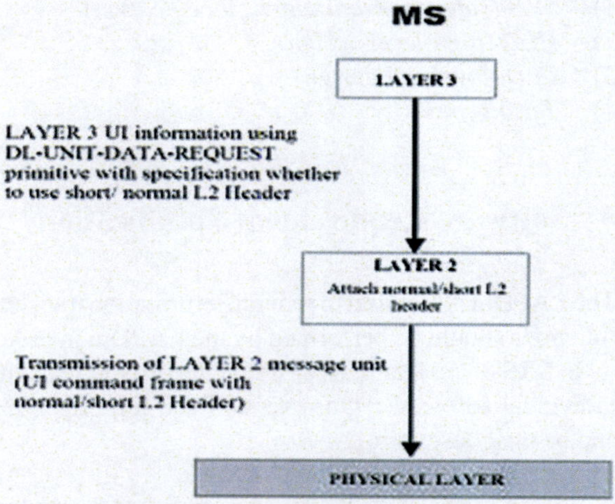

Fig. 5 Unacknowledged information transfer [7]

5 HLD (High Level Design)

The HLD of the LAPDm protocol explains in detail the flow of primitives across layers and their relationships.

The high-level design provides the detailed functionality of the LAPDm protocol to realize the functions specified in the software requirement analysis. The high-level design is prepared for the unacknowledged Information Transfer (Fig. 5).

6 LLD (Low Level Design)

The comprehensive implementation model of LAPDm is given by LLD. A brief introduction on LAPDm procedures in the MS is provided in the preceding sections [8–10]. A theoretical design arrangement of a send queue, a send buffer and a transmit buffer for illustrating procedures are exhibited in Fig. 6 as assumed. For ease of understanding of procedures in MS, it is presumed that for transferring any message from layer 3 to layer 2, it must first be sent via SAPI 0 in DL-DATA-REQUEST or DL-UNIT-DATA primitive before it is placed in the send queue. A FIFO (first in-first out) mechanism is followed by the layer 2 entity to identify messages and placing them into send buffer [11]. It must be noted that a send buffer cannot hold multiple messages of layer 3 at any given instance of time so a UI frame is developed from the contents of the send buffer. These frames that are later placed in a transmit buffer are created one at a time. The frame is then transmitted.

LAPDm is implemented as two different processes LAPDm UP and LAPDm Down. LAPDm accepts primitives from layer 3 along with the attached protocol development unit (PDU), prepares frame and passes them for transmission to layer 1. Layer 1 stub is also implemented in two parts. One part is used as sink for the

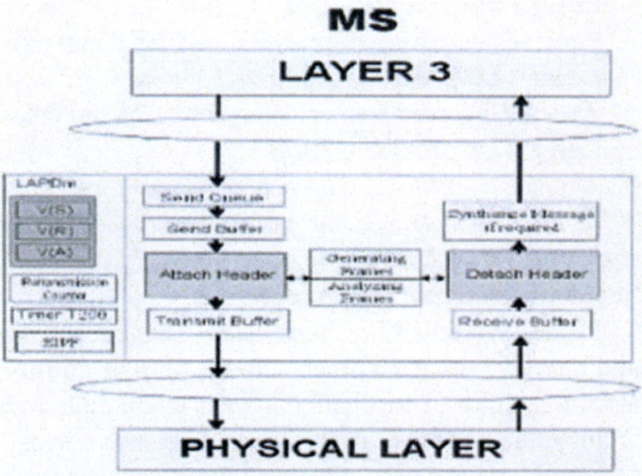

Fig. 6 Detailed implementation model of LAPDm protocol

frames received from layer 2. As the implementation of layer 1 is not the focus of the work, it is implemented to simply accept the prepared frames from layer 2 and displaying them assuming that layer 1 is working perfectly and frames have been transmitted successfully over the radio interface.

Another part of layer 1 is implemented as layer 1 Up which accepts frames transmitted by peer entity and indicates about the same to the layer 2. Practically received frames are passed to upper layer only when the received data is found error free [12]. Assuming that the error free reception has occurred through the dummy frames, which exactly matches with the actually received response/data frames in practical scenario, are prepared and passed to layer 2 by attaching them with an indication primitive.

LAPDm Up process accepts the primitive from the layer 1 Up process detaches the received frames, analyzes frame header and takes necessary action, for example if the received frame is a response frame used for providing acknowledgement then the timer corresponding to the related procedure in LAPDm Down process will be stopped. If the received frames are data frames then after detaching the header the messages are prepared by performing synthesizing/ fragmentation and passed to the upper layer by attaching them with the suitable indication primitive. Layer 3 process accepts the received message and displays them on the same screen. In summary, there are six processes implemented in order to demonstrate and test the functionality of LAPDm but in our case we use only down processes.

Layer 3 Down process – L3

Prepare message and pass that as PDU to layer 2 for transmission.

Layer 2 Down process – L2

Accepts layer 3 request and follows necessary steps by taking the services from lower layer.

Layer 1 Down process – L1

Accepts frames from L2 and transmits them bit by bit over radio interface.

Layer 1 Up process – L1up

Receives data from peer entity, verifies it and passes the same to layer 2.

Layer 2 Up process – L2up

Accepts received frames, analyzes them and takes necessary action.

Layer 3 Up process – L3up

Act as a sink for L3 data received from the peer entity.

As explained above, the various processes implemented needs to interact with each other. The concept of Inter Process Communication (IPC) is used to realize interaction between these processes.

In the layered architecture, various layers are implemented as separate process and interact with each other with the help of primitives. The IPC concept serves as a very useful tool for realizing such interaction amongst various layers in layered architecture [13–15]. There are various ways using which this interaction can be realized. We have used message queues and shared memory for this purpose.

7 Testing of Protocol Stack

Protocol stack testing is done to ensure the proper and desired functionality of the implemented protocol as per the specifications. It is always recommended to test a protocol stack before actually placing it for operation in real world [16].

Development of each layer takes place independently to the development of other layers. In order to check for the proper functionality of each layer, testing is done at the layer level. Once the testing of all layers at individual level is done, then all the layers are integrated to form a protocol stack. The testing of complete protocol stack, fully responsible to realize communication between two entities, is entirely different then the testing done at each layer module partially contributing in the whole process of communication between the peer entities [17–19].

As the focus of the work is at specific layers, here the testing methodology suggested are applicable for the testing of DLL functionality instead of testing of the complete protocol stack. In order to test the functionality of a layer module, stub testing is proposed here: - Stub **Testing** As shown in Fig. 4, the LAPDm has been implemented as two different processes named, LAPDm down Process (LDP) and LAPDm up Process (LUP) [20, 21]. LDP accepts primitives from upper layer, takes necessary action and passes primitives to lower layer for transmission of prepared frames. Similarly, LUP accepts primitives along with the received frames from the physical layer, analyzes and detaches header and passes the received messages after necessary processing to the layer 3. As the two processes, LDP and LUP, correspond to the same layer, regular interaction among them is required. Concept of shared memory is used to realize interaction between these two processes. In Stub testing, a stub of upper layer and lower layer is created which passes test primitives to the LDP and LUP respectively. With the help of stub testing the desired functionality of implemented layer can be tested.

8 Conclusion

The focus of the work is on unacknowledged mode LAPDm protocol development at MS side of GSM network. The implemented work has been done by taking LAPDm as a process in C and the Interprocess communication (IPC) concept is used in order to realize communication with upper and lower layer of GSM protocol stack. In order to test the validity of the implementation stubs of testing of information transfer from one layer to of layer 3 and layer 1 have been used. In addition, the procedure described in paper can be used for development of any protocol of a layered architecture protocol suite.

References

1. 3GPP TS 04.05. Data Link Layer General Aspects, version 8.0.2, Release 1999. www.3gpp.org/ftp/Specs/html-info/0405.htm
2. 3GPP TS 04.06. MS-BSS interface: Data Link Layer specification, version 8.2.1, Release 1999. www.3gpp.org/ftp/Specs/html-info/0406.htm
3. Heine G (1999) GSM networks: protocols, terminology and implementation. Artech House, pp 10, 43–44, 55–86, 101–106, 296
4. Rappaport T (2001) Wireless communications: principles and practice, 2nd ed., PHI
5. Sharma R (2007) Implementation Strategy for LAPDm protocol at MS side of GSM Network. M. Tech thesis, Jaypee Institute of Information Technology, University, Noida
6. Bhargava N (2004) Protocol stack implementation using IPC as shared memory. Linux Gazette web J.
7. Lee CYW (1989) Spectrum efficiency in cellular (radio). IEEE Trans Veh Technol 38(2):69–75
8. Lee WCY (1988) Spectrum efficiency and digital cellular. In: 38th IEEE Vehicular Technology Conference. IEEE
9. GSM Recommendation 04.03. MS-BSS Interface: Channel Structures and Access Capabilities
10. GSM Recommendation 05.01. Physical Link Layer on the Radio Path (General Description)
11. GSM Recommendation 05.02. Multiplexing and Multiple Access on the Radio Path
12. Conference Proceedings. Digital Cellular Radioconference, Hagen FRG (1988)
13. GSM Recommendation002.02. Bearer Services Supported bya PLMN
14. GSM Recommendation 09.01. General Aspects on PLMN Inter-working
15. GSM Recommendation 03.04. Signaling Requirements Related to Routing of calls to Mobile Subscribers
16. GSM Recommendation 08.02. 855-MSC Interface-Interface Principles
17. GSM Recommendation 08 04. BSS-MSC Layer 1 Specifications
18. GSM Recommendation 08.06, " Signaling Transport Mechanisms for BSS-MSC Interface."
19. GSM Recommendation 09.02. Mobile Application Part (MAP) Specifications
20. GSM Recommendation 08.08. 855-MSC Layer 3 Specifications
21. GSM Recommendation 04.08. Mobile Radio Interface-Layer 3Specifications

Wide Band Vivaldi Antenna Design by Using SIW

Chanchala Kumari and Neela Chattoraj

Abstract Over the last two decades, tapered Slot antenna have effective performance for wideband applications. The Vivaldi antenna is one type of planer antenna having the property of easy design, lower cost with light weight. We have designed a Vivaldi antenna with microstrip to SIW feed different parameters are studied and analyse. CST software (microwave studio suit) was used for simulation and optimization in Time domain analysis. The Antenna resonates at 10 GHz. It is designed for wideband application Design methodology consist is to divide the Vivaldi antenna into two parts i) microstrip-to-SIW transition ii) tapered curve part. The objective of this proposed design model is to compute the far-field pattern and impedance bandwidth of the structure. Good matching is observed over a wide frequency band for single element.

The gain of the single element was found to be not so good. The result for impedance band width radiation pattern (E field and H field are obtained, to prevent the back radiation a conducting plate was fixed in to the back side of antenna.

Keyword SIW · Vivaldi antenna · Impedance bandwidth

1 Introduction

Vivaldi antenna is light weight, small size planar antenna. It has been used in various research and technological fields. It was introduced by Gibson in 1979 [1]. The first formulas was formulated in 1986, in which simple TSA without a substrate was inspected and later on more productive methods were gradually introduced.

Lewis [2] found that a TSA as a broad band stripline array which is multi octave bandwidth. And it was followed by Gibson in 1979, he introduced Vivaldi antenna which is exponentially tapered slot antenna. He analysed that it has sufficient gain in frequency range below 2 GHz and above 40 GHz and it is linearly polarized.

C. Kumari · N. Chattoraj (✉)
Department of ECE, Birla Institute of Technology Mesra Ranchi, Jharkhand, India

V. Nath and J. K. Mandal (eds.), *Proceeding of Fifth International Conference on Microelectronics, Computing and Communication Systems*, Lecture Notes in Electrical Engineering 748,
https://doi.org/10.1007/978-981-16-0275-7_35

As asymmetric one sided microstrip to slotline transition was also designed on alumina plate it was served fairly in the range 8–40 GHz.

Three different types of Tapered Slot Antenna was compared by Yngvesson [3] Gibson's Exponential tapered slot antenna, (LTSA) Linearly tapered slot antenna and constant width slot antenna (CWSA). He analysed that CWSA and LTSA had largest side lobes as compared to Vivaldi antenna and beamwidth of Vivaldi was widest and in case of CWSA, it was narrowest. He also probed that how the length of antenna and the thickness of substrate effect on the beamwidth.

By using the substrate of low dielectric that is cuclad = 2.45 in place of alumina had been taken by Gazi [5] and he studied about the antipodal slotline transition. It was designed by tapering the microstripline through parallel strip to an asymmetric double sided slotline. He observed that in this type of structure available bandwidth was wider which was not available in case of traditional Vivaldi antenna. But one problem with this designed was that having high cross polarization, further investigation done by Langley [5] given good results in term of cross polarisation as well as wideband.

To make easy fabrication a circular stubs put in microstrip slotline transition had investigated by Schuppert [8] further in place of circular a radial stubs was used by Sloan [9] given the better bandwidth. And by using both, that is radial and circular to design a stripline feed investigated by Schaubert [10] conducting plates on both side on the Vivaldi antenna.

By investigating the last few papers we conclude that the bandwidth of the stripline feeding have better performance than microstrip feeding. Further we also conclude that arrays of Vivaldi antenna can also use this type of feeding structure by the concept of power splitter and the microstrip impedance matching. Since its structure is lossy, an alternate solution for that was studied by Kazemi [13] inbuilt by using SIW.

2 Antenna Design

The structure of the proposed antenna is the like traveling wave antenna.it designed by etched the thin metal to make the shape of tapered by using substrate and the back side of the substrate is metal plated and followed by feeding structure. The main property of Vivaldi antenna is that having lightweight, higher bandwidth, higher gain and lower side lobs. And also produce a symmetrical end-fire beam. First step to design the antenna is that identified the suitable feeding techniques. For that we need to understand the basic characteristics of Vivaldi antenna it will help to design antenna as well as suitable feeding techniques. As we know that Vivaldi has wider bandwidth, higher directivity and having symmetrical end fire radiation pattern. These basic features are available into all types of exponentially flared structure, like as we know the traditional TEM horn antenna where also as length of the antenna increases flares also increases, so we can say that the Vivaldi is the planer form of horn antenna. Here wave guided structure is a printed slotline that is tapered exponentially. The basic tapered design equation [5] given below for tapered part is given by

$$y = Ae^{r*x} + B \tag{1}$$

$$A = \frac{Ws + W1}{2} * \left(e^{r*Lt} - 1\right)^{-1} \tag{2}$$

$$B = \frac{(Ws + W1)e^{r*Lt}}{2} * \left(e^{r*Lt} - 1\right)^{-1} \tag{3}$$

Due to the tapered shape and planer characteristics of Vivaldi antenna it is compatible with printed circuit board. It is actually the class of travelling wave antennas. As the structure is flare the waves travel along its structure. When the distance between the metal is small as compared to the free space wavelength, the electromagnetic waves are closely and firmly bounded with each other as the distance increases the bond becomes steadily weaker due to that radiated waves get away from antenna. The tapered-slot antennas behave as a travelling takes place along the flare structure, because the phase velocity V_{ph} is less than the velocity of light in free space or $V_{ph} < c$ as well as the limiting case when $V_{ph} = c$) [23]. In this structure the end fire radiation takes place at the wider end. when $V_{ph} = c$ then we consider antenna with air as the dielectric. In this case the beamwidth and side lobs are much high as compared to when we are using dielectric materials. Thickness and dielectric constant effects on the phase velocity and guide wavelength are significantly. And it is also affected by tapered shape.

The basic property of these types of antenna is that having the infinite bandwidth of operation because at different frequency antenna radiates in different area. Where the radiating part is constant in wavelength. We can say that it is frequency independent. With the variation in wavelength, it scaled down in size in proportional to the wavelength with same shape. Table 1 shows the comparative data for the proposed antenna with the some other Vivaldi antennas.

Table 1 .

Ref	Substrate	ε_r	Loss tangent	Thickness	Size (mm^2)	RL (dB)	Gain (dB)	Fr. (GHz)
[19]	FR 4	4.4	0.02	1.6	80 × 60	−30	2.8	3.1–10.6
[20]	F4BM265	2.65	0.001	1	130 × 80	−27	2.27–14	1.13–12
[21]	RO3006	6.15	0.002	1.28	45 × 60	−30	2.2–6.8	1.7–11
Proposed design	**Ro4003**	**3.55**	**0.002**	**0.183**	**24 × 12.5**	**−23**	**1.7–4.2**	**8–12**

3 Micro Strip SIW Transition To

A. *Substrate Integrated Waveguide (SIW)*

A new perspective to design planer waveguide structure is SIW for the application in microwave and millimetre-wave systems. It is the combination of waveguide and microstrip circuit which is smartly replace the traditional waveguide. Issues to using traditional waveguide is that the large structure and stocky. To make it compatible with planer circuit we need extra supporting structure. Due to advantage of SIW over a non-planer rectangular waveguide, it can be easily fabricated with planer antenna. The basic design structure for SIW is that dielectric substrate having two parallel lines of via-holes in place of metallic wall in traditional waveguide structure. The main feature of using SIW is that the passive components like filter coupler diplexers mixer etc., active element like amplifier, oscillator and antenna can be fabricated on the same substrate by planer transmission lines. Planer nature of SIW makes it easy to fabricate on the PCB and other microstip line and planer filter with antenna can be easily inbuilt on single planer structure.

B. *Design Technique For SIW*

For Rectangular SIW important parameters to be calculated are the width and height. Because the cutoff frequency SIW depends on them. As we know that for rectangular waveguide TE_{10} mode is dominant mode. The spacing between the vias "p" known as pitch and the diameter of vias "D" are the important parameter to control the radiation loss and return loss whereas the "W" waveguide width examine the cut-off frequency and propagation constant. "Lw" is the length of the SIW. There are two design rule given by Eq. (4) and (5).

The microstrip transition consists two parts that is series combination of quarter-wave tapered transformer and a 50 Ω microstrip line as shown in Fig. 1.

$$D < \lambda g/5 \tag{4}$$

$$P \leq 2D \tag{5}$$

Where λg is the guided wavelength in the SIW.

Due to these transformations a good impedance matching can be built for the 50 Ω microstrip impedance. The length og tapered line and width of tapered shape are the main parameter for effective performance.

Fig. 1 Microstrip to SIW transition

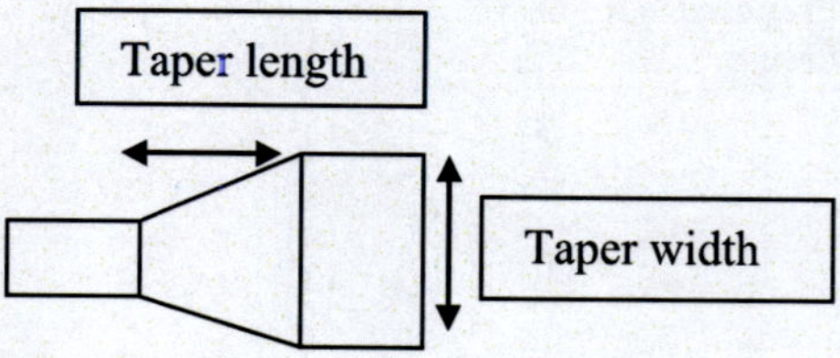

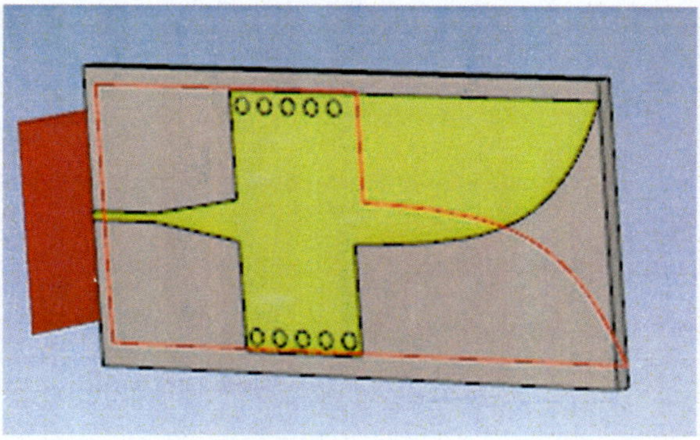

Fig. 2 Final single antenna modeled by CST

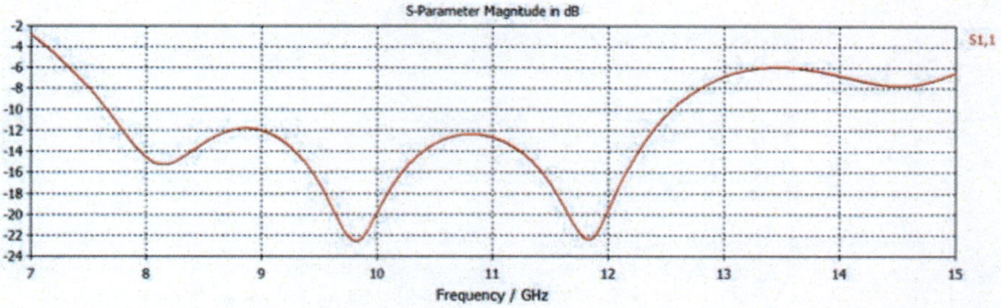

Fig. 3 Final single antenna modeled by CST

4 Numerical Results And Discussions

A. *Single Element Vivaldi Antenna*

The final Vivaldi element is designed for the frequency range 8–12 GHz at the central frequency $f_o = 10$ GHz, it is implemented on **Rogers 4003(lossy)** substrate with dielectric constant $\varepsilon r = 3.55$ and thickness $T = 0.813$ mm, loss tangent $\eth =$ 0.0027 all design. All design parameter are given below in Table 2.

1. *Impedence Bandwidth:*

The impedance matching is considered as in term of return loss S_{11}. The Fig. 3 shows the result for impedance matching for single optimal Vivaldi antenna from 8–12 GHz. From figure it shows that there is good impedance matching for entire band, the S11 response is below −12 dB from 8 to 12 GHz (Fig. 2)

2. *Radiation Pattern in E Plane and H Plane*

Radiation pattern at frequency 8, 10 and 12 GHz is plotted with the 2D plots in the Fig. 4(a) 4(b) and 4(c) respectively, Which shows that the Vivaldi antenna is an

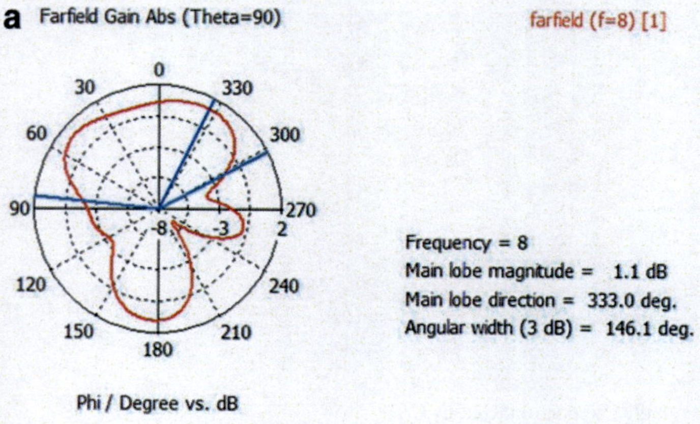

Fig. 4(a) Radiation pattern E plane at 8 GHz

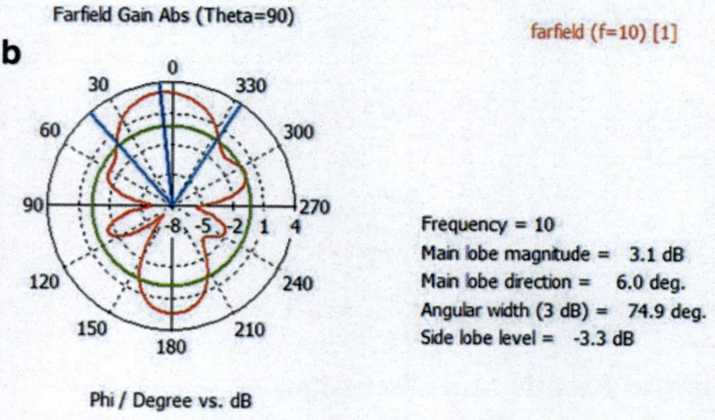

Fig. 4(b) Radiation pattern E plane at 10 GHz

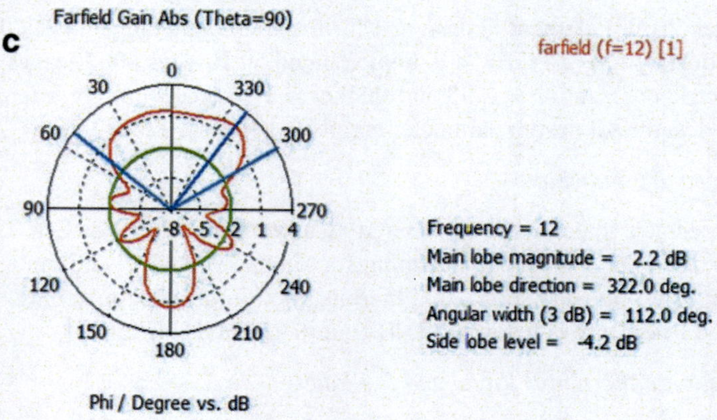

Fig. 4(c) Radiation pattern E plane at 12 GHz

Table 2 .

Parameter	Dimension (mm)	Description
Fw	2.5	Feeding width
Lt	16.8	Antenna tapered shape length
Lw	Ni*p	Total length of SIW
Mt	0.02	Metal thickness
N	1.8	Via centre point from Width W
T	0.813	Substrate thickness
W	12.5	Width of tapered shape
W1	1.88	Tunning width for Vivaldi
d	0.8	Diameter of via for SIW
Mst	4	Microstrip length for transition part
mtap	5.5	Tapering length for transition part
n1	5	No. of cylinder to design the SIW
p	N*d	Centre to centre distance of via (PITCH)
r	0.38	Tapering rate
t1	0.45	Width of microstrip line for 100 Ω

end fire radiating type antenna. We can see that in case of E Plane and H plane, the main lob that is the direction of radiation is the plane contain E vector and passing through the origin in that case theta = 90° E plane.

3. *Microstrip to SIW transition simulation*

Using Eq. (4) and (5), the width of taper comes out be 4.5 mm for 0.831 mm height of the substrate while, 4.1 mm is the tapered length for n = 5. Based on the above formula back to back microstrip to SIW transition design and simulation results shown in Fig. 5(a), 5(b) and 5(c), in Fig. 5(b) we can see that the result showing for S11 below −55 db at 9.9 GHz. And in Fig. 5(c) S12 and S21 shows that nearly about zero dB.

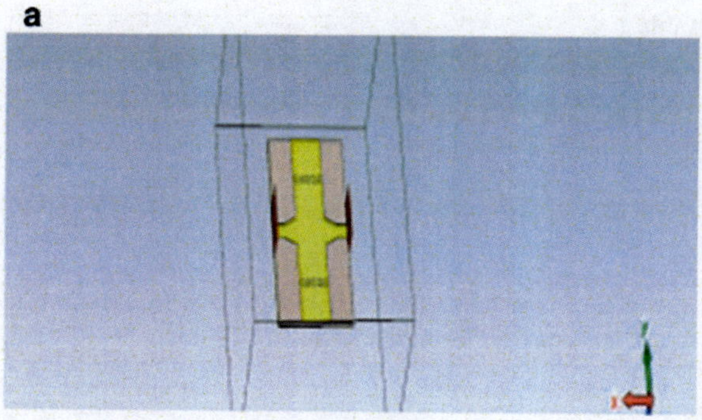

Fig. 5(a) Back to back transitions simulation model

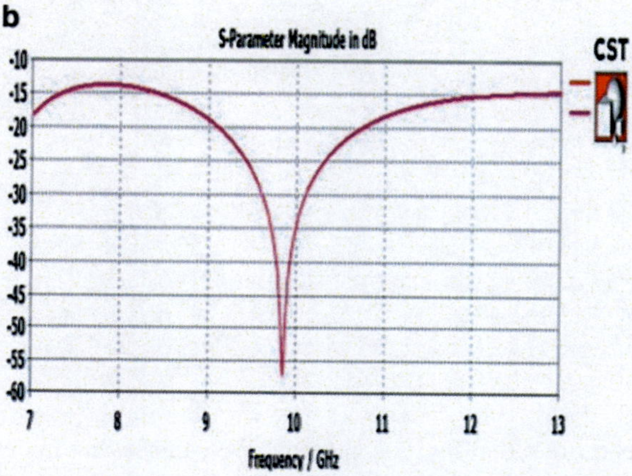

Fig. 5(b) Return loss for S11 and S22

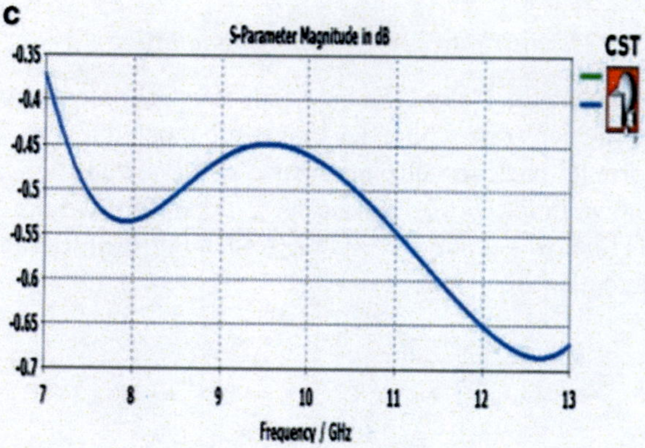

Fig. 5(c) Transmission loss S12 and S21

5 Conclusion

Single element Vivaldi antenna is designed and optimized for X band having centre frequency 10 GHz by using microstrip to SIW feeding technique, shows good impedance matching which results S11 (return loss) dB below −55 dB. Transmission loss S12 and S21 both are found approximate 0 dB. Band width is around 8–12 GHz. As the gain of single element is very low it can be increased by array design. This design can be easily modified, improved and effectively utilized for wide band applications.

References

1. Gibson PJ The Vivaldi aerial. In: Proceedings of the 9th European
2. Microwave Conference (1979), pp 103–105
3. Lewis LR, Fasset M, Hunt J (1974) A broadband stripline array. In: IEEE A P-SSynip, p. 335, June 1974
4. Yngvesson KS, Korzienowski TL, Kim YS, Kollberg EL, Johansson JF (1985) Endfire tapered slot antenna on dielectric substrates. IEEE Trans. Antennas Prop AP-33:1392–1400
5. Gazit E (1988) Improved design of a Vivaldi antenna. In: IEEE Proceedings H, pp 89–92, April 1988
6. Langley JDS, Hall PS, Newham P (1993) Novel ultrawide-bandwidth Vivald antenna with low cross polarization. Electr Lett 29(23):2004–2005
7. Langley JDS, Hall PS, Newham P Balanced Antipodal Vivaldi
8. Antenna for Wide Bandwidth Phased Arrays. In: IEEE Proceedings of Antennas and Propagation, vol 143(2), pp 97–102, April 1996
9. Kim SG, Chang K (2004) A low cross-polarized antipodal Vivaldi antenna array for wideband operation. In: Proceedings of IEEE International AP-S Symposium, Monterey, CA, pp 2269–2272, June 2004
10. Schuppert B (1988) Microstrip/Slotline transitions: modeling and experimental investigation. IEEE Trans Microw Theory Tech 36(8):1272–1282
11. Sloan R, Zinieris MM, Davis LE (1998) A broadband microstrip to slotline transition. Microwave Opt Technol Lett 185:339–342
12. Schaubert DH, Shin J (1999) A parameter study of stripline-fed vivaldi notch antenna arrays. IEEE Trans Antennas Propag 47:879–886
13. Adamiuk G, Zwick T, Wiesbeck W (2010) Compact, dual-polarized UWB antenna, embedded in a dielectric. IEEE Trans Antennas Propag 58(2):279–286
14. Kazemi R, Fathy AE, Sadeghzadeh RA (2012) Dielectric rod antenna array with substrate integrated waveguide planar feed network for wideband applications. IEEE Trans Antennas Propag 60(3)
15. Deng C, Xie Y-J (2009) Design of resistive loading vivaldi antenna. IEEE Antennas Wirel Propog Lett 8
16. Guillanton E, Dauvignac JY, Pichot C, Cashman J (1998) A new design tapered slot antenna for ultra-wideband applications. Microw Opt Technol Lett 19(4):286–289
17. Mohammadi M, Kashani FH, Libafan JG A Compact Planar Monopulse Combining Network At W-Band, Electrical Engineering Department, Ira University of Science and Technology (IUST), Tehran
18. Skolnik MI (2008) Radar Handbook, 3rd edn. McGraw-Hill, New York
19. Deslandes D, Wu K (2002) Design consideration and performance analysis of substrate integrated waveguide components. In: Microwave Conference 2002, 32nd European

20. Deslandes D (2010) Design Equations for Tapered Microstrip-to-Substrate Integrated Waveguide Transitions. In: IMS 2010
21. Schuppert B (1988) Microstrip/slotline transitions: modeling and experimental investigations. IEEE Trans MTT-36:1272–1282
22. Knorr JB (1974) Slotline transitions. IEEE Trans MTT-22:548–554
23. Bozzi M, Xu F, Deslandes D Ke WU (2007) Model and design considerations for substrate integrated waveguide circuits and components. In: TELSIKS 2007, Serbia, Nis, 26–28 September
24. Pozar DM Microwave Engineering. Fourth edition. Wiley publication

Issuing and Verifying of Blockchain Based Certificates

Yousuf Nizam, Tanvi Tadimeti, Prarthana, Ch. Jayanth Babu, and R. Padmavathy

Abstract The issuance of fraud certificates has become an unrestrained and unchecked problem around the globe, because of its inevitable simplicity. A fake academic certificate can be defined as a counterfeit degree from a legitimate or even an illegitimate university. As a result, companies end up spending millions of dollars to simply verify the legitimacy of the certificates and transcripts of the applicants. Blockcerts, MIT Lab project, presented a solution, but their design was based on the Bitcoin chain which limited the scope of the project and left several functionalities unaddressed. Our approach implements a technique that uses the Blockchain (Ethereum) technology that can implement functional logic, to solve this issue. The two major advantages of Blockchain are that the verification can be done through an application and authentication or verification can occur instantaneously once the certificate holder approves the access. Also, the records cannot tamper as changes to individual blocks can be done only with the approval of all the parties processing the block. For years now, we have been relying on just a password to be the guardian of our most confidential information. In today's modern world, this has become the greatest threat to our security. Passwords are very easily decodable and can be cracked in just a matter of seconds, while users are being subjected to phishing scams with alarming regularity. A way to reduce this security threat is two factor-based authentication. Two-factor authentication (2FA) is a type of multi-factor authentication method which confirms a users claimed identity by utilizing a combination of two different components, instead of one. This paper is on issuing certificates, verification and proposed decentralized two factor-based authentication feature using the Ethereum smart contract, which makes the entire authentication process even more secure.

Keyword Blockchain · 2-factor authentication · Ethereum · Smart Contract

---

Y. Nizam · T. Tadimeti · Prarthana · Ch. J. Babu (✉) · R. Padmavathy
Department of Computer Science and Engineering, National Institute of Technology, Warangal, India
e-mail: jayanthbabu_786@student.nitw.ac.in

R. Padmavathy
e-mail: rpadma@nitw.ac.in

V. Nath and J. K. Mandal (eds.), *Proceeding of Fifth International Conference on Microelectronics, Computing and Communication Systems*, Lecture Notes in Electrical Engineering 748,
https://doi.org/10.1007/978-981-16-0275-7_36

1 Introduction

The process of issue and validation of academic certificates has remained unchanged from centuries, regardless of the information revolution from the last couple of decades. Very few institutions maintain the digital counterpart of the certificate, and internal records. Therefore many of the problems listed below remain unresolved.

(a) Loss of Certificates: In assumption, a recipient damages or destroys a certificate, then a revival of the certificate becomes a highly cumbersome process. It deals with extensive paperwork where the recipient is expected to apply to the institution forre-issue. In the best-case scenario, assuming that everything is in order, a new certificate is issued and sent to the recipient. This process is generally very complex requiring sanction from numerous records, signatures from multiple parties and voluminous databases of student records. It is slow, extensive and burdensome but all in all, it is impossible if the institution ceases to exist.

(b) Certification Validation: The certificate is sent from the recipient to the potential employer, who is required to validate the legitimacy of the certificate and ensure that it was in no way tampered with. Similar to the above-mentioned problem, this process can take days or weeks but is also impossible without the existence of the institution. This is logistical problem as keeping track of old records is generally a manual process.

(c) *Internal Records Lost or Modified:* Even with the existence of the institution, it is possible that the records of internal certification are lost, destroyed or damaged prohibiting validation and re-issuing of a certificate. They may also manipulated by human intervention.

In the year 2008, blockchain [1], the technology underlying Bitcoin [3], gained popularity as a technology whose features could be used for more than just crypto currencies. Regardless of it being public or private, the immutability of the records of the blockchain-based distributed ledgers provides certitude. Only data that is determined to be legitimate is validated and added to the chain of blocks, in the form of a block. Once added to the remainder of the chain, records are impossible to chain, thus resulting in unarguable content in records and permit complete transparency. This has tremendous implications for the origin of credentials and certificates, along with the speed with which the third party (e.g. an employer) can certainly verify their accuracy. While universities and issuing institutions can now ensure protection for their brand name, businesses and HR departments can now screen through an enormous volume of data or credentials with speed and ease.

Currently available methods of certificate verification is very time consuming and costly. An easy and quick method is required for the same not only by the educational institutions but also by companies during the recruitment and many other institutions. Ethereum [2] uses 2FA, but the second factor, it uses is Google authenticator which is not completely secure.

In this paper, a solution is proposed to resolve the issues of certificate issuing and verification, including, utilizing a two-factor decentralized Blockchain-based authentication scheme [6] for authentication.

The rest of the paper is organized as: Sect. 2 discusses the related work in this domain. Next, Sect. 3 covers the basics of the blockchain and related terminology. Section 4 consists of a detailed explanation of the proposed methodology. The results achieved from the proposed methodology covered in Sect. 5 and Sect. 6 covers the conclusion of this paper.

2 Related Work

In the current digital world, the hacker becoming more strong day by day. So, the blockchain-based applications need to be secured by more powerful authentication methods like Two factor-authentication system, not by simple user-name/password like methodologies. As blockchain technology is decentralized, every part that related to security needs to implement in decentralized approach. Therefore, there is a need of decentralized 2FA based authentication approach.

There are many approaches designed and implemented based on the blockchain technologies. The approach introduced in 2015 by Gipp, Meushke, and Gernandt to address these needs. This approach showed crypto currencies such as Bitcoin or Etherium can be used as ledger of type Decentralized Trusted Time stamping (DTT) [10]. The internal architecture and implementation of DTT is explained in this paper along with the integration of multiple blockchain types.

The approach [10] is improved and it is called Origin-Stamp [11] implemented based on blockchain technologies for securing intellectual property rights. It is a trusted time stamping, web-based service, runs based of the decentralized Bitcoin blockchain to store digital content. The additional feature of this approach is the fingerprint of the document is kept in cryptocurrency. The fingerprint used in this approach is SHA 256 checksum based of RFC6234 [12]. This approach is similar to decentralized 2FA to some extent because the person has to prove his identity who knows the secret.

Later, the decentralized naming services came into existence, such as Blockstack [13], Namecoin [14] and Cert-coin [15]. Blockstack is like new internet, where trust points are removed in the network, so that user doesn't need to trust remote servers. But, security of critical data bindings is handled by blockchains. Blockstack provides services like identity, storage and discovery and it handles underlying block chain failures.

Namecoin [14] is another naming service based on the concepts of the peer-to-peer (P2P) based cryptocurrency bitcoin. This service is provided without the need of the trusted authority. The distributed Namecoin blockchain is a publicledger for all executed Namecoin transactions. The Namecoin domains are stored in ledger, and regular domain name system (DNS) are not integrated directly. It requires additional extension/addition for browser.

The Certcoin [15] is also public ledger of domains with associated public keys. Thefeature of optimized Certcoin is accessible to limited storage capacity devices, such as smartphones. The optimizations tools used for Certcoin are distributed hash tables and cryptographic accumulators.

The Two-factor authentication (2FA) is implemented based on the Time-based One-Time Password algorithm (TOTP) [16]. A Time-based One-time Password Algorithm (TOTP) is an algorithm uses current time and shared secret key to computes a one-time password (OTP), which is used in many 2FA systems. In this, both client and server generate the token, then server verifies the clients token with the token locally generated at server.

The 2FA integration with the client is provided by Google Authenticator [17] and 1 Password [18]. Google Authenticator [17] is an authenticator app that generates the verification code, sends it smartphone. The user has to pass the two step verification process to use it. This Google authenticator application scans the QR code displayed on the screen. Then, if generates a counter based code or time based code via mail,text or voice call that should be entered to get the access.

1 Password [18] is a password manager app developed by AgileBits Inc. This application is a container of sensitive information called virtual vault, where we can store secret passwords, software licenses, and other secret information. The virtual vault is locked with a PBKDF2-guarded master password. This password protected vault provided for the company and it is charged monthly. Generally, this secure virtual vault stored in server system of the company.

Two-Factor Authentication on the Bitcoin protocol [19] uses a two-party signature scheme compatible with ECDSA (Elliptic Curve Digital Signature Algorithm) [20]. This protocol used the smartphone/mobile for the second authentication factor. The problem with this protocol is, the smart-phone that generates the second factor need to communicate directly with computer/PC where transaction occurs. Generally, this direct communication not possible. The generation of the transaction depends on the network configuration of the computer. Therefore, this approach should be generalized.

A. Motivation-Blockcerts-MIT labs project

In the year 2017, Massachusetts Institute of Technology Media lab created a project named Block-certs. Block-certs isan open standard that is used for building applications for issuing and verifying blockchain-based official records. The scope of certificates here can be defined as civic records, professional licenses, work-force development, academic credentials and much more. It consists of open-source libraries, tools, and mobile applications that enable a decentralized, standards-based, recipient-centric ecosystem, enabling trustless verification through blockchain technologies. Blockcerts utilizes and encourages consolidation of open standards.

Blockcerts is pledged to the self-defined identity of all participants and enabling recipient control of their claims through easy-to-use tools such as the certi_cate wallet (mobile app). Along with this, blockcerts is pledged to the availability of credentials in the absence of single points of failure. These open-source repositories may be utilized by other research projects and commercial developers. It contains components for creating, issuing, viewing, and verifying certificates across any blockchain. These

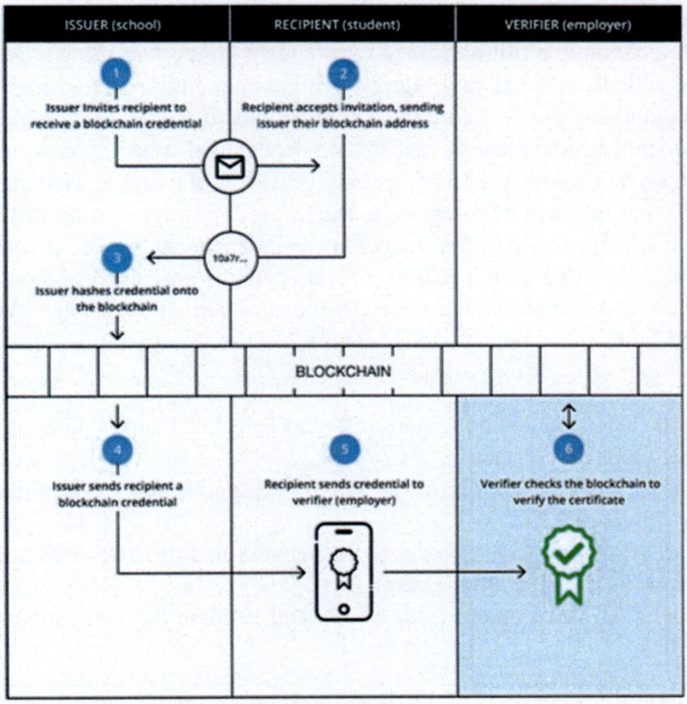

Fig. 1 Basic structure of blockcert

components form all the parts needed for a complete ecosystem. Figure 1 illustrate the structure of the blockchain.

2FA in Ethereum: Ethereum combines the user's password to a dynamic token generated automatically by the application Google Authenticator, available for iOS and Android.

Pros:

- Two-factor authentication software can be utilized in absence of cellular coverage or internet.

Cons:

- Smartphone or a similar device is necessary
- Application is hackable
- The battery of the smartphone may discharge
- The smartphone reset/ lost, or authenticator applicationis deleted, then token would be lost and its recovery ishard.

2FA Decentralized Protocol

This crypto-economic primitive aimed to provide an important infrastructure piece that can be added to the existing set of tools that help engineers and organizations globally build and deploy decentralized applications. It is vital to consider that the suggested identity authentication mechanism itself can also be implemented on any smart contract technology platform (including EOS, NEO, Stellar, Hedera, Dfinity) and provide that infrastructure service to the respective ecosystem. In this authentication mechanism, a session ID (provided by an independent 3rd party) is encrypted with the public key of a user. The user must decrypt the encrypted session ID using its private key. If the session ID provided by the third party is the same as the session ID declared by the user, then a user's identity has been successfully authenticated. The economic actors in this game are:

- *Users (U):* Consumers who require different levels of identity authentication to purchase goods and services
- *Third-Party Service (3PS)*: Firms looking to sell goods and services that require different levels of KYC
- Session ID Providers (SIP): Nodes in the network that provide session IDs (used to authenticate the identity of Users)
- Validators (V): Stake tokens within SIPs and validate the authenticity of their session IDs.

After a thorough case study of the Block-certs project, we found certain problems in the project:

Ethereum Blockchain: [2] The project was built on a Bitcoin blockchain [3], which requires the maintenance of thousands of participants in the cryptocurrency ecosystem. It is not to be predicted that the Bitcoin would work well continuously in the future because a wide range of stakeholders influences blockchain ecosystem or business model. Hence, we will adopt the Ethereum blockchain to eliminate factors of instability. Ethereum allows one to add business logic whilst using smart contracts (a computer code running on top of a blockchain containing a set of rules under which the parties to that smart contract agree to interact with each other).These contracts are capable of tracking performance in real-time and can bring tremendous cost savings.

Partial Disclosure of Information: Block-certs project has no proper support for partial disclosure of information. Certificates are issued separately for the high level information and detailed personal information. But, fully-featured selective disclosure within a single certificate is possible theoretically by spreading the contents of a document across a Merkle tree. This is what we aim to achieve in our approach.

No Proper Structure for Revocation: The Block-certsproject does not have proper structure for revocation. Serious authentication defects: The current project has some serious authentication defected in recipients side and the certificate issuing organization side.

B. Currently Used 2FA in Etherem

Ethereum presently uses Google Authenticator as the second factor, which makes it attack prone and serve as a bottleneck for the entire method. Although using 2-Factor authentication Ethereum has overcome the issues associated with single password authentication but still, Google Authenticator that Ethereum uses as the second factor of authentication is notcompletely rid of security threats, can be breached easily and is centralized. Ethereum is used for several crucial transactions including monetary transaction for which a strong authentication system is necessary. This security threat can be overcome by the decentralization of the second factor used for authentication.

3 Blockchain Basics

A blockchain is a public digital ledger that is distributed and decentralized that is used to record transactions across many computers. After a single party initiates the process of a transaction by creating a block, it is verified by millions of computers distributed over the internet. This verified block is then stored on the chain creating a unique record and having a unique history. Falsifying a single record would imply falsification of the entire chain in millions of instances. A blockchain database is managed independently using a peer-to-peer network. A peer-to-peer network consists of peersor computer systems that are connected over the internet. There is no necessity of a central server and the peers can directly communicate and share _les, acting as both servers and clients. Blockchain also has a distributed time stamping server. This kind of design ensures a strong workflow where the participants very rarely doubt the security of the data (Fig. 2).

Ethereum
Ethereum is a programmable blockchain. Ethereum allows the user to create their operations of any complexity, instead of giving them an already existing set of operations (e.g. bitcoin transactions). This is how it acts as a platform for cryptocurrencies along with a multitude of decentralized blockchain applications.

In a general sense, Ethereum refers to a set of protocols that define a platform for decentralized applications. In it is crux, the Ethereum Virtual Machine (EVM) executes code of arbitrary algorithmic complexity. Computer scientists refer it as "Turing complete" and can create applications that run on the machine using simple programming languages modeled on already defined languages such as Python and Java Script Like all other blockchains, Ethereum uses a peer-to-peer network protocol, whose database is maintained and updated by multiple nodes connected to the network. Each node on the network runs the virtual machine to maintain consensus across the blockchain and executes the same set of instructions [4]. Decentralization property of the ethereum provides high levels of fault tolerance, zero downtime, and makes datastored on the blockchain forever unchangeable and resistant to censorship.

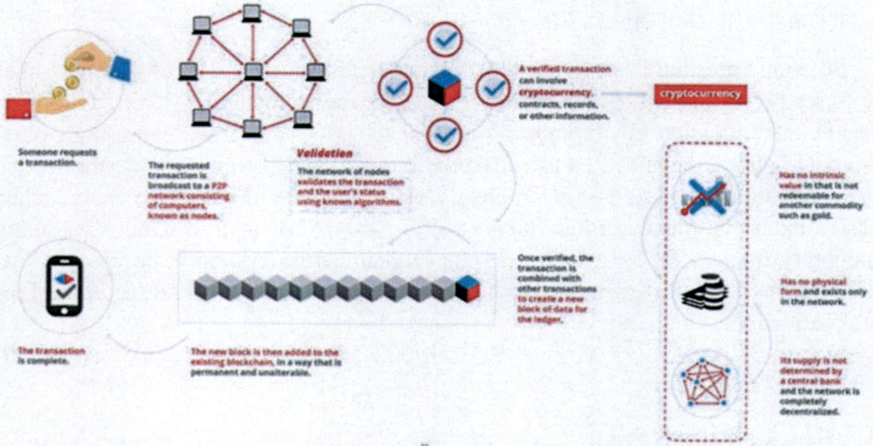

Fig. 2 Basic structure of blockchain

Smart Contracts

Ethereum is a platform specifically designed to create smart contracts. Smart contracts help you to exchange value in a clear, error-free way while avoiding a middleman's services.

The language is Turing-complete, implying that a wider range of computational protocols is supported. Smart contracts by definition are the digitized version of legal contracts. They are deployed, stored and executed using the EVM. They can store data and be used to record information needed to implement the logic for real world contracts. Smart contracts are like object oriented classes, and it can tell other smart contracts to create and use objects of another class. Think of the smart as a small program consists of functions. We can create an instance of the smart contract and use functions of logic to view and update contract data. Example functions are:

- As a multi-signature accounts, so that funds are spent only after a minimum percentage of people accept
- As an agreement manager between users
- As a utility provider to other contracts (similar to how a software library works)
- As an information store for applications, such as domain registration information or membership record.

2-Factor Authentication

Two-factor authentication (2FA) is used to get the rights to access any resource or information. As compared to the one-factor authentication like login-password pair, the 2FA is much more trustworthy. Nowadays there are a variety of innovative methods, from social engineering to distributed brute force techniques, to hack and to evade password authentication [5]. Some users use the same password for multiple login accounts making it easier for the hacker to access the protected information and private transaction. The main edge that 2FA has over other authentication schemes is the high login security. However, it has two major disadvantages: increased time of entry into the system and risk of losing, which is the physical media has to pass authentication step.

4 Proposed Methodology

The proposed work has been categorized into two distinct sections. The former deals with the creation and verification of certificates while the latter involves in adding authentication for the issuing of certificates and solving the issues related with present authentication schemes.

A. Creating a Model for the Issuing of Certificates

The digital certificate created by issuing authority is signed cryptographically and transaction for the certificate is registered on blockchain. The blockchain is a sequential chain of blocks with each building upon the last. It is immutable and is a distributed store of transactions. A receipt is generated when the certificate is issued. Issuing of the certificate creates data compression into a one-way hash and logs it on the blockchain. It helps for the purpose of veri_cation process that include shashing of certificate and comparing with certificate on the blockchain. This ensures that the data cannot be replicated.

In the solution, it is a provision to have logical grouping of recipients called batches that are not expected to change, e.g. Batch of 2017. The issuer can cancel certificates in case of mistakes and issue another batch. The issuing applications are in charge of certificate issuing which forms the primary business logic. It merges the certificates hash in a Merkletree and sends the Merkle root to the blockchain by APIs.

Hashes - Hash of the local certificate is computed. The whole approach is based on document hashes, in our case using an algorithm called SHA-256 (the same used in the Bitcoin protocol).

Root Hash - There are hundreds of students every graduating batch. Each of the students will be given a certificate. However, we want only one hash that represents all the certificates. We achieve this by using a Merkle tree structure. In essence certificates hashes are combined and then re-hashed to create a tree-like structure which represents the whole set of certificates and any change to any of the certificates will result in a completely different root (top)hash.

Blockchain record-The hash of the document is then entered into the field of *OP_RETURN* in an unspendable.

Bitcoin transaction to serve as the permanent record under-pinning the overall approach we are taking. The transaction identifier is recorded for further reference.

Self-Verification - Every instruction is required for self-verification process. The self-verification of certificate leads to interesting chicken-and egg problem because we can't say about bitcoin transaction hash until creation of the hash. But, hash will be changed once the transaction entered into the certificate.

This issue is resolved in the proposed scheme by keeping the metadata after issuing in the blockchain. The metadata stored doesn't change the _le structure so that is is easy to verify. This metadata contains Markle root hash, Markle proof, transaction identifier, public address and issuing institution (Figs. 3 and 4).

B. Creating a Model for Verification of Certificates

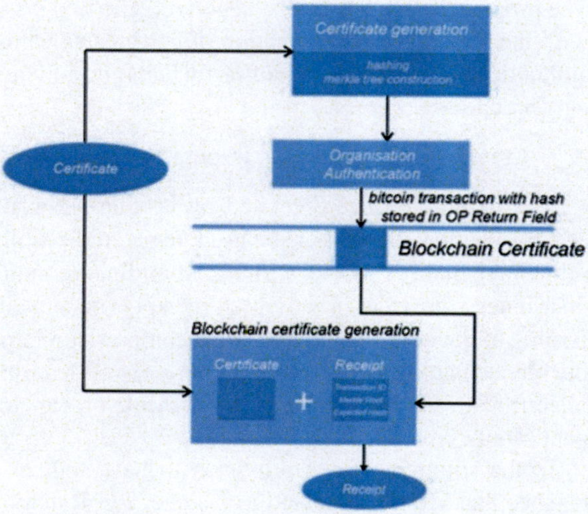

Fig. 3 Old issuer proposed by blockcerts

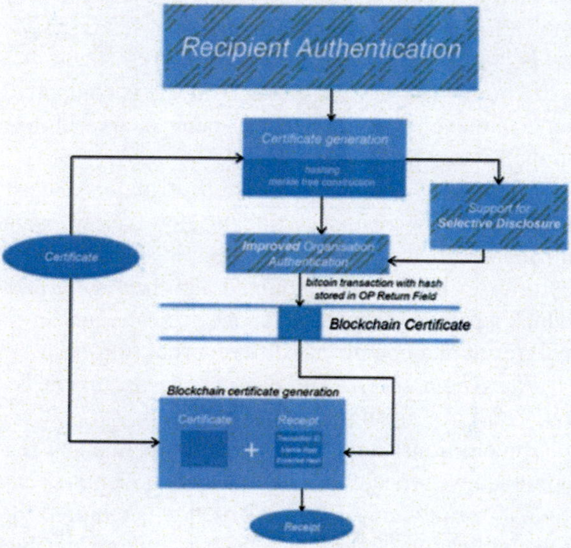

Fig. 4 Proposed issuer

The process of certificate verification should ensure that the certificate issued was not tampered, not expired, not revoked and is authentic. The verification process is responsible for checking the integrity and legitimacy of the certificate [9]. The solution that proposed provides a link to the additional inputs needed to be verified and the blockchain certificate.

Hashes - Hash of the local certificate is computed. The whole approach is based on document hashes, in our case using an algorithm called SHA-256 (the same used in the Bitcoin protocol). Hashing is an algorithm that takes as input any arbitrary data (in our cases, it would be the PDF documents of the certificates) and produce a series of unique numbers and letters. You cannot recreate a document from a hash, but you can recreate the hash from a document.

Fetching the hash in OP_RETURN *field* - When a certificate is issued, a hash of the document is then entered into the *OP_RETURN* field in an unspendable Bitcoin transaction to serve as the permanent record. This hash is fetched.

Comparing the Local and Blockchain Hashes - The verification to confirm if the hash value is in the Merkle tree. The verification to confirm if the hash value of the Merkle treeroot is on the blockchain. Above three steps make sure that the certificate is not tampered (integrity test) and verifies the validity of the certificate (to avoid the revoked certificate). The verification of the issuer is shown Fig. 6 (Fig. 5).

C. Proposed 2-Factor Authentication Scheme

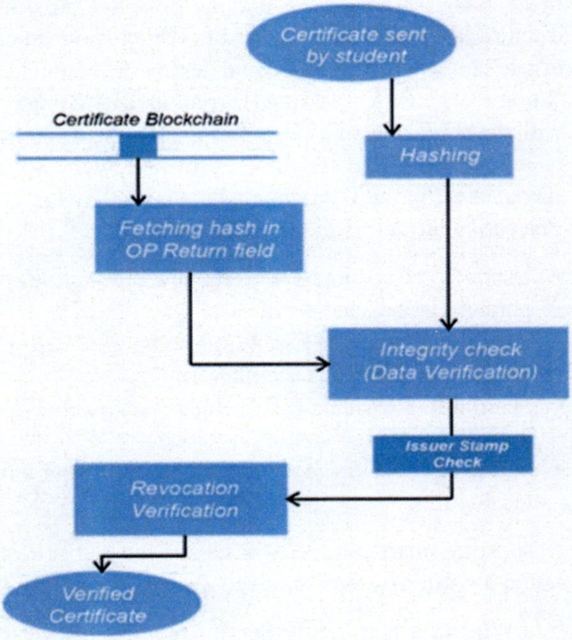

Fig. 5 Old verifier proposed by blockcerts

Fig. 6 Proposed verifier

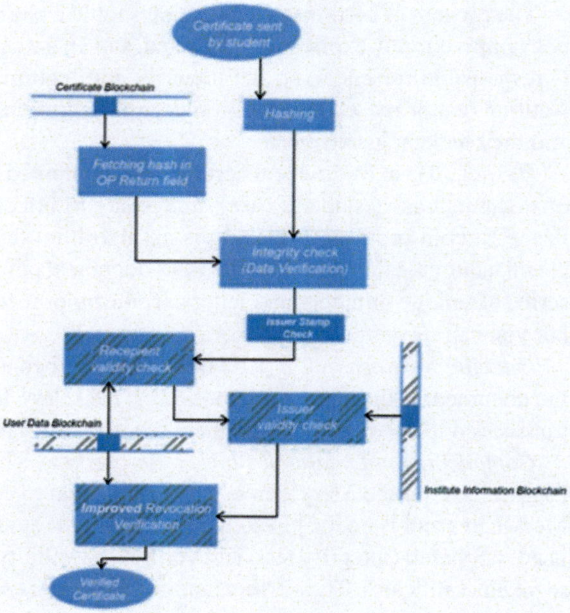

The 2-factor authentication framework built on top of the Ethereum public blockchain. This crypto-economic primitive aims to provide an important infrastructure piece that can be added to the existing set of tools that help engineers and organizations globally build and deploy decentralized applications. In this authentication mechanism, a session ID (provided by an independent 3rd party) is encrypted with the public key of a User. The User has to decrypt the encrypted session ID using its private key. If the session ID provided by the 3rd party is the same as the session ID declared by the User, then a User's identity has been successfully authenticated. The economic actors in this game are:

- Users (U): Consumers who require different levels of identity authentication to purchase goods and services
- 3rd Party Service (3PS): Firms looking to sell goods and services that require different levels of KYC authentication
- Session ID Providers (SIP): Nodes in the network that provide session IDs (used to authenticate the identity of Users)
- Validators (V): Stake tokens within SIPs and validate the authenticity of their session IDs.

A proxy smart contract will handle the authentication requests and the authentication algorithm. Steps followed during authentication.

1. Whenever a 3rd Party Service requires a User to authenticate his or her transaction, the 3rd Party Service can call a function within a proxy contract with a fixed address.

2. This function publishes a message to the distributed ledger (a network of nodes) indicating a 3rd Party Service is requesting an identity authentication, this message also contains: the public key of the User, encrypted with 3rd Party Services private key (who will be authenticated), the public key of the 3rd Party Service.
3. A Session ID Provider node generates a random session ID (X) and publishes it to the proxy contract.
4. The User uses Recursive Length Prefix encoding to sign the session ID (X) on the proxy contract.
5. Validator decrypts the signature on the session ID using user's public key.
6. Validator then validates the Users (identity) private key by checking if the decrypted signature matches the session ID (X) in the proxy contract.

The following are functions used in this process are:

1. *requestAuth*: Is called by the 3rd Party Service and deploys an authentication request message and its respective parameters (Q, M).
2. *requestSign*: Is called by the SÎP providing the authentication for the 3rd Party Service. This method deploysa randomly generated session ID to the proxy contract.
3. *proofSign*: Is called by the User, where the Users private key is used to deploy Recursive Length Prefix encoded signature to the proxy contract.
4. *validateMessage:* Is called by a Validator, the no devalidates the authentication using HMAC (the session ID, public key of User and signature) (Fig. 7).

5 Results and Discussion

A. Certificate Issuing and Verification

The Ethereum blockchain setup is made on local node which uses network with built-in RPC. The transaction which we make is recorded in it. The setup of the network shown in Fig. 8.

Instead of developing the application against the live blockchain, we have used in-memory blockchain (think of it as a blockchain simulator) called ganache.

Certificate Issuer: We have front end in which we will type values of certificate and issue certificate. When "Issue Certificate" is clicked, two things will be happened, One thing is a json is generated and hashed whose values are sent to blockchain transaction then you will generate a transaction ID correspond to json hash. another thing is, the json certificate is stored and downloaded with the transaction ID to local node.

Certification Verification: In this, We will upload the json by keeping the transaction ID in it that fetches the transaction in the blockchain. The two hashes are compared, if both are not equal it considered as the certificate issued is not valid certificate else it is valid.

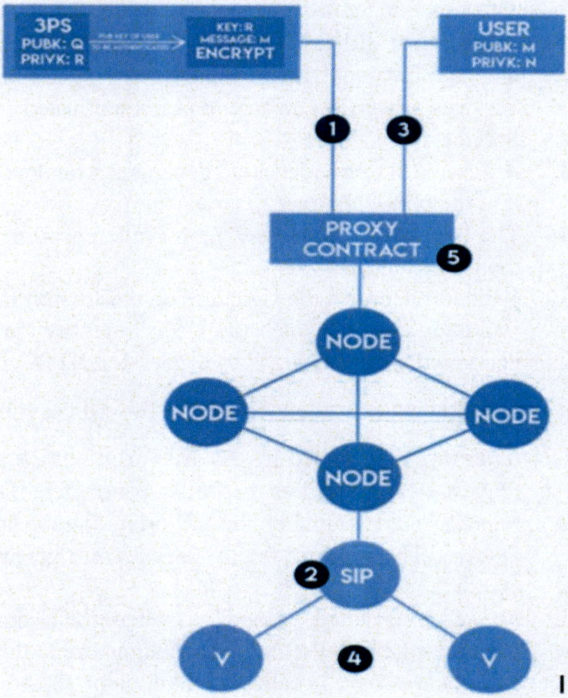

Fig. 7 Proposed authentication scheme

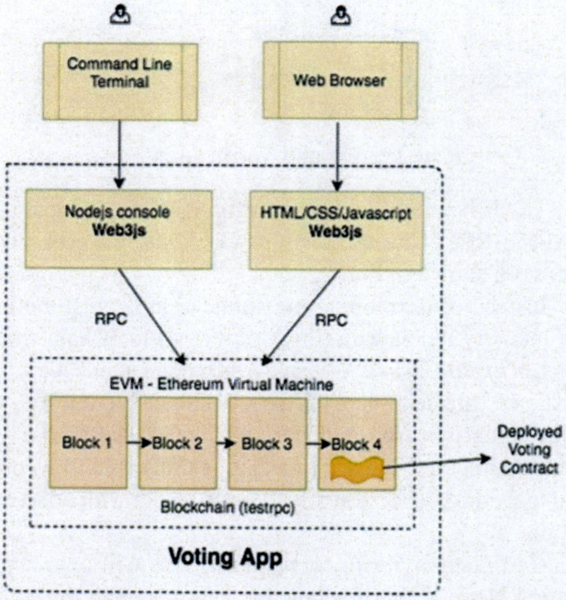

Fig. 8 Setting up of the network

The issuing of certificate is also possible by unauthorized party so it need to be authenticated with the efficient de centralization. We proposed a 2-Factor decentralized authentication scheme for blockchain security which provides additional benefits for the smart certificate issue and verification process.

A. 2FA Decentralized Authentication based on Blockchain-Security Analysis

We designed the 2-factor based decentralized Authentication protocol which solves the problem of single password- based authentication and the problem with the presently used method for 2FA in Ethereum and did a security analysis of the same. This section analyzes proposed schemes security property [7, 8].

Replay Attack:
In the proposed scheme, We use a session ID provider generates a random number, SID and sends it to the contract.

The SID is then encrypted by the user who is to be authenticated. Every time, a new number is generated by the Session ID provider and hence it cannot be reused to authenticate any other user for the second time. The adversary cannot guess the value of SID every time and thus won't be able to compute an encrypted SID for verification. Additionally, the adversary cannot change the value of the SID in the smart contract as it is accessible to only the User to be authenticated, the third-party software, the Session ID provider and the validators. This is achieved by restricting the access of the smart contract to only the public keys of the nodes involved in the authentication process. Therefore, the proposed scheme with stands a replay-attack.

Insider Attack Resistance:
The proposed scheme is decentralized, as it spread over various nodes. An insider does not have access to smart contract as its access is restricted to only the nodes involved in the authentication of a user. Thus no other node will be able to eavesdrop or make changes to the data stored in the smart contract. The SID required for authentication is randomly generated thus making it difficult to be guessed by the malicious user. Therefore, the insider attack can be restricted by the proposed scheme.

Mutual Authentication:
In the initial part of authentication, the third party system provides a message to the smart contract. The message includes the public key of the user which is encrypted using the private key of its own. The user then decrypts the message in the smart contract using the public key of the third party software. The adversary cannot forge a message without knowing the private key of the third party software. Therefore, mutual authentication is provided by the proposed scheme.

Perfect-Forward Secrecy:
In the proposed algorithm, the Session ID is generated by the session ID provider. The session ID is in no way related to the private key of the user or the third-party software. Even if the two private keys are compromised the adversary cannot

compute the value of the previously used session ID. Thus, proposed algorithm achieved perfect-forward secrecy.

Withstand Impersonation Attack:
In our algorithm, the authentication request is sent by the third-party software which starts the whole process of authentication. The public key of the user to be authenticated is provided by the third-party software. Even if a particular adversary tries to impersonate the user corresponding to the public key provided by the third-party software, it will not be able to compute the encrypted message that is to be verified by the verifier without knowing the Users private key, the user to be authenticated. User's private key in the Ethereum decentralized network is secret and known only to the user itself. Thus, no other person in the blockchain has access to it. Therefore, the proposed algorithm can withstand an impersonation attack.

User Anonymity:
Our algorithm is built on the ethereum blockchain. When a transaction is made on a blockchain it is recorded in a block along with the public key of the user that issues the transaction which is available to everyone on the blockchain. Thus the public key of the user to be authenticated is publically available on the blockchain making it not anonymous during its authentication. Every node in the decentralized network knows the identity of the user which is getting authenticated. Therefore, in the proposed algorithm the user is not anonymous.

6 Conclusion and Future work

Conclusion

We redesigned the model of Blockcerts, MIT Labs project, which was developed on the Bitcoin chain using Ethereum blockchain. The new issuer and verifier design is implemented. There were some more issues to be solved in the design. One of the major issue that was found after the analysis was that the authentication of issuers was not looked into. To solve this issue we designed a 2-Factor decentralized authentication scheme to overcome the problems faced by single password and centralized authentication schemes.

Future Work

1. To add a certificate revocation feature to the certificate system.
2. To integrate the 2-Factor decentralized scheme proposed with the issuing module of the certificate system.
3. To do other formal security analysis on the authentication protocol developed and solve the issues highlighted during this analysis.
4. Implement Burrows Abadi Needham(BAN) logic and do analysis using Vista tools.
5. Make the project more scalable and secure.

References

1. Zheng Z, Xie S, Dai H, Chen X, Wang H (2017) An Overview of Blockchain Technology: Architecture, Consensus, and Future Trends. IEEE 2017, pp. 1–8
2. Wood G (2014) (Founder Ethereum and Parity) Ethereum: A secure decentralized generalized transaction ledger, pp. 1–32
3. Nakamoto S (2018) Bitcoin: A Peer-to-Peer Electronic Cash System, pp 1–9 (2018)
4. Hyperledger White Paper An Introduction to Hyperledger, pp 1–32
5. HYDRO Raindrop: (2018) Public Authentication on The Blockchain - June 2018, pp 1–25
6. Cresitello-Dittmar B (2016) Application of the Blockchain for Authentication and Verification of Identity, pp 1–9, 30 November 2016
7. Altinkemer K, Wang T (2011) Cost and benefit analysis of authentication systems. Dec Support Syst 51:394–404
8. Mo J, Hu Z, Lin Y (2018) Remote user authentication and key agreement for mobile client server environments on elliptic curve cryptography. J Supercomput 74:5927–5943
9. University of Birmingham Blockchain based Academic Certificate Authentication System Overview, pp 1–16
10. Gipp B, Meuschke N, Gernandt A (2015) Decentralized Trusted Timestamping using the Crypto Currency Bitcoin, p 16
11. Hepp T, Schoenhals A, Gondek C, Gipp B (2018) Origin Stamp: A blockchain-backed system for decentralized trusted time stamping, it Information Technology, vol. 60, pp 273–281 (2018)
12. IETF. RFC 6234 US Secure Hash Algorithms b (SHAand SHA-based HMAC and HKDF) (2011)
13. Nelson J, Shea R, Freedman MJ, Ali M (2017) Blockstack: A New Internet for Decentralized Applications (2017)
14. Loibl A (2014) Namecoin:Seminars FI/IITM SS2014, Network Architectures and Services, August 2014. https://doi.org/10.2313/NET-2014-08-114
15. Fromknecht C, Velicanu D (2014) CertCoin: A Name Coin Based Decentralized Authentication System 6857 Class Project (2014)
16. M'Rahi, D, Machani S, Pei M, Rydell J (2011) TOTP: Time-Based One-Time Password Algorithm. RFC 6238, pp 1–16 (2011)
17. https://en.wikipedia.org/wiki/Google Authenticator
18. https://en.wikipedia.org/wiki/1Password
19. Mann C, Loebenberger D (2017) Two-factor authenticationfor the Bitcoin protocol. Int J Inf Secur 16:213–226. https://doi.org/10.1007/s10207-016-0325-1
20. Lpez J, Dahab R (2000) An Overview of Elliptic Curve Cryptography. Technical Report IC-0-10, State University of Campinas 2000

Spectrum Sensing Techniques in Cognitive Radio Networks: Challenges and Future Direction

Neha Chaudhary and Rashima Mahajan

Abstract Cognitive radio being a key factor to fulfill demand of high data transmission and has improved the utilization of available frequency spectrum for next generation technology. For dynamic allocation of frequency hole to unlicensed user/secondary user, cognitive radio (CR) network requires detection of licensed user/primary user. Spectrum sensing phenomenon is used to estimate the unutilized frequency band in the available spectrum area. This is one of the important functions and challenging task in cognitive radio. This paper lists the comprehensive literature review of the spectrum sensing technique and presents various research areas of spectrum sensing (SS) in cognitive radio networks (CRN). An attempt to summarize currently existing research issues with spectrum sensing technique is also made through this paper. This paper also includes the classification of different spectrum sensing techniques, their comparison and finally presents future scope of this research.

Keywords Cognitive radio (CR) · Spectrum sensing technique (SS) · Primary user (PU) · Secondary user (SU) · Wireless communication

1 Introduction

Cognitive radio networks (CRN) have been implemented to increase the access of available radio spectrum by addressing the latter's limitation of spectrum scarcity. The survey report by Federal Communications Commission (FCC) indicates, some of the frequency bands in radio frequency spectrum are not utilized or partially utilized leading to spectrum scarcity. Opportunistically radio frequency environment sensing and detection of spectrum holes takes place by using this intelligent programmable

N. Chaudhary (✉)
Manav Rachna International Institute of Research and Studies, EEE Department, Faridabad, Haryana, India
e-mail: nehachaudhary.fet@mriu.edu.in

R. Mahajan
Manav Rachna International Institute of Research and Studies, CSE Department, Faridabad, Haryana, India

V. Nath and J. K. Mandal (eds.), *Proceeding of Fifth International Conference on Microelectronics, Computing and Communication Systems*, Lecture Notes in Electrical Engineering 748,
https://doi.org/10.1007/978-981-16-0275-7_37

radio [1]. Cognitive Radio (CR) has two main parameters. One is their cognitive capabilities, which defines the capability to observe the environment to subsequently create, plan and orient itself. The other is re-configurability property, which defines ability of cognitive radio to be programmed with dynamism for each range of the radio environment [2, 6]. Spectrum sensing (SS), spectrum decision (SD), spectrum sharing (SS) & spectrum mobility (SM) are four main functions of CR network. Figure 1 shows the general model used for spectrum sensing.

The objective of this paper is to review the spectrum sensing technique and find the latest trends & future scope of spectrum sensing (SS). The Sect. 2 consists of detailed explanation about the spectrum sensing techniques (SST) followed by issues in spectrum sensing techniques as discussed in Sect. 3.

2 Categorization of Spectrum Sensing Techniques

Numerous classifications of spectrum sensing techniques (SST) are available [14, 15]. Figure 2 shows classification of spectrum sensing technique.

(A) Non-coherent Detection Spectrum Sensing Technique: This technique is also called as signal processing technique. In this technique, the channel co-ordination is not present for sensing between the cognitive radio networks [20].

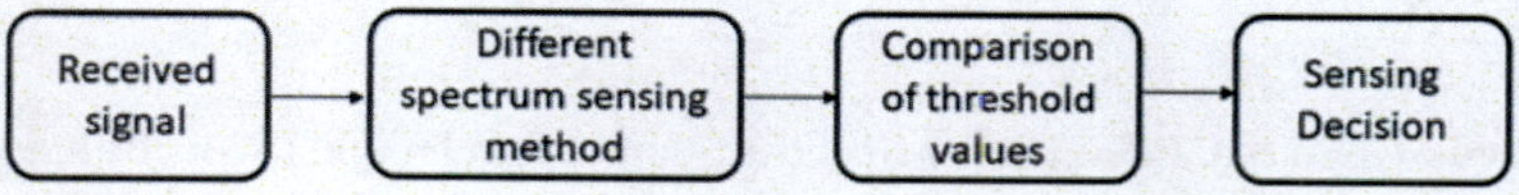

Fig. 1 Spectrum sensing model

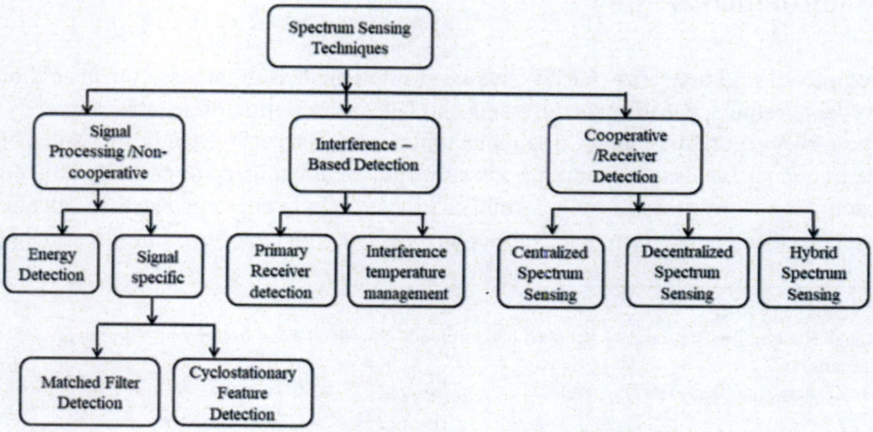

Fig. 2 Classification of spectrum sensing techniques

(i) Energy Detection (ED): In this spectrum sensing technique, no prior information of the primary user signal is required [3, 4]. Detection of primary user takers place by comparing the output of energy detector (E) with set value of threshold (λ). Let us consider received signal r(n) represented by Eq. (1)

$$r(n) = t(n) + g(n) \tag{1}$$

Where g(n) is the gaussian noise and t(n) is the signal to be detected.

Energy of the received signal is calculated by Eq. (2)

$$E = \sum_{(n=0)}^{N} r(n)^2 \tag{2}$$

Where N is the number of samples. By comparing this calculated energy to the set value of threshold, detection of primary user takes place by using the following hypothesis equation:

r(n) = g(n) : H_0 (absence of primary user).
r(n) = t(n) + g(n) : H_1 (presence of primary user).

Two performance parameters are used to analyze the performance of energy detector. First is probability of detection (P_d)which indicates existence of primary user and is represented by:$P_d = Pr(E > \lambda | H1)$. The other one is probability of false alarm (P_f) which shows incorrect or false existence of primary user, represented by: $P_f = Pr(E > \lambda | H0)$. For better utilization of spectrum, P_f should be low and P_d should be high. Performance of this technique is better for high signal to noise ratio [11]

(ii) Matched Filter Detection (MFD): In matched filter detection, received signal is projected towards the pilot signal and estimates its value by using test statistics given in Eq. (3)

$$T_m = \sum_{n=0}^{N_s} r(n) x_p(n) \tag{3}$$

where r(n) is the received signal, x_p(n) is pilot signal and T_m is the test statics. This estimated value is compared with set threshold to give decision. Performance parameters (P_d *and* P_f) are expressed in Eqs. (4) and (5)

$$P_d = \mathrm{Q}\sqrt{\frac{\lambda s - Ep}{Ep\sigma_p^2}} \tag{4}$$

$$P_f = \mathrm{Q}\sqrt{\frac{\lambda s}{Ep\sigma_p^2}} \tag{5}$$

Where λs represents the sensing threshold is Ep is the energy of the primary signal and σ_p^2 represents the standard deviation of the primary user signal. Sensing threshold can be calculated by using the formula given in Eq. (6):

$$\lambda s = Q^{(-1)}(P_f)\sqrt{Ep\sigma_p^2} \tag{6}$$

This technique requires minimal sensing time as compared to other sensing technique and it is suitable for all signal to noise ratio ranges [7, 8].

(iii) Cyclostationary Feature Detection: This spectrum sensing technique is based on introducing periodic redundancy into signal either by modulation or sampling method. For detection of primary user signal, cyclostationary feature detector observes the mean & autocorrelation of the received signal. Autocorrelation function of the given signal m(t) is represented in Eq. (7)

$$R_m(\tau) = \int_{-\infty}^{\infty} m(t) * m(t-\tau)dt \tag{7}$$

According to the distribution of autocorrelation function, sensing decision will take place. For presence of noise, small lag of the autocorrelation function is considered and for presence of signal, significant large lag value of the autocorrelation is considered.

$$\text{if lag0} > \text{lag1}: \text{ indicates PU absent.}$$
$$\text{if lag0} \approx \text{lag1}: \text{ indicates PU present.}$$

This technique is able to differentiate signal and noise, also well performed at low signal to noise ratio.

(iv) Waveform Detection: In this technique pattern of the transmitted signal is known prior and this includes the information of the preambles (known sequence transmitted before each slot) & midambles (known sequence transmitted in middle of each slot) of the transmitted signal [13]. By correlating the received signal with the known pattern, detection of the primary user takes place. This technique is suitable for all SNR ratio and required less time for sensing.

(v) Eigen value Detection: Step 1: is to calculate the number of samples in the received signal and after that formation of Hankel matrix which is called the catalecticant matrix. Step 2: includes the decomposition of the given matrix and factorization of the given matrix can be calculated by using singular value- decomposition (SVD). Summation of the diagonal matrix results in calculation of the eigen values and by arranging the maximum (λmax) and minimum (λmin) eigenvalues threshold value can be detected [5]. If the detected threshold value is greater than the set value of threshold, PU is considered present else absent.

B) Cooperative Spectrum Sensing (CSS): In this technique, cognitive radio shares its information with other secondary users (SU) and combines the findings of all users to give final decision. This technique provides the solution of the problems arises by the fading, shadowing and noise uncertainty [9, 12]. In centralized spectrum sensing (CSS) technique, one central unit is present that is called cluster head to control overall network. If the primary user is present then first this information is given to cluster head. By broadcast method, cluster head forward this information to the unused spectrum bands and accordingly cognitive devices will take decision. In distributed spectrum sensing technique, cluster head is not present, individual cluster nodes share information with each other and take final decision. In hybrid technique, cognitive radio independently detects & vacates the channel and there is no need to inform other nodes for detection process.

(C) Interference Based Detection: In this, detection of primary user takes place by considering one important measuring parameter i.e. interference. Primary receiver detection (PRD) and interference management detection (IMD) are the two classification of this technique. Management of interference is important to maintain quality of service [10].

2.1 Performance Parameters

Following are the performance parameters used to analyze spectrum sensing:

(a) Probability of Detection ($\mathbf{P_d}$): Defines as the correct detection of primary user (PU) w.r.t the total sensing trials. The formula for P_d is given in Eq. (8).

$$P_d = Q\left(\frac{TED - N_s\left(\sigma_n^2 + \sigma_p^2\right)}{\left(\sigma_n^2 + \sigma_p^2\right)\sqrt{2N_s}}\right) \tag{8}$$

where Q function $= \frac{1}{2\pi}\int_X^{\infty} \exp(\frac{-u2}{2})$du, σ_n and σ_p are representing the standard deviation (SD) of noise & primary user signal,N_s represents total number of samples and TED (Total energy detection) is calculated by the given relation:

TED $= \sum_{n=1}^{N_s}(y[n]^2)$, Where y[n] is received samples.

(b) Probability of False-Alarm (P_f): Defines as the falsely detection of primary user (PU) w.r.t the total sensing trials. The formula for P_f is given in Eq. (9).

$$P_f = Q\left(\frac{\lambda_{det} - N_s\sigma_n^2}{\sigma_n^2\sqrt{2N_s}}\right) \tag{9}$$

where λ_{det} is detection threshold, obtained by Eq. (10)

$$\lambda_{\text{det}} = \sigma^2(Q^{(-1)}(P_f a)\sqrt{2N_s} + N_s) \quad (10)$$

where σ = Standard deviation of the noise.

N_s = Number of samples.

P_f = Probability of false alarm.

Q^{-1} is the inverse Q-function (This special function represents the complement of (SNDF) standard normal distribution function and gives the information of the Gaussian distribution (GD)).

Probabilities of sensing decision (P_r) obtained by comparing TED to λ_{det} and this is given by:

$P_d = P_r(\text{TED} > \lambda_{det})$: Primary user present.

$P_f = P_r(\text{TED} > \lambda_{det})$: Primary user absent.

For better spectrum sensing technique, the value of P_d should be high and P_f should be low [16].

3 Challenges and Future Scope

This section presents the various challenges associated with each technique:

(i) Channel Uncertainty: Due to shadowing and fading effects, false detection of primary user (PU) takes place [17, 18]. Hence, cognitive radio (CR) must have to differentiate shadowed or faded signal from an available spectrum hole. (ii) Selection of threshold: In case of energy detection spectrum sensing technique, selection of threshold is a difficult task as tradeoff is done between the probability of miss-detection (P_d) and the probability of false alarm (P_f). By setting the constant false alarm rate (CFAR) the level of threshold can be increased or decreased [19]. (iii) Sensing throughput trade-off: Probability of detection (P_d) & probability of false alarm (P_f) are the two main parameters which affect the performance of sensing. Also, the irregular arrivals & departures of primary users affect the sensing throughput. Hidden primary user, security, duration of sensing and frequency time, efficient information sharing, sensing interference limit and hardware platform are the others challenging factors in spectrum sensing [16, 21].

Various spectrum sensing techniques are available in cognitive radio network and each technique has its own advantages and disadvantages in different aspects. In future, diverse investigation is required, to find issues associated with different spectrum sensing technique (SST). Enhancement in detection capability, accuracy and detection in multi-user network, required more future research.

4 Conclusions

Demand of upcoming wireless technologies can be fulfilled by the efficient utilization of spectrum and this is proven that cognitive radio technology has great impact on this. To overcome various challenges associated with spectrum sensing, researchers have proposed various solutions to deal these issues effectively. For future research, concept of dual stage sensing (DSS) is a great scope to improve the detection parameters by combining the effect of different spectrum sensing techniques (SST). Also, the performance of spectrum sensing (SS) can be improved by the combination of best sensing techniques. Hence, spectrum sensing is still an open research area.

References

1. Amrutha V, Karthikeyan KV (2017) Spectrum sensing methodologies in cognitive radio networks: a survey. In: Proceedings of IEEE International Conference on Innovations in Electrical and Electronics, Instrumentation and Media Technology (ICIEEIMT), pp 306–310
2. Ali A, Hamouda W (2016) Advances on spectrum sensing for cognitive radio networks: theory and applications. IEEE Commun Surv Tutor 19(2):1277–1304
3. Arjoune Y, Mrabet Z, Ghazi H, Tamtaoui A (2018) Spectrum sensing: enhanced energy detection technique based on noise measurement. Electr Eng Syst Sci, 1–6
4. Ahmed O, Salam A, Ray E, Saleh R, Araji A, Mezher K, Nasir Q (2019) Adaptive threshold and optimal frame duration for multi-taper spectrum sensing in cognitive radio. Sci Direct ICT Exp 5(1):31–36
5. Awin FA, Abdel-Raheem E, Ahmadi M (2016) Designing an optimal energy efficient cluster-based spectrum sensing for cognitive radio networks. IEEE Commun Lett 20(9):1884–1887
6. Amjad M, Rehmani MH, Mao S (2018) Wireless multimedia cognitive radio networks: A comprehensive survey. IEEE Commun Surv Tutor 20(2):1056–1103
7. Di Benedetto M-G, Cattoni AF, Fiorina J, Bader F, De Nardis L (eds) (2015) Cognitive Radio and Networking for Heterogeneous Wireless Networks. Springer , Cham
8. Deng M, Hu BJ, Li X (2017) Adaptive weighted sensing with simultaneous transmission for dynamic primary user traffic. IEEE Trans Commun 65(3):992–1004
9. Ejaz W, Hattab G, Cherif N, Ibnkahla M, Abdelkefi F, Siala M (2018) Cooperative spectrum sensing with heterogeneous devices: hard combining versus soft combining. IEEE Syst J 12(1):981–992
10. Geetha Priya C, Sowmiya Sri S, Renganayagi M, Varsha TM (2019) Average information based spectrum sensing for cognitive radio. In: IEEE International Conference on Communication and Signal Processing, p 876–880
11. Hu H, Zhang H, Liang YC (2016) On the spectrum-and energy-efficiency tradeoff in cognitive radio networks. IEEE Trans Commun 64(2): 490–501
12. He Y, Xue J, Ratnarajah T, Sellathurai M, Khan F (2018) On the performance of cooperative spectrum sensing in random cognitive radio networks. IEEE Syst J 12(1):881–892
13. Illanko K, Naeem M, Anpalagan A, Androutsos D (2016) Energy efficient frequency and power allocation for cognitive radios in television systems. IEEE Syst J 10(1):313–324
14. Javed JM, Khalil M, Shabbir A (2019) A survey on cognitive radio spectrum sensing: classifications and performance comparison. In: International Conference on Innovative Computing.
15. Karimi M, Sadough SMS (2018) Efficient transmission strategy for cognitive radio systems under primary user emulation attack. IEEE Syst J 12(4):3767–3774

16. Yang K, Martin S, Quadri D, Wu J, Feng G (2017) Energy-efficient downlink resource allocation in heterogeneous OFDMA networks. IEEE Trans Veh Technol 66(6):5086–5098
17. Lopez OLA, Sanchez SM, Mafra SB, Fernandez EMG, Brante G, Souza RD (2018) Power control and relay selection in cognitive radio ad hoc networks using game theory. IEEE Syst J 12(3):2854–2865
18. Zheng M, Chen L, Liang W, Yu H, Wu J (2017) Energy-efficiency maximization for cooperative spectrum sensing in cognitive sensor networks. IEEE Trans Green Commun Netw 1(1):29–39
19. Manesh MR, Kaabouch N (2018) Security threats and countermeasures of MAC layer in cognitive radio networks. Elsevier Ad Hoc Netw 70:85–102
20. Suseela B, Sivakumar D (2015) Non-cooperative spectrum sensing techniques in cognitive radio-a survey. In: IEEE International Conference on Technological Innovations in ICT for Agriculture and Rural Development
21. Torabiand M, Nerguizian C (2017) Adaptive transmission in spectrum sharing systems with AlamoutiOSTBC under spatially correlated channels. IEEE Trans Veh Technol 66(4):3131–3142

Scaling Blockchain by Autonomous Sidechains

Abhishek Vispute, Siddhesh Patel, Yuvraj Patil, Shubham Wagh, and Mahesh Shirole

Abstract Blockchain is one of the most promising technologies of the future. But till date, it is mostly used for transactions that involve mostly cryptocurrencies. To expand the scope of blockchain beyond cryptocurrency, it is really important to rethink the structure of existing blockchain architecture. One of the major factors which hold blockchain back from being an option to host real-world applications is its lack of Scalability. This paper presents the solution to scale blockchain using the concept of Autonomous Sidechains based on the Plasma framework. Instead of forcing the implementation of all transactions to the mainchain, we propose to create a new sidechain for every decentralized application. The sidechains created are autonomous with respect to each other, with mainchain acting as a backbone of the entire network. This concept increases the scalability of the blockchain by increasing the number of transactions network can process at a time.

Keywords Blockchain · Scalability · Plasma · Sidechains · DApps

A. Vispute (✉) · S. Patel · Y. Patil · S. Wagh · M. Shirole
Department of Computer Engineering and Information Technology, Veermata Jijabai Technological Institute, Matunga, Mumbai, India
e-mail: asvispute_b16@it.vjti.ac.in

S. Patel
e-mail: snpatel_b16@it.vjti.ac.in

Y. Patil
e-mail: yapatil_b16@it.vjti.ac.in

S. Wagh
e-mail: srwagh_b16@it.vjti.ac.in

M. Shirole
e-mail: mrshirole@it.vjti.ac.in

V. Nath and J. K. Mandal (eds.), *Proceeding of Fifth International Conference on Microelectronics, Computing and Communication Systems*, Lecture Notes in Electrical Engineering 748,
https://doi.org/10.1007/978-981-16-0275-7_38

1 Introduction

When Bitcoin [1] was invented, "Decentralized ledger for Cryptocurrency" was seen as the only use case of blockchain. Then Ethereum [2] came, with the vision of becoming the world's decentralized computer. It expanded the reach of blockchain with the help of smart contracts [3] to cover a wide range of use cases. This enabled developers to build decentralized applications (DApps) [4] using programming languages like solidity [5]. One of them was CryptoKitties, which is an online game that uses Ethereum as a backend. The game debuted on Nov. 28, 2017 and then the problem in Ethereum was seen by a larger audience. As more people started using Cryptokitties, it clogged the network, leading to slower transaction (TX) time for all users of the blockchain [6]. As more incidents like this arise, the blockchain community realized that if they want to see blockchains replace the centralized systems they need to match their Transaction throughput. Currently, Ethereum can process roughly 15–20 Transactions per Second (TPS), while in comparison Visa processes approximately 24,000 TPS [7].

The power of blockchain is a trust-less environment in peers. This trust-less environment originates from immutability and network protocol they use. The same protocol is the reason behind the scalability problem. Protocol says, all the systems should have the same state. To enable this single global state, all nodes should simultaneously process every TX happening on the network. So blockchain can't move to the next state until all nodes have reached the previous state.

This puts a fundamental limit on public blockchain's transaction throughput, it cannot be higher than what an individual node can process [8].

The problem of scalability comes with the trilemma which states that you can have only two of the three things shown in Fig. 1.

1) *Decentralization:* This refers to the Dynamics of Power in the System. The system is called decentralized when every peer of the network exercise same

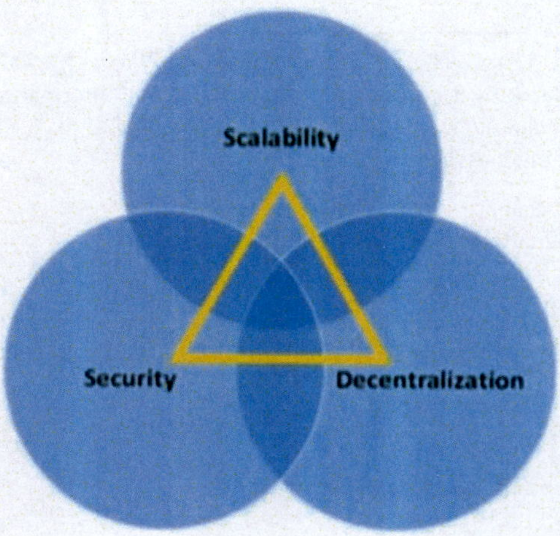

Fig. 1 The blockchain scalability trilemma [9]

amount of power, anyone can participate in network. No single peer or group of peers have complete autonomy on the system.

2) *Security:* As we discussed in the decentralization, everyone should exercise same amount of power, which includes fraudulent peers too, so the Security refers to how other ethical peers can guard the system from frauds to maintain the integrity of the system.
3) *Scalability:* This refers to ability of the system to handle a growing amount of TX and Users without increasing load on the system.

The objective of this paper is to find a solution that increases scalability without sacrificing Security or Decentralisation. The design choices of Bitcoin [1] and Ethereum favour decentralization and security while making a sacrifice in scalability [9]. Solutions proposed by community, fall into three categories: on-chain solutions (Layer 1) like Sharding [10], off-chain solutions (Layer 2) [11] like State channels [12] and Sidechains, and Switching to efficient consensus protocols like Proof of Stake (PoS) [13].

One promising solution is a side-chain solution with Plasma (Layer 2) [14]. In Plasma, instead of putting all transactions in the main chain, Plasma allows anyone to create side chains and bond side chains into the global blockchain. The main idea behind Plasma is to take advantage of increased scalability of sidechains backed with the security of the mainchain. Plasma is a framework and there are various implementations of it [15]. **Our proposed system is implementation of Plasma Framework, only.** Forward in this paper, we propose our system in Section II. We analyse the results and compare our solution with the existing plasma solutions in Section III. At last we conclude in Sect. 4.

2 Proposed System

We propose to create an Autonomous sidechain per DApp, with one parent chain (Mainchain) as an anchor. They are autonomous in the sense that two sidechains are independent of each other. Each sidechain user must also be part of the mainchain. Any user who wants to deploy DApp will create a new sidechain. Other users who want to use that DApp will join that particular sidechain. Figure 2 shows the Basic architecture. as you can see in the Figure, each sidechain has its own set of users. Users on a particular sidechain don't have to maintain other sidechains. All users are part of Mainchain.

By this architecture, users don't need to maintain all Transactions happening on the network. Users will maintain the mainchain and only sidechains they are part of. Thus, we counter the reason explained in Introduction, as now every transaction need not be processed by every single node in the network. So, there are two advantages here, Firstly, you have to maintain only a part of the total blockchain network, so physical hardware requirements are low, Secondly, as you divide resources as per DApp, blocks are added more frequently, so TPS increases. **But... what about the**

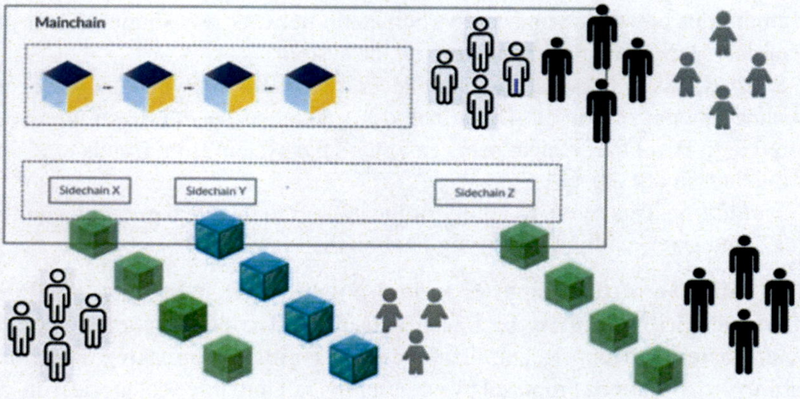

Fig. 2 Proposed architecture: basic

security concerns as each sidechain only has users of that particular DApp, the number of users on a sidechain will be less than the total number of users in the complete system. We know that this will increase the probability of fraud on the sidechain. To solve this problem, we are using the mainchain as an anchor, how we do this is mentioned ahead in this section. Regarding the currency model, each sidechain can have its local currency, backed by mainchain's global currency. how we secure the valuation of currency is also explained in this section.

2.1 Architecture in Detail

In this section, firstly we will define the terms used in architecture and then we will see architecture from global perspective.

1) Blockchain: Defines how blocks are linked.
   - Sidechain: There is one sidechain per DApp. All TX for respective DApp are maintained in that sidechain. Once a block is mined by validators in that sidechain, its block header is committed to the mainchain.
   - Mainchain: It maintains the ledger for TX containing main-chain cryptocurrency and block headers from sidechain.

2) Account: Wallet on chain
   a) User Accounts
      - Master Account: Each user has one Master Account on Mainchain. The funds in the master account are frozen, i.e. the funds in the master account are not directly accessible to a user.
      - Transaction Account: Each user has a Transaction Ac-count on Mainchain and on the sidechains they are part of.

The funds in TX account are liquid, i.e. they are accessible to the user directly.

b) Exit Account: Each sidechain has its own exit account. There is only one exit account per sidechain. The funds transferred to exit account are not accessible to anyone as no one owns this account.

3) Currency: Define the medium of exchange.

- Global currency: Currency of Mainchain
- Local currency: Each sidechain has its own local currency which is backed by Global currency. Global currency = Local Currency * Conversion factor. The local currency is inter-convertible with global currency on the basis of this conversion factor.

4) Consensus: Describes how nodes agree on global truth.

- Main Chain: Main chain nodes have to reach a consensus on TX of the main ledger (Main chain global currency TX) as well as have a choice to verify TX in side chain from block header. Proof of work (PoW) is used as a consensus algorithm for mainchain.
- Sidechain: To speed up the process of block generation, consensus algorithm on sidechain are kept as Delegated Proof of Stake (DPoS) [16] and Proof of authority (PoA).

There is one more reason why we are using DPoS or PoA on Sidechain other than speed and efficiency. Consider, we have PoW on the sidechain. In PoW case the intelligent miner will join the sidechains with lesser number of users, as there is less competition there. So, it creates unfairness to sidechains with more number of users. Plus, In PoW case, there is a mining reward, so essentially we are creating new coins, which disturbs the inter-convertible connection between Mainchain global currency and Sidechains local currency. In case DPoS and PoA our problem is solved as there is no mining reward or race here, Validator gets TX fees only [16].

5) Blocks

- Main chain blocks: Block size is kept fixed
- Side chain blocks: Block size is kept Variable Reason behind this approach: As A DApp developer you need variable block size for flexibility, But if you have variable block size, the size of the block can go to such height, where users only with high computing power can sync the block, this progresses to Centralization. The same thing is happening with ethereum [17]. In our case,As mainchain don't run any DApps so, we can keep it as fixed for decentralization and can keep Variable size for sidechain to give flexibility for DApps.

6) Transaction model: Account-based model [2].

7) P2P communication: It's done through Channel. There is a different channel dedicated to mainchain and each sidechain.

In our prototype we used Pub sub protocol [18].

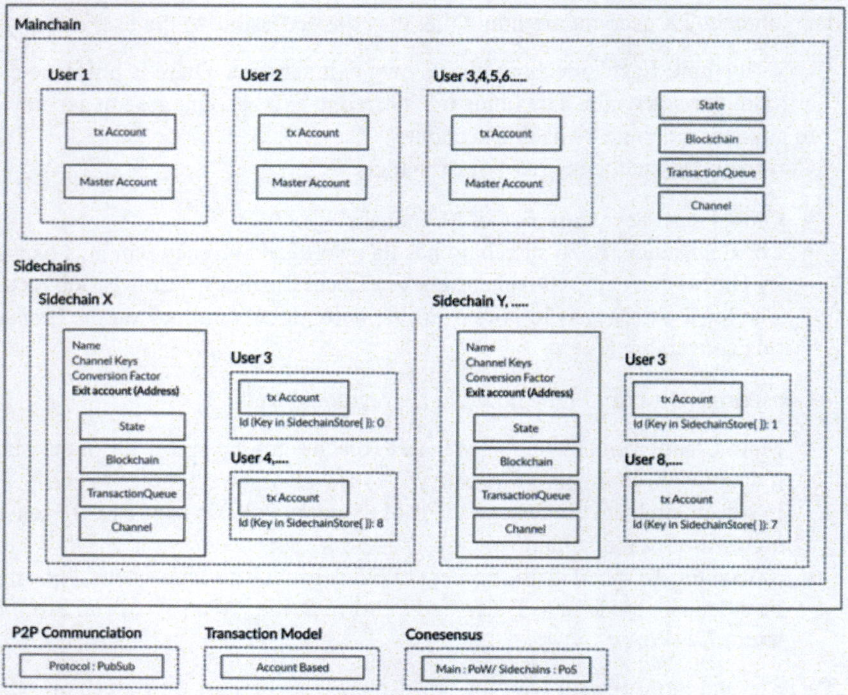

Fig. 3 Architecture: global view

8) Users

   a) Any user of the mainchain can create sidechain, and any user of the mainchain can join it.
   b) When a user joins sidechain an account on that sidechain is created.
   c) Each User account has a local Sidechain Store. It is a Key-value pair map with key as the local id of sidechain specific to the user and value is sidechain instance.

9) Transaction Queue contains the TX which are not yet included in the block. State is current information of all nodes [2]. Their Implementation is same in Sidehcains and Mainchain. **Please refer Fig. 3 for Architecture from global view.**

2.2 *Methods*

In this section, we explain the methods in our system.

1) Create Sidechain

a) User transfers “freezing amount” in user’s master account. This amount in the master account will not be released or accessible until user leaves that sidechain.
b) The new TX account is created in sidechain with balance as an equivalent amount, this equivalent is calculated by multiplying conversion factor into freezing amount.
c) The Exit account of that sidechain is also created at the time of sidechain creation.
d) The new sidechain is stored in user’s local sidechain store
With id = previous sidechain’s id + 1.

2) Join Sidechain

a) User fetches the sidechain blocks from other peers.
b) The state gets synced in the background
c) The next process is same asthe method
Create Sidechain’s a) b) and d).

3) Exit Sidechain

a) User starts exit challenge by submitting proof of ownership of their funds. This exit challenge is on for a specified time.
b) Sidechain users can submit a challenge to counteract the proof submitted in an exit challenge, if user submitted proof is fraudulent.
c) If a successful challenge is received within the specified time, all the exit amount is burned [19].
d) If no challenge is received within specified time, then
    i) Burn: the amount is transferred to the exit account
    ii) Release:Amount equal to *BurnAmount/Conversionfactor* is transferred from user’s master account to TX account.
    iii) Delete: The sidechain from the local sidechain store is deleted.

4) Sidechain Block-header Verification

a) The sidechain validator submits block header to the mainchain, to be verified and added to the mainchain blockchain.
b) The block header is submitted every time a block is mined. These periodic commitments are necessary to maintain the state of the blockchain.
c) They also lead to enhanced security, as any sidechain user can submit Fraud Proofs against a fraudulent block, within DTF of that block.
d) Each submitted block header has a DTF timestamp. DTF (Dispute Time Frame) is the time for which fraud proofs will be accepted for a block.

The verification algorithm is given in **Algorithm 1**.

Algorithm 1 Sidechain Block-Header Verification

Precondition: Sidechain store for every transaction will have a *fraudProof* map, with *key* = *blockHash* and *value* = *FraudProof* Array.

```
Sidechain validator publishes block header with merkle root
Block header added in mainchain transaction queue
if currentDateTime ≥ BlockHeader.DTF then
    Select Block Header for mining
end if
for Every block header do
    Miner gets all fraud proofs from sidechain store
    TotalValidProof ← 0
    for Each fraud proof do
        if Fraud Proof is Valid then
            TotalValidProof ← TotalValidProof + 1
        end if
    end for
    if TotalValidProof ≥ TotalFraudProof/2 then
        The block header is discarded
    else
        The block header is added to the block
    end if
end for
```

5) Publishing Fraud Proof [20]

a) Sidechain users can submit fraud proofs for one of the 3 types mentioned in the Fraud Proof (point 6) below, within the block's DTF.

b) If these fraud proofs are verified by the mainchain user, then the sidechain block producer will be penalized.

c) A Fraudulent transaction proof will have
   i) Merkle path of fraudulent TX
   ii) Merkle path of TX in current block
   iii) The transaction details

d) A Fraudulent block hash proof should submit
   i) Previous block header
   ii) If previous block header is added in mainchain block, then that mainchain block number must be provided.

e) A Fraudulent block merkle root proof will have
   i) Block Merkle root
   ii) All transaction hashes in that block

6) Validating Fraud proof

Currently we have identified 3 means of potential fraud, which can be described as:

a) Invalid transaction by double spend
b) Invalid block by wrong block hash
c) Invalid block by wrong merkle root

The invalid block cases will be considered valid only if Fraud proofs are submitted by a considerable number of people. The case of invalid transaction by double spend is explained in the **Algorithm 2**.

Algorithm 2 Validate Fraud Proof

```
Precondition: fraudProof object has all the required variables
procedure VALIDATEFRAUDPROOF (fraudProof)
    proofType ← fraudProof.proofType
    if proofType == DoubleSpendTx then
    if Double spend TX in same block then
        Compare Merkle path of both TX hashes
        if Both paths valid and different then
            Return True
        end if
    else if Double spend TX in different block then
        HeaderA ← Block header containing current block TX
        HeaderB ← Block header containing other TX
        if HeaderA precedes HeaderB then
            Return False
        end if
        Evaluate Merkle path of both TX hashes
        if currentDateTime ≥ HeaderB.DTF then
            HeaderB should exist in mainchain TX queue
        else
            Mainchain block No. containing HeaderB to be given
            if Block number not provided then
                Return False
            end if
        end if
        Return True
    end if
  end if
end procedure
```

7) TX on Sidechain

When TX on sidechain is validated by sidechain validator two special TX are broadcast on the mainchain channel. These TXs are special in the sense that they are not validated by mainchain miners.

When peers on the mainchain receive them it directly affects the state.

Suppose the TX is Transfer x from user 1 to 2.

Two special TXs are.

a) First: Transfer the equivalent amount from user 1′s master account to user 2′s master account
b) Second: Transfer TX fees from User 1's Master account to Sidechain Validator's master account.

The reason behind this approach is *we consider TXs are "valid" if it is validated by sidechain validator, so we don't validate it again on the mainchain.* But, if the sidechain validator goes rogue, the fraud proofs are used to secure the network.

3 Results

In this section at first, we will see the observations from testing we conducted on the prototype. We will derive the theoretical throughput mathematically. We will compare our system with existing solutions. At last we will see how we solved scalability trilemma.

3.1 *Observations Derived from Testing*

We tested 1000 TX in batch of 100 TX on prototype [21] we built, and recorded time required for the batch to get added in global state. We started with 0 sidechains and then incremented from 1 to 5 [22].

1) *Direct Observation:* As we can see in Fig. 4, we have drawn a graph between No of transactions vs. Time required to execute them (micro-seconds). Each line in the graph represents the combination of Mainchain and no of sidechains [0–5]. As we can see as the number of sidechains are increasing the time required for 1000 tx execution decreases. Thus, we can conclude that,

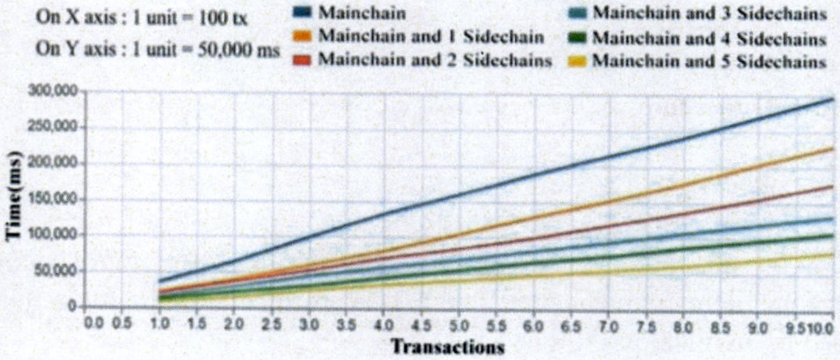

Fig. 4 Transaction time graph

Table 1 TPS mapped with no of sidechains

Number of sidechains	Transactions per second
0	3372
1	4384
2	5711
3	7685
4	9425
5	12478

$$Time\,required\;for\,'X'\,TX \propto \frac{1}{number\;of\;sidechains}$$

2) *Throughput:* We calculated the TPS for each case, as shown in the Table 1.

Percentage increase in Transaction Per Second per sidechain is **29.98%**. Thus, we can conclude that,

$$TransactionsPerSecond \propto Numberofsidechains \tag{2}$$

3.2 Theoretical Throughput

Theoretical throughput of the system can be calculated as follows: Consider the mainchain transaction throughput as,

$$m_{tps} = \frac{\text{mainchainBlockSize}}{\text{mainchainTransactionSize} \times \text{mainchainBlockGenTime}} \tag{3}$$

The total transactions in the sidechain block are,

$$transactionPerBlock_{sidechain} = \frac{\text{sidechainBlockSize}}{\text{sidechainTransactionSize}} \tag{4}$$

Thus, overall throughput of system can be calculated as,

$$system_{tps} = m_{tps} \times transactionPerBlock_{sidechain} \tag{5}$$

Note: We are calculating the maximum achievable tps (theoretical), so the general assumptions are as follows:

1) Mainchain blocks contain only sidechain block headers as transactions. Therefore,

$$mainChainTransactionSize = sideChainBlockHeaderSize \tag{6}$$

2) For the above point to be true, the sidechain blocks must already be mined when the mainchain block is to be mined. So we conclude,

$$mainChainBlockGenTime > sideChainBlockGenTime \quad (7)$$

3) Since our sidechains have variable block size, upto a certain limit, we assume the maximum possible block size for the sidechains.

$$sideChainBlockSize = maxSideChainBlockSize \quad (8)$$

Considering our variable values as,

1) mainChainBlockSize = 1 MB
2) sideChainBlockHeaderSize = 1 KB
3) maxSideChainBlockSize = 10 MB
4) sideChainTransactionSize = 1 KB
5) mainChainBlockGenTime = 20 s

Thus, using the above values, we get total transaction throughput to be,

$$OverallTheoriticalThroughput = 5 * 100,000tps \approx 10^5 tps \quad (9)$$

Table 2 Comparison with existing plasma solutions

Parameters	Autonomous sidechains	Loom sidechain [23]	Matic network [24]
Root chain	Our Mainchain	Bitcoin, Ethereum, Binance	Ethereum
Consensus model	POA, Delegated POS	Delegated POS	Delegated POS
Transaction model	Account based	UTXO	Account based
Fungibility[1]	Yes	No	Yes
Block size	Variable	Variable	Variable
Theoretical throughput	$10^5 tps$	NA[2]	$2^{16} tps$

[1]Fungibility is the property of a token or fund whose individual similarly valued units are essentially interchangeable
[2]Theoretical Throughput of Loom is not mentioned in their documentation, but it's supposed to be in range of our protocol and Matic only

3.3 Comparison with Existing Plasma Solutions

Following research compares the existing plasma solution with our proposed system. The existing solutions we chose to compare are Loom sidechain [23] and Matic network [24].

a) *Comparison with Loom sidechain:* In Loom sidechain, every time a user deposits some funds on the mainchain, a unique token is created on the sidechain [23]. Hence loom tokens are not fungible. The user only has to track their token. However, over time, the size of one token gets very large, which results in new set of problems [25]. Since we have an Account based transaction model, these problems are out of the question.
b) *Comparison with Matic sidechain:* Matic sidechains are deployed on ethereum. But due to variable block size, ethereum is moving towards centralization, as mentioned in [17], which can compromise the Matic network's security. Since our mainchain has a fixed block size, this problem won't arise in our solution.

The other properties of Loom and Matic are compared in the Table 2.

3.4 Achievement of Objective

Our objective was to increase the scalability of the blockchain without sacrificing Decentralization or Security. Let's see if we have achieved it.

1) *Scalability:* As we saw in "Observations derived from Testing" [Section III-A], and in "Theoretical throughput" [Section III-A2], when we increased the no of sidechains, the throughput of the system has also increased. We agree that throughput is one of the measurements to measure scalability, but it's the most scrutinized in the community.
2) *Security:* As per mainchain is concerned, all users are part of the mainchain, it has PoW as consensus, so to execute fraud, a user or group of users need have 51% of the current computing power of the chain [1]. For sidechain, we agree that due to the lesser number of users and comparatively less secure consensus (DPoS/PoA) it's not secure on its own [26]. But the catch is sidechains are only autonomous with respect to each other, and are safeguarded by mainchain. The users of the mainchain secure the network by provisions such as Block header verification and Fraud proofs [20]. Despite of this provisions, if user of the sidechain feels that the current sidechain network is fraudulent, they can exit the network by Exit sidechain method.
3) *Decentralization:* Any user can join the mainchain. If user is part of the mainchain they can join any sidechain. So there is no censorship. Consensus has major role in decentralization, It should be such that all peers should have equal right in deciding the global truth. On the mainchain we have PoW consensus so it's quite immune to Centralization [16]. Coming to Sidechains, they have

DPoS/PoA consensus protocol [16], these protocols are not as decentralized as PoW. But as we back our sidechains with Mainchain, the decentralization in the mainchain, secure the sidechains.

4 Conclusion

Our proposed solution creates a new sidechain for each DApp thereby reducing the number of transactions per sidechain. Result shows the percentage increase in TPS for system is **29.98%** per Sidechain. Thereby we can say that we have increased the throughput which reflects in increased scalability of the blockchain. So, the proposed concept of Autonomous Sidechains extends the scope of blockchain technology beyond, to create real-world applications capable of handling a growing number of users.

References

1. Nakamoto S (2009) Bitcoin: a peer-to-peer electronic cash system. https://www.bitcoin.org/bitcoin.pdf.
2. Buterin V, et al (2014) Ethereum white paper. https://github.com/ethereum/wiki/wiki/White-Paper
3. Lauslahti K, Mattila J, Seppala T (2017) Smart contracts–how will blockchain technology affect contractual practices? Etla Rep 68
4. Cai W, Wang Z, Ernst JB, Hong Z, Feng C, Leung VCM (2018) Decentralized applications: the blockchainempowered software system. IEEE Access 6:53019–53033
5. Dannen C (2017) Solidity programming. In: Introducing Ethereum and Solidity, pp 69–88. Springer, Heidelberg
6. Kharif O (2017) Cryptokitties mania overwhelms ethereum network's processing. https://www.bloomberg.com/news/articles/2017-12-04/cryptokittiesquickly-becomes-most-widely-used-ethereum-app.
7. Rieth Y (2018) Payment systems: Visa vs. bitcoin, decenter. https://decenter.org/en/paymentsystems-visa-vs-bitcoin
8. Stark J (2018) Making sense of ethereum's layer 2 scaling solutions: State channels, plasma, and truebit. https://medium.com/l4-media/making-senseof-ethereums-layer-2-scaling-solutions-state-channels-plasmaand-truebit-22cb40dcc2f4
9. Brian W (2019). Ethereum blockchain performance and scalability. https://blog.bybit.com/research-and-analysis/ethereum-blockchainperformance-and-scalability
10. Gao Y, Nobuhara H (2017) A proof of stake sharding protocol for scalable blockchains. Proc Asia-Pacific Adv Netw 44:13–16
11. Casey MJ (2018) layer 2' blockchain tech is an even bigger deal than you think. https://www.coindesk.com/layer-2blockchain-tech-even-bigger-deal-think
12. Dziembowski S, Faust S, Hostakova K (2018) Foundations of state channel networks. IACR Cryptol ePrint Arch 2018:320
13. Johannes H (2018) Scaling solutions on ethereum explained. https://medium.com/coinmonks/scaling-solutions-onethereum-explained-d970b66e28e5
14. Poon J, Buterin V (2017) Plasma: Scalable autonomous smart contracts. In: White paper, pp 1–47

15. Aiden (2018) Plasma world map - the hitchhiker's guide to the plasma. https://ethresear.ch/t/plasma-world-map-the-hitchhiker-s-guide-to-the-plasma/4333
16. Anwar H (2018) Consensus algorithms: The root of the blockchain technology. https://101blockchains.com/consensus-algorithms-blockchain
17. Stop And Decrypt (2018) The ethereum-blockchain size has exceeded 1tb, and yes, it's an issue. https://medium.com/hackernoon/the-ethereum-blockchainsize-has-exceeded-1tb-and-yes-its-an-issue-2b650b5f4f62
18. Krishnamachari B, Wright K (2017) The publish-process subscribe paradigm for the internet of things. USC ANRG Technical Report
19. Kuznetsov N (2019) Token burning, explained. https://cointelegraph.com/explained/token-burningexplained
20. Al-Bassam M, Sonnino A, Buterin V (2018) Fraud and data availability proofs: Maximising light client security and scaling blockchains with dishonest majorities. arXiv preprint arXiv:1809.09044
21. Vispute A, Patel S, Patil Y (2020) Autonomous sidechains prototype. https://github.com/abhishekvispute/AutonomousSidechains
22. Vispute A, Patel S, Patil Y (2020) Autonomous sidechains performance analysis. https://github.com/abhishekvispute/AutonomousSidechainsBasic/blob/master/Testing-Report%2030042020.pdf
23. Konstantopoulos G (2019) Plasma cash: Towards more efficient plasma constructions. arXiv preprint arXiv:1911.12095
24. Kanani J, Nailwal S, Arjun A (2018) Matic whitepaper, version 1.1. https://whitepaper.matic.network/#whitepaper
25. Fichter K, et al (2018) Plasma cash, shorter proofs. https://www.learnplasma.org/en/research/#plasmacash
26. Li C, Palanisamy B (2020) Comparison of decentralization in dpos and pow blockchains. arXiv preprint arXiv:2002.02082

Blockchain Interoperability Using Hash Time Locks

Snoviya Dcunha, Srushti Patel, Shravani Sawant, Varsha Kulkarni, and Mahesh Shirole

Abstract The recent escalation of individual, independent blockchain networks has engendered the need for a robust solution to communication between the various token networks. The solution must conform to industry standards and must be universally endorsed. This research paper explores various studies and implementations conducted under the topic of interoperability between different blockchain networks. Discovering various aspects of a blockchain system, comparing existing solutions and proposing a solution for communication between any two token blockchain networks are the steps effectuated. The proposed solution allows for a secure transfer of tokens between homogeneous blockchains using the concept of hash time locked contracts. Involving a self-enlisted middleman in the transaction eliminates the need for an external interface or a centralised authority for transaction, and reinforces trust. As a result, a safe and reliable transfer of tokens without the involvement of a centralised third party takes place. The solution warrants a mass adoption for further testing, however, it executes in a stable environment between two homogeneous token blockchain networks.

Keyword Blockchain · Interoperability · Smart contracts · HTLC - hashed time lock contracts

---

S. Dcunha (✉) · S. Patel · S. Sawant · V. Kulkarni · M. Shirole
Department of Computer Engineering and Information Technology, Veermata Jijabai Technological Institute, Matunga, Mumbai, India
e-mail: srdcunha_b16@it.vjti.ac.in

S. Patel
e-mail: sspatel_b16@it.vjti.ac.in

S. Sawant
e-mail: smsawant_b16@it.vjti.ac.in

V. Kulkarni
e-mail: vvkulkarni_b16@it.vjti.ac.in

M. Shirole
e-mail: mrshirole@it.vjti.ac.in

V. Nath and J. K. Mandal (eds.), *Proceeding of Fifth International Conference on Microelectronics, Computing and Communication Systems*, Lecture Notes in Electrical Engineering 748,
https://doi.org/10.1007/978-981-16-0275-7_39

1 Introduction

Blockchain is simply a chain of blocks, where each block contains validated transactions. These blocks are usually of fixed size and they store information about transactions like the date, time, amount, information about who is participating, and information (hash values) that distinguishes them from other blocks. The transactions are done by account holders of that particular blockchain and the validation is done by miners and stakeholders based on the consensus mechanism [22]. Blockchain is based on the three pillars of immutability, transparency, distributed ledger technology [3, 14, 16].

There are many blockchains in the world today - 9,150 by the year 2019 [4]. Currently, one blockchain does not know information that might exist in a different blockchain. For instance, the Bitcoin (BTC) blockchain exists fully independently of the Ethereum (ETH) blockchain and vice versa [19]. As the several networks carry different kinds of units, there will be requirements for exchange between them. Currently, no choice of exchange, even among homogeneous chains is provided. This is where interoperability in blockchain comes into picture.

Interoperability in blockchain [8] is a technical word for exchanging information or value across different chains in different networks, without restrictions. To increase the growth in business or evolve further, interaction is required and interoperability eases that process, maintaining the major pillars of blockchain intact. Thus, blockchain interoperability can significantly increase scalability, speed, and extensibility of blockchains. The primary challenge [18] to face here is how to cross the interface barrier, keeping in mind the issues faced by previous solutions as well as the original blockchain solutions, and provide for a solution to meet the compatibility and security standards to guarantee efficient delivery of messages and sound reliability of the system.

Rest of the paper is organised as follows: Sect. 2 discusses related work, Sect. 3 gives an overview of the basic concepts used in this paper, Sect. 4 describes the proposed solution and the alternate flow. Section 5 provides an overview of the implementation of the proposed system and Sect. 6 gives results. Finally, Sect. 7 concludes the paper [6].

2 Related Work

There have been several attempts at implementations to blockchain interoperaility, and while solving some problems, they have also created others. We studied a history of interoperability solutions, summarised as follows.

Tendermint

Tendermint is mainly a consensus protocol based on Proof-of-Stake consensus mechanism as well as an open source software for programming in blockchain. It uses a Byzantine Fault Tolerant engine for peer-to-peer architectural networking. The problem with the consensus is that it can get centralized towards the participants with higher stake in the chain [7].

Polkadot

Polkadot uses a relay chain mechanism. It acts as the hub through which the parachains connect to and coordinates the consensus as well as transferring of messages and data between the parachains. One of the major concerns in polkadot is scalability, which will only be tested with mass adoption in future [5].

Cosmos

Cosmos, like Tendermint, uses Byzantine Fault Tolerant mechanisms for consensus and networking. It is essentially a group or cluster of independent, parallel blockchain networks. It uses IBC (Inter Blockchain Communication) protocol for operations between heterogeneous tokens in blockchains. The problem with cosmos is that while it is developer-friendly, it is not user friendly for now [9, 11].

XClaim

XClaim is a framework for interoperability. It is trustless and uses cross-chain swaps. It allows to create assets that are backed 1:1 by existing cryptocurrencies. The drawback of xclaim is that the actual swap takes minutes to be executed, which might not be acceptable once it is adopted widely [21].

3 Basic Concepts

The solution proposed requires basic understanding of few concepts mentioned below:

A. Smart Contract

If you want to verify, or enforce the negotiation or performance of a contract digitally then you need Smart Contract which is a computer protocol.. It will help you to exchange money, property, shares, or anything of value in a transparent, conflict-free way while avoiding the services of a third party [15].

How it works:

1 The terms, rules and the conditions of the agreement are established by all the counter parties and translated into code that is written in blockchain.
2 The parties involved in the contract are generally anonymous and the code is on public blockchain.

3 If the events specified in the conditions occur then the code is automatically executed.
4 Once executed the terms of the contract will automatically transfer the value to the relevant parties.
5 The transfer of value will be recorded on the blockchain then.

The advantages include trust, backup, no intermediary, speed, accuracy and autonomy. At the same time the disadvantage may be include assumption of ideal actors and scenario which is not the case in real life, enforceability and also issues like coding error or system malfunction [17].

B. HTLC - Hash Time Locked Contracts

Hash Time Locked Contract (HTLC) refers to a special feature that is used to create smart contracts that are able to modify payment channels.

There two key elements which distinguish HTLC from standard cryptocurrency transactions, which are:

1 Hashlock: a function that restricts the spending of funds until a certain piece of data is disclosed (as a cryptographic proof). The pre-image is simply the piece of information that is used to generate the hashlock, and to later unlock its funds.
2 Timelock: is a function that restricts the spending of funds until a specific time in the future [2] (time bound transactions basically).

The recipient is required to acknowledge payment by generating cryptographic proofs within a certain time frame leading to completion of the transaction. Failing to do so would lead to cancellation of transaction taking place. Advantages include time sensitivity, validation, existence of trustless system and transfer that can take place between multiple blockchain [12].

4 Proposed System

The diagram for the Proposed system is shown in Fig. 1. The Process is as mentioned:

STEP 1. *Participant C*, who has an account on two token contracts – *Blockchain 1* and *Blockchain 2* - registers in the application as a verified middleman, and includes their charges in the declaration.

STEP 2. *Participant A* of *Blockchain 1*, who wishes to transfer tokens to *Participant B* of *Blockchain 2*, initiates the process by opening the list of Middlemen available.

STEP 3. *Participant A* views the list of Middlemen available, and chooses *Participant C*.

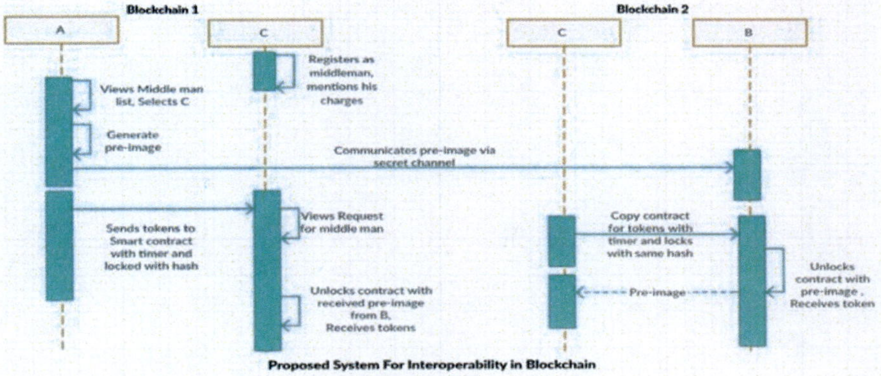

Fig. 1 Sequence diagram of proposed solution

STEP 4. *Participant A* generates a pre-image based on random nonce and a hash value based on the generated pre-image value.

STEP 5. *Participant A* communicates the pre-image value over to *Participant B* via a secret channel that does not involve *Participant C*. The pre-image is known only to *A & B*, and *Participant C* will not be involved in this transaction. Thus, confidentiality is ensured and attack by a potentially harmful middleman is prevented by this method.

STEP 6. *Participant A* now send tokens[1] to the smart contract with hash time locked, pre-determined timestamp until which the transaction would be valid, and after which the transaction would be nullified and go back to STEP 1.[2]

STEP 7. *Participant C* receives request for their services as Middleman. They initiate a transaction of the same amount as received from *Participant A* in *Blockchain 1* to be sent to *Participant B* in *Blockchain 2*.

STEP 8. In *Blockchain 2*, *Participant C* copies the contract for tokens with the relevant hash time locked with the same hash, and sends it over to *Participant B*.

STEP 9. *Participant B* unlocks the contract they received from *Participant C* in *Blockchain 1*, with the pre-image they received from *Participant A* initially via the secret channel, and thus receives the tokens sent by *Participant A*.

[1] No new tokens can be created using this protocol, only existing tokens can be transferred from one blockchain network to another.

[2] Locktime(C) = Locktime(A) - 2*(Average time of transaction in blockchain).

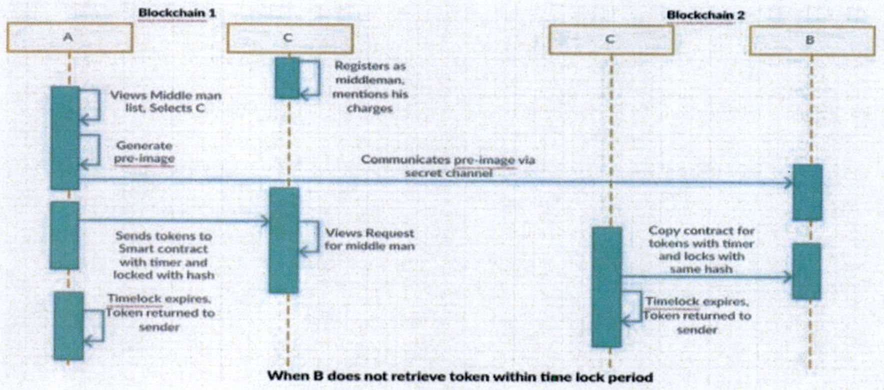

Fig. 2 Sequence diagram of alternate flow

STEP 10. *Participant B* sends *Participant C* in *Blockchain 2* the pre-image. *Participant C* unlocks the contract with the received pre-image, and thus gets the tokens.

A. Alternate Flow - When Receiver Does Not Retrieve Tokens Within Pre-set Time Lock

The diagram for the Alternate Flow is as shown in the Fig. 2. Step 1 to Step 8 are same as mentioned above and later steps are as follows:

STEP 9. The timelock set by *Participant A* expires, and the tokens that were supposed to be sent to *Participant B* will be initiated to be returned back to sender.

STEP 10. The expired tokens are sent back to *Participant A* of *Blockchain 1* and *Participant C* of *Blockchain 2* and both the token networks return to initial states, with no transaction performed ultimately.

5 Implementation

A. Data Structures

For the implementation of Interoperability in Blockchain using HTLC, we have used the following data structures.

```
struct Middleman{
  string name;
  string charges;
  bytes32 password;
  bool exists;
  bool free;
}

struct Request{
  bytes32 hashkey;
  uint256 releaseTime;
  uint256 amount;
  address receiver;
  address sender;
}

struct Transaction{
  address sender;
  address middleman;
  string preimage;
  bytes32 hashkey;
  uint256 releaseTime;
  uint256 amount;
  address receiver;
}
```

Mapping of addresses to structs help in quick fetching of data from smart contract. Following mappings are used:
mapping(address =>Middleman)middlemen;
mapping(address =>Transaction)transactions;
mapping(address =>Request)requests;
mapping(address =>address)middlemanReceiver;

B. Algorithms

Following are some of the algorithms of our implementation:

Algorithm 1: sendTokens

Input: account_address, amount, receiver_address, middleman_address, preimage, locktime(in hours, minutes, seconds)

Output: -

Calculate releaseTime = block.timestamp+h*1 hours+m*1 minutes+s*1 seconds

Calculate SHA256 hash of preimage

if *middleman address != 0* **and** middleman exists **and** middleman is free **then**

  Create a transaction with sender address, middleman address, preimage(null), hash of preimage, releaseTime, amount and receiver address

  Push the transaction into transactionList

  Create a request for the middleman using sender address, receiver address, amount, releaseTime, hash

  Push the request in requestList

end

The sendToken algorithm corroborates the data to be validated before a transaction can be considered for execution, and initiates the process of transmission. It first validates whether the current time is within the predefined releaseTime. The SHA256 hash of the input preimage, known only to the sender and receiver, is calculated. Then, testifying whether the middleman is existing, available and has a valid address is performed. A new transaction is created comprising the requisite information for the process, i.e. sender address, middleman address, preimage (null), hash of preimage, releaseTime, amount, and receiver address and pushed onto the transactionList, for enhanced tracking of transactions. A request to include the chosen middleman has to be created next, using sender address, receiver address, amount, releaseTime, and hash and is pushed onto requestList for the middleman to keep track of.

Algorithm 2: copyContract

Input: account_address, receiver_address, hash, amount, timestamp
Output: -
Calculate middleman releaseTime = timestamp-2*avgTime
if *releaseTime >block.timestamp* **and** receiver is currently free **then**
  Create a Transaction with sender as middleman address, hash, releaseTime, receiverAddress, amount and preimage(as null) and middleman(as null)
  Add a lookup entry for middleman and receiver
end

The copyContract algorithm is called by the middleman on the receiver blockchain after they view the pending transactions, which is unique to each middleman. It then calculates the release time for this contract based on the formula mentioned. The average time is set to 15 s by default [1]. This acts as the timelock of the contract between middleman and receiver. Next it checks if the release time is greater than the current time, and if the receiver is available for transaction and if both these conditions are true then it invokes the create transaction function where the middleman becomes sender, with the hash value and receiver address remaining the same as in the transaction mentioned above, the release time as timestamp and amount to be transferred. The middleman sets the preimage and middleman fields as empty. Following this step, a lookup entry for the middleman and the receiver is added into the data.

Algorithm 3: withdraw

Input: account_address, preimage
Output: -
Store address as recipient's address and convert it into payable.
if *the hash of given preimage = original hash* **and** timestamp <release time set by the sender **then**
  Transfer the amount to the recipient address
  Store the key as a preimage and reset the release time
end

The withdraw algorithm is utilised by the receiver to withdraw tokens from the smart contract. Initially, the transaction is fetched using the receiver's address. The transaction release time is examined, and if it has not exceeded then the tokens are transferred to the receiver's wallet following the entry of a valid preimage. The preimage is confirmed by computing its SHA256 hash value and comparing it with the hash value stored in the transaction structure. The preimage is saved in the smart contract for transaction records and the release time is reset.

[Note: If the receiver is a middleman and accessing the function from his middleman account then the same steps are conducted followed by setting the middleman status as free].

Algorithm 4: release

Input: account_address

Output: -

1 Store address as recipient's address and convert it into payable.

2 **if** *releaseTime != 0* **and** releaseTime has exceeded **and** amount >0 **then**

3 Transfer the amount to payable address

4 Reset releaseTime, set middleman as free, reset amount to 0 i.e. paid in requests

end

The release algorithm is invoked when the receiver fails to withdraw the tokens within the time defined during development. First, the sender address is converted to a payable address. Next, required conditions for transferring are tested i.e. releaseTime to be exceeded and amount to be greater than zero. Once these conditions are satisfied, the amount mentioned is all set to return to the payable address i.e., sender address. Lastly, the algorithm resets the releaseTime and the amount to zero and frees the middleman.

6 Results

After proposing and implementing the solution mentioned above, we came up with the following result and analysis as shown in Table 1 [9, 10, 13, 20]:

The proposed system thus ensures that there is no central party or an external third party in the transfer process. It does have a middleman who is elected by the sender's free will and no one has control over it. In addition to security advantages, the sender gets to set his timelock. A small downside is that the sender has to pay a little extra (middleman charges) and that a single token transfer process involves four transactions hence execution cost is more.

Table 1 Comparison with existing interoperability solutions

Parameters	Cosmos	Polkadot	Our solution
Consensus Algorithm	Tendermint BFT	Delegated PoS	Consensus Algorithm of participating blockchain
Interoperability technology	Notary scheme	Relaychain and Parachain/Multichain of Parachains	Hash-locking
Participating chain conditions	Member chains require compatibility. Non-compatible chains can use Pegzones	Member chains require compatibility. Non-compatible chains can use Bridgechains	Member chains must support smart contracts and some members of the blockchain must have accounts on other blockchains to act as middlemen
Validators	Required	Required	Not required
Native tokens	ATOM	DOTs	Tokens of participating blockchain
Security	Shared security solution for member chains but not for Peg-zones using chains	Shared security solution for member chains but not for chains using Bridgechains	Secure transfer of tokens using hash time locks
Reach	n(any ledger)	n	2

7 Conclusion

A universally interoperable system would have great benefits, and only then might the domain of cryptocurrency benefits, and only then might the domain of cryptocurrency begin to replace fiat currency, a dream held distantly by many blockchain advocates. However, no one can know if such a system might be able to exist fully functionally in the future. We realised during our study of blockchain development that while there might be a host of technologies and platforms available for both a programmer and a user to choose from, there scarcely exists any fully operating mechanism by which interoperability within two kinds of blockchain networks is made possible. This led us to form the problem statement of a homogenised interoperability solution. We followed this up with implementation, which followed the design and architecture of the system as detailed previously.

8 Future Work

While far from being a perfectly homogenised solution to the existing problem of communication between blockchain networks, our proposed system aims to be a standard protocol to be adopted as a step towards a democratic, distributed resolution to the issue. If a denial-of-service attack occurs because of a middleman failing to perform benevolently, we can place them at the bottom of the middleman list or display number of transactions successfully completed by the middleman sorted in descending order. Apart from this, there is need for both optimisation as well as standardisation of the solution in theory as well as design, and a need to embrace scalability formulae once the solution is put into mass usage. For the purposes of this paper, however, the solution is fit to be adopted into existing blockchain network systems and tested for efficacy.

References

1. Anonymous. Ethereum Average Block Time Chart (2020). https://etherscan.io/chart/blocktime
2. Anonymous. Hashed TimeLock Contract (HTLC). 2017–20. https://www.binance.vision/glossary/hashed-timelock-contract
3. Anonymous. Interoperability Protocols – The Current Landscape (2018). https://ieeexplore.ieee.org/document/8343163
4. CoinMarketCap. Kryptowaehrung Marktkapitalisierun- gen (2018). https://coinmarketcap.com/de/.
5. Polkadot Community. Polkadot Documentation (2019). https://polkadot.network/technology/
6. Darisi M, Salva J, Shirole M, Bhirud S (2019) STEM: Secured Token Exchange Mechanism. https://link.springer.com/chapter/10.1007/978-981-15-2693-0_15
7. Interchain Foundation (2017) Consensus Compare: Tender-mint BFT vs. EOS dPoS. https://blog.cosmos.network/consensus-compare-tendermint-bft-vseos-dpos-46c5bca7204b
8. Grayblock (2018) Interoperability - The Holy Grail of Blockchain. https://medium.com/int-chain/interoperability-the-holy-grail-of-blockchain-eac8e689d50d
9. Kajpust D (2018) Blockchain Interoperability: Cosmos vs. Polkadot. https://medium.com/@davekaj/blockchain-interoperability-cosmos-vs-polkadot48097d54d2e2
10. Koens T, Poll E (2019) Assessing Interoperability Solutions for Distributed Ledgers Extended version. https://www.ingwb.com/media/2667864/assessing-interoperability-solutions-for-distributed-ledgers.pdf
11. Kwon J, Buchman E (2015) Cosmos Whitepaper - A Network of Distributed Ledgers. https://cosmos.network/resources/whitepaper
12. Ometoruwa T (2019) An Introductory Guide to Hashed TimeLock Contracts. https://btcmanager.com/an-introductory-guide-to-hashed-timelock-contracts/
13. Specs Rationality (2018) Interoperability Protocols – The Current Landscape. https://spec-rationality.com/interoperability-landscape/
14. Reiff N (2020) Blockchain: Everything You Need to Know. https://www.investopedia.com/terms/b/blockchain.asp
15. Rosic A (2016) Smart Contracts: The Blockchain Technology That Will Replace Lawyers. https://blockgeeks.com/guides/smart-contracts/
16. Rosic A (2016) What is Blockchain Technology? A Step- by-Step Guide for Beginners. https://blockgeeks.com/guides/what-is-blockchain-technology/
17. Sharma R (2018) How Are Disputes In Smart Contracts Resolved? https://www.investopedia.com/news/how-are-disputes-smart-contracts-resolved/
18. Tasatanattakool P, Techapanupreeda C (2018) Blockchain: Challenges and applications. https://ieeexplore.ieee.org/document/8343163
19. Bitcoin Wiki. Lightweight node - Bitcoin Wiki (2018). https://en.bitcoin.it/wiki/Lightweight_node

20. Wood G (2019) Polkadot Documentation. https://polkadot.network/PolkaDotPaper.pdf
21. Zamyatin A, Harz D, Lind J, Panayiotou P, Arthur G, Knottenbelt W (2019) XCLAIM: Trustless, Interoperable, Cryptocurrency-Backed Assets https://ieeexplore.ieee.org/document/8835387
22. Zheng Z, Xie S, Dai H, Chen X, Wang H (2017) An Overview of Blockchain Technology: Architecture, Consensus, and Future Trends. https://ieeexplore.ieee.org/document/8029379

Automotive Ethernet Physical Optimization and IEEE 1588 Implementation

Yash Agarwal and D. R. Shilpa

Abstract The use of Ethernet protocol in automotive industry for in vehicle communication has seen rapid rise in recent years. The Ethernet provides with higher data rates/bandwidth which is essential in time critical communications. The Ethernet physical is a device that comes in and facilitates the communication between the MAC layer and network layer of Ethernet implementation [1]. The device can be programmed by a set of configuration registers. The modern automotive applications such as Advanced Driver Assistance Systems (ADAS), requires further optimization of Ethernet physical devices deployed in the various Electronic Control Units (ECU's) in the car. The paper estimates the power and area of the configuration registers and implementation of the IEEE 1588 Precision Time Protocol (PTP) in Ethernet physical.

Keywords Ethernet physical · Advanced driver Assistance systems · Configuration registers · Precision Time Protocol

1 Introduction

With the advances in the field of automotive industry, a rise in the deployment of electronics in automobiles has been observed. Various government regulations about emissions require efficient engine control. The advent of self-driving cars and Advanced Driver Assistance systems (ADAS) require sophisticated state of the art electronic modules. The development of low cost per performance solid state digital electronics further encourages the electronics deployment in electronics. Modern age automobiles consist of a plethora of electronically controlled components or units. These are more formally called Electronic Control Units (ECU's). These ECU's can be deployed in engine control, infotainment system, doors and windows control, safety units, etc. These various deployed units need to communicate and exchange data with each other in various situations [2].

Y. Agarwal (✉) · D. R. Shilpa
R.V. College of Engineering, Mysore Road, Bengaluru 560059, India
e-mail: yashagrawal.ec16@rvce.edu.in

V. Nath and J. K. Mandal (eds.), *Proceeding of Fifth International Conference on Microelectronics, Computing and Communication Systems*, Lecture Notes in Electrical Engineering 748,
https://doi.org/10.1007/978-981-16-0275-7_40

Ethernet has emerged as the main protocol of interest for automotive applications. The modern age and upcoming automobiles requires higher bandwidth communications between various modules. Ethernet is very much capable of providing such high data rates in an automotive environment. Also the use of Ethernet in automobiles has been recognized as a future proofing measure. Therefore there is an eminent need for optimizing such Ethernet devices. By comparing with the work done in [2] and [3] it can be observed that the major areas of improvement and optimization in these devices are in terms of power and area consumption of the product. Also there can be various such Ethernet nodes deployed in automobiles. These nodes will constitute a communication network which can be used for performing some processing or computation among multiple nodes. But in order to ensure smooth processing and communication across this network, the local clocks of the modules should operate in synchronization with each other with certain degree of precision and accuracy. To serve this purpose the IEEE standard 1588 protocol can be implemented. It is precise clock synchronization protocol for Networked measurement and control systems. This protocol generates a master slave hierarchy among the clocks in the given system [3]. In [4–6] there is a mention of Network Time Protocol (NTP) which has been prevalent for synchronization purposes. After comparing it with Precise Time Protocol (PTP) it can be observed that NTP provides accuracy in microsecond range and is not suitable for applications requiring precision and accuracy in sub microsecond range. Also it is not administration free. This further creates the need for PTP implementation.

2 Configuration Register Implementation

A typical Ethernet physical device has a total of 12,000 flip flops out of which 3,000 of them implement configuration registers [7–9]. Here we have considered a comparative analysis between the regular flip flop implementation and latch implementation of configurable registers. The Fig. 1 gives the flip flop implementation of configuration registers. Here we considered a module consisting of various control signals (including a clock signal). The module consists of 200 configuration registers of 16 bits each that can be implemented via 3,200 D-type flip flops. An 8 bit address bus specifies the address of the register to be read or written. Whether we want to read or write a configuration registers depends on the read and write enable signals. The data to be written is provided by the 16 bit input data bus din and the result of a read operation is available on a 16 bit output dout. The specified module is then synthesized using genus synthesis solution for power and area estimation.

Figure 2 deals with the latch implementation of the configurable registers. Here also just like the previous case we have considered 200 configurable registers of 16 bits each, which are implemented via 3,200 D type latches. Here we have an additional signal, a reset signal which sets all the registers to zero. The specified module is then synthesized using genus synthesis solution for power and area estimation.

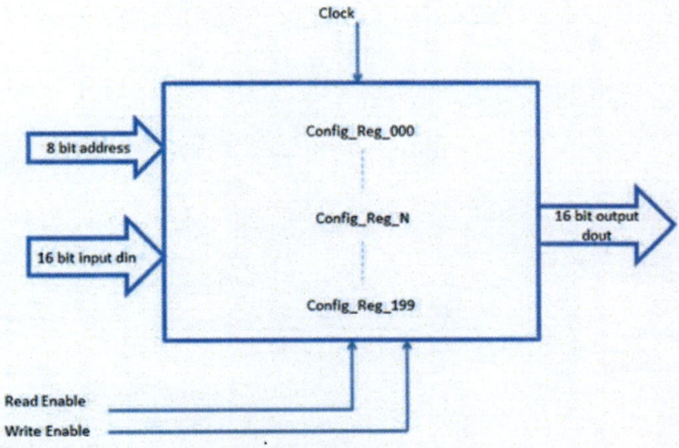

Fig. 1 Flip flop implementation of configurable registers

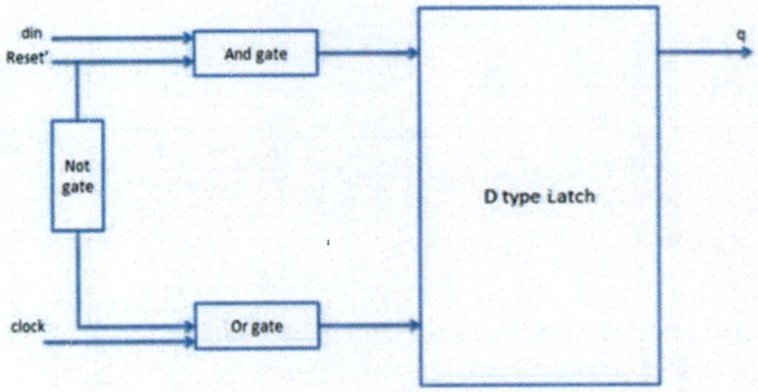

Fig. 2 Latch implementation of configurable registers

3 IEEE 1588 Implementation

IEEE 1588 standard defines a protocol enabling precise synchronization of clocks in measurement and control systems implemented with technologies such as network communication, local computing, and distributed objects. The protocol generates a master – slave hierarchy among the clocks in the system. All clocks ultimately derive their time from a clock known as the grandmaster clock. In its basic form, this protocol is intended to be administration free.

Measurement and control applications are increasingly using distributed system technologies such as network communication, local computing, and distributed objects. Without a standardized protocol for synchronizing the clocks in these devices, it is unlikely that the benefits will be realized in the multivendor system component market [3].

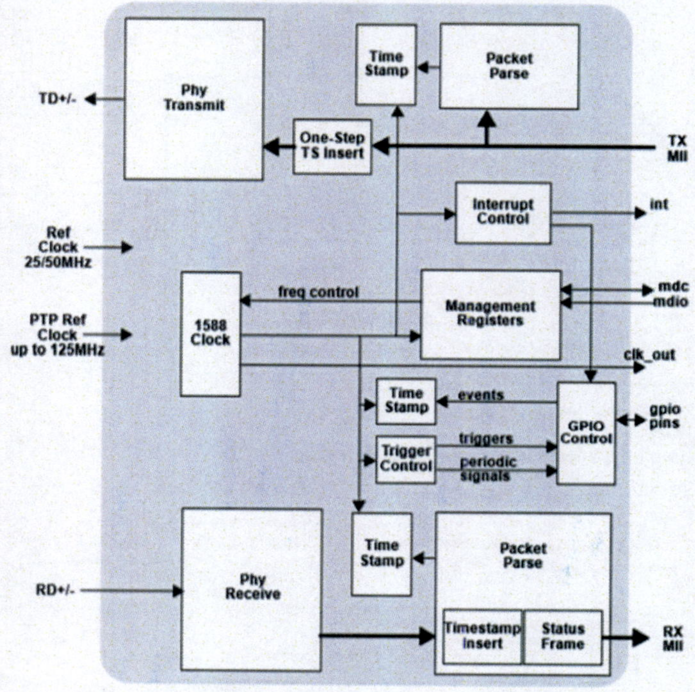

Fig. 3 Ethernet physical with IEEE 1588 core

The protocol referred to as Precision Time Protocol (PTP) supports system-wide synchronization accuracy in the sub-microsecond range with minimal network and local clock computing resources. The default behavior of the protocol allows simple systems to be installed and operated without requiring the administrative attention of users [3, 10–12]. The Fig. 3 gives the IEEE 1588 digital core implementation of Ethernet physical.

The implementation of various units of the IEEE 1588 is discussed in further sections.

4 Transmit Packet Parser and Timestamp Unit

The transmit packet parser and timestamp unit monitors the transmit packet to detect IEEEz1588zversion 1 and versionz2 event messages. Upon detection of a PTP event message, the transmit timestamp is captured and is provided to the software through the PTP Transmit Timestamp Register. The software knows the order of the packet transmission. Therefore only the timestamp is required. The device will indicate the availability of the timestamp by setting a certain bit in PTP Status and Control register. The transmit timestamp operations are controlled through PTP Transmit Configuration registers [13]. The timestamp unit can be programmed to detect messages encoded in the form of three main types of packets:

1. UDP/IpV4
2. UDP/IpV6
3. Layer 2 Ethernet

The transmitter unit can be programmed to operate in one step mode. In this mode for Sync event messages; the transmitter unit can automatically insert the timestamp information in the outgoing packet. This eliminates the need of a separate Follow_up packet for transmitting timestamp information [4].

5 Receive Packet Parser and Timestamp Unit

This unit on detection of a PTP event message, will capture the receive timestamp and provide the receive timestamp value to the software via PTP Receive Timestamp register. In addition to the timestamp the unit will capture the 16zbit sequence id, 4xbit message type field. The device will also indicate that the timestamp is available by setting a certain bit in PTP status register.

The IEEE 1588 receive timestamp unit can provide timestamp value to the software by inserting it in the receive packet. This gives a simple method to deliver the packet to the software without being concerned about how to match the timestamp to the correct packet. This eliminates the need to provide the software with sequence id, message type and source information [4].

6 IEEE 1588 Triggers

The device can be programmed to generate a trigger signal on an output pin based on IEEE 1588 clock value. The trigger can be programmed to generate a onetime rising or falling edge or a single pulse of programmable width or a periodic signal. The device supports up to 8 trigger signals which can be output on any of the available GPIO signal pins. The triggers can be configured via PTP Trigger Configuration registers. There are various configuration options available for the triggers. They are mentioned below:

1. **Single/Periodic Control:** Indicates whether a single edge/pulse or a periodic signal will be generated.
2. **Pulse/Edge Control:** Indicates whether the signal is a pulse or an edge.
3. **Trigger if Late Control:** allows immediate trigger if the start time is earlier than the current clock time.
4. **Notify:** Generate status on trigger completion or trigger error.
5. **Toggle mode:** Toggle from current state.
6. **GPIO Pin Select:** Indicates which I/O pin is used for output signal.

7 Support for 1 Gigabits Data Rate

The paper has described the design of Ethernet physical device based on the DP83640 device from Texas Instruments Inc. The device supports data rates of 10 and 100 Megabits. Here an attempt is made to further up the data rate to 1 Gigabits. The data is transmitted between the physical and mac layers by Media Independent Interface/Reduced Media Independent Interface. This interface has a number of pins or control signals for transmit and receive process [14–17]. For transmission we have:

- A nibble wide data bus, TXD [3:0].
- Transmit enable control signal, TX_EN.
- Optional transmit error control, TX_ER.
- A transmit clock, TX_CLK.

Similarly for receiving we have:

- A nibble wide data bus, RXD [3:0].
- Receive data valid flag, RX_DV.
- Receive error control, RX_ER.
- A receive clock, TX_CLK.

Apart from all these there are two special pins available across the MII/RMII interface:

- COL: Collision Detect.
- CRS: Carrier Sense.

The existing physical devices support 10 and 100 Megabits data rates over the MII/RMII interface (the difference between the two interfaces is MII operates over 4 bit data bus and RMII operates over 2 bit data bus). For supporting 10 Megabits the 4 bit data bus, along with the operating frequency of 2.5 MHz is utilized. Similarly for 100 Megabits the 4 bit data bus, at operating frequency of 25 MHz was utilized. Now for design for 1 Gigabits data rate, we proposed an 8 bit data bus over the MII/RMII interface which when operated at 125 MHz will give a data rate of 1 Gigabits. But the challenge here was ensuring we have all available data rates, i.e. 10 and 100 Megabits also. For this purpose we introduced a control signal **'Bus_Select'.** This control signal switches the data bus between 4 and 8 bits according to our data rate requirement. The operating clock frequency required for the given data rate (2.5, 25, 125 MHz) can be output on the TX_CLK and RX_CLK of the MII/RMII interface.

The Fig. 4 describes the Ethernet physical device's IEEE 1588 digital core implementation for 1 Gigabits data rate.

The Fig. 4 clearly describes the various modules of the IEEE 1588 digital core, some of which are already discussed. For the purpose of switching the data bus between 4 and 8 bits we are using Bus_Select signal. This signal is fed to as an input to 8 And logic gates, 4 for the receiving bus and 4 for the transmission bus. When the Bus_Select signal is low the And gates block the upper 4 bits (Most Significant Bits) and allow the lower 4 bits of the 8 bit data bus. When the signal is high it allows

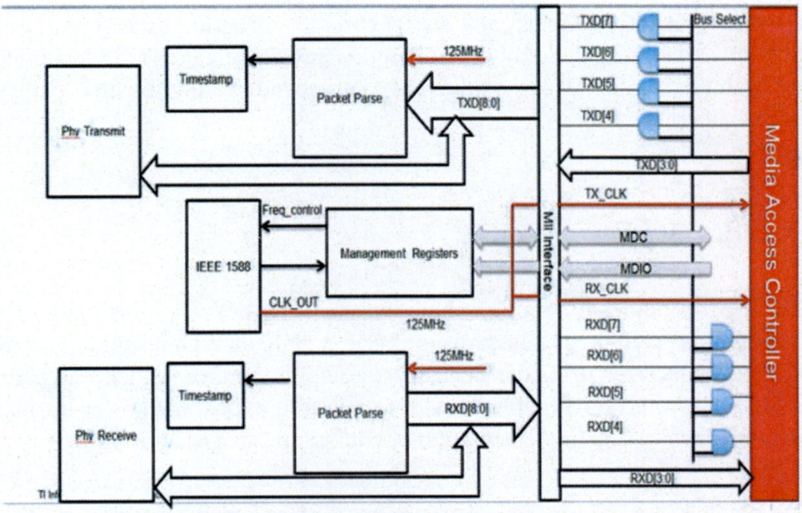

Fig. 4 1 Gigabits IEEE 1588 Digital Core

all the 8 bits of the data bus over the MII/RMII interface to facilitate 1 Gigabit data transfer.

Apart from this modification all other modules needed changes to accommodate for the 1 byte long data bus. The transmit and receive packet parser and timestamp units have the ability to detect the frames by means of Start of Frame Detect (SFD). It is in the beginning of the frame along with the preamble. After this SFD only the 16 bit data word is transmitted. For the Ethernet protocol the SFD sequence is fixed at **01011101**. Also in these frames in order to detect for the errors in transmission we make use of 32 bit Cyclic Redundancy Check bits (CRC). For Ethernet there is a fixed generator polynomial:

$$\mathbf{x}^{32} + \mathbf{x}^{26} + \mathbf{x}^{23} + \mathbf{x}^{22} + \mathbf{x}^{16} + \mathbf{x}^{12} + \mathbf{x}^{11} + \mathbf{x}^{10} + \mathbf{x}^{8} + \mathbf{x}^{7} + \mathbf{x}^{5} + \mathbf{x}^{4} + \mathbf{x}^{2} + \mathbf{x}^{1}$$

This generator polynomial is used for generating the code word which is appended with the transmitted frame. The code word is checked again for the received frame. If there is a mismatch it will indicate there is an error in the received signal.

The mechanisms for both of these functionalities were initially designed to support nibbles of data. In order to accommodate for 8 bit data bus for 1 Gigabit data rate changes have to be made. The SFD checker mechanism which initially used to check for SFD sequence in 2 nibbles first 1101 and then 0101, could now check the entire 1 byte long SFD sequence 01011101 in one go. Also for the CRC calculation mechanism which earlier had the provision for the code word calculation using the generator polynomial using a nibble of data could now calculate it using a byte of data [18–21].

All other units of the IEEE 1588 digital core, like event timestamp unit, trigger control unit and interrupt control doesn't require any modifications to accommodate for 1 Gigabit data rate. All these other units can operate at any required frequency as per the data rate.

8 Results and Discussions

The describes modules for configurable registers using both flip flop and D- Latch implementation were programmed using Verilog Hardware Description Language (HDL) and synthesized using the Genus synthesis solution from cadence to perform power and area analysis. Figures 5 and 6 represent the area and power estimation respectively for flip flop implementation of configurable registers.

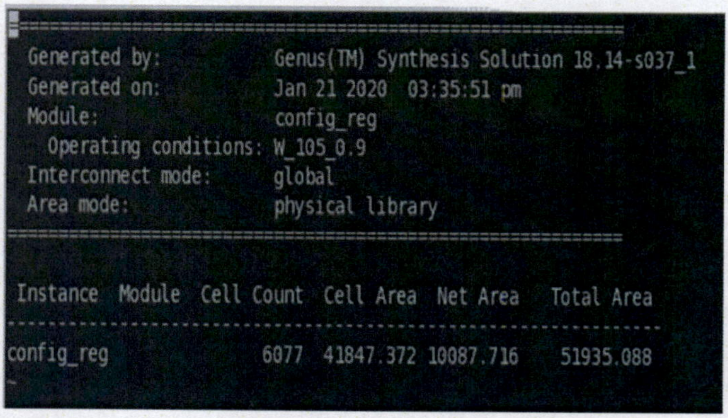

```
Generated by:            Genus(TM) Synthesis Solution 18.14-s037_1
Generated on:            Jan 21 2020  03:35:51 pm
Module:                  config_reg
  Operating conditions:  W_105_0.9
Interconnect mode:       global
Area mode:               physical library

Instance  Module  Cell Count  Cell Area  Net Area   Total Area
--------------------------------------------------------------
config_reg              6077  41847.372 10087.716    51935.088
```

Fig. 5 Area in micro meter sq.

```
Generated by:            Genus(TM) Synthesis Solution 18.14-s037_1
Generated on:            Jan 21 2020  03:36:00 pm
Module:                  config_reg
  Operating conditions:  W_105_0.9
Interconnect mode:       global
Area mode:               physical library

             Library        Leakage    Dynamic     Total
Instance     Domain Cells   Power(nW)  Power(nW)   Power(nW)
-----------------------------------------------------------
config_reg      0     6077 335122.419 604924.001 940046.420
```

Fig. 6 Power in nano watts

Figures 7 and 8 represent the area and power estimation respectively for latch implementation of configurable registers.

From the obtained results it is quite evident that the latch implementation is more area efficient but at the same time less power efficient than the flip flop implementation. This comparative analysis can be taken into consideration while selecting the implementation type based on the application. The synthesis results of the latch and flip flop implementations are tabulated in Table 1.

Fig. 7 Area in micro meter sq.

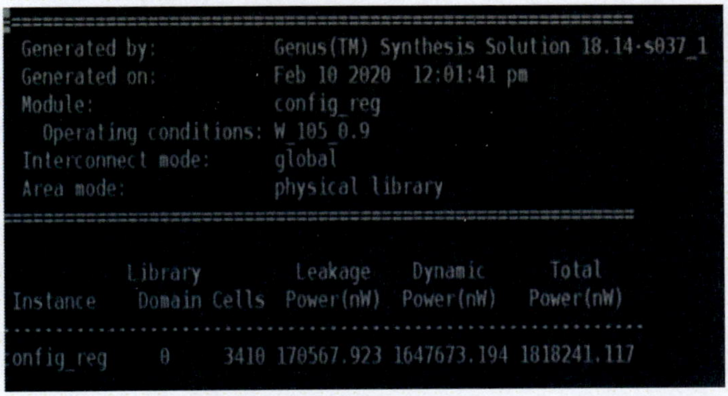

Fig. 8 Power in nano watts

Table 1 Synthesis results comparison

Sl no.	Implementation	Area (micro meter sq.)	Power (Nano watts)
1.	Flip flops	51935.088	940046.420
2.	Latches	21926.602	1818241.117

```
============================================================
  Generated by:            Genus(TM) Synthesis Solution 18.14-s037_1
  Generated on:            Mar 26 2020  11:40:33 am
  Module:                  ptp_top
    Operating conditions:  W_105_0.9
  Interconnect mode:       global
  Area mode:               physical library
============================================================

     Instance           Module      Cell Count  Cell Area  Net Area   Total Area
--------------------------------------------------------------------------------
ptp_top                                  23018  91147.644 34550.498   125698.142
  ptp_status_pkt_1 ptp_status_pkt         2702  13390.524  3847.060    17237.584
  event_tsu_1      ptp_event_tsu          2513  12416.796  3305.446    15722.242
  ptp_registers_1  ptp_registers          1687   7303.666  1860.163     9163.828
  tx_parse_1       ptp_tx_parse           1217   4151.398  1704.019     5855.416
  ptp_clock_gen_1  ptp_clock_gen           136    717.948   121.469      839.417
```

Fig. 9 Area in micro meter sq.

```
============================================================
  Generated by:            Genus(TM) Synthesis Solution 18.14-s037_1
  Generated on:            Mar 26 2020  11:40:49 am
  Module:                  ptp_top
    Operating conditions:  W_105_0.9
  Interconnect mode:       global
  Area mode:               physical library
============================================================

           Library         Leakage     Dynamic      Total
Instance   Domain  Cells  Power(nW)   Power(nW)   Power(nW)
-------------------------------------------------------------
ptp_top       0    23018 839344.299 3122761.062 3962105.361
```

Fig. 10 Power in nano watts

Similar to this various units of the IEEE 1588 core (Transmit packet parser and Timestamp, Receive Packet Parser and Timestamp and Trigger control unit) were programmed using Verilog to be implemented for Ethernet physical device. The core was synthesized to obtain power and area estimates. Figures 9 and 10 represent the area and power estimation respectively for IEEE 1588 core.

9 Conclusion

The emergence of Ethernet protocol in automotive applications is undeniable. It provides with higher data rates and bandwidths for time critical operations. Therefore there is a need for optimizing the Ethernet devices. Here we discussed about the Ethernet device at the physical layer. The comparison between latch and flip flop implementation of configuration registers of the Ethernet physical device was performed. It was observed that the latch implementation is more area efficient (21926.602 micro meter square) and the flip flop implementation is more power efficient (940046.420 nano watts). So depending on the application the type of implementation can be selected. The automotive environments involve multiple Electronic Control Units (ECU's) which deploy multiple Ethernet physical devices. Therefore there is a need of synchronizing the clocks in these devices by implementing IEEEz1588 protocol. The various units of thexIEEEz1588 core were programmed using Verilog and synthesized using Genus synthesis solution for power and area estimation.

References

1. Reyes C Anatomy of Ethernet Physical Layer Transceivers. Texas Instruments – Precision Labs (Ethernet)
2. Hank P (2013) Automotive ethernet: in-vehicle networking and smart mobility. In: Design, Automation and Test in Europe conference & Exhibition (DATE)
3. Technical committee on sensor technology (2019) IEEE Standard for a Precision Clock Synchronization Protocol for Networked Measurement and Control Systems, IEEE Instrumentation and Measurement Society
4. National Semiconductors (2008) AspenPhy Digital core: 10/100 Ethernet Trans receiver Design Development Specification
5. Liu Z (2016) A universal method for implementing IEEE 1588 with the 1000 M Ethernet Interface. In: IEEE Autotestcon
6. Erikkson H (2001) VLSI implementation of CRC-32 for 10 gigabit ethernet. In: The 8th IEEE International Conference on Electronics, Circuits and Systems
7. Diaz JC (2004) Clock management in a gigabit ethernet physical layer transceiver circuit. In: Design, Automation and Test in Europe conference & Exhibition (DATE)
8. Neuhaus R (2005) A beginner's guide to ethernet 802.3. In: Engineer-to-Engineer Note EE-269
9. Lajmi H, Alimi AM (2013) Using ethernet technology for in-vehicle's network analysis. In: Fifth International Conference on Computational Intelligence, Communication Systems and Networks
10. Rahmani M, Pfannenstein M, Steinbach E, Giordano G, Biersack E (2009) Wireless media streaming over IP-based in-vehicle networks. In: IEEE International Conference on Communications Workshops 2009, pp 1–6
11. Bogenberger R, Rahmani M, Hillebrand J, Hintermaier W, Winckler A (2008) Design and realization of an IP-based In-car network architecture. In: The First Annual International Symposium on Vehicular Computing Systems, Dublin, Ireland
12. Zeng H, Di Natale M, Ghosal A, Sangiovanni-Vincentelli A (2011) Schedule optimization of time-triggered systems communicating over the Flexray static segment. IEEE Trans Ind Informat 7(1):1–17
13. Wolfgang B (2012) MOST in driver assistance. In: The Special Issue for the 4th MOST Forum
14. Road Vehicles—Interchange of Digital Information—Controller Area Network for High-Speed Communication (1994) ISO 11 898
15. Zinner H, Noebauer J, Oallner T, Seitz J, Waas T (2011) Application and realization of gateways between conventional automotive and IP/Ethernet based networks. In: Proceedings ACM/EDAC/IEEE. DAC 2011, pp 1 – 6
16. Lajmi H, Alimi AM (2013) Optimized packet processing architecture for ethernet trafic monitoring and analysis in the automotive field. International Journal of Vehicle Information and Communication Systems (IJVICS) (in press)

17. Diaz JC, Parodi G (2002) Static timing analysis of the everest MA1110 gigabit PHY transceiver. In: Proceedings XVII DCIS Conference, pp 513–516
18. The Institute of Electrical and Electronics Engineers (October 2000) IEEE 802.3 Standard, 2000 Edition
19. Skeie T, Johannessen S, Holmeide O (2001) Highly accurate time synchronization over switched ethernet. In: Proceedings of 2001 8th IEEE International Conference on Emerging Technologies and Factory Automation, pp 195–204
20. Lv S, Lu Y, Ji Y (2010) An enhanced IEEE 1588 time synchronization for asymmetric communication link in packet transport network. IEEE Commun Lett 14:764–766
21. Burch J, Green K, Nakulski J, Vook D (2009) Verifying the performance of transparent clocks in PTP systems. In: 2009 International Symposium on Precision Clock Synchronization for Measurement, Control and Communication, pp 1–6

Design and Implementation of a Small Hand Held Camera Detector on FPGA

N. Shylashree and H. M. Supritha

Abstract Nowadays, abuse of camera is high due to the rapid development technology. In a video surveillance system, wireless camera is commonly used. Even though, privacy issues related to illegal videos have recently increased. Due to accuracy limitation, conventional technique is not recommended for practical real-time implementations. In this paper, we propose A Small Hand Held Hidden Camera detector to detect the Hidden cameras in public places like trial rooms, hotels, and restrooms. This device uses a camera to scan the areas to capture the lights emitted by the components of the hidden camera that maybe the LED indicator or maybe the IR light emitted by the camera lenses. In this project, an adaptive thresholding method-based segmentation algorithm will segment the abnormal portion of the image and send it to the monitor via a Video Graphics Array (VGA) port. A hidden camera detector prototype is implemented using a Field Programmable Gate Array (FPGA) and Raspberry Pi Board. Around 98% of classification accuracy is obtained using proposed method which is high when compared to the existing techniques.

Keywords Hidden camera detection · Infrared camera · Field Programmable Gate Array (FPGA) · Video Graphics Array · Hand-Held device · Embedded system · Convolutional neural network · Portable hidden camera detector

1 Introduction

With the rapid increase in the number of wireless video cameras, the financial rate of attacking confidentiality using spy cameras has dropped significantly, causing serious concern. In particular, Due to portable size, wireless camera can be simply

N. Shylashree (✉)
Department of Electronics and Communication Engineering, RV College of Engineering, Bengaluru 560059, Karnataka, India
e-mail: shylashreen@rvce.edu.in

H. M. Supritha
VLSI Design and Embedded Systems, RV College of Engineering, Bengaluru 560059, Karnataka, India

V. Nath and J. K. Mandal (eds.), *Proceeding of Fifth International Conference on Microelectronics, Computing and Communication Systems*, Lecture Notes in Electrical Engineering 748,
https://doi.org/10.1007/978-981-16-0275-7_41

installed wherever in a personal location or on everyday objects such as smoke detectors, USB chargers, watches or sockets, without having to rely on cables to connect to a network [1]. Nowadays, hidden cameras are extensively available in the market. Despite growing concerns about confidentiality, the discovery of wireless spy cameras is currently an open source of serious problems [2].

Masked camera identification is important, still a consistent clarification has not yet been developed. Several applications for android and apple mobile applications can be found in google and apple stores [3]. These applications use any one of the following techniques: based on reflected light, detects small the camera surface or EMI technique [4]. Also, a small device which will detects the leakage of RF signal in the surrounding area for detect the RF based hidden camera [5].

In this work, a prototype for hidden camera detection is proposed to detect both infrared cam and general can in a simple device. The infrared-based camera detection process is implemented in the FPGA environment. Mobile camera detection is performed in the raspberry pi board using a machine learning algorithm such as a convolutional neural network. This system can be implemented in a small area such as the shape of a handheld device. Due to the usage of high-speed FPGA and embedded processors, this can be work in a real-time environment without any delay.

The organization of this paper follows: In Sect. 2, the general information about the background and requirements of the hidden camera detection process. Section 3 explains the proposed using FPGA and embedded environment with more detailed block-level and circuit-level schematics. Section 4 discusses the experimental results to demonstrate the performance and characteristics. In the 5th section, conclusion and future work are explained.

2 Literature Survey

Various techniques are proposed previously to detect the hidden can in a real-time environment. In this section, some of the previous works which are performed to detect the hidden camera are explained. Most of the design is concentrating infrared cameras because it can be detected easily using normal cameras. Similarly, streaming cameras can be detected by using various RLC circuits based on the interference which is generated by the cameras. A detailed review of some of the work which is done previously to perform the hidden camera detection is explained below:

And Kevin et al. Surveillance has been introduced to identify cameras that broadcast user videos [6]. This can be achieved by using a well-known camera and recording medium in a situation such as mobile phone camera. Meanwhile, the network interface enters monitoring mode, recording nearby data communications and registering the number of bytes each wireless device transmits at each stage. He applies measurements of similarity between a known record of data synchronization and each network device.

And Sindhu et al. Offer a portable hidden camera and interference detector to find a spy camera [7]. The proposed device has an RF signal detector and silencer. At that moment, when the RF transmission signal detector is detected from the activated camera, it starts warning with the LED indicator on the device.The LED is on until a signal is transmitted. The broadcast frequency of a mobile phone's camera or camera ranges from 0.05 to 3 GHz. The high frequency GHz detection circuit is used to detect the camera.

T Liu et al. Specific practical systems, called Flickr and Flickr, allow you to exactly recognize packet streams of hidden cameras [8]. Blink is an android application that can be organized on ready-made mobile devices to detect hidden cameras in live mode. The inexpensive and portable circuit flicker used by LEDs is used daily to stimulate wireless spy cameras with a hidden light flicker. Due to the lack of local storage, most miniature and implanted wireless spy cameras transfer images as soon as they are taken, so they respond to light stimuli in real time.

Vaishali et al. analyzes the magnetic activity in the proposed area, and if it is similar to the camera, the spy issues an alarm for further investigation to gain access to the camera [9]. This mobile application can detect the hidden camera in trial room, private hotel room, etc. This application can identify infrared rays. It similarly helps you to easily share your location with friends.

Cheng, Yushi, Xiaoyu Ji et al. implemented a spy cam detector to find the hidden cameras present around us [10]. This app is a tool to help you find a hidden camera lens that you suspect is in the room. Similar to mobile phones with a camera detector, it tries to detect cameras using the magnetic sensor of an Android device. If magnetic activity seems similar to that of a camera in the room, our app will beep and raise alarm, so that you can further investigate.

3 Proposed Method

The proposed work mainly concentrates to implement both IR based cameras and normal cameras in the same device. This need both FPGA and embedded devices to implement IR and general camera detection respectively. IR camera detection using FPGA improve performance in terms of computational time. Similarly, a trained CNN based vision model is developed to detect the general type of cameras in the raspberry pi environment. Both systems can be integrated in a single device to perform the detection process. The detailed information about the proposed hidden camera detection is explained in the following section.

The Block diagram for the hidden camera detection technique process which is proposed in this work can be shown in Fig. 1. Here we have used both FPGA and embedded processor to perform the camera detection process separately. In FPGA, the thresholding and segmentation process is performed to detect the IR based cameras. Also, in the embedded side, a trained CNN is used to detect the camera as shown in the figure. Detailed information about the separate modules is shown in the following sections.

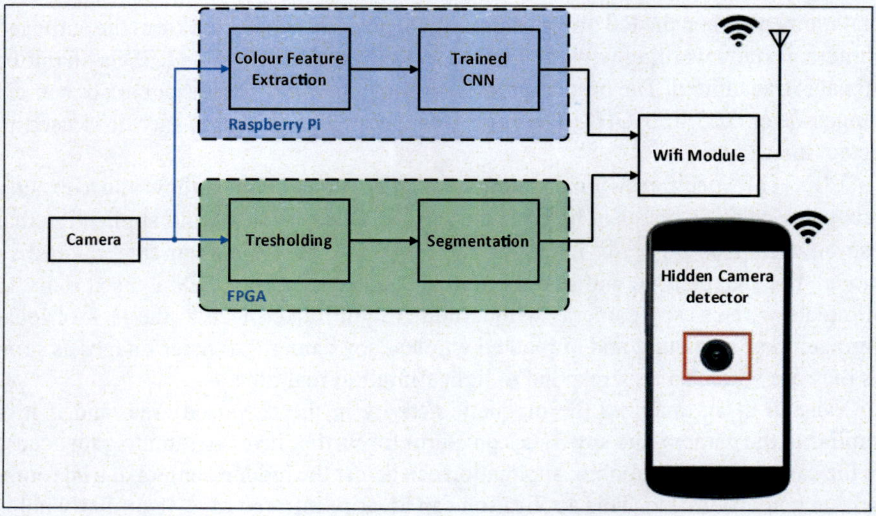

Fig. 1 Block diagram for proposed hidden camera detection technique

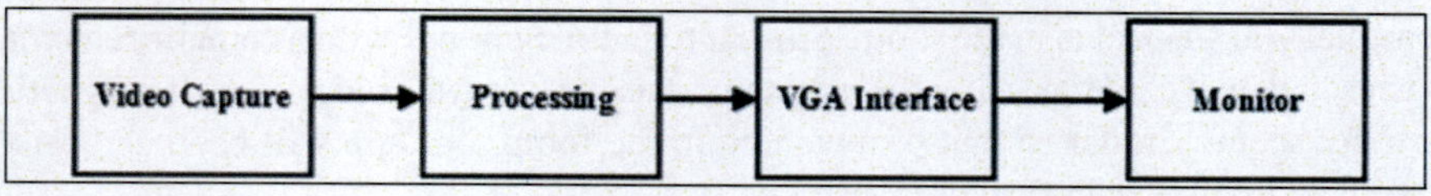

Fig. 2 FPGA based hidden camera detection

3.1 FPGA Based Detection Process

This device uses a camera to scan the areas to capture the lights emitted by the components of the hidden camera that maybe the LED indicator or maybe the IR light emitted by the camera lenses. In this project the segmentation algorithm will segment the abnormal portion of the image and send it to the monitor via Video Graphics Array (VGA) port Hidden camera detector is implemented using FPGA [11, 12]. Figure 2 shows the FPGA based hidden camera detection method with block diagram.

a) Thresholding
Thresholding is performed to improve the quality of the detection process [13]. In this stage, the pixels are converted into the unsigned 8-bit format and finalized. The threshold value will be selected based on the pixel information. Generally,the average of the pixel matrix will be considered as the threshold value [14]. Since the input images can be taken at different times of the day, a certain limit may not work for different lighting conditions. For this reason, an adaptive threshold determination method based on discriminant analysis is included, where the image is divided into two classes which can be categorized

as objects and background, respectively. The optimal range is obtained by maximizing the difference between classes. The threshold image is inverted to show abnormal areas in the form of white blocks on a black background. That image is displayed on the monitor via VGA port. If any abnormal region is found, then the segmentation algorithm will highlight the portion.

b) The Canny Edge Detection technique.
In general, purpose of edge detection is to expressively reduce the volume of information in the image while retaining the fundamental features used for additional image processing [15]. There are a number of algorithms, this worksheet is by John F. Wright. It focuses on the proposed algorithm developed by Kani. Although very old, it has converted one of the standard techniques of finding the edge and is still used in most applications [16].

The algorithm works in 5 different steps:

- Smoothing: Smooth the pixel based on 3×3 size window.
- Finding gradients: Estimate the gradient of the image by calculating horizontal and vertical differences
- Non-maximum suppression: To calculate the edge based on the local maximum intensity.
- Double thresholding: Based on the threshold, the final decision will be taken for decision

Generally, images will be affected by noise while capturing due to sensitivity in circuit and environmental effect. The sound should be reduced so as not to confuse the edges of the sound. The cyan component will be extracted to perform the gaussian filtering to remove the noise present. The following equation is used to perform the filtering using 5×5 mask.

$$B = \frac{1}{159} \cdot \begin{bmatrix} 2 & 4 & 5 & 4 & 2 \\ 4 & 9 & 12 & 9 & 4 \\ 5 & 12 & 15 & 12 & 5 \\ 4 & 9 & 12 & 9 & 4 \\ 2 & 4 & 5 & 4 & 2 \end{bmatrix} \tag{1}$$

Caney's algorithm traces the most changing edges of the grayscale image intensity. These regions are found by defining image gradients. The first step is to approximate the gradient in the x and y directions, respectively, using the kernels shown in Eqs. (2) and (3). The gradients at respectively pixel of the smooth image are determined by the so-called canny operator.

$$H_x = \begin{bmatrix} -1 & 0 & 1 \\ -2 & 0 & 2 \\ -1 & 0 & 1 \end{bmatrix} \tag{2}$$

$$H_y = \begin{bmatrix} 1 & 2 & 1 \\ 2 & 0 & 2 \\ -1 & -2 & -1 \end{bmatrix} \tag{3}$$

Gradient values (also known as boundary values) can be defined as the Euclidean distance measure using the Pythagorean law, can be shown in Eq. (3). To minimize complexity, Manhattan distance measurement technique is used based on the Eq. (4).

$$\begin{aligned} |G| &= \surd\left(G_x^2 + G_y^2\right) \\ |G| &= \left|G_x\right| + \left|G_y\right| \end{aligned} \tag{4}$$

Gradients of x and y directions are represented as G_x and G_y respectively shown in Eq. (5). It is clear from Fig. 3 that the gradient values often indicate the edges. The edges are usually wide, so they do not indicate accurately where the edges are. In order to be able to define an edge, the route of the edges must be defined and preserved.

$$\theta = \arctan\left(\frac{|G_y|}{|G_x|}\right) \tag{5}$$

The edges can be used further processed to perform the camera location in the video frames.

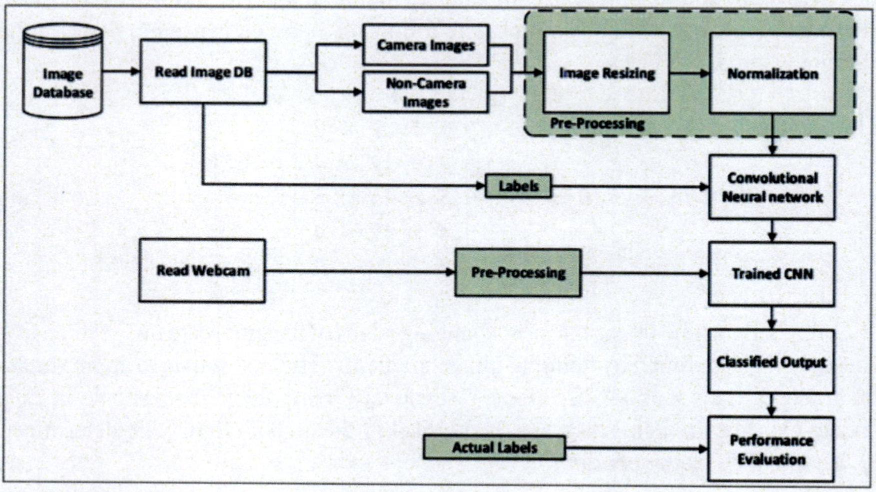

Fig. 3 CNN based hidden camera detection for Raspberry Pi

3.2 *CNN Based Hidden Camera Detection*

By using IR based hidden camera detection, only the infrared cameras can be detected. Also, this process should be performed in a dark environment. The usage of IR cameras very less nowadays due to lower quality and complex management. CNN based hidden cam detection and detect both IR and general type of camera in the complex color environment. Here pixel intensity value is considered as a color feature for both training and testing processes for the convolutional neural network. Figure 3 represents the overall block diagram for the CNN based hidden camera detection method.

Convolutional Neural Network

Conventional neural network is a category of deep neural networks that is often used to examine graphic images [17]. Based on their architecture with common weights and non-translational characteristics, they are also called immutable or spatially invariant artificial neural networks (SIANNs). CNN is most commonly used for classification related applications in image processing signal processing communication engineering etc. [18].

Design

The CNN having various layers such as input layer, output layer, hidden layers etc. Usually, the activation function is an RELU layer, followed by additional convolutions such as combining layers, fully connected levels, and normalization levels called hidden levels, because their inputs and p outputs hide the activation function and the final convention [19]. CNN's hidden layers usually contain an array of convolutional layers that are folded using multiplication or another point product [20]. The final evolution involves back propagation to more accurately determine the weight of the final product. Some of the layers in the CNN is given below:

Convolutional
ReLU layer
Loss layer
Pooling
Fully connected layer
Pooling layer

Weights

Individually a single neuron of a neural network estimates the production value by relating a particular function which is coming to the approachable field at the previous level. The function that applies to the input values is determined by the weight vector and the offset. A distinctive feature of CNN is that many neurons can use the same filter. This reduces the occupied memory, because one offset and one vector of weights are used in all accessible fields sharing this filter, in contrast to each receptive field that has its own offset and weight. Figure 4 shows the CNN structure. The study of the neural network is in progress by making repetitive variation the

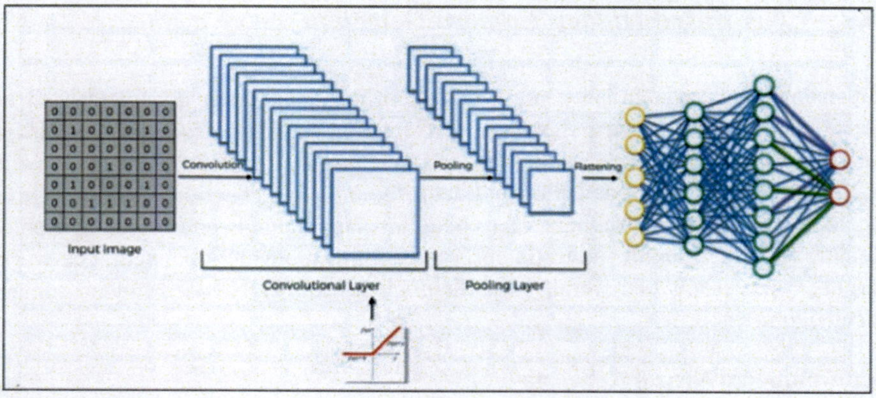

Fig. 4 CNN architecture

weights. The masses and the offset are called filters and represent specific features of the input data.

4 Results and Discussion

The proposed prototype is implemented using FPGA and Raspberry pi board. In this work, the database is created by using a camera to train and test the CNN. CNN algorithm is implemented using the python language and simulated in Google colab. Also, FPGA based IR camera detection process is implemented using Verilog programming language in Xilinx software. Performances of the proposed technique are measured in terms of various quality evaluation techniques and computation time.

Figure 5 shows some of the camera images which are used to perform the CNN training. These sting and validation process. Images are captured by using a normal camera to perform testing. Images are converted into128 $\times$ 128 size to reduce the computational time and complexity. CNN is trained by giving labels 0 and 1 for camera and no-camera images. Similarly, no-camera images are captured for CNN training. These images also reduced to 128 $\times$ 128 format to reduce the computational complexity of the system. These images can be shown in Fig. 6. Figure 7 shows the testing results for the CNN based hidden camera detection technique. The first frame has a mobile with a hidden camera. That can be visible in a small spot format. Here, a red color indication showing there is a hidden camera present on that frame.The second image section not having any hidden camera and there is no indication.

Fig. 5 Non-camera images used for CNN training

Figure 8 shows the segmentation input and output of FPGA based IR hidden camera detection process. Here, thresholding and segmentation are performed to detect the hidden cam. In the first image, the white spot shows the real IR hidden cam in the dark environment. The second image shows the segmented output concerning the input. Table 1 shows the Performance for Hidden camera detection with conventional techniques.

Fig. 6 Camera images used for CNN training

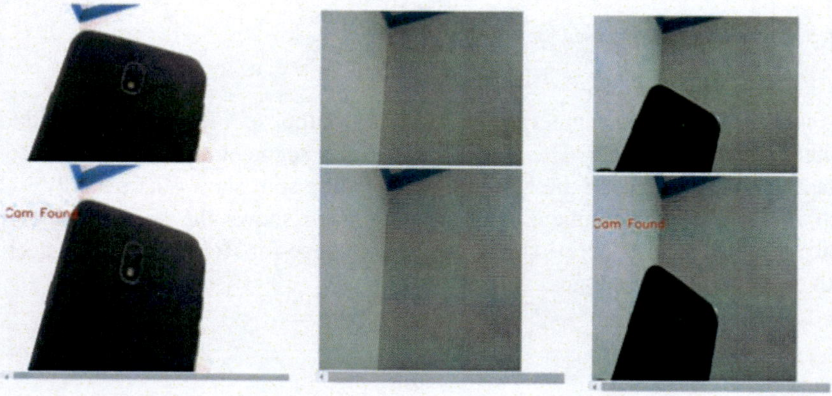

Fig. 7 Testing results for camera and Non-camera images

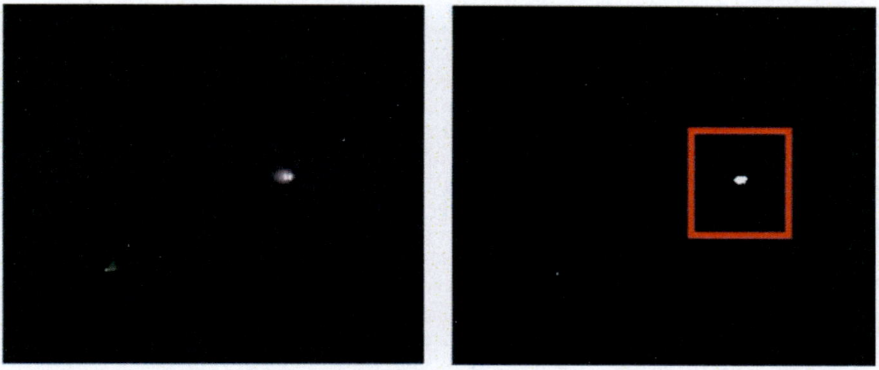

Fig. 8 Testing results for camera and Non-camera images

Table 1. Performance for Hidden camera detection

	Accuracy	Precision
Proposed method	98.2	95.6
[6]	96.55	84.384
[1]	87.304	83.611
[10]	96.7	94.7

5 Conclusion

In this work, a prototype model for hidden camera detection is implemented and tested. FPGA based infrared camera detection is implanted to detect the IR type hidden camera. Also, the general type of hidden camera detection is implemented by using CNN by performing separate training and testing in the Raspberry pi board using the python programming language. To perform the detection process, the database is separately collected by capturing a sample camera and non-camera images. The size of the image is reduced to 128×128 to reduce the computational complexity and hardware usage. Proposed method increases the accuracy by 3% and precision around 11.3% it is best when compared to the conventional techniques.

References

1. Wu K, Lagesse B (2020) Detecting hidden webcams with the delay-tolerant similarity of simultaneous observation. Pervasive Mob Comput 65:101154
2. Gayathri N, Sivasakthi T (2016) Presence of active mobile phones and hidden camera detection. Int J Comput Commun Inform Syst India 8(1):01–07
3. Meghana S, Magrabi SAR Detection of hidden spy cams in the public/private areas using IoT based mobile application

4. Karthikeyan R, Sasirekha K (2018) Identification of hidden camera using mobile RF signal. Int J Pure Appl Math 119(16):1915–1918
5. Pragna BR et al (2018) Women safety devices and applications. Int J Eng Res Technol 7(07):2395–2456
6. Wu K, Lagesse B (2019) Do you see what i see? detecting hidden streaming cameras through similarity of simultaneous observation. In: 2019 IEEE International Conference on Pervasive Computing and Communications. (PerCom. IEEE)
7. Liu T et al (2018) Detecting wireless spy cameras via stimulating and probing. In: Proceedings of the 16th Annual International Conference on Mobile Systems, Applications, and Services
8. Sindhu K et al (2018) A women safety portable hidden camera detector and jammer. In: 2018 3rd International Conference on Communication and Electronics Systems (ICCES). IEEE
9. Bhuvaneshwari D, RibhuVats \Emissary camcordrevealer : Spy cam detector
10. Cheng Y, Ji X, Lu T, Xu W (2019) On detecting hidden wireless cameras: a traffic pattern-based approach. IEEE Trans Mob Comput 19(4):907–921
11. Gandhare S, Karthikeyan B (2019) Survey on FPGA architecture and recent applications. In: 2019 International Conference on Vision towards Emerging Trends in Communication and Networking (ViTECoN). IEEE
12. Rose J, El Gamal A, Sangiovanni-Vincentelli A (1993) Architecture of field-programmable gate arrays. Proc IEEE 81(7):1013–1029
13. Asari KV, Srikanthan T (2002) Segmenting endoscopic images using adaptive progressive thresholding: a hardware perspective. J Syst Architect 47(9): 759–761
14. Asari KV et al (1999) A pipelined architecture for image segmentation by adaptive progressive thresholding. Microprocess Microsyst 23(8–9): 493–499
15. Xu Q, Chakrabarti C, Karam LJ (2011) A distributed canny edge detector and its implementation on FPGA. In: 2011 Digital Signal Processing and Signal Processing Education Meeting (DSP/SPE). IEEE
16. Liu Z et al (2019) Implementation of a FPGA-ARM-based canny edge detection system. In: 2019 Chinese Control Conference (CCC). IEEE
17. Chauhan R, Ghanshala KK, Joshi RC (2018) Convolutional Neural Network (CNN) for image detection and recognition. In: 2018 First International Conference on Secure Cyber Computing and Communication (ICSCCC). IEEE
18. Albawi S, Mohammed TA, Al-Zawi S (2017) Understanding of a convolutional neural network. In: 2017 International Conference on Engineering and Technology (ICET). IEEE
19. Van den Oord A, Dieleman S, Schrauwen B (2013) Deep content-based music recommendation. In: Advances in Neural Information Processing Systems
20. Kölsch A et al (2017) Real-time document image classification using deep CNN and extreme learning machines. In: 2017 14th IAPR International Conference on Document Analysis and Recognition (ICDAR). vol 1. IEEE

DC-DC Converter for Energy Efficient DC Drive System

Y. Jeyashree, Y. Sukhi, and A. Sahaya Ponrekha

Abstract In case of converters used for power conversion, pulse width modulation is used. These converters use switches which operate in switched mode. The load current flows through the switches during its switching operation. Therefore switches are having high stresses and losses which are directly proportional to the frequency of operation. This also results in high value of dv/dt and di/dt due to EMI. In order overcome these disadvantages, LC circuits are used to provide zero current switching (ZCS) and/or zero viltage switching. This circuit overcomes the problems associated with high switching frequency. These soft switched converters have the advantages of better efficiency, compact; less weight, reduced stresses on components and low electromagnetic interference. The series-parallel resonant converters are the preferred topologies for dc-dc power conversion. This type of resonant converters is designed and simulated using resistive load to analyze their performance. Experimental setup has been done for series parallel resonant converter to prove its application to DC motor. Efficiency is calculated. Switching losses is calculated and comparison is made with the conventional system.

1 Introduction

Power processing can be done in Pulse width modulation and resonance circuit schemes [1–5]. During the converter operation using these circuits, the switches are operated under full load conditions. Due to this, the switching losses and stress on the switching devices increases. This varies linearly with switching frequency [6, 7].

Y. Jeyashree (✉) · A. Sahaya Ponrekha
EEE Department, SRMIST, Kattankulathur, Tamilnadu, India
e-mail: jeyashry@srmist.edu.in

A. Sahaya Ponrekha
e-mail: sahayapa@srmist.edu.in

Y. Sukhi
EEE Department, R.M.K. Engineering College, Kavaraipettai, Tamilnadu, India
e-mail: ysi.eee@rmkec.ac.in

V. Nath and J. K. Mandal (eds.), *Proceeding of Fifth International Conference on Microelectronics, Computing and Communication Systems*, Lecture Notes in Electrical Engineering 748,
https://doi.org/10.1007/978-981-16-0275-7_42

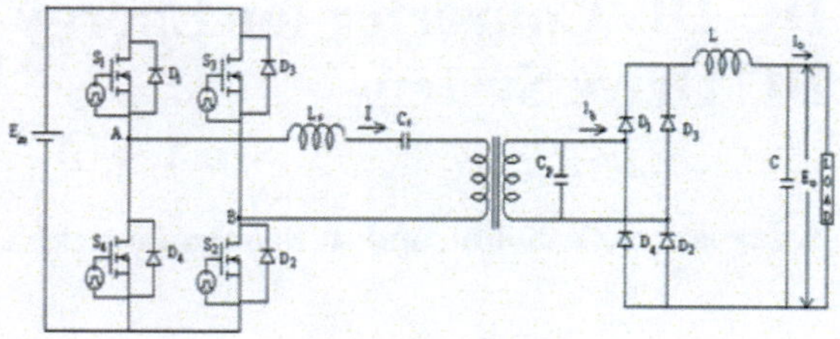

Fig. 1 Basic circuit diagram of proposed converter

This method of power processing also results in high value of di/dt and dv/dt due to electromagnetic interference [8–10]. In order to have stable operation with high efficiency for variable loads, resonant converter design plays an important role. The switching frequency is varied to obtain the required output voltage. However the losses increase all loading conditions due to the increase in switching frequency. The constant frequency operation is discussed in literature survey. In this method duty cycle is varied to control the voltage. Due to the LC resonant elements, zero switching losses can be obtained with resonant converters [13, 14].

2 Circuit Description

The functional diagram of the proposed DC-DC converter using resonant elements is shown in Fig. 1. Switches S_1–S_4 are devices having base turnoff capability. Diodes D_1–D_4 are connected across switches. This can also be the internal diode of MOSFET. If they have slow recovery characteristics, external fast recovery diodes have to be used. Bidirectional switch is formed by MOSFETs and its anti parallel diodes. Inductor and capacitor in the circuit form the resonant elements. In this circuit to make use of leakage inductance of high frequency transformer (HFT), the parallel resonant capacitor placed on the secondary of the transformer. The output voltage V_{AB} of High Frequency Bridge Inverter (HFBI) is applied to the tank circuit. The diodes used in high frequency bridge rectifier should have fast recovery characteristics. Load is connected across the filter output.

3 Operation of Circuit

As explained duty ratio 1 is used for mode 1 operation with the condition that impedance across AB is capacitive reactance. The inductor current *i(t)* lead the voltage applied to the tank thus converter operates in leading power factor mode. Figure 2 shows waveforms of this operating mode. Initially, S_1 & S_2 are made on at $t = 0$. The tank voltage becomes +Ein immediately. When the current through the resonant elements becomes zero at $t = t_2$, S_1 & S_2 are naturally commutated.

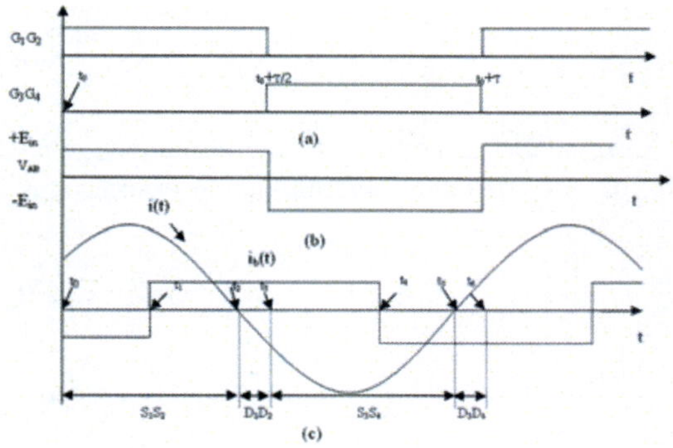

Fig. 2 Typical wave form in mode-1(a) gate signals (b) V_{AB} (t) (c) i(t) and i_b(t)

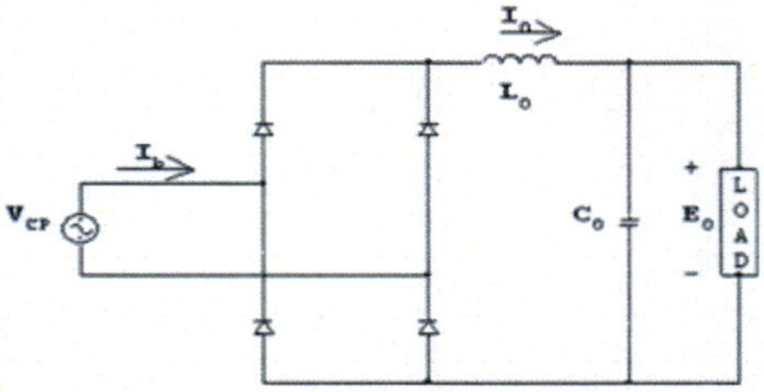

Fig. 3 Output circuit of the converter

Therefore, zero current turn off is realized. As the resonant current reversus at t = t_2, with the tank connected to $+E_{in}$, D_1 & D_2 turn on allowing a part of the tank energy to return to the source during the time interval t_2–t_3. At the instant t = t_3, switches S_3 & S_4 are turned on and V_{AB} changes from $+E_{in}$ to $-E_{in}$. At t = t_5 when *i(t)* becomes zero and begins to reverse forcing the diodes D_3 & D_4 to conduct. The interval t_2–t_3 & t_5 & t_6 are called regenerative interval (RI) and the intervals t_0–t_2 and t_3–t_5 are called power intervals. Thus in this mode there are two RI & two PI (Fig. 3).

As the switches are on at a finite current and at finite voltages, this results in turn on switching power loss. If the diodes have slow recovery characteristics, at t = t_0 large reverse current may flow through D_3 S_1 & D_4 S_2. In order to avoid these current and to minimize the diode turn off losses, the freewheeling diodes should have fast recovery characteristics. It is also evident that mode 1 requires lossy snubber & di/dt limiting inductors.

4 Mathematical Analysis Under Steady State

The D.C. voltage across capacitor is obtained as below,

$$E_0 = \frac{1}{\pi}\int_{0}^{\pi} \sqrt{2}V_{cp} \sin \omega_s t d(\omega t) \tag{1}$$

$$E_0 = \frac{2\sqrt{2}}{\pi} V_{cp} \tag{2}$$

$\omega_s = 2\pi f_s$ and f_s is the switching frequency.
Diode current is obtained using Fourier analysis as

$$I_b = \frac{2\sqrt{2}}{\pi} I_0 \tag{3}$$

Using Eq. 1 and Eq. 3, the resistance at input of the rectifier bridge is given by

$$R_{ac} = \frac{V_{cp}}{I_b} = \frac{\pi^2}{8} R_L \tag{4}$$

δ and D are related by

$$\delta = \pi D \tag{5}$$

The Voltage across A and B is given by

$$V_{AB} = \frac{2\sqrt{2}E_{in} \sin \delta/2}{\pi} \tag{6}$$

The ratio of output to input voltage is

$$\frac{E_0}{E_i} = \frac{\sin \delta/2}{\frac{\pi^2}{8}\left(\frac{m+1}{m}\right)(1 - y^2) + jQ\left(y - \frac{1}{(m+1)y}\right)} \tag{7}$$

where

$$m = C_s/C_p,\ Q = w_o L/R_L = 1/w_o C R_L,\ y = w/w_o$$

In the optimum design of the converter, the losses in the switches, losses due to magnetic circuit elements are also considered. Maximum efficiency occurs at a operating frequency close to resonant frequency. For reliable operation the normalized frequency *f/fr* should be in between 1.05 to 1.15. There are three important parameters to be considered for optimum design. They are efficiency, KV A rating of the converter per KW output and peak device; current. The efficiency should be maximum, the peak current and KVA/KW of the converter should be minimum. These three parameters do not occur at the same operating frequency. Therefore a compromise should be made for these three values to active at the choice of operating frequency.

The next important consideration is the losses of the circuit. The losses take place in the inverter section, the transformer and the rectifier-filter section. The inverter losses consist of switching losses during on and off, conduction losses including on-state loss and dynamic saturation losses of the switching devices such as BJT or MOSFETs, the base drive circuit losses, the losses of the inductor and the conduction losses of the anti parallel diodes. Let C_1, C_2, C_3, C_4 represents snubber capacitors across devices 1, 2, 3 and 4 respectively. The turn-on losses of the switches operating above resonant frequency is negligibly small because the switches are made on when the current through the anti parallel diodes reaches zero.

Losses occur in the switches during turn-off in this mode of operation. It is known that the turn-off period consists of the storage time and fall time. During the storage time the transistor is still in saturation and the voltage across it is the saturation voltage of the device. The loss during this period is the conduction loss and it should be included in it. However, the storage time is important in selecting the maximum operating frequency. The fall time commences after the storage period, and current begins to fall and voltage starts to rise.

Referring to Fig. 1, let the switches S_1 and S_4 be turned off and current through them begins to fall off, and current begins to flow through D_3 and D_4. The snubber capacitors play an important role in calculation of switching losses. Let the capacitor across D_3 and D_4 be charged fully to voltage equal to V_I before the commencement of the fall time and during the fall time the current in the devices S_1 and S_2 falls linearly to zero. That is

$$i_{S1} = i_{S2} = I\left(1 - \frac{t}{t_{off}}\right) \tag{8}$$

where I = link current before turn off

t_{off} = turn off time of Switches.

As soon as the current through S_1 and S_2 begins to fall, the capacitors C_3 and C_4 begin to discharge, and the discharge current through the snubber capacitors rises linearly to the switch current at the beginning of the fall time. Thus, it may be assumed that the total current through (C_3, C_4) and (S_1, S_2) is equal to the link current. The snubber capacitors at all the four devices are equal.

Since the capacitor current is assumed to rise linearly, the voltage would fall in a linear way. Thus, equation of capacitor voltage (C_3. C_4) may be written as

$$V_{C3} = V_{C2} = V_1 - \frac{1}{C}\int_0^{t_f} I\frac{t}{t_{off}}dt = V_1 - \frac{It^2}{2C_2 t_{off}} \tag{9}$$

The voltage across S_1 and S_2 are the difference of the input voltage and the capacitor voltages V_{C1} and V_{C2} respectively, hence voltage across S_1 and S_2 rises as

$$V_{S1} = V_{S2} = \frac{I_0 t^2}{2C_2 t_{off}} \tag{10}$$

The loss during turn off is

$$P_{S1(toff)} = f\int_0^{t_f} i_{Sj}\cdot V_{S1}dt = \frac{I_0^2 f t_{off}^2}{24C_2}\omega = P_{S2(toff)} \quad (11)$$

Similar amount of power take place in S_3 and S_4. Thus the total losses in the bridge circuit is

$$P_{T(toff)} = 4\cdot\frac{I_0^2 f t_{off}^2}{24C_2} \quad (12)$$

Thus

$$C_2 = \frac{I t_{off}}{2V_1} \quad (13)$$

Thus the total device turn off losses in the resonant bridge circuit is

$$P_{T(toff)} = \frac{4}{12} I_0 f t_{off} V_1 \quad (14)$$

For MOSFET's, conduction loss per device is

$$P_{con} = I_{rms}^2 R_{ds(on)} \quad (15)$$

P_T = total device loss

$$P_{in} = P_{out} + P_T \quad (16)$$

Efficiency of converter is expressed as

$$\%\eta = \frac{P_{out}}{P_{in}} \times 100 \quad (17)$$

5 Variation of Converter Gain (E_o/E_{in}) Versus Normalized Switching Frequency (y)

Switching frequency variation is plotted against voltage gain with quality factor Q as a varying parameter with the use of Eq. 7. The curve is plotted for quality factor varies between 1 and 6 and the value of capacitance ratio is taken as 1 as shown in Fig. 4. The converter gain increases up to a certain frequency ratio and then decreases. The maximum gain is obtained at y = 0. 707 for large values of Q. The maximum gain

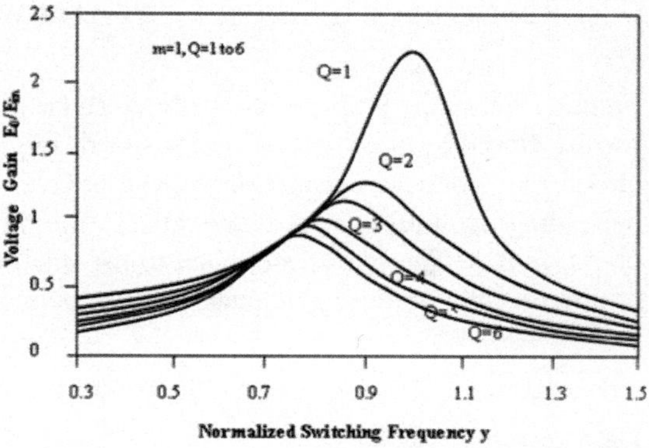

Fig. 4 Voltage gain versus normalized switching frequency

occurs at frequency greater than resonant frequency for lower values of Q. During loading conditions the resonant peak occurs at resonant frequency when y = 1.0.

Design of Converter

Applied voltage in the input side $E_{in} = 30$ volts
Load voltage $E_o = 24$ volts
Load Current = 1.5 Amps
Switching frequency $f_s = 50$ kHz
Capacitor ratio = $C_s/C_p = 1$,
Quality factor = 5,
Frequency ratio y = 1.1
Load resistance $R = V_0/I_0 = 16\Omega$

$$\sqrt{\frac{L}{C}} = Q \times R = 5 \times 16 = 80$$

Resonant frequency f_o is given by
$f_o = f/y = 50{,}000/1.1 = 45.45$ kHz
But

$$f_0 = \frac{1}{2\pi\sqrt{LC}}$$

$$\frac{1}{\sqrt{LC}} = 2\pi \times 45.45 \times 10^3$$

The values of L & C are L = 261μH &C = 0.0237 μF
Since Cs = Cp, Cs = Cp = 2C = 0.0474μF

6 Experimental Setup of Converter with Motor Load

Figure 5 shows the proposed converter with the microcontroller and the control unit for the closed loop control of motor speed. Figure 6 shows the flow chart for the closed loop controls. A variable voltage is given to the analog to digital converter depending upon the required gate pulses. The ADC converts the analog voltage to digital pulses. The pulses are given to the microcontroller. The microcontroller is programmed to produce gate pulses. The converter output is varied by changing the gate pulses.

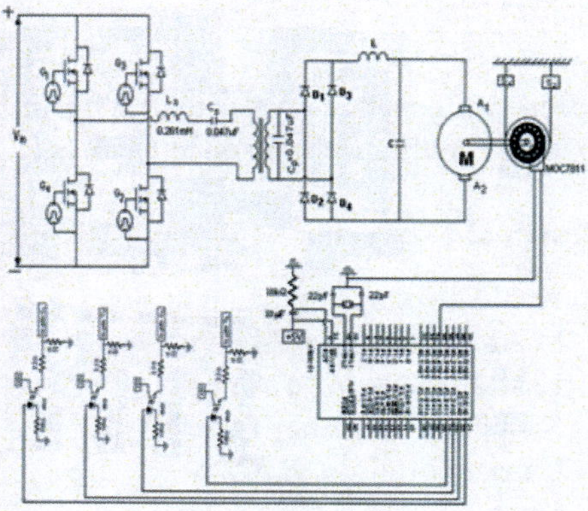

Fig. 5 Experimental circuit diagram of series parallel resonant converter with motor load under closed loop

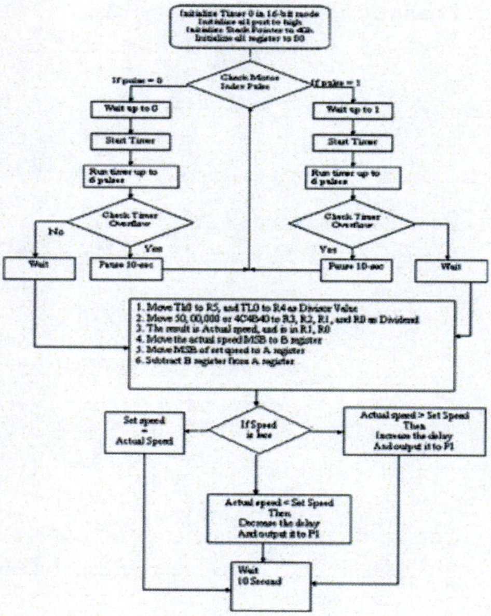

Fig. 6 Flow chart of proposed converter for closed loop with motor load

The experimental results are given to verify the theoretical prediction. Figure 7 to Fig. 10 show the experimental outputs obtained for the series inductor voltage V_{LS}, the series capacitor voltage V_{cs}, the parallel capacitor voltage V_{cp} and the load voltage V_0 with filter elements are obtained from the experimental set up. The experimental results are obtained under closed loop condition. The experimental results are obtained for various values of load and with the variation in duty ratios to maintain the speed constant. The Table 1 gives the calculated results of Converter for motor load under closed loop.

The Table 2 gives the experimental results of proposed Converter for motor load under closed loop. Due to high value of armature resistance, as the value of load current decreases, the output voltage also decreases even for constant speed. The series inductor voltage, series capacitor voltage and parallel capacitor voltage decreases as the load current decreases. The Table 3 gives the efficiency obtained with conventional method. As the load is decreased, the output voltage increases. The duty ratio is reduced and input current, output current and output voltage are noted (Figs. 8, 9).

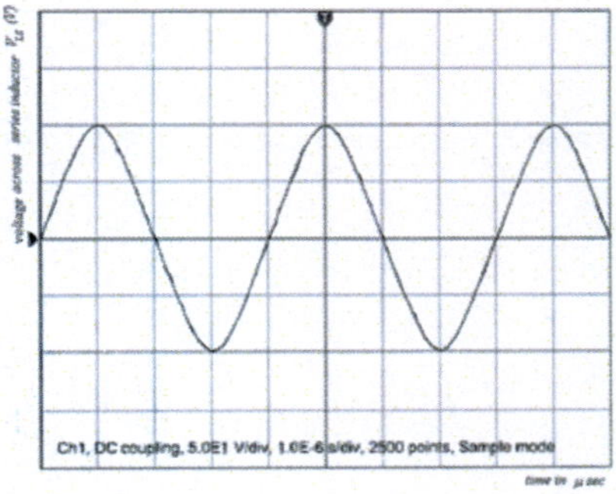

Fig. 7 Experimental results of V_{LS}

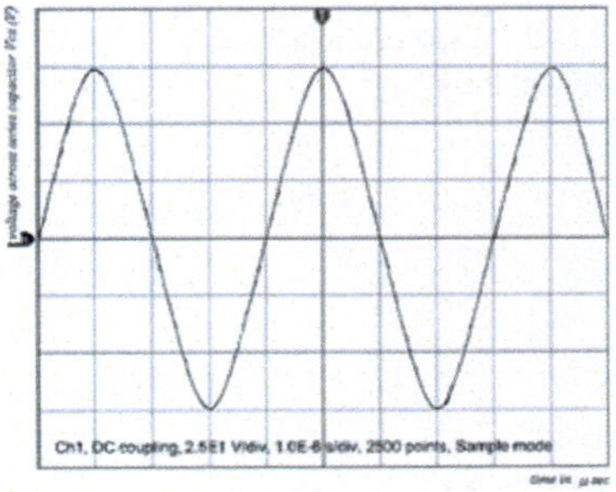

Fig. 8 Experimental results of Vcs

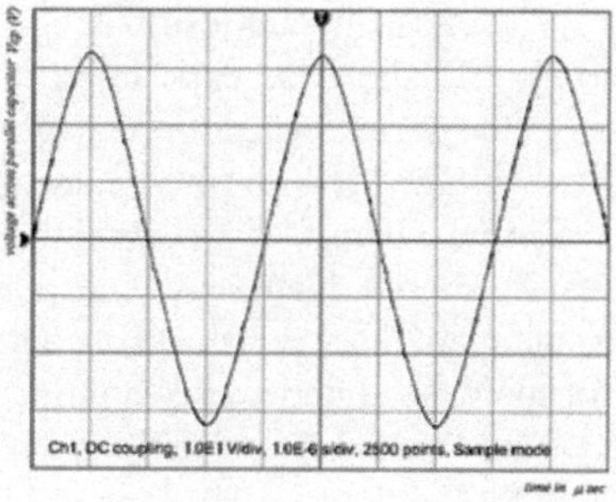

Fig. 9 Experimental results of Vcp

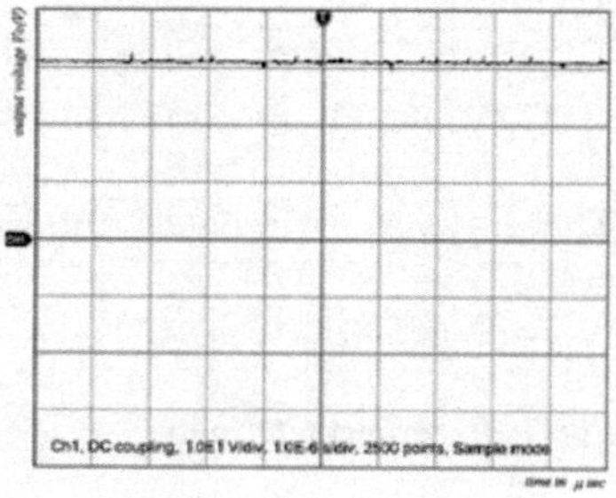

Fig. 10 Experimental results of V_0

Table 1 Calculated results

S. No	Duty Ratio	Peak Current through Series Inductor (amps)	Peak Voltage across Series Inductor (volts)	Peak Voltage across Series Capacitor (volts)	Peak Voltage across Parallel Capacitor (volts)	Output Current (amps)	Output Voltage (volts)
1	0.9	2.35	156.25	127.19	36.99	1.5	35.27
2	0.85	2.1	121.53	96.56	35.26	1.24	33.62
3	0.83	1.9	95.47	72.53	32.13	1.05	30.53
4	0.81	1.4	78.32	54.81	31.45	0.74	28.14
5	0.78	0.9	64.34	40.76	29.42	0.46	27.88

Table 2 Experimental results closed loop

S. No	Duty Ratio	Input DC Current (amps)	Peak Voltage across Series Inductor (volts)	Peak Voltage across Series Capacitor (volts)	Peak Voltage across Parallel Capacitor (volts)	Output Voltage (volts)	Output Current (amps)	Efficiency (%)
1	0.91	1.9	159.4	129.8	37.9	36.25	1.5	95.1
2	0.86	1.5	125.1	99.2	35.9	34.34	1.25	94.6
3	0.82	1.14	98.6	74.9	32.7	31.17	1.04	94.2
4	0.8	0.77	81.7	58.7	31.2	29.74	0.73	93.8
5	0.78	0.47	67.1	43.1	30.1	28.56	0.46	93.4

Table 3 Efficiency details (conventional)

S. No	Duty Ratio	Input Current (amps)	Input Voltage (volts)	Output Current (amps)	Output Voltage (volts)	Efficiency %
1	1	1.64	36	1.5	34.2	86.63
2	0.92	1.31	36	1.2	32.6	83.17
3	0.86	0.95	36	0.9	29.7	80.84
4	0.79	0.62	36	0.6	28.4	76.29

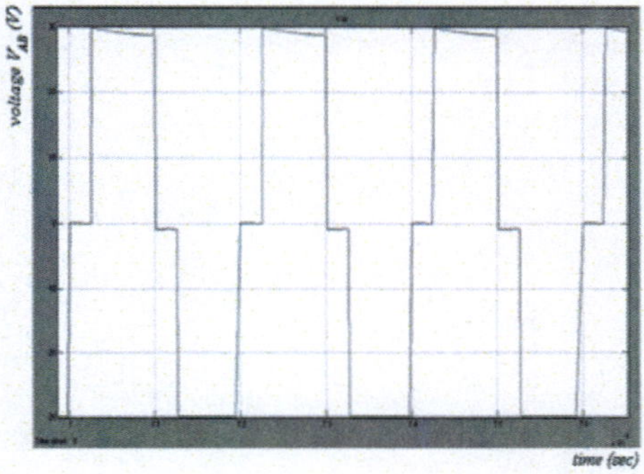

Fig. 11 Simulation results of V_{AB}

7 Simulation Results of Proposed Converter

A simulation has been implemented for motor load. Figure 11 to Fig. 21 show the simulation output obtained for motor load under closed loop keeping constant speed with motor load having converter output current of 0.8 A and D = 0.81 with variation in the duty cycle. The voltage input to the resonant circuit, the voltage across series capacitor, the voltage across inductor, the voltage across parallel capacitor, the current through the series inductor, the input current, shows the current through switch1, the current through switch2, the current through load and the output voltage with filter are shown. The Table 4 Simulation results of Series Parallel Resonant Converter for motor load under constant speed. As the load value decreased, the load current decreases. The decrease in load current in turn results in reduced current through capacitor and inductor. In order to reduce the output voltage with decreased load, the duty ratio is reduced. The Table 5 gives the comparison the between the hard switching and soft switching. The soft switching reduces the switching losses due to zero voltage switching. This results in increased efficiency of the converter (Figs. 12, 13, 14, 15, 16, 17, 18, 19, 20 and 22).

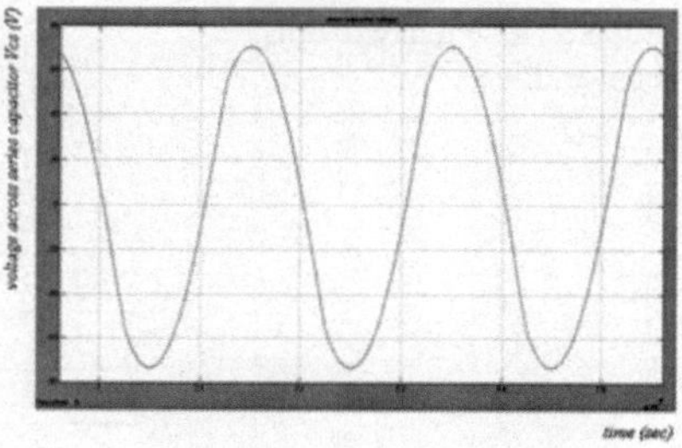

Fig. 12 Simulation results of v_{cs}

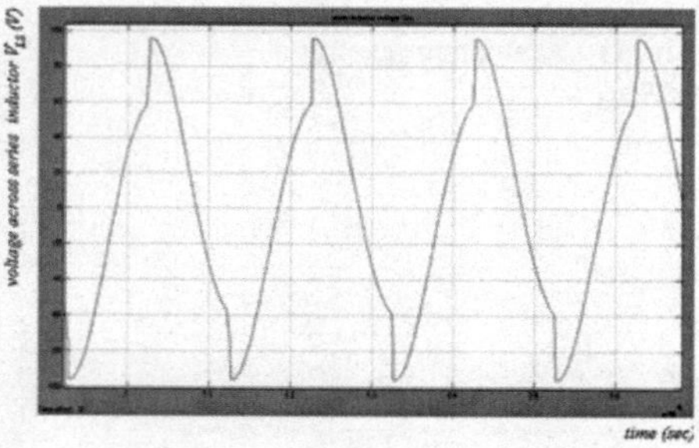

Fig. 13 Simulation results of V_{LS}

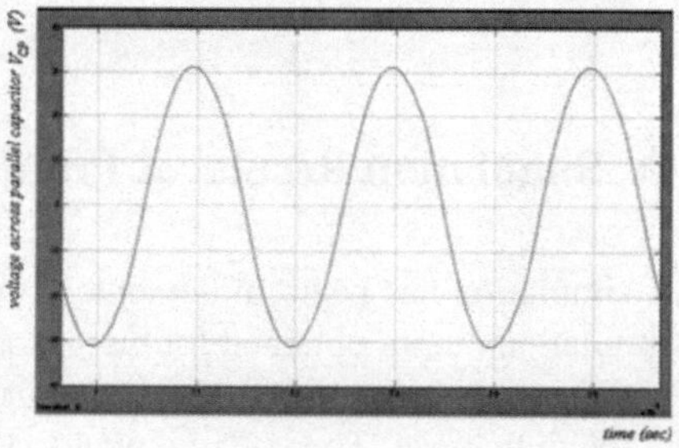

Fig. 14 Simulation results of V_{cp}

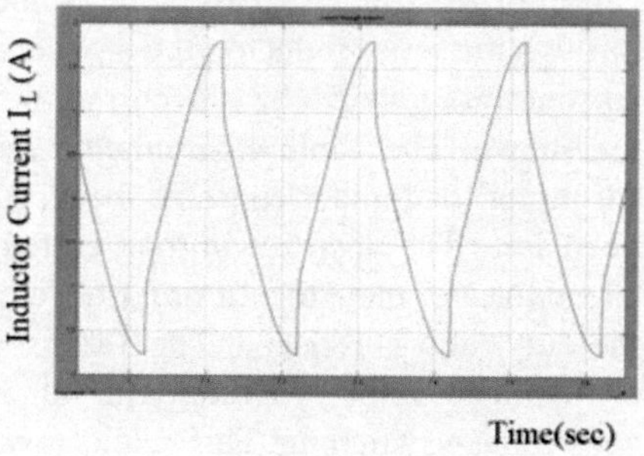

Fig. 15 Simulation results of I_L

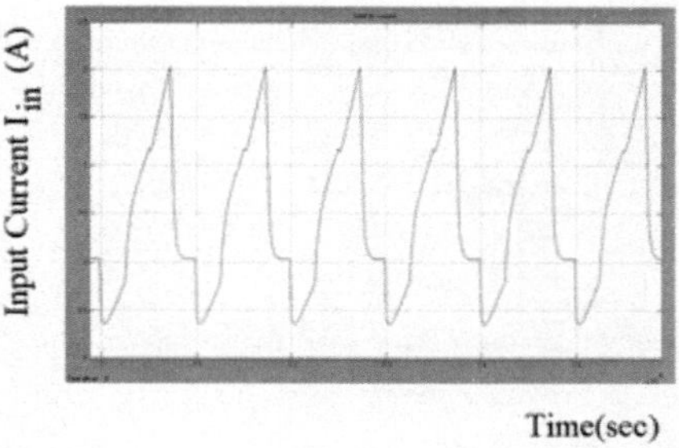

Fig. 16 Simulation results of I_{in}

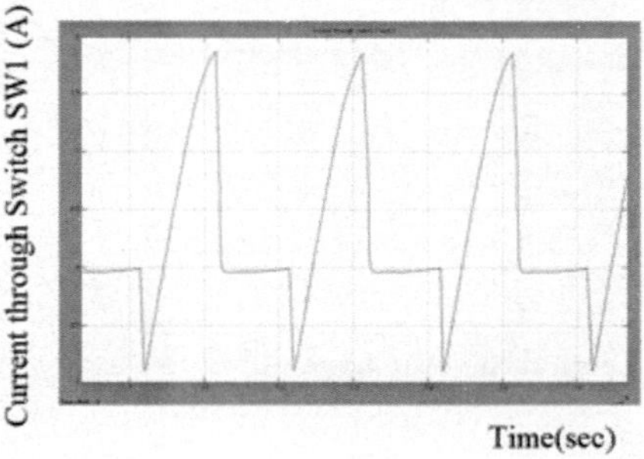

Fig. 17 Simulation results of I_{sw1}

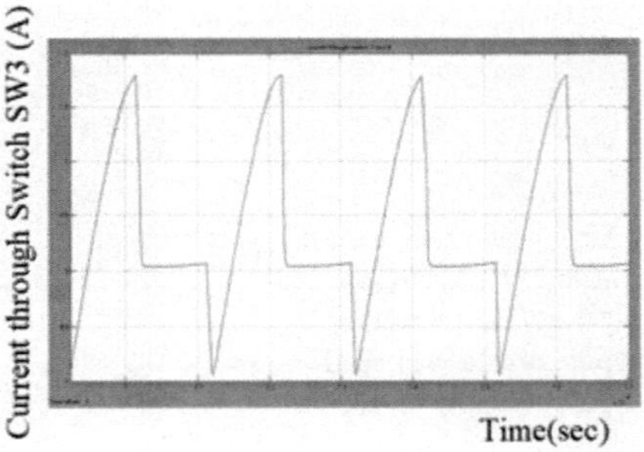

Fig. 18 Simulation results of I_{sw3}

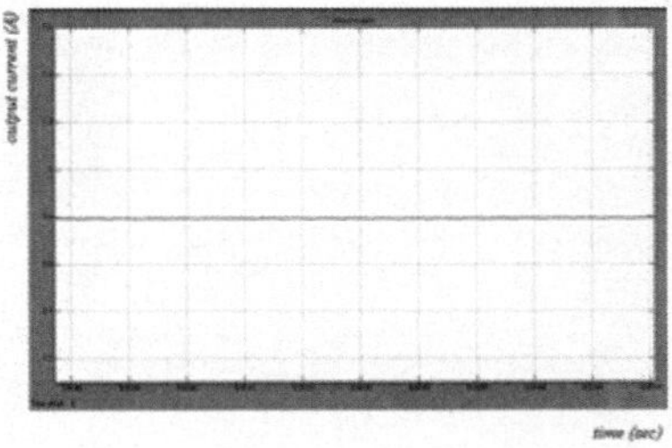

Fig. 19 Simulation results of I_0

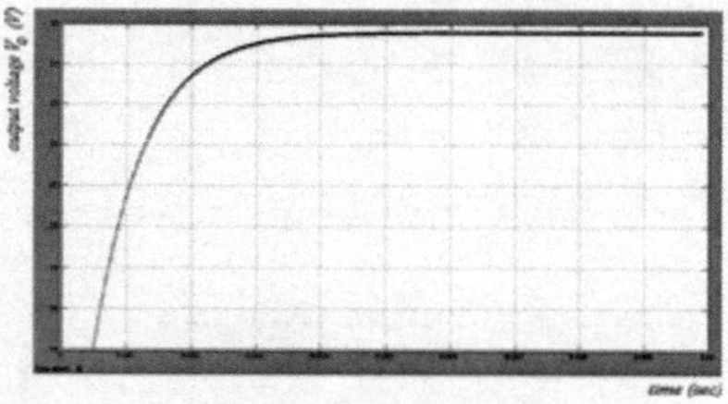

Fig. 20 Simulation results of V_0

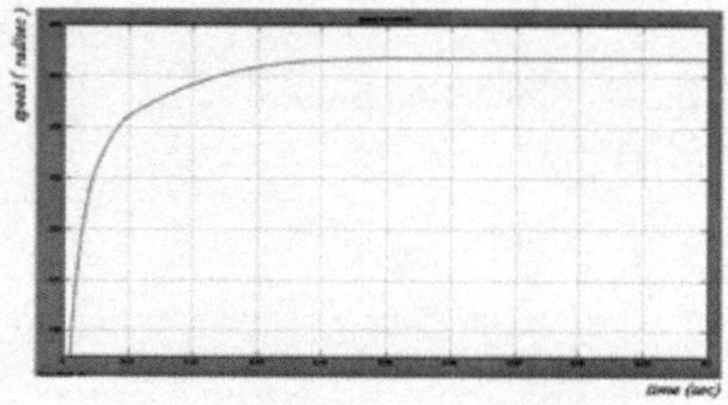

Fig. 21 Simulation results of motor speed

Table 4 Simulation results

S. No	Duty Ratio	Input DC Current (amps)	Peak Input DC Current (amps)	Peak Voltage across Series Inductor (volts)	Peak Current through Series Inductor (amps)	Peak Voltage across Series Capacitor (volts)	Peak Voltage across Parallel Capacitor (volts)	Output Voltage (volts)	Output DC Current (amps)	Efficiency (%)
1	0.9	1.79	2.5	157.1	2.4	127.5	36.1	34.5	1.5	96.4
2	0.85	1.49	2.2	140.8	2.0	11238	35.4	33.1	1.3	96.1
3	0.83	1.24	2.0	123.4	1.8	97.4	34.7	32.4	1.1	95.7
4	0.81	0.83	1.5	96.7	1.3	72.6	31.4	29.7	0.8	95.4
5	0.78	0.48	0.96	65.2	0.8	41.2	29.2	27.6	0.5	95.1

Table 5 Comparison between hard switched topology and proposed topology under closed loop

Type of loss	formulae used	hard switched topology	proposed topology
Switching loss	$P_{sw}=\frac{4}{12}\cdot I_1\cdot f\cdot t_{off}+t_{on}\cdot V_{in}$	$\frac{4}{12}\cdot 1.61\cdot 50000\cdot 100\cdot (10^{-9}+10^{-9})\cdot 30=1.6W$	$\frac{4}{12}\cdot 0.6\cdot 50000\cdot 100\cdot (0+10^{-9})\cdot 30=0.3W$
Conduction loss	$P_{con}=I_{rms}^2 R_{ds(on)}$	$1.5^2\cdot 0.4=0.9W$	$1.34^2\cdot 0.4=0.718W$
Diode conduction loss	$P_d=V_F\cdot I_D$	$0.8\times 0=0W$	$0.8\times 0.024=0.0192$
% Efficiency	$\%\eta=\frac{P_{out}}{P_{out}+P_{loss}+P_d}\times 100$	$\frac{51.75}{51.75+1.61+0.9}\times 100=95.37$	$\frac{51.75}{51.75+0.3+0.718+0.0192}\times 100=98.03$

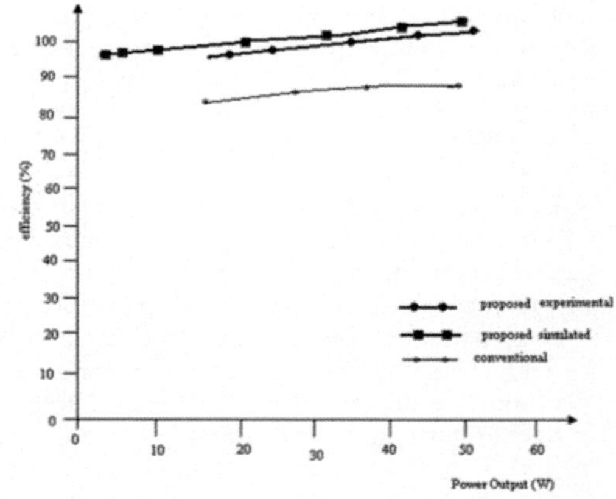

Fig. 22 Efficiency of the converter

8 Conclusion

The design of dc-dc series parallel resonant converter topology has been proposed. Due to the simple control method, phase shifted pulse width modulation is used in high power applications. The converter operates under zero-voltage switching (ZVS) for lagging power factor. The current stress on the switches is low and the output current is in continuous conduction. Experiments are carried out with a load rating having voltage levels above and below the input. The leakage inductance of the transformer is utilized. Delay time is varied to get desired operation as mentioned. Figures depict experimental and simulation results obtained.

ReferenCES

1. Zhao C, Wu X, Meng P, Qian Z (2009) Optimum design consideration and implementation of a novel synchronous rectified soft-switched phase-shift full-bridge converter for low-output-voltage high-output-current applications. IEEE Trans Power Electron 24(2):388–397
2. Da Silva ES, Coelho EAA, de Freitas LC, Vieira Jr JB, Farias VJ (2004) A soft-single-switched forward converter with low stresses and two derived structures. IEEE Trans Power Electron 19(2):388–396
3. Canales F, Barbosa P, Lee FC (2002) A zero-voltage and zero-current switching three-level DC/DC converter. IEEE Trans Power Electron 17(6):898–904
4. Hey HL, de Oliveira Stein CM, Grundling HA, Pinheiro H, Pinheiro JR (2004) Zero-current and zero-voltage soft-transition commutation cell for PWM inverters. IEEE Trans Power Electron 19(2):396–404
5. Chung HS-H, Tse KK, Ron Hui SY, Mok CM, Ho MT (2003) A novel maximum power point tracking technique for solar panels using a SEPIC or cuk converter. IEEE Trans Power Electron 18(3):717–724
6. Grundling HA, De Oliveira Stein CM, Pinheiro H, Pinheiro JR, Hey HL (2004) Zero-current and zero-voltage soft-transition commutation cell for PWM inverters. IEEE Trans Power Electron 19(2):396–404

7. Burdio JM, Monterde F, García JR, Barragan LA, Martínez A (2005) A two-output series-resonant inverter for induction-heating cooking appliances. IEEE Trans Power Electron 20(4):815–822
8. Carr JA, Rowden B, Balda JC (2009) A three-level full-bridge zero-voltage zero-current switching converter with a simplified switching scheme. IEEE Trans Power Electron 24(2):329–338
9. Huber L, Hsu K, Jovanovic MM, Solley DJ, Gurov G, Porter RM (2006) 1.8-MHz, 48-V resonant VRM: analysis, design, and performance evaluation. IEEE Trans Power Electron 21(1):79–89
10. Pilawa-Podgurski RCN, Sagneri AD, Rivas JM, Anderson DI, Perreault DJ (2009) Very-high-frequency resonant boost converters. IEEE Trans Power Electron 24(6):1654–1665
11. Song TT, Huang N (2005) A novel zero-voltage and zero-current-switching full-bridge PWM converter. IEEE Trans Power Electron 20(2):286–294
12. Chen W, Ruan X, Zhang R (2008) A novel zero-voltage-switching PWM full bridge converter. IEEE Trans Power Electron 23(2):793–803
13. Xiong Y, Sun S, Jia H, Shea P, Shen ZJ (2009) New physical insights on power MOSFET switching losses. IEEE Trans Power Electron 24(2):525–531
14. Lu Y, Cheng KWE, Ho SL (2008) Quasi current mode control for the phase-shifted series resonant converter. IEEE Trans Power Electron 23(1):353–358

Chest X-Ray Analysis and Tuberculosis Detection Using CNN

Rahul Singh, Arnab Chakravarty, and Biswajit Ghosh

Abstract Looking for anomalies in a Chest X-ray (CXR) is a tedious and demanding job, usually carried out by a radiologist as a sequence of predefined steps. We present a framework to help reduce a radiologist's time in examining a CXR by pre-computing some of the more mechanical aspects of the examination while highlighting some of the features, for example, the view position, patient's gender, and critical conditions are observable, using deep-learning (DL), neural networks. Our framework helps radiologists to focus their time and attention more towards the critical areas of the examination rather than the more mechanical aspects. It also reduces the possibility of missing important anomalies in CXRs. We integrated a set of eight Convolutional Neural Networks (CNN) into our framework that operates in a hierarchical order to collaboratively predict the features and metadata of a CXR. Our framework focuses on training lighter convolutional models to predict a single feature each rather than training deep Neural Networks to predict all the features at once, this makes each model lighter and easier to train.

Keywords CNN · Keras · Machine learning · Tuberculosis · X-ray

1 Introduction

Radiologists examining a Chest X-ray (CXR) undertake a large number of sequential evaluations in a very short period [1]. This necessary but tedious manual process occasionally leads to inaccuracies and omissions. These sequential steps can be better performed by a computer using advanced machine learning techniques like Convolutional Neural Network (CNN) [2] working under the supervision of the radiologist.

Our framework is presented to radiologists as a Progressive Web Application (PWA) [3] that assists in diagnosing a CXR image. The PWA is cloud-hosted, interactive, and easy to use, which can be easily accessed by a radiologist without going

R. Singh · A. Chakravarty (✉) · B. Ghosh
IT Department, Future Institute of Engineering and Management, Kolkata 700150, WB, India

V. Nath and J. K. Mandal (eds.), *Proceeding of Fifth International Conference on Microelectronics, Computing and Communication Systems*, Lecture Notes in Electrical Engineering 748,
https://doi.org/10.1007/978-981-16-0275-7_43

through the hassle of environment setup and running code scripts to get computerized predictions and make use of convolutional neural networks.

Many health conditions require patients to take a CXR and thus radiologists often have to conduct routine examinations of a large number of normal CXRs, then being able to afford detailed examination of critical ones. By pre-computing the more mundane aspects of the examination with this framework, we intend to divert a radiologist's time more towards the critical cases. Therefore making Radiologists our target audience. Algorithms developed using deep convolutional neural networks [4, 5] have proven to be quite good at making diagnostic predictions better than the radiologists themselves. Techniques such as transfer learning and fine-tuning equip us with the ability to make use of pre-trained models as to our needs [6].

The main objective of this framework is to assist a radiologist in diagnosing a chest x-ray image and to provide the radiologist with the metadata and predicted state of the image, the final decision, however, is of the radiologist to accept or reject the assistance of our framework.

We have developed a framework consisting of a hierarchical sequence of 8 CNN models to automatically evaluate and suggest several steps in the radiologists' workflow. The models used are relatively small and easier to train when compared with using deep CNNs for CXR images [7]. The datasets used for the training of these models were obtained from the National Institutes of Health (NIH) chest x-ray datasets available via Box [8] and some other publically available datasets [9–12]. Our framework also makes use of bootstrap learning during its training to maximize the use of the data we have.

In this paper our main motivation to use CNNs for tuberculosis detection rather than any other image processing techniques such as Contrast-Limited Adaptive Histogram Equalization (CLAHE) as used by Gabriella, I., Kamarga, S. A., and Setiawan, A. W. [13], optimized gray level co-occurrence matrix as used by Junaedi, I., Yudaningtyas E., and Rahmadwati, R. [14], contour models for feature selection as used by Afzali, A., Mofrad, F. B. and Pouladian, M. [15], localization algorithms as used by Song, Y. and Yang, Y. [16], etc., is that CNNs can automatically detect the important features that make the best prediction without any human intervention. We aim to leverage this specialty of CNNs by using a hierarchical sequence of 8 CNN models to automatically evaluate and predict the most important features in each hierarchical step of our framework's workflow.

The rest of this paper is organized as follows. Section 2 presents the Network Architectures of the models used. In Sect. 3 the dataset used is described. Section 4 consists of the Algorithm and workflow of the framework. In Sect. 5 the training of the models is discussed. Section 6 provides the results we obtained Finally, Sect. 7 concludes the paper.

2 Network Architecture

To train our CNN models there are two network architectures that we have created using Keras and TensorFlow. One is for classification (Architecture 1) and the other is for predicting the mask (Architecture 2).

A. Architecture 1

This network architecture (Architecture-1) is used to train all the classification models in our framework. This architecture consists of only three convolutional layers each having a kernel size of 5 as shown in [Fig. 1].

The first convolutional layer consists of 32 neurons with a kernel size of 5 and padded such that the layer's outputs will have the same spatial dimensions as its inputs. This layer uses the relu activation function and has an input size of 64 × 64.

The second convolutional layer consists of 50 neurons with a kernel size of 5 and padded such that the layer's outputs will have the same spatial dimensions as its inputs. This layer also uses the relu activation function and has an input size of 32 × 32. The input to this layer happens after a max-pooling from the previous layer.

The third convolutional layer is of 80 neurons with a kernel size of 5 and padded such that the layer's outputs will have the same spatial dimensions as its inputs. This layer uses the relu activation function and has an input size of 16 × 16. The input to this layer happens after a max-pooling from the previous layer.

Next, we used is a fully connected layer with 512 neurons and relu activation function having 41,472 trainable parameters.

The last layer that we have is also a fully connected layer with 2 neurons and softmax activation function having 1026 trainable parameters. The network uses categorical cross-entropy as the loss function and RMSprop optimizer with a learning rate of 0.0001 and decay of $1e^{-6}$.

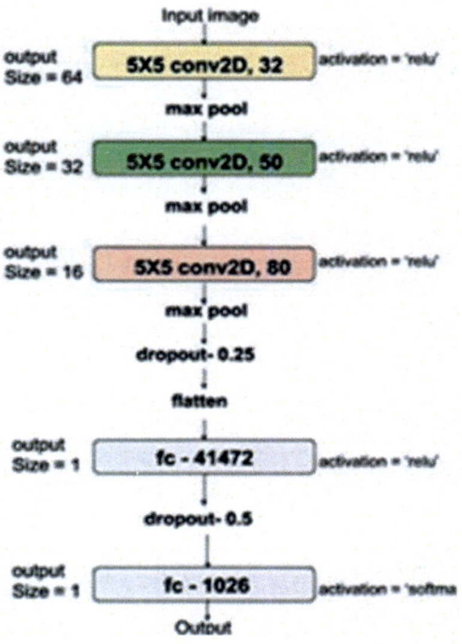

Fig. 1 Architecture-1; Network architecture for our classification models

B. Architecture 2

This network architecture (Architecture-2) is used to train our mask model in this framework. This Architecture is not our own we have taken this from a public Github repository (Fig. 2).

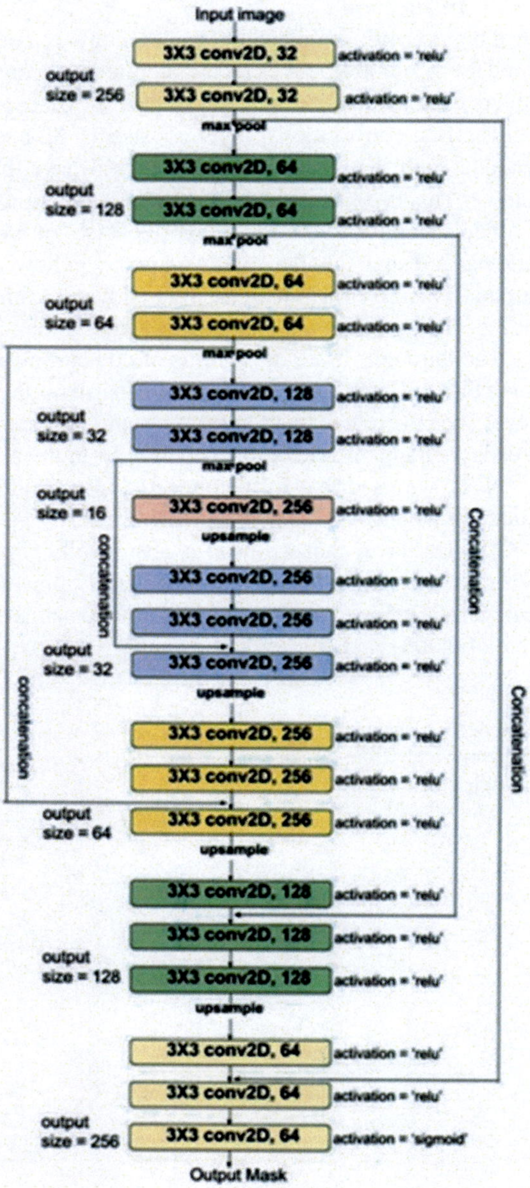

Fig. 2 Architecture-2; Network architecture for our mask prediction model

3 Dataset Description

We have used a total of 5 datasets to train our convolutional networks. Each of the datasets used specializes in training the network for a particular feature.

The first dataset that we have is built by us and consists of 94 different hand-picked images. Which are divided among two classes lateral and not-lateral. This dataset specializes in training the network in determining whether an X-ray image's view position is lateral or not lateral. The second dataset is from the National Institutes of Health (NIH) chest x-ray dataset which is hosted in Kaggle [9]. This set consists of 5606 images. This dataset contains information about the view position of the image, patient gender, and diseases. This dataset has been used to train the network to determine whether a CXR image's view position is Anteroposterior (AP) or Posteroanterior (PA), whether the patient's gender is male or female and if the status is normal or not normal. This framework deals with only tuberculosis and other diseases if detected will be shown as abnormal. The third dataset is also obtained from the National Institutes of Health (NIH) chest x-ray dataset [8] by web scraping. This set consists of 7470 CXR images which are divided into two categories Normal or Not_Normal. The fourth dataset is also downloaded from Kaggle and consists of 662 CXR Images [10]. This dataset consists of information about the patient's gender, view position of the image, age, and whether it has tuberculosis or not. This dataset specializes in training the network in determining whether an X-ray image might have tuberculosis or not. The fifth dataset consists of 247 Chest X-ray images in.img format along with the Chest mask for each in Run-Length Encoding (RLE). This dataset was downloaded from JSRT [11] and it specializes in predicting the lung mask in a Chest X-ray image. The sixth dataset consists of 10,675 dicom images and their corresponding RLE mask for pneumothorax. This dataset was also downloaded from kaggle [12]. We have used this dataset specifically to test some of the models that we have trained using the above datasets.

4 Algorithm

The algorithm used can be divided into two steps, Pre-processing, and Training.

A. Preprocessing Algorithm

The unknown chest x-ray entered by the user is first analyzed for its format whether it is.png,.jpeg,.jpg, or.dcm if it is found to be a dicom image then it is first converted into a.png format. Then the image is converted into grayscale and resized using OpenCV. After that, the image is converted into a 2D matrix using numpy (Fig. 3).

B. Training Algorithm

- The pre-processed x-ray is initially passed to the model_L to confirm if the input Chest X-ray is of lateral view position or not. If the view position is lateral, then we discard the image as our models are not built for lateral chest x-rays. If the

Fig. 3 This figure demonstrates the workflow of the pre-processing step in our framework

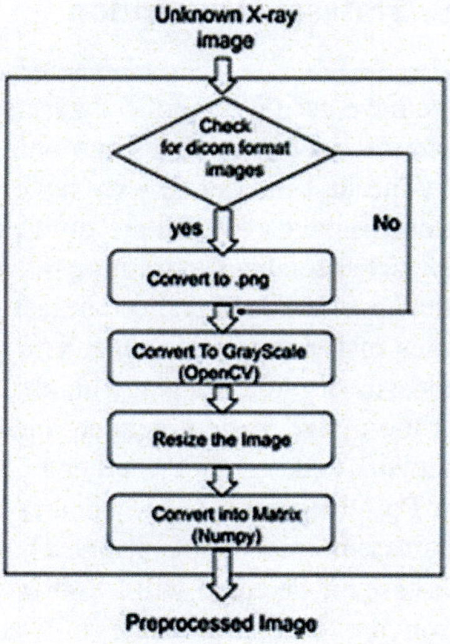

model_L gives a false output on analyzing the chest x-ray, then we can be sure that the unknown x-ray is either taken from the front (AP) or the back (PA).

- After model_L we pass the x-ray image to the model_AP_PA to classify if the image is Anterior–Posterior (taken from the front) or Posterior-Anterior (taken from the back). We need to know whether our unknown x-ray image is AP or PA as our following models are built separately depending on the x-ray view positions.
- Next in line are the models model_AP_G and model_PA_G which determines the gender of the unknown X-rays for AP view position and PA view position respectively. After analyzing the x-ray these models can classify if the patient whose x-ray was taken was male or female.
- Simultaneous to model_AP_G and model_PA_G, we pass the unknown x-ray through another model model_D_T which determines if the x-ray has tuberculosis or not.
- In the end, if the X-ray image is found not to have tuberculosis then it is passed to model_AP_D and model_PA_D according to the image's respective view position AP and PA to determine any other anomalies due to some other disease.
- The X-ray image is also passed through a masking model model_mask in parallel to the above steps to predict the lung mask in the image. Which is then displayed in the application (Fig. 4).

5 Training

In this project, 8 CNN Models have been trained using 5 different datasets collected from the internet (Fig. 5).

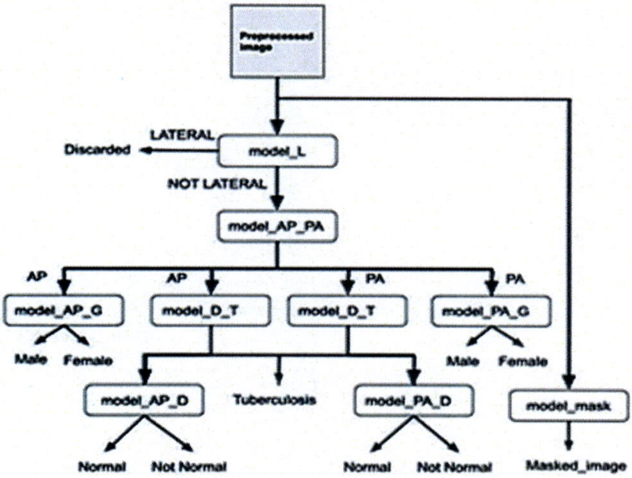

Fig. 4 This figure demonstrates the prediction workflow of our framework

Fig. 5 This figure is a visual demonstration of the training algorithm of all of our models

A. Training of the First Model

First CNN model that we have trained is model_L. This model is trained on 64 samples and validated on 30 sample x-ray images to differentiate between lateral and non-lateral view positions of an x-ray image. The training is done over 60 epochs and the learning curve is shown in Fig. 6.

B. Training of the Second Model

The second model that we have trained is model_ AP_PA. This model is trained on 1500 image samples and validated on 71 x-ray image samples taken from the second data set we have from the Kaggle data set consisting of 5606 images. This

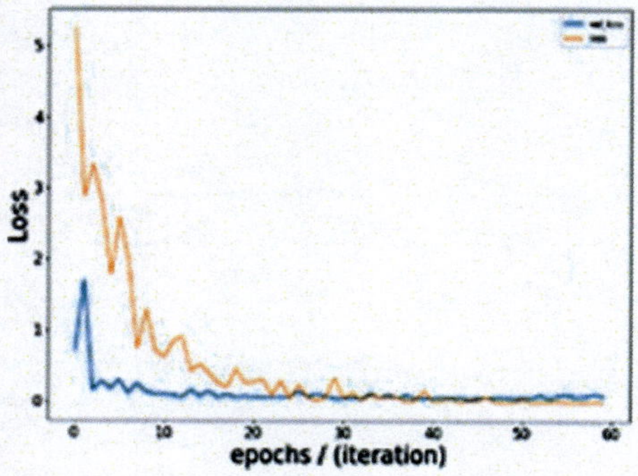

Fig. 6 This graph demonstrates the learning curve for model_L with the number of epochs in the x-axis and loss in the y-axis. The blue line represents the validation loss and the orange line represents the training loss

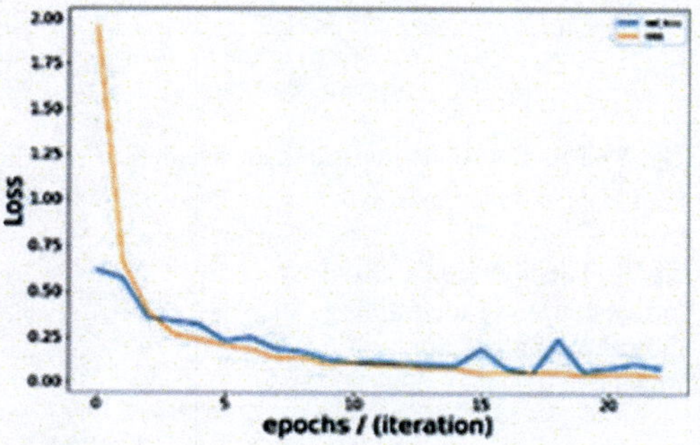

Fig. 7 This graph demonstrates the learning curve for model model_AP_PA with the number of epochs in the x-axis and loss in the y-axis. The blue line represents the validation loss and the orange line represents the training loss

model decides whether the image is AP or PA if it is found to be not lateral by the previous model (model_L). This model is trained for 23 epochs and the learning curve is shown in Fig. 7.

C. Training of the Third Model

The third model that we have trained is model_AP_G. This model is trained in 1989 images and validated on 224 images from the second Kaggle data set of 5606 images. This model is trained only with the AP images contained in the data set. This model predicts the gender of the patient that is male or female if the image is found to have view position AP by the previous model (model_AP_PA). The model was trained for 26 epochs and the learning curve is shown in Fig. 8.

D. Training of the Fourth Model

The fourth model that we have trained is model_PA_G. This model is trained on 3011 images and validated on 382 images from the second Kaggle data set of 5606 images. This model is trained only with the PA images contained in the data set. This model predicts the gender of the patient that is male or female if the image is found to have view position PA by the previous model (model_AP_PA). The model was trained for 31 epochs and the learning curve is as shown in Fig. 9.

E. Training of the Fifth Model

The fifth model that we have trained is model_D_T. This model is trained with 600 images and validated on 62 images that we have downloaded from the third data

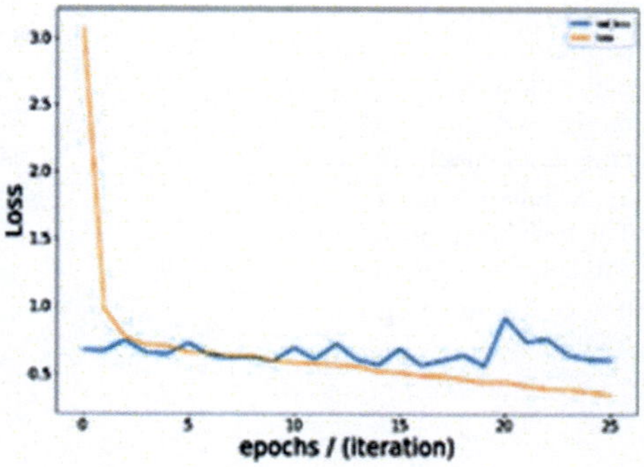

Fig. 8 This graph demonstrates the learning curve for model model_AP_G with the number of epochs in the x-axis and loss in the y-axis. The blue line represents the validation loss and the orange line represents the training loss

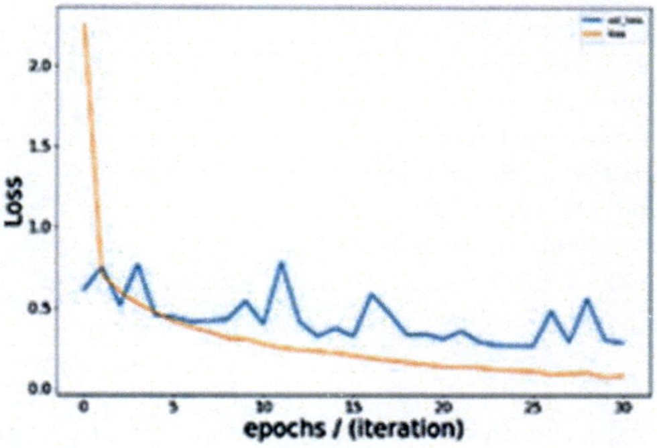

Fig. 9 This graph demonstrates the learning curve for model model_PA_G with the number of epochs in the x-axis and loss in the y-axis. The blue line represents the validation loss and the orange line represents the training loss

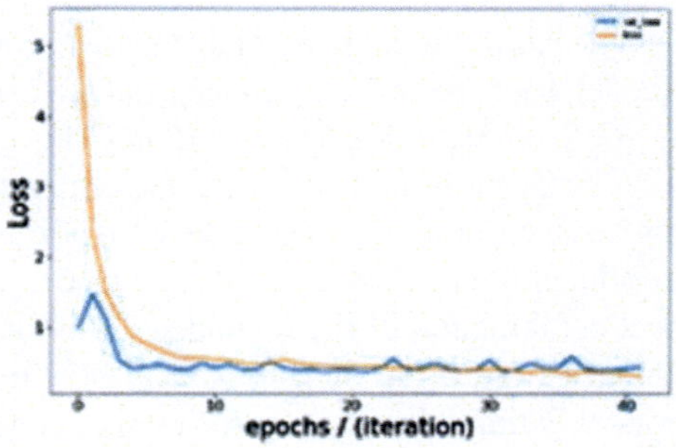

Fig. 10 This graph demonstrates the learning curve for model model_D_T with the number of epochs in the x-axis and loss in the y-axis. The blue line represents the validation loss and the orange line represents the training loss

set from Kaggle especially for tuberculosis. This model predicts whether a patient has tuberculosis or not. The model was trained for 42 epochs and the learning curve is as shown in Fig. 10.

F. Training of the Sixth Model

The sixth model that we have trained is model_AP_D. This model is trained by bootstrap training. The data set that this model is trained using is the predicted results of the models model_L and model_AP_PA. The data set used for this model is a combination of the second set of 5606 images and the fourth data set of 7470 images from these images only the AP images are used to train the model. This model is

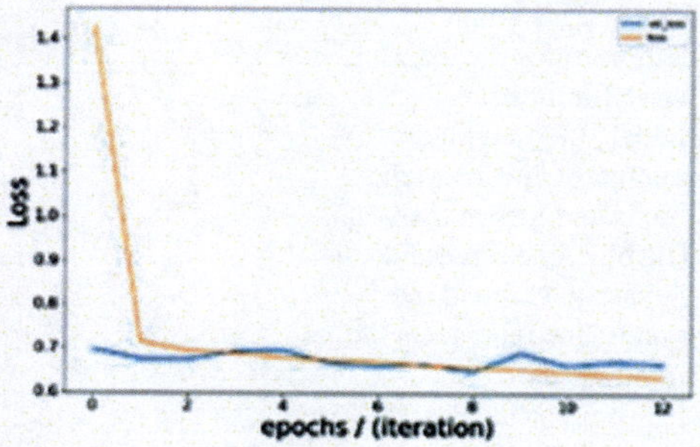

Fig. 11 This graph demonstrates the learning curve for model model_AP_D with the number of epochs in the x-axis and loss in the y-axis. The blue line represents the validation loss and the orange line represents the training loss

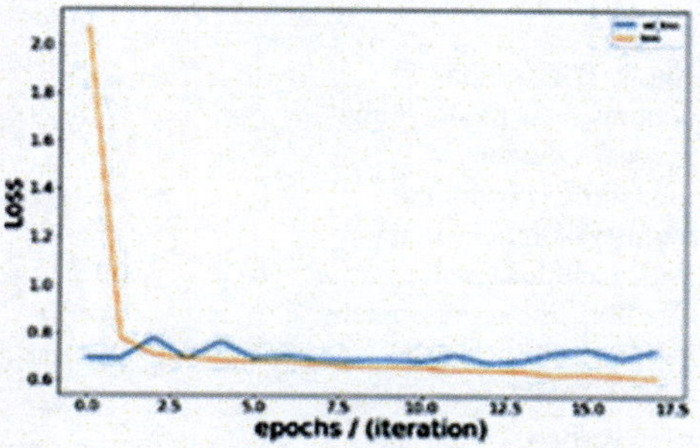

Fig. 12 This graph demonstrates the learning curve for model model_D_T with the number of epochs in the x-axis and loss in the y-axis. The blue line represents the validation loss and the orange line represents the training loss

trained with only AP images contained the data set which counts to 5599 images and validated on 224 images. This model is only used if the patient is found not to have Tuberculosis. This model predicts that if other than tuberculosis there are any other abnormalities in the x-ray image if so this model predicts abnormal else normal. The model was trained for 13 epochs and the learning curve is as shown in Fig. 11.

G. Training of the Seventh Model

The seventh model that we have trained is model_PA_D this model is trained by bootstrap training. The data set that this model is trained using is the predicted results of the models model_L and model_AP_PA. The data set used for this model is a combination of the second set of 5606 images and the fourth data set of 7470 images from these images only the PA images are used to train the model. This model is trained with only AP images contained the data set which counts to 3498 images and validated on 382 images. This model is only used if the patient is found not to have tuberculosis. This model predicts that if other than tuberculosis there are any other abnormalities in the X-ray Image if so this model predicts abnormal else normal. The model was trained for 18 epochs and the learning curve is as shown in Fig. 12.

H. Training of the Eighth Model

The eighth model that we have is model_mask. This model is trained by JSRT dataset which contains 247 images in.img format. This model predicts the lungs region in the image and masks it. This model is only trained for AP and PA images.

6 Results

Some basic features of a chest x-ray can easily be predicted accurately with a small convolution network shown (Fig. 1). We achieved an accuracy of 98% as shown in (Table 1) with our model model_L which only determines whether the view position of a Chest X-ray image is Lateral or not, the roc curve for the model is shown in (Fig. 13a). Using the same architecture we classified whether the view position if not lateral is Anteroposterior (AP) or Posteroanterior (PA) with our model model_AP_PA and achieved an accuracy of 96% as shown in (Table 1), the roc curve for the model is shown in (Fig. 13b).

We also achieved good accuracy in predicting the patient's gender from an unknown chest x-ray image knowing it's view position with the same network architecture shown in (Fig. 1) with our models model_AP_G and model_PA_G as shown in (Table 1). The roc curves for the models model_AP_G and model_PA_G are shown in (Fig. 13c and Fig. 13d) respectively.

The workflow of this framework primarily focused on feature extraction techniques that are letting the model be aware of the view position and patient gender beforehand to detect abnormalities in a chest x-ray with a small convolution network rather than using a deep convolutional network [17]. Small convolution networks (Architecture 1) (Fig. 1) do not perform as well as the deep convolutional networks

Table 1 This table demonstrates the score of each of our classification model

Models	Acc.	Precision	F1	Recall	AUC	Conf. Mat.
Model_L	0.98	1.0	0.977	0.965	0.978	$\begin{pmatrix} 55 & 0 \\ 2 & 43 \end{pmatrix}$
Model_AP_PA	0.966	0.953	0.973	0.993	0.958	$\begin{pmatrix} 178 & 15 \\ 2 & 305 \end{pmatrix}$
Model_AP_G	0.778	0.766	0.822	0.886	0.758	$\begin{pmatrix} 133 & 78 \\ 33 & 256 \end{pmatrix}$
Model_PA_G	0.906	0.899	0.908	0.917	0.906	$\begin{pmatrix} 220 & 26 \\ 21 & 233 \end{pmatrix}$
Model_AP_D	0.571	0.5	0.495	0.489	0.561	$\begin{pmatrix} 81 & 47 \\ 49 & 47 \end{pmatrix}$
Model_PA_D	0.560	0.634	0.503	0.417	0.571	$\begin{pmatrix} 129 & 49 \\ 119 & 85 \end{pmatrix}$
Model_D_T	0.855	0.862	0.847	0.833	0.854	$\begin{pmatrix} 28 & 4 \\ 5 & 25 \end{pmatrix}$

Fig. 13 These figures show the ROC curves for our models with the true positive rate on the y-axis and false positive rate on the x-axis

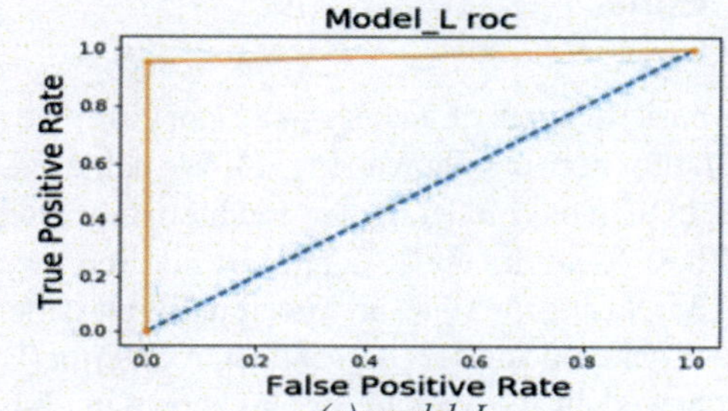

(a) model_L

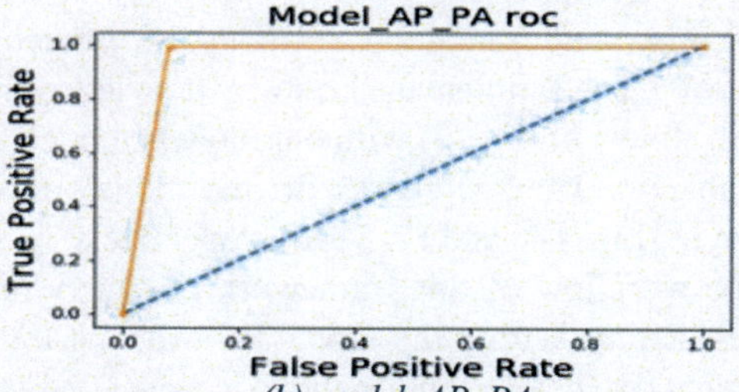

(b) model_AP_PA

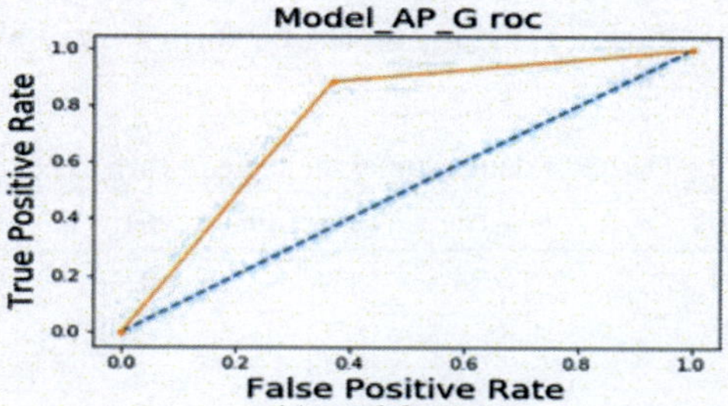

(c) model_AP_G

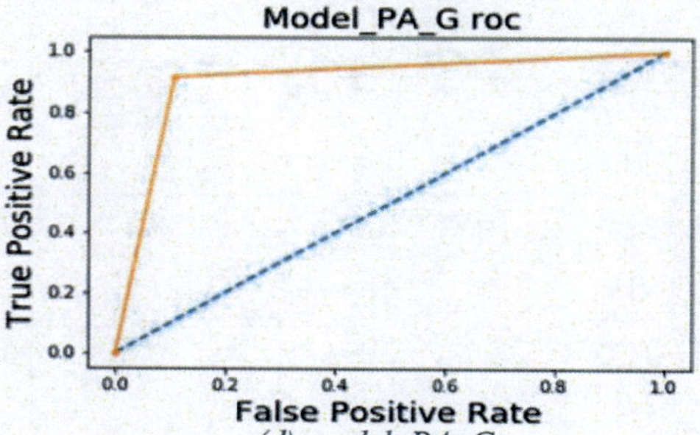

(d) model_PA_G

Fig. 13 (continued)

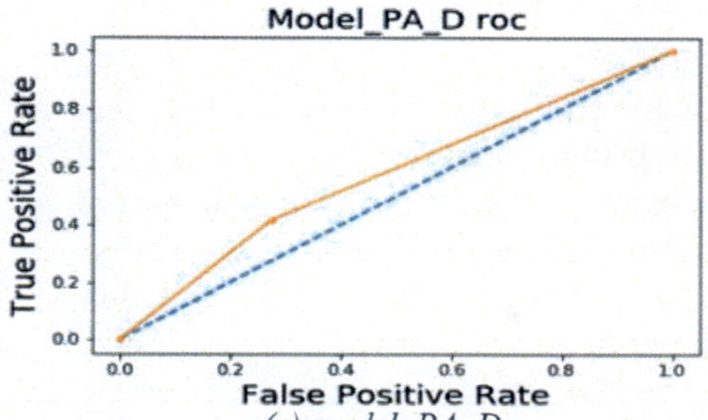

(e) model_PA_D

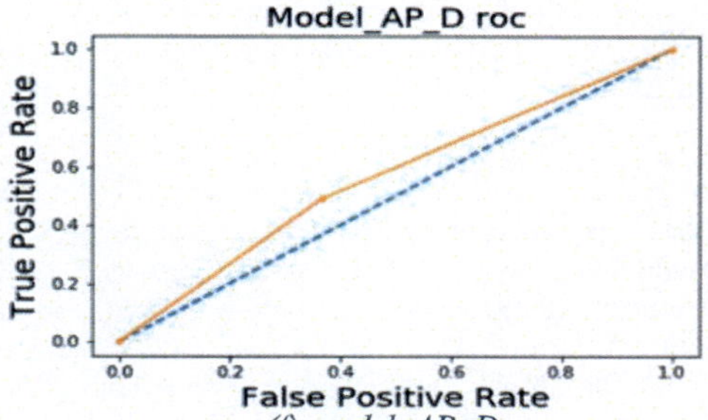

(f) model_AP_D

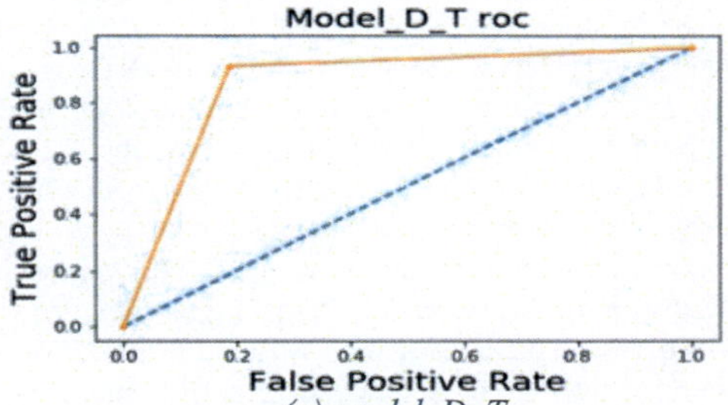

(g) model_D_T

[18–21] in predicting all the abnormalities in a chest x-ray at once as shown by the models model_AP_D and model_PA_D in (Table 1), the roc curve for the models model_AP_D and model_PA_D as shown in (Fig. 13e, Fig. 13f). On the other hand, our network (Architecture 1) (Fig. 1) did seem to do a pretty good job of predicting a single disease with good accuracy. In our case, we predicted tuberculosis with our model model_D_T with an accuracy of about 86% as shown in (Table 1), the roc curve for the model model_D_T is shown in (Fig. 13 g). This shows that small convolution networks if used with proper feature extraction techniques may also be as useful as the deep convolutional networks architectures such as vgg-16 [19], resnet [18], and others [17, 20, 21] to detect tuberculosis.

7 Conclusion

The framework that we have developed integrates all of the above-mentioned models in an easy to use interactive web application. By using this framework, Radiologists can save a lot of their time and accurately be assisted with the metadata of an unknown chest x-ray image resulting in a better diagnosis. The workflow architecture of our framework demonstrates that small convolutional networks by making use of feature extraction techniques, may also be efficient in making accurate predictions like the deep convolutional networks and provide us with satisfactory results.

References

1. Dixon A, Jones J (2018) Chest X-ray: ABCDE (summary). www.radiopaedia.org/articles/chest-x-ray-abcde-summary. Accessed on Jul 2018
2. Yamashita R, Nishio M, Do RKG, Togashi K (2018) Convolutional neural networks: an overview and application in radiology. Insights Imaging 9:611–629
3. Biørn-Hansen A, Majchrzak T, Grønli T (2017) Progressive web apps: the possible web-native unifier for mobile development. In: Proceedings of the 13th International Conference on Web Information Systems and Technologies, vol 1, pp 344–351
4. Abiyev RH, Ma'aitah MKS (2018) Deep convolutional neural networks for chest diseases detection. Journal of Healthcare Engineering, vol 2018
5. Rajpurkar P, Irvin J, Zhu K, Yang B, Mehta H, Duan T, Ding DY, Bagul A, Langlotz CP, Shpanskaya KS, Lungren MP, Ng AY (2017) CheXNet: Radiologist-Level Pneumonia Detection on Chest X-Rays with Deep Learning. ArXiv, abs/1711.05225
6. Shin H, Roth HR, Gao M, Lu L, Zu X, Nogues I, Yao J, Mollura D, Summers RM (2016) Deep convolutional neural networks for computer-aided detection: CNN architectures, dataset characteristics, and transfer learning. IEEE Trans Med Imaging 35(5):1285–1298
7. Bindiya HM, Chethana HT, Pavan Kumar SP (2018) Detection of anomalies in fetus using convolutional neural networks. MECS Press IJITCS, vol 10, no 11
8. National Institute of Health and Clinical Center (2017) NIH Chest X-ray Dataset. Retrieved Jun 2018 from https://nihcc.app.box.com/v/ChestXray-NIHCC
9. National Institute of Health and Clinical Center (2017) Sample of NIH Chest X-ray Dataset, Version 4. Retrieved Jun 2018 from https://www.kaggle.com/nih-chest-xrays/sample
10. Quantitative Imaging in Medicine and Surgery (2014) Two public chest X-ray datasets for computer-aided screening of pulmonary diseases. Retrieved Jun 2018 from https://www.ncbi.nlm.nih.gov/pmc/articles/PMC4256233/
11. Japanese Society of Radiological Technology (2006) Digital Chest X-ray Image Database. Retrieved Jun 2018 from https://db.jsrt.or.jp/eng.php
12. See—(2019) SIIM_TRAIN_TEST, Version 4. Retrieved Jul 2019 from https://www.kaggle.com/seesee/siim-train-test/metadata
13. Gabriella I, Kamarga SA, Setiawan AW (2018) Early detection of tuberculosis using chest X-Ray (CXR) with computer-aided diagnosis. In: 2nd International Conference on Biomedical Engineering (IBIOMED) 2018, pp 76–79. https://doi.org/10.1109/IBIOMED.2018.8534784
14. Junaedi I, Yudaningtyas E, Rahmadwati R (2019) Tuberculosis detection in chest X-Ray images using optimized gray level co-occurrence matrix features. In: 2019 International Conference on Information and Communications Technology (ICOIACT), pp 95–99. https://doi.org/10.1109/ICOIACT46704.2019.8938584
15. Afzali A, Mofrad FB, Pouladian M (2020) Feature selection for contour-based tuberculosis detection from chest X-Ray images. In: 2019 26th National and 4th International Iranian

Conference on Biomedical Engineering (ICBME), pp 194–198. https://doi.org/10.1109/ICBME49163.2019.9030395
16. Song Y, Yang Y (2010) Localization algorithm and implementation for focal of pulmonary tuberculosis chest image. In: 2010 International Conference on Machine Vision and Human-machine Interface, pp 361–364. https://doi.org/10.1109/MVHI.2010.180
17. Rajaraman S, Antani SK (2020) Modality-specific deep learning model ensembles toward improving TB detection in chest radiographs. IEEE Access 8:27318–27326
18. He K, Zhang X, Ren S, Sun J (2016) Deep residual learning for image recognition. In: 2016 IEEE Conference on Computer Vision and Pattern Recognition (CVPR), pp 770–778
19. Simonyan K, Zisserman A (2015) Very deep convolutional networks for large-scale image recognition. CoRR, abs/1409.1556
20. Rajaraman S, Kim I, Antani SK (2020) Detection and visualization of abnormality in chest radiographs using modality-specific convolutional neural network ensembles. Peer J 8:e8693
21. Nguyen QH, Nguyen BP, Dao SD, Unnikrishnan B, Dhingra R, Ravichandran SR, Satpathy S, Raja PN Chua MCH (2019) Deep learning models for tuberculosis detection from chest X-ray Images. In: 2019 26th International Conference on Telecommunications (ICT), pp 381–385. https://doi.org/10.1109/ICT.2019.8798798

Defect Analysis in Memory Unit

Mitali Mutkekar and M. Govinda Raju

Abstract 3D NAND flash memory is a better alternative to 2D NAND as it offers higher speed, capacity and better performance. 3D NAND technology is rapidly evolving with ever increasing number of layers in devices. With increasing layers, the design and test complexity of memory devices increases rapidly. With these advances in NAND flash technology comes a host of challenges in regards to the ensuring the reliability and quality of memory devices reaching the end customer. Thus, analysis and study of the 3D NAND flash failure mechanisms is of prime importance. This paper focuses on defect and failure analysis of prominent 3D NAND flash failures using NanoNT tool to understand the nature of the failures, their potential causes and suggests on method to diagnose and screen these failures. The test flow is implemented in Advantest tester, improved yield of 98.7% and test time reduction of 6.63% is achieved.

Keywords 3D NAND flash · 2D NAND flash · Defect analysis · Reliability · Flash memory · Failure analysis

1 Introduction

With the advent of the zettabyte age the need for capable data storage solutions has rapidly increased. To meet the diverse needs of the semiconductor memory market many different types of memory technologies have evolved. One such popular technology widely in use today is the flash memory technology. Flash memory technology has evolved from Electrically Erasable Programmable Read Only Memory (EEPROM) technology. There are two main types of flash memory based on the structure of the core memory array, NAND flash and NOR flash. Both categories

M. Mutkekar (✉) · M. Govinda Raju
Department of Electronics and Communication Engineering, RV College of Engineering, Bengaluru, India

M. Govinda Raju
e-mail: govindarajum@rvce.edu.in

V. Nath and J. K. Mandal (eds.), *Proceeding of Fifth International Conference on Microelectronics, Computing and Communication Systems*, Lecture Notes in Electrical Engineering 748,
https://doi.org/10.1007/978-981-16-0275-7_44

have their own distinct advantages and usage. In NOR flash the core memory array consists of cells connected in parallel bearing close resemblance to that of NOR logic gate. Series connected cells resembling a NAND logic gate structure form the core NAND flash memory array. The series array structure of NAND flash memory allows for higher bit density.

NAND flash technology offers considerable benefits over other storage technologies including SRAM, DRAM, NOR Flash etc. Ever since NAND flash memory entered the market in 1987, its compact size, low cost, speed and performance have made it a popular choice [1, 13, 21]. The memory technology landscape has been ever-evolving as a result of the continuous thrust to enhance memory capabilities. Memory manufacturers are under constant pressure to increase the performance, density and speed of memory devices. Planar 2D NAND technology based on floating gate transistors has been the mainstay of storage industry for many years. In planar NAND technology the memory cells are connected to each other in a horizontal string. Increased storage density in 2D (planar) NAND flash is achieved by increasing the memory cell bit density and miniaturization of feature size. This miniaturization has led to worsening of physical issues in 2D NAND degrading the endurance, reliability and performance [2, 21].

2D NAND technology has approached its limit at the 15nm/14nm node leading to the emergence of 3D NAND flash technology, the successor to 2D NAND flash technology. The benefits of 3D NAND flash technology include better performance with lower power consumption, higher speeds, higher density and lower cost per gigabyte. On the flip side the processes in manufacturing 3D NAND flash memory are significantly more complex requiring stacking of multiple layers closely on top of each other (see Fig. 1) and high aspect ratio etch often leading to manufacturing defects like shorts or open causing failures [15–17]. Consequently, the task of failure and defect analysis in advanced technology nodes becomes increasingly complex [8]. To enhance the reliability and performance of memory devices an exhaustive understanding of various failure modes and mechanisms arising in its life cycle are supremely important.

The two main factors impacting the reliability of NAND flash are data retention and endurance. Endurance is the ability of cell to sustain program and erase cycling without significant degradation. Data retention is the duration for which data written

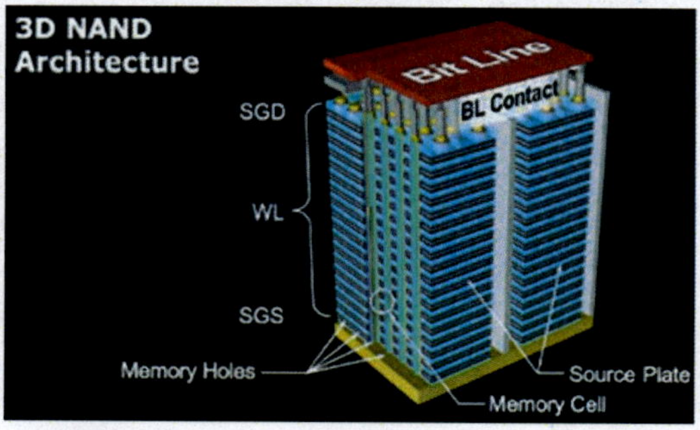

Fig. 1 3D NAND flash memory

to a cell is maintained [2–4, 10]. NAND flash reliability issues are caused due to degradation of tunnel oxide, charge de-trapping, stress induced leakage current and manufacturing process related defects [6]. NAND flash memories have a central core array with a peripheral circuitry providing access to each storage cell. There are certain failure modes common in NAND flash memory devices they are be classified as hard defect-based errors, write errors, errors due to disturbance phenomenon and data retention errors at cross temperatures [2, 6, 12, 23].

With advances in technology and fabrication processes memory design and architecture gets increasingly intricate and complex. These intricate structures often give rise to a number of defects and failure modes in advanced memory nodes [15]. Thus, failure analysis in advanced memory devices is of prime importance to have reliable and high performing memory devices. Failure signature are analysed by electrical failure analysis (EFA) [8]. The underlying failure mechanism is diagnosed and corrective action is suggested. The operation unit of NAND flash memory is per page and that of DRAM, NOR flash is bit unit thus March algorithms need to be modified for NAND flash memory. March-like modified algorithms along with different addressing modes and data patterns can be used for test and detection of the faults in NAND flash memory [9–12]. This paper focuses on the electrical failure analysis of prominent failure mechanisms and suggests the testing method that can be implemented to screen out these defects for 3D NAND flash memory devices. Yield enhancement and test time reduction methods are implemented to reach the product yield goal and test time requirements.

2 NAND Flash Operation

2D and 3D NAND flash memory operate on the principle of floating gate (FG) technology and charge trap (CT) technology [14, 19, 20]. Basic flash cell is as shown in the Fig. 2 and 3. Flash memory is programmed by the process of Fowler-Nordheim (FN) tunnelling to store electrons on the storage layer using incremental step pulse program (ISPP) technique [4, 7]. The program cycle begins with the selected wordline at the high positive programming voltage (18–22 Volts) as shown in Fig. 4 this attracts the electrons towards the storage layer changing its threshold voltage. The corresponding cells on the same wordline are deselected by biasing their bit lines

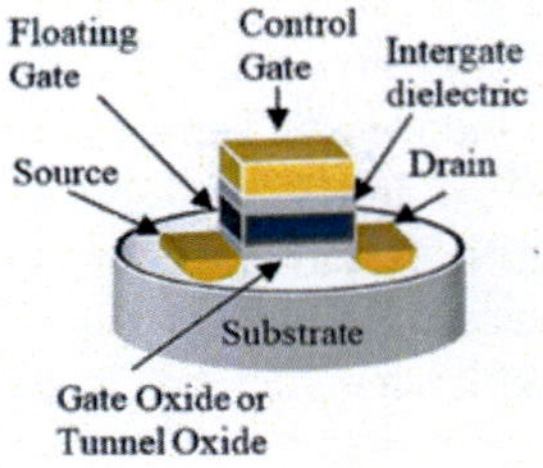

Fig. 2 2D NAND flash memory cell [3]

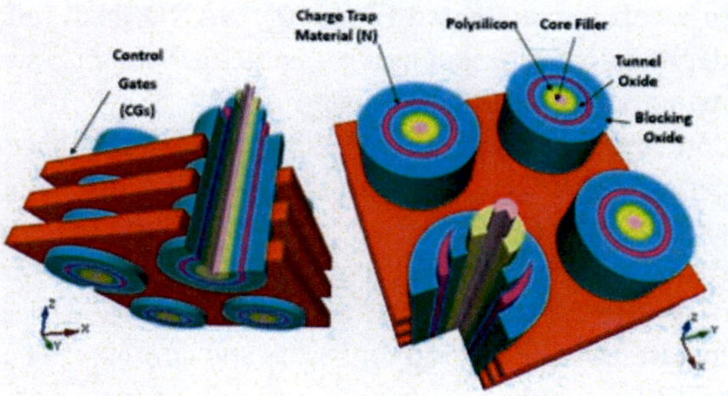

Fig. 3 3D NAND flash cell [14]

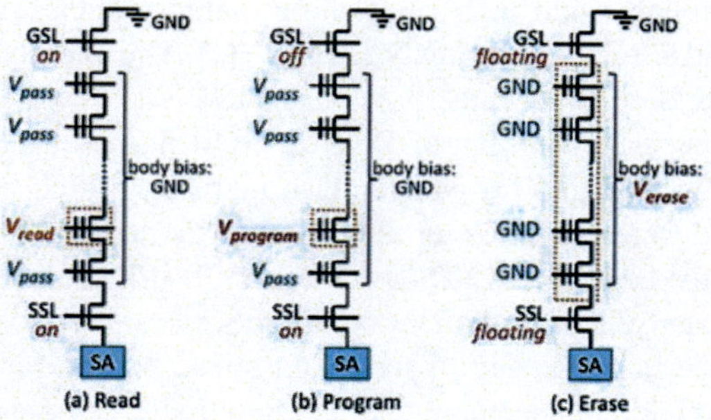

Fig. 4 Flash memory operation [4]

(2 Volts). ISPP technique consists of applying step programming pluses to inject electrons into the storage layer and iteratively verifying whether the cell has reached desired threshold voltage (V_t).

ISPP only allows for an increase in voltage thus flash memory can be reprogrammed only after erase. Flash memory is erased by the process of hole injection or electron de-trapping from the storage layer by biasing the substrate to a high positive erase voltage (18–22 Volts) [7] and by setting the control gate to ground [4]. Since the substrate is common for a block, erase operation occurs at block level. The cell threshold voltage is sensed by biasing the control gate to a read reference voltage. The number of reference voltages required to sense the cell state depends on number of bits stored [4, 7]. The sense amplifier senses the cell data this is aided by biasing rest of the cells on that bit line to a pass voltage to provide access to the selected cell being read on a particular wordline [4].

NAND flash cells are classified as single level cell (SLC), multi-level cell (MLC) and triple level cell (TLC) on the basis of number of bits of data stored. The SLC flash stores single bit of data per flash memory cell and has two threshold voltage windows (V_{th}), erase (ER) and one program state (P1) as shown in Fig. 5. Similarly, MLC and TLC flash store two and three bits per cell respectively. They have four (ER, P1-P3) and eight cell states (ER, P1-P7) [4]. Reliability margin is the window space between two threshold voltage states. With increasing bits per cell, reliability with which the data is interpreted reduces [1].

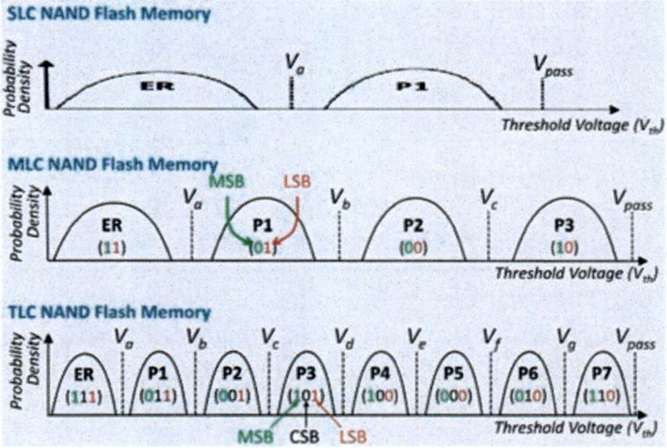

Fig. 5 SLC, MLC and TLC flash memory threshold voltage distribution [4]

3 NAND Flash Reliability

Building quality memory products requires a fine balance between design complexity, cost and reliability [5]. There are certain failure modes common in NAND flash memory devices they are write errors, data retention errors, hard-defect-based faults and errors due to disturbance [2, 6, 12, 18].

Write errors occur due to incorrect placement of cell threshold voltage levels during program and verify operation. As a result, data programmed in the cell is erroneous leading to loss of information. When the cell is programmed with data by ISPP and foggy-fine programming the threshold voltages on the partially programmed cells get disturbed due to erratic tunnelling as a result of program noise on adjacent cells and tunnel-oxide defects [2, 4].

Data retention errors are caused due to leakage of charge over time. Due to smaller feature sizes in smaller technology nodes the gain or loss of electrons significantly shifts the threshold voltage. The phenomena causing data retention errors are charge de-trapping from the gate oxide and trap-assisted tunnelling (TAT) which increases the conductivity of the tunnel oxide [2, 4]. With increasing retention age, the threshold voltage distributions become wider and the threshold voltages of the cells shift in opposite directions [4, 22]. These effects are exacerbated by repeated program and erase operations.

Hard-defect-based faults are caused by manufacturing process related issues [6]. With smaller feature sizes requiring higher precision these defects are becoming more common. These types of defects are of two types, shorts and opens between bit lines, wordlines, source select lines and drain select lines. The fault pattern can be characterised by the stuck-at fault model [11, 12]. These types of defects are managed by column repair and bad-block handling [12].

Errors due to disturbance phenomenon are caused by a shift in threshold voltage of the affected cell due to erase, program, read (EPR) operations on the neighbouring adjacent cells with the same bit line, word-line or source-line [2, 9, 12, 24, 25]. The main causes of these errors are undesired injection of electrons to the floating from the channel and cell-to-cell electrostatic interference [2, 4].

4 Methodology

The existing March algorithms aren't ideally suited for 3D NAND flash memories. Various March-like modified test algorithms are proposed for testing NAND flash memories [9–12]. Due to the varied nature of different failure signatures in 3D NAND flash memory it is better to design individual tests to screen out each failure as shown in Table 1. These directed test screens provide better fault diagnosis than a unified algorithm, which is often more complex and time consuming.

The initial design and definition of the high-level test strategy was started with the study of the past test flow design and the failure modes in the previous NAND technology nodes. The test starts with initialization of the memory followed by various contact tests to check the connectivity. The memory is then subjected to stresses and the functionality is evaluated by applying known set of inputs and checking for the corresponding output. Failure modes in 3D NAND flash memory are identified by conducting failure and defect analysis in NanoNT tool. The failure signatures are obtained by the measurement of electrical parameters related to threshold voltage distribution and leakage current.

The timing parameters like program time, erase time, read time and delays are monitored. The root cause of the failures is identified by careful examination of the failure signatures. The details of the failures observed is utilized to suggest corrective action in the test and screening methodology. The test flow is evaluated for redundant test blocks and methods are employed to enhance the yield and reduce the test time.

Table 1 Fault coverage of various tests

Test method	Failure type							
	Wordline-wordline leakage failure		Peripheral data latch failure		Erase disturb failure		Cluster block failure	
Memory	2D	3D	2D	3D	2D	3D	2D	3D
March-FT [9]	Yes	–	–	–	Yes	–	–	–
March-FTE [9]	Yes	–	–	–	Yes	–	–	–
March flash-FD [10]	Yes	–	–	–	Yes	–	–	–
Flash MLC march [12]	Yes	–	–	–	Yes	–	–	–
Proposed individual ad-hoc tests	–	Yes	–	Yes	–	Yes	–	Yes

5 Failure Analysis

The reliability of NAND flash memory depends on a variety of process dependent factors. Failure modes and mechanisms with their cause is identified by the process of failure analysis. 3D NAND flash memory is subjected to initial testing and an extensive study is conducted of the failures reported. Prominent failure modes are chosen for detailed analysis.

A. *Wordline-Wordline Leakage Failure*

NAND memory cells are programmed in the form of threshold voltage distribution. Each voltage loop corresponds to a memory state and provides details of the data stored. This stored data is sensed by reading the cell state at a particular reference voltage. TLC programming of NAND cells corresponds to seven distinct program loops.

On programming data into the NAND memory, the threshold voltage distribution was noticed to have out of place memory cells as shown in Fig. 6. This noisy distribution gives a clear indication of the existence of a failure. It is observed that all cells on the corresponding word-line fail program operation.

On further analysis of the failure signature it is noticed that the cells on the failing wordline fail to reach programming voltage. Nature of the defect is identified to be wordline to wordline leakage failure. The root cause is identified as a short between wordlines. Test is designed by biasing alternate wordlines, drain and source select lines to different voltages such that a short between adjacent wordlines causes a leakage due to potential difference which is detected.

B. *Peripheral Data Latch Failure*

NAND flash peripheral circuitry consists of peripheral data latches, data path, input and output pads. The data to be written into the NAND is transferred by peripheral circuitry to the core NAND array. The data latches are a temporary buffer to store the data before it is written into the NAND array. Since all the data being written

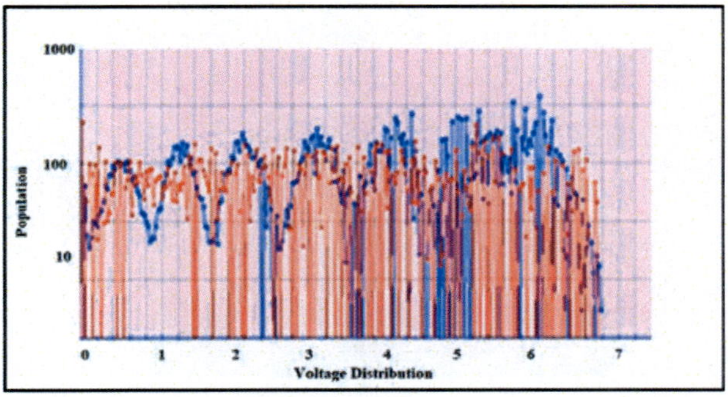

Fig. 6 Wordline-wordline leakage failure

to the NAND passes through this circuitry the correct functioning of peripheral data latches is of prime importance.

The failure presented as an increase in the number of bad columns. The failure was identified to be a data latch failure since the data loaded into the peripheral data latch was not matching with the data being clocked out as shown in Fig. 7. The designed test loads and compares data for various patterns. The patterns written are a sequence of ones and zeros with varying positions for each pattern to achieve better coverage by exercising all the parts of the periphery.

C. *Erase Disturb Failure*

Erase disturb failure is observed as a shift in threshold voltage distribution on the failing block. The threshold voltage of the victim block shifts upon erase and reprogram of a neighbouring block. During failure analysis two adjacent blocks are observed to erase disturb each other where the performance of erase operation on either of these blocks disturbs the other block. Figure 8 shows the V_t distribution on the failing block for initial erase and reprogram. This distribution is clear and as expected.

Erasing other non-adjacent blocks doesn't have an impact on the failing block distribution. Upon erasing the adjacent block an erase disturb fail occurs on the victim block as shown by the voltage distribution in Fig. 9 this is due to the blocks sharing the same outer decoder. Test is designed with a decremental program, erase, read sequence, where the upper block is erased and the immediately lower block is read to check for possible disturb to screen out the failure.

D. *Cluster Block Failure*

During memory testing the failing blocks are marked as bad-blocks. These blocks fail for various status operations such as read, erase, program. If there exists a long cluster of bad-blocks these failing blocks might impact the voltage distribution on

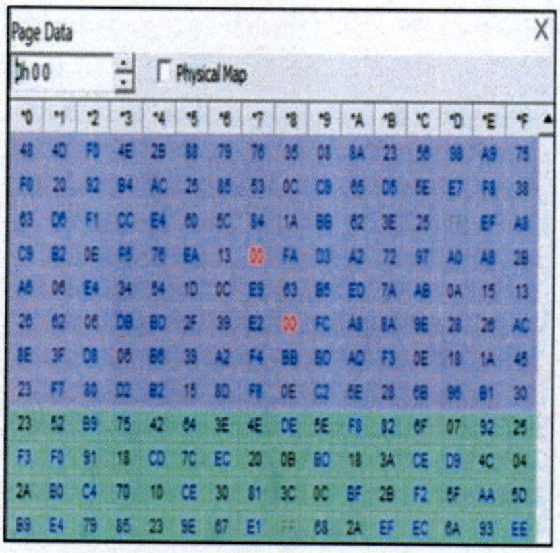

Fig. 7 Faulty data read from peripheral data latch

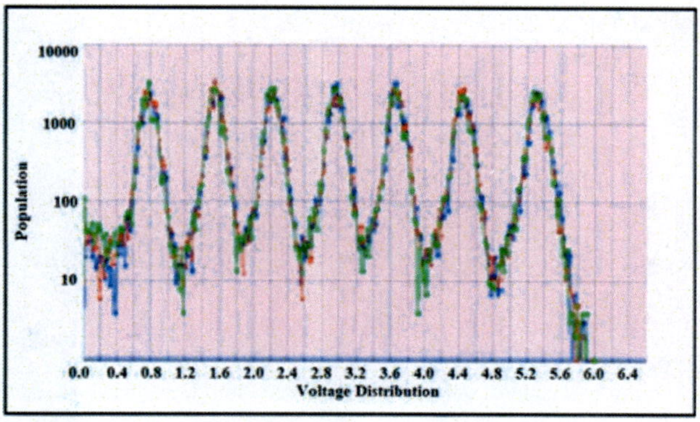

Fig. 8 Voltage distribution post erase and reprogram

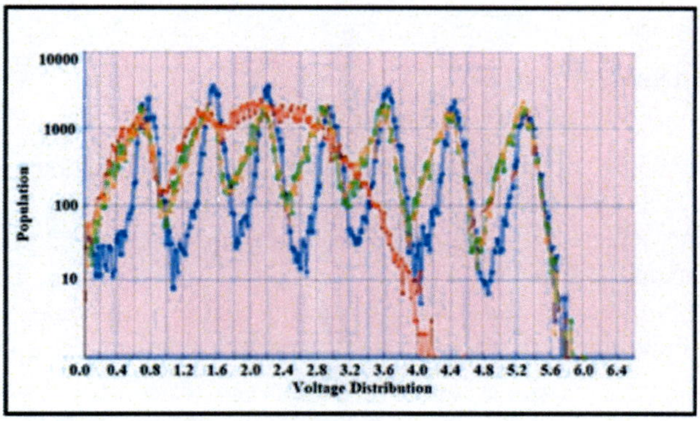

Fig. 9 Voltage distribution post erase of adjacent block

the adjacent good blocks and cause them to fail. The failing blocks were found in consecutive address locations forming a cluster of bad-blocks.

For cluster block fails upon further EPR cycling a growth in bad-blocks is observed as the voltage distribution shift disturbs the neighbouring good blocks causing them also to fail. This indicates the need for a guard band to protect the data integrity. The bad-blocks are observed in consecutive locations with a cluster size of thirty-one bad-blocks. The addresses of blocks in the cluster are odd thus they lie on consecutive locations in plane one (3E7-423) as shown in Fig. 10.

Test is designed with bad-block cluster size as 25, which checks if twenty-five or more adjacent blocks consecutively fail status read operations and applies a guard-band padding set to 3. Hence, three good blocks above and below the cluster are marked as bad thus isolating the bad-blocks and preventing the good blocks from failing.

Fig. 10 Bad-block cluster

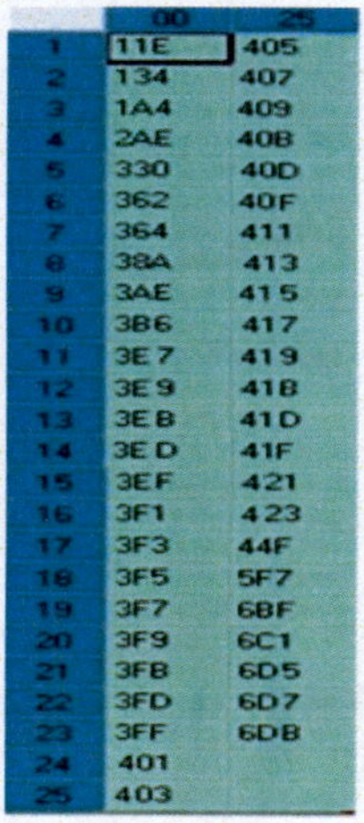

	00	25
1	11E	405
2	134	407
3	1A4	409
4	2AE	40B
5	330	40D
6	362	40F
7	364	411
8	38A	413
9	3AE	415
10	3B6	417
11	3E7	419
12	3E9	41B
13	3EB	41D
14	3ED	41F
15	3EF	421
16	3F1	423
17	3F3	44F
18	3F5	5F7
19	3F7	6BF
20	3F9	6C1
21	3FB	6D5
22	3FD	6D7
23	3FF	6DB
24	401	
25	403	

6 Yield Enhancement

Yield is the percentage of non-defective products of the total manufactured products. The product reaching the end customer has to be of prime quality ensuring that it doesn't develop defects during its lifecycle. To ensure that the end product is non-defective the memory devices are passed through extensive testing. The main goal of this testing is to ensure that all the failure mechanisms are detected and while ensuring that there the tests are not too stringent thus causing false fails for good units.

The yield of a product should be maintained at acceptable limits to reduce loss of good quality material and to thus keep the production costs optimum for the product to be profitable. The yield can be improved by designing the test flow such that it caters to the product requirements by choosing only those test blocks which are needed. Another way of reducing the fail percentage for a test block is by relaxing the pass/fail criteria pertaining to the bad-block and bad-columns.

Figure 11 shows yield goal for the memory is 98.5%. The monitored yield was observed to be 96.3%, which doesn't meet the yield goal. The test flow was evaluated for test blocks which were causing the yield loss. The SLC Program (Pattern A) was found to be causing a yield loss of 2.4%.

Upon failure analysis and manual validation, it was observed that the units failing for SLC Program (Pattern A) weren't valid fails but were false fails. This test block in the initial test flow was a screen which rejected failing units thus causing yield loss. The test block was kept as monitor to collect pass/fail data without rejecting any failing units. The resulting yield after this change was 98.7% which meets the set yield goal.

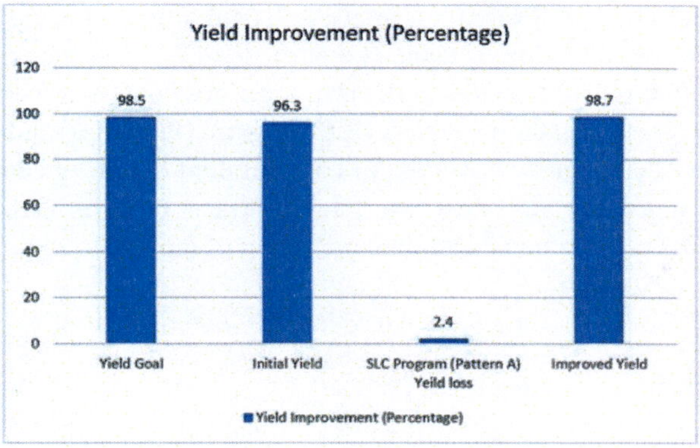

Fig. 11 Yield improvement

Fig. 12 Major time-consuming test blocks

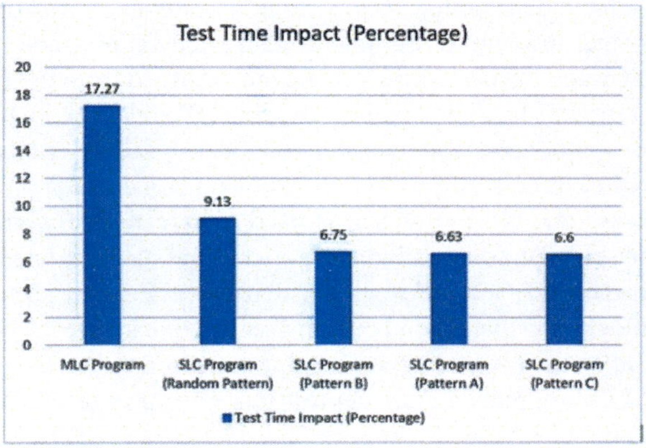

7 Test Time Reduction

The goal of testing is to ensure product performance and quality by ensuring all the failure modes are detected. The trade off in ensuring greater test coverage is the resulting test time increase. In order to optimize the overall impact on the test time various test time reduction schemes can be utilized such as reduction in EPR operation delays, use of faster operating modes, sample testing rather than full array testing and removal of redundant test blocks.

The test flow is studied and the various test time reduction methods are considered for their overall impact and effectiveness for test time reduction. The most effective means are noted to be the removal of redundant test blocks and usage of sample testing. The entire test flow is evaluated for the topmost time-consuming test blocks as shown in Fig. 12.

The test block for SLC Program (Pattern A) is a monitor, thus the main purpose of this test block is collection of data. The purpose of this test block is to program and verify whether the all the cells in the array can store data effectively.

The SLC Program (Random Pattern), SLC Program (Pattern B) and SLC Program (Pattern C) provide adequate coverage in ensuring that the array can store various data patterns. Hence, the SLC Program (Pattern A) test block is redundant and can be removed. This helps achieve test time reduction of 6.63%.

8 Conclusion

This paper analyses the fault signatures of prominent failure modes in the NAND flash memory. The Wordline-Wordline leakage failure caused by a short between two adjacent wordlines is identified by electrical failure analysis of the noisy threshold voltage distribution. Test screen is designed with stripe pattern voltage biasing to detect the failure. Peripheral data latch failure mode is identified by a mismatch between the data expected and the data read from the latch. The designed test block programs four data patterns to read and compare data by exercising the various parts of the peripheral circuitry. Erase disturb failure is observed as a threshold voltage distribution shift on the failing block upon erase of an adjacent block. Decremental erase and read sequence is used to screen the failure. Cluster block failure was observed in consecutive address locations on plane one forming a cluster of thirty-one bad blocks. Test is designed with bad-block cluster size as 25 or more adjacent failing blocks and guard band padding set to 3 blocks. Hence prominent failure mechanisms are studied with diagnosis and test method for each suggested. The units failing for SLC Program (Pattern A) test block were false fails causing a yield loss of 2.4% and the test block was made a monitor. The resulting yield of 98.7% meets the set yield goal. Since SLC program (Random Pattern), SLC program (Pattern B) and SLC program (Pattern C) provide adequate coverage for the memory array. The SLC program (Pattern A) is redundant and can be removed. Thus, test time reduction of 6.63% is achieved. The future scope of this work can be identification, diagnosis and test of other novel failure modes and mechanisms in NAND flash memory.

References

1. Li Y, Quader KN (2013) NAND flash memory: challenges and opportunities. Computer 46(8):23–29. https://doi.org/10.1109/MC.2013.190
2. Monzio Compagnoni C, Spinelli AS (2019) Reliability of NAND flash arrays: a review of what the 2-D–to–3-D transition meant. IEEE Trans Electron Dev 66(11):4504–4516. 10.1109/ted.2019.2917785
3. Mielke NR, Frickey RE, Kalastirsky I, Quan M, Ustinov D, Vasudevan VJ (2017) Reliability of solid-state drives based on NAND flash memory. Proc IEEE 105(9):1725–1750. https://doi.org/10.1109/JPROC.2017.2725738

4. Cai Y, Ghose S, Haratsch EF, Luo Y, Mutlu O (2017) Error characterization, mitigation, and recovery in flash-memory-based solid-state drives. Proc IEEE 105(9):1666–1704. https://doi.org/10.1109/JPROC.2017.2713127
5. Zambelli C, Micheloni R, Olivo P (2019) Reliability challenges in 3D NAND flash memories. In: 2019 IEEE 11th International Memory Workshop (IMW), Monterey, CA, USA, pp 1–4. 10.1109/IMW.2019.8739741
6. Muroke P (2006) Flash memory field failure mechanisms. In: 2006 IEEE International Reliability Physics Symposium Proceedings, San Jose, CA, pp 313–316. 10.1109/RELPHY.2006.251235
7. Cai Y, Haratsch EF, Mutlu O, Mai K (2012) Error patterns in MLC NAND flash memory: measurement, characterization, and analysis. In: 2012 Design, Automation & Test in Europe Conference & Exhibition (DATE), Dresden, pp 521–526. 10.1109/DATE.2012.6176524
8. Li SX (2011) Advanced failure analysis of memory devices. In: 18th IEEE International Symposium on the Physical and Failure Analysis of Integrated Circuits (IPFA), Incheon, pp 1–8. https://doi.org/10.1109/ipfa.2011.5992799
9. Martirosyan S, Harutyunyan G, Shoukourian S, Zorian Y (2017) An efficient testing methodology for embedded flash memories. In: 2017 IEEE East-West Design & Test Symposium (EWDTS), Novi Sad, pp 1–4. 10.1109/EWDTS.2017.8110074
10. Martirosyan S, Harutyunyan G (2019) An efficient fault detection and diagnosis methodology for volatile and non-volatile memories. In: 2019 Computer Science and Information Technologies (CSIT), Yerevan, Armenia, pp 47–51. 10.1109/CSITechnol.2019.8895189
11. Hsu M-H, Hsing Y-T, Yeh J-C, Wu C-W (2006) Fault-pattern oriented defect diagnosis for flash memory. In: 2006 IEEE International Workshop on Memory Technology, Design, and Testing (MTDT 2006), Taipei, p 6. https://doi.org/10.1109/mtdt.2006.13
12. Cha J, Kim I, Kang S (2011) New fault detection algorithm for multi-level cell flash memroies. In: 2011 Asian Test Symposium, New Delhi, pp 341–346. 10.1109/ATS.2011.13
13. Kim H, Ahn S, Shin YG, Lee K, Jung E (2017) Evolution of NAND flash memory: from 2D to 3D as a storage market leader. In: 2017 IEEE International Memory Workshop (IMW), Monterey, CA, pp 1–4. 10.1109/IMW.2017.7939081
14. Micheloni R, Aritome S, Crippa L (2017) Array architectures for 3-D NAND flash memories. Proc IEEE 105(9):1634–1649. https://doi.org/10.1109/JPROC.2017.2697000
15. Aritome S (2015) Challenges of three-dimensional nand flash memory. In: NAND Flash Memory Technologies, IEEE, pp 335–379. https://doi.org/10.1002/9781119132639.ch8
16. Maejima H et al (2018) A 512 Gb 3b/Cell 3D flash memory on a 96-word-line-layer technology. In: 2018 IEEE International Solid - State Circuits Conference - (ISSCC), San Francisco, CA, pp 336–338. 10.1109/ISSCC.2018.8310321
17. Kim C et al (2018) A 512-Gb 3-b/Cell 64-Stacked WL 3-D-NAND flash memory. IEEE J Solid-State Circ 53(1):124–133. https://doi.org/10.1109/JSSC.2017.2731813
18. Grossi A, Zambelli C, Olivo P (2016) Reliability of 3D NAND flash memories. In: Micheloni R (ed) 3D Flash Memories. Springer, Dordrecht
19. Feng W, Deng N (2015) Study on cell shape in 3D NAND flash memory. In: 2015 IEEE International Conference on Electron Devices and Solid-State Circuits (EDSSC), Singapore, pp 387–390. https://doi.org/10.1109/edssc.2015.7285132
20. Aritome S (2015) Three-dimensional nand flash cell. In: NAND Flash Memory Technologies, IEEE, pp 273–334. https://doi.org/10.1002/9781119132639.ch7
21. Lee S (2016) Technology scaling challenges and opportunities of memory devices. In: 2016 IEEE International Electron Devices Meeting (IEDM), San Francisco, CA, pp 1.1.1–1.1.8. https://doi.org/10.1109/iedm.2016.7838026
22. Liu W et al (2019) Characterizing the reliability and threshold voltage shifting of 3D charge trap NAND flash. In: 2019 Design, Automation & Test in Europe Conference & Exhibition (DATE), Florence, Italy, pp 312–315. https://doi.org/10.23919/DATE.2019.8714941
23. Zambelli C, Crippa L, Micheloni R, Olivo P (2018) Cross-temperature effects of program and read operations in 2D and 3D NAND flash memories. In: 2018 International Integrated Reliability Workshop (IIRW), South Lake Tahoe, CA, USA, pp 1–4. https://doi.org/10.1109/IIRW.2018.8727102

24. Zou X, Jin L, Jiang D, Zhang Y, Chen G, Huo Z (2018) Investigation of cycling-induced dummy cell disturbance in 3D NAND flash memory. IEEE Electron Dev Lett 39(2):188–191. https://doi.org/10.1109/LED.2017.2785843
25. Zou C, Zhao Y, Chien W-TK, Tang J (2017) Fail mechanism of program disturbance for erase cells VT positive shift in NAND flash technology. In: 2017 China Semiconductor Technology International Conference (CSTIC), Shanghai, pp 1–3. https://doi.org/10.1109/CSTIC.2017.7919835

UVM Based Design Verification of Interconnect Block

G. Madhura and Ravishankar Holla

Abstract System on Chip (SOC) designs involves the integration of intellectual property (IP) cores, each separately designed and verified. Most important is the method by which the IP cores are connected i.e., Interconnect. Verification of new Interconnect block design with a uniform communication bus is discussed in this paper, which can be reused among many projects. This block is combination of different bus protocols, removing signals that are chip or block specific. UVM based framework is developed to verify the new Interconnect block. Random constraint stimulus vectors generated using UVM are driven automatically to the Design Under Test (DUT) to obtain higher functional coverage. The new design and its UVM verification approach developed makes it easy to reuse the Interconnect block in different SOC architectures at both design and verification levels.

Keywords Communication bus · Interconnect · Universal Verification Methodology (UVM) · Design Under Test (DUT) · Design verification · Coverage

1 Introduction

The SOC modules typically contains several IP blocks. All the SOC modules need to communicate with each other for the proper operation of system. Interconnects are used to do the communication between them. As technology is improving day by day, the complexity of SOC blocks increases. There is need for new Interconnect block which has unified communication bus that can be reused across different SOC blocks and across different projects [1, 2]. As the design itself is moving towards a reusable environment, the verification environment should also move toward a reusable environment, so that verification does not become the bottleneck of the design which consumes up to 70% of the total design time. So, a reusable, stand-alone verification environment is needed to decrease the time needed for verification

G. Madhura · R. Holla (✉)
Department of Electronics and Communication, R V College of Engineering, Bengaluru, 560059, India
e-mail: ravishankarholla@rvce.edu.in

V. Nath and J. K. Mandal (eds.), *Proceeding of Fifth International Conference on Microelectronics, Computing and Communication Systems*, Lecture Notes in Electrical Engineering 748,
https://doi.org/10.1007/978-981-16-0275-7_45

[3–5]. This paper focuses on the design verification of new Interconnect block [6]. UVM based framework is developed to help the reuse of verification testbench along with design. Random constraint stimulus generation [7] is performed, and coverage metrics are analyzed [8–10].

Design and implementation of new approach of interconnect block, UVM based testbench framework, verification and test plan with table of testcase is described in Sect. 2. Results, waveforms of testcase and coverage report is presented in Sect. 3 and this paper is finally concluded in Sect. 4.

2 Design and Implementation of Interconnect Block

SOC designs comprises the integration of various individually designed and verified intellectual property (IP) cores. The process by which the IP cores are connected is the most important issue. SOC interconnect architectures are of types Network—on chip (NOC) and Bus architectures. Switch based interconnects used in SOC are referred to as NOC.

2.1 Basic Interconnect Architecture

The general architecture of an interconnect is as shown in the Fig. 1.

There are two types of interconnects, On-chip interconnects and off chip interconnects. There are many IP blocks (processors) in the SOC module. Cache, data or instruction storage are numerous types of on chip memory present with addition to these IP blocks. Graphics, processors, video codecs and network control units are some other application specific IP blocks integrated in SOC. For the proper operation

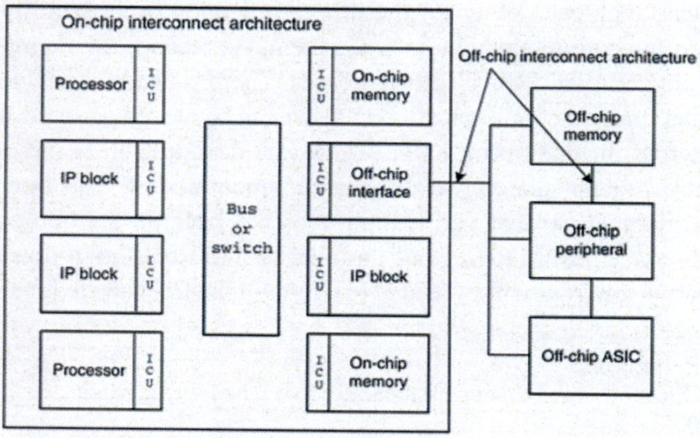

Fig. 1 Block interconnect architecture

of system communication with each other is needed by all the above SOC modules. For the communication interconnects are used between the modules.

2.2 *Design of Interconnect Block*

The UVM based verification test bench framework architecture is as shown in Fig. 2. The new Interconnect design block consists of combination of different communication protocols as shown in Fig. 1. The Interconnect block has 7 masters and 7 slaves per master for data transmission. It has sideband transmitter (tx) and receiver (rx) signals for general purpose controls. The block also includes sensor placed at uniform distance for monitoring and scan signals for DFT purpose. UVM is used for verification of this interconnect block with combination of all these different signals. A general UVM testbench architecture with single agent and multiple agents is shown in Figs. 4 and 5 respectively. The tool used for design verification is Cadence Incisive and Simvision.

The block diagram of interconnect block with signal combination of different communication protocols is shown in Fig. 3.

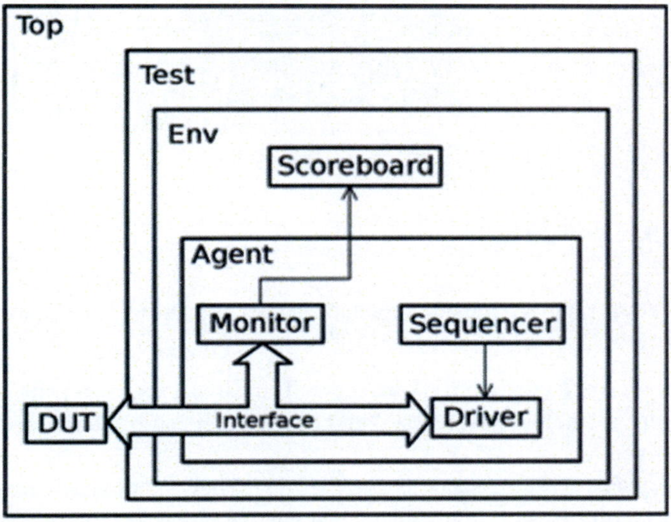

Fig. 2 Typical UVM testbench

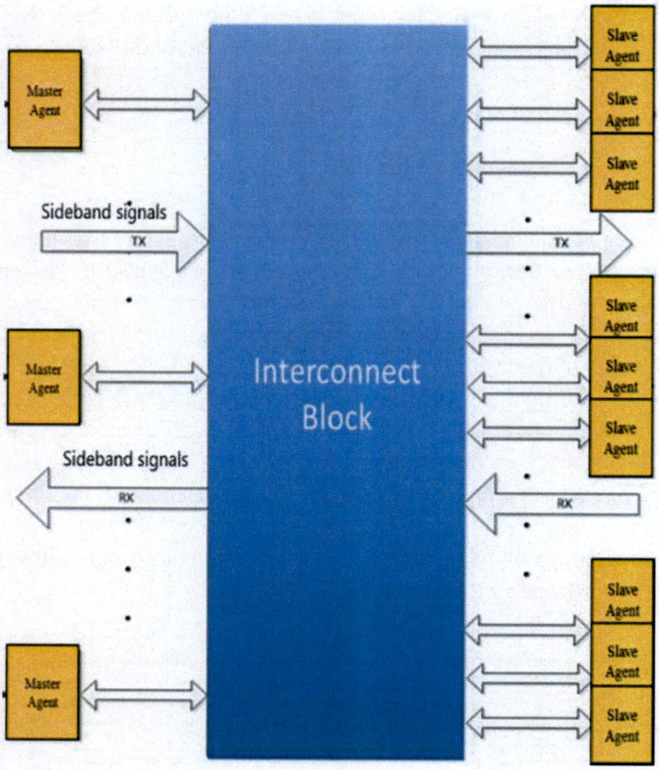

Fig. 3 Block diagram of interconnect block

2.3 Verification Plan

The Verification plan for the Interconnect design for each UVM components is explained below.

1) Agent—It is extended from UVM Agent. It consists of sequencer, master_driver, slave_driver, monitor, scoreboard and coverage. It can be configured as active or passive.
   Active agents drive transactions into the DUT. They can be configured as active in 2 ways: Master Agent and Slave agent. Passive agents observe the transactions happening on the bus and don't drive any transactions to the DUT.
2) Sequencer—It controls the flow of request and response of sequence items between sequence and the driver. Sequencer and driver use TLM Interface to communicate transactions. In this VIP, sequencer is used to drive transaction from sequence to master driver.
3) Master_driver—It is in control for communicating at the transaction level with the sequence via TLM communication with the sequencer and converting

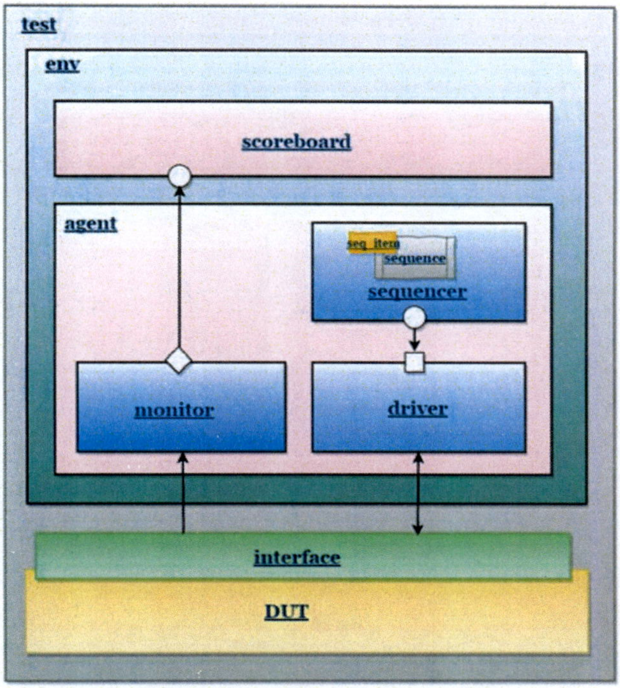

Fig. 4 Block diagram of UVM testbench with single agent

between the sequence_item on the transaction side and pin-level activity in communicating with the DUT via a virtual interface. It typically gets a sequence_item and uses this information to drive signals to the DUT. It also sends the response back to sequence_item for READ type command.

4) Slave_driver—It communicates with the DUT at pin-level via a virtual interface. It typically gets a WRITE/READ type of transaction on the pins and uses this information to drive signals to the DUT.

It does not follow typical UVM sequence-based architecture as Slave in our test-bench framework doesn't initiate the transaction as per the design protocol. Thus, we don't require a sequencer to drive sequence items to slave_driver.

In case of WRITE, it latches the address and data to memory model present in config class.

In case of READ, it sends back the data already available in memory model in config class to the corresponding pins connected through virtual interface.

5) Monitor—It communicates with DUT signals through a virtual interface and contains code that recognizes AHB transaction in the signal activity. Once a transaction is recognized, it builds an abstract transaction model representing that activity, and broadcasts the transaction to any interested components such as scoreboard in our case.

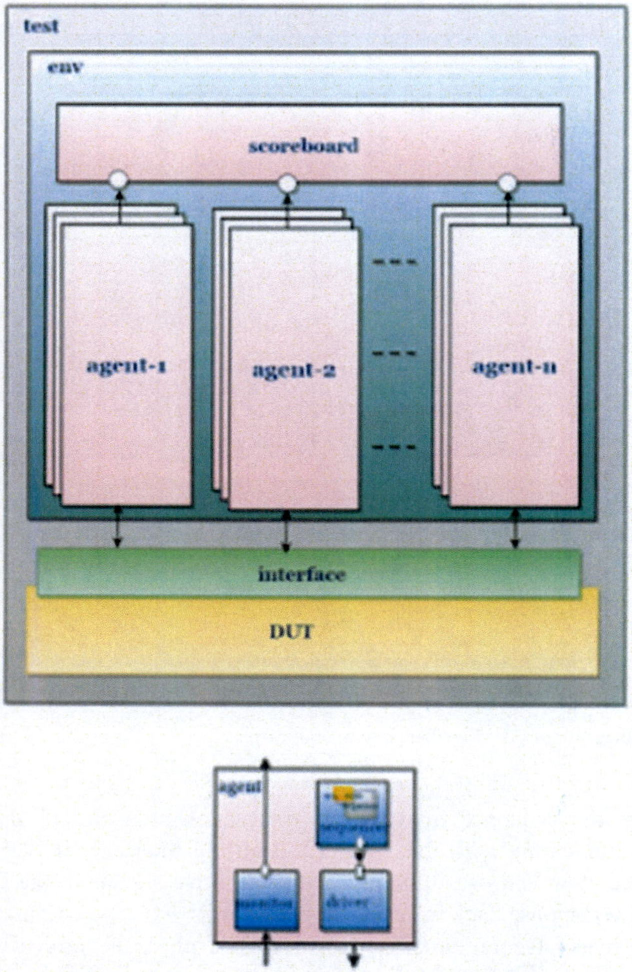

Fig. 5 Block diagram of UVM testbench with multiple agents

6) config—It has multiple functions:

a. Configures the agent. It consists of two switches: active/passive and master/slave.

Active switch controls whether the agent is in active mode or passive mode. As described above, in the active mode—sequencer and master_driver will be created or slave_driver will have instantiated based on type of agent. In the passive mode, no sequencer or driver will be created. Master switch controls whether the agent is instantiated as master agent or slave agent. monitor is instantiated always irrespective of the switch value configurations.

Based on the values for both switches, the agent will be created as one of the three possible configurations shown below:

1. agent in ACTIVE Mode as MASTER
2. agent in ACTIVE Mode as SLAVE
3. agent in PASSIVE Mode with only MONITOR

b. Configures the different types of transfers from MASTER agent.

7) sequence—It contains two kinds of items:

   a. write_trans—This is instantiated when the transfer type is WRITE. We do not wait for the response from this type of transfer.
   b. read_trans—This is instantiated when transfer type is READ. We wait for the response from this transaction and next sequence is allowed after we get response from DUT.

8) sequence_item—It contains different signal-based variables such as address, data, transfer type, size, burst type. Object of item is passed from sequence with user-defined values via sequencer to the driver and later captured by monitor and sent to scoreboard.
9) Scoreboard—The correctness of the DUT will be checked by comparing the expected values with the DUT output. It's a vital component and part of the agent. It compares the actual transactions received from monitor via analysis port with expected transactions received from memory model implemented inside config and reports the success or failure of the comparison.
10) Coverage—It will collect a copy of transactions forwarded to scoreboard from monitor. It contains cover groups sampled per transaction basis and between transactions to check whether agent completely exercises all possible configurations.

2.4 *Implementation*

The design verification flow chart is shown in Fig. 6. The design specification specified are studied and verification plan for all UVM components and test plan is prepared. Based on the verification plan of the components, environment is developed using UVM concepts.

Tests, functional coverage models are developed to ensure that the block is bug free. Regressing the tests and analyzing code coverage. Upon any failures, holding detailed discussion with the designer and modifying tests if required. Finally ensuring 100% coverage after exclusions.

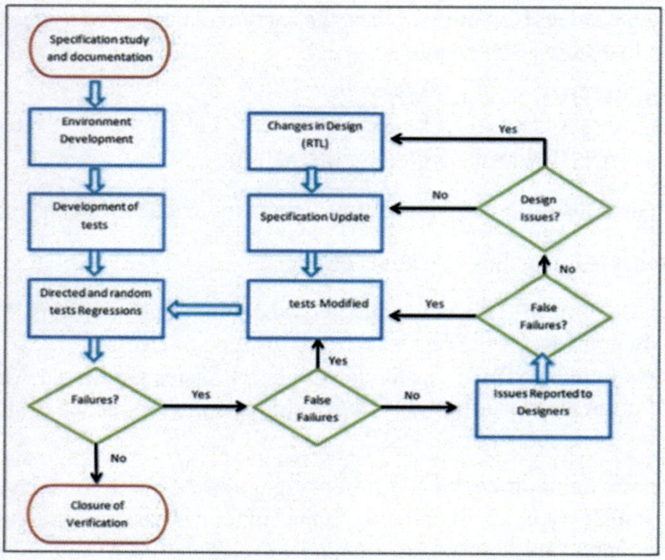

Fig. 6 Design verification flow

2.5 Testplan

Based on design specifications testcases are developed. The testcases defined for verification of few major features of this block is shown in Table 1. Every testcase addresses a particular design feature and aims for verification of that. Each testcase has randomization or stimulus strategy and self-checking strategy.

Finally, the UVM components are coded based on all the testcases along with verification plan which verifies the design.

Table 1 Verification testplan

Testcases	Stimulus strategy
sanity_test	Basic test for clock and resets up
sensor_test	Sensor connectivity check
gpc_test	To drive gpc inputs, sample gpc outputs
data_test	Issue series of random write/read across all possible legal address map ranges
scan_test	To enable and toggle scan signals for coverage

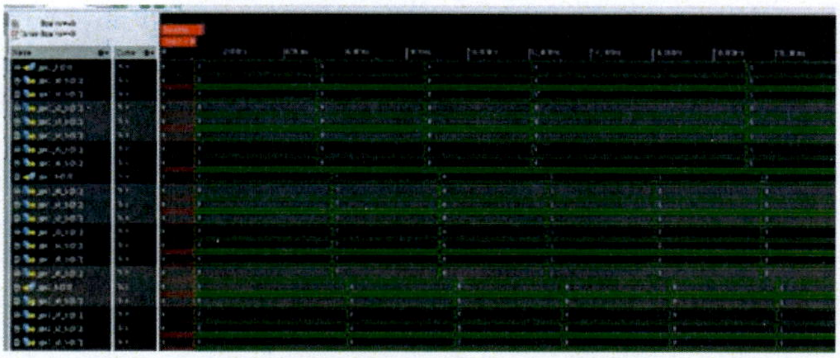

Fig. 7 Sideband signal test output waveform

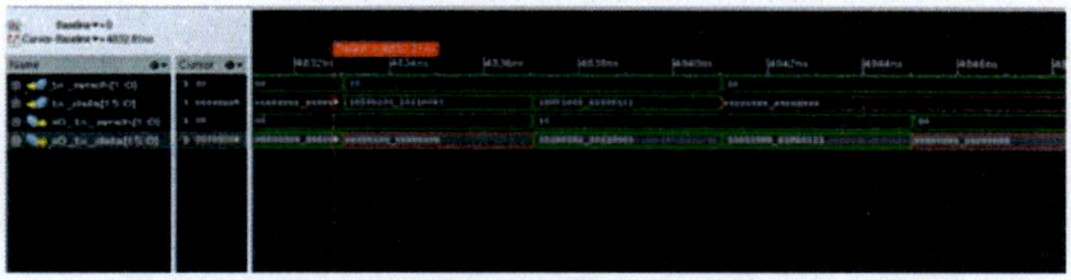

Fig. 8 Data packet test output waveform

3 Results and Discussion

Design functionality of the Interconnect block specified in Sect. 2.1, is verified using the UVM based framework designed in Sect. 2.2. The simulation was done in Cadence Incisive as well as runsim and the output waveform is are obtained using Cadence simvision.

3.1 Simulation Waveforms

The waveform of DUT behavior for gpc test is shown in Fig. 7.

Waveform displaying DUT behavior for random write transaction in data packet test is as shown in Fig. 8. It shows that random data is written for random address from the master agent is issued to right slave.

3.2 Coverage Metrics

Functional coverage was calculated using coverage model [10] and could achieve good coverage result as per verification plan. The coverage report generated for the interconnect DUT is shown in Fig. 9. The coverage includes code, all inter blocks, expression and toggle. All possible cover-groups allowed by UVM and different tests

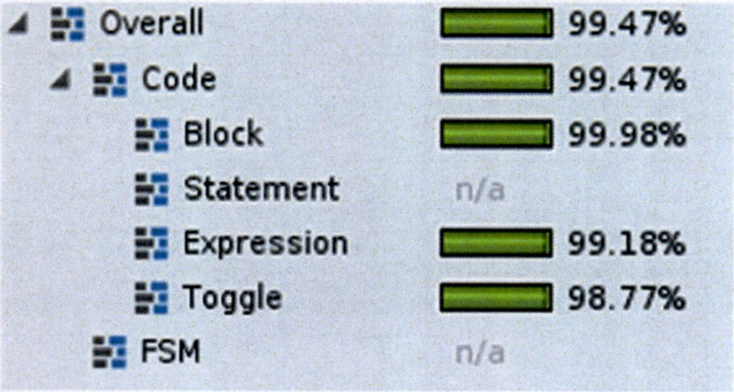

Fig. 9 Code coverage reports without exclusions

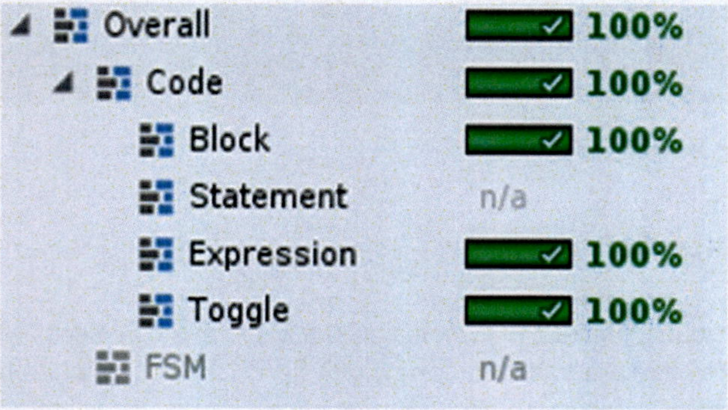

Fig. 10 Code coverage reports with exclusions

are written to remove the coverage holes in verification. Toggle coverage checks are done for signals like scan signals to check for the transition of signals from 0 to 1 and vice versa.

There are few corner cases like the code that gets covered only in SCAN mode verification or dead code that never gets hit during verification. In such cases, an exclusion list is written after getting a signoff from designer to achieve 100% coverage as shown in Fig. 10.

4 Conclusion

In this paper, design of Interconnect block with a unified communication bus with combination of different bus protocols is outlined. A UVM based framework is developed for design verification of block specified.

As the SOC design complexity is increasing day by day, the interconnect block with unified communication channel is useful for reuse across different projects. Also, UVM helps in reuse of testbench across entire project. UVM framework which has strong base class with many inbuilt features and system Verilog concepts helps in reducing the time spent on verification.

The UVM environment framework developed has master and slave agent for data packets transmission and sideband agents for general control signals. Various design features are verified providing different input stimuli based on testcases defined. It also has coverage collector component with coverage group defined for all possible configurations to enhance and achieve 100% coverage.

Acknowledgements I would like to thank my guide Ravishankar Holla, Assistant Professor, Dept. of Electronics & Communication Engineering, RV College of Engineering for his help, tremendous support and timely advice throughout the course of the project work.

References

1. Saleh R, Wilton S (2006) System-on-chip: reuse and integration. Proc IEEE 94(6):1050–1069
2. Ho WH, Pinkston TM (2006) A design methodology for efficient application-specific on-chip interconnects. IEEE Trans Parallel Distrib Syst 17(2):174–190
3. Georgoulopoulos N, et al (2018) UVM-based verification of a mixed signal design using systemverilog. In: 28th international symposium on power and timing modeling, optimization and simulation (PATMOS)
4. Krishnan KY, Bhakthavatchalu R (2019) Block Level SoC verification using systemverilog. In: IEEE conference proceedings of the third international conference on electronics communication and aerospace technology, ICECA 2019
5. Samanta P, Chauhan D, Deb S, Gupta PK (2014) UVM based STBUS verification IP for verifying SOC architectures. In: Proceedings IEEE VLSI design and test, 18th international symposium, Coimbatore. https://doi.org/10.1109/ISVDAT.2014.6881037
6. Drechsler R, Chevallaz C, Fummi F, Hu AJ, Morad R, Schirrmeister F, Goryachev A (2014) Panel: future SoC verification methodology: UVM evolution or revolution? In: Design, automation & test in Europe conference & exhibition (DATE), pp 1–5
7. Zhou Z, Xie Z, Wang X, Wang T (2012) Development of verification envioronment for SPI master interface using SystemVerilog. In: 2012 IEEE 11th international conference on signal processing, Beijing, pp 2188–2192
8. Salah K (2014) A UVM-based smart functional verification platform: concepts, pros, cons, and opportunities. In: 2014 9th international design and test symposium (IDT), Algiers, pp 94–99
9. Ni W, Wang X (2015) Functional coverage-driven UVM-based UART IP verification. In: IEEE international conference on ASIC (ASICON)
10. Cheng AC, Yen CC, Jou JY (2012) A formal method to improve SystemVerilog functional coverage. In: 2012 IEEE international high level design validation and test workshop (HLDVT), Huntington Beach, CA, pp 56–63

Adaptive Feedback Mechanism for Silicon Bug Detection

B. P. Ashish and Namita Palecha

Abstract Silicon validation is the most time-consuming process in the VLSI design flow. It mainly involves pre-silicon validation and post silicon validation. Post-silicon validation has limited visibility and control of internal signals and is the most time-consuming task. To help in narrowing down the features responsible for failure and reproduce the bug, the power of machine learning is leveraged. Based on the data provided, machine learning is used to provide the feature weights that control the failure of the device, thereby leading to faster reproduction of the failure. The framework that stresses the CPU requires manual intervention for providing test vectors and hence to have a faster response, the framework is automated with a reset mechanism so that it can include tests that are expected to fail on the device. Based on sample data, the machine learning model using supervised learning algorithms is developed. Results show that the model is accurately able to predict the features responsible for failure when applied on real time chip data. Due to automation, there is about 66% saving in time when compared to the manual process, thereby giving more time for more experiments to be run on the device and faster time to market.

Keywords Post silicon validation · Machine learning · SoC · Automation

1 Introduction

Chips are the silicon brains of mobiles, computers and servers that contain millions and millions of transistors. Correct functionality and operations of systems needs to be ensured despite rising complexity levels of chips according to Moore's law. Increased advances in theory and practice of logic and design verification pre and post manufacture of digital systems are required. Validation mainly contains two

B. P. Ashish (✉)
VLSI Design and Embedded Systems, Rashtreeya Vidyalaya College of Engineering, Bengaluru 560059, India

N. Palecha
Department of Electronics and Communication Engineering, Rashtreeya Vidyalaya College of Engineering, Bengaluru 560059, India

V. Nath and J. K. Mandal (eds.), *Proceeding of Fifth International Conference on Microelectronics, Computing and Communication Systems*, Lecture Notes in Electrical Engineering 748,
https://doi.org/10.1007/978-981-16-0275-7_46

phases, pre-silicon and post silicon. Pre-silicon verification consists of techniques that are mainly emulation, simulation and formal verification based. The limitation of pre-silicon verification is that the current technologies and tools are not sufficient for full system level verification. Hence, Post-silicon validation is required for full system functional validation.

Post silicon validation methodologies mainly focus on detection and fixing or bypassing functional issues of systems after manufacture. High design complexity makes it difficult to detect all functional bugs before manufacture, which escape in pre-silicon verification stage. Post silicon validation has four major steps. Initially, bugs are detected by running test suites containing test vectors which are combinations of features, on system under test (SUT) until the system fails. Then, accordingly, the system failure or bug is localized to a region of the system. The root cause of the bug is identified. According to the root cause, the problem is fixed or bypassed if possible. Else it is given back to designer to make changes required changes in the design.

Initial stages of validation process are slow even though testing is conducted with high speeds, due to the huge amount of data that needs to be collected and highly intensive tests that stress the device are to be executed on each chip. Once data is collected and fed back, understanding the data, relating that with chip functionalities, root-causing and fixing the problem if any requires large amounts of time. Also, data and measurements are easily impacted by environmental and process variations, hence it is a challenge to obtain effective models between data and chip functionalities. Therefore, it is highly desirable to have innovative and efficient methodologies to handle such complex and data-intensive tasks at both the pre-and post-silicon stages to help in accelerating chip designs.

Recently, great advances are made in the field of machine learning due to its unique feature of extracting structure from intensive data. Various literature proposes promising approaches for handling complex problems varying from pre-silicon to post-silicon validation and tuning. Machine learning algorithms use statistical models, from learning data and are then used to predict output with respect to new samples. For example, fraud detection algorithms are trained using the normal behavior of a customer and then monitored for fraudulent behavior and flagged if found so. Anomaly detection algorithms have similarities to bug detection in digital designs. Hence n this project, Machine learning is used, and an adaptive feedback mechanism is utilized in order to help in narrowing down of silicon bugs to speed up the validation process.

Machine learning is a concept that uses a statistical approach, in which a particular problem is modelled by using input and output data sets instead of actual system equations and response. In most real-world problems system dynamics cannot be known and modelling the ideal system model may not be easy. Machine learning techniques eases the task modelling as it does not depend on the internal system equations. Most of the machine learning algorithms are directly inspired by nature. Fuzzy logic uses human thinking, neural networks utilize the working of brains, simulated annealing uses processes of metallurgy, genetic algorithms adopt the concept of survival of fittest concept. Machine learning algorithms are classified on the basis of treatment of inputs. It is broadly classified as supervised and unsupervised. Machine learning algorithms are used in areas like computer vision, recommendation systems, medical diagnosis, stock market analysis, data mining, search engines. Post silicon validation field has been slower to utilize the power of modern machine learning techniques as compared other fields.

The validation algorithm discussed in [1], useful server processor feature validation. It is adaptive to the learning's of previous validation. The algorithm of [2] applies anomaly detection based on machine learning similar to that used in detection of credit card fraud. It focuses on intermittent bugs, where multiple executions of the same test lead to varying outcomes. In paper [3] application cases of machine learning in pre-silicon and post silicon validation such as variation extraction, bug localization and hotspot detection are discussed. [4] proposes the use of machine learning in post silicon validation for automation of trace dump diagnosis and aiding in bug localization. [5] discusses bug localization in processors using bug localization graphs. [6] talks about the opportunities, challenges and advances in post silicon validation. In [7], a novel method for volume diagnosis is proposed using techniques of machine learning instead of the conventional cause effect analysis. In [8], a wafer fault detection system based on machine learning is proposed. In [9], it is discussed that fault detection techniques help semiconductor manufacturers by reducing test wafer usage, equipment uptime and reduce scrap. In [10], the challenges of intermittent bugs are discussed. These are bugs that cause multiple executions of the same test to produce varying outcomes. These are usually caused due to asynchronous events on the chip or due to electrical effects. In [11], the challenges faced during post silicon debug are discussed.

From the literature review it is found that machine learning can be used in both pre-silicon and post silicon validation phases. Literature shows previous works that have implemented the same. Different algorithms like anomaly-based detection techniques are used for bug localization. Trace dump analysis to find bugs is also used in literature. Various works discuss of the model, but the training process is done using historical static data. For iteratively training the data of new chipsets, automating this process and detecting feature weightage, there would be a need of feedback to the validation framework which hasn't been discussed in literature.

Present work proposes an adaptive feedback mechanism that uses machine learning to learn the feature weights of the features responsible for failure. The automation is done on an existing stress test framework that is executed on the chip. The reset mechanism is implemented using Lauterbach Trace32 software which allows programming the debugger to which the chip is connected. The machine learning model is developed using Python on Spyder tool. Results show significant saving of time due to automation when compared to the manual process and the developed machine learning model is verified to be accurate in its prediction.

The present paper is organized as follows. Section 2 deals with the fundamentals of machine learning and supervised learning algorithms. Section 3 contains the implementation details of the proposed work. The results obtained are presented in Sect. 4. Finally, Sect. 5 concludes the paper.

2 Overview of Machine Learning

Application of artificial intelligence is Machine Learning. It enables systems to learn automatically and improvise from past experience without actually being programmed. Machine learning is mostly used for developing computer programs for accessing data and using it learn for themselves. The learning process involves data or observations, such as direct experience, instruction, and examples to look for and find patterns in data that can be used to make better predictions on the future data based on the learnings and observations provided. The major goal of machine learning is to allow computers to automatically learn without manual intervention or help and make changes to their actions accordingly.

Machine learning algorithms are basically categorized as supervised or unsupervised. Supervised algorithms are those which learn from labelled data and apply the same to new incoming data using learnt examples for predicting future events. The training dataset which consists of examples and observations is used by the learning algorithm to infer an inherent function that can be used to make predictions about the output values. Once sufficient training is done, the model is able to predict outputs of any new incoming data. Comparison with the intended output of the predicted output helps the learning algorithm to find errors and modify accordingly.

In contrast, unsupervised machine learning algorithms are those which are used when unlabeled or classified information is present. Unsupervised learning algorithms try to determine system function that can be used to identify a hidden structure from data that is not labeled. The model doesn't know the correct output but can explore and draw conclusions from datasets to find out hidden patterns or structures that define the system function. Semi-supervised machine learning algorithms are those that utilize both labeled and unlabeled data for training the model and hence fall in between supervised and unsupervised algorithms. Usually, a large amount of unlabeled data and small amounts of labeled data is used for the purpose of training the model. Examples show that this type of mixed learning models show high learning accuracy. This type of machine learning algorithm is used when the labeled data requires extra skills and more resources for learning and training the model and unlabeled does not require extra sources.

Reinforcement learning is another machine learning method whose main feature includes interacting with the environment by performing certain actions and evaluating these actions as correct or wrong using rewards or punishments. Optimum accuracy is obtained using trial and error methods and by providing rewards for good predictions and punishments for bad or wrong predictions. It allows for the model to maximize its performance by determining the optimum relations between input and output parameters for a given problem. A reward feedback called the reinforcement signal is used to make the model understand that the action performed is best. In a similar way punishment are awarded for bad actions.

As the problem of narrowing down features involves labeled data, supervised learning algorithms would be the most efficient way to approach the problem. Linear regression is a supervised learning algorithm which assumes that the input variables and output variables have a linear relationship with each other. To elaborate, the output variable can be easily calculated by using a combination of the input variables in a linear fashion.

Decision Tree is another important supervised learning algorithm under machine learning. CART, stands for classification and regression trees, is used for modeling applications involving classification and regression. The CART model is represented using a binary tree. The familiar binary tree used in data structures and various algorithms is used here. Each input variable is represented by each node in the binary decision tree. The output variables are represented by the leaf nodes of the decision tree. The leaf nodes are used to make the final prediction.

Naive Bayes is a simple and powerful supervised learning algorithm used in predictive modeling. Often in statistics and machine learning, the main goal is to select the best hypothesis that suits the data. For problems involving classification, the hypothesis can be taken as the label or the class to which a new sample is to be assigned. The best suited hypothesis can be selected based on the given data using past experience or prior knowledge. Bayes' Theorem is a mathematical equation that helps in calculating the probability of an outcome or a hypothesis given the prior knowledge or past history.

The K-Nearest Neighbor algorithm is another supervised learning model that uses the training data for its model representation. As there is no requirement of modeling in KNN, learning is not required. Data structures like k-d trees can be used for making look up tables and for the matching procedure of new patterns thereby making the overall prediction mechanism quite efficient. The KNN model directly makes predictions using training data. When a new sample comes in, the model searches in the existing data set for the most similar neighbors and provides an output based on the search.

3 Implementation

This section discusses the implementation details of the adaptive feedback mechanism. For implementation of the machine learning model, Python scripting on Spyder tools is used. Machine learning libraries including scikit-learn, pandas, numpy, and matplotlib are used. The existing framework stress tests the CPU based on low power modes, frequency and voltage scaling and other such workloads. Framework is written in C and runs over the APSS (Application Processor Sub System) of SoC. It dynamically takes user input and performs required test in an infinite loop. The tests can be changed on the y. The framework runs on the APSS and hence if there is a fail scenario and APSS is hung, the framework also is hung until reset. As manually checking for each test case and waiting for results for each case can be tedious, an input file containing all tests to be run is created. The framework is modified so as to take each row of the input CSV (Comma Separated Values) file as a test vector and within the loop perform the tests one row after another. For implementing the automation of input vectors, advanced file handling operations are utilized.

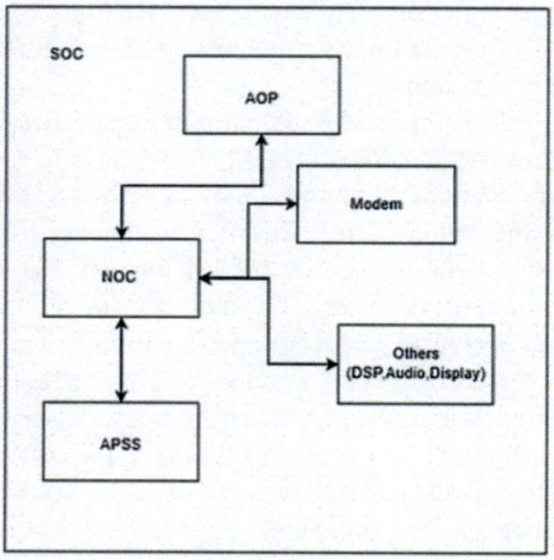

Fig. 1 Basic block diagram of SoC

Feedback loop is implemented using reset mechanism. As framework stress tests the APSS Proc, if there is a fail scenario, the framework is hung and requires a reset. In an octacore CPU, the framework can be run on any of the cores. But in a fail scenario if the CPU is hung, a manual reset would be required. In existing SoC architecture there exists another Proc called AOP (Always on Processor) which is basically another processor external to the CPU which is used for resource power management as shown in Fig. 1. A timeout period is used to ensure failure has occurred and hence reset the APSS Proc. Trace32 is used to run the framework. In order to communicate between processors, cmm (Trace32) script is used.

In order to account for the test cases that result in a failure, a reset mechanism is used. The reset mechanism makes use of the AOP to monitor a timer and reset the APSS processor it there is no response from the APSS within the time out period. A file containing the experiment and iteration count values are monitored. If after a maximum time required to run the test cases, the experiment and iteration numbers are not updated, it is assumed that the chip has failed for the particular test and that the stress test framework is hung and requires a reset.

The reset mechanism flow is as shown in Fig. 2. The reset mechanism is explained as follows. Initially reset variable is made 1 and the device is reset. AOP instructs the APSS proc to run the validation suite through JTAG Debugger. A while loop comparing current experiment to previous experiment is checked for. If current experiment is equal to previous experiment it means that updation has not taken place and there is a hang detected. Else the loop is entered and iteration counters are initialized. Current iteration is compared with previous iteration, if not equal (running), previous iteration variable is updated with current iteration value and current iteration variable is updated with incremented value. Between updation of iteration variables, a timeout is set so that reset is not activated unnecessarily. If iteration values are equal, a hang is detected and reset variable is made 1. The mechanism is repeated for all experiments in the input file. Once last experiment is reached, the script exits

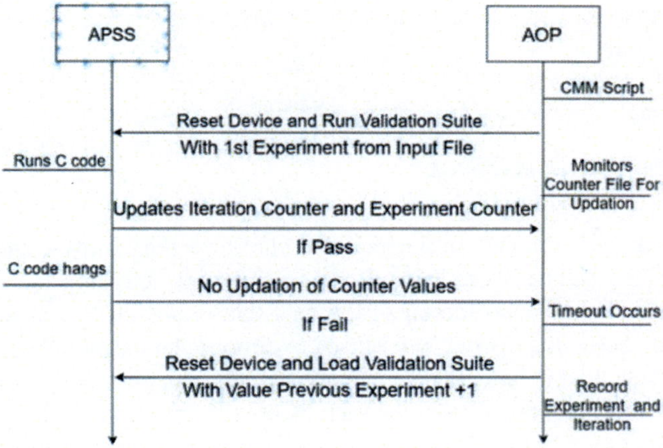

Fig. 2 Reset mechanism flow

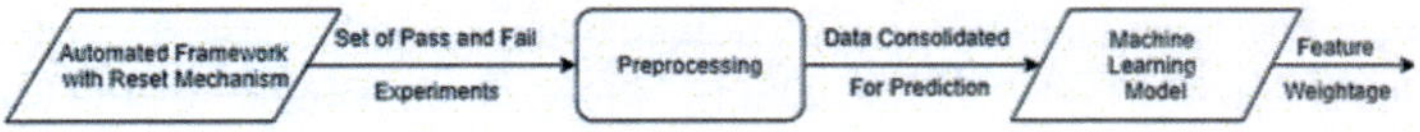

Fig. 3 Overall adaptive feedback mechanism flow

from the loop and ends the execution. In case a failure has occurred, reset is made 1 and the experiment number and iteration at which the experiment failed is recorded. The iteration number gives a notion of time, as in, in a fail scenario, if the same experiment ran for more number of iterations when compared to another which ran for few iterations, we can conclude that the experiment that ran for more iterations took longer to fail. Along with the experiment number and the iteration number the symptoms of failure, detectable if any are recorded in a .txt file. The output of the automation of test vector input to the framework, along with the reset mechanism gives the set of experiments that have failed and the iteration at which the experiment failed gives a notion of time. Using this information, data is pre-processed before providing it to the machine learning model to detect feature responsible or failure.

Initially when test vectors are executed on the chip, the results of whether the chip will pass or fail the given test is not known. In order to find out which feature in the experiment is responsible for failure, it is required to know which experiment when executed, fails. From the output of the automated framework, with the reset mechanism, the set of failing experiments are obtained. Before providing this data to the model, the set of passes and failing data must be provided in a specific format that enables prediction by the machine learning model. Hence, preprocessing the data is required. Figure 3 shows the overall process flow. Python script is used to post processes the data and formulate it in a cleaner fashion indicating pass or fail, so that the machine learning model can work on the data. The output of the model is weightage of the corresponding parameters thereby giving an idea of which parameter is causing failure.

4 Results and Discussions

Initially, for the development of the model to check the feasibility, a sample dataset was used. The machine learning model is developed by utilizing the sample data which resembles the actual format of the data that is obtained from the real time chip. Python code using Spyder tool is used to develop the machine learning model. As the data involved is labeled data, supervised learning algorithms are used. The major supervised learning algorithms such as linear regression, logistic regression, CART, Naive Bayes and KNN are compared using testing and training data split, cross validation technique and accuracy of prediction.

For automating the test vector input, the framework was altered using file handling operations. The input file is basically a CSV file that holds all possible values for each feature and workload. The input values 1 and 0 indicate whether the feature or workload is enabled or disabled. The iterations column indicates the number of times the current experiment is to be run as the test may fail only after being run for more than once. At each iteration, the framework is made to take the complete row of workload and features and execute it on the chip and log its results. Once the experiment is completed, the next experiment is taken and so on.

Once the test vector input file is loaded onto the automated framework, each test is executed and the output of the test is logged. More specifically, the experiments that fail are logged onto another.txt file. The information that is logged consists of the experiment number, the iteration number at which the experiment failed, the symptom of the failed experiment which is the state of the CPU which could be a hang, may be a data abort, may be a memory abort or may not show any symptom. Once the logging process is over, the chip is automatically reset and the next experiment in the input vector file is run. The process again repeats until all experiments in the test vector file have been exhausted. The output of the automated framework is as shown in Fig. 4.

The reset mechanism works on a time out basis. It waits for a file counter to be updated within the specified time, failing upon which, the chip is automatically reset. The file containing the symptoms and failed experiments information is preprocessed before providing as input to the developed machine learning model.

The preprocessing stage is used to format the data in order to provide to the machine learning model for prediction of feature weights. It is a python script that consolidates the experiments executed and the failed experiments. The actual real time data given to the chip is as shown in Fig. 5. The present scenario consists of 25 experiments that need to be run on the chip. The results of the experiments are not known before hand and hence all possible combination to stress test the CPU are incorporated into the file. Once the framework executes all the experiments of the input vector file, the output obtained would be the set of failed experiments and

```
Experiment Number:1,Failed at iteration number:2
With Symptom:


Experiment Number:2,Failed at iteration number:3
With Symptom:
Core 6 hang


Experiment Number:3,Failed at iteration number:4
With Symptom:
Core 6 hang


Experiment Number:4,Failed at iteration number:2
With Symptom:
Core 0 hang
Core 1 hang
Core 2 hang
Core 3 hang
Core 4 hang
Core 5 hang
Core 6 hang
Core 7 hang


Experiment Number:6,Failed at iteration number:2
With Symptom:
Core 6 hang
```

Fig. 4 Output of automated framework

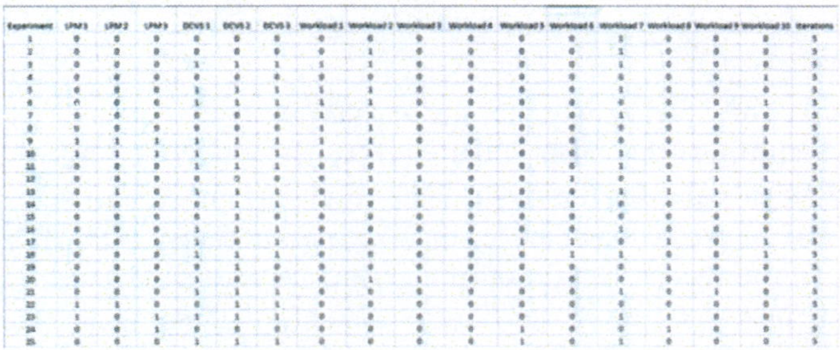

Fig. 5 Real time input vector to chip

their iterations along with their symptoms. As symptoms are used later on, another file stores the experiment numbers and their iterations as shown in Fig. 6.

The input test vector file, along with the failed experiments file is given as the input to the preprocessing stage. The preprocessing stage consolidates the two data files and formats it as shown in Fig. 7 so that it can be given to the machine learning model for prediction. The output consists of all experiments performed and whether they passed or failed as indicated with a 0 if fail and a 1 if pass.

Experiment	Iteration
3	3
5	1
10	1
11	1
13	3
16	1
18	1
21	1
23	1
25	1

Fig. 6 Failed experiments and their iterations

```
[[0 0 0 0 0 0 0 0 0 0 0 0 1 0 0 0 1]
 [0 0 0 0 0 0 0 1 0 0 0 0 1 0 0 0 1]
 [0 0 0 1 1 1 1 1 0 0 0 0 1 0 0 0 0]
 [0 0 0 0 0 0 0 0 0 0 0 0 0 0 0 1 1]
 [0 0 0 1 1 1 1 1 0 0 0 0 1 0 0 0 0]
 [0 0 0 1 1 1 1 1 0 0 0 0 0 0 0 1 1]
 [0 0 0 0 0 0 1 1 0 0 0 0 1 0 0 0 1]
 [0 0 0 0 0 0 1 1 0 0 0 0 0 0 0 0 1]
 [1 1 1 1 1 1 1 1 0 0 0 0 0 0 0 1 1]
 [1 1 1 1 1 1 1 1 1 0 0 0 1 0 0 0 0]
 [0 0 0 1 1 1 0 0 0 0 0 0 1 0 1 0 0]
 [0 0 0 0 0 0 0 1 1 0 0 1 0 1 1 1 1]
 [0 1 0 1 1 1 0 0 1 0 0 0 1 1 1 1 0]
 [0 0 0 0 1 1 0 0 1 0 0 1 0 0 1 1 1]
 [0 0 0 0 1 0 0 0 0 0 0 0 0 0 0 0 1]
 [0 0 0 1 1 1 0 0 0 0 0 0 1 0 0 0 0]
 [0 0 0 1 0 1 0 0 0 0 1 1 0 1 0 1 1]
 [0 0 0 1 1 1 0 0 0 0 0 1 1 0 0 1 0]
 [0 0 0 0 1 0 0 0 0 0 0 0 0 1 0 0 1]
 [0 0 0 1 0 0 0 1 1 0 0 0 0 0 0 0 1]
 [0 0 0 1 1 1 0 0 0 0 0 0 1 0 1 0 0]
 [1 1 0 0 1 1 0 0 0 0 0 0 0 0 0 0 1]
 [1 0 1 1 1 1 0 0 0 0 0 0 1 1 0 0 0]
 [0 0 1 0 0 0 0 0 0 0 1 0 0 1 0 0 1]
 [0 0 0 1 1 1 0 0 0 0 1 0 1 0 0 0 0]]
```

Fig. 7 Output of preprocessing stage

Once the data of experiments performed and the failed experiments is consolidated in the required format that enables the machine learning model for prediction, it is provided as an input for the model. The results of the comparison of various algorithms and the prediction of the feature weights is as shown in Fig. 8. The CART algorithm performs better than the other algorithm as seen from its accuracy score. The last row shows the feature weight responsible for failure. As observed, the 6th feature has 28.5% and the 13th feature has 71.42% control over the outcome. This is again verified by running experiments from the database that have feature 6 and 13.

The main objective of this work is to eliminate manual intervention in order to enable faster bug resolution through root cause analysis. If the whole process is done manually, the average execution time required to run an experiment with all features enabled for onetime would take 30 min. For dynamically loading the next experiment it would take at least 15 min to skim through required features, and in case there is a failure, time for resetting the device manually would be around 15 min. Analysis of symptoms by looking at each symptom register would take another 15 min. Thereby, for 1 iteration of an experiment, it would take 35 min in case of a pass and 60 min in case of a fail in order to move to the next experiment. Through automation, in

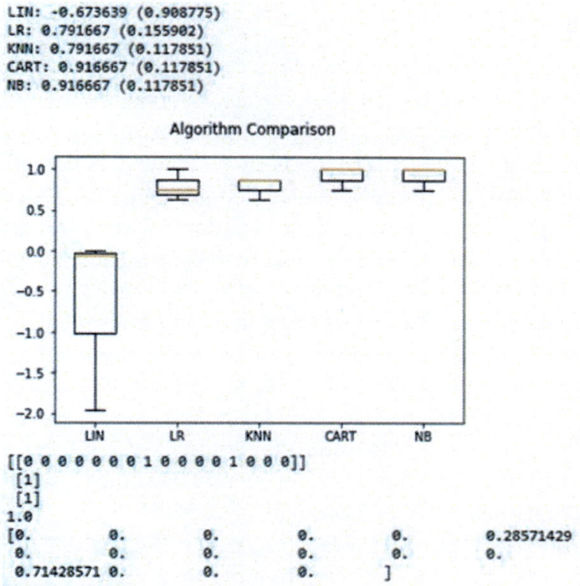

Fig. 8 Comparison and prediction by the model

Table 1 Comparison without automation

Method	Max time per experiment iteration	Max time for 25 experiments with 5 iterations each
Without automation	75 min	6 days
With automation	30 min	2 days

case of a passing scenario, there is not much significant gain as only time required to skim through features and load the experiment is saved. But in the case of a failing scenario, along with time required to skim through features, time for reset and time or analyzing symptoms is saved, thereby saving up to 45 min per iteration. Consider a case where 25 experiments known to fail, each with 5 iterations are present. The overall time required without automation would be approximately 6 days. Whereas with automation, the same is done in approximately 2 days, which is equal to the actual run time of the tests, saving time of about 4 days which is immensely significant in post silicon validation. Per iteration, about 2/3rd or about 66% of the time is saved. The time saved using automation is summarized in the Table 1.

5 Conclusion

This work describes an adaptive feedback mechanism that is used in post silicon bug detection. The power of machine learning techniques involving supervised learning algorithms are leveraged to help in faster root cause analysis. Automation of the stress test framework is done using file vector input. In order to account for experiments that cause failure of the device, a reset mechanism is implemented and before the data is provided to the machine learning model, it is preprocessed and made in the required format. Finally, the developed machine learning model is used on the processed data and feature weights of each feature responsible for failure is predicted.

Proposed work is unique as compared to other works in literature, as it utilizes a feedback mechanism, along with automated test vector input and analysis of symptoms. The feedback mechanism is capable of handling experiments that lead to both pass and fail. The automation of the framework provides significant saving of time. On comparison with the manual process, the results show the advantages of automation. Without automation, tasks such as dynamic entry of features to enable or disable for each experiment, resetting of the device in case of failure and analysis of symptoms by reading of system registers require more time. A significant time of about 45 min per iteration is saved using automation which is about 2/3rd or about 66% of the time. With the saved time, more number of experiments can be run allowing faster root cause analysis and faster time to market of the chips.

References

1. Pridhiviraj P, Tomar Dheerendra S, Muralidhar P (2015) Adaptive post-silicon server validation using machine learning. Int J Appl Inf Syst 9(1):24–32
2. DeOrio A, Li Q, Burgess M, Bertacco V (2013) Machine learning-based anomaly detection for post-silicon bug diagnosis. In: Design, Automation & Test in Europe Conference & Exhibition (DATE)
3. Zhuo C, Yu B, Gao D (2017) Accelerating chip design with machine learning: from pre-silicon to post-silicon. In: 30th IEEE International System-on-Chip Conference (SOCC)
4. Mandouh EE, Wassal AG (2017) Application of machine learning techniques in post-silicon debugging and bug localization. J Electron Test 34(2):163–181
5. Park S, Bracy A, Wang H, Mitra S (2010) Blog: post-silicon bug localization in processors using bug localization graphs. In: Proceedings of the DAC
6. Mitra S, Seshia SA, Nicolici N (2010) Post-silicon validation opportunities, challenges and recent advances. In: Proceedings of the DAC
7. Wang S, Wei W (2009) Machine learning-based volume diagnosis. In: Proceedings of the DATE
8. Kim D, Kang P, Cho S, Lee H, Doh S (2012) Machine learning-based novelty detection for faulty wafer detection in semiconductor manufacturing. Exp Syst Appl 39(4):4075–4083
9. He QP, Wang J (2007) Fault detection using the k-nearest neighbor rule for semiconductor manufacturing processes. IEEE Trans Semicond Manuf 20(4):345–354
10. DeOrio A, Khudia DS, Bertacco V (2011) Post-silicon bug diagnosis with inconsistent executions. In: IEEE/ACM International Conference on Computer-Aided Design (ICCAD)
11. Basu K, Mishra P (2013) Rats restoration-aware trace signal selection for post- silicon validation. IEEE Trans Very Large-Scale Integr (VLSI) Syst 13(4):605–613

12. Keshava J, Hakim N, Prudvi C (2010) Post-silicon validation challenges: how EDA and academia can help. In: Design Automation Conference
13. Josephson D (2006) The good, the bad, and the ugly of silicon debug. In: 43rd ACM/IEEE Design Automation Conference
14. Bishop CM (2007) Pattern Recognition and Machine Learning
15. Mitchell T (1997) Machine Learning
16. Rahmani K, Mishra P (2017) Feature-based signal selection for post-silicon debug using machine learning. IEEE Trans Emerg Top Comput 8:907–915
17. Rahmani K, Ray S, Mishra P (2017) Postsilicon trace signal selection using machine learning techniques. IEEE Trans Very Large Scale Integr (VLSI) Syst 25(2):570–580
18. Youn S, Gu C, Kim J (2014) Probabilistic bug localization via statistical inference based on partially observed data. In: Proceedings of the DAC, pp 1–6
19. Yang H, Su J, Zou Y, Yu B, Young EFY (2017) Imbalance aware lithography hotspot detection: a deep learning approach. In: Proceedings of the DAC, pp 62:1–62:6
20. Capodieci L (2017) Data analytics and machine learning for design-process-yield optimization in electronic design automation and IC semiconductor manufacturing. In: Proceedings of the CSTIC, pp 1–3
21. De Paula FM, Gort M, Hu AJ, Wilton SJE, Yang J (2008) Backspace: formal analysis for post-silicon debug. In: Formal Methods in Computer-Aided Design
22. Wagner I, Bertacco V (2008) Reversi: post-silicon validation system for modern microprocessors. In: IEEE International Conference on Computer Design

Efficient Use of DDR Bandwidth and Space for Rendering 3D Graphics

Sagar Bamashetti, Himanshu Govil, Pavan Kumar Akkaraju, and Sudha R. Karbari

Abstract Latest mobile device processors like Snapdragon have multicore CPU and a mobile GPU embedded on a single chip, which enables implementation of high-end 3D graphics application into the mobile devices. As a result, the emerging demand in on rendering on power and memory budget. Texture Processing is a DDR intensive operation Therefore, it becomes important to find ways of reducing DDR BW usage by Texture processors especially in mobile devices. In this paper we propose a novel method to effectively use the available DDR space on a mobile device by calculating the bounding box for all the textures and loading the texture grids which has maximum coverage percentage and storing those texture in the memory. Manhattan benchmark was used for graphics simulation and for a single frame the average texture coverage for top 5 APIs was found to be 62.81% by using texture gridding logic. Next, 30 consecutive frames were used to analyze the texture reusage in multiple frames using PSNR value and we proposed a novel algorithm based on PSNR value to effectively store the texture in DDR.

Keywords 3D graphics · Adreno · PSNR · Texture mapping · Texture processor · Texture gridding

1 Introduction

Mobile gaming is advancing at an unprecedented rate, with vast amounts of money being invested for large commercial gain. It is expected that the gaming industry will be worth over 250 billion rupees by 2024, marking a marvelous potential in gaming

S. Bamashetti (✉)
VLSI Design and Embedded Systems, RV College of Engineering Bengaluru, Bengaluru, India

H. Govil · P. K. Akkaraju
Qualcomm Pvt. Ltd. India, Bengaluru, India

S. R. Karbari
Department of Electronics and Communication Engineering, RV College of Engineering Bengaluru, Bengaluru, India

V. Nath and J. K. Mandal (eds.), *Proceeding of Fifth International Conference on Microelectronics, Computing and Communication Systems*, Lecture Notes in Electrical Engineering 748,
https://doi.org/10.1007/978-981-16-0275-7_47

sector. In recent years we can see that most of the gaming companies are bringing computer games to mobile devices. Fortnite, PUBG and Call of Duty are recent examples of this porting. This insurgence has put lot of demand on mobile GPU capabilities because people want to play PC like games on mobile now. Because of this, mobile GPUs need to improve their performance and power to enable these new games. Though the performance of mobile phones has improved by a large number of times it still can't be compared with the desktop PC's in terms of CPU frequency, DDR, battery life.

There are many effective techniques in synthesizing high quality image, Texture mapping is one such effective technique. Hence study of texture mapping is important because it is applicable throughout the image processing and computer graphics. Texture Processing is a DDR intensive operation because it needs to read data from input texture images which are stored in DDR. Therefore, it becomes important to find ways of reducing DDR BW usage by Texture processors especially in mobile devices. Saving DDR BW will also result in lower DDR power and hence, will help in improving both performance and power. Also, the DDR will be shared between multiple application on the device, operating system's kernel background applications, updating the software. The overall performance of the mobile device depends on the Texture processing methods adopted in the gaming application.

Consider a 16,384 $\times$ 16,384 texture at 32 bits per pixel it would take 1 GB of memory, and if a single frame uses multiple textures for rendering, the size of the memory will increase drastically. Also, the available cache memory is very limited within the GPU. Loading the entire texture to DDR is not feasible, this will increase the fetch time for the texel and in turn affect the overall performance of the GPU. To avoid we can load & unload only a certain portion of the texture which is currently being used by drawing a bounding box, so that it can reduce the overall memory consumption.

In this paper we will discuss a framework to extract the coordinates for textures and determine the bounding box for various mipmaps. Also suggest a method to decide which texture must be stored in DDR to improve the efficiency. Manhattan benchmark from the GXFBench was used to carry out the simulation.

2 Literature Review

User's perspective model to predict the performance of mobile GPUs particularly on the Qualcomm Adreno GPUs was proposed in [1]. They proposed a Unified Shader (US) based model for calculating the performance they used US's instruction per second (USI) to represent the core graphics performance. The proposed model had an average error rate of 3.32% in predicting the performance on the Adreno GPUs when compared with the performance parameters from the Snapdragon profiles. In [2] they proposed a more accurate and practical model to measure the GPU performance by modifying the model presented in [1]. Along with the US they took texture mapping performance also as a parameter to predict the GPU performance. They measured the

GPU efficiency by varying the number of texture mapping units. They also proposed a performance prediction model based on frames per second (FPS). They derived an equation based on the FPS to predict the GPU performance. Texture based micro Benchmarks were used to conduct the experiments. The proposed model had an average error rate of 5.77%. All the experiments were carried out on the Qualcomm Adreno GPUs and they claim the proposed model can be applied to GPUs other than Adreno. In [3] they propose a framework for rendering on a power budget for graphics applications. This framework used optimal power and maximizes the visual quality with minimum power consumption. They disabled as many external CPUs as possible to reduce the CPUs impact on mobile GPUs.

To reduce the computational complexity in determining the partial derivatives while calculating the position of the pixels, in [4] they proposed Heckbert method which reduce the number of addition and multiplication by caching the intermediate values of computation. Storing of the intermediate values improves the performance of filtering significantly. In [5] they propose an algorithm to recover missing regions in an image. The proposed algorithm has 2 stages to reconstruct the images. A design framework for embedded game on mobile devices was proposed in [6] which implements a highly productive mobile billiards game in mobile devices. They also discussed a high-performance collision detection algorithm based on mobile device characteristics. In [7] they propose a method for filling in blocks of lost data of images in transmission. They aimed to rebuild the missing data using correlation between the lost block and its neighbours. New algorithm for adapting dictionaries in order to accomplish sparse signal representation was proposed in [8]. A new technique for texture synthesis using optimization by defining Markov Random Field based similarity matric with respect to the given sample was discussed in [9]. In [10] they present a model to implement a sparse description of the images content based on the texture.

In [11] a complete pipeline for 3D modelling from terrestrial image data, different approaches are considered, and various steps involved were analyzed. In [12, 13] new and effective texture-based methods were discussed for detecting the moving objects. New texture classification method was proposed in [14]. A detailed review on the fundamentals of texture mapping, filtering in texture mapping was detailed in [15]. Methods to reconstruct the images in the transmission were discussed in [16, 17].

Details of the sparse texture implemented in latest Apple bionic chips was detailed in [18]. These textures control the storage of the texture at fine granularity. These textures are not completely resident in the memory they are stored on a special memory called 'Spare Heap'. In sparse texture the textures are split into units called 'Sparse tiles'. In these chips unlike other chips the granularity can be maintained at tiles level instead of the mipmap level.

3 Background

3.1 Adreno GPU

Adreno is a GPU from Qualcomm embedded in its Snapdragon processors. Adreno GPU is specially designed for mobile devices taking into consideration various mobile device constraints. Latest Adreno GPUs support various APIs like Vulkan, OpenGL, OpenVG, Directx and Direct3D. Adreno has various special features like Multiple texture: uses more than one texture for a single polygon. Video textures: real time video is applied to the surface as a texture. Texture compression: reduce the memory usage and bus bandwidth required. Early Z rejection: It helps in rejection of unwanted or hidden pixels and avoiding render passes for these objects. Tile-based rendering: scene is broken down into multiple smaller rectangular regions for rendering. Scalar architecture: supports scalar component natively. In terms of hardware usage this is more efficient as a full vector component is not used to process a single scalar [19].

3.2 Texture Mapping

Computer graphics is commonly used today in various fields like mobile gaming, videos, medical investigation etc. A computer graphics can be a 2D or 3D. In synthesizing high-quality images texture mapping is effective method.

In Texture mapping a texture which is a 2D surface is wrapped around an object which is either a 3D object. Thus, these 3D object gets a surface texture like that of the texture or 2D surface. A simple texture mapping involves processes as in shown in Fig. 1, where a sphere-shaped 3D object is applied a texture to create a globe.

3D is basically all about representations of shapes in a 3D space, with a coordinate system which is used to calculate the position of shape. The x axis is towards the right, the y axis is towards the up, and the z axis points out of the screen. Different objects are build using vertices. Texture, normal, Position and colour are few of the parameters by which each vertex is defined. All the GPUs used have a specific

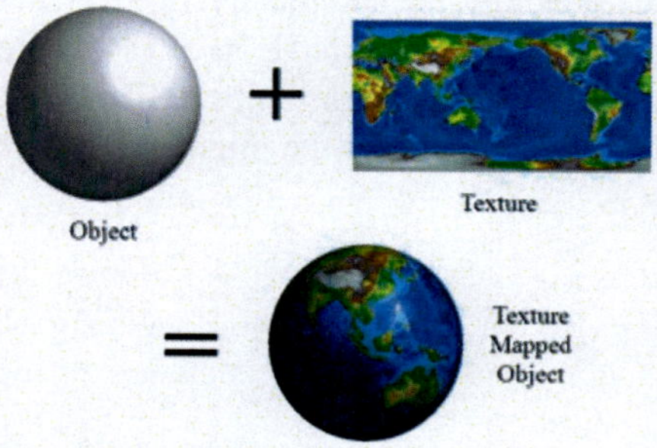

Fig. 1 Texture mapping

rendering pipeline by which the images are prepared and displayed on the screen. There are various steps involved with in this pipeline like *Vertex Processing*: where the primitives are formed from the vertex by setting there coordinates in the 3D space. *Rasterization*: this converts the primitives to a set of fragments which are the prediction of the pixels on the display with respect to the 3D object it is basically determining which pixels belongs to which objects. *Fragment Processing*: calculates the final colour based on the given parameters and it mainly focuses on the textures. *Output merging:* all the fragments of the primitives from the 3D space are converted into 2D grids of pixels and displayed on the screen.

3.3 Texture Processor

Texture processor (TP) is a unit in the modern GPUs. It performs various operations like rotation, resizing a bitmap image before placing the texture on the 3D objects. It acts as an independent. independent unit in most of the modern GPUs. Before applying the texture to any object's, they must be addressed based on the pixel address and apply various filters as per the requests. TP works with the vertex and pixel shader to complete the texture mapping. Performance of any GPUs can be measured by taking the number of TPs into consideration [20]. It is obvious that the processor with a greater number of TPs will be faster at processing the texture mapping. TP unit contains many sub modules like Texture Address (TA) unit, Texture filtering (TF) units and Texture Cache (TC) unit texture decompressor units. To map the texel's to pixels TA unit is used [21]. There are various filtering modes which can be applied to the textures the filtering operation in the TF is handled by the TF unit.

There are various filtering methods supported by the TP. Nearest-neighbour interpolation: is the simplest filtering method. It just uses the colour of the texel closest to the pixel centre to colour the pixel. Nearest neighbour with mipmapping: uses nearest neighbour interpolation but also adds mipmapping. Bilinear filtering: four adjacent texel's to the centre of a pixel are sampled, and their colours are combined by weighted average according to distance. Trilinear filtering: bilinear filtering performed on two nearest mipmap levels. Anisotropic filtering: non-square shape sampling of the textures.

4 Implementation

4.1 Generation of Interface Files

Standard graphics benchmarks, Manhattan from GFXBench was simulated on the software modules of the Adreno GPUs and various necessary interface files which contain the coordinates of each pixel request and the dimensions of the textures were extracted from the simulation results. Further these interface files where studied for the necessary values and these values were extracted as descried in the Sect. 4.2. Manhattan is a complex test based on OpenGL ES 3.0. This test contains a visual content of a city during night. Figure 3(b) shows a scene from the Manhattan scene.

4.2 Extraction of Texture Dimensions

TP will receive individual pixel request. These requests can be handled individually or can be grouped depending on the architecture of the TP. The interface file received by the TP unit contains all the necessary information required of the individual pixel requests, this includes the U and V coordinates of the pixel, filter type, LOD of the texture etc.

A single texture can be accessed by many of the pixel's requests. We grouped the request being made to the same texture based on the individual texture ID's. The portion of the texture which is accessed by these requests may be spread all over the texture. By finding the maximum and minimum coordinates in both U and V direction we drawn bounding box to each of the textures which represent the portion of the texture being used. Consider a texture of dimension of 1024 × 1024 pixels as shown in Fig. 2, if there are total 25,000 requests made to this texture and the region accessed are scattered across various part of the texture. One API is accessing a portion of 100 × 100 pixels with 10,000 requests. Similarly, other API's are accessing a portion of 150 × 150 pixels and a 50 × 50 pixels.

In the interface file received by the TPU unit each of these requests will have the coordinates in terms of U and V, we must iterate through each request individually to find the final coordinates corresponding to Umin, Vmin, Umax and Vmax. Finally, we took get the original dimensions (width, height) of the texture from the appropriate interface files. Python script were written to extract all the required values from interface files and 'Pyglet' library was used for visualization of the textures and bounding boxes.

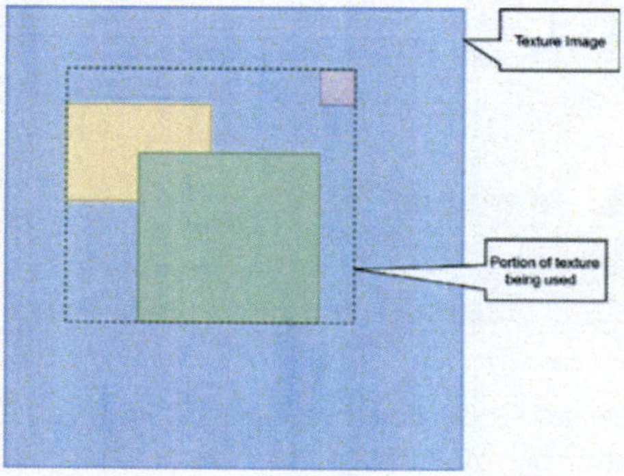

Fig. 2 Texture vs Bounding box

(a)

(b)

Fig. 3 **(a)** Input texture used by the TP to render an output image. **(b)** Single frame output generated by the TP [22].

4.3 Input Texture vs Rendered Output

To illustrate how the input textures are used to render the 3D graphics we considered few input textures and output images.

Multiple input textures are used in rendering a single frame here we have taken a single frame. Figure 3(b) is an output frames for the Manhattan benchmark. From these figures we can see there are multiple texture used to render the complete output. The major textures are used to render the roads, billboards, buildings, steps and vehicles. Portion of the Fig. 3(b) is highlighted which uses the input texture shown in Fig. 3(a) From the Fig. 3(b) we can see that there is no need to load the entire textures to render this output scene if the required texture is just a small portion of the texture. Just by loading the grids which contains the required textures we can still render complete output.

Fig. 4 Texture gridding

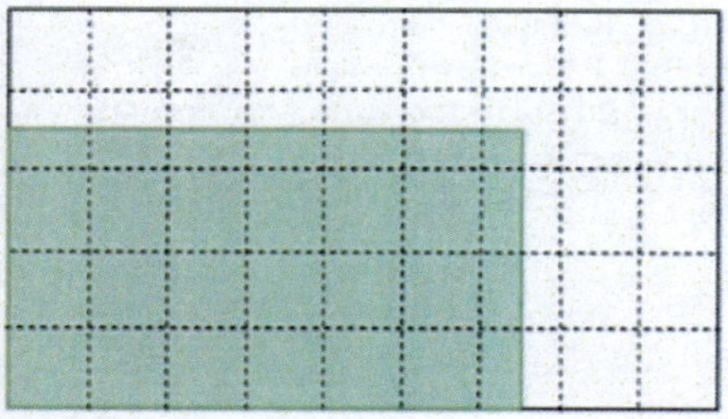

Fig. 5 Matrix representation of the texture gridding

0	0	0	0	0	0	0	0	0
1	1	1	1	1	1	1	0	0
1	1	1	1	1	1	1	0	0
1	1	1	1	1	1	1	0	0
1	1	1	1	1	1	1	0	0

4.4 *Texture Gridding*

To differentiate between the portion of the texture used and portion of the texture unused, gridding logic was applied to the texture. Depending on the dimension of the texture it was divided into many smaller tiles. The logic was implemented so that the minimum number of tiles is 16, staring from tile size of 64×64 pixels. If the number of tiles for a given texture are smaller than 16.

The size of tiles is reduced to 32×32 pixels and so on till 4×4. As shown in Fig. 4 the total texture is divided into 45 tiles, the bounding box corresponding to this matrix is also shown. Out of total 45 tiles 28 tiles are under the bounding box which means the usage of this texture is around 62%. Therefor loading or storage of the rest 38% of the texture is redundant and effects the memory usage.

To represent this gridding, matrix representation of the texture with rows equal to the number of horizontal tiles columns equal to the number of vertical tiles in the texture. As shown in Fig. 5 matrix elements corresponding to the tiles which are covered by the gridding box are represented as 1 and the tiles not covered by the gridding box are represented as 0.

4.5 *Multi Frame Data*

In previous section we concluded that just by loading the grids of the textures which are accessed more we can render output scenes. In this section we discuss how the

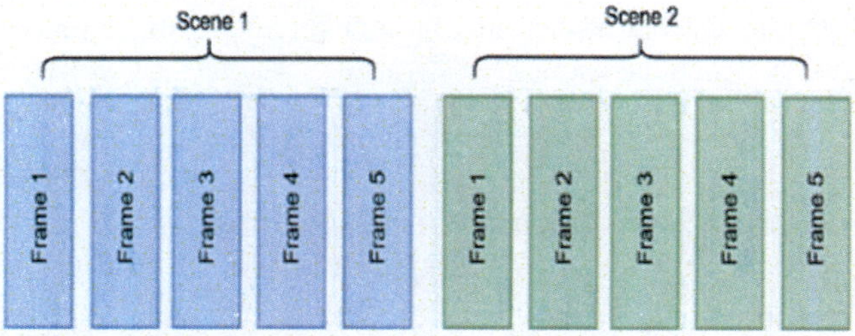

Fig. 6 Multi frame data

grids loaded to the memory can be re-used to render multi frames within a scene. As shown in the Fig. 6 each scene in any graphics is composed of multiple frame and the frame within the scene mostly use similar texture. So, choosing the texture with maximum coverage for storing in the memory and the texture which is reused in most of the frame will improve the memory efficiency by reducing the chances of page fault.

4.6 PSNR Comparison for Multi Frame Data

To illustrate that the adjacent uses the same textures we use Peak Signal to Noise Ratio (PSNR) value between the consecutive frames. PSNR is used as a quality measurement between two images to find the similarities between them. Higher the value of the PSNR implies the given two images are similar. Mean-square error (MSE) is used to compute the PSNR. MSE is the sum of squared error between the two images. The equation to calculate PSNR and MSE are as mentioned in Eq. 1 and Eq. 2 respectively.

$$MSE = \frac{1}{\text{mn}} \sum_{i=0}^{m-1} \sum_{j=0}^{n-1} (I_1[i, j] - I_2[I, J])^2 \tag{1}$$

$$PSNR = 10 \log 10\left(\frac{MAX_i^2}{MSE}\right) \tag{2}$$

where MAXi is 255 for 8-bit pixel representation. PSNR is also applicable for the coloured images which have 3 components (RGB). The definition for the PSNR for the coloured images remains the same except the calculation of the MSE changes, which is now the sum over all the squared value difference of the three components divided by the image size (width * height) and by three.

The PSNR values for the 30 consecutive frames were calculated. PSNR values for first 10 consecutive frames with respect to the first frame in the scene. If the textures used in the adjacent frames are the same there will not be much variation in the PNSR value between the frames.

4.7 *Algorithm to Store Texture in the Memory*

To carefully load the input textures, we propose the algorithm which reduces the number of page faults. In the proposed algorithm we set a threshold for the page fault values based on which the content of the memory can be reloaded. In the proposed algorithm we load the entire texture to the memory for the first frame of every scene. Once the textures are loaded into the memory the grid counters are initialized which keeps track on the number of times each grid of the textures as accessed. After the completion of the first frame we have the total number of times each of the grids are accessed, based on this counter values we can load the textures for the next frames. For the second frame onwards, we propose to load only the grids which are accessed more based on the grid counter values instead of the entire texture and avoid loading the grids which are never used. Loading only the grids will improve the memory efficiency as the space used to store the grids will be less compared to the space needed to store the entire texture. The space left unused can be used to load other texture or can be used by the system.

Algorithm 1:

1.**while(frames to render){**
2. **if(the current frame = first frame in the scene)**
3. Load entire texture required to render the graphics.
4. Initialize the texture grid counter.
5. Update the counters by number of times the grid is accessed.
6. else:
7. Load only the grids with maximum counter value if the texture was loaded in the previous frame else load the entire texture.
8. Calculate the page fault.
9. if (page fault value > threshold value)
10. Clear the texture grid counter.
11. Consider the next frame as the first frame of the scene.
}

5 Results and Discussions

Figure 7 represents Texture vs bounding box visuals for Manhattan benchmark for a single frame. We can see from the figure that for most of the textures, portion used is just a small percentage of the overall texture area. All the textures dimensions are scaled down by a factor of five.

The rectangle filled with green colour are the dimensions of the bounding boxes of their respective textures. Texture with different surfaces are filled with different colours to differentiate the textures. Visual contents are shown only for top three mipmap levels. Bounding boxes are drawn separately for each APIs in the processor.

Five textures with maximum number of API request are considered for discussion. Total coverage of these textures is discussed. Texture dimensions, grid sized used and the percentage coverage of the textures are listed in the Table 1.

Average texture coverage is 62.81% so loading the rest of the texture will have no effect on the rendered output. Thus, DDR efficiency can be increased by storing only the textures which have maximum coverage, and which are currently used in the scene.

The PSNR values for 30 consecutive frames were calculated. Figure 8(a) shows the PSNR values for first 10 consecutive frames with respect to the first frame in the scene. We can see from the graph the value of the PSNR drops slightly as the

Fig. 7 Texture vs Bounding boxes for a single frame from Manhattan benchmark

Table 1 Coverage details of top five textures

Texture dimension	Grid size	No. of grids	Grids used	% coverage
1920 × 1080	64 × 64	480	276	57.50%
120 × 67	16 × 16	28	28	100%
1920 × 1080	64 × 64	480	360	75.00%
1024 × 1024	64 × 64	256	132	51.56%
1024 × 1024	64 × 64	480	144	30.00%

Fig. 8 Texture vs bounding boxes for a single frame from Manhattan **(a)** frame 1–10 **(b)** frame 11–20 **(c)** frame 21–30

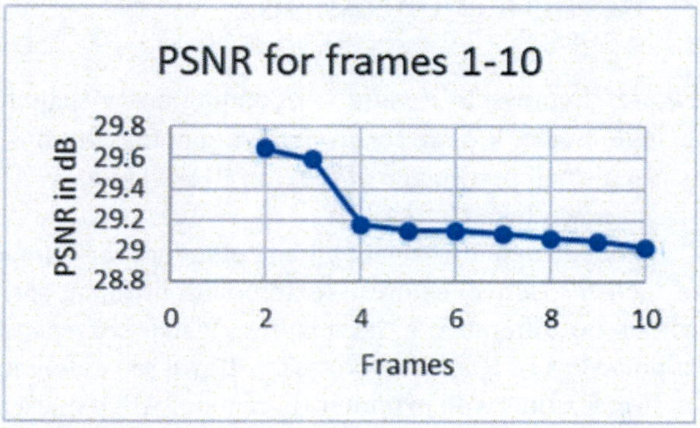

(a)

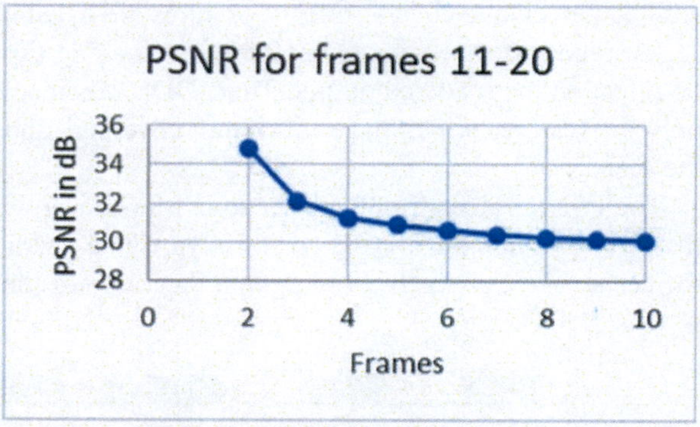

(b)

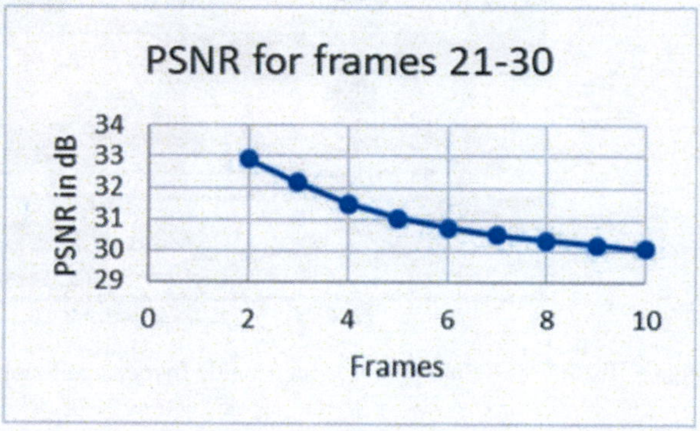

(c)

frame changes the PSNR between first and second frame is higher as compared to the first and tenth frame. From observing the PSNR values of consecutive frames we can conclude that the nearby frames use the same texture to produce the output.

This is as per the discussed algorithm that the texture used in the current frames changes when there is a scene change. Figure 8(a) and 8(c) are the PSNR values for the frame 11 to 20 and frame 21 to 30 respectively. In these graphs also we can see similar characteristics as shown in Fig. 8(a). By resetting the texture grid counter and reloading the textures to the memory we can achieve high memory hit rate whenever the scene changes and efficiently use the memory.

6 Conclusion

In this paper we built a framework to compare the percentage coverage of the textures during Texture Mapping in the Texture processors. Standard graphics benchmark Manhattan was used to conduct the simulation and it was found that for the top five textures with maximum number of requests in a single frame the average texture coverage was 62.81%. From the results we can conclude that by carefully choosing the textures with maximum texture coverage and storing these textures in DDR we can improve memory efficiency. Further we analysed how the input textures are used to render the graphics and used the PSNR values for multiple frames to show that similar textures are used to render multiple frames in the scene. From the experimental values we concluded that similar textures are used in adjacent frames. We also proposed an algorithm to efficiently store the textures grids in the DDR to improve the memory usage.

Future work can implement the proposed gridding technique to manage grid level granularity instead of mipmap granularity. Which allows more details of a texture and improve the overall efficiency of the GPU. Also, a texture access counter can be used to determine how often a grid of texture is accessed this can be used to priorities texture loading grid wise as the grid accessed more will be visible in the current frame.

References

1. Yun J, Lee J, Kim CG, Lim Y-K, Nah J-H, Kim Y, Park W-C (2018) A novel performance prediction model for mobile GPUs. IEEE Access 6:16235–16245. https://doi.org/10.1109/ACCESS.2018.2816040
2. Yun J et al (2019) A practically applicable performance prediction model based on capabilities of texture mapping units for mobile GPUs. IEEE Access 7:102975–102984
3. Wang R, Bowen Yu, Marco J, Tianlei Hu, Gutierrez D, Bao H (2016) Real-time rendering on a power budget. ACM Trans. Graph. 35(4):1–11
4. Romanyuk AN, Dudnyk OO (2017) Ways to improve performance of anisotropic texture filtering. In: Proceedings of International Siberian Conference on Control and Communications (SIBCON), Astana, pp 1–6. https://doi.org/10.1109/SIBCON.2017.7998589
5. Lakshman H, Köppel M, Ndjiki-Nya P, Wiegand T (2010) Image recovery using sparse reconstruction-based texture refinement. In: Proceedings of IEEE International Conference on Acoustics, Speech and Signal Processing, Dallas, TX, 2010, pp 786–789. https://doi.org/10.1109/ICASSP.2010.5494974

6. Yang B, Zhang Z (2010) Design and implementation of high-performance mobile game on embedded device. In: 2010 International Conference on Computer Application and System Modeling, ICCASM 2010, Taiyuan, pp V8-196–V8-199. https://doi.org/10.1109/ICCASM.2010.5619347
7. Rane SD, Sapiro G, Bertalmio M (2013) Structure and texture filling-in of missing image blocks in wireless transmission and compression applications. IEEE Tran Image Process 12(3):296–303
8. Aharon M, Elad M, Bruckstein A (2006) *K-SVD*: an algorithm for designing over complete dictionaries for sparse representation. IEEE Trans. Sig. Process. 54(11):4311–4322
9. Kwatra V, Essa I, Bobick A, Kwatra N (2007) Texture optimization for example-based synthesis. IEEE Trans. Image Process 5(11)
10. Peyre G (2008) Sparse modeling of textures. J Math Imaging Vis 34:17–31. https://doi.org/10.1007/s10851-008-0120-3
11. Remondino F (2006) Image based 3d modelling: a review. Photogram Rec 21(115):269–291
12. Heikkila M, Pietikainen M (2006) A texture-based method for modeling the background and detecting moving objects. IEEE Trans. Pattern Anal Mach Intell 28(4):657–662
13. Lazebnik S, Schmid C, Ponce J (2005) A sparse texture representation using local affine regions. IEEE Trans Pattern Anal Mach Intell 27(8):1265–1278. https://doi.org/10.1109/TPAMI.2005.151 inria-00548530, 2008
14. Xie J, Zhang L, You J, Zhang D (2010) Texture classification via patch-based sparse texton learning. In: Conference paper in international conference on image processing. https://doi.org/10.1109/ICIP.2010.5651387
15. Heckbert PS (2011) Survey of texture mapping. In: Graphics Interface
16. Efros AA, Leung TK (2009) Texture synthesis by non-parametric sampling. In: IEEE International conference on computer vision, Corfu, Greece, September 2009
17. https://developer.apple.com/videos/play/tech-talks/608/
18. https://developer.qualcomm.com/qfile/34706/80-nb295-7_adreno_vulkan_developer_guide.pdf
19. Robinson JA (1997) Efficient general-purpose image compression with binary tree predictive coding. IEEE Trans Image Process 6:601–607
20. https://www.iskmogul.com/amd-working-on-their-own-ray-tracing-technology/
21. https://gfxbench.com/company.jsp
22. Heeger DJ, Bergen JR (2005) Pyramid-based texture analysis/synthesis. In: Proceedings of ACM SIGGRAPH, August 2005, pp 229–238
23. Wei L-Y, Levoy M (2013) Fast texture synthesis using tree-structured vector quantization. In: Proceedings of the 27th annual conference on computer graphics and interactive techniques. Addison-Wesley, New York, Reading, pp 479–488. Assoc. Comput. Mach.
24. Shirani S, Kossentini F, Ward R (2000) Reconstruction of baseline JPEG coded images in error prone environments. IEEE Trans Image Process 9(7):1292–1299. https://doi.org/10.1109/83.847842
25. Debevec PE, Taylor CJ, Malik J (2008) Modeling and rendering architecture from photographs: a hybrid geometry and image-based approach. In: SIGGRAPH 1996, August 2008, pp 11–20
26. Krishnamurthy V, Levoy M (2009) Fitting smooth surfaces to dense polygon meshes. In: Proceedings of SIGGRAPH 96, New Orleans, Louisiana, August 2009, pp 313–324. ISBN 0-201-94800-1
27. Segal M, Korobkin C, van Widenfelt R, Foran J, Haeberli P (1992) Fast shadows and lighting effects using texture mapping. ACM SIGGRAPH Comput Graph 26(2):249–252. https://doi.org/10.1145/142920.134071
28. www.statista.com/statistics/739508/number-of-mobile-gamers-india/
29. www.statista.com/statistics/235850/value-of-the-gaming-industry-in-india/

Retentive True Single Phase Clock 18T Flip-Flop with SVL Technique

Pooja Patil and K. B. Sowmya

Abstract In this paper, we propose an 18-transistor true single-phase-clock (TSPC) flip-flop (FF) by employing SVL technique with static data retention based on two forward-conditional feedback loops, without increasing the clock load. Power dissipation mainly occurs due to leakage currents of different forms such as Sub threshold & Gate leakage etc. The main reason for data loss in Flip Flop is leakage current so we need to reduce leakage current in Flip Flop. This work mainly aims at the design of Low power Retentive TSPC Flip Flop with improved Self Controllable Voltage Level circuits. The performance of proposed FF is compared with c2mos M-S FF, Topologically Compressed FF, and recently proposed retentive TSPC FF. It is shown that both the delay time and power consumption are reduced considerably in the proposed Retentive TSPC FF.

Keywords Improved SVL (I -SVL) · Low power · Leakage current · Self Controllable Voltage Level (SVL) · Upper SVL · Lower SVL · Flip flop

1 Introduction

Most important storage device in digital circuit is Flip flop. The flip flop can store either zero or one bit. Low and high are two stable states in Flip Flop. It will be instate of infinite till it sent to change to other state. Power, performance and area are the three major factors play vary important role in VLSI design. Light weight batteries and portable devices are most required in industries. In order to achieve that complexity of circuit and area are reduced in circuit. In this work attempt is made to full fill these above requirements by applying different power reduction technique. As technology grows the area of the circuit and chip are reduced. TSPC FF is designed so that it reduces the less power consumption as it works on both positive and negative clock pulses.

P. Patil (✉) · K. B. Sowmya
Department of ECE, RVCE College Bangalore, Bangalore, India

V. Nath and J. K. Mandal (eds.), *Proceeding of Fifth International Conference on Microelectronics, Computing and Communication Systems*, Lecture Notes in Electrical Engineering 748,
https://doi.org/10.1007/978-981-16-0275-7_48

As the number of transistor increases in circuit the power consumption and area is increases in circuits. Without effecting the working of flip flop have to reduce area and power dissipation in circuits. Size and power dissipation play very important role in deciding area and size of the chip.

Application of flip flop in digital circuits are registers, memory, Bounce switch elimination, latch, data storage etc. Power dissipation are increases complexity of design, leakage current, transition count and switching activity. Power dissipation are of 2 types static and dynamic power dissipation.

(a) Static Power Dissipation-It is the power consumed by the circuit during no switching activity. Sub-threshold leakage current flows when the device is in an inactive state. It is represented by the equation.

$$\text{P stat} = \text{V}_{\text{DD}}^{*}\text{I leakage} \tag{1}$$

(b) Dynamic Power Dissipation-The circuit consumes power when the switching activity is present and short circuit path. Short circuit current flows when there is a closed path between supply and ground. It is represented by the equation.

$$\text{P Dyn} = \alpha \text{CV}_{\text{dd}}^{2}\text{f} \tag{2}$$

Now the paper is presented as follows: in Sect. 2 it defines the sort of different architectures of TSPC FF by advantages and adversities. In the last section the proposed Retentive TSPC is clarified. Section 3 includes simulations and results. Simulation result shows that the proposed Retentive TSPC uses less power compared to other TSPC architectures.

2 True-Single Phase Clock Flip-Flop

2.1 C2MOS TSPC Flip Flop

Conventional clocked C2MOS consists of 2 stages like master and slave stages. This FF works on both positive and negative edge of clock is the advantage of it. For generation of CPI and CPN additional 8 transistors are connected to clock signal. Which increases the complexity of the circuit. Due to this there will be power dissipation in circuit and clock uncertainty increases.

Schematic of C2MOS is shown in Fig. 1. This FF works on both low and high clock pulses. The C2MOS works as follows.

1. CP = 0 slave latch enters into transparent mode and master latch enters into activated. CPN becomes low and CPI becomes high. Data is reached to second stage and data will be stored in master latch.

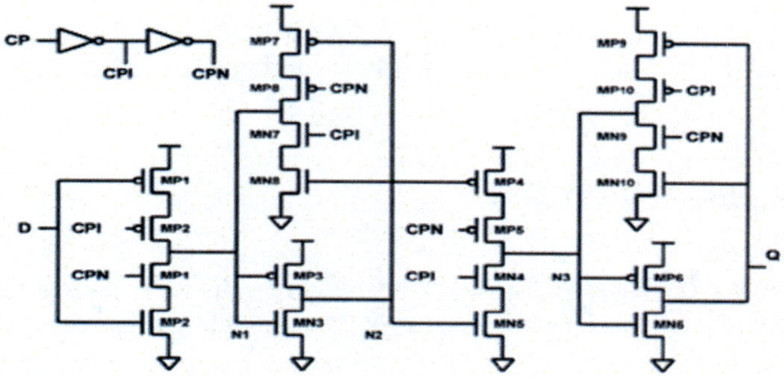

Fig. 1 Schematic structure of conventional C2 mos M-S FF architecture [1].

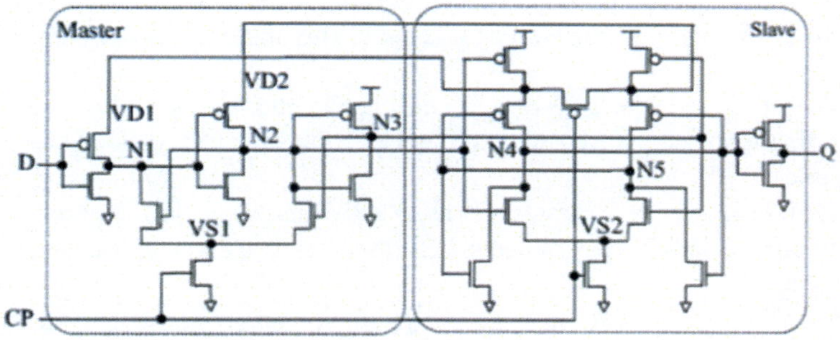

Fig. 2 Schematic structure of topologically compressed FF [2]

2. CP = 1 slave latch enters into activated and master latch enters transparent. Stored data at master latch at master latch will be transmitted to output of slave.

2.2 Topologically Compressed Flip Flop

Topologically compressed FF is shown in Fig. 2. In this topologically compressed Flip flop the transistors of identical operation are combined. Initially, with in the latches generation of equivalent nodes are done and conversion of flip flop is done into AND-OR complex gates instead of TG gates. Topologically compressed Flip flop consists of 21 transistors by lowering transistors connected to clock pulse.

Working of TCFF is done when CP = 0 and CP = 1. When CP = 0 sampling of input data is done by master stage and master stage is acts as series of 3 inverters. By holding previously sampled input data the slave stage acts as cross coupled inverter. Integrity of signal is stored by utilizing power supply direct PMOS and NMOS transistors.

When CP = 1 the input data which is sampled during negative phase of clock pulse is hold by master stage and master stage is in hold state. Sampled data is processed to slave stage and then passed to slave stage. NMOS transistor which is there in input stage is used to cutoff input data which changes during positive phase of clock.

2.3 Retentive TSPC FF

The architecture of Retentive TSPC is shown in Fig. 3 includes master, slave stage, and 2 conditional feedback paths. Because of 2 conditional feedback loop data retention is done. It has reduced clock load with only 4 transistors compared to other architectures of FF.

Data transfer from D to Q takes place as fallows.

A) When D = 0, cp = 0 master will be in active state and slave will be in active. During this data get transmitted to N2 node. This phase is called as data retention phase.
B) When D = 0, cp = 1 slave will be in active and master will be in inactive. During this data get transmitted to node Q. This phase is called Data transfer phase.
C) When D = 1, cp = 0 master will be in active state and slave will be in active. During this data get transmitted to N2 node. This phase is called as data retention phase.
D) When D = 1, cp = 1 slave will be in active and master will be in inactive. During this data get transmitted to node Q. This phase is called Data transfer phase.

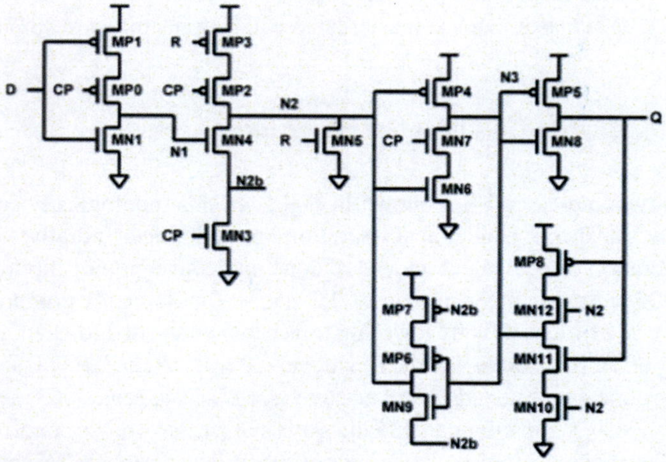

Fig. 3 Schematic structure of retentive TSPC FF [1]

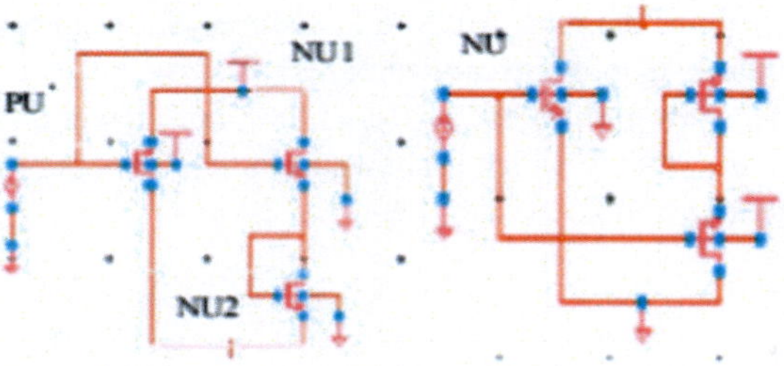

Fig. 4 I-USVL & I-LSVL circuits [3]

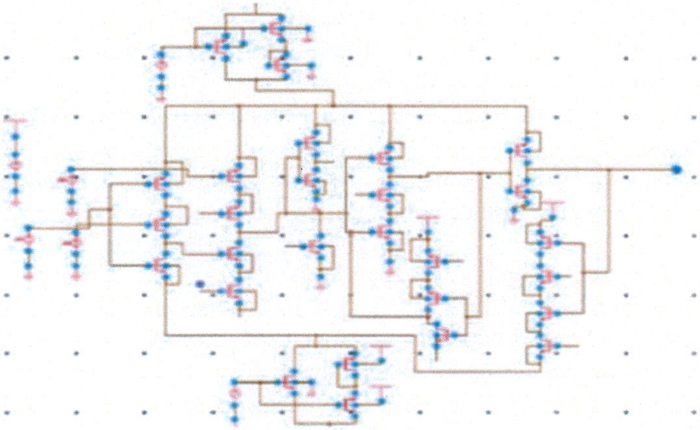

Fig. 5 Schematic of retentive TSPC Flip Flop with upper SVL and lower SVL scheme.

TSPC FF suffers from data loss and main reason for data loss is leakage current and power dissipation is circuit. Power dissipation in circuit can be reduced by applying low power technique.

2.4 Improved Retentive TSPC with SVL Technique

Improved upper SVL and lower SVL circuit is shown in Fig. 4. There are two types in SVL technique Upper SVL and Lower SVL technique which uses three additional transistors. Upper SVL is connected to VDD and FF, Lower SVL is connected to FF and Gnd. During the operation of TSPC FF, some transistors will be active mode and some will be inactive mode. By application of the SVL technique load circuit is supplied with full supply voltage in active mode and by this, we can achieve less gate leakage current in the circuit. Hence in standby mode, a provision is made to incorporate slightly lower voltage and relatively higher voltage to load using ON switch, such that Vds of the OFF MOSFETs reduces Vsub, in turn, Vth increases consequently as sub-threshold current decreases.

In order to reduce power dissipation, Upper SVL and Lower SVL techniques are applied to Retentive TSPC FF which is shown in Fig. 5. The operation of this circuit is the same as retentive TSPC.

3 Simulation and Results

The performance of proposed FF is compared with C2MOS M-S FF, Topologically Compressed FF, and recently proposed retentive TSPC FF. The proposed Retentive TSPC with SVL technique is simulated in the Cadence virtuoso using 45 nm technology and CMOS device with supply voltage 0.9 V.

Figure 6 shows the transient response of the Proposed Retentive TSPC with SVL technique.

The results are performed using the Cadence virtuoso ADEL window and the transient analysis was performed with transient analysis = 100 ns.

The obtained wave forms as shown in Fig. 6 has parameters input as data and clock and output as Q and power dissipation.

The comparison of Retentive TSPC with SVL technique, Retentive TSPC FF, Topologically Compressed Flip Flop, C2MOS TSPC Flip Flop is shown in Table 1.

The performance parameters i.e. number of transistors, clock to Q delay, average power dissipation at 10% activity factor, and the power delay product of Retentive TSPC with SVL technique, Retentive TSPC FF, B.Topologically Compressed Flip Flop, C2MOS TSPC Flip Flop in Table 1.

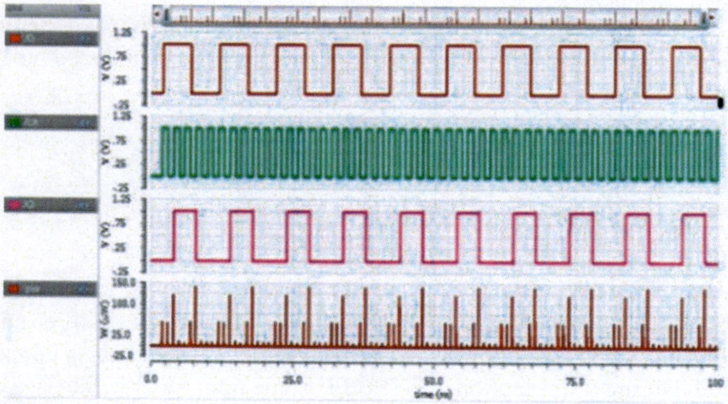

Fig. 6 Output waveform of proposed retentive TSPC Flip Flop with SVL technique

Table 1 Comparison of implemented designs

	No. of transistors	Power dissipation	Clk to Q delay	D to Q delay
C^2MOS TSPC FF	24	54.15 uW	46.42 ns	30.17 ns
Topologically compressed TSPC FF	21	70.85 uW	93.78 ns	19.89 ns
Retentive TSPC FF	18	639.4 nW	283.99 ps	450.64 ps
Improved Retentive TSPC FF SVL technique	24	583.6 nW	271.18 ps	487.19 ps

4 Conclusion

This paper presents the delay, power analysis and Retentive TSPC with SVL technique shows better performance compared to other FF. The power consumption and delay of Retentive TSPC FF with SVL technique is less compared to other TSPC FF s.

References

1. Stas F, Bol D (2017) A 0.4-V 0.66-fJ/cycle retentive true-single-phase-clock 18T flip-flop in 28-nm fully-depleted SOI CMOS. IEEE Trans Cırc Syst I Regul Pap Exp Briefs 56(11):810–814 (2017)
2. Savanth A, Ultra-low power 18-transistor fully static contention-free single-phase clocked flip-flop in 65-nm CMOS. In: IEEE J. Solid-State Circ Syst I Reg Papers **61**(4), 961–975 (2019)
3. Hemanth kumar CS, Kariyappa BS (2017) Analysis of low power 7T SRAM cell employing improved SVL (ISVL) technique. In: 2017 International Conference on Electrical, Electronics, Communication, Computer and Optimization Techniques (ICEECCOT), vol. 12, no. 5, pp. 477–484, September 2017
4. Zhao P, Darwish T, Bayoumi M (2004) High-performance and low-power conditional discharge flip-flop. IEEE Trans Very Large Scale Integr (VLSI) Syst 12(5):477–484. https://doi.org/10.1109/TVLSI.2004.826192
5. Phyu MW, Kangkang Fu, Goh WL, Yeo K-S (2011) Power-efficient explicit-pulsed dual-edge triggered sense-amplifier flip-flops. IEEE Trans Very Large Scale Integr (VLSI) Syst 19(1):1–9. https://doi.org/10.1109/TVLSI.2009.2029116
6. Bhargava B, Sharma PK, Akashe S (2014) Comparative analysis of self-controllable voltage level (SVL) and stacking power gating leakage reduction techniques using in sequential logic circuit at 45nanometer regime. Int J Intel Electron Syst 8(1):8–13. https://doi.org/10.18000/ijies.30133
7. Karthikeyan S, Sivakumar S (2013) Leakage current reduction on different SRAM cells used adaptive voltage level technique. Int J Adv Inf Sci Technol (IJAIST) 2. ISSN 2319-268
8. Lin J-F, Sheu M-H, Hwang Y-T, Wong C-S, Tsai M-Y (2017) Low-power 19-transistor true single-phase clocking flip-flop design based on logic structure reduction schemes. IEEE Trans Very Large Scale Integr (VLSI) Syst 25(11):3033–3044. https://doi.org/10.1109/TVLSI.2017.2729884
9. Kawai N (2014) A fully static topologically-compressed 21-transistor flip-flop with 75% power saving. IEEE J. Solid State Circ 49(11):2526–2533

10. Abdollahi R, Hadidi K, Khoei A (2016) A simple and reliable system to detect and correct setup/hold time violations in digital circuits. IEEE Trans Circ Syst I Regul Pap 63(10):1682–1689. https://doi.org/10.1109/TCSI.2016.2582239
11. Nikolic B, Oklobdzija VG, Stojanovic V, Jia W, Chiu JK-S, Leung MM-T (2000) Improved sense-amplifier-based flip-flop: design and measurements. IEEE J Solid-State Circ 35(6):876–884
12. Stojanovic V, Oklobdzija VG (1999) Comparative analysis of masterslave latches and flip-flops for high-performance and low-power systems. IEEE J Solid-State Circ 34(4):536–548
13. Partovi H, Burd R, Salim U, Weber F, DiGregorio L, Draper D (1996) Flow-through latch and edge-triggered flip-flop hybrid elements. In: ISSCC (IEEE Dig.), September 1996, pp 138–139
14. Klass F et al (1999) A new family of semidynamic and dynamic flip-flops with embedded logic for high-performance processors. IEEE J Solid-State Circ 34(5):712–716
15. Tschanz J, Narendra S, Chen Z, Borkar S, Sachdev M, De V (2001) Comparative delay and energy of single edge-triggered and dual edge triggered pulsed flip-flops for high- performance microprocessors. İn: Proceedings of the ISPLED 2001, pp 207–212
16. Kong B-S, Kim S-S, Jun Y-H (2001) Conditional-capture flip-flop for statistical power reduction. IEEE J Solid-State Circ 36(8):1263–1271
17. Anghel L, Alexandrescu D, Nicolaidis M (2000) Evaluation of a soft error tolerance technique based on time and/or space redundancy. In: SBCCI 20000, p 237
18. Krueger D, Francom E, Langsdorf J (2008) Circuit design for voltage scaling and SER immunity on a quad-core itanium processor In: ISSCC, 2 February 2008, pp 94–95

Verification of SerDes Design Using UVM Methodology

K. Arpitha Nagesh and D. R. Shilpa

Abstract There has always been a need to transmit data since the computer's inception. When data is sent out of the network, it will require cable sizes to carry the data. To ease the data transmission protocols, cabling scheme, and serial communication was introduced. Numerous issues arise while transmitting data parallely such as skew, cross talk, cost and board space of System On Chip (SOC). Hence SerDes can be a great solution in moving large data from point A to Point B within the system, between two different systems or indeed between systems in different places. SerDes allows data to be transmitted at a higher rate and is less expensive. In this paper, design and verification of SerDes has been proposed. Verilog HDL was used in the design of SerDes and verification was carried out using Universal Verification Methodology (UVM) as it provides a reusable test bench and hence significantly reducing time to market.

Keywords Universal Verification Methodology (UVM) · Finite State Machine (FSM) · Design Under Test (DUT) · Transaction Level Modeling (TLM) · System On Chip (SOC)

1 Introduction

The VLSI industry is developing quickly and demands systems which perform at high speed, use less power and do not affect the performance of the device [1–3]. A high speed parallel bus consumes significant power and creates complications in routing and also suffers from timing issues. System architects frequently have issues in keeping up with skew between parallel lines and problems elevate as frequency and distance increases. Also serial connections minimize cost reliably due to the lowered

K. A. Nagesh (✉) · D. R. Shilpa
Department of Electronics and Communication Engineering, R V College of Engineering, Bengaluru, India

D. R. Shilpa
e-mail: shilpadr@rvce.edu.in

V. Nath and J. K. Mandal (eds.), *Proceeding of Fifth International Conference on Microelectronics, Computing and Communication Systems*, Lecture Notes in Electrical Engineering 748,
https://doi.org/10.1007/978-981-16-0275-7_49

pin count. Hence serial communication is preferred over parallel communication. Ethernet, PCI express and numerous other applications use SerDes architecture [4]. The design needs to be thoroughly verified for errors and simple test cases wouldn't ramp up to verify complex systems reasonably which necessitate the use of robust verification models [5]. To correct this errors in simulations SOC and IP engineers work together to identify the error. Universal Verification Methodology has grown as a well-known framework for verification of SOC. The highlights of UVM are reusability of code, checkers, assertions, automation and many more [6]. Universal Verification Methodology supports a layered verification environment and UVM components like driver, monitor, agent can be reused in different Design Under Test (DUT). Object-oriented concepts along with TLM are the attributes of UVM that help in the abstraction level of modeling. Flexibility is one of the important features of UVM [7, 8]. In this paper, Serdes is designed and verification is carried out by writing a verification plan to carefully verify the functionality of design. A UVM test bench is used for this purpose and the connection from test bench to DUT is made with the help of GMII interface which supports high speed operations [15]. Comparative Analysis of research papers led to an understanding that SerDes is a key component of a physical layer that is accountable for information transmission in media and defining a reusable test bench such as UVM serves the purpose of reduced market time as 70% of product cycle corresponds to verification. Therefore both design and verification of Serdes is carried out in this work.

2 Methodology

The work is carried out by understanding the design functionality and carefully deciding the method to carry out the project. Methodology and design is explained in detail. The top design module includes Encoder/decoder, TX/RX FSM, Serializer and Deserializer. These modules are coded using Verilog HDL. The proposed design is shown in Fig. 1. Serializer and Deserializer blocks operate at 1.25 GHz while the rest blocks in the design work at 125 MHz. Hence a clock divider circuit is used to divide the clock by 10. The input to the test bench comes from a UVM test bench through the GMII interface. The design of each block in the top level module along with the importance of each module is explained in detail in this section.

A. *8b/10b Encoder/Decoder*

This technique is largely used in high speed serial communication mainly because the design scheme has neutral dc and limitation to highest run length possible. Run length can be defined as the number of continuous zeros or ones. These two features ensure that the data is properly recovered in the clock and data recovery circuit at the receiver end as this circuit needs data transitions in order to recover data. The coding scheme splits the eight bit input information in two parts i.e. three bit MSB and five bit LSB. The design involves the use of conversion tables, signals for error if incorrect control character is requested, and signal for illegal code received. A LUT

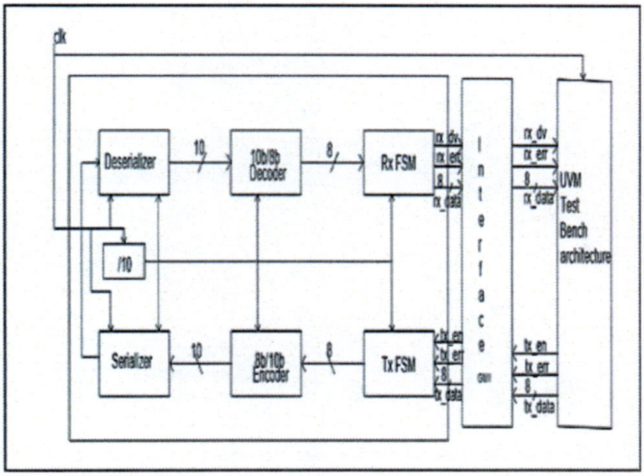

Fig. 1 Proposed design

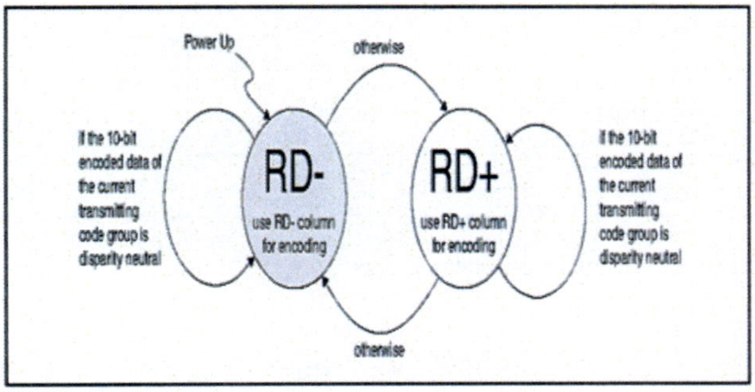

Fig. 2 Running disparity state machine

in design maps the three bit and five bit data to four bit and six bit encoded data. Similarly decoder splits the data in 10 bit into 6 bit and 4 bit data and uses the LUT to decode the data to 8 bit. In order to ensure DC balance, a running disparity state machine is used as shown in Fig. 2. The flowof the design is shown in Fig. 3.

B. *Serializer and Deserializer*

Serilalizer/Deserializer functional blocks perform conversion of parallel data to serial data and vice-versa. Serialization of parallel data is necessary as it reduces skew, crosstalk, cost incurred in transmitting parallel data by a great amount.

C. *GMII Interface and TX/RX FSM*

GMII is a standard interface. The letters GMII stands for Gigabit Media Independent Interface. The aim of the interface is to offer affordable, simple and convenient to implement connectivity to the physical layer where Serdes are placed [9, 10]. Six

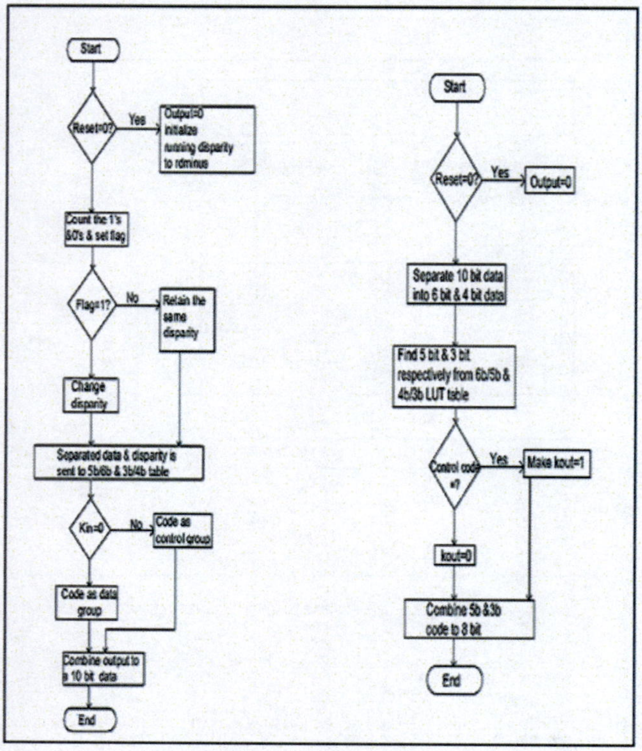

Fig. 3 Flowchart for the design of encoder/decoder

GMII signals are used with 3 each in transmitter and receiver side. The signals at transmitter side indicate the conditions under which data or error is transmitted from the test bench and the signals at receiver side indicate the correctness of received data. Transmitter and receiver use FSM to achieve these conditions and transmit/receive data accordingly. The Table 1 shows conditions for transmitter and receiver FSM respectively. The interface facilitates 1000 Mbps operations [9]. A state machine is designed to decide the conditions at which data transfer can happen. Similarly, it decides the information and control signals that have to be sent to MAC from PHY. Based on the design specification, a verification plan is made to verify the design features and the test cases are defined as shown in Table 2. UVM test bench is created and the interfaces are passed to the wrapper module which can be considered as the top design module. GMII and SerDes agents are built with their interfaces i.e. GMII interface defines the control signals as per specification and the SerDes interface holds reset and DUT clock definitions. Both interfaces operate at different clock frequencies. A UVM test bench that is linked via GMII interface to the top design module. The DUT is linked via driver and monitor to the test bench. The general description of a UVM Test bench for the design is shown in Fig. 4.

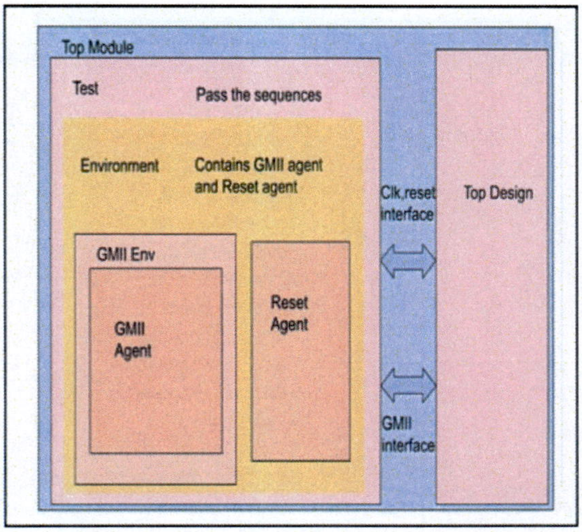

Fig. 4 General information of UVM test bench

3 Results

The design of SerDes is written in Verilog HDL. The simulations are carried out using cadence tools. The description of module hierarchy is shown in Fig. 5.

D. *Simulation Results*

The 8bit data is converted to one of 256 data characters or one of 12 control characters based on kin input. Output to indicate the error in coding of control character is also shown in Fig. 6 of encoder simulation waveform and the test cases are shown in Table 2.

The 10bit data is decoded into 8bit data in the decoder. A signal to indicate whether decoded data is data character or control character is shown in Fig. 7.

Serializer transforms encoded data into serial data at transmitter side and deserializer transforms serial data back to 10 bit data. This is indicated in Fig. 8. And the final output of top level design which includes GMII signals, FSM at transmitter and receiver side along with SerDes is verified by writing UVM test bench is shown in Fig. 9.

Transmitter FSM converts GMII signals coming from test bench to 8bit code and Receiver FSM performs the reverse operation i.e. converts 8bit code group into GMII signals. The output of the states of FSM as indicated in Table 1 is shown in Fig. 10.

```
Design hierarchy summary:
                                 Instances  Unique
        Modules:                        11      10
        Interfaces:                      2       2
        Verilog packages:                6       6
        Registers:                   15915   10489
        Scalar wires:                   55       -
        Vectored wires:                 21       -
        Named events:                    4      12
        Always blocks:                  22      21
        Initial blocks:                329     167
        Parallel blocks:                27      28
        Cont. assignments:               2       2
        Pseudo assignments:             43      43
        Assertions:                      2       2
        SV Class declarations:         208     322
        SV Class specializations:      441     441
        Simulation timescale:          1ps
Writing initial simulation snapshot: worklib.top:sv
```

Fig. 5 Description of module hierarchy

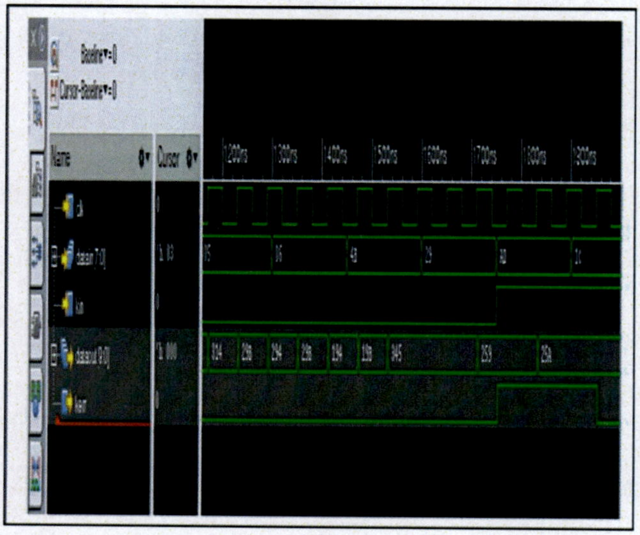

Fig. 6 Encoder output waveform

Fig. 7 Decoder output waveform

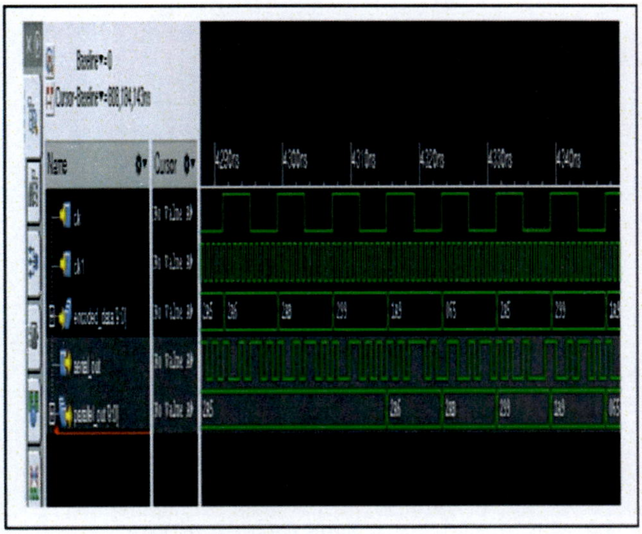

Fig. 8 Waveform of Serializer/Deserializer

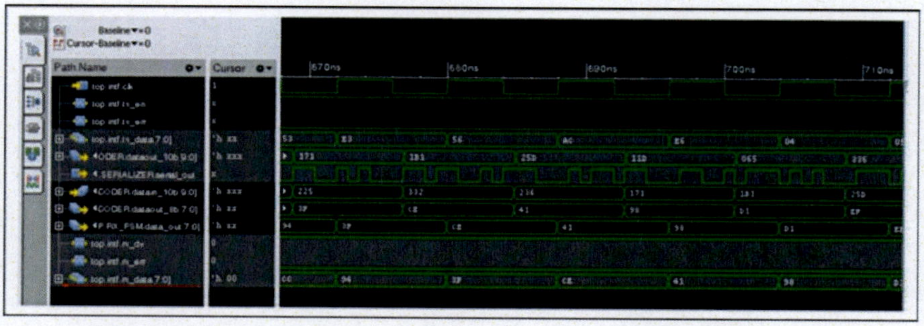

Fig. 9 Top module output

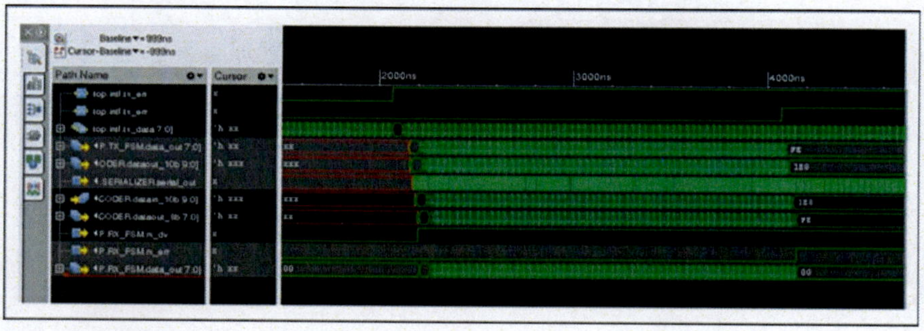

Fig. 10 Simulation result of FSMs

Table 1 Description of FSM [14]

TX FSM

TX_EN	TX_ER	Description
0	0	Normal inter-frame
1	0	Normal Data Transmission
1	1	Transmit Error Propagation

RX FSM

RX_DV	RX_ER	Description
0	0	Normal inter-frame
1	0	Normal Data Reception
1	1	Data Reception Error

Table 2 Test case definition

Functionality	Test Name
Reset Test	basic_reset
Negative Tests	Invalid_preamble, Invalid_SFD,idle_condition
Data Transfer Test	transfer_test
Error Condition	err_condition_test
Error Propagation	err_prop_test
Encoding/Decoding	enc_dec_test
Data transfer and reset test	data_reset_test

4 Conclusion

The paper discusses the design of SerDes and verification using UVM. The design and verification is carried out using Cadence tools. One of the essential advantages of UVM is that the technique provides a set of rules to be followed in order to create a test bench for verification. This will guarantee uniformity between various verification teams. There are standardized handshake mechanisms, a reporting mechanism that saves a lot of time while creating test bench and reusability offers improved efficiency. The design modules include FSMs that decide the conditions under which transmission and reception of information have to happen and the SerDes block that is wrapped to form the top design module.

Encoder/Decoder helps in shaping data streams suitable for serialization. Continuous zeros and ones are not suitable for transmitting over serial channels. Hence encoder encodes incoming parallel data into 10 bit data with run-length less than 5 [11–13]. This encoded data is serialized using serializer and sent over the channel. At the receiver end, the deserializer transforms serial information into parallel information and passes it on to the decoder. The parallel data comes to DUT from UVM test bench through GMII interface.

Acknowledgements Our heartfelt thanks to Dr K S Geetha, Professor and Head, ECE, R V College and Ignitarium Technology Solutions Pvt Ltd. for their generous assistance.

References

1. Lear JA (2009) Digital and mixed-signal verification differences. In: 2009 10th International Workshop on Microprocessor Test and Verification, pp 91–94
2. Salah K (2014) A UVM-based smart functional verification platform: concepts, pros, cons, and opportunities. In: 2014 9th International Design and Test Symposium (IDT), Algiers, pp 94–99

3. Ustaoğlu B, Bağbaba AÇ, Örs B, Erdem İ (2015) Creating test environment with UVM for SPI. In: 2015 23nd Signal Processing and Communications Applications Conference (SIU), Malatya, pp 2373–2376
4. Maadi M (2014) An 8b/10b encoding Serializer/Deserializer (SerDes) circuit for high speed communication applications using a DC balanced, partitioned-block, 8b/10b T. Int J Electr Electr Eng 3(2):144–148
5. Ni W, Zhang J (2015) Research of reusability based on UVM verification. In: 2015 IEEE 11th International Conference on ASIC (ASICON), Chengdu, pp 1–4
6. Pavithran TM, Bhakthavatchalu R (2017) UVM based testbench architecture for logic subsystem verification. In: 2017 International Conference on Technological Advancements in Power and Energy (TAP Energy)
7. Hussien A et al (2019) Development of a generic and a reconfigurable UVM-based verification environment for SoC busses. In: 2019 31st International Conference on Microelctronics (ICM), Cario, Egypt, pp 195–198. https://doi.org/10.1109/ICM48031.2019.9021657
8. Verificationacademy.com. 2020. Verification Academy - The Most Comprehensive Resource For Verification Training. Verification Academy. https://verificationacademy.com/
9. Georgoulopoulos N, Giannou I, HAtzopoulos A (2018) UVM – based verification of a mixed signal design using system verilog. In: 2018 28th International Symposium on Power and Timing Modeling, Optimization and Simulation (PATMOS), pp 97–102
10. IEEE Standard for Information technology-Telecommunications and information exchange between systems-Local and metropolitan area networks-Specific requirements Part 3: Carrier sense multiple access with collision detection (CSMA/CD) access method and physical layer specifications. In: IEEE Std 802.3-2002 (Revision of IEEE Std 802.3, 2000 edn.), pp 1–1550, 8 March 2002
11. Quesada-Martínez A, Aparicio-Morales J, Campos-Araya J, Chacón-Rodríguez A, Garcia-Ramirez R, Rimolo-Donadio R (2020) Evaluation of 8b/10b FPGA encoder implementations for SerDes links. In: 2020 IEEE 11th Latin American Symposium on Circuits & Systems (LASCAS), San Jose, Costa Rica, pp 1–4. https://doi.org/10.1109/LASCAS45839.2020.9069001
12. Samanta S, Dastidar A (2015) Implementation of 10bit SerDes for Gigabit Ethernet PHY. In: 2015 International Conference on Man and Machine Interfacing (MAMI), pp 112–117
13. Takeda M, Yamasaki N, (2019) Finding the maximum number of symbols for the 4b/10b line code with error correction. In: 2019 Seventh International Symposium on Computing and Networking Workshops (CANDARW), Nagasaki, Japan, 2019, pp 460–464. https://doi.org/10.1109/CANDARW.2019.00087
14. IEEE Standard for Ethernet. In: IEEE Std 802.3–2018 (Revision of IEEE Std 802.3–2015), pp 1–5600, 31 August 2018. https://doi.org/10.1109/IEEESTD.2018.8457469
15. IEEE Standard for Universal Verification Methodology Language Reference Manual. In: IEEE Std 1800.2–2017, pp 1–472, 26 May 2017. https://doi.org/10.1109/IEEESTD.2017.7932212

Clock Power Reduction Using NDR Routing

Shridhar Alure, N. Ramavankateswaran, Ramcharan Buddi, and Varinder Kumar

Abstract The recent advancement in nanotechnology over a different scope of industries and an expanded microelectronics market demand for low power, high performance and complexity on System on Chip (SoC) have made Electronic Design Automation (EDA) teams to investigate and innovate at all stages involved in the design development procedure. Mass production of ICs at lower technology nodes have been introduced by some foundries and have directed many industries to use this technology for their upcoming devices. This kind of technology has bought many challenges for EDA vendors, Physical designers due to the physical constraints and design rules so as to meet the foundry prerequisites. Non Default Rule (NDR) routing is a technique applied during clock routing in order to reduce the Clock power and hence the overall power dissipation of the subsystem. Basically Clock Tree in a design is built to reduce the skew and the latency, so in this process of reducing skew and latency the tool adds Buffers/Inverters. By applying NDR technique the tool add lesser number of Buffer/Inverters to maintain skew and latency. Before performing clock routing Floorplan, Placement and Clock Tree Synthesis (CTS) is performed on the design. The execution of the Subsystem design to which NDR is to be applied is done in Place and Route Synopsis tool.

Keywords System on chip · Non Default rule · Clock tree synthesis · ICs · Clock power · ED · Electromigration · Resistance · Coupling capacitance · Node · Slew · Skew · Latency · Power

S. Alure (✉) · N. Ramavankateswaran · R. Buddi · V. Kumar
VLSI Design and Embedded Systems, Department of Electronics and Communication Engineering, RV College of Engineering, Bengaluru 59, India
e-mail: shridharalure.lvs18@rvce.edu.in

Department of Electronics and Communication Engineering, RV College of Engineering, Bengaluru 59, India

V. Nath and J. K. Mandal (eds.), *Proceeding of Fifth International Conference on Microelectronics, Computing and Communication Systems*, Lecture Notes in Electrical Engineering 748,
https://doi.org/10.1007/978-981-16-0275-7_50

1 Introduction

Clock distribution in number of clock tree levels plays a major role in latency, skew and have major impact on Integrated circuit performance [1]. For Nanoscale technology interconnect width is plays a major role in defining the slew i.e. for larger interconnect width the resistance will be less and hence slew will be less as compared to small width interconnect [2]. For clocking structure in CTS H-tree and clock mesh planning is performed for high performance computing [3]. While performing CTS, tool insert/size the Inverters/Buffers that are specified in the library file to prevent signal degradation and meet the slew constraint [4]. The regular or symmetrical clock tree for high performance chips help in improving the timing results such skew in the design [5]. The 3D gated clock tree approach in the design shown the improvement in the clock skew and power reduction as compared to 2D state of art [6].

Due to usage of FinFET devices [7] for technology node below 16 nm in the design has introduced lot of challenges and merits as well such as drive strength capabilities of the transistor which drives 25% high current than 20 nm node transistor [8].

Due to less spacing between the two metal layers after routing the design causes the coupling capacitance and Electromigration effect. Coupling Capcitance effect causes the logic failure and degradation in timing, Electromigration i.e. accumulation of charges on metal layer which leads to hillock and hollow in the layer and hence damaging the device. Also lesser the width of the metal layer leads to higher resistance which directly degrades the slew.

Our paper adopt the application of Non Default Rule routing in CTS stage which introduce higher the metal width and spacing in the design causing reduction in skew, latency, clock cell count, clock Power.

2 Concept of Clock Tree Synthesis

The CTS consist of creating a distributed clock network to all synchronous circuits of the IC based on prerequisites of the design. H-Tree or X-tree topology is followed for distributing the clock from root node to leaf pin of the circuit. H-tree topology implies the clock tree is built in H shape fashion, similarly the X tree topology. These two tree topologies are considered best layouts for maintaining uniform clock distribution over entire chip.

In case of Multimode MultiCorner design the Clock tree is built based on the corner (Slow, Fast, Typical) specified by the designer. CTS will apply the default slow corner in case when no corner is specified.

During CTS the tool inserts the buffer in clock network to maintain the acceptable skew that is provided in the CTS specification. Tool also tries to minimize the latency, area.

Skew is defined as the difference in the delay of the maximum and the minimum path.

Latency is defined as time difference between the clock at the register (definition point) and the clock source point such as PLL (Phase Locked Loop).

The below Fig. 1 shows the demonstration of the term skew.

Skew = t2–t1.

2.1 Need of Applying Clock Tree Synthesis

The below Fig. 2 shows the reason for applying clock tree and buffering the clock the clock network. Due to signal degradation the functionality of the design might change, hence to avoid this situation we buffering of the clock network which avoids signal degradation.

2.2 Concept of NDR in Clock Routing

The below Fig. 3 explains the nature of NDR technique applied for H tree clock structure and also explains how the width for metal layers is being defined during clock routing.

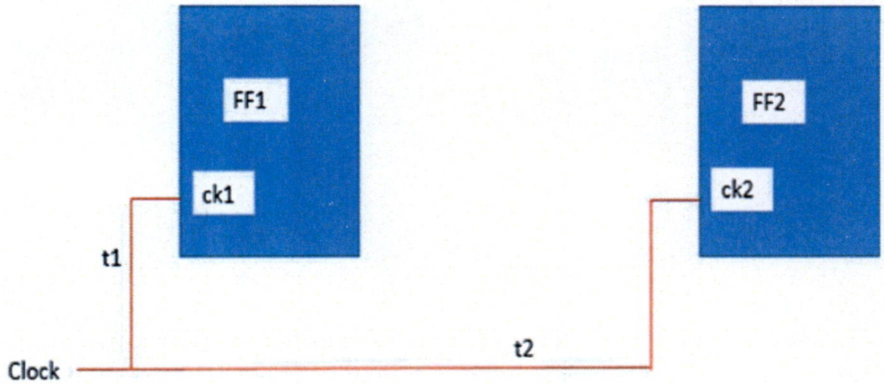

Fig. 1 Skew definition

Fig. 2 Demonstration of signal degradation

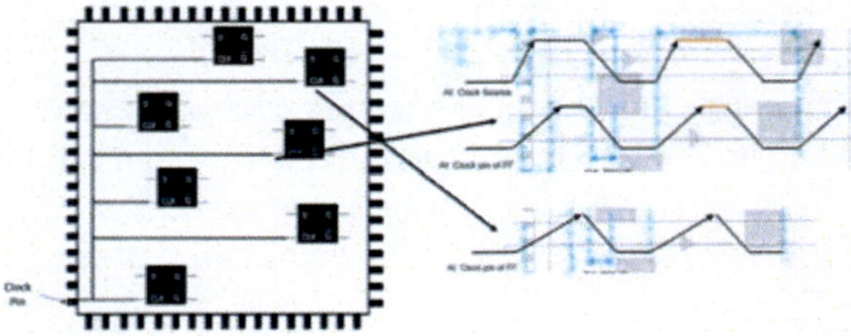

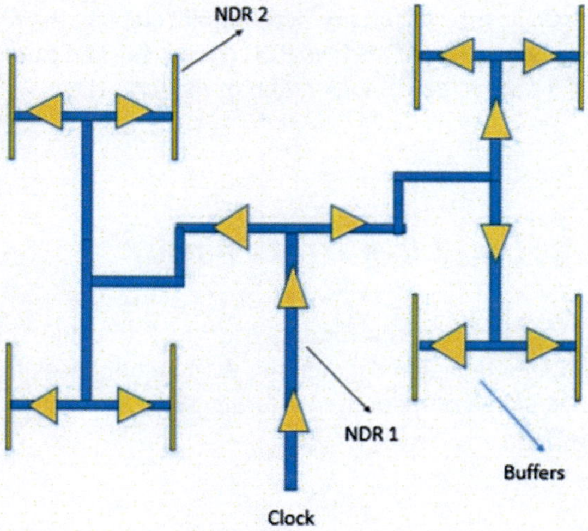

Fig. 3 Concept of NDR in clock routing

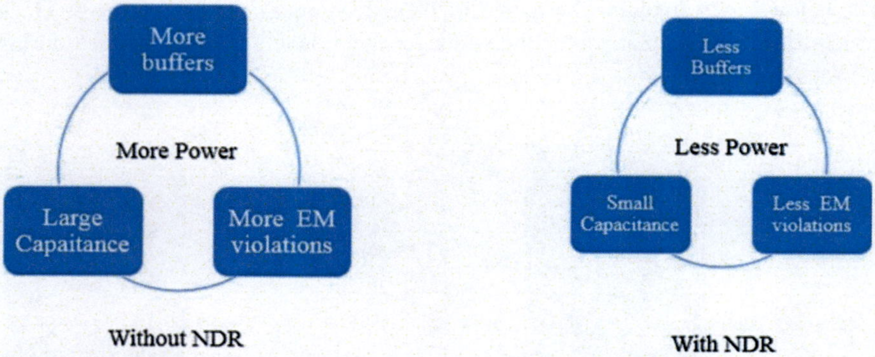

Fig. 4 Effect of NDR on SoC

Let suppose 'x' is some um in width then NDR 2 is x um and NDR 2 will be 2×, 3× etc.

2.3 *Effect of NDR on System on Chip*

The below Fig. 4 explains the comparison between the effect of with and without NDR on SoC. It shows that with NDR design can insert less number of Buffers in CTS stage since slew is not getting much degraded, Coupling Capacitance effect and Electromigration is less due to maintenance of more space between the layers.

2.4 Methodology

The below Fig. 5 depicts the methodology adopted for implementing the project. In the physical design domain for any project to implement first step would be performing the floorplan stage. Second stage will be Placement stage, then next comes CTS stage in which NDR needs to be applied with different width of highest metal layer and spacing between the two metal layers.

A) **Floorplan Stage:** In this stage all the macros are placed in the core area of the chip according the connectivity of the cells/macros with ports or other macros. Hard/Placement blockages are placed over macros and in between channels of the macros such that standard cells won't get overlapped with the macros and also avoids the placing of standard cells in the channels near to the macros. After placing macros, ports placement is done then Power ground mesh is created. Below Fig. 6 shows the completed Floorplan stage of the block.

B) **Placement Stage:** In this stage placement of all standard cells is done and the command "Legalization" is performed so that no two cells overlap.

C) **Clock Tree Synthesis with NDR:**
Steps involved CTS stage:

  i. Create scenarios (Multimode multicorner)
  ii. Apply CTS exceptions (Ignore pin, Stop pin etc.)
  iii. Set CTS settings (NDR)
  iv. Synthesize clock tree (Build clock tree)

Below Fig. 7 shows the command to creating and applying the NDR

Fig. 5 Methodology

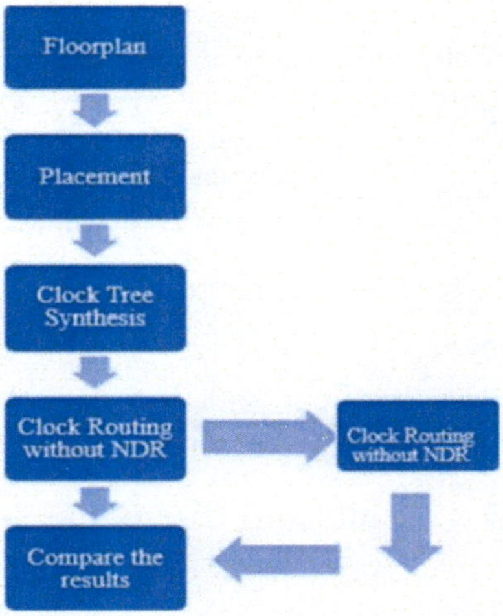

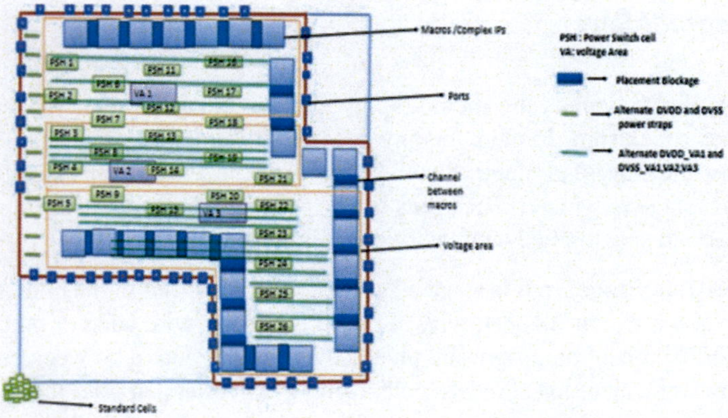

Fig. 6 Floorplan stage

```
create_routing_rule  CLK_M5_M7  -multiplier_width  1.0  –spacing 2.0
create_routing_rule  CLK_M5_M7 { -default reference rule
                        -width {M5  0.04  M6 0.04  M7 0.04}
                        -spacing {M5  0.04  M6 0.04  M7 0.04}
                        }
Set_routing_rule CLK_M5_M7 -Min_routing_layer  -Max_routing_layer –net type root
Set_routing_rule CLK_M5_M7 -Min_routing_layer -Max_routing_layer –net type internal
```

Fig. 7 Command to apply NDR.

Table 1 Complete list of values, result for all parameters

NDR/parameters	Clock cell count	LVT cells count	Utilization (%)	Skew (ns)	Latency (ns)	VT variant (LVT)	Leakage power (nw)
Default NDR (1w2s)	2496	72,147	58.53	0.14	0.904	2390	3.6
2w4s	2425	72,080	58.47	0.112	0.758	2319	3.53
2w3s	2417	72,062	58.5	0.105	0.757	2311	3.52
3w3s	2108	71,749	58.46	0.130	0.746	2002	3.06

3 Results and Discussion

The below Table 1 proves that by applying NDR technique of different width and spacing power (leakage) reduction can be achieved. This paper shows the leakage power reduction of approximately 15% with 3w3s NDR as compared to default

NDR. Also there is 0.1% improvement in utilization, 7.1% improvement in Skew and 17.4% improvement in Latency.

4 Conclusions

Though there is a good improvement of 15% of leakage power, 0.1% improvement in utilization, 7.1% improvement in Skew and 17.4% improvement in Latency but the congestion increases in the design due to increase in spacing between the metal layers from 2 to 3 s there would be scarce in availability of metal tracks to route the design. Hence reducing congestion due to this effect of NDR can be considered as future scope or study.

References

1. Avdalyan N, Petrosyan K (2019) Processing an effective method for clock tree synthesis. In: 2019 IEEE east-west design & test symposium (EWDTS), Batumi, Georgia, pp 1–5
2. Liu W, Sitik C, Salman E, Taskin B, Sundareswaran S, Huang, B (2019)SLECTS: slew-driven clock tree synthesis. IEEE Trans Very Large Scale Integr (VLSI) Syst **27**(4):864–874
3. Chou Y, Wu C, Tsai C (2019)Clocking for HPC design: challenges and experience sharing. In: 2019 international symposium on VLSI design, automation and test (VLSI-DAT), Hsinchu, Taiwan, pp 1–2
4. Choi M, Oh D, Kim J (2018) Slew-aware fast clock tree synthesis with buffer sizing. In: 2018 international conference on electronics, information, and communication (ICEIC), Honolulu, HI, pp 1–4
5. Liu M, Zhang Z, Sun W, Wang D (2017)Obstacle-aware symmetrical clock tree construction. In: 2017 IEEE 60th international midwest symposium on circuits and systems (MWSCAS), Boston, MA, pp 515–518
6. Lin M, Sun H, Kimura S (2016) Power-efficient and slew-aware three dimensional gated clock tree synthesis. In: 2016 IFIP/IEEE international conference on very large scale integration (VLSI-SoC), Tallinn, 2016, pp 1–6
7. Mei S., Raghavan N, Bosman M, Pey KL (February 2018) Stochastic modeling of FinFET degradation based on a resistor network embedded metropolis Monte Carlo method.IEEE Trans Electron Devices **65**(2), 440–447
8. Yang Y, Chou W, Lin M, Kang P, Oates AS, Peng Y (2017)On-die 16nm metal critical peak current test methodology with 100ps pulse width. In: 2017 symposium on VLSI technology, Kyoto, 2017, pp. T130–T131
9. Wang T, Huang S, Cheng W, Chou Y (2016)Top-level activity-driven clock tree synthesis with clock skew variation considered. In: 2016 IEEE international symposium on circuits and systems (ISCAS), Montreal, QC, 2016
10. Fan X, et al (July 2016)Frequency-domain optimization of digital switching noise based on clock scheduling. IEEE Trans. Circuits Syst Regular Papers **63**(7):982–993
11. Xie F, Wan P (2016) Clock tree synthesis and optimization in BES1300 IC Smart Card. In: 2016 13th IEEE international conference on solid-state and integrated circuit technology (ICSICT), Hangzhou, 2016, pp 846–848.
12. Riahi Alam M, Salehi Nasab ME, Fakhraie SM (2015) Power efficient high-level synthesis by centralized and fine-grained clock gating. IEEE Trans. Comput. Aided Des Integr Circuits Syst **34**(12), pp 1954–1963, Dec. 2015.

13. Seo H, Kim J, Kang M, Kim T (2015) Synthesis for power-aware clock spines. In: 2015 IEEE/ACM international conference on computer-aided design (ICCAD), Austin, TX, 2015, pp 126–131
14. P. P. Saha, S. Saha and T. Samanta, "Rectilinear Steiner Clock Tree Routing Technique with Buffer Insertion in Presence of Obstacles," 2015 28th International Conference on VLSI Design, Bangalore, 2015, pp. 447–451
15. Liu W, Wang Y, Chen G, Ma Y, Xie Y, Yang H (2015)Whitespace-aware TSV arrangement in 3-D clock tree synthesis. IEEE Trans Very Large Scale Integr (VLSI) Syst **23**(9):1842–1853
16. Deng C, Cai Y, Zhou Q (2015)Fast synthesis of low power clock trees based on register clustering. In: Sixteenth international symposium on quality electronic design, Santa Clara, CA, pp 303–309
17. Roy S, Mattheakis PM, Masse-Navette L, Pan DZ (2015) Clock tree resynthesis for multi-corner multi-mode timing closure. IEEE Trans Comput-Aided Des Integr Circuits Syst**34**(4):589–602
18. Ewetz R, Koh C (2015) Cost-effective robustness in clock networks using near-tree structures. IEEE Trans Comput Aided Des Integr Circuits Syst 34(4):515–528
19. M. P. Lin, C. Hsu and Y. Chen, "Clock-Tree Aware Multibit Flip-Flop Generation During Placement for Power Optimization," in IEEE Transactions on Computer-Aided Design of Integrated Circuits and Systems, vol. 34, no. 2, pp. 280–292
20. Park H, Kim T (2015) Synthesis of TSV fault-tolerant 3-D clock trees. IEEE Trans Comput Aided Des Integr Circuits Syst 34(2):266–279
21. Joo D, Kim T (2015) Managing clock skews in clock trees with local clock skew requirements using adjustable delay buffers. In: 2015 International SoC Design Conference (ISOCC), Gyungju, 2015, pp 137–138
22. Cai Y, Deng C, Zhou Q, Yao H, Niu F, Sze CN (2015) Obstacle-avoiding and slew-constrained clock tree synthesis with efficient buffer insertion. IEEE Trans Very Large Scale Integr (VLSI) Syst **23**(1):142–155
23. Sharma V (2015)Minimum current consumption transition time optimization methodology for low power CTS. In: 2015 design, automation & test in europe conference & exhibition (DATE), Grenoble, 2015, pp:412–416
24. Puneeth L, Murty NS (2015) Low power clock Optimized Digital De-Skew Buffer with improved duty cycle correction. In:2015 IEEE International Conference on Computational Intelligence and Computing Research (ICCIC), Madurai, pp 1–5
25. Lu T, Srivastava A (2014)Gated low-power clock tree synthesis for 3D-ICs. In: 2014 IEEE/ACM international symposium on low power electronics and design (ISLPED), La Jolla, CA, 2014, pp 319–322
26. Deng C, Cai Y, Zhou Q (2014) A register clustering algorithm for low power clock tree synthesis. In:2014 IEEE international symposium on circuits and systems (ISCAS), Melbourne VIC, 2014, pp 389–392
27. Roy S, Mattheakis PM, Masse-Navette L, Pan DZ (2014)Evolving challenges and techniques for nanometer SoC clock network synthesis. In: 2014 12th IEEE international conference on solid-state and integrated circuit technology (ICSICT), Guilin, 2014, pp 1–4
28. Joo D, Kim T (2014) A fine-grained clock buffer polarity assignment for high-speed and low-power digital systems. IEEE Trans Comput Aided Des Integr Circuits Syst 33(3):423–436
29. Shih X, Hsu T, Lee H, Chang Y, Chao K (2013) Symmetrical buffered clock-tree synthesis with supply-voltage alignment. In:2013 18th Asia and South Pacific design automation conference (ASP-DAC), Yokohama, 2013, pp 447–452
30. Lim K, Joo D, Kim T (2013) An optimal allocation algorithm of adjustable delay buffers and practical extensions for clock skew optimization in multiple power mode designs. IEEE Trans Comput Aided Des Integr Circuits Syst 32(3):392–405
31. Trifković M, Raič D, Strle D (2013) Optimal deployment of shadowed registers in systems with serial clock distribution. In: Proceedings of the 2013 9th conference on Ph.D. research in microelectronics and electronics (PRIME), Villach, 2013, pp 161–164
32. Shang Y, Zhang C, Yu H, Tan CS, Zhao X, Lim SK (2013) Thermal-reliable 3D clock-tree synthesis considering nonlinear electrical-thermal-coupled TSV model. In: 2013 18th Asia and South Pacific design automation conference (ASP-DAC), Yokohama, 2013, pp 693–698

33. Kahng AB, Lin B, Nath S (2013) High-dimensional metamodeling for prediction of clock tree synthesis outcomes. In: 2013 ACM/IEEE international workshop on system level interconnect prediction (SLIP), Austin, TX, 2013, pp 1–7
34. Park H, Kim T (2013) Comprehensive technique for designing and synthesizing TSV Fault-tolerant 3D clock trees. In: 2013 IEEE/ACM international conference on computer-aided design (ICCAD), San Jose, CA, 2013, pp 691–696
35. Hang G, Yang Y, Zhao P, Hu X, You X (2013) A clocked differential switch logic using floating-gate MOS transistors. In: 2013 IEEE 10th International Conference on ASIC, Shenzhen, pp 1–4
36. Kim T, Kim T (2013) Resource allocation and design techniques of prebond testable 3-D clock tree. IEEE Trans Comput Aided Des Integr Circuits Syst 32(1):138–151
37. Chong AB (2013) ASIC clock tree estimation in design planning. In: 2013 4th International Conference on Intelligent Systems, Modelling and Simulation, Bangkok, 2013, pp 619–626
38. Lung C, Su Y, Huang H, Shi Y, Chang S (2013) Through-silicon via fault-tolerant clock networks for 3-D ICs. IEEE Trans Comput Aided Des Integr Circuits Syst 32(7):1100–1109
39. Liu W, et al (2013) TSV-aware topology generation for 3D clock tree synthesis. In: International symposium on quality electronic design (ISQED), Santa Clara, CA, 2013, pp. 300–307
40. Yuce B, et al (2013) Synthesis of clock trees for sampled-data analog IC blocks. In: East-West Design & Test Symposium (EWDTS 2013), Rostov-on-Don, 2013, pp. 1–4
41. Li X, Liu W, Du H, Wang Y, Ma Y, Yang H (2013) Whitespace-aware TSV arrangement in 3D clock tree synthesis. In: 2013 IEEE computer society annual symposium on VLSI (ISVLSI), Natal, 2013, pp 115–120
42. Sai MPD, Yu H, Shang Y, Tan CS, Lim SK (2013) Reliable 3-D clock-tree synthesis considering nonlinear capacitive TSV model with electrical–thermal–mechanical coupling. IEEE Trans Comput Aided Des Integr Circuits Syst 32(11):1734–1747
43. Giraud B, Noel JP, Abouzeid F, Clerc S, Thonnart Y (2013) Robust clock tree using single-well cells for multi-VT 28nm UTBB FD-SOI digital circuits. In: 2013 IEEE SOI-3D-subthreshold microelectronics technology unified conference (S3S), Monterey, CA, 2013, pp 1–2
44. Chakrabarti P (2012) Clock tree skew minimization with structured routing. In:2012 25th international conference on VLSI design, Hyderabad, 2012, pp 233–237
45. Shih X, Chang Y (2012) Fast timing-model independent buffered clock-tree synthesis. IEEE Trans Comput Aided Des Integr Circuits Syst 31(9):1393–1404
46. Lu J, Chow W, Sham C (2012) Fast power- and slew-aware gated clock tree synthesis. IEEE Trans Very Large Scale Integr (VLSI) Syst **20**(11):2094–2103
47. Chen F, Hwang T (2012) Clock tree synthesis with methodology of re-use in 3D IC. In:DAC design automation conference 2012, San Francisco, CA, 2012, pp:1094–1099
48. Shelar RS (2012) A fast and near-optimal clustering algorithm for low-power clock tree synthesis. IEEE Trans Comput Aided Des Integr Circuits Syst 31(11):1781–1786
49. Parthibhan N, Ravi S, Mallikarjun KH: Clock skew optimization in pre and post CTS. In: 2012 International Conference on Advances in Computing and Communications, Cochin, Kerala, 2012, pp 146–149

Clock Domain Crossing—Design, Verification and Sign-Off

Krishna B. Pandit and Sudha R. Karbari

Abstract As the technology nodes are getting smaller and smaller the design of a SoC is becoming more complex day by day. Modern day SoCs are a collection of individual intellectual properties [IP]. Each IP has its own clocking regime and each clock belongs to the separate clock domains. This leads to the challenge in design and verification of SoCs. Designing of the SoCs involves proper synchronization schemes to be employed to synchronize signals crossing clock domains. The functional verification of SoCs is becoming more and more difficult due to the data transferred between individual IP modules. When data transfers between two IPs the data crosses clock domains leading to clock domain crossing [CDC]. For a SoC to be free of all CDC errors, verification is paramount along with meeting the timing requirements. The present work focuses on reduction of waivers and improving the QoR and the coverage of the design. This work also focuses on the verification of functionality of the synchronizers by simulation and assertion-based verification. Then finally to close the gap from verification to sign-off, the assertions are converted to equivalent timing checks and are validated through timing analysis. The results show that the methodology shows a reduction of waivers by 45% and improved the design coverage to 93%.

Keywords Clock domain crossing · Verification · System Verilog Assertions · Verification · Sign-off

1 Introduction

Modern day SoCs are extremely complex in their composition. Each SoC is composed of smaller individual modules called as intellectual properties. Each IP block operates on its own clock and clock domains. When data transfers in-between the IPs, the data crosses clock domains giving raise to clock domain crossing [CDC]. When transfers from one clock domain to another, it is called as clock domain crossing. CDC usually

K. B. Pandit (✉) · S. R. Karbari
Electronics and Communication Engineering, RV College of Engineering, Bangalore, India

V. Nath and J. K. Mandal (eds.), *Proceeding of Fifth International Conference on Microelectronics, Computing and Communication Systems*, Lecture Notes in Electrical Engineering 748,
https://doi.org/10.1007/978-981-16-0275-7_51

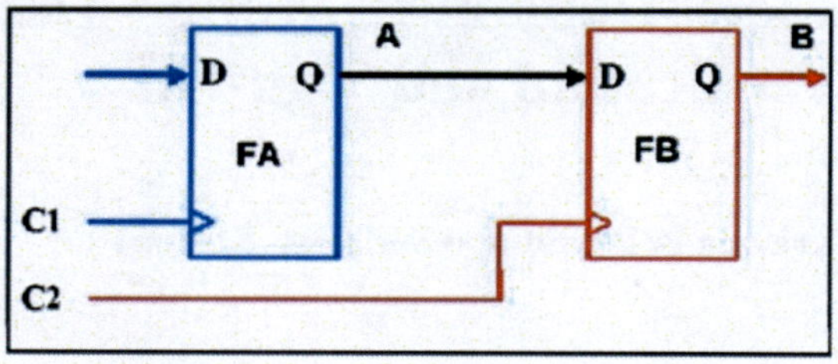

Fig. 1 Clock domain crossing

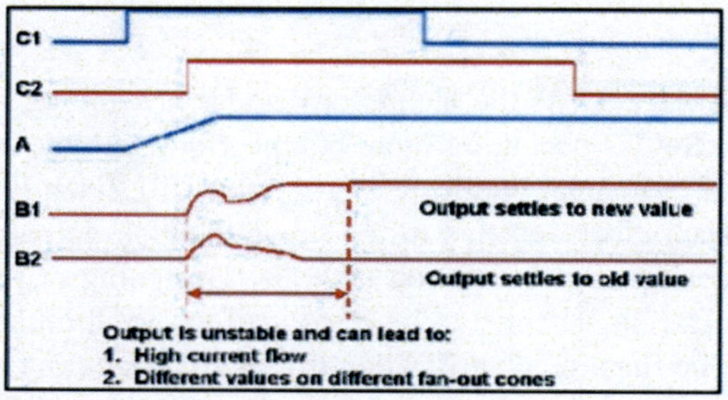

Fig. 2 Metastability

happens when the sending and the receiving clocks are asynchronous w.r.t to each other [1].

In Fig. 1, flip-flop F_A and flip-flop F_B are clocked by two clocks C_1 and C_2. C_1 and C_2 are asynchronous w.r.t each other. Data is said to cross from C_1 domain to C_2 domain. The major issues of CDC are metastability, data loss and data incoherency. Metastability is the major error of CDC. Metastability is the state of the flip-flop when the sampled data is in the unknown state of either 1 or 0 (Fig. 2).

When the transition on signal A happens very close to the rising edge of C_2, violates the setup time or the hold time of the flip-flop clocked by C_2 i.e. F_B goes metastable. The value sampled by the flip-flop may settle to either B_1 or B_2 and is completely unknown. Metastability leads to data loss and data incoherency. Two of the most common problems of CDC. Data loss whenever the destination flip-flop captures source data, if each transition is captured, then data is not lost. Data loss happens in a serial transmission of data bits where each transmitted bit may settle at either 0 or 1 due to metastability owing to data loss. To prevent data loss, the sending end data is to be held stable for a minimum period of 1.5×clock period of the receiving clock. This includes the setup-hold margin and the single clock cycle latency in the receiving domain [2] and [3].

Figure 3 represents data loss. Here instead of capturing 1011 the flip-flop captures 0011 losing the first bit. Data incoherency is the third most common error of CDC. When multi-bit data cross clock domains, due to metastability each bit transferred may settle at a different logic state compared to the input. The data sent and the data received will not match and hence are incoherent. To solve this problem handshaking and asynchronous FIFO based synchronizers are employed [4].

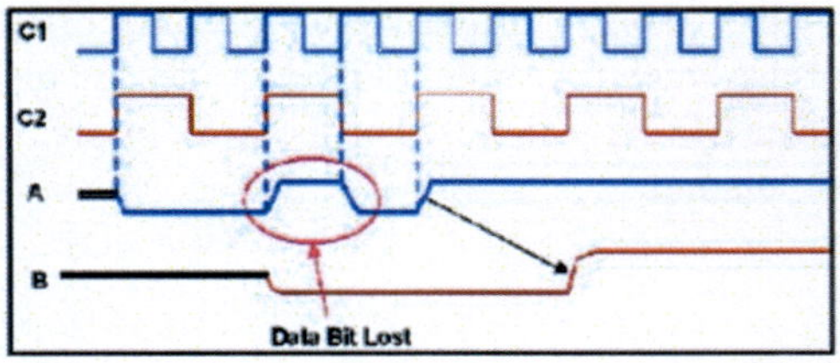

Fig. 3 Data loss

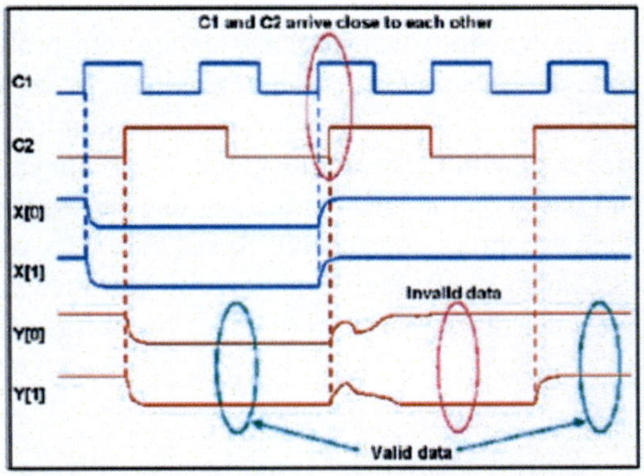

Fig. 4 Data incoherency

Figure 4 represents data incoherency. The data being sent is X[0] and X[1] and the data received is Y[0] and Y[1]. Due to metastability, Y[1] does not match with X[1] leading to data incoherency [5].

2 Synchronization Reliability

The reliability of a type of synchronizer is based on the probabilistic value. This value is known as mean time between failures [MTBF]. MTBF value gives the mean time between two successive failures of the synchronizers. MTBF is given by the equation:

$$\text{MTBF} = \frac{\exp\left(\frac{s}{\tau}\right)}{\text{Tw} * \text{Fc} * \text{Fd}} \tag{2.1}$$

Equation 2.1 represents the MTBF where,
S - the synchronization window where the data must be synchronized
τ - the time constant of the flipflop
Tw - Timing Window, sum of the setup and hold times of the flip-flop
$T_W = T_{SETUP} + T_{HOLD}$
Fc - Clock frequency
Fd - Data toggling rate

The 2FF synchronizer has an MTBF of 10^6 years. MTBF gives the application of the synchronizer i.e. MTBF decides the type of the synchronizer to be used for a given CDC signal. Based on the type of synchronizer used, SVAs are verified through simulation.

3 Verification and Static Sign-Off

As known from the previous section, the problems of CDC are enormous when gone unnoticed. It may even lead to functional and electrical failure of the chip. To combat this problem, synchronizers are employed across clock domains to synchronize data with the receiving clock. There are various types of synchronizers present and based on the data being sent, one can employ different types of synchronizers. Each synchronizing scheme comes with a set of built in System Verilog Assertions [SVAs] to verify their functionality and their usage. Practices from the past have made use of this part to verify synchronizers and declare a chip as CDC bug free [6].

4 Clock Domain Crossing Synchronizers

There are various types of synchronizers used for CDC synchronization. Here are the most common schemes along with their SVAs.

A. *2 Flip-flop synchronizer*
The basic of all the type of synchronizer is the 2 flip flop synchronizer. It consists of 2 back-to-back flip-flops. The 2 flip-flop synchronizer resolves metastability issues (Fig. 5).
The major checks for 2FF synchronizer are the metastability check and data stability check when it comes to verification. Metastability check checks

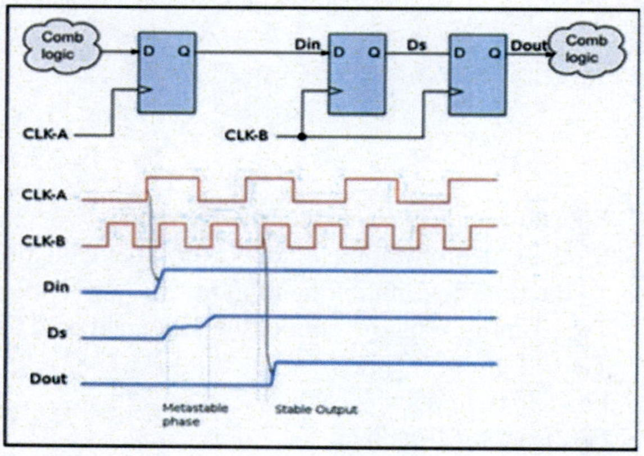

Fig. 5. 2 flip-flop synchronizer

whether there are glitches in the data path and the data stability check checks whether the data is stable for a minimum of 1.5 x the receiving clock cycles [7]. The SVAs for the two checks are given below.

```
property p_stability;
@(posedge clk-B)
!$stable(Din) |= > $stable(Din) [*2];
endproperty: p_stability
property p_no_glitch;
logic data;
@(D_in)
(1, data = !Din) |= > @(posedge clk-B)
(Din == data);
endproperty: p_no_glitch
assert property(p_stability);
assert property(p_no_glitch);
```

The above properties describe the SVAs for a 2 flip-flop synchronizer. SVAs must pass in the simulation for the design to be functionally correct [8].

B. *Handshaking Synchronizer*

Handshaking synchronizer works on the protocol of request and acknowledge. There are two types of handshaking synchronizers-2phase and 4phase. The major functional criteria for a handshaking synchronizer are—each request must get an acknowledge, each acknowledge must be preceded by a request and when the request is asserted, the data sent must be stable [9] (Fig. 6).

The sample SVAs for a handshake synchronizer are:

```
sequence Data_tx;
@(posedge clk)
req ##1 !req [*1:max] ##0 ack;
endsequence
property Req_G_Ack;
@(posedge clk)
req |- > Data_tx;
endproperty
```

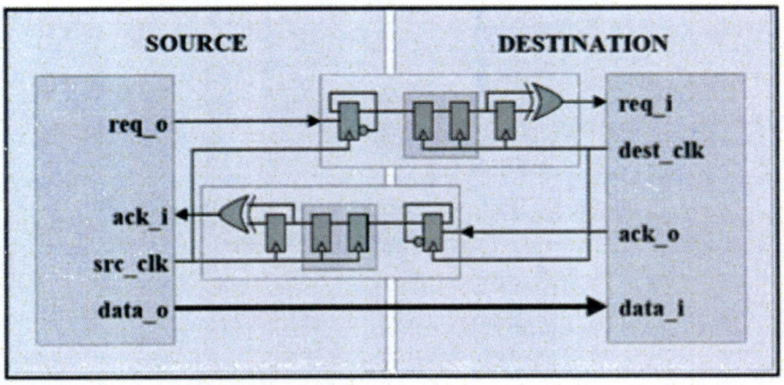

Fig. 6. Handshaking synchronizer

```
property Ack_H_Req;
@(posedge clk)
ack |- > Data_tx.ended;
endproperty
property Data_Stability;
@(posedge clk)
req |= > $stable(data) [*1:max] ##0 ack;
endproperty
assert property(Req_G_Ack);
assert property(Ack_H_Req);
assert property(Data_Stability);
```

C. *Asynchronous FIFO Based Synchronizer*

The most commonly used synchronizer is dual clock asynchronous FIFO based synchronizer. It is composed of a memory element a block RAM or a dual port SRAM, a 2 flip-flop synchronizer, binary to gray and gray to binary counters write and read control logics separated by asynchronous read and write clocks. This synchronizer is used to synchronize multibit data crossing clock domains such as address bus [10].

Figure 7 shows the basic block diagram of asynchronous FIFO synchronizer. The basic checks for a FIFO synchronizer are – never write a full FIFO, never read to an empty FIFO, binary to gray checks and 2 flip-flop synchronizer checks. The sample SVAs of a FIFO based synchronizer is given below.

```
property Access;
@(posedge clk)
inc |- > !flag;
endproperty
property Gray_Code
@(posedge clk) disable iff (!rst_n)
!$stable(data) |- > $onehot(data ^ $past(data));
endproperty
```

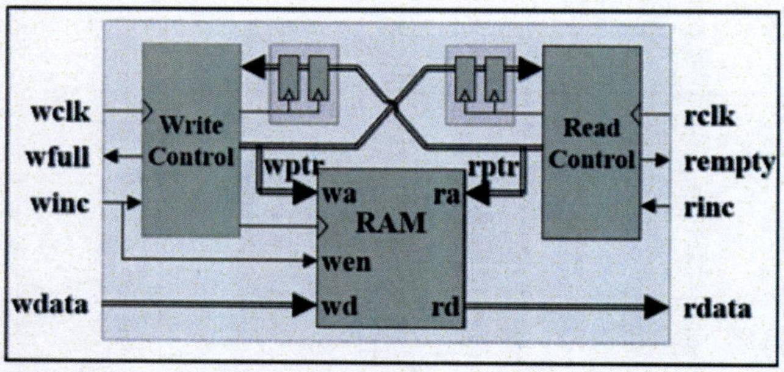

Fig. 7 Asynchronous FIFO synchronizer

The property Access is used for both the write and read domains. The property Gray_Code checks whether the consecutive pointers have a single bit change between them.

5 Sign-Off

For the proper functioning of SoC, not only should it be functionally correct but also it must meet the timing requirements. The timing requirements are checked at the synthesis stage. There are a variety of timing checks that are performed on the SoC for timing sign-off. They are:

A. *Setup Check [max-delay check]*
Setup checks are done on the receiving flip-flop to check whether the incoming data meets the set-up time of the flip-flop. For positive edge triggered launch and capture flops, setup checks are called as single cycle checks. Since the data launched in the present posedge is checked for setup violation in the next posedge. Setup checks consider the maximum delays of the data and clock paths. Hence called as max-delay checks (Fig. 8).

B. *Hold Checks [min-delay checks]*
Hold checks check if the hold time of the flip-flop is met by the incoming data. For flip-flops triggered on the same clock edge, the hold checks are termed as zero cycle checks. Hence hold checks are not dependent on the frequency of the clock triggering the flip-flop [11] and [12].

C. *Data Setup and Hold Checks*
The normal setup and hold checks are done on the data signal with the clock being the reference. Data setup and hold checks are performed keeping one of the data signals as a reference. The signal on which the check is being performed is called the constrained signal and the signal which acts as the reference is known as reference signal. Data set-up time is defined as the amount of time the constrained data signal must be stable w.r.t the reference signal. Data set-up also the time by which the constrained signal must arrive before the reference signal. Data hold time is defined as the amount of time the constrained signal must be stable after the toggling of the reference signal. The data - to data checks

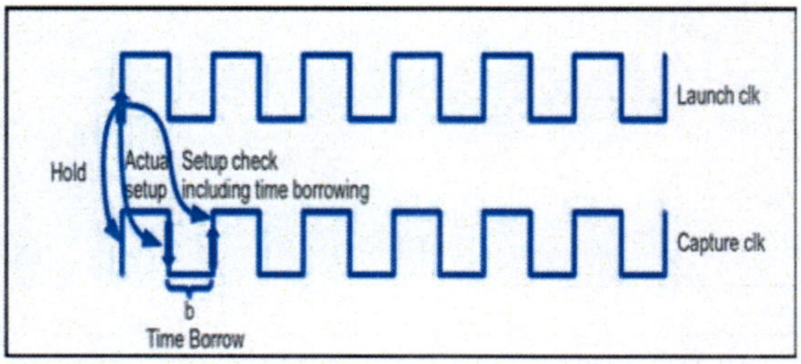

Fig. 8 Setup check hold check

verify the arrival times of the two data signals under consideration. Hence data to data checks are known as skew checks.

6 Constraints Driven CDC Methodology

Constraints driven CDC methodology is based on writing the constraints in tcl format for a given design. The methodology focuses on the reduction of waivers and increase of constraints to make sure that the design is free of all CDC violations.

A. *Waivers*
Waivers are commands given to the tools to ignore certain violations present in the design. Waivers are detrimental to the design. Waivers mask the actual violations to ensure the RTL is clean. Waivers also reduce the quality of design. Reduction of waivers by the addition of equivalent constraints improve the design and the QoR ensuring the design is bug error free (Fig. 9).

B. *Methodology*
The major factor of this methodology is constraints of the design in the SDC format. Figure 9 shows the block diagram of constraint driven CDC methodology.

Step 1
The base of all the constraints is the clock definitions and the clock relationships between two given clocks in the design. It is of paramount importance to identify this in the design.
Step 2
The second step of this methodology is to identify all the CDC paths present in the design, the from and to clocks and the number and signals crossing clock domains.

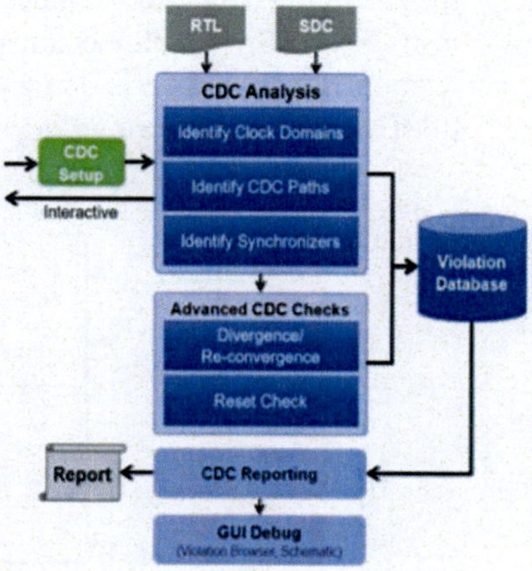

Fig. 9 Constraints driven CDC methodology

Step 3

Upon knowing the number and type of signals crossing clock domains, the third part is to identify the type of synchronizers used to synchronize data at the CDC boundary. The correct instantiation of the synchronizer cells and the correct constraints are necessary for this step.

Step 4

Ensuring the above 3 steps are properly taken care of, the fourth step is to find out and debug actual CDC bugs present in the design. The major type of bugs waived are reset violations and the reconvergence violations. To remove reset violations, the constraint of asynchronous reset de-assertion does the job and to remove reconveregence violations, if the reconverging signals are independent of each other based on timing, then such signals can be declared as timing independent instead of waiving the violations.

7 Verification to Sign-Off

Verification to sign-off methodology is an extension of constraint driven CDC methodology where verification and sign-off are not treated as separate entities. Previous works in this domain have treated verification and sign-off as two separate entities in the design of an SoC. The verification checks verify the SoC based on functionality whereas the timing sign-off checks verifies the SoC based entirely on timing. The functional pre-synthesis verification does not consider the net and the cell delays and the timing sing-off checks post synthesis, do not take into account the functionality of the chip. For a proper and reliable functioning of the chip both in terms of logical functionality and meeting the timing requirements, there must be an intersection between verification and sign-off.

A. *Methodology*

The verification to sign-off methodology bridges the gap between verification and static sign-off first manually and then the process is automated. In the manual process the methodology is as follows (Fig. 10):

Step 1:

For the closing the gap between verification and sign-off, it is paramount to find out the number of asynchronous paths present in the given design. This is done by running CDC on RTL on the tool spyglass. This tool verifies the each asynchronous crossing and the presence of synchronization scheme for that particular crossing. This step requires the inputs to the tool about the design in the form of tcl constraints and based on the design, waivers.

Step 2:

After a successful CDC run, with resolution of all the CDC violations by writing proper constraints and waivers, the tool generates the SVAs for each waiver present in the waiver file and each constraint in the constraint file. These SVAs are verified

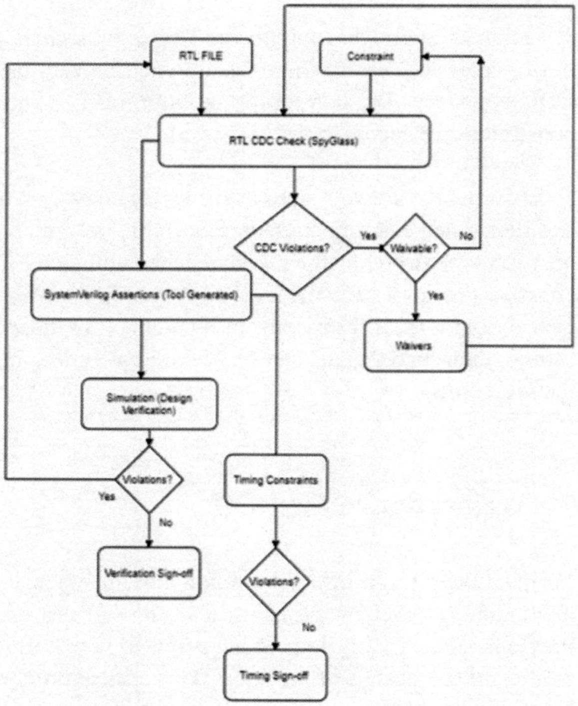

Fig. 10 Verification to sign-off methodology

along with the SVAs written for each type of synchronization cell in the simulation of the design.

Step 3:

After verifying the SVAs for constraints, waivers and synchronization schemes, we need to detect the presence and the number of untimed paths present in the design. Untimed paths are basically paths where we cannot put a timing check. All asynchronous paths present in the design are untimed paths along with multi-cycle paths and false paths. Here to perform the timing checks on these paths, the constraints are written by the conversion of SVAs for the synchronization cell to timing check. This timing check verifies the timing of that given data path along the synchronization cell. Hence, we're verifying the functionality through SVAs and the timing by converting those SVAs to timing checks for the timing sign-off.

Step 4:

As this is a manual and time-consuming job, scripts were written to find out the number of asynchronous paths and to find the type of synchronization scheme used and the number of a synchronization scheme present in the design.

Step 5:

The manual mapping of SVAs to the timing checks were put in a file for each of the synchronization scheme present in the design. This gives the necessary timing check for a given SVA for a given synchronizer present in the design.

8 Results

This section describes the various results obtained at the time of conducting this project. The first section of the results describe the number of violations and the waiver count reduced in the form of a table and the type and number of synchronizers present in the design along with the simulation of the SVAs for the basic 2FF synchronizer. The second section describes the conversion of SVAs for 2FF synchronizer into equivalent tcl script and the timing report upon running the script.

A. *Design Results*

Table 1 describes the total number and the types of synchronizers present in the design. This table gives the inputs to the type of SVAs to be simulated for various types of signals to verify the functionality of the synchronizers.

Table 2 lists the number of CDC violations present in the design. The CDC violations are divided into 4 major categories. The most important violations are clock and reset violations. The table shows the actual violations that waivers were masking in the design. The total waiver count for the design was found to be 15,863. Due to so many number of waivers, the coverage of the design was at 80%.

Table 1 List and number of synchronizers

Serial no.	Type of synchronizer	Number of synchronizer
1	2 flip-flop	150
2	4-phase handshaking	27
3	Asynchronous FIFO	17

Table 2 List of all CDC violations with and without the waivers

Type of CDC violations	Violations with waivers	Violations without waivers
Clock violations	1,353	1,35,387
Reset violations	1,856	3,45,786
Data violations	386	30,583
Reconvergence violations	35	583

Table 3 List of CDC violations after updating the constriants

Type of violations	Violations with updated constraints	Violations with updated constraints and waivers
Clock violations	0	0
Reset violations	5	0
Data violations	2	0
Reconvergence violations	1	0

Table 3 shows the list of violations after updating the necessary constraints. This reduced the waiver count by 45% and improved the design coverage to 92%.

B. *Simulation Results*

Figure 11 shows the simulation graph of SVAs of 2FF synchronizer. The graph shows the passing of the SVAs. The SVAs are converted into equivalent timing script to validate timing.

C. *Timing Results*

Figure 12 shows the timing report of a data paths of synchronized by the 2FF synchronizer. The sending end clock is Aux_clk and the receiving end clock is the Main_clk where both the clocks are asynchronous to each other. The report shows the slack is met for all the paths. This validates the SVAs in terms of timing ensuring the design is CDC error free.

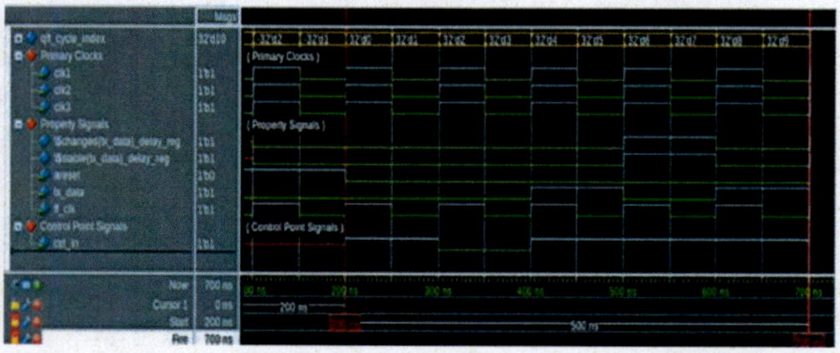

Fig. 11 Simulation results of SVAs of 2FF synchronizers

Fig. 12 Timing report of the SVAs of 2FF synchronizer

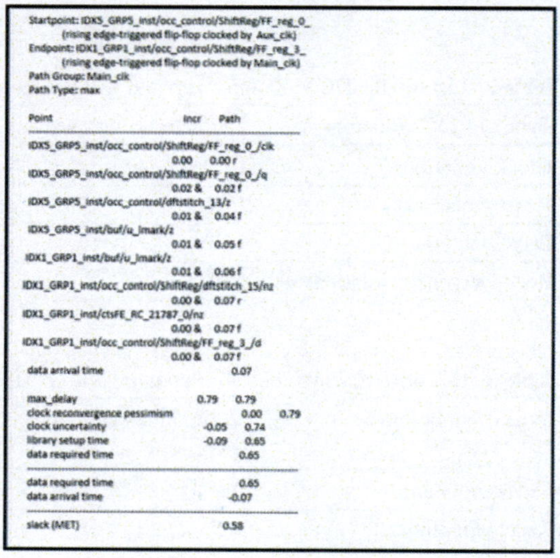

```
Startpoint: IDX5_GRP5_inst/occ_control/ShiftReg/FF_reg_0_
        (rising edge-triggered flip-flop clocked by Aux_clk)
Endpoint: IDX1_GRP1_inst/occ_control/ShiftReg/FF_reg_3_
        (rising edge-triggered flip-flop clocked by Main_clk)
Path Group: Main_clk
Path Type: max

Point                                       Incr     Path
-----------------------------------------------------------
IDX5_GRP5_inst/occ_control/ShiftReg/FF_reg_0_/clk
                                            0.00     0.00 r
IDX5_GRP5_inst/occ_control/ShiftReg/FF_reg_0_/q
                                            0.02 &   0.02 f
IDX5_GRP5_inst/occ_control/dftstitch_13/z
                                            0.01 &   0.04 f
IDX5_GRP5_inst/buf/u_lmark/z
                                            0.01 &   0.05 f
IDX1_GRP1_inst/buf/u_lmark/z
                                            0.01 &   0.06 f
IDX1_GRP1_inst/occ_control/ShiftReg/dftstitch_15/nz
                                            0.00 &   0.07 r
IDX1_GRP1_inst/ctsFE_RC_21787_0/nz
                                            0.00 &   0.07 f
IDX1_GRP1_inst/occ_control/ShiftReg/FF_reg_3_/d
                                            0.00 &   0.07 f
data arrival time                                    0.07

max_delay                                   0.79     0.79
clock reconvergence pessimism                        0.00     0.79
clock uncertainty                           -0.05    0.74
library setup time                          -0.09    0.65
data required time                                   0.65
-----------------------------------------------------------
data required time                                   0.65
data arrival time                                   -0.07
-----------------------------------------------------------
slack (MET)                                          0.58
```

9 Discussion and Future Scope

This project combines two of the most important part of VLSI design flow i.e. verification and synthesis. The transition from verifying the design functionally into the sign-off of the design w.r.t timing has a huge impact on how future SoCs are designed. The verification to sign-off methodology removes the human dependency in the type and manner of checks that are supposed to be done on a multi clock complex SoC. This project does not consider the resets present in the design. The major factor for resets is the de-assertion of resets if it is close w.r.t the clock, it requires a reset synchronizer and gives raise to extra complexities. This can be used as a reference for the future works on resets.

References

1. Bartik, M (June 2018) Clock domain crossing-an advanced course for future digital engineers. In: MECA BUDVA Montenegro
2. Verma, S, Dabare, AS (Dec 2007) Understanding Clock Domain Crossing Issues. EE Times, India
3. Hatture, S, Dighe, S (2015) Open loop and closed loop solutions for clock domain crossing faults. In: GCCT
4. Tarawneh, G, Mokhov, A, Yakovlev, A (2016) Formal verification of clock domain crossing using gate-level models of metastable flip-flops. In: DATE, pp 1060–1065
5. Beer S, Ginosar R (2014) Eleven ways to boost synchronizer. IEEE Trans VLSI Syst 23(6):1040–1049
6. Hatture, S, Dighe, S (2015) Multi clock domain synchronizers. In: 2015 International Conference ON Computation of Power, Energy, Information AND Communication, pp 403–408
7. Mjumdar, A, Jayadev, B (2018) Handling clock-domain crossings in dual clock-edge logic for DFx features. In: 2018 IEEE 27th Asian Test Synposium
8. Litterick, M (2006) Pragmatic simulation-based verification of clock domain crossing signals and jitter using systemverilog assertions. In: DVCon 2006
9. Ginosar, R (2011) Metastability and synchronizer a tutorial. VLSI Systems Research Centre, IEEE
10. Ashar, P, Vishwanath, V (2019) Closing the verification gap with Static sign-off. In: 20th International Symposium on Quality Electronic Design. IEEE
11. Ashar, P, Sachdeva, V, Vishwanath, V (2017) Failures and verification solutions related to untimed paths in SoCs. In: ISQED 2017, pp 460–465
12. Litterick, M (2016) Full Flow Clock Domain Crossing – From Source to Si. Verilab, Munich

Flame Sensor Based Autonomous Firefighting Robot

Nikhil Agarwal and Yogesh Rohilla

Abstract Firefighting is a challenging job, and it comes with a risk of loss of life. With some technical advancements, this risk can be reduced. This paper explains the making of a firefighting robot with the help of flame sensors and Arduino UNO microcontroller board. The robot is made to detect and extinguish the fire. A detailed explanation of components used for making the firefighting robot has been provided. Closed loop control diagram and flow chart are provided for better understanding of principle and control process of firefighting robot, respectively. Every action of the robot has been explained using Proteus 8 Professional simulation environment. Finally, the step by step process of hardware connections has been provided.

Keywords Arduino UNO · Flame sensors · Closed loop control system · Firefighting robot · Proteus 8

1 Introduction

What causes fire to breakout? There are many possible ways for that to happen. It may be accidental or human-made. It could be a short-circuit, gas leakage or any other such activity [1]. Precautions could be and have been taken for these cases, but they still happen. Firefighters risk their lives in extinguishing and mitigation of fire. They try to save lives and the property from damage. Firefighting is a risky job, and sometimes firefighters lose their lives in performing their duty.

Since 60 AD, when the firefighting corps was first established, the firefighters have saved many people from fire incidents. But many people have lost their life too. Lack of firefighters is reported from time to time [2]. Generally, firefighters work in two teams, one saves the lives, and other extinguishes the fire. There is always

N. Agarwal · Y. Rohilla (✉)
Electrical and Electronics Engineering, Institute of Engineering and Technology, JK Lakshmipat University, Jaipur, India
e-mail: yogeshrohilla@jklu.edu.in

N. Agarwal
e-mail: nikhilagarwal@jklu.edu.in

V. Nath and J. K. Mandal (eds.), *Proceeding of Fifth International Conference on Microelectronics, Computing and Communication Systems*, Lecture Notes in Electrical Engineering 748,
https://doi.org/10.1007/978-981-16-0275-7_52

a risk of loss of human resources. But what if all the firefighters could focus only on saving the people and leave the job of extinguishing the fire to someone else? A robot, maybe?

It is possible using a firefighting robot. The job of a firefighting robot could be to detect the fire and extinguish it or to save the people lives. The earlier option seems to be suitable for the firefighting robot, and later option should be left to firefighters. Using such a robot, the firefighters could focus on saving the people and have less worry about the extinguishing the fire. It helps to increase the number of members in their team so that they could cover more ground and save more people. Another benefit of having more members is that if by accident any member of the team got injured, they would have someone to help them too.

Many firefighting robots are reported in the literature. These are based on PID control [3], Bluetooth control [4], innovative algorithms [5, 6], mechanical advancements [7], sensors [8], and Arduino and IoT devices [9, 10].

In this paper, a flame sensor based autonomous firefighting robot is made using Arduino UNO. The firefighting robot works using three flame sensors, that are spread out in three directions to cover more area, to detect fire. Once sensors detect the fire, they then send a signal to the Arduino UNO microcontroller board. Arduino UNO acts as the brain of the robot. The Arduino commands the two DC motors to move the robot in the direction of the fire. After reaching the location of the fire, the water pump starts working on extinguishing the fire. To increase the fire extinguishing range, the pump is attached to a servo motor, which can change the direction of the pump accurately.

The paper arrangement follows the introduction in Sect. 1; components and their descriptions in Sect. 2; concept, working, and control mechanism of the robot in Sect. 3, simulation circuits in Sect. 4; hardware connections in Sect. 5; and finally results and discussion in Sect. 6.

2 Components

Many components have been used in this project to make the firefighting robot, as given in Table 1. Main components are discussed here.

- Arduino UNO
  Arduino boards can be called the brain of a project. It takes the inputs from sensors and then generates outputs based on a stored control program in it. Outputs may either be a display on screens or controlled power supply which can activate the actuators or can blink LEDs [15–17].
  The Arduino board that is being used in this project is Arduino UNO. It is a microcontroller board based on the Microchip ATmega328P [11]. Pinout diagram of Arduino UNO microcontroller board is shown in Fig. 1. Description of its pins are as follows:
- *Pin Description:*

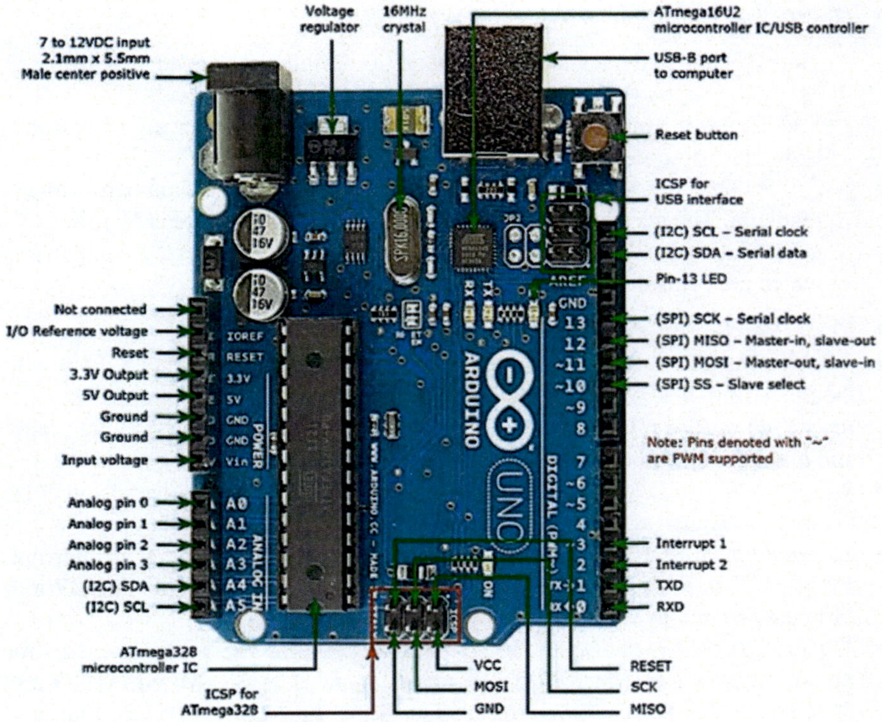

Fig. 1. Pinout diagram of Arduino UNO microcontroller board

Table 1. List of components used to make the firefighting robot

Sr. No	Name	Quantity
1	Arduino UNO board	1
2	IR fire sensor	3
3	Servo motor	1
4	L293D motor driver module	1
5	Small breadboard	1
6	Connecting wires	-
7	Robot chassis	1
8	5 V DC Motors	2
9	Small Can	1
10	Battery 9 V	1
11	MOSFET and resistor	1-1

- *Power (V_{in}, 3.3 V, 5 V, GND):*

  Vin: This pin provides an input voltage to Arduino whenever it is using an external power source.
  5 V: This is the regulated power supply that is used to power microcontroller and other components on the board.
  3.3 V: This pin provides 3.3 V supply generated by an onboard voltage regulator. The maximum current that can be drawn from here is 50 mA.
  GND: This pin is used to ground Arduino and other connected components which are controlled by Arduino.

- *Reset (Reset): This pin is used to reset the micro-controller.*
- *Analog Pins (A0 – A5):* This set of pins is used to provide an analog input with the range of 0–5 V.
- *Input/Output Pins (Digital Pins 0–13):* This set of pins can be used as both input and output pins.
- *Serial {0(Rx), 1(Tx)}:* The serial pins are used to receive and transmit TTL (Transistor-Transistor Logic) serial data.
- *External Interrupts (2, 3):* The external interrupt pin is used to trigger an interrupt.
- *PWM (3, 5, 6, 9, 11):* These pins are used to take an 8-bit PWM (Pulse Width Modulation) output.
- *SPI {10 (SS), 11 (MOSI), 12 (MISO) and 13 (SCK)}:* These pins are used for Serial Peripheral Interface (SPI) communication. Here SS, MOSI, MISO and SCK stands for Select Slave, Master Out Slave In, Master In Slave Out, and Serial Clock respectively.
- *Inbuilt LED (13):* This pin is used to turn on the LED that is inbuilt in the Arduino.
- *TWI {A4 (SDA), A5 (SCA)}:* These two pins are used for TWI (Two Wire Interface) communications. Here, SDA stands for Serial Data, and SCA stands for Serial Clock.
- *AREF (Analogue Reference):* This pin is used to provide a reference voltage to the input voltage.
- *Flame Sensor*

  A flame sensor is a sensor that is used to detect fire. This sensor consists of an infrared (IR) module or photodiode that detects the IR rays emitted from the fire. It then uses an operational amplifier to check the voltage difference, if any, in the sensor. If there is a fire, then the voltage output would be LOW, otherwise HIGH. The flame sensor finds its applications in petroleum gas lines, fire alarms, hydrogen stations, furnaces, generators, metal fabrication, storage tanks, and many more [12].

  An IR flame sensor module is shown in Fig. 2. The flame sensor consists of three pins; DO, GND, and VCC. VCC pin supplies power to the sensor, GND pin provides the grounding to the module, and DO sends the sensed signal to the connected location.
- *Servo Motor*

  A servo motor is an electrical device, used for movement. It uses gears for high torque. These are used in industries to shift an object from one place to another

Fig. 2. The infrared flame sensor

Fig. 3. The servo motor and three wires

with accuracy. Here, servo motor shifts the direction of the water pump while it extinguishes the fire.

Servo motor consists of three wires, brown, red and orange, as shown in Fig. 3. The brown wire is for grounding, red wire powers the motor, and orange wire takes PWM signal to drive the motor.

- *DC Motors*
  A DC motor is the most commonly used type of actuator whose speed of rotation can be easily controlled. The DC motors used in the paper are 5 V DC motors. DC motors are preferred here due to their properties of quick starting, stopping, and control.
- *Motor Driver Module*
  A motor driver module is an IC that is used to control the DC motors and stepper motors. The L293D driver module is used in this paper to make autonomous firefighting robot. It includes four quadruple half-H bridges which can control two motors simultaneously. The L293D motor driver IC based module and IC's pin layout is shown in Fig. 4. L293D IC is a 16 pin IC, and a description of its pin is shown in Table 2 [13]. It has been used to control two DC motors in this paper.
- *Breadboard and connecting wires*
  A breadboard is a plastic board using which prototype circuits can be made in an early stage before the final version. Connecting wires connect one component with other, supply power and ground, send signals and help in communication between components. Current flows through these wires. Figure 5 shows a DC motor, a breadboard, and connecting wires.

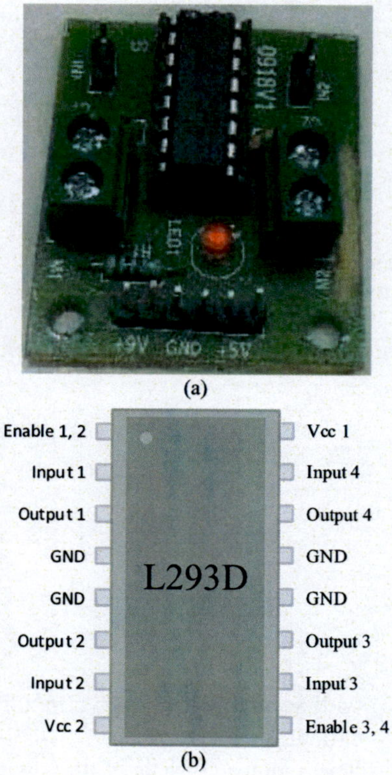

Fig. 4. The motor driver module. (a) Motor driver module with L293D IC and other components arrangement, (b) Pin layout of L293D motor driver IC

3 The Firefighting Robot

The firefighting robot has fire sensors interfaced in its control hardware which detects the fire and make the responsive move as needs are. The robot is intended to recognize the intensity of the fire, and work first, a place where the intensity of the fire is more. It is a programmed robot and doesn't work from any remote control. One just needs to convey the robot in a fire inclined zone, and the robot consequently starts activity once it recognizes a fire breakout. This robot discovers its applications in rescue tasks during fire mishaps, where the opportunities for administration men to enter the fire inclined territories is poor.

The control hardware of the robot is based on Arduino UNO. There are three fire sensors interfaced in the control hardware in the forward, left and right half of the robot, Fig. 6. A water pump motor is appended on the robot which reproduce the working of a water siphon. Aside from the segment's interfacing in the circuit, the primary noteworthiness is of the Arduino board running on the controller circuit. It is the Arduino microcontroller which gives the product the knowledge to detect fire utilizing fire sensors, move the robot towards the high-intensity fire, and speed up

Table 2. Pin description of L293D motor driver IC

Pin number	Pin name	Pin description
1	Enable 1-2	It is a master control pin for the left part of the IC. Left part of IC works when this pin is HIGH
2	Input 1	Pin 2 is used to allow the current to flow through the Output 1 terminal
3	Output 1	It is connected to any one of the terminals of the motor
4, 5	GND	Ground
6	Output 2	It is connected to second terminal of the motor whose first terminal is connected to pin 3
7	Input 2	Pin 7 is used to allow the current to flow through the Output 2 terminal
8	Vcc 2	It is used to power the IC
9	Enable 3-4	It is a master control pin for the right part of the IC. Right part of IC works when this pin is HIGH
10	Input 3	Pin 10 is used to allow the current to flow through the Output 3 terminal
11	Output 3	It is connected to any one of the terminals of the motor
12, 13	GND	Ground
14	Output 4	It is connected to second terminal of the motor whose first terminal is connected to pin 10
15	Input 4	Pin 15 is used to allow the current to flow through the Output 4 terminal
16	Vcc 1	It is used to power the IC

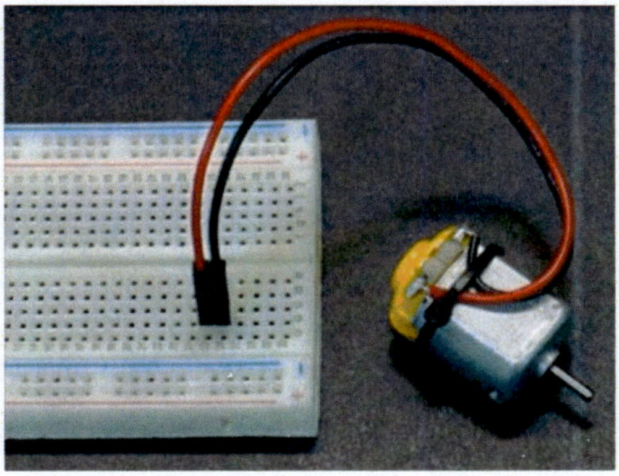

Fig. 5. A representation of DC motor, breadboard, and connecting wires

the water siphon. The Arduino board is composed and aggregated utilizing Arduino IDE.

The robot works by identifying the fire utilizing the three fire sensors. After coming in fire range, the siphon begins pouring water on the fire. The water stops as

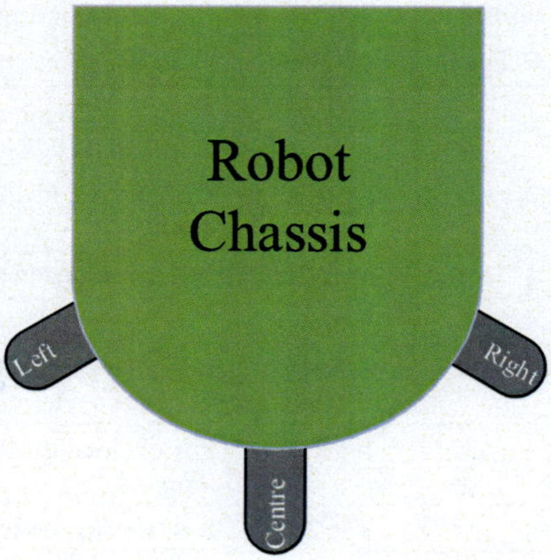

Fig. 6. Location of three flame sensors on robot chassis. They are located at front-side centre, left, and right

soon the fire sensors quit recognizing the fire. The fundamental working of this robot depends on the Arduino and the three flame sensors. The sensor module includes an infrared sensor that senses the fire in the vicinity. Fire emits infrared rays which are sensed by the infrared sensor. It then uses an operational amplifier to check if there is any change in the voltage over the flame sensor. Now, if it detects any fire, then the output pin of the sensor gives 0 V (LOW), and if not, then it gives 5 V (HIGH) [14].

Three flame sensors have been used to detect the fire in three different directions, left, right, and centre. By using these sensors, the robot then determines the path to the fire and then uses the DC motors to move towards the fire with the help of the L293D motor driver module. When the robot reaches near the fire, it then uses the 5 V water pump to through the water on the fire with the help of the servo motor.

The firefighting robot follows a closed loop control system framework to take its decision independently, Fig. 7. The system aim is to extinguish the fire and finally, make sure that there is no fire. The input of the system is the status of fire detected by the flame sensors. Input and output are compared using the Arduino UNO microcontroller. Based on the difference between input and output, microcontroller issues the control commands to the actuators. Actuators are DC motors, servo motor, and water pump motor. Actuators run the whole plant, i.e. firefighting robot. Plant processes and try to extinguish the fire if present, or take no action. The output of the plant again feeds back to the controller and compares with the input. This process repeats and alternatively, make the firefighting robot autonomous.

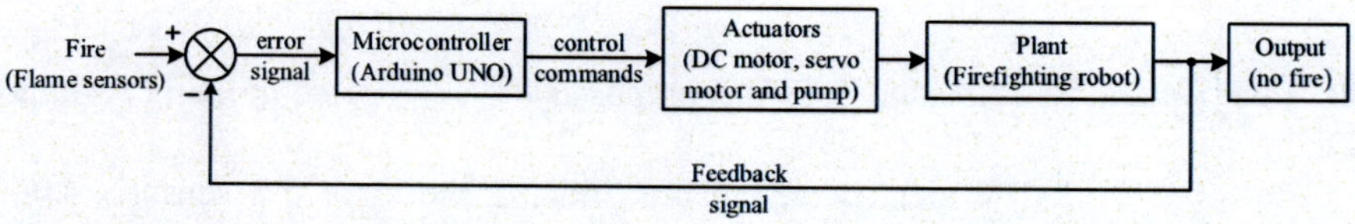

Fig. 7. Closed loop control system for firefighting robot

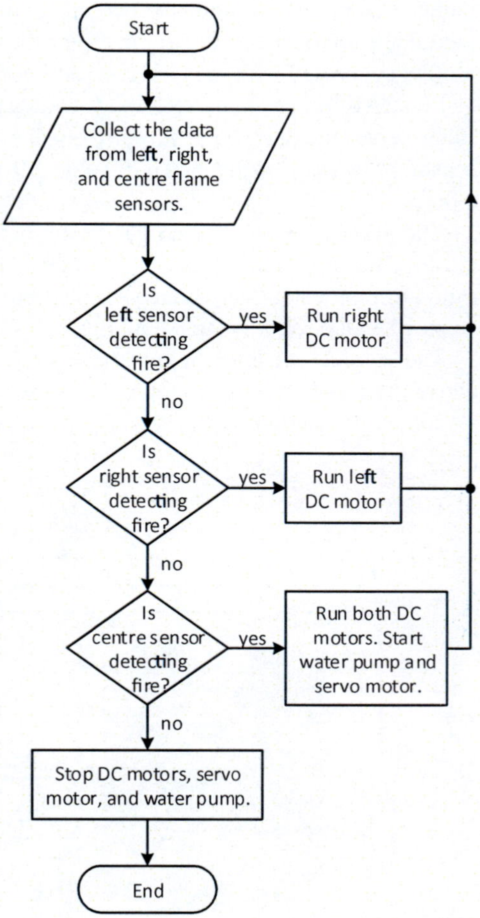

Fig. 8. Flowchart for the control actions of the autonomous firefighting robot

The control commands generated by the Arduino microcontroller depend on the coding algorithm fed in it. Coding algorithm is written in Arduino IDE software which is based on C++ language, and it is then uploaded using USB D type port on Arduino UNO. A flow chart explaining the coding algorithm is shown in Fig. 8.

4 Simulation Validation

Autonomous firefighting robot has been simulated using Proteus 8 Professional software. The simulation shows the working of the DC motors along with the pump based on the direction of the fire, which is detected using three flame sensors. Figure 9 shows the simulation circuit of the system. It includes three flame sensors, denoted by RIGHT_S, LEFT_S, and STRAIGHT_S; three logicstates connected with each flame sensor; one Arduino UNO board denoted by ARD1; one L293D motor driver module denoted by U1; one servo motor located above the Arduino Board; two DC motors connected with motor driver; and one pump motor shown between LEFT_S and STRAIGHT_S flame sensors. It shows the idle condition of the robot. Logic-states serves the purpose of fire. If the value of logicstate is 1, that means the flame sensor is detecting no fire, and if its value is 0, that means the flame sensor is detecting a fire.

The first case is when the right sensor detects fire. The first step for the robot is to move towards the fire, which is achieved by running the left DC motor with the right DC motor stationary. In result, the robot moves around its axis. This movement stops when the robot is facing the fire.

The second case is when the left sensor is the one detecting the fire. Just like the above case, only the right motor runs in the forward direction with the left motor stationary. In result, robot moves left and face towards the fire.

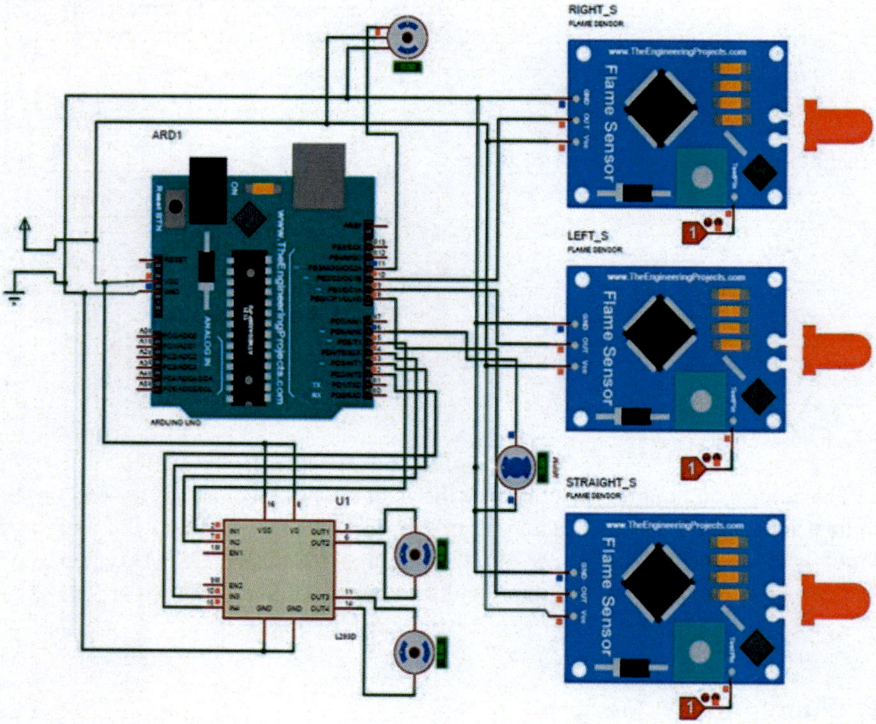

Fig. 9. Simulation of autonomous firefighting robot. There are three flame sensors, denoted by RIGHT_S, LEFT_S, and STRAIGHT_S; three logicstates connected with each flame sensor; one Arduino UNO board denoted by ARD1; one L293D motor driver module denoted by U1; one servo motor located above the Arduino Board; two DC motors connected with motor driver; and one pump shown between LEFT_S and STRAIGHT_S flame sensors

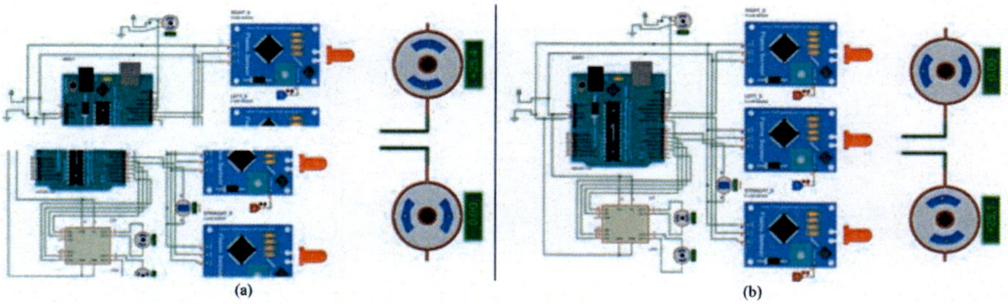

Fig. 10. Operation of autonomous firefighting robot. (a) Case 1: When only the right flame sensor, RIGHT_S, is sensing the fire, shown by logicstate position 0. The left DC motor is running, shown by +25.4 rpm speed, whereas the right DC is at a standstill. (b) Case 2: When only the left flame sensor, LEFT_S, is sensing the fire, shown by logicstate position 0. The right DC motor is running, shown by +25.1 rpm speed, whereas the left DC motor is at a standstill

Both of the above cases are shown in the simulation, as given in Fig. 10. In the first case, only the right sensor detects a fire, and in response, the robot needs to move in the right direction to face towards the fire. It is shown by the movement of the left DC motor in the forward direction, i.e. + 25.4 rpm, keeps the right DC motor at a standstill, Fig. 10(a). As the left motor runs and the right motor is at a standstill, in response robot turns around on its axis till it faces fire. Similarly, in the second case, the robot needs to turn in the left direction because the left sensor is detecting fire. This was achieved by running the right DC motor in the forward direction, i.e. + 25.1 rpm, and leaving the DC motor at a standstill, Fig. 10(b).

In both of these cases, where either a right or a left flame sensor detects a fire, the robot rotates around its axis to face towards the fire. As it rotates, side sensors stop detecting the fire, and only the centre sensor starts detecting it. This is the third case where only the centre sensor is detecting the fire, Fig. 11(a). In this case, both DC motors run in a forward direction at the same speed to reach the fire location, Fig. 11(b). After reaching the location, DC motors slow down to stop. Water pump starts throwing water on the fire, and servo motor adjusts the position of the pump to cover the surroundings, Fig. 11(c) and Fig. 11(d). It throws the water till centre flame sensor stops detecting the fire. This completes the simulation of the autonomous firefighting robot.

5 Hardware Connections

In this section, a step by step process of making the hardware connections of all the components to design an autonomous firefighting robot is being discussed. The process of connecting the components are given in such a manner so that anyone can make these connections by following the process. As given in Table 1, the components required for making the firefighting robot are, one Arduino UNO board, three IR flame sensors, one L293D motor driver IC, two DC motors, one servo motor, one

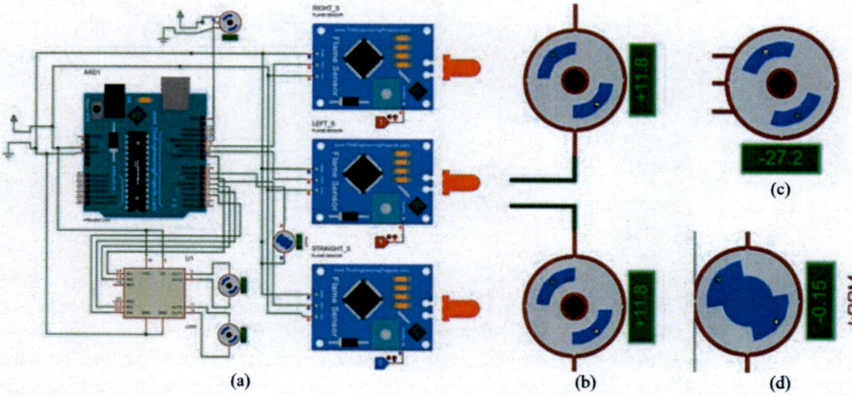

Fig. 11. Operation of autonomous firefighting robot. (a) Case 3: When only centre flame sensor, STRAIGHT_S, is sensing fire, shown by logicstate position 0. (b) Both DC motors are running at same speed in forward direction, shown by +11.8 rpm speed. (c) Movement of servo motor in degrees. Negative sign shows movement in clockwise direction. (d) Water pump motor is running at a speed of 150 rpm

water pump motor, one breadboard, one n-channel MOSFET, one 10 kΩ resistor, and some connecting wires. All the connections are shown in Fig. 12, and step by step process for these connections is as follows:

- Place all the components, as shown in Fig. 12.
- Make a common bus ground (Bus GND) and a common bus positive (Bus + ve) on the breadboard.
- Connect Arduino UNO 5 V to Bus + ve and its GND to Bus GND.
- Connect L293D Output 1 and Output 2 to first DC motor Coil 1 and Coil 2 respectively. Similarly, connect L293D Output 3 and Output 4 to second DC motor Coil 1 and Coil 2 respectively.
- Connect L293D Vcc 2 with Arduino UNO Vin and pump motor Coil 2, and its Vcc 1 to Bus + ve.
- Connect Enable 1, 2 and Enable 3, 4 of L293D to Arduino UNO 5 and 6 respectively.
- Connect all the GND pins of L293D to Bus GND.
- Connect Input 1, Input 2, Input 3, and Input 4 of L293D to Arduino 2, 3, 4, and 7 respectively.
- Connect pump motor Coil 1 to MOSFET D.
- Connect servo motor's brown wire to Bus GND, red wire to Bus + ve, and orange wire to Arduino 8.
- Connect G pin of MOSFET with one end of the resistor and Arduino 9, and its S pin with Bus GND.
- Connect VCC of all the flame sensor to Bus + ve and their GND to Bus GND. Connect DO pins of first, second and third flame sensor to Arduino A3, A4 and A1 respectively.

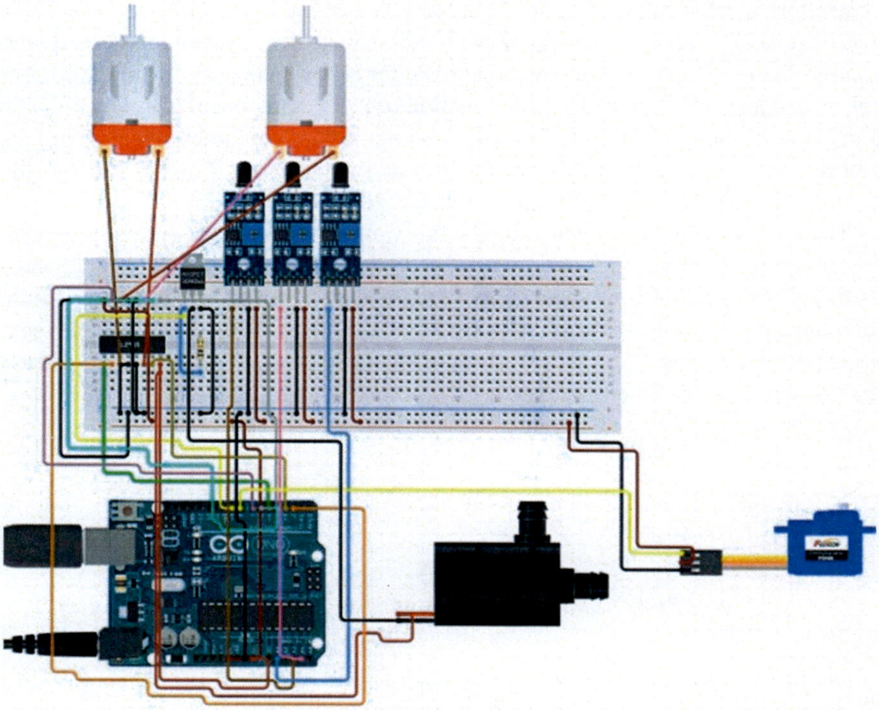

Fig. 12. Hardware connections of different components used for making an autonomous firefighting robot

- Supply power with laptop or PC. Connect the USB cable and upload the coding algorithm to Arduino microcontroller.
- Test each component and troubleshoot for any issues.
- Shift the power supply to the battery.

After following all these steps and testing the working, all the connections need to shift on chassis to make the final version of the firefighting robot. Two wheels connected with the shaft of DC motor supports the structure with the help of one free-moving wheel fixed at the front side of the chassis. A metal or plastic Can help in storing the water. Pump motor can be installed at any appropriate location, and water can be transferred with the flexible PVC pipes.

6 Results and Discussion

A detailed process of making an autonomous firefighting robot has been given in this paper. The focus has been given on every detail related to components selection, controlling mechanism, simulation, and hardware connections. Every section has a good amount of details which equip everyone to make their own robot. Closed loop control diagram and flowchart has been given for better understanding of the control concept of a robot. A physical representation of every component has shown which remove any ambiguity in the selection of component. Simulation diagrams and its descriptions help in testing the concept before implementing. Finally, step by step process of hardware connections has been provided.

There is ample scope of extending the functioning of this robot. A remote control function can also be added to control it remotely overriding its predefined commands. With the inclusion of a webcam, a live feed can be telecast and stored for future reference. Fire extinguishing gases can also be included side by side with the water. This robot can run on flat surfaces only. Including the ability to move on the staircase may be a reasonable extension of this robot.

References

1. Ritchie, H, Roser, M (2018) Causes of Death. Our World in Data 14 Feb 2018
2. Jha, A, Gupta, S, Zaidi, ZH, Gupta, A, Ramaswamy, S, Sharma, B: India Risk Survey 2017. India2018. https://ficci.in/Sedocument/20416/India-Risk-Survey-2017-Report.pdf. Accessed 22 June 2020
3. Rakib, T, Sarkar, MR (2016) Design and fabrication of an autonomous fire fighting robot with multisensor fire detection using PID controller. In: 2016 5th International Conference on Informatics, Electronics and Vision (ICIEV), pp 909–914. IEEE
4. Zaman, HU, Khan, TA, Falgunee, SR, Rashid, GS, Talukder, FH (2018) Autonomous fire-fighting robot with optional bluetooth control. In: 2018 4th International Conference on Computing Communication and Automation (ICCCA), pp 1–4. IEEE
5. Kim, J-H, Keller, B, Lattimer, BY (2013) Sensor fusion based seek-and-find fire algorithm for intelligent firefighting robot. In: 2013 IEEE/ASME International Conference on Advanced Intelligent Mechatronics, pp 1482–1486. IEEE
6. Hidayatullah, AA, Handayani, AN, Fuady, MJ (2017) Performance analysis of A∗ algorithm to determine shortest path of fire fighting robot. In: 2017 International Conference on Sustainable Information Engineering and Technology (SIET), pp 53–56. IEEE
7. Jia, Y-Z, Li, J-S, Guo, N, Jia, Q-S, Du, B-F, Chen, C-Y (2018) Design and research of small crawler fire fighting robot. In: 2018 Chinese Automation Congress (CAC), pp 4120–4123. IEEE
8. Gao, S, Zhang, Z, Zhao, Z, Jamali, MM (2018) Vision and infra-red sensor based fire fighting robot. In: 2018 IEEE 61st International Midwest Symposium on Circuits and Systems (MWSCAS), pp 873–876. IEEE
9. Pan, T, Zhu, Y (2018) A fire-fighting robot using arduino. In: Designing Embedded Systems with Arduino: A Fundamental Technology for Makers Singapore, pp 189–203. Springer, Singapore. https://doi.org/10.1007/978-981-10-4418-2_7
10. Kanwar, M, Agilandeeswari, L (2018) IOT based fire fighting robot. In: 2018 7th International Conference on Reliability, Infocom Technologies and Optimization (Trends and Future Directions) (ICRITO), pp 718–723. IEEE
11. Arduino Uno Rev3 | Arduino Official Store (2020). https://store.arduino.cc/usa/arduino-uno-rev3
12. A Guide to the Applications of Flame Detectors (2013). https://www.azosensors.com/article.aspx?ArticleID=335

13. L293D datasheet | TI.com (ed) (2020)
14. LM393 datasheet | TI.com (ed) (2020)
15. Parihar V, Rohilla Y, Kumari K (2020) Ultrasonic sensor based smart cap as electronic travel aid for blind people. In: 2020 Third International Conference on Smart Systems and Inventive Technology (ICSSIT), pp. 873–877. IEEE
16. Patel D, Rohilla Y (2020) Infrared sensor based self–adaptive traffic signal system using arduino board. In: 2020 12th International Conference on Computational Intelligence and Communication Networks (CICN), pp. 175–181
17. Shekhawat AS, Rohilla Y (2020) Design and control of two-wheeled self-balancing robot using Arduino. In: 2020 International Conference on Smart Electronics and Communication (ICOSEC), pp. 1025–1030. IEEE

Design of Band Pass Filter and Main Amplifier for Biomedical Application Using 180 nm Technology

Lavesh Jain, P. P. Bansod, D. K. Mishra, and Rupali Jarwal

Abstract This paper presents a band pass filter and a main amplifier for biomedical applications. Biomedical signals have very low frequency and low amplitude, So signal filtering is a crucial stage in biomedical system and due to very low amplitude of signal amplification is also important stage in the biomedical system. By cascading high pass filter, buffer and low pass filter, Band pass filters designed. The main amplifier is designed by combining two single stage operational amplifiers in parallel. These circuits are designed by Semi-conductor Laboratory 180 nm Technology in Cadence Virtuoso Analog Design Environment. The signal coming from previous stage to filter is the output signal of amplifier stage contains noise and unwanted frequency component, so filtering of signal is important. After filtering of the signal it is sent to the main amplifier stage to provide more gain to low amplitude signal.

Keywords Biomedical signal · Band pass filter · Main amplifier · Low pass filter · High pass filter

1 Introduction

The biomedical signals have very low amplitude and low frequency. Signals are taken from the human body, so recording of the signal is great challenge because the biomedical signals have low frequency and very low amplitude, therefore high CMRR is required in the Biomedical Signal Front End Design to sense the very weak biomedical signals because the common mode signal might affect the biomedical signal. So the Instrumentation amplifier is used as the pre-amplifier. It has high differential gain as well as high CMRR. The common biomedical signals are Electroencephalography (EEG), Electromyography (EMG), and Electrocardiography (ECG).

A certain amount of charge is transferred to the microchip through the human body these phenomena is known as electrostatic discharge (ESD). Consequence of these phenomena, a huge amount of current can pass in a very small-time duration.

L. Jain (✉) · P. P. Bansod · D. K. Mishra · R. Jarwal
Electronics and Instrumentation Engineering, Shri Govindram Seksaria Institute of Technology and Science, Indore, India

V. Nath and J. K. Mandal (eds.), *Proceeding of Fifth International Conference on Microelectronics, Computing and Communication Systems*, Lecture Notes in Electrical Engineering 748,
https://doi.org/10.1007/978-981-16-0275-7_53

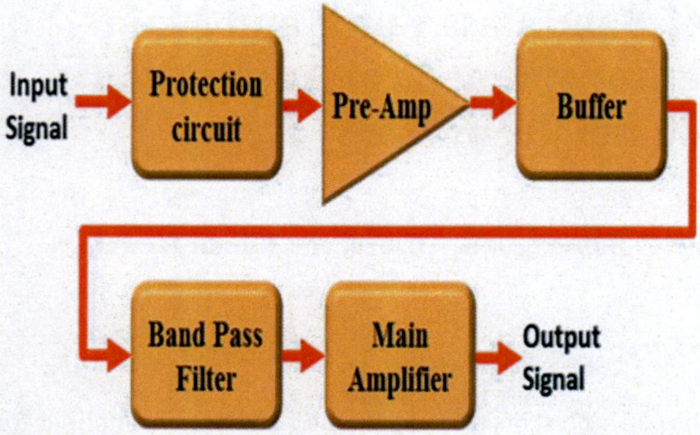

Fig. 1 Block diagram of biomedical signal monitoring system

Therefore, it is one of the most Common hazards to the electronic circuits and integrated circuits (ICs) [4].

The biomedical signals are distorted by the noise caused by the electrosurgical RF units (ESUs). The distortion came into the signals scales from KHz to MHz a RC band pass filter (B.P.F) is used to filter the signal from noise which is designed by cascading of a high pass filter (H.P.F), buffer and low pass filter (L.P.F) [2].

The signal coming from filter output contain very low amplitude, main amplifier stage is used to provide extra gain to the signal by which signal can be detectable at the output of the system.

Protection circuit is used to protect machine and patient from the large amount of current or voltage. Biomedical signals are applied to the protection circuit which is taken from the human body by the leads as shown in Fig. 1. Instrumentation amplifier block and the main amplifier block is used to amplify the signal by providing high gain to it. The amplified signal contains some distortion like noise due high value resistance used in Instrumentation amplifier which is filtered out by the band pass filter.

2 Realization of Low Pass Filter

A 3-dB low pass butter worth filter of 2nd order is designed using Sallen-key topology. This filter is designed to remove noise present in the signal [9]. Biomedical signals have very low frequency range as shown in Table 1, so a low pass filter frequency cut-off of 354.81 Hz is designed with 0 dB gain. It rejects the higher frequency component then 354.81 Hz. The total power consumed by this circuit is 115 μW (Fig. 2).

Butterworth filter is preferred over other filters because it provides more flat pass band and stop band as compared to other filters.

Table 1 Biomedical signals

Biomedical signals	Frequency range	Amplitude range
ECG	0.05 to 150 Hz	1 to 5 mV
EEG	0.5 to 40 Hz	0.5 to 100 μV
EMG	5 to 450 Hz	0 to 10 mV
EOG	0.05 to 41 Hz	50–3500 μV

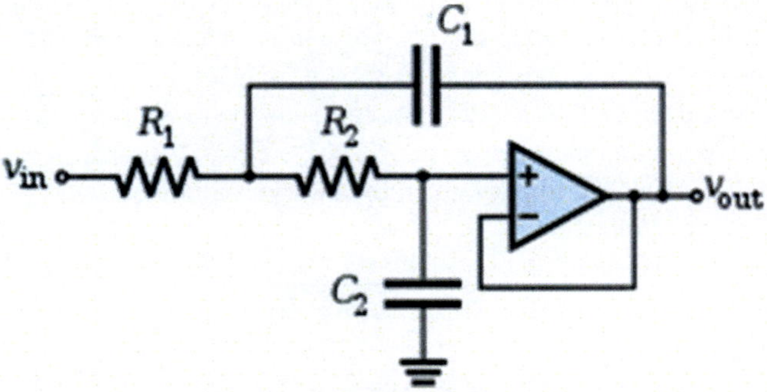

Fig. 2 Architecture of the proposed low pass filter

IInd order system transfer function is given by.

$$\frac{Vout(s)}{Vin(s)} = \frac{A}{S^2 + 2\xi\omega_n S + \omega_n^2} \tag{1}$$

A = overall block gain
ξ = damping of system
ωn = natural frequency of oscillations
Frequency cut-off of the filter is given by natural frequency

$$\omega_H = \frac{1}{\sqrt{R_1 R_2 C_1 C_2}} \tag{2}$$

$$f_H = \frac{1}{\sqrt{2\pi R_1 R_2 C_1 C_2}} \tag{3}$$

3 Realization of Buffer

A negative feedback configuration operational amplifier of unit gain is implemented. The output terminal of the operational amplifiers connected with the inverting terminal of the operational amplifier. Buffer is used to pass same signal from previous stage to the next stage as original level by impendence matching between two phases.

4 Realization of High Pass Filter

Of A3-dB high pass butter worth filter of 2nd order is designed using Sallen-key topology. This filter is designed to remove noise present in the signal [9]. Biomedical signals have very low frequency range as shown in Table 1, so frequency cut-off of filter is 0.05 Hz is designed with 0 dB gain. It rejects the lower frequency component then 0.05 Hz (Figs. 5 and 6).

Frequency cut-off of the filter is given by natural frequency

$$\omega_L = \frac{1}{\sqrt{R_1 R_2 C_1 C_2}} \tag{4}$$

$$f_L = \frac{1}{\sqrt{2\pi R_1 R_2 C_1 C_2}} \tag{5}$$

5 Realization of Band Pass Filter

This filter is designed by the cascading of high pass, filter buffer and low pass filter shown in Figs. 3 and 4 these three blocks are connected in the cascading form (Figs. 7 and 8).

The proposed band pass filter has the lower cut-off frequency 0.05 Hz and higher cut-off frequency of 316 Hz [12]. The band pass filter takes advantage of the high pass configuration as well as the low pass configuration. Band pass filter allows a range of frequency called pass filter. Signals below the lower cutoff frequency (fl) are truncated in the same way as higher cutoff frequency (fh) signals are above (Fig. 9).

$$\omega_H = \frac{1}{\sqrt{R_1 R_2 C_1 C_2}} \tag{6}$$

$$f_H = \frac{1}{\sqrt{2\pi R_1 R_2 C_1 C_2}} \tag{7}$$

$$\omega_L = \frac{1}{\sqrt{R_1 R_2 C_1 C_2}} \tag{8}$$

$$f_L = \frac{1}{\sqrt{2\pi R_1 R_2 C_1 C_2}} \tag{9}$$

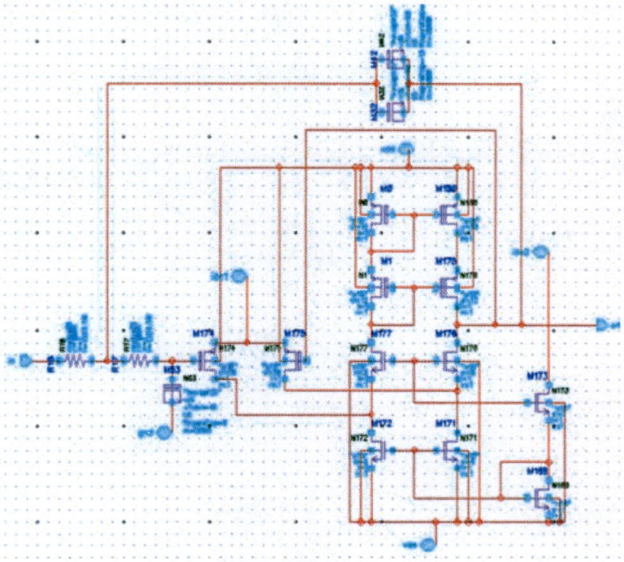

Fig. 3 Schematic design of the proposed low pass filter

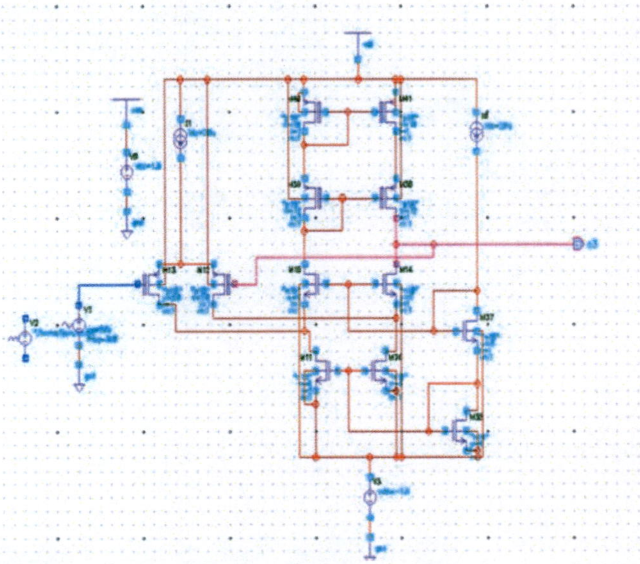

Fig. 4 Schematic design of the proposed buffer

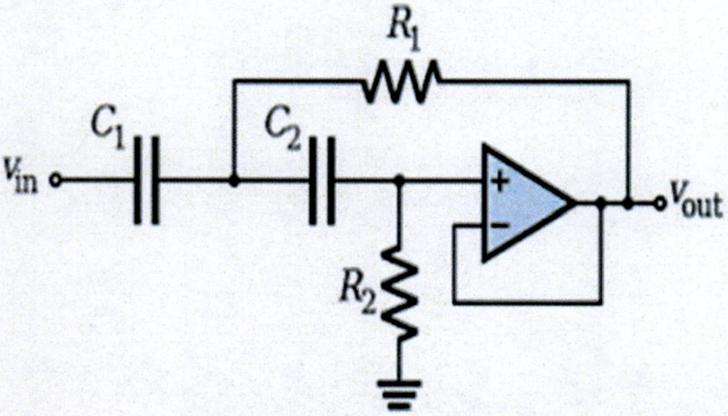

Fig. 5 Architecture of the proposed high pass filter

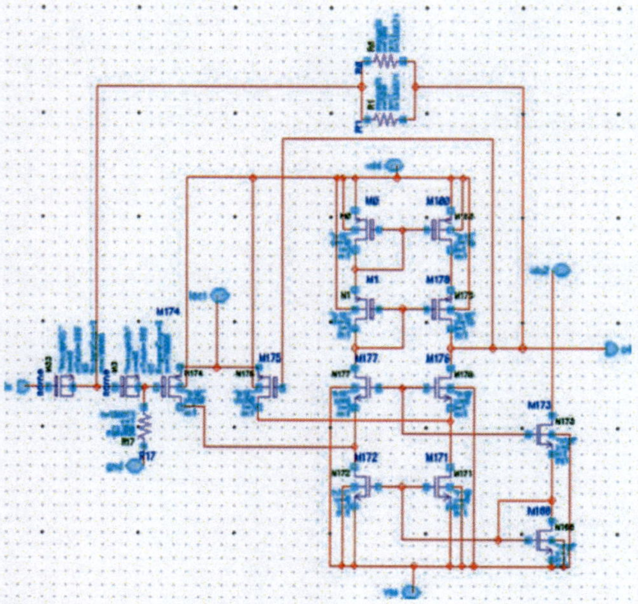

Fig. 6 Schematic design of the proposed high pass filter

Fig. 7 Block diagram of the proposed band pass filter

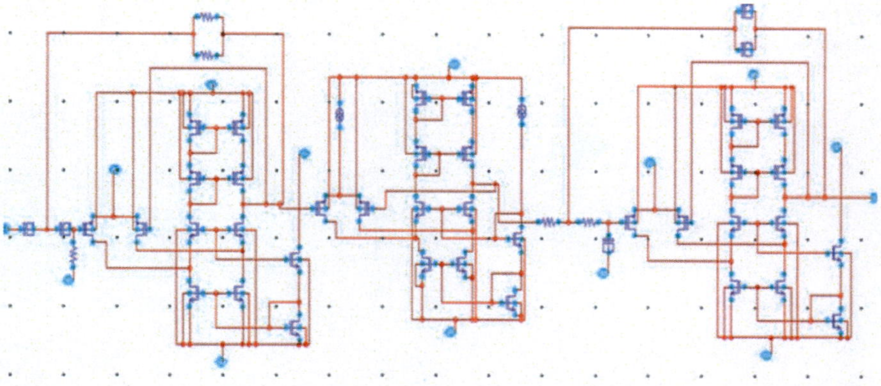

Fig. 8 Schematic design of the proposed band pass filter

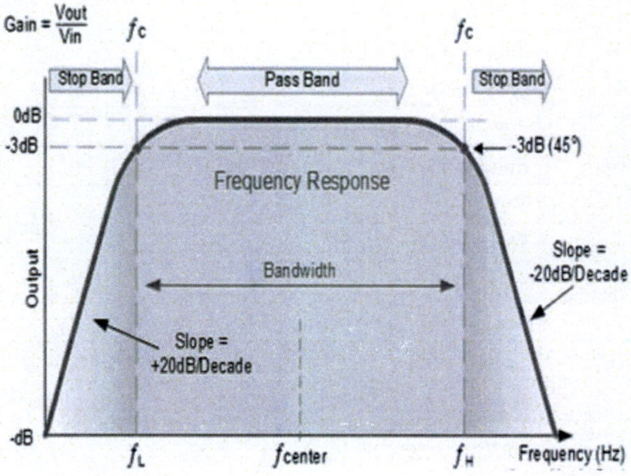

Fig. 9 Frequency response of the band pass filter

Bandwidth is given by the equation

$$BW = f_H - f_L \tag{10}$$

6 Realization of Main Amplifier

The proposed main amplifier schematic is shown the Fig. 10. The main amplifier is designed using single stage differential amplifier. Single stage differential amplifier is one of the most commonly used analog circuits[1, 3, 18], because it provides better CMRR and highest output voltage swing [8]. The main amplifier is used to provide extra gain to biomedical signal after the filtering process of the signals. The

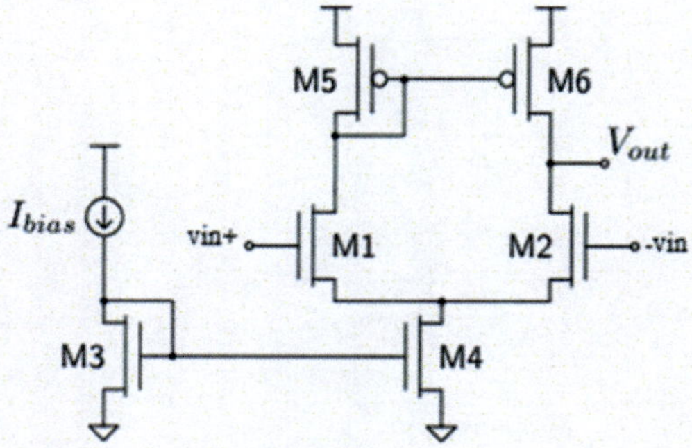

Fig. 10 Single stage differential amplifier

Table 2 Comparison result of this work with previous results

S. No.	Simulated results of band pass filter				
1	Source	[11]		[12]	This work
2	Technology (nm)	180		180	180
3	Supplied power (V)	1.8		1.8	1.8
4	Gain (dB)	0		0	0
5	Power dissipation (μW)	–		–	378.40
6	Bandwidth (Hz)	100		0.05–150	0.05–354
7	Topology	Butterworth		Switched capacitor	Sallen-Key
	Simulated results of main amplifier				
8	Source	[9]	[11]	[12]	This work
9	Gain (dB)	66	67.49	85.57	49
10	CMRR (dB)	–	116	112.52	101
11	Topology	Two stage	Cascode	Two stage	Parallel
12	Power dissipation (μW)	558	–	–	655

biomedical signals are very weak signals, so it requires a high gain and high CMRR for good extraction of signal at the output. Only pre-amplifier stage is not sufficient to provide enough gain to signals that's why need main amplifier stage. Amplifier is designed using a current mirror circuit (Table 2).

To use this circuit as an amplifier, all the transistors should operate in saturation region [8]. Condition for transistor to operate in saturation region is given by.

$$V_{DS} \geq V_{GS} - V_t \tag{11}$$

Drain current in saturation region is given by

$$I_D = \frac{1}{2}\mu_n c_{ox}\frac{w}{l}(V_{GS} - V_t)^2 \tag{12}$$

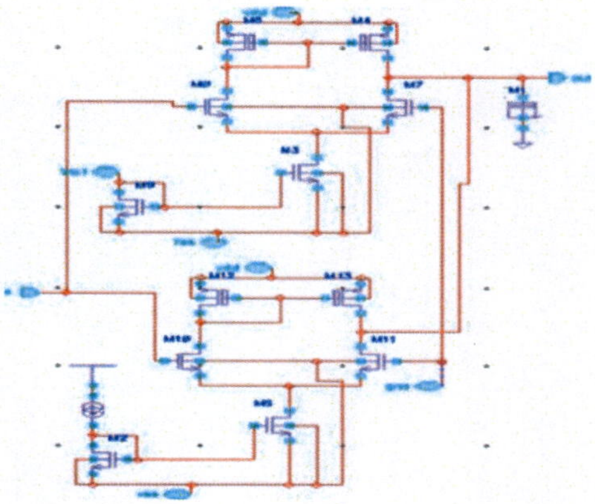

Fig. 11 Schematic design of the proposed main amplifier

Gain and Gain Bandwidth Product is given by

$$Gain = \frac{-g_m}{2\pi f C_l} \tag{13}$$

$$Gain * BW = \frac{-g_m}{2\pi C_l} \tag{14}$$

W/L ratio of all transistors is set by below equation

$$\frac{W}{L} = \frac{g_m^2}{2 I_D \mu_n C_{ox}} \tag{15}$$

This amplifier is designed by using two single stage amplifier connected parallel with each other. The inverting terminal of both are connected together and non-inverting amplifier are connected with each other. The differential gain of this amplifier is 49 dB (Fig. 11).

7 Analysis of Circuit

The Fig. 12 (a), (b), (c) are simulated results of L.P.F, H.P.F and B.P.F its shows the cut-off frequency of the respectively circuit. L.P.F have cut-off frequency of 354 Hz, H.P.F have cut-off frequency of 0.05 Hz and cut-off of B.P.F have range of (0.05–354) Hz its desirable results for the signals.

The Fig. 13 (a), (b) shows the transient analysis of the biomedical signals ECG and EEG through band pass filter the input and output signals are same which shows that the output signal contain zero distortion in the signal.

The Fig. 13 shows the gain of the main amplifier which is 49 dB which provides strength to weak biomedical signals.

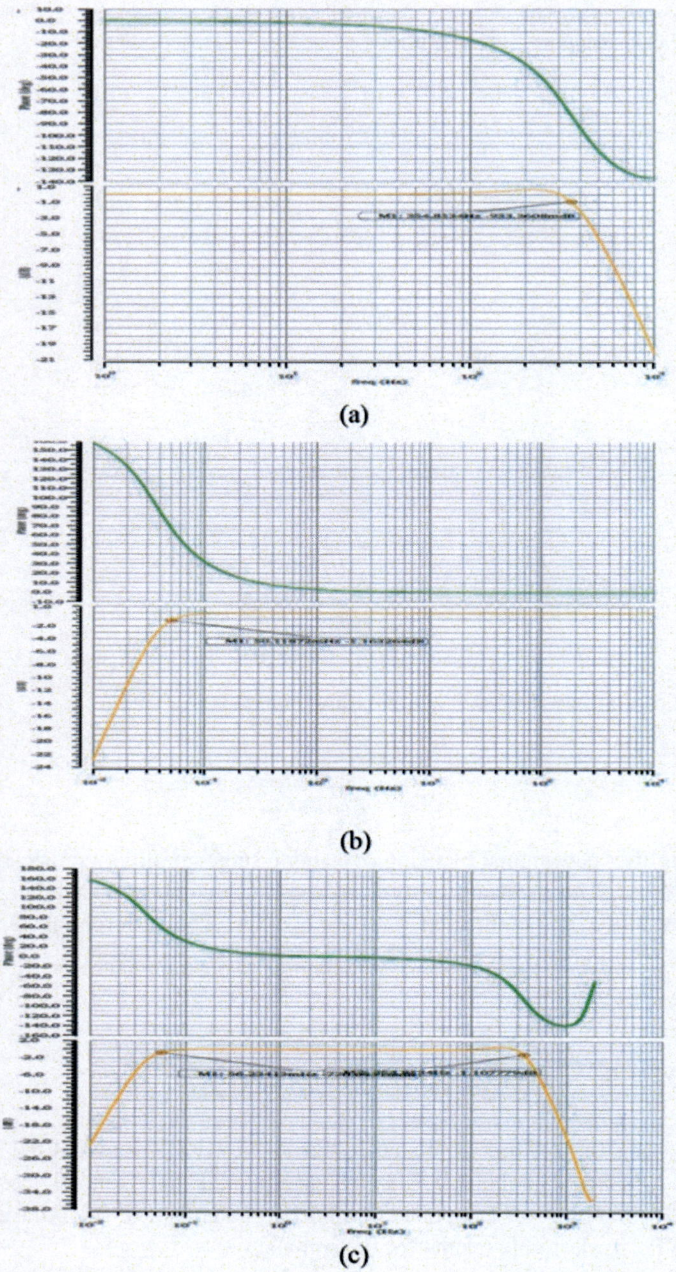

Fig. 12 **a** Frequency cut-off representation of low pass filter **b** Frequency cut-off representation of high pass filter **c** Frequency cut-off representation of Band pass filter

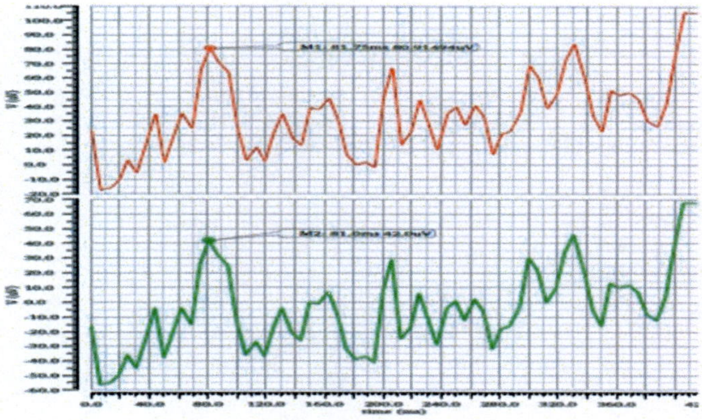

(a)

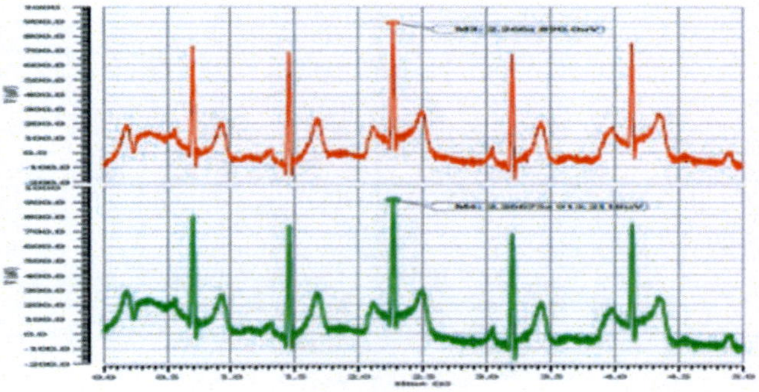

(b)

Fig. 13 **a** Transient response waveform representation of EEG signal through Band Pass filter **b** Transient response waveform representation of ECG signal through Band Pass filter

The Fig. 14 (b), (c) shows ECG and EEG signal respectively they are in mV and μV the main amplifier provides extra gain to change signal in volts at output.which makes signal in practical analysis form.

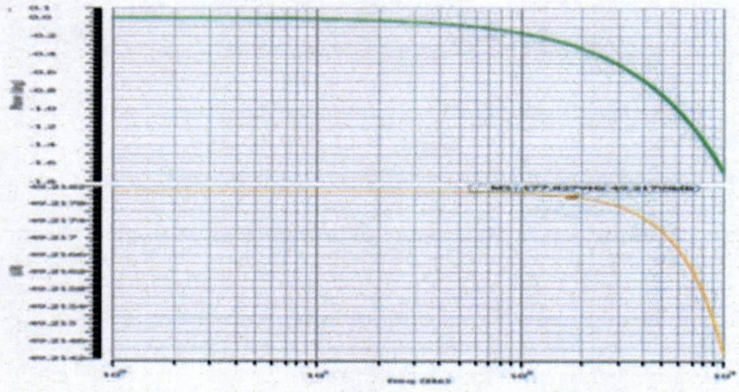

(a)

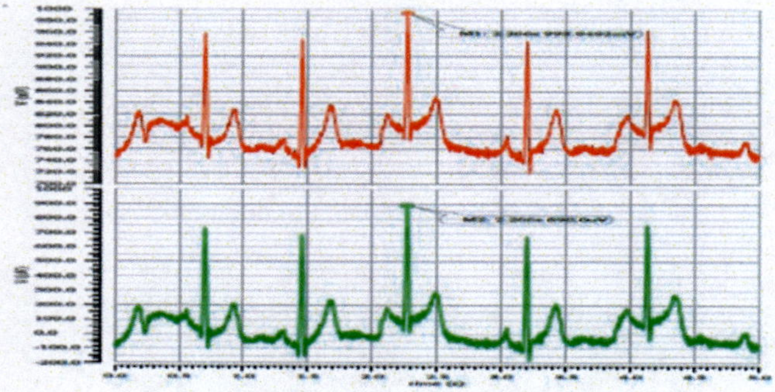

(b)

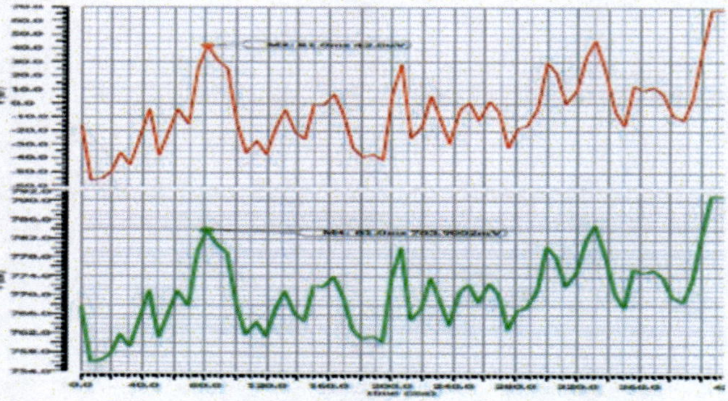

(c)

Fig. 14 **a** Gain of main amplifier is shown in graphical form **b** Transient response waveform representation of ECG signal through main amplifier **c** Transient response waveform representation of EEG signal through main amplifier

8 Conclusion

Filtering of the noisy biomedical signals (ECG and EEG) are done by proposed band pass filter as shown in Fig. 13 (a), (b). The proposed filter has sharp cut-off. Another circuit of main amplifier presented in this paper is used to provide extra gain and CMRR. Main amplifier is used to amplify biomedical signal and to reduce distortion present in it. The amplified biomedical signals (ECG and EEG) are shown in Fig. 14 (b), (c). These circuits are designed by Semi-conductor Laboratory 180 nm Technology. The power supply used is 1.8 V. The auto correlation analysis is done on MATLAB the correlation factor of input and output wave is 0.95 which shows the output and input waveform are almost similar.

References

1. Johns D, Martin K (1997) Analog Circuit Design. John Willy & Sons, New York
2. Sedra AS, Smith KC (2015) Microelectronics Circuits, 7th edn. Oxford University Press, Oxford, pp 1300–1319
3. Razavi B (2002) Design of Analog CMOS Integrated Circuits. Tata McGraw Hill Education, New York
4. Harrison RR, Charles C (2003) A low-power low-noise CMOS amplifier for neural recording applications. IEEE J Solid-State Circuits 38(6):958–965
5. Raghav HS, Maheshwari S (2013) Design of low voltage OTA for biomedical application. IEEE
6. Wen-Chung SW (1994) Digital-compatible high-performance operational amplifier with rail-to-rail input and output ranges. IEEE J Solid-State Circuits, SC-29(1), 63–66
7. Replacing the original Physio Bank web services introduced in 2000, the ATM offers a variety of tools for visualizing PhysioBank data and converting them to text, EDF, or. mat files, and Available: https://phisionet.org/cgi-bin/atm/ATM
8. Fisher JA, Koch R (1987) A highly linear CMOS buffer amplifier. IEEE J Solid-State Circuits, SC-22(3), 330–334
9. Gupta A, Singh S (2018) Design of two stage CMOS Op-Amp with high slew rate and high gain in 180 nm. In: 2018 2nd International Conference on I-SMAC (IoT in Social, Mobile, Analytics and Cloud) (I-SMAC)I-SMAC (IoT in Social, Mobile, Analytics and Cloud) (I-SMAC), 2018 2nd International Conference On. https://doi.org/10.1109/i-smac.2018.8653750
10. Kumngern M, Dejhan K (2009) Voltage-mode low-pass, high-pass, band-pass biquad filter using simple CMOS OTAs. In: IEEE Conference on International Instrumentation and Measurement Technology (12MTC'09), May 2009, pp 924–927
11. Singh KS, Bansod PP, Gamad RS (2019) ECG monitoring system using 180 nm technology. In: 2019 3rd International Conference on Electronics, Materials Engineering & Nano-Technology (IEMENTech)
12. Singh N, Bansod PP (2017) Switched-capacitor filter design for ECG application using 180 nm CMOS technology. In: 2017 International Conference on Recent Innovations in Signal Processing and Embedded Systems (RISE). https://doi.org/10.1109/rise.2017.8378195
13. Barsaiyan M, Bansod PP (2017) Protection circuit design for electrocardiograph (ECG) with input filtering. In: 2017 8th International Conference on Computing, Communication and Networking Technologies (ICCCNT). https://doi.org/10.1109/icccnt.2017.8204168
14. Gualsaqui Miranda MV, Vizcaino Espinosa IP, Flores Calero MJ (2016) ECG signal features extraction. In: 2016 IEEE Ecuador Technical Chapters Meeting (ETCM). https://doi.org/10.1109/etcm.2016.7750859

15. Ahmed W, Khalid S (2016) ECG signal processing for recognition of cardiovascular diseases: a survey. In: 2016 Sixth International Conference on Innovative Computing Technology (INTECH). https://doi.org/10.1109/intech.2016.7845089
16. Hsiao CL, Wei HC, Huang RB, Tan KY (2013) A fully integrated switched-capacitor filter design for ECG application. In: IEEE 2nd Global Conference on Consumer Electronics (GCCE), pp 247–248
17. Allen PE, Hollberg DR (2002) CMOS Analog Circuit Design, 2nd edn. Oxford University Press, Oxford
18. Sedra AS, Smith KC Microelectronic Circuits 6th edition
19. Hamzah MH, Jambek AB, Hashim U (Apr 2014) Design and analysis of a two-stage CMOS Op-amp using Silterra's 0.13 μm technology. In: IEEE Symposium on Computer Applications & Industrial Electronics (ISCAIE), pp 55–59
20. Hosticka BJ (1979) Improvement of the gain of MOS amplifiers. IEEE J Solid-State Circuits, SC-14(6), 1111–1114

Design of Instrumentation Amplifier and Buffer for Biomedical Applications Using 180 nm Technology

Vaibhav Pardhi, P. P. Bansod, D. K. Mishra, and Rupali Jarwal

Abstract This paper presents a high-performance Op-Amp based CMOS Instrumentation Amplifier (IA) and a Buffer for biomedical signal applications. This Instrumentation Amplifier is constructed using three two stage operational amplifiers. To achieve the desirable gain for this instrumentation amplifier the aspect ratios and passive resistances of op-amps are selected very carefully. To obtain a required common mode rejection ratio (CMRR) and gain, the appropriate selection of the transistor sizing and op-amp circuit topology were essential for the design of IA for biomedical applications. A voltage buffer amplifier is cascaded with this IA. Voltage follower or buffer circuit is generally used to isolate different stages from each other. The gain and CMRR of two stage amplifier are 60.917 and 81.012 dB respectively. The CMRR and overall differential gain of this IA are 111.21 and 67.41 dB respectively. The total power absorbed by this IA is 368 μW.

Keywords Instrumentation Amplifier · Buffer · Gain · CMRR · Two stage op-amp · Biomedical signals

1 Introduction

The biomedical signals are taken from the human body, so recording of the signal is great challenge because the biomedical signals have low frequency and very low amplitude, therefore high CMRR is required in the biomedical signal front end design to sense the very weak biomedical signals because the common mode signal might affect the biomedical signal. So the IA is used as the pre-amplifier. It has high differential gain as well as high CMRR [9, 10]. The common biomedical signals are ECG, EEG, EMG and EOG. Table 1 shows different Biomedical Signals and the ranges of their frequencies and amplitudes.

The block diagram of physiological monitoring is depicted in Fig. 1. The biomedical signals taken from the leads are applied to protection circuit. Main function of

V. Pardhi (✉) · P. P. Bansod · D. K. Mishra · R. Jarwal
Electronics and Instrumentation Engineering, Shri Govindram Seksaria Institute of Technology and Science, Indore, India

V. Nath and J. K. Mandal (eds.), *Proceeding of Fifth International Conference on Microelectronics, Computing and Communication Systems*, Lecture Notes in Electrical Engineering 748,
https://doi.org/10.1007/978-981-16-0275-7_54

Table 1 Standard biomedical signals

Biomedical Signals	Frequency Range (in Hz)	Amplitude Range (in mV)
ECG	0.05 - 150	1 - 5
EEG	0.5 - 40	0.0005 - 0.1
EMG	5 - 450	0 - 10
EOG	0.05 - 41	50 - 3.5

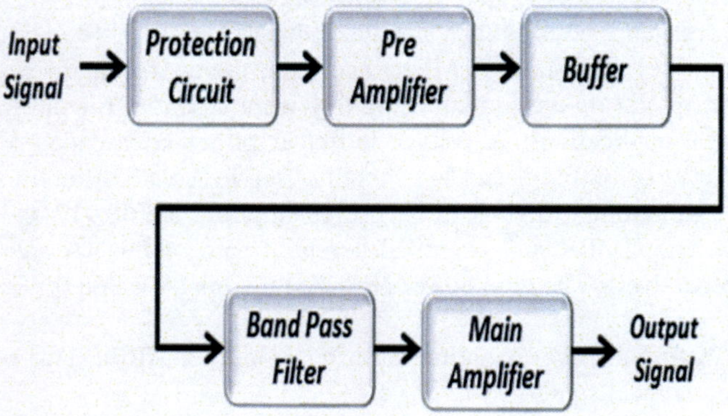

Fig. 1 Block diagram of biomedical signal monitoring system

the protection circuit is to protect the patient and latter stages of the system from high current and voltages. In the next stage the signal is applied to IA where it is amplified and the interferences generated due to common mode signals are rejected [12, 13]. In next stage the signal is passed through a BPF by which the signal is band-limited to a desired frequency. Later this signal is passed through a main amplifier to strengthen the biomedical signals and to achieve the desired gain.

In applications like low power physiological monitoring system, biomedical IA plays a very important role. As the biomedical signals are very weak, the interference caused by DC offset and the artifact signals is very large therefore, for biomedical applications IA should have high differential gain and high CMRR [2, 6].

A voltage buffer amplifier is cascaded with the IA. The reason for using a voltage buffer between two stages of a system is that if a circuit having large output impedance is directly connected to another circuit of low input impedance, the desired functionality of the latter stages of the system will be affected drastically.

In some situations source impedance can be larger than load impedance. In such scenario if source is directly connected to the load there might have a problem of signal attenuation. Hence using buffer or voltage follower, individual stages of the circuit are linked together without mismatching of impedances.

2 Circuit Design of Instrumentation Amplifier and Buffer

An amplifier with the property of high CMRR is required to detect biomedical signals because there might be some noise at the input terminals of the amplifier in the form of common mode voltages [1, 3]. The magnitudes of these interferences are very small therefore to reject these unwanted signals amplifier should have very high CMRR [7]. A two stage op-amp based IA configuration is used for this application. This IA is designed using a fully differential first stage followed by a differential amplifier. Figure 2 shows a conventional IA using op-amps and passive resistances. Two cascaded stages of differential amplifier are used in this IA. The first stage comprises two op-amps 1 and 2 and their corresponding resistances, and the second stage includes Op-Amp 3 and its four corresponding resistances. The Op-Amps of first stage are configured in non-inverting mode and amplify both the input signals V_1 and V_2 [4]. The outputs of these Op-Amps are applied to the input terminals of difference amplifier of second stage. In second stage the difference amplifier amplifies the difference signals i.e. $(V_2 - V_1)$ of previous stage and provides a differential gain as:

$$Gain(A_d) = \frac{R_3}{R_2}\left(1 + \frac{2R_1}{R_{gain}}\right) \tag{1}$$

If $R_3 = R_2$ the differential gain of this IA is given as:

$$Gain = 1 + \frac{2R_1}{R_{gain}} \tag{2}$$

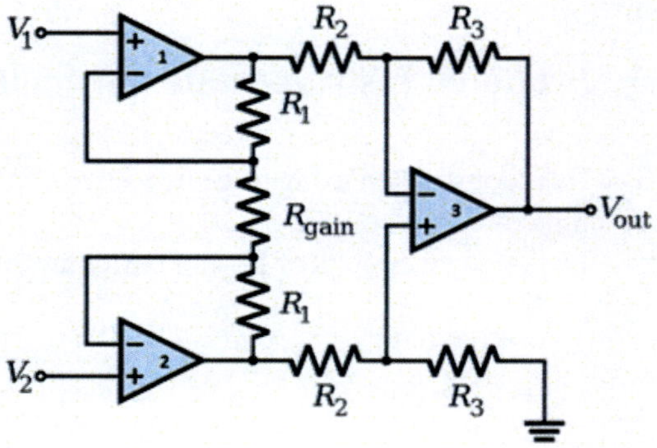

Fig. 2 Standard IA using two stage op-amps

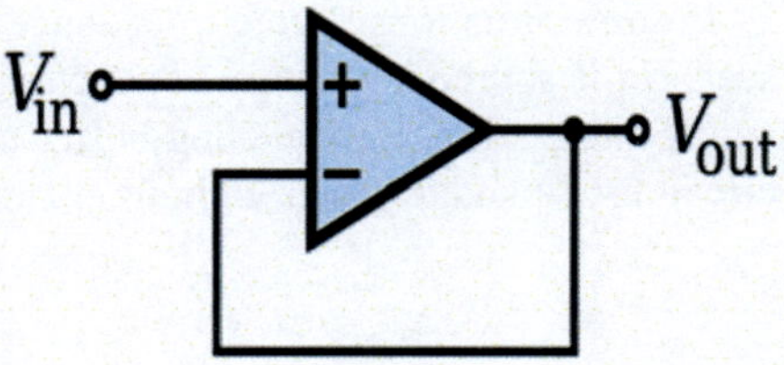

Fig. 3 Voltage follower op-amp or buffer

Input signals are taken from the leads and applied to both the terminals of the difference amplifier and the difference of the two applied signals are amplified and the signals which are common to both the inputs are rejected by the amplifier. Differential amplifier amplifies entirely the difference of applied input signals i.e. $V_d = V_2 - V_1$ and rejects the common mode signal represented by V_{cm}, practical circuits will have an output voltage V_o which is given by

$$V_o = A_d V_d + A_{cm} V_{cm} \tag{3}$$

where the differential mode gain of IA is denoted by A_d and the common mode gain of the IA is denoted by A_{cm}. The differential mode gain of a differential amplifier must be as minimum as possible and ideally zero.

The CMRR is a measure of ability of the differential amplifier to reject the common signals that appear concurrently on both the inputs, which is defined as:

$$CMRR = \frac{|A_d|}{|A_{cm}|} \tag{4}$$

$$CMRR = 20log\frac{|A_d|}{|A_{cm}|}dB \tag{5}$$

A voltage buffer is a unity-gain amplifier. This shows that the buffer circuit does not provide any amplification to the applied signal but it is mainly used to transform the impedance. A buffer circuit is generally used to isolate different stages from each other. The buffer circuit using op-amp is depicted in Fig. 3.

3 Proffered Instrumentation Amplifier and Buffer

The IA is constructed using three two stage CMOS op-amp. The proposed two stage op-amp circuit is shown in Fig. 4.

The preferred circuits of IA and buffer are shown in Fig. 5 and 6 respectively.

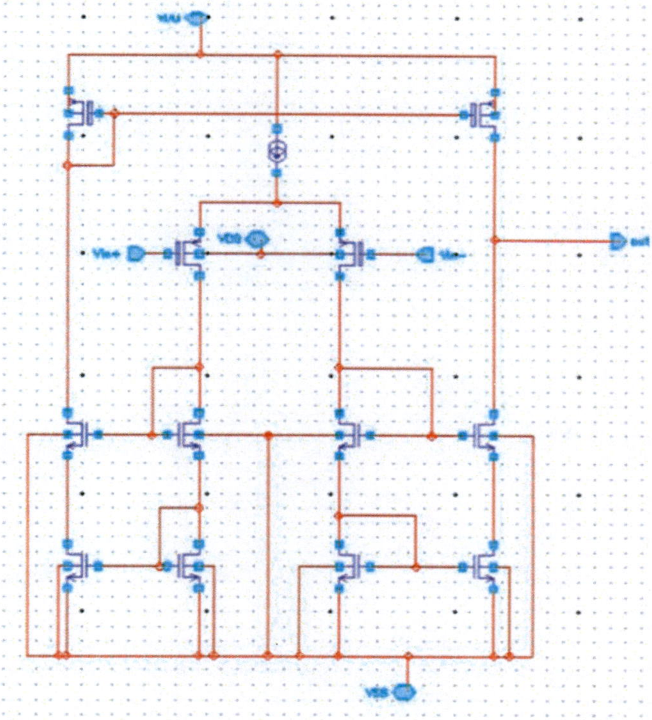

Fig. 4 Two stage operational amplifier

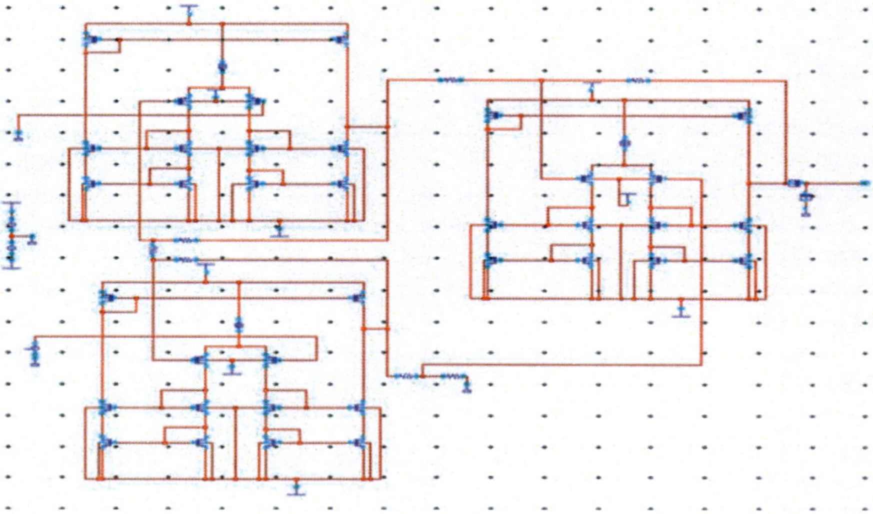

Fig. 5 Proposed schematic of IA

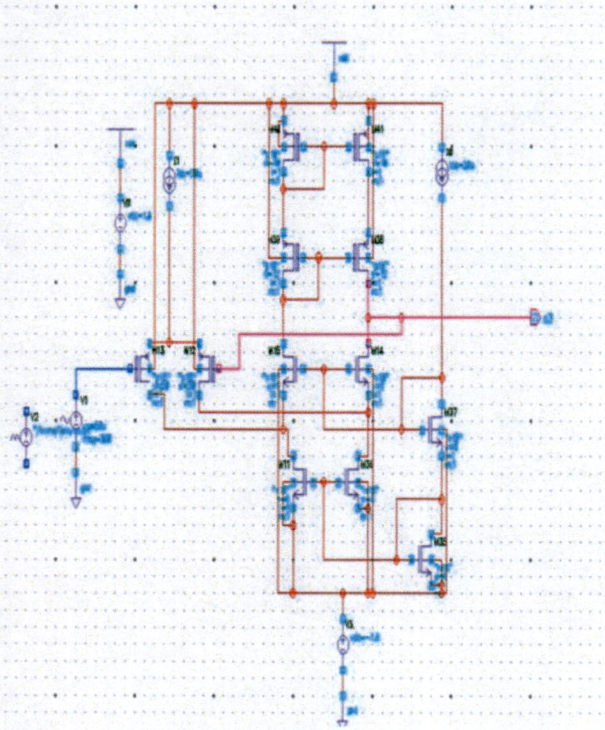

Fig. 6 Proposed schematic of buffer

4 Results of Simulation

The Cadence Virtuoso tool with SCL 180 nm technology is used for simulation of the proposed IA and buffer circuits. The achieved differential gain and CMRR for this IA are 67 and 111 dB respectively. The total power absorbed by this circuit is about 368 μW. The plot of differential mode gain and common mode gain with respect to frequency of this IA are depicted in Fig. 7 and in 8 respectively. The transient response of this IA for ECG and EEG signals are depicted in Fig. 9 and 10 respectively.

The gain, CMRR, power dissipation and other parameters of this IA is shown in Table 2.

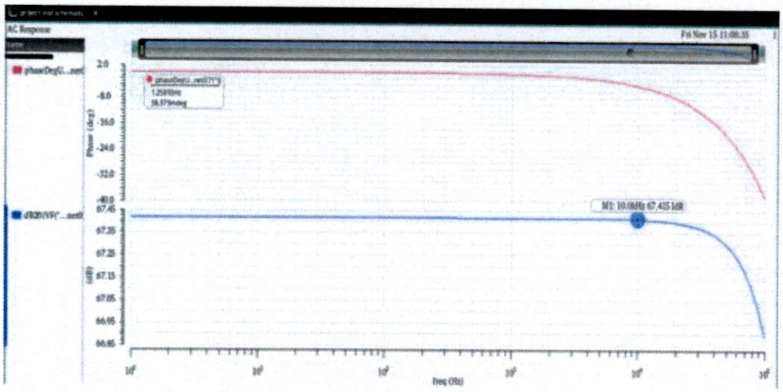

Fig. 7 Differential gain of IA

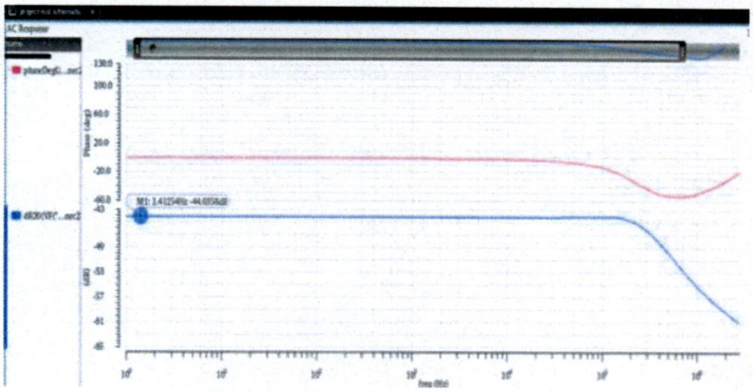

Fig. 8 Common mode gain of IA

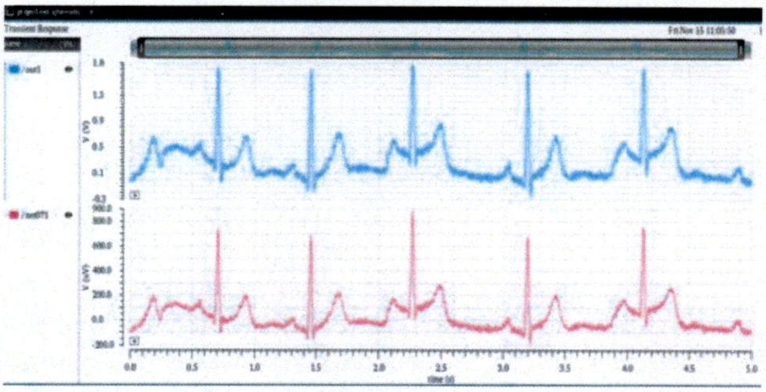

Fig. 9 Transient response of ECG signal

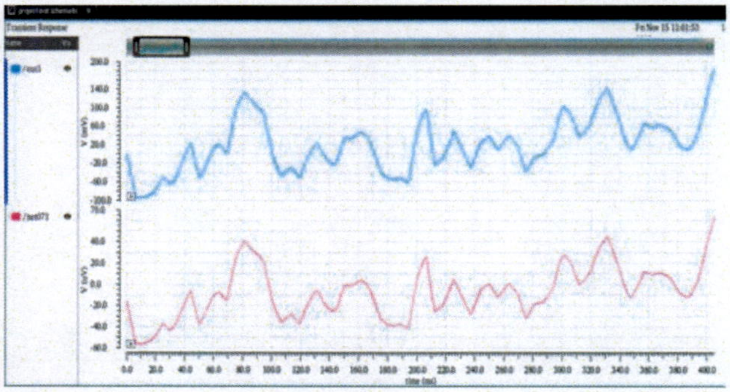

Fig. 10 Transient response of EEG signal

Table 2 Comparison with other IAs

S. No	Parameters	This Work	[3]	[5]	[8]	[11]
1	Technology (μm)	0.18	0.5	0.18	0.18	0.18
2	Power Supply (V)	1.8	–	1.8	–	1
3	Gain (dB)	67.41	45	67.7	79.16	59.8
4	CMRR (dB)	111.21	75	92	98	105
5	Power Dissipation (μW)	368	280	263	409.14	–

5 Conclusion

An Instrumentation Amplifier designed using CMOS is presented in this paper which has the property of high gain and is able to reject most of the common mode noise because it has high CMRR. A voltage follower or buffer circuit is also designed using CMOS. The circuits are implemented and simulated using Cadence Virtuoso tool in 180 nm SCL technology.

References

1. Chou C-J, Kuo B-J, Chen L-G, Hsiao P-Y, Lin T-H (2010) A 1-V low noise readout front end for biomedical applications in 0.18μm CMOS. In: Proceedings of the international symposium on VLSI-DAT, Hsin Chu, April 2010, pp 295–298
2. Yen C-J, Chung W-Y, Chi MC (2004) Micro-power low offset instrumentation amplifier IC design for bio-medical system applications. IEEE Trans Circuits Syst- I Regul Pap 51(4):691–699
3. Goswami M, Khanna S (2011) DC suppressed high gain active CMOS instrumentation amplifier for biomedical application. In: Proceedings of the IEEE international symposium on circuits and systems (ISCAS'04), vol 4, pp 9–12, September 2011
4. Sedra AS, Smith KC (2015) Microelectronics circuits, 7th edn. Oxford University Press, Oxford, pp 82–83

5. Goel A, Singh G (2013) Novel high gain low noise CMOS instrumentation amplifier for biomedical applications. In: IEEE international conference on machine intelligence and research advancement, pp 392–396
6. Dangi J, Gurjar RC (2014) An ECG instrumentation amplifier with improved CMRR and gain using 180nm technology. In: IEEE conference, pp 92–95, June 2014
7. Chow H-C, Wang J-Y (2007) High CMRR instrumentation amplifier for biomedical aplications. In: Proceedings of the 9th international symposium IEEE ISSPA, pp 1–4
8. Sharma BP, Mehra R (2016) Design of CMOS instrumentation amplifier with improved gain & CMRR for low power sensor applications. In: 2nd international conference on next generation computing technologies (NGCT), pp 72–76
9. Qian Y-Y, Wang Z-G, Liu Y-K, Zhou Z-J (2019) A high gain and high CMRR instrumentation amplifier for biomedical applications. In: IEEE 4th international conference on integrated circuits and microsystems (ICICM), pp 61–64
10. Nagulapalli R, Raparthy S, Venkatareddy A (2017) A low offset instrumentation amplifier for biomedical applications. In: International conference on circuits, controls, and communications (CCUBE), pp 191–193
11. Zhou Y, Zhao M, Dong Y, Wu X, Tang L (2018) A low-power low-noise biomedical instrumentation amplifier using novel ripple-reduction technique. In: IEEE biomedical circuits and systems conference (BioCAS), pp 1–2
12. Sanabria EH, Palacio JA, Herrera HH, Van Noije W (2017) A design methodology for an integrated CMOS instrumentation amplifier for bioespectroscopy applications. In: CHILEAN conference on electrical, electronics engineering, information and communication technologies (CHILECON), pp 1–6
13. Paraskevopoulou SE, Eftekhar A, Kulasekeram N, Toumazou C (2015) A low-noise instrumentation amplifier with DC suppression for recording ENG signals. In: 37th annual international conference of the IEEE engineering in medicine and biology society (EMBC), pp 2693–2695

Analysis of Resonance Matching Zig Zag Fat Arrow Shaped Slot Antenna for 60 GHz Short Distance Applications

Srinivasa Nookala and Kammara Bharath Kumar

Abstract Today High data rate components are required for wireless communication to full fill the needs of society in Indoor and outdoor applications. These applications require good connectivity and high bandwidth. In this article zig-zag unequal slotted antenna analyzed, how reflection coefficient and directivity change by inserting unequal slots in the structure. The proposed structure is given good results compared with existing structures. Fat shaped unequal slotted antenna gives max reflection coefficient S_{11} as 42.83 dB at 60 GHz, bandwidth of 16.37 GHz, directivity as 5.8 dBi with the efficiency of 55%. Dimensions of the Zig-Zag fat arrow-shaped unequal slotted antenna are $6 \times 4 \times 0.508$ mm^3. Due to its dimensions and bandwidth this Zig-Zag fat arrow-shaped unequal slotted antenna is more fluxiable and suitable for Femtocell, Microcell and macrocell applications.

Keywords Femto Access Point (FAP) · Mobile management entity · Service gate way · Microcell · Fixed wireless access system

1 Introduction

Now a day high data rate components are required for wireless networks system to full fill the needs of today's society in indoor and outdoor short-distance applications. For these applications high security, good connectivity, compact size, fewer losses, less fabrication cost, good directivity and high bandwidth microwave components are playing a vital role in communications systems. These requirements point view short distance communication systems are comes on the picture. These short-distance communication systems full fill all the needs of today's world. In short distance communication systems femto access point (FAP) technology is placed in

S. Nookala (✉)
Electronics and Communication Engineering, Matrusri Engineering College, Hyderabad, India
e-mail: srinivasa.nookala@gmail.com

K. Bharath Kumar
Electronics and Communication Engineering, CMR Technical Campus, Hyderabad, India

V. Nath and J. K. Mandal (eds.), *Proceeding of Fifth International Conference on Microelectronics, Computing and Communication Systems*, Lecture Notes in Electrical Engineering 748,
https://doi.org/10.1007/978-981-16-0275-7_55

the Heterogeneous Radio Access Network (Het-RAN), which is placed in the base station of the communication network. This is shown in Fig. 1. This Het-RAN consists of different elements like mobility management entity (MME), Service Gateway (S-GW), microcell and femtocells. This mobility management system is used for mobile management of user equipment. Gateway is used for sending data to the user equipment. S1 is used as an interfering component in- between macro cell and MME/S-GW. HeNB Gateway is used for connecting femtocell and MME/S-GW. X2 is an interface device [1] in between small cells and femtocells. This X2 is also used for handover the user equipment to the different cells.

Due to FAP, the overall efficiency of communication systems is improved like compact size, performance and less cost [2]. This FAP mainly used in fixed wireless systems (FWS). Different types of FWS are shown in Fig. 2.

The main future of this femtocell covers the gap between different local area networks. In FAP high bandwidth, compact size [3] and high directivity [4, 5] antennas are placed. To design these types of antennas rectangular waveguide i.e.

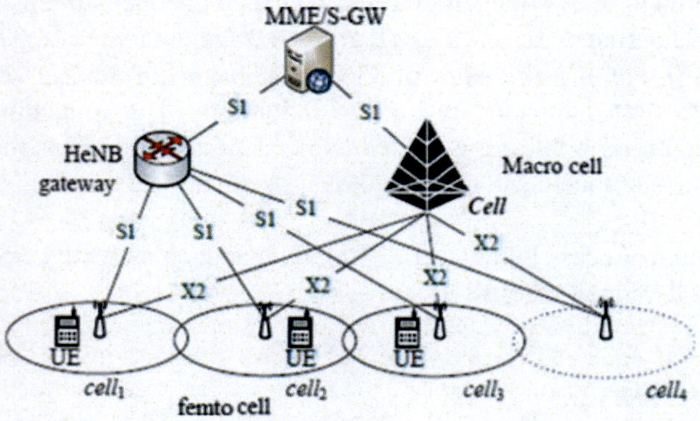

Fig. 1 The architecture of Het-RAN in the wireless network [1]

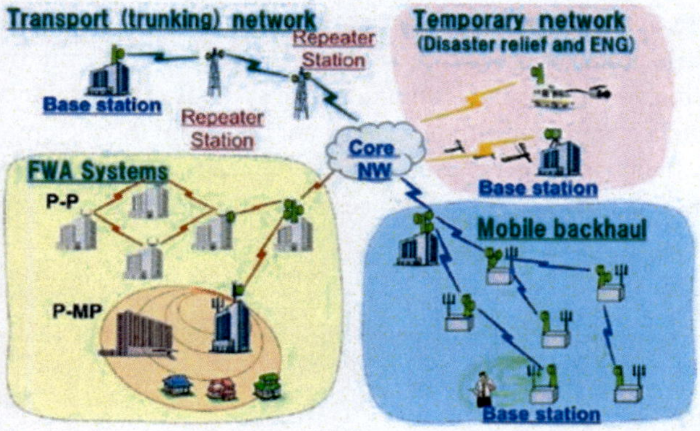

Fig. 2 Different types of Fixed Wireless System [2]

Non-planer structure is more suitable for developing systems, because these structures provide high power handling capability, fewer losses and high quality factor. Drawbacks of this structure are bulky in size. So these non-planer structures i.e. waveguides are replaced by Rectangular Substrate Integrated Waveguide (RSIW). RSIW is compact in size, the fabrication cost is less and power handling capability is moderate. Federal Communication Commission (FCC) announced 64 GHz to 57 GHz i.e. 7 GHz as unlicensed for short-distance applications [6]. So the antenna is designed using RSIW technology at 60 GHz [7, 8]. For evolution purposes Grounded Coplanar Waveguide (GCPW) transition is used because this transition provides good impedance between the components [8]. Previously different types of antenna are developed based on their applications and requirements [9–12]. At 10 GHz the first slotted four by four antenna array was designed [13]. Recently suppress the back lobe in SIW slotted antenna structure many topologies are proposed. But shortened quarter wavelength microstrip slotted SIW antenna structure gives the best solution for suppressing the back side radiation of the antenna [14]. Folded corrugated stub slotted SIW antenna also suppresses the back lobe [15]. To increase the bandwidth of the RSIW slotted antenna structure discussed in [16]. Unequal slots are introduced in the RSIW structure to increate the bandwidth of the structure. Introducing the unequal slotted in RSIW antenna structure and planar RSIW slotted array antenna structure [17, 18].

Next part of this article discussed design of zig zag fat- arrow slotted antenna. Theird part of this article explains the results and compression of the existing structure. Finally section gives the conclusion of zig-zag fat-shaped RSIW antenna.

2 Analysis, Results and Discussions

A good Antenna improve the overall performance of the wireless network. To developh an antenna in V-band RSIW technology is used. RT duriod 5880 substrate material is used to design of RSIW slotted antenna [8]. The widths between the two

Fig. 3 Proposed zigzag fat arrow-shaped slotted RSIW antenna structure

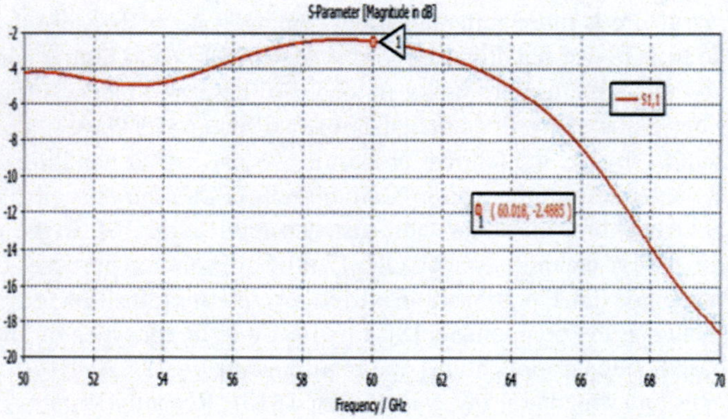

Fig. 4 Proposed zigzag fat arrow-shaped slotted RSIW antenna simulation result

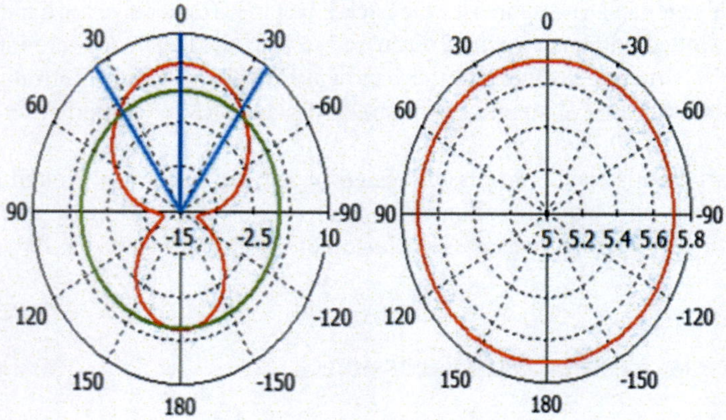

Fig. 5 2D Radiation patterns of proposed zigzag fat arrow-shaped slotted RSIW antenna a) E-plane b) H-plane

arrays of row vias are 3.45 mm. GCPW dimensions are the center conductor width is 1.008 mm and spacing between the center conductor and grounded layer on the top side is 0.1 mm [19]. The developh the basic RSIW fat shaped antenna structure is shown in Fig. 3. 6 mm as length and is 4 mm as width of the top side layer of SIW antenna.

The proposed structure is simulated in the CST EM simulator. The CST simulation graph is shown in Fig. 4.

In the above simulation graph frequency resonance point is not occur because perfect impedance is not matching. The radiation pattern is shown in shown Fig. 5.

To improve the resonance frequency point in the simulation result created unequal slots in the proposed antenna structure and observe the performance in simulation result as well radiation pattern. Step by step insertion of unequal slots are shown in Fig. 6. To get a good resonance frequency point in the reflection coefficient graph

Fig. 6 Step by step implementation of the proposed zigzag fat arrow- shaped unequal slotted RSIW antenna structure

Table 1 S11 of each step analysis of proposed zigzag fat arrow-shaped unequal slotted RSIW antenna structure

	S11 (dB)	Fr (GHz)	Directivity (dBi)	Bandwidth (GHz)	Quality factor
Fat arrow	–	–	5.83	–	1117
Slot1	−37.94	58.92	5.2	16.2	1236
Slot2	−29.25	60.98	5.3	16.79	718
Slot3	−42.83	59.9	5.8	16.37	742

unequal slots are inserted below the zigzag fat arrow-shaped. In this structure each slot reduced the width by 0.4 mm. The first slot length is maintained by 0.3 mm and the next slots are created by reducing slot length as 0.05 mm. the entire is analyzed.

The each step changes of reflection coefficient (S_{11}) of the proposed zigzag fat arrow-shaped slotted RSIW antenna structure results are shown in below Fig. 7.

In the above simulation result graph red color indicates a reflection coefficient (S_{11}) of the zigzag fat arrow-shaped antenna. In the graph no resonance peak present. But this structure provides a high quality factor which is given in Table 1. To improve the resonance frequency point in the result of first slot is created. Due to the first slot reflection coefficient results are changed. The changed graph is with the orange color shown in Fig. 7. This graph changed by creating an unequal slot in the structure because the entire structure impedance changed. Due to the first slot resonance peak occurs at 58.92 GHz with reflections coefficient of–dB. This structure gives 16.2 GHz of bandwidth. The next second slot is created to shift the resonance frequency point towards 60 GHz. Due to second slot resonance frequency point shifted to 60.99 GHz with reflections coefficient of –29.67 dB. This structure provides 16.8 GHz of bandwidth, shown in Fig. 7 with green color. But we need resonance at 60 GHz. The third slot is created in the structure to shift the resonance peak. Due to third slot resonance peak occurs at 29.9 GHz with a reflections coefficient of –42.7 dB. This structure provides 16.37 GHz of bandwidth. Each step analysis of proposed antenna results are given in Table 1.

Proposed fat arrow-shaped slotted RSIW antenna structure polar plot graphs are shown in Fig. 8. This proposed structure gives 3 dB as gain and 5.8 dBi as directivity.

In Fig. 9 gives the total proposed and proto type of antenna structure. In this figure c figure indicates measurement of prototype structure. Figure 10 indicates the simulation and mesurement results of proposed antenna structures.

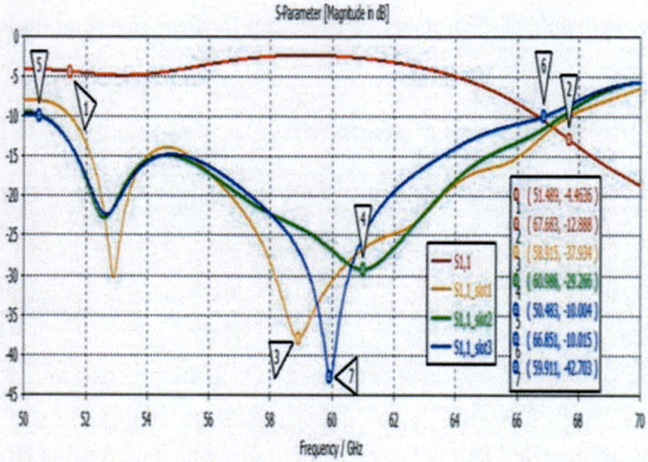

Fig. 7 Each step analysis of reflection coefficient of the proposed zigzag fat arrow-shaped unequal slotted RSIW antenna structure

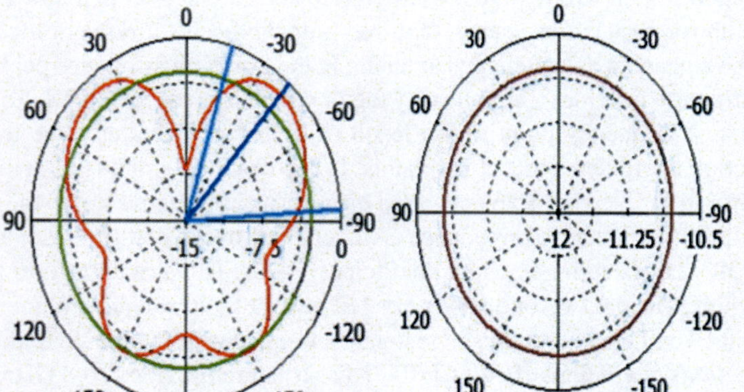

Fig. 8 2D polar plot graphs of proposed zigzag fat arrow-shaped unequal slotted RSIW antenna a) E-plane b) H-plane

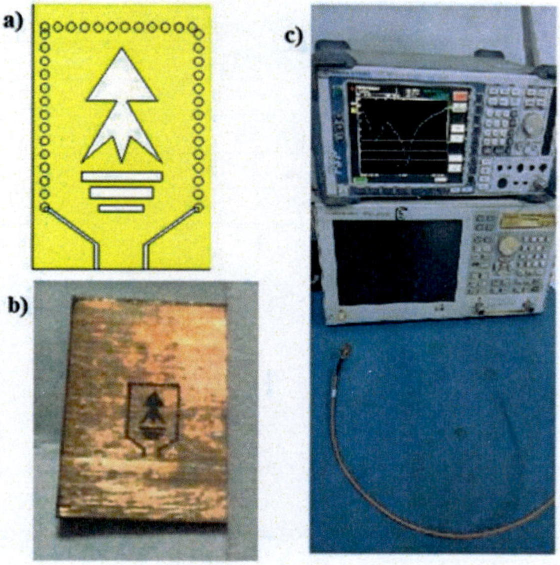

Fig. 9 a) Total proposed structure b) fabricated structure c) Measurement setup of proposed structure

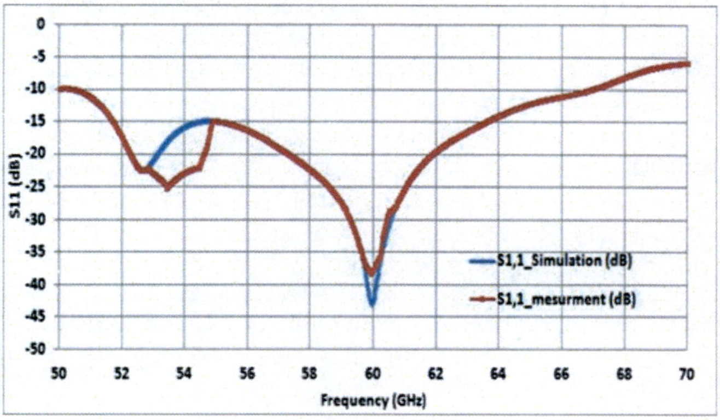

Fig. 10 Reflection coefficient of simulation and mesurment result of the proposed zigzag fat arrow-shaped antenna structure

In Fig. 10 red colour indicates measurement results and blue colour indicate simulation results of reflection co-efficient of proposed zigzag fat arrow-shaped antenna structure. Here small diviation occurs in simulation and fabrications results. Maximum reflection co-efficient occurs at 59.9 GHz in simulation result but in mesurment peak result occurs at 60 GHz. This dividion occurs due to creating of vias in fabrication process.

Table 2 Comparisons of existing and proposed zigzag-shaped antenna

Parameter	Ref. [20]	Ref. [21]	Ref. [22]	Proposed	
				Simul.	Measur
Freq.(GHz)	59.9	58	60	59.9	60
B.W (GHz)	4.028	9.5	12	16.37	16.4
S_{11} (dB)	41	37.5	43	42.83	37.5
Size (mm^3)	8 × 8 × 1.6	7.5 × 9.5 × 0.508	6 × 4 × 0.508	6 × 4 × 0.508	6 × 4 × 0.508
Gain (dBi)	5.48	8.2	5.31	5.8	-

3 Comparison

Table 2 gives an overall comparison between existing antennas with zigzag fat arrow-shaped slotted RSIW antenna structure.

In the above comparisons table proposed zigzag fat arrow-shaped slotted RSIW antenna structure gives good results when Compared with existing structures [20–22]. $S_{11,}$ bandwidth and gain are improved with lower dimensions. Generally for femtocell applications point of view greater than 5 dBi directivity is requires for a single element of the antenna. The proposed zigzag fat arrow- shaped structure fulfills the required bandwidth and directivity of femtocell, micro and macro cell applications.

4 Conclusion

The proposed zigzag fat arrow-shaped slotted RSIW antenna is develophead for FAP in indoor applications. Zigzag fat arrow-shaped slotted RSIW antenna structure operated from 64 to 57 GHz bands. Zigzag fat arrow-shaped slotted RSIW antenna gives 5.8 dB as gain and 7 dBi as directivity. Due to its results like dimensions, bandwidth and directivity of the Zigzag fat arrow-shaped slotted RSIW antenna this structure of antenna is most suitable to developh the FAP in outdoors applications. Zigzag fat arrow-shaped slotted RSIW antenna structure covers the entire V-band particularly for all RADAR applications.

References

1. Yeh EH, Lin P, Lin YB, Lee CP (2016) A connection-driven mechanism for energy saving of small-cell networks. In: IEEE conference. ISBN 978-1-5090-2279-3
2. Series of ITU-R Reports (2015) Fixed service use and future trends. Electronic Publication, Geneva

3. Astuti DW, Rahardjo ET (July 2019): Size reduction of substrate integrated waveguide cavity backed u-slot antenna. In: IEEE Indian conference on antennas and propogation (InCAP), Hyderabad, India. ISBN 978-1-5386-7061-3
4. Chen Z, Liu H, Yu J, Chen X (2018) High gain, broadband and dual-polarized substrate integrated waveguide cavity-backed slot antenna array for 60 GHz band. IEEE Access 6:31012–31022. ISSN 2169-3536
5. Pourahmadazar J, Denidni TA (Nov 2015) High gain substrate integrated waveguide resonant slot antenna array for mm-wave band radio. In: 2015 IEEE international conference on ubiquitous wireless broadband (ICUWB), Montreal, QC, Canada. ISBN 978-1-4673-6554-3
6. Zhu Q, Ng KB, Chan CH, Luk KW (2017) Substrate-integrated-waveguide-fed array antenna covering 57–71 GHz band for 5G applications. In: IEEE Trans Antennas Propag 65:6298–6306. ISSN 1558-2221
7. Hosseini-Fahraji A, Mohammadpour-Aghdam K, Faraji-Dana R (Mar 2017) Millimeter wave rectangular waveguide to grounded CPW transition on multi-layer substrate. In: 2016 fourth international conference on millimeter-wave and terahertz technologies (MMWaTT), Tehran, Iran. ISBN 978-1-5090-5415-2
8. Kumar KB, Shanmuganantham T (2018) Four dielectric substrate analyses for millimeter-wave applications. J Mater Today 5(4):10771–10778. Elsevier, Part 3
9. Kumar KB, Shanmuganantham T (Mar 2018) Blue-tooth shaped SIW antenna for RADAR applications. In: 2017 IEEE international conference on circuits and systems (ICCS), Thiruvananthapuram, India. ISBN 978-1-5090-6481-6
10. Wali MR, Tahir FA, Khan MU, Hussain R (Jan 2019) A broadband SIW antenna for 5G applications. In: 2018 IEEE international symposium on antennas and propagation & USNC/URSI national radio science meeting, Boston, MA, USA. ISBN 978-1-5386-7103-0
11. Deckmyn T, Reniers AC, Smolders AB, Rogier H, Ginste DV, Agneessens S (Dec 2018) Band substrate integrated waveguide antenna for next-generation mmWave wireless systems. In: 12th European conference on antennas and propagation (EuCAP 2018) IET, London, UK. ISBN 978-1-78561-815-4
12. Kim YM, Ha KH, Lee JS, Kang HJ, Jung IH, Cho CS (Jan 2020) 60 GHz substrate integrated waveguide balun. In: IEEE conference on antenna measurements & applications (CAMA), Kuta, Bali, Indonesia, Indonesia. ISBN 978-1-7281-2397-4
13. Yan L, Hong W, Hua G, Chen J, Wu K, Cui TJ (2004) Simulation and experiment on SIW slot array antennas. IEEE Microwave Wireless Component Letter 14(9):446–448
14. Wei J, Chen ZN, Qing X, Shi J, Xu J (2013) Compact substrate integrated waveguide slot antenna array with low back lobe. IEEE Antenna Wirel Propag Lett 12:999–1002
15. Cho D, Lee HY (2013) Folded corrugated SIW (FCSIW) slot antenna for back lobe suppression. IEEE Antenna Wirel Propag Lett 12:1276–1279
16. Mbaye M, Hautcoeur J, Talbi L, Hettak K (2013) Bandwidth broadening of the dual-slot antenna using substrate integrated waveguide. IEEE Antenna Wirel Propag Lett 12:1169–1172
17. Hosseininejad SE, Komjani N (2013) Optimum design of traveling wave SIW slot array antennas. IEEE Trans Antenna Propag 12:1971–1975
18. Hosseininejad SE, Komjani N, Mohammadi A (2015) Optimum design of traveling wave SIW slot array antennas. IEEE Antenna Wirel Propag Lett 14:261–264
19. Kumar KB, Shanmuganantham T (2018) Design of hut shaped SIW antenna on analog integrated circuits for 60 GHz applications. Int J Comput Appl 16(1):165–169. ISSN 0973-5704
20. Saini J, Agarwal SK (2017) Design a single band microstrip patch antenna at 60 GHz millimetre wave for 5G application. In: IEEE international conference on computer, communications and electronics (comptelix). ISBN 978-1-5090-4708-6
21. Liu H, He Y, Wong H (2013) Printed U-slot patch antenna for 60 GHz applications. In: IEEE international workshop on electromagnetic, applications and student innovation completion, pp 153–155
22. Kumar KB, Shanmuganantham T (Dec 2018) Analysis of impedance matching star-shaped slot antenna for 60 GHz femtocell applications. In: IEEE Indian conference on antenna and propagation. ISBN 978-1-5386-7060-6

Design of Radial Reentrant Cavity for V-Band Vacuum Microwave Devices

M. Santosh Kumar, Chaitali Koley, Santigopal Maity, A. K. Bandyopadhyay, Vatsav Kolluru, and Debashish Pal

Abstract The V-band frequency range becomes popular in the context of mm-wave wireless communication. RF amplifiers with medium output power and bandwidth may be used to provide high data rate wireless communication over a wide area. In this paper, cavities operating in the V-band frequency range have been investigated. These cavities may be used in the RF section of a medium powered V-band klystron. Three types of radial reentrant cavities on V-band have been considered. Three-dimensional electromagnetic simulations of these cavities were carried out using commercially existing simulation software (CST MW Studio) and different cavity parameters have been compared.

Keywords Radial Reentrant Rectangular Cavity (RRRC) · Radial Reentrant Cylindrical Cavity (RRCC) · Radial Reentrant Square Cavity (RRSC) · V-Band · R/Q · CST MW studio

1 Introduction

The Klystron amplifier is considered as the most favorable device used in several high gain and power applications like mm-wave communication, radar, etc. There are several unique features of the Klystrons, which include high output power, high efficiency, high gain, and excellent phase and amplitude stability [1–3]. However, there is a significant decrease in the output power of the Klystron amplifier in the

M. Santosh Kumar (✉) · C. Koley · S. Maity · V. Kolluru
Department of ECE, NIT Mizoram, Aizawl, India
e-mail: santu.1503@gmail.com

C. Koley
e-mail: chaitali.ece@nitmz.ac.in

A. K. Bandyopadhyay · D. Pal
Microwave Devices Area, CSIRCEERI, Pilani, Rajasthan, India
e-mail: akb@ceeri.res.in

V. Nath and J. K. Mandal (eds.), *Proceeding of Fifth International Conference on Microelectronics, Computing and Communication Systems*, Lecture Notes in Electrical Engineering 748,
https://doi.org/10.1007/978-981-16-0275-7_56

mm-wave regime, due to its physical restrictions such as the voltage breakdown and the amount of available beam current [4].

Klystron amplifiers offer high gain and efficiency but they have narrow operational bandwidth. To overcome this narrow operational bandwidth, reentrant RF cavities [5–8] are introduced in place of RF cavities. In general, reentrant RF cavities are used as RF interaction structure of klystrons. The reentrant cavity structures considered in this article are shown in Fig. 1. These cavities are a re-entrant metallic structure where a gap is formed at the middle through which the electron beam passes. This geometry structure of the cavity is optimized to support the electromagnetic field at the operating frequency [9, 10]. The performance of an RF cavity is characterized by several parameters [5] like the resonant frequency of the operating mode, quality factor (Q), and shunt impedance to quality factor ratio (R/Q) [11, 12].

Radial reentrant cavities operating in the Ka-Band frequency range have been reported in [13, 14]. In the present work, different configuration of radial reentrant cavities (Fig. 1) operating in the V-band frequency range [15–17] is considered. V-Band frequency range may be used for high data rate mm-wave transmissions using the unlicensed spectrum. The large available bandwidth allows wide channels through which high data rate communication is possible. However, to compensate high losses at this frequency band, the transmitted RF power should be high – which may be achieved by the usage of vacuum electronic devices. In this work, the RF cavities of a klystron working in V-band have been simulated using three-dimensional electromagnetic simulation software [18]. Cavity parameters corresponding to the operating frequency, unloaded quality factor, and characteristic impedance of the cavity (R/Q) values have been estimated from the simulations.

This paper is arranged as follows: The designs of the V-band RRC structures are presented in Sect. 2. The simulated and observed results are given in Sect. 3. And the conclusion is in Sect. 4.

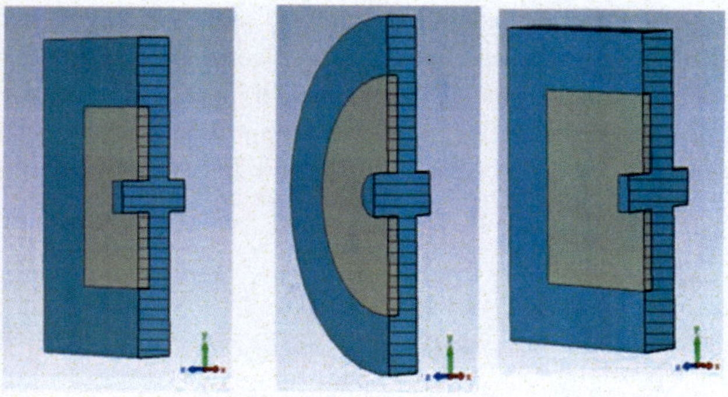

(a) (b) (c)

Fig. 1 Designed R F cavities are: a) Radial Reentrant Square Cavity, b) Radial Reentrant Cylindrical Cavity, and c) Radial Reentrant Rectangular Cavity

2 Design

The main feature of a reentrant cavity is its ability to concentrate a strong electric field in the interaction gap space. Although the RF cavities are narrowband, the operational bandwidth of a klystron can be widened employing stagger tuned RF cavities. As the interaction gap widens the operating frequency and shunt impedance of the cavity increases. The shunt impedance represents the interaction of the electric field with the electron beam at the cavity gap. For a cavity with low R/Q value, the interaction of the electric field with the electron beam degrades. In such cases, it might not be able to establish the required level of beam modulation to achieve the desired gain and output power specifications.

Here we have considered three different resonant cavities namely Radial Reentrant Square Cavity (RRSC), Radial Reentrant Cylindrical Cavity (RRCC), and Radial Reentrant Rectangular Cavity (RRRC). Figure 2 show a schematic diagram of the designed radial reentrant cavity. In the sectional view, we have represented w, g, c, b, and a as the width of the cavity, interaction gap of the cavity, beam tunnel diameter, reentrant boundary height, and height of the cavity respectively.

The geometrical cavity parameters used in the simulation of the RRRC, RRCC, and RRSC are presented in Table 1.

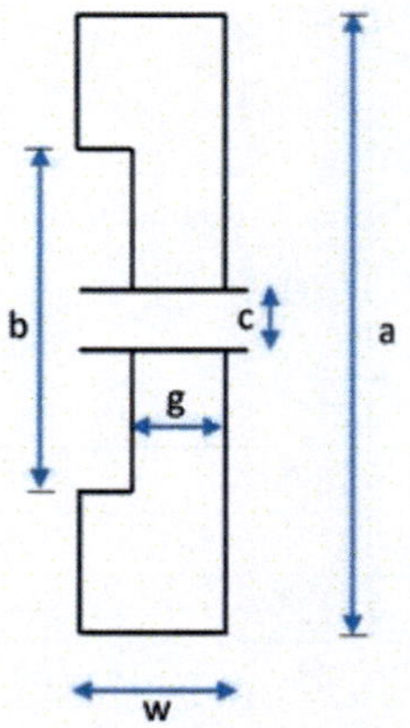

Fig. 2 Schematic figure of the Radial Re-entrant Cavity

Table 1 Designed cavities measurements and simulated cavity parameters. (Cavity Measurements in mm)

	a_x (mm)	a_y (mm)	b_x (mm)	b_y (mm)	c (mm)	g (mm)	w (mm)	f (GHz)	R/Q (Ω)	Q
RRRC	3	2.6	2.7	1.6	0.3	0.325	0.5	60.6	60	1137
RRCC	1.65	1.65	1.1	1.1	0.2	0.325	0.5	60.4	69	1159
RRSC	3	3	1.7	1.7	0.3	0.325	0.5	60.9	72	1219

(RRCC is cylindrical cavity, where a_x, a_y, b_x, b_y and c are treated as radius)

In Table 1 the RRC dimensions are height in the direction of X (a_x) and direction of Y (a_y), width (w), reentrant boundary height in the direction of X (b_x) and direction of Y (b_y), beam tunnel gap (c) and interaction gap (g) and all cavity dimensions are in mm.

3 Simulation Results

The radial reentrant cavities (RRRC, RRSC, and RRCC) operate in TM_{010} mode and geometrical parameters of the simulation models are optimized using the Eigen-mode solver of the electromagnetic simulation software, CST EM Studio. The comparative study between RRRC, RRCC, and RRSC with varying interaction gap of each cavity is given in Table 2. The operating frequency, R/Q value, and quality factor of the cavity are shown in Table 2.

The product of the maximum achievable gain of a klystron and its operational bandwidth is proportional to the R/Q value of the RF cavity. Therefore, RF cavities with higher R/Q values are more suitable for being used in the RF section of the klystron.

The variations of the cavity parameters like operating frequency, R/Q value, and quality factor as functions of the interaction gap are presented in the following sections.

Table 2 Simulated cavity parameters for different cavities

Interaction gap (g) mm	RRCC	RRRC	RRSC	RRCC	RRRC	RRSC	RRCC	RRRC	RRSC
	Frequency (f) GHz			(R/Q) value			Quality Factor (Q)		
0.275	56.7	57	57	58	50	61	1060	1041	1131
0.2875	57.6	57.9	58	61	54	65	1085	1066	1154
0.3	58.6	58.9	59	63	55	67	1109	1090	1177
0.3125	59.5	59.8	60	66	60	68	1134	1113	1199
0.325	**60.4**	**60.6**	**60.9**	**69**	**61**	**72**	**1159**	**1137**	**1219**
0.3375	61.2	61.5	61.8	72	64	76	1184	1160	1242
0.35	62.1	62.3	62.6	75	65	77	1210	1185	1263
0.3625	62.9	63	63.4	78	71	85	1236	1209	1284
0.375	63.7	63.8	64.3	81	72	89	1264	1234	1306

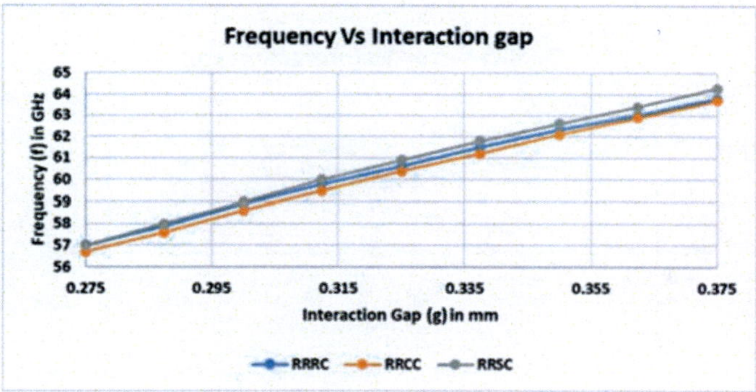

Fig. 3 Interaction gap (g) vs. Frequency (f) plot

3.1 *Interaction Gap Vs Frequency*

In Fig. 3 shows the plot between the variation of frequency with the change in the interaction gap of the cavity. We observe that for RRC's, the operating frequency increases with an increase in the interaction gap (g).

3.2 *Interaction Gap Vs (R/Q) Value*

Figure 4 shows the variation of (R/Q) value with change in the interaction gap of the cavity. We observe that for RRC's, the characteristic impedance increases with an increase in the interaction gap (g). Based on the plot, RRSC provides better characteristic impedance.

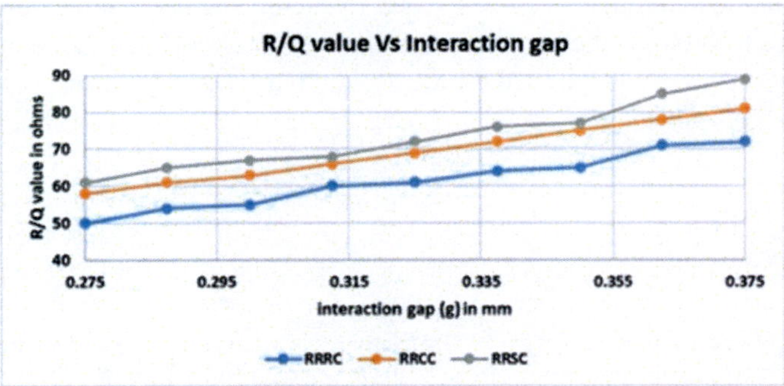

Fig. 4 Interaction gap (g) vs. Characteristic Impedance (R/Q) plot

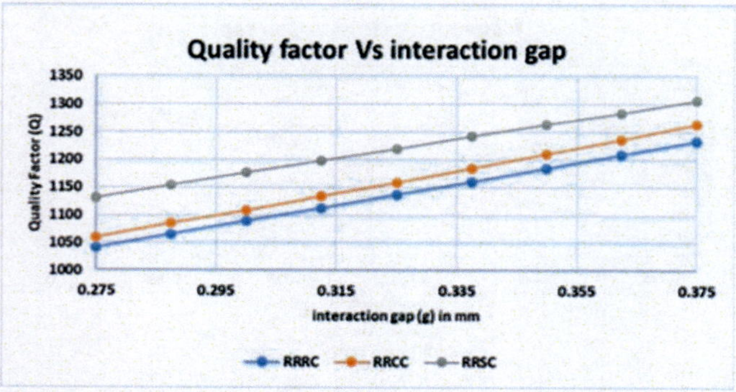

Fig. 5 Interaction gap (g) vs. Quality Factor (Q) plot

3.3 *Interaction Gap Vs Quality Factor*

In Fig. 5, the variation of the quality factor with the change in the interaction gap of the cavity is shown. We observe that for RRC's, the quality factor increases with an increase in the interaction gap (g). Based on the plot, RRSC provides a better quality factor.

The comparative plots Fig. 3, 4, and 5 indicate that among three types of radial reentrant cavities, square RRC exhibits better performance.

4 Conclusion

Radial reentrant cavities operating in the V-band frequency range were designed and simulated by using simulated software (CST MW Studio). Cavity parameters namely the operating mode frequency, unloaded quality factor, and R/Q values were obtained and compared. Based on the simulation result, it is concluded that the square-shaped radial reentrant cavity can be used for the design of a V-band klystron amplifier with moderate output power.

References

1. Lau YY, Friedman M, Krall J et al (1990) Relativistic klystron amplifiers driven by modulated intense relativistic electron beams. IEEE Trans Plasma Sci 18(3):553–569
2. Wu Y, Xie HQ, Li ZH et al (2013) Gigawatt peak power generation in a relativistic klystron amplifier driven by 1 kW seed-power. Phys Plasmas 20(11):
3. Zhao YC, Li SF, Huang Z et al (2015) Analysis and simulation of a multigap sheet beam extended interaction relativistic klystron amplifier. IEEE Trans Plasma Sci 43(6):1862–1870
4. Friedman M, Pasour J, Smithe D (1997) Modulating electron beams for an X band relativistic klystron amplifier. Appl Phys Lett 71(25):3724–3726

5. Bansiwal A, Raina S, Datta S, Vinoy KJ, Datta SK (2019) A simple method for estimating the quality factor of cylindrical re-entrant cavity of Klystrons. J Electromagn Waves Appl 33(8):1082–1091
6. Bansiwal A, Raina S, Datta S, Vinoy K (2018) Calculation of equivalent circuit parameters of a rectangular reentrant cavity for klystron. Int. J. Microw. Opt. Technol. 13(6):487–492
7. Bansiwal A, Raina S, Vinoy KJ, Datta SK (2018) A ridge loaded rectangular re-entrant cavity structure for multiple beam klystron tube. Indian Patent Appl. 2018:11034857
8. Bansiwal A, Raina S, Datta S, Vinoy K (2019) A broadband rectangular reentrant cavity for multiple-beam klystron. IEEE Trans Electron Devices 66(7):3168–3170
9. Samuel YL (2003) Microwave Devices and Circuits. India, Pearson Education, Inc. and Dorling Kindersley Pvt. Ltd., pp. 353–363
10. Gilmour AS Jr (1994) Klystrons, Traveling Wave Tubes, Magnetrons, Crossed-Field Amplifiers, and Gyrotrons. Artech House, Norwood, pp 281–310
11. Chang Z, Meng L, Yin Y, Wang B, Li H, Rauf A, Ullah S, Bi L, Peng R (2018) Circuit design of a compact 5-kV W-band extended interaction klystron. IEEE Trans Electron Devices 65(3):1179–1184
12. Bansiwal A, Raina S, Datta S, Vinoy K (2019) Equivalent circuit analysis of a rectangular double-reentrant cavity with circular cylindrical ferrule for klystrons. IEEE Trans Electron Devices 66(11):4952–4956
13. Santosh Kumar M, Chaitali, K, Santigopal M, Bandyopadhyay AK, Pal D (2019) Design and simulation of Radial Reentrant Cavities For Ka band Vacuum Microwave Devices. In: IEEE Conference TENCON-2019, pp. 1908–1909
14. Santosh Kumar M, Chaitali K, Santigopal M, Bandyopadhyay AK, Pal D (2019) Design and simulation of high gain RRC For Ka-band vacuum microwave device. In: Vacuum Electronic Devices and Applications (VEDA) Conference
15. Tianxiang Z, Hu Y (2015) Design of a Novel High Power V-Band Helix-Folded Waveguide Cascaded Traveling Wave Tube Amplifier", Hindawi Publishing Corporation, Active and Passive Electronic Components, vol. 2015, Article ID 846425, 9pps
16. Acosta R, Nessel J, Simons R, Zemba, M, Morse J, Budinger J (2012) W/V-Band RF Propagation Experiment Design, NASA Glenn Research Center https://ntrs.nasa.gov/search.jsp?R=20120016067
17. Hannachi C, Hammou D, Djerafi T, Ouardirhi Z, Tatu SO (2013) Complete Characterization of Novel MHMICs for V-Band Communication Systems", Hindawi Publishing Corporation. Journal of Electrical and Computer Engineering, vol. 2013, Article ID 686708, 9pps
18. CST (2018) User's Manual, CSST GmbH. www.cst.com

Implementation of Discrete Sine Transform Realization Though Systolic Architecture

Anamika Jain and Neeta Pandey

Abstract In this paper, an efficient design approach is used for implementation of discrete sine transform (DST). A new algorithm for DST for $N = 4^n$ has been suggested and based on an appropriate formulation of transform a systolic architecture is presented. Design uses lesser hardware and outperform in terms of delay thus the architecture well suited for VLSI implementation.

Keywords Discrete sine transform · Discrete cosine transform · Systolic array · Recursive technique

1 Introduction

The discrete sine transform (DST) is well known discrete trigonometric transform. DST has application in interference reduction of multiple code CDMA, signal processing such as data compression [1, 2] and filtering the speech signal, transform–domain filtering, [3], under water target recognition and used for generation of waveform in frequency Division Multiplexing (OFDM) System [4]. Ramadan et al. [5] found DST shows superior performance as compare to Discrete Fourier transform (DFT) for MIMO orthogonal OFDM. Also, DST generates better results for noise estimation compared with the Discrete Cosine Transform [6] and the DFT. In high efficient video coding, for achieving efficiency, DCT-II and DST VII are being used [7]. Moreover, some fast computation algorithms have been presented to compute the DST/DCT including sparse-matrix factorization, fast recursive algorithm as well as recursive algorithm etc. [8–10]. Perera [11] has proposed signal flow graph of fast, efficient, completely recursive and forward stable DST I-IV algorithms that gives sparse, scaled orthogonal and rotation-reflection matrices and provide speed

A. Jain (✉)
Department of Electronics and Communication Engineering, Maharaja Agrasen Institute of Technology (MAIT), Rohini Sector-22, Delhi 110086, India

N. Pandey
Department of Electronics and Communication Engineering, Delhi Technological University (DTU), Rohini Sector-17, Delhi, India

V. Nath and J. K. Mandal (eds.), *Proceeding of Fifth International Conference on Microelectronics, Computing and Communication Systems*, Lecture Notes in Electrical Engineering 748,
https://doi.org/10.1007/978-981-16-0275-7_57

improvement. Fast DST Algorithms based on butterfly structures can reduce the number of multiplications, but result to irregularity in the structure with complicated data routing and require large design time. Also, the successive truncation involved in the recursive decomposition decline the accuracy for a fixed point implementation. Computation of DST is an intensive process so it is necessary to develop effective, high speed dedicated hardware to use them in real time application through VLSI technology. For efficient implementation of any algorithm in VLSI, data flow plays an important role. Thus, there is a requirement of regular and modular structure for realizing the algorithm in VLSI. Many algorithms and their architectures for implementation of DST/DCT/DFT has been suggested in the literature [12–15]. Meher [16] proposed a common computational algorithm for computing DCT as well as DST and then design was implemented through DA based parallel computing structure. A new VLSI algorithm implementation approach through parallel short pseudo-cycle convolution has been suggested by Chiper et al. [17] and is realised through two linear systolic arrays that have a low hardware complexity. In this paper, a new algorithm is developed for efficient computation of DST. The suggested algorithm is realized though systolic architecture and It is found that the suggested architecture provides a superior performance in terms of the hardware complexity, speed, I/O Costs, in addition to features such as regular, modular, pipeline capability and local connectivity which make architecture well suited for are suitable for VLSI implementation. The work is organized as follows: Sect. 2 deals with new algorithm for DST, corresponding architecture is described in Sect. 3, comparison of the proposed work with the existing work in the literature is done in Sect. 4. Finally, the proposed work is summarized in Sect. 5.

2 A New Algorithm for DST

The Discrete Sine Transform (DST)

$$X[K] = \sum_{n=0}^{N-1} 2x[n]\mathrm{Sin}\left(\frac{\pi k(2n+1)}{2N}\right) \quad k = 1, 2, ..., N \tag{1}$$

Dividing Eq. (1) into four groups.

After mathematical manipulation Eq. (2) reduces to

$$\begin{aligned} X[K] &= \textstyle\sum_{n=0}^{\frac{N}{4}-1} x[n]Sin\left(\frac{\pi k(2n+1)}{2N}\right) \\ &+ \textstyle\sum_{n=\frac{N}{4}}^{\frac{N}{2}-1} x[n]Sin\left(\frac{\pi k(2n+1)}{2N}\right) \\ &+ \textstyle\sum_{n=\frac{N}{2}}^{\frac{3N}{4}-1} x[n]Sin\left(\frac{\pi k(2n+1)}{2N}\right) \\ &+ \textstyle\sum_{n=\frac{3N}{4}}^{N-1} x[n]Sin\left(\frac{\pi k(2n+1)}{2N}\right), k = 1, 2, ..., N \end{aligned} \tag{2}$$

Solving Eq. (2), we get

$$X[K] = \sum_{n=0}^{\frac{N}{4}-1} \left\{ \begin{array}{l} x[n]Sin(\theta_k) \\ +x\left[n + \frac{N}{4}\right]Sin\left(\theta_k + \frac{\pi K}{4}\right) \\ +x\left[n + \frac{N}{2}\right]Sin\left(\theta_k + \frac{\pi K}{2}\right) \\ +x\left[n + \frac{3N}{4}\right]Sin\left(\theta_k + \frac{3\pi K}{4}\right) \end{array} \right\} \quad (3)$$

Let

$$x[n] = A, \quad x\left[n + \frac{N}{4}\right] = B,$$
$$x\left[n + \frac{N}{2}\right] = c, x\left[n + \frac{3N}{4}\right] = D$$

And $\theta_k = \frac{\pi K(2n+1)}{2N}$
Equation 3 can be written as

$$X[K] = \sum_{n=0}^{\frac{N}{4}-1} \left\{ \begin{array}{l} A\ Sin(\theta_k) + B\ Sin\left(\theta_k + \frac{\pi K}{4}\right) \\ +C\ Sin\left(\theta_k + \frac{\pi K}{2}\right) + D\ Sin\left(\theta_k + \frac{3\pi K}{4}\right) \end{array} \right\} \quad (4)$$

$$X[K] = \sum_{n=0}^{\frac{N}{4}-1} \left\{ \begin{array}{l} Sin\ (\theta_k)\left[\begin{array}{l} A + B\ Cos\left(\frac{\pi K}{4}\right) \\ +C\ Cos\left(\frac{\pi K}{2}\right) + D\ Cos\left(\frac{3\pi K}{4}\right) \end{array}\right] \\ +Cos(\theta_k)\left[\begin{array}{l} B\ Sin\left(\frac{\pi K}{4}\right) + C\ Sin\left(\frac{\pi K}{2}\right) \\ +D\ Sin\left(\frac{3\pi K}{4}\right) \end{array}\right] \end{array} \right\} \quad (5)$$

Let $X[K] = X[4p+q]$
Where p = 0 to N/4–1, and q = 1, 2, 3, 4

$$\theta_{p,q} = \frac{\pi(2n+1)(4p+q)}{2N}$$

Case 1: q = 1

$$X[4p+1] = \sum_{n=0}^{\frac{N}{4}-1} \left\{ \begin{array}{l} Sin\left(\theta_{p,1}\right)\left[\begin{array}{l} A + B\ Cos\left(\frac{\pi}{4}(4p+1)\right) \\ +C\ Cos\left(\frac{\pi(4p+1)}{2}\right) \\ +D\ Cos\left(\frac{3\pi(4p+1)}{4}\right) \end{array}\right] + \\ Cos\left(\theta_{p,1}\right)\left[\begin{array}{l} B\ Sin\left(\frac{\pi(4p+1)}{4}\right) \\ +C\ Sin\left(\frac{\pi(4p+1)}{2}\right) \\ +D\ Sin\left(\frac{3\pi(4p+1)}{4}\right) \end{array}\right] \end{array} \right\} \quad (6)$$

i.e.

Equation 6 reduces to

$$X[4p+1] = \sum_{n=0}^{\frac{N}{4}-1}\left\{Sin(\theta_{p,1})\left(A + \frac{(-1)^p}{\sqrt{2}}(B - D)\right) + Cos(\theta_{p,1})\left(C + \frac{(-1)^p}{\sqrt{2}}(B+D)\right)\right\} \quad (7)$$

Similarly.
Case2: q = 2

$$X[4p+2] = \sum_{n=0}^{\frac{N}{4}-1}\left\{\begin{array}{l} Sin(\theta_{p,2})(A - C) \\ +(-1)^p Cos(\theta_{p,2})(B - D) \end{array}\right\} \quad (8)$$

And Case 3: q = 3

$$X[4p+3] = \sum_{n=0}^{\frac{N}{4}-1}\left\{\begin{array}{l} Sin(\theta_{p,3})\left(A - \frac{(-1)^p}{\sqrt{2}}(B - D)\right) \\ -Cos(\theta_{p,3})\left(C - \frac{(-1)^p}{\sqrt{2}}(B + D)\right) \end{array}\right\} \quad (9)$$

Case 4: q = 4

$$X[4p+4] = \sum_{n=0}^{\frac{N}{4}-1}\{Sin(\theta_{p,4})((A+C) - ((-1)^p(B+D)))\} \quad (10)$$

Therefore, for even value of q = 2,4 generalized equations:

$$X[4p+q] = \sum_{n=0}^{\frac{N}{4}-1}\left\{Sin(\theta_{p,q})\left(A + (-1)^{q/2}C\right) + (-1)^p Sin\left(\theta_{p,q} + \frac{q\pi}{4}\right)\left(B + (-1)^{q/2}D\right)\right\} \quad (11)$$

Similarly generalized equation for odd values of q = 1, 3 is:

$$X[4p+q] = \sum_{n=0}^{\frac{N}{4}-1}\left\{Sin(\theta_{p,q})\left(A + \frac{(-1)^{\frac{q+3}{2}+p}}{\sqrt{2}}(B - D)\right) + (-1)^{\frac{q+3}{2}} Cos(\theta_{p,q})\left(C + \frac{(-1)^{\frac{q+3}{2}+p}}{\sqrt{2}}(B + D)\right)\right\} \quad (12)$$

3 Block Diagram of the Proposed Algorithm

Cost-effectiveness and Fast solutions are always main concern while designing any specific system. Fast computations can be achieved by using fast algorithms and costs can be reduced by the use of suitable architectures. Great saving can be achieved by decomposing a structure into a few simple substructures, which are used repetitively with simple interfaces. This is especially true for VLSI designs where a single chip comprises hundreds of thousands of components. Systolic architectures are concurrent structures-that can map high-level computations into hardware structures. Figure 1 shows a basic block diagram of the proposed architecture. Pre-processing units comprises of adders and Subtractors only. Output generated by the pre-processing unit act as input for processing elements (PE).

The idea for pre-processing unit for N = 16 can be seen from Fig. 2. There are N/4 pre–processing unit in stage 1(PP1, PP2, PP3 and PP4) where all the inputs are added and subtracted to provide input to the next stage of pre-processing (PP5).

O/P of PP-5 are I0 to I11,where.

I0-I7:- (An + Cn) ± (Bn + Dn), n = 0 to N/4–1.

(I0, I1,I2, and I3 shows addition of the input samples and I4,I5,I6,I7shows subtraction between the input samples).

Output of the function {(An-Cn) + (B((N/4)-1-n)-D((N/4)-1-n))} obtained at I8 to I11 for n = 0 to N/4–1.

Similarly, output of PP-7unit are I12 to I27 which is the results of equations.

I12–I19:-[An ± (Bn-Dn)/$\sqrt{}$2] = [An ± (1/$\sqrt{}$2 of O/Ps from PP-3 and PP-4)],

I20-I27:-:[Cn ± (Bn + Dn)/$\sqrt{}$2] = [Cn ± (1/$\sqrt{}$2 of O/Ps from PP-3 and PP-4)],

Eg:- for n = 0, I12 = A0 + (B0–D0)/$\sqrt{}$2,I13 = A0–(B0–D0)/$\sqrt{}$2, for n = 1, I14 = A1 + (B1–D1)/$\sqrt{}$2.

I15 = A1-(B1-D1)/$\sqrt{}$2, I16, I17 for n = 2, I18, I19 for n = 3, I20 = C0 + (B0 + D0)/$\sqrt{}$2, I21 = C0–(B0 + D0)/$\sqrt{}$2 and so on.

Systolic Array of PE: Processing Elements (PE's) consist of an adder and a multiplier. All the PE's are arranged in row and column fashion so that the entire network is a pipelined network arrangement of cells. It is a specialized form of parallel computing, where cells compute the data which is coming as input and store them independently and each cell shares the information with its neighbour immediately after processing. If we assume m_t and a_t be the time for performing a multiplication and adding operation respectively in a PE unit. The (m_t + a_t) time unit elapse in one processing elements. Thus, ignoring the data loading operating in the PE unit.

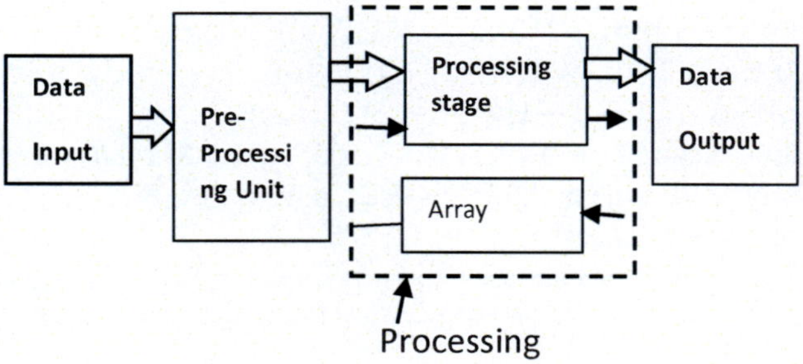

Fig. 1 Block diagram of the proposed DST algorithm

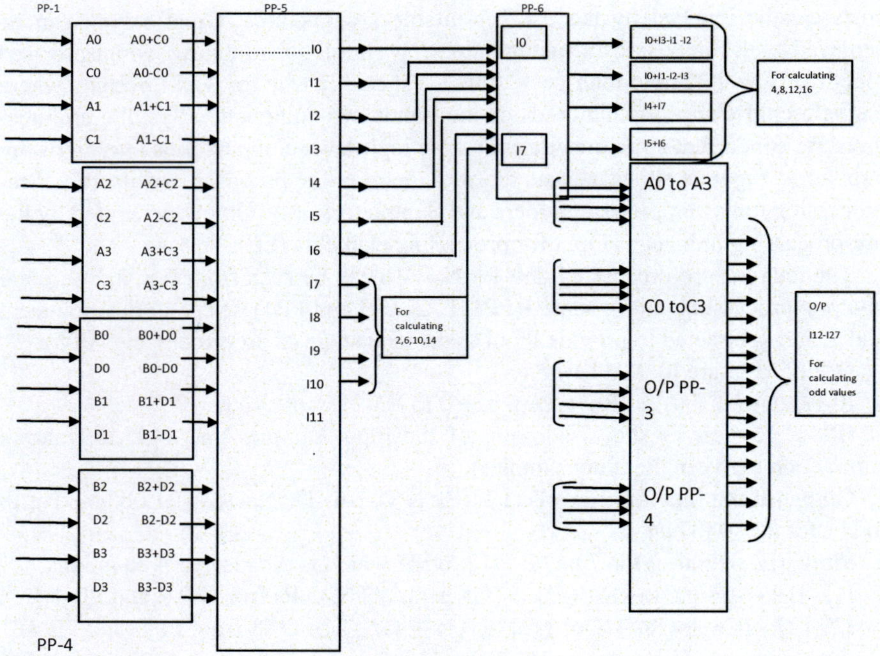

Fig. 2. Preprocessing unit for N = 16

Coefficients X[8] and X[16] are obtained after ($1*m_t + 0*a_t$) time units as can be seen from Fig. 3. As the output from one PE is added with other PE therefore, X[4] and X[12] are obtained in ($1*m_t + 2*a_t$) time units.

For even DST coefficients: X[k] (k = 4,8,12,16) = > X[4p + q] (q = 4).

X[4] = (I4 + I7)* α + (I5 + I6)*γ, X[8] = (I0 + I1−I2−I3)* β, X[12] = (I4 + I7)* γ -(I5 + I6)* α, X[16] = (I0 + I3-I1-I2)*1.

Here α = 0.3826, β = 0.7071, γ = 0.9238.

Remaining even coefficients X(2, 6, 10, 14),here q = 2 are obtained as shown in Fig. 4.

X[2] = (I8*(a) + I9*(b) + I10*(c) + I11*(d)),

X[6] = (I8*(b) + I9*(d) + I10*(a) + I11*(−c)),

X[10] = (I8*(c) + I9*(a) + I10*(-d) + I11*(b)),

X[14] = (I8*(d) + I9*(-c) + I10*(b) + I11*(−a)).

Multiplication factors:-a = 0.1950, b = 0.5555, c = 0.8314, d = 0.9807.

Systolic array of PE (one multiplier is used in these Pes, Fig. 4).

Now for obtaining odd coefficients, X (1, 3, 5, 7, 9, 11, 13, and 15) a systolic array of 8 × 4 processing elements are used and it is shown in Fig. 5, where outputs are:-

X[1] = m0*I12 + m1*I14 + m2*I16 + m3*I18 + m7*I20 + m6*I22 + m5*I24 + m4*I26.

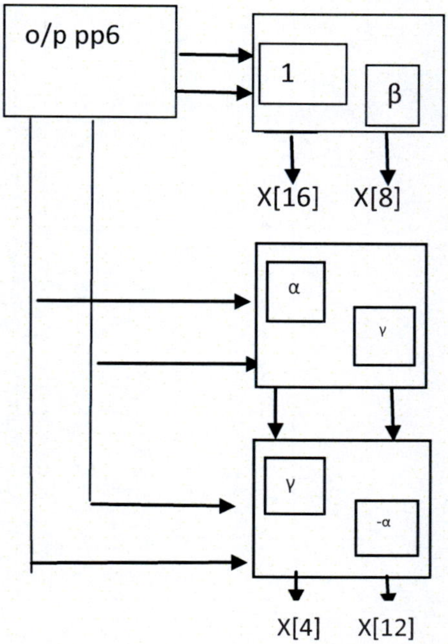

Fig. 3 Shows the systolic array unit for computation of even coefficients

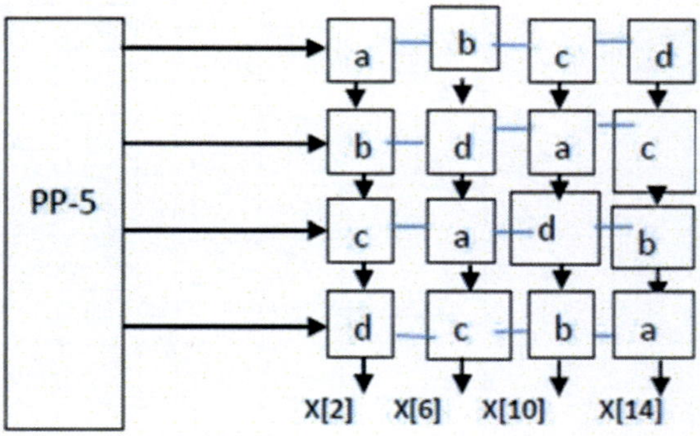

Fig. 4 Processing element for computation of even coefficients of X[K]

X[5] = m2*I13 + m7*I15 + m3*I17 − m1*I19 + m5*I21 + m0*I23 − m4*I25 − m6*I27.

X[9] = m4*I12 + m2*I14 − m6*I16 − m0*I18 + m3*I20 − m5*I22 − m1*I24 + m7*I26.

X[13] = m6*I13 − m3*I15 + m0*I17 + m2*I19 + m1*I21 − m4*I23 + m7*I25 − m5*I27.

X[3] = m1*I13 + m4*I15 + m7*I17 + m5*I19 − m6*I21 + m3*I23 + m0*I25 − m2*I27.

X[7] = m3*I12 − m5*I14 − m1*I16 − m7*I18 − m4*I20 + m2*I22 + m6*I24 − m0*I26.

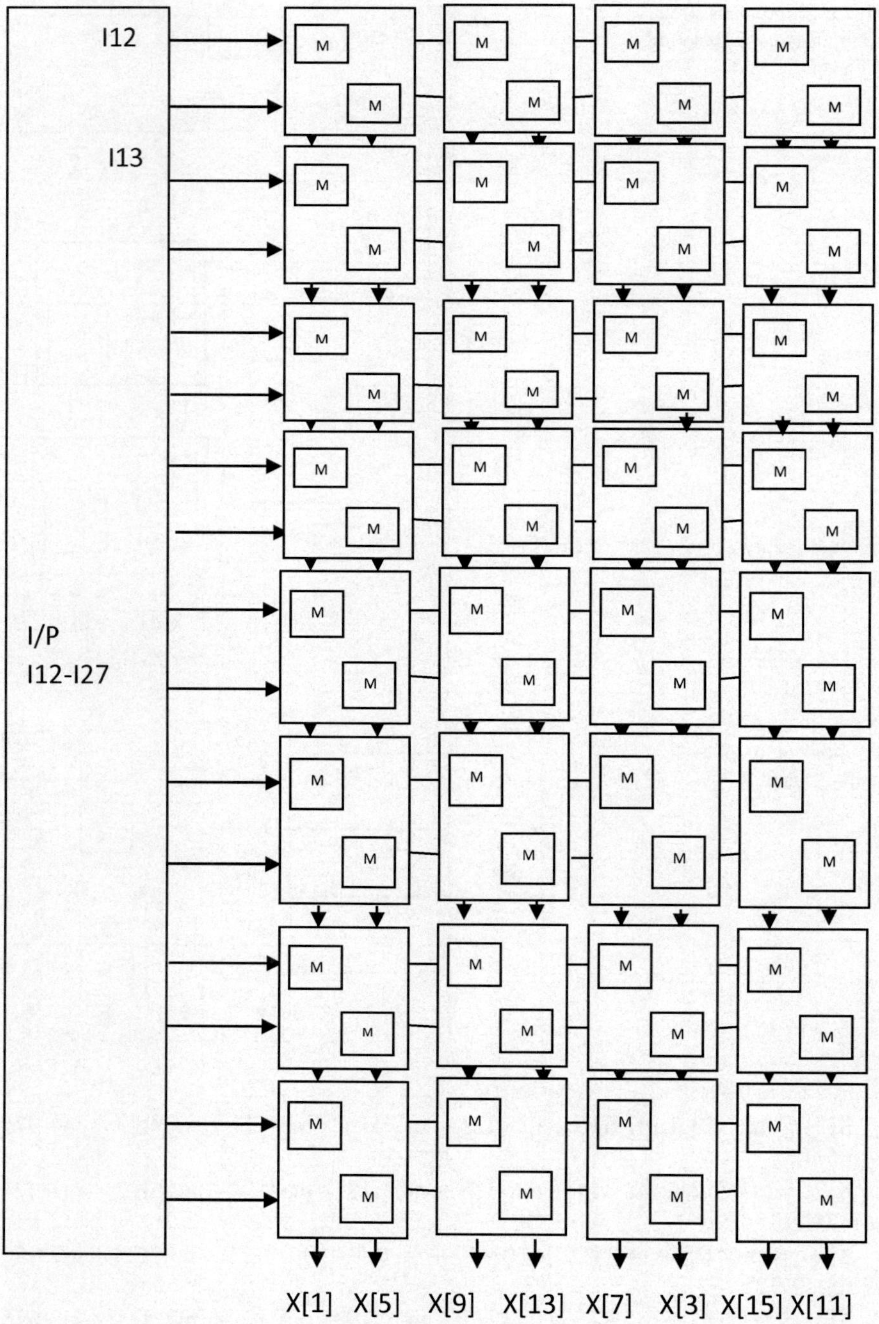

Fig. 5. Systolic array of PE for computing odd DST coefficients

X[11] = m5*I13 − m0*I15 − m4*I17 + m6*I19 − m2*I21 + m7*I23 − m3*I25 − m1*I27.

X[15] = m7*I12 − m6*I14 + m5*I16 − m4*I18 − m0*I20 + m1*I22 − m2*I24 + m3*I26.

Multiplication factors:-

m0 = 0.0980, m1 = 0.2902, m2 = 0.4713, m3 = 0.6343, m4 = 0.7730, m5 = 0.8819, m6 = 0.9569, m7 = 0.9951.

4 Comparison

As can be seen from Eq. (11) and (12) for computation of any coefficient K = 1,2,…,N only 2(N/4) multiplications has to perform instead of N therefore, for computing all coefficients total multiplications are $N(N^2/2)$.There are certain redundancy as multiplication with 1 so using the algorithm the number of multiplications for N = 16 are 86, that is nearly reduced to 1/3 of the N^2 (no of multiplications required in general method). Total 43 PE's are needed to calculate all the DST coefficients of a sequence N = 16. Parallel computing of even and odd coefficients leads to a latency (time required to compute the DST computations) of ($1*m_t + 8*a_t$) as it needs maximum time to compute the odd coefficients. In [18], a systolic architecture for DST computation was presented where DST has been expressed in terms of DFT and number of multiplication are $3N^2 - 5$ N (for N = 16 = > 688) also lot of control circuitry is involved in implementation. Chang et al.[19] presented a unified systolic structure for DCT and DST that require N-1 PE's but for the pipelining, the computation of the whole transformed sequence needs (2 N − 2)(m, + a,) units of time[for N = 16 = > 30(m_t + a_t) unit times]. Chiper [17] presented a systolic architecture where direct ROM-based implementation is used for DST. In, direct ROM based implementation, the multipliers used for multiplication of input values with the fixed transform kernel coefficients, are replaced by a ROM-based look-up table (LUT) of size 2^L, where 'L' is the word length, in which each of the ROM tables contains the precomputed product values for all possible values of input samples. This technique is used to implement the DCT and the DST in linear systolic arrays, after converting the transform in to a circular convolution or convolution-like form [15]. Besides, it also requires the transform length to be a prime number to make it possible to convert the DCT or the DST into circular-convolution structure. Hence, proposed architecture is efficient as it requires less number of processing elements as well as it is faster as its latency is less. A Comparison Table 1. Shows that the proposed systolic architecture out forms the other structure available in the literature.It can seen in [20] that inverse DST could be computed through systolc architercture and this work can be done in future.

Table 1 Comparison of different systolic designs for N = 16

Designs	No. of multiplication	No. of PE units	Latency
[18]	688	16	$62(m_t + a_t)$
[19]	240	15	$30(m_t + a_t)$
Proposed	86	43	$(1m_t + 8a_t)$

5 Conclusion

A new algorithm for computing DST and its VLSI implementation using parallel and pipelined network of processing elements is proposed in the paper. The array of processing elements works in systolic fashion which makes the computation fast.

References

1. Choi JW, Kim NU, Lim SC, Kang J, Kim HY, Lee YL (2017) Shuffled discrete sine transform in inter-prediction coding. ETRI J 39(5):672–682
2. Malini S, Moni RS (2014) Use of discrete sine transform for a novel image denoising technique. Int J Image Process (IJIP) 8(4)
3. Tseng CC (2016) Closed-form design of fir frequency selective filter using discrete sine transform. In: IEEE Asia Pacific conference on circuits and systems (APCCAS)
4. Kus Y, Kavlak M, Elmagoz U, Ozen A (2018) A novel waveform design based on discrete sine transform for 5G and beyond In: International conference on telecommunications and signal processing (2018)
5. Ramadan K, Fiky AS, Dessouky MI, Fathi E, Samie AE (2019) Equalization and carrier frequency offset compensation for UWA-OFDM communication systems based on the discrete sine transform. J Digit Sign Process 90:142–149
6. Dhamija S, Jain P (2011) Comparative analysis for discrete sine transform as a suitable method for noise estimation. Int J Comput Sci 8(5)
7. Choi JW, Kim NU, Lim SC, Kang J, Kim HY, Lee YL (2017) Shuffled discrete sine transform in inter –prediction coding. J ETRI 39(5):672–682
8. Jain P, Kumar B, Jain SB (2007) Discrete sine transform and its inverse realization through recursive algorithms. Int J Circ Theory Appl 36(4):441–449
9. Rama Murthy N, Swamy MNS (1994) On the real-time computation of DFT and DCT through systolic architectures. IEEE Trans Signal Process 42(4):988–991
10. Dhaiya P, Jain P (2019) Realization of recursive algorithm for DCT transform. Int J Circ Syst Sign Process
11. Perera SM (2017) Signal flow graph approach to efficient and forward stable DST algorithm. J Linear Algebra Appl
12. Vinitha CS, Sharma RK (2019) An efficient LUT design on FPGA for memory-based multiplication. Iran J Electr Electron Eng 15(04):462–476
13. Kumar GG, Sahoo SK, Meher PK (2019) 50 Years of FFT algorithms and applications. Int J Circ Syst Sign Process 38(12):5665–5698
14. Chiper DF, Swamy MNS, Omair Ahmad M (2005) Systolic algorithms and a memory-based design approach for a unified architecture for the computation of DCT/DST/IDCT/IDST. IEEE Trans Circ Syst 52(6):1125–1137
15. Meher PK, Swamy MNS (2007) New systolic algorithm and array architecture for prime-length discrete sine transform. IEEE Trans Circ Syst II Exp Briefs 54(3):262–266

16. Meher PK (2005) Unified DA-based parallel architecture for computing the DCT and the DS. In: 5th international conference on information communications & signal processing
17. Chiper DF, Cracan A (2017) A novel algorithm and architecture for a high throughput VLSI implementation of DST using Short Pseudo-cycle convolutions. In: International symposium on signals, circuits and systems (ISSCS)
18. Ramakrishnan K, Ling N (1993) An efficient systolic array design for the discrete sine transform. In: Proceedings of 36th midwest symposium on circuits and systems
19. Chang LW, Wu MC (1991) A unified systolic array for discrete cosine and sine transforms. IEEE Trans Sign Process 39(1):192–194
20. Florin CD (2017) A new systolic array algorithm and architecture for the VLSI implementation of IDST based on a pseudo-band correlation structure. Adv Electr Comput Eng 17(1):75–80

Anamika Jain B.E., M.E. in Electronics and Communication Engineering, Assistant Professor, Maharaja Agrasen Institute of Technology (MAIT) Affiliated to Guru Gobind Singh Indraprastha University, ECE department. More than 18 years of teaching experience. Her field of interest is VLSI, DSP.

Prof. Neeta Pandey M. E. (Microelectronics) Ph.D. Professor in Delhi Technological University (DTU) with approximately 30 years teaching and research experience in Electronics and Communication Engineering. More than 250 international journal publications. Her area of interest are Analog and Digital VLSI Design, Current mode ADC Design.

Design of OTA Based Band Pass Filter Using 180 nm Technology

Shailesh Jakodiya, R. C. Gurjar, and R. S. Gamad

Abstract This paper presents a fully integrated operational transconductor amplifier (OTA) based band pass filter (BPF). This design has been implemented using simple CMOS based current mirror circuits. The overall analysis and simulation implemented in UMC 180 nm technology. The band pass filter depends on MOS based inductor that is constructed from Operational Trans-conductor amplifiers (OTA). Centre frequency and quality factor depends upon the capacitor values. This circuit offers malleable common-mode ranges. The 3 dB lower cutoff and the upper cutoff frequency is 1.5 kHz and 6.9 kHz respectively. The centre frequency can be changed by varying capacitor value. The total use of the DC current is 5.8 mA @ 1.8 V supply. This circuit also gives the DC gain nearly 22 dB. And overall power dissipation is nearly 2.4 μW .

Keywords Operational trans-conductance amplifier (OTA) · Band pass filter (BPF) · Active inductor

1 Introduction

Wireless systems require future performance in terms of electronic performance as well as reducing manufacturing time and costs [1, 4, 10]. Much attempt has been built to improve the chip area and power usage and frequency [3, 4]. The trend of expansion of microwave filters has gained great recognition. The resulting wireless technology made it possible to fulfill the design precondition [4, 8]. The large bandwidth attributes make for various receiver and transmitter systems to use of a similar device which improves high speed and less size consumption and also low-cost design. The interest in large bandwidth and high-speed performance is rapidly

S. Jakodiya (✉) · R. C. Gurjar · R. S. Gamad
Department of Electronics and Instrumentation Engineering, SGSITS, Indore, Madhya Pradesh, India
e-mail: sjakodiya@gmail.com

R. C. Gurjar
e-mail: rgurjar@sgsits.ac.in

V. Nath and J. K. Mandal (eds.), *Proceeding of Fifth International Conference on Microelectronics, Computing and Communication Systems*, Lecture Notes in Electrical Engineering 748,
https://doi.org/10.1007/978-981-16-0275-7_58

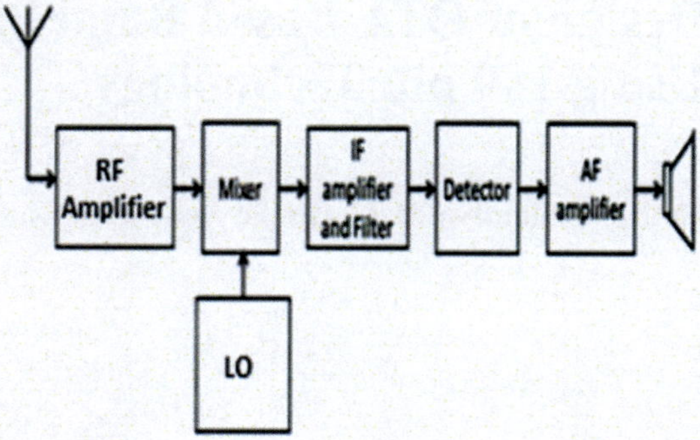

Fig. 1 Super heterodyne type receiver block diagram

increased recently. The wideband systems give high transmission rate with very low power dissipation [11–16].

After antenna, the Radio Frequency is the first juncture of any receiver system which also consists of low noise amplifier circuit [6]. The RF part of receiver system chooses the puny signals within the required band frequency, amplifies the signal for the next processing and also minimizes the noise [2, 6].

There are a lots of methodologies to design a CMOS based fully integrated inductor. But in nowadays, OTA based concept is largely validated in grounded active inductor as well as in floating realization [5]. The basic building block diagram representation of a super heterodyne receiver is shown in the below figure. It is more admire structure of commercial RF receiver that is utilized effectively in TV and various applications like radios. Intermediate frequency section is used to filter the results of mixer for amplifying the voltage increases to that level where it can be demodulated properly for gaining information and obtaining required frequency ranges signal. This stage consists of a Band pass filters which provides amplification and selectivity for obtaining proper gain [6].

The Intermediate frequency (IF) block contain of band pass filters focused at the IF frequency gives amplification for required gain (Fig. 1).

For designing an active inductor the OTA circuit is to be used [5]. This paper consists of following sections: - Sect. 2 consists of fundamentals of OTA circuit. Section 3 deals the active inductor based band pass filter. Section 4 describes the analysis and simulation results of a band pass filter.

2 Ota Basics

OTA is a flexible active device used in large number of applications die to its large tunability and less complex structure [9]. The basic implementation of OTA is a bias current source followed by MOSFET [7]. Voltage controlled current source [VCCS] based an ideal OTA has very large input and output impedance as well as bandwidth. Common drain and common gate configuration behaves as a positive trans-conductance [7]. In the below figure shows the circuit and block diagram of Operational trans-conductance amplifier.

Figure 2 shows the differential OTA. In high speed, the differential signals have many desired ascribes like CMRR from the steady-state signal and matched response of filter, parasitic effect of components and even harmonic rejection [10]. However, most filter design, software and methodology all are for single-ended [10] (Fig. 3).

This section consists of a design of active inductor using OTA.

For Band pass filter design, the current mirror cascade OTA topology has been used. It provides a pliable input common-mode because of two diode-connected elements below the pair of input side [6]. The basic OTA consist of perfectly matched and mirrored MOS transistor which operating in saturation mode and single current source (Fig. 4).

Fig. 2 Differential type OTA

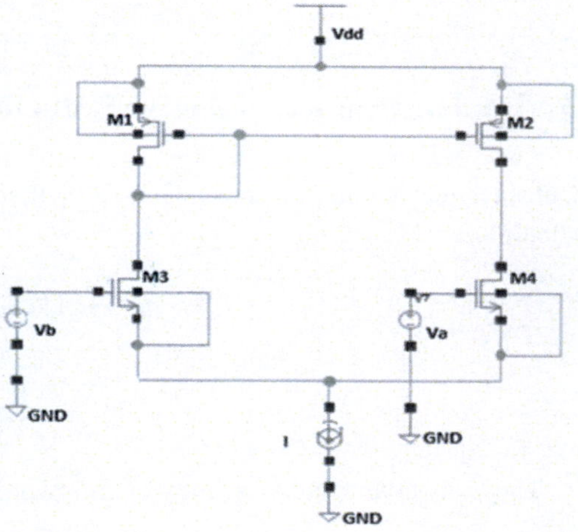

Fig. 3 Basic symbolic OTA

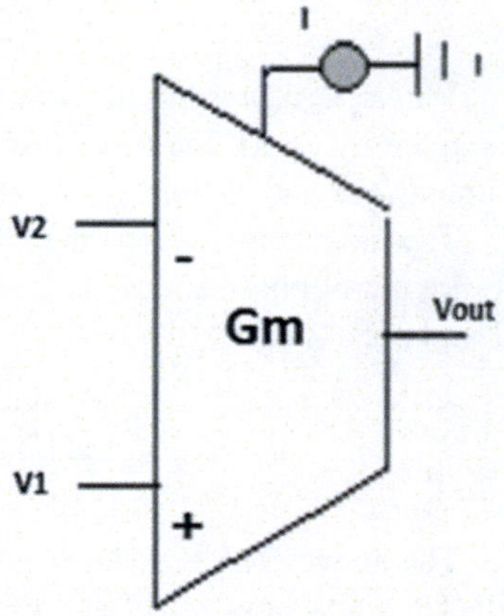

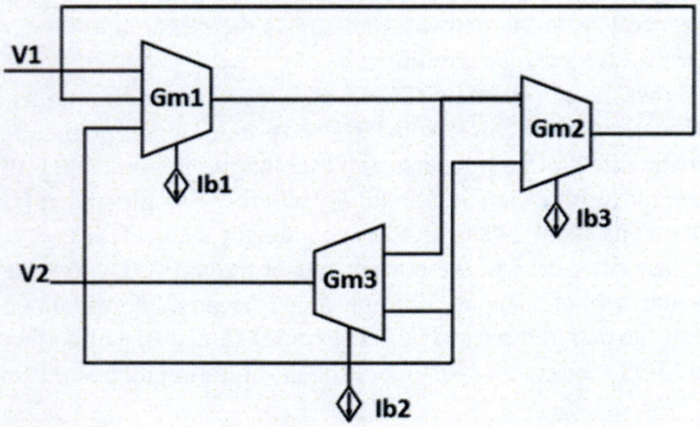

Fig. 4 Architecture of active inductor

3 Implementation of Active Inductor

Consider the output current of Gm_1 is I_1 then I_1 can be written as below in the equation

$$I_1 = Gm_1[V_1 - V_2]$$

$$Gm_1 = \frac{I_b}{2Vc}$$

The trans-conductance is given by following equation [4]:-

$$g_m = \sqrt{2k_p I_{bias}\frac{W}{L}}$$

Where K_p represents the Boltzmann's constant, Width is W and length of transistor is given by L. All the transistors are working in saturation mode and consist same dimensions and characteristics (Fig. 5).

The inductance is controlled by capacitor value as well as bias current. The I_1 and I_2 are out of phase and same in magnitude. The impedance evaluated between node 1 and node 2 is

$$Z12 = \frac{4v_t^2 sC}{I_{b1} * I_{b2}} = \frac{sC}{g_m^2} = sL$$

The above OTA based circuit consists of basic current mirror-based NMOS and PMOS transistors in differential pairs. This topology provides a malleable input

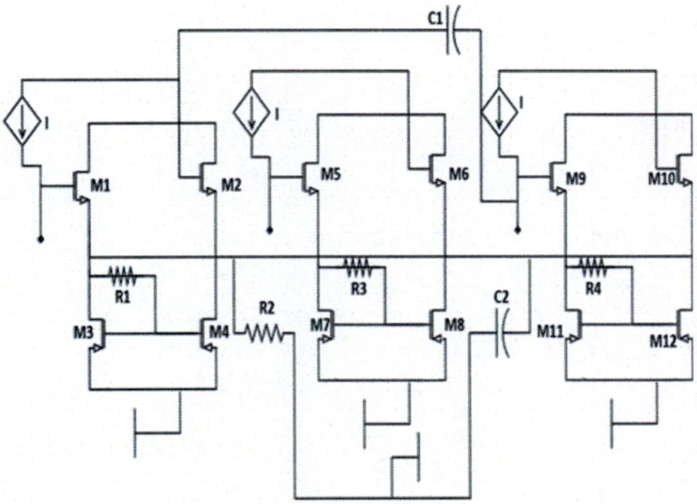

Fig. 5 OTA based band pass filter

Fig. 6 Equivalent of above circuit

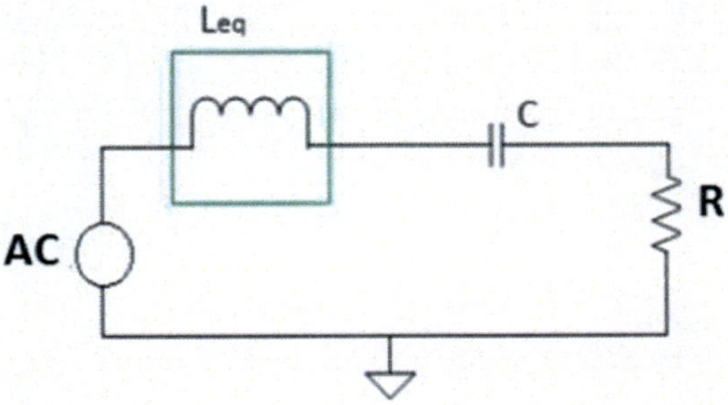

common-mode for the reason that of two diode-connected components above the input side configuration [6] (Fig. 6).

4 Analysis and Simulation Results

From below figure shows the transient response of band pass filter and it is clear that the current lag voltage by 90°. It means that our design behaves as an inductor (Figs. 7, 8, 9 and 10).

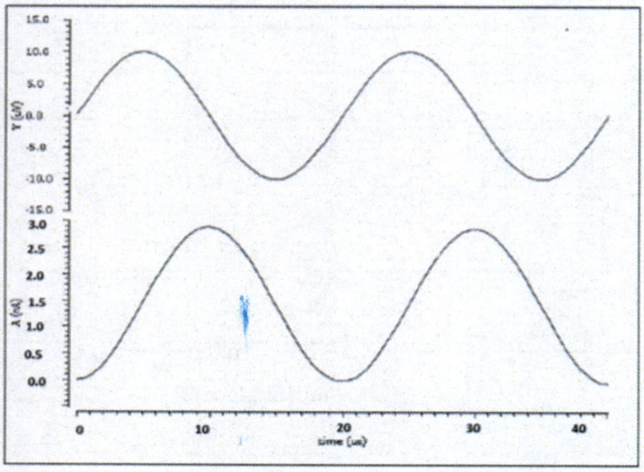

Fig. 7 Transient analysis

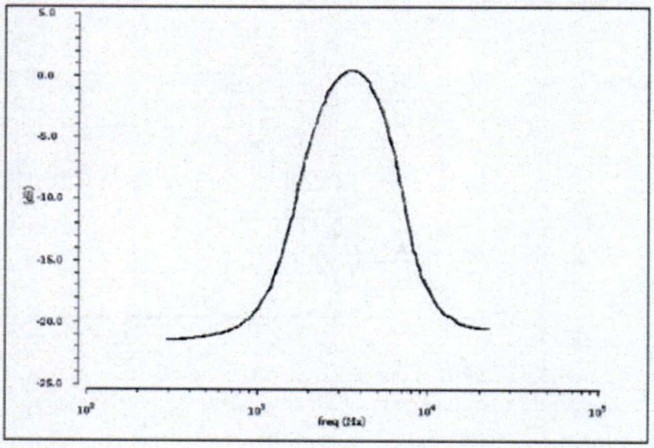

Fig. 8 AC analysis

By Z parameter analysis at input side we can plot the graph between inductance and frequency. From above graph we can say that circuit behaves inductive up to 80 kHz and above it. For Band pass filter analysis, at load resistance side we must have to perform AC analysis. Resonant frequency 3.3 kHz is obtained by theoretically. By AC analysis of Band Pass filter, we obtained that the 3 dB lower cutoff and the upper cutoff frequency is 1.5 kHz and 6.9 kHz respectively. The centre frequency can be changed by varying capacitor value (Table 1).

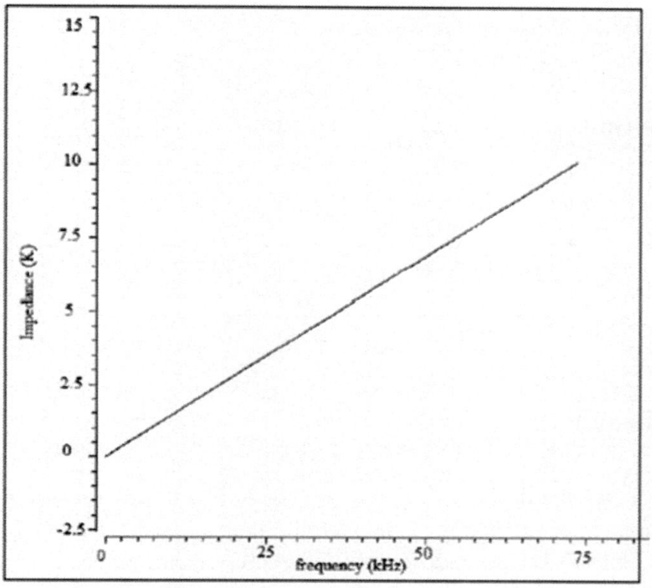

Fig. 9 Inductance VS Frequency plot

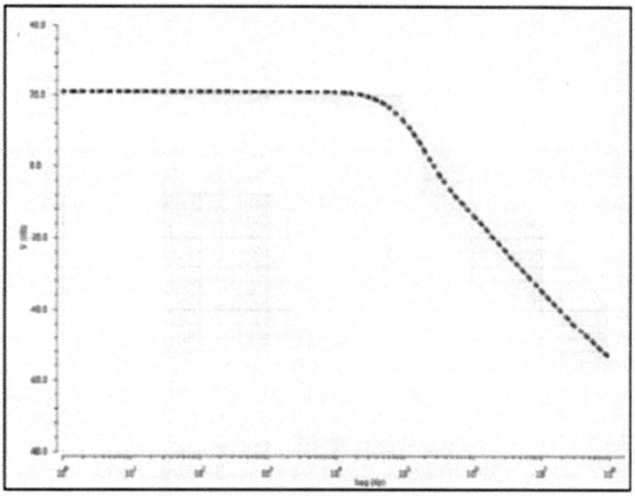

Fig. 10 DC gain

Table 1 Performance summary and comparison

Source	Technology	Cut-off frequency	Power dissipation	Bandwidth	Supply voltage(V)
This work	180 nm	3.3 kHz	2.4 μW	1.5 kHz−6.9 kHz	1.8
[19]	180 nm	0.28 Hz	0.8 μW	0.2 Hz−5.8 Hz	1.2
[20]	0.35 μm	1 Hz(Lower) 5 Hz(Upper)	4.2 μW	4 Hz	3
[6]	–	>100 kHz	–	82 MHz	3
[17]	0.35 μm	2 K Hz	303 nW	2	–
[18]	0.8 μm	900 Hz	2.5 μW	17.20	1.25

5 Conclusion

This paper shows the design of OTA based easily tuned, less transistor Band pass filter with high frequency range. The overall circuit consists of two capacitors and three OTA. This can be more advantage that Quality factor and center frequency of Band pass filter can changed by desired value by capacitors. This circuit can be used in IF stage of the receiver system.

References

1. Prommee P, Wongprommoon N, Khateb F, Manositthichai N (2019) Simple structure OTA-C elliptic band-pass filter. In: 16th International Conference on Electrical Engineering/Electronics, Computer, Telecommunications and Information Technology (ECTI-CON)
2. Kim T, Lee D, Kwon K (2020) CMOS channel-selection low-noise amplifier with high-Q RF band-pass/band-rejection filter for highly integrated RF front-ends. IEEE Microwave Wirel Compon Lett. https://doi.org/10.1109/lmwc.2020.2966361
3. Mehral R, Kumar V, Islam A (2018) Reliable and Q-enhanced floating active inductors and their application in RF bandpass filters. IEEE Access 6: 48181–48194
4. Jarjar M, El Ouazzani N (2017) Design of an OTA-based microwave active bandpass filter. In: 2017 International Conference on Wireless Technologies, Embedded and Intelligent Systems
5. Pramanik A (2015) Design of bandpass filter using operational transconduction amplifier based active inductor. In: 2015 International Conference on Pervasive Computing (ICPC)
6. Hashim NZB, Jamuar SS (2014) Gm-C based band pass filter. In: 2nd Internation Confrence on Electronic Design (ICED), Panang Malaysia
7. Asto T, Ito T (2013) Design of low distortion active inductor and its applications. Analog Integer Circ Sig Process 75:245–255
8. Jarjar M, El Ouazzani N (2016) Synthesis of a microwave CMOS active inductor using simple current mirror-based operational transconductance amplifier (OTA). Int J Sci Eng Res 7(8): 2229–5518
9. Petrashin P, Peretti G, Romero E (2010) OBT implementation on an OTA-C band-pass filter. In: 2010 11th International Symposium on Quality Electronic Design (ISQED). https://doi.org/10.1109/isqed.2010.5450427

10. Rao MS, Mahalakshmi I, Mourougayane K (2015) Elliptical function based differential band pass filter for IF signal processing and EMI applications. In: 2015 13th International Conference on Electromagnetic Interference and Compatibility (INCEMIC)
11. Gao Z, Ma J, Yu M, Ye Y (2008) Afully integrated CMOS active bandpass filter for multiband RF front-ends. IEEE Trans Circuits Syst II Exp Briefs 55(8): 718–722
12. Wu B, Chiu, Y (2015) A 40 nm CMOS derivative-free IF active-RC BPF with programmable bandwidth and center frequency achieving over 30 dbm IIP3. IEEE J Solid-State Circuits 50(8): 1772–1784
13. Kumar V, Mehra R, Islam A (2017) A 2.5 GHz low power, high-Q, reliable design of active bandpass filter. IEEE Trans Device Mater Rel 17(1): 229–244
14. Van Liempd B et al (2016) A +70-dBm IIP3 electrical-balance duplexer for highly integrated tunable front-ends. IEEE Trans Microw Theor Tech 64(12):4274–4286
15. Hsiao Y-C, Meng C, Yang S-T (2016) 5/60 GHz 0.18 μm CMOS dualmode dual-conversion receiver using a tunable active filter for 5-GHz channel selection. IEEE Microw Wireless Compon Lett 26(11):951–953
16. Barzgar A, Oraizi H (2017) A low profile band pass DGS filter for harmonic suppression. 2017 Sensors Networks Smart and Emerging Technologies (SENSET). https://doi.org/10.1109/senset.2017.8125003
17. Peng S-Y, Lee Y-H, Wang T-Y, Huang H-C, Lai M-R, Lee C-H, Liu L-H (2018) A Power-Efficient Reconfigurable OTA-C Filter for Low-Frequency Biomedical Applications. IEEE Trans Circuits Syst I Regul Pap 65(2):543–555. https://doi.org/10.1109/tcsi.2017.2728809
18. Rodriguez-Villegas E, Yufera A, Rueda A (2004) A 1.25-V micropower Gm − C filter based on FGMOS transistors operating in weak inversion. IEEE J Solid-State Circuits 39(1):100–111
19. Liu L, Zou X, Goh WL, Ramamoorthy R, Dawe G, Je M (2012) "800 nW 43 nV Hz neural recording amplifier with enhanced noise efficiency factor. Electron Lett. 48(9): 479–480
20. Aziz JNY, et al (2009) 256-channel neural recording and delta compression microsystem with 3D electrodes. IEEE J Solid-State Circuits 44(3): 995–1005

Quantifying apt of RNN and CNN in Image Classification

Vani A. Hiremani and Kishore Kumar Senapati

Abstract This paper presents a comparative study of understanding the apt of Recurrent Neural Network (RNN) and feed forward artificial neural network like Convolutional Neural Network (CNN) algorithms from deep learning area in the image classification process. We have considered MNIST handwritten digits dataset for image classification using RNN and CNN classification techniques from deep learning to understand the nature of their working in classification. The results of these two classification techniques have shown their own sway with set of inputs. The pros and cons of each technique are recorded during the process. This work edifies the researchers the strength of different classification techniques and supports them in selecting suitable algorithm.

Keywords RNN · CNN · Deep neural network · Feature extraction and image classification

1 Introduction

Image classification is process of segregating images among classes based on n number of similar features. This process plays an important role in handling umpteen numbers of miscellaneous images and fastens the accessibility. It has applications in industry 4.0 era and social platform ranging from predicting defective product [1] out of good products to classifying spam mails based on bag of words method [2] respectively. The state of the art has elucidated Multilayer Perceptron Neural Network (MLPNN), Convolutional Neural Network (CNN) and Recurrent Neural Network (RNN) like deep learning approaches [3] to pursue image classification and these approaches have touched upon manufacturing industries, smart phone

V. A. Hiremani (✉)
Department of Computer Science and Engineering, Amity University, Jharkhand, India

K. K. Senapati
Department of Computer Science and Engineering, Birla Institute of Technology, Mesra, Ranchi, India
e-mail: kksenapati@bitmesra.ac.in

V. Nath and J. K. Mandal (eds.), *Proceeding of Fifth International Conference on Microelectronics, Computing and Communication Systems*, Lecture Notes in Electrical Engineering 748,
https://doi.org/10.1007/978-981-16-0275-7_59

companies, stock market prediction, sentiment analysis, recommendation engines, healthcare, cyber security, forensic science and crime investigation. This work has used MNIST data focused on image classification using the CNN and RNN deep learning approaches to quantify the favorable method for image classification.

Of late Gmail has launched new setting to ease email writing by predicting supposed to write text in next. This magic happened by learning the predictive patterns and it is the work of a specific type of neural network called a Recurrent Neural Network (RNN). The difference between an RNN and a Feed-Forward Neural Network i.e., CNN is that as illustrated in Fig. 1 the previous state of hidden layer is stored i.e., state of the hidden layer at the previous pattern presentation. RNN uses Backpropagation algorithm to train the model.

It eventually makes copy of hidden layer which is referred as the context layer. Context layer pretends as extension to input layer, feeding signals of previous network state to hidden layer. Hence the input vector in Eq. (1) below is combination of actual input and context units.

$$Z = \left(\underbrace{I_i, \ldots\ldots, I_{i+1}}_{\text{Actual inputs}}, \underbrace{I_{i+2}, \ldots\ldots I_{i+1+j}}_{\text{Context units}} \right) \quad (1)$$

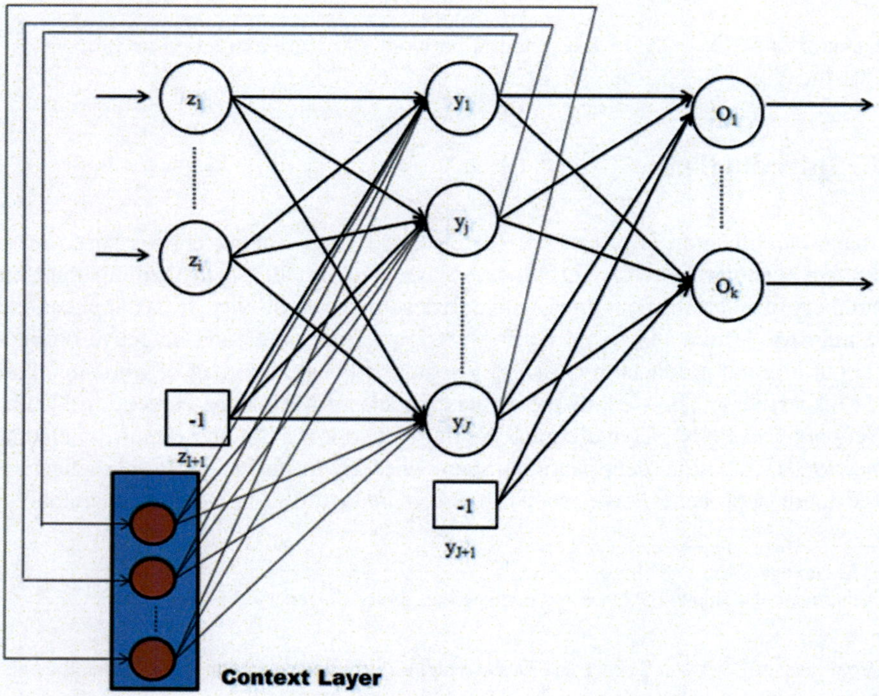

Fig. 1 Recurrent neural network

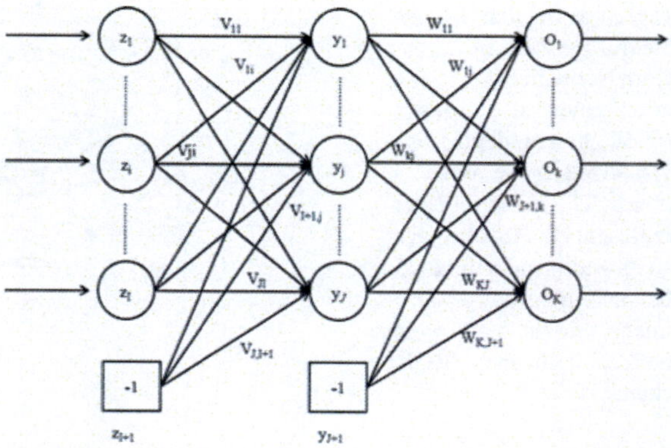

Fig. 2 Feedforward neural network

Context units $I_{i+2}, \ldots\ldots, I_{i+1+J}$ are fully interconnected with all hidden layers. The connections from each hidden layer K_n (for n = 1,…., n) to its corresponding context layer I_{i+1+J} have a weight W = 1 so as the activation value K_n is copied to I_{i+1+J}. Equation (2) shows calculation of each output units activation function.

$$O_{k,p} = f_{ok}\left(\left(\sum\nolimits_{j=1}^{J+1} .w_{kj} f_{yj}\left(\sum\nolimits_{i=1}^{i+1+J} .v_{ji} z_{i},_{p}\right)\right)\right) \quad (2)$$

Where $(I_{i+2,p,\ldots\ldots}, Ii_{+1+J,p}) = \left(y_{1},_{p}(t-1), \ldots., y_{J},_{p}(t-1)\right)$.

Whereas in feed forward artificial neural net shown in Fig. 2 NN receives external signals and keep propagating these through further layers to obtain the end result without feedback connection to previous layers.

CNN is a type of such network in which the connectivity pattern between neurons/nodes is inspired by animal visual cortex. The core of CNN is that, the values of images pixels and filters are convolved to reduce the size of image to feed to next layer. The basic architecture of CNN is shown in Fig. 3.

The boom of CNN lies in its learning local patterns of images as compared to densely connected layers which learn global patterns. The key characteristics of CNN labeled are 1) The patterns are translation invariant and 2) They learn spatial hierarchies of patterns this allows convnets to efficiently learn increasingly from abstract to complex features layer by layer. In CNN the neuron in a layer is only connected to small portion of layer ahead of it this reduces the number of computing parameters. CNN compares images piece by piece i.e., features. The two key parameters of CNN observed are 1) Size of patches extracted from input image typically 3 × 3 or 5 × 5 matrices and 2) Depth of the output feature map i.e., the number of filters computed by convolution.

CNNs are regularized versions of Multilayer Perceptron's (MLP's). MLP's are fully connected networks where single neuron in one layer has connection with every neuron in the next layer. This "fully-connectedness" of MLP's makes them prone to overfitting data. Typical ways of regularization include adding some form of magnitude measurement of weights to the loss function. CNNs performs regularization

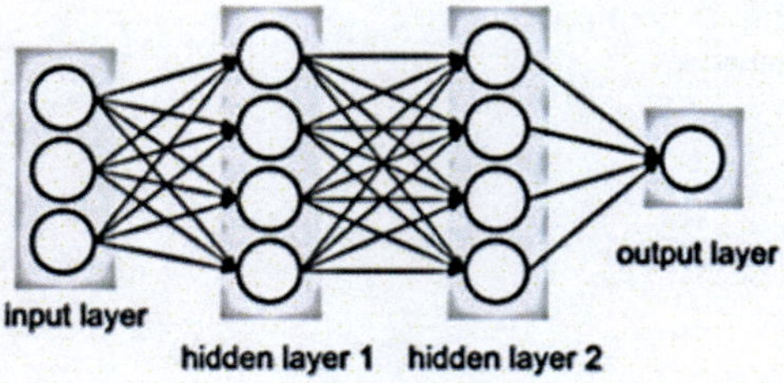

a)

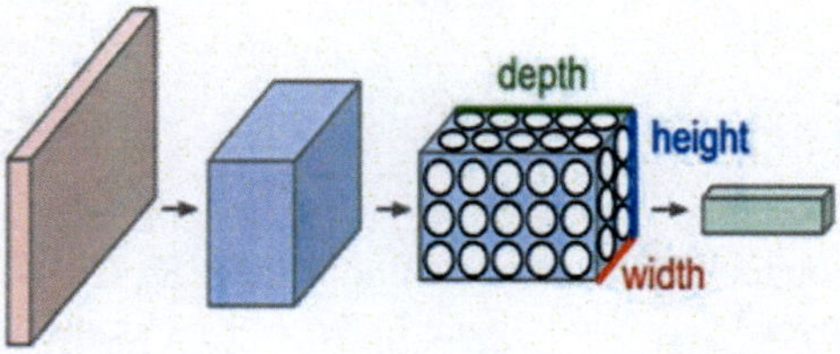

b)

Fig. 3 a) Regular 3-layer Neural Network b) A ConvNet with 3D arrangement of neurons (width, height, depth), as visualized in one of the layers. The red input layer represents the input image, so its width and height would be the dimensions of the image, and the depth would be 3 color channels (RGB channels)

using spatial hierarchies of patterns this allows convnets to efficiently learn increasingly from abstract to complex features layer by layer. Therefore, on the measure of connectedness and complexity, CNNs are on the lower extreme.

The major advantage of convolutional neural network is that it doesn't hold back memory. This means each input provided at an instance is processed independently, with no reference to previous inputs. So to process a new sequence of data we need to show the entire sequence to the network at once in terms of single vector. In contrast with simple recurrent networks since they retain information about the inputs seen long timestamp before, practically such long term dependencies become unviable to learn due to the vanishing gradient problem [4]. In this work we have used Long Short-Term Memory (LSTM) layer to elude this problem. As a variant to simple RNN LSTM layer saves information in parallel to ongoing sequential processing meanwhile preventing older signals from vanishing gradually.

This comparative study of RNN and CNN helps in selection among these neural networks for relative computer vision problems like recognition, analysis and classification. These models are resilient to fit other existing data sets. For comparative study reason being we have considered the MNIST handwritten digits dataset.

2 Background

Few state of the art works explaining the utilization RNN and CNN are summarized below:

In [5] Chiun-Li Chin et al. has built two CNN models to classify good and unhealthy skin quality for aging product recommendation. First model with 2 convolutional layers, 2 pooling layers and 3 fully connected layers and the other with 3 convolution layers, 3 pooling layers, and 4 fully connected layers. Their experimental results shown having deeper network improves accuracy as compared to shallow one. In [6] Vera Wati et al. have used CNN model to classify the facial expressions like pleasure, sadness, surprise, disgust, neutral, anger and fear from real time face captured for non-verbal communication. In [7] Farhana Sultana et al. have explained different CNN models from classification varying from LeNet-5 to SENet model which will be ease for researchers finding suitable method for classification. In [8] Xiangrong Zhang et al., have proposed a novel local space sequential method used in recurrent neural network (LSS-RNN) to extract local and semantic information for hyper spectral image classification. The low level features from hyper spectral images are extracted including texture and differential morphological profiles (DMP) feature. The local spatial sequential (LSS) features are built by combining the low level features together. Recurrent neural network and LSS feature used to train the network for the systematic parameters. Then final classification is carried by feeding softmax layer with the high level semantic feature generated by RNN. In [9] Danni Cheng et al., have proposed the combined model of RNN and CNN in diagnosing Alzheimer's disease using PET images, the hierarchical 2D CNNs are used for intra-slice feature extraction while inter-slice features extracted by the gated recurrent unit (GRU) of RNN for final classification. In [10] Lichao Mou et al., have proposed a RNN model to analyze Hyperspectral pixels as sequential data and then determine information categories via network reasoning. They have proposed parametric rectified tanh (PRetanh) activation function for Hyperspectral sequential data analysis instead of tanh a rectified linear unit. In [11] B. Chandra et al., have used two independent RNN model in parallel to overcome limitation of flow of gradient and the output of these computed as the mean of the modulus. In [12] Nishank Singhal et al., have proposed a comparative study on CNN and RNN neural networks to predict whether the candidate is confident and attentive or unconfident and inattentive during interview based on his movement of eyes, face pose, body movement and activeness to judge the performance.

The state of the art has shown the RNN and CNN models strength in individual and together in many sectors like health care, product recommendation, Hyperspectral image classification and etc. The learning from this survey encourages to, make use of these classification techniques in underlying comparative classification study.

3 Comparative Study

Many recent works have shown quantitative comparative study of exploring best suitable deep neural network architecture. The Table 1 presents the comparison study we performed with few state of art work.

Table 1 Comparative study table

Title of paper	Dataset	Models	Overview	Accuracy%	Observations
Comparative Study of CNN and RNN for natural language processing [13]	Sentiment classification (SentiC), Relation classification (RC), Textual entailment (TE) and answer selection (AS)	CNN, LSTM, GRU	Authors have compared the three DNNs – CNN, GRU and LSTM – with respect to various datasets for solving semantic classification problem under NLP	CNN – 82 GRU – 86 LSTM- 86	RNN has shown good performance but limited to recognition task as compare to CNN but optimizing hidden layer size and batch size improved performance of CNNs and RNNs
Comparative study of singing voice detection based on deep neural networks and ensemble learning [14]	Jamendo	CNN, LSTM, CLSTM, and capsule net	Authors have compared four CNN and RNNs models on features like MFCC, spectrogram/raw PCM samples	CNN- 91.8 LSTM – 84 CLSTM- 85 Capsulenet- 74.9	Post classification voting ensemble method increases accuracy along with multiple network structures
Exploring convolutional, recurrent, and hybrid deep neural networks for speech and music detection in a large audio dataset [15]	Google audioSet dataset	CNN, RNN, Hybrid Deep Neural Network (C2-LSTM)	Authors have evaluated variants of neural networks to for speech and music event detection	C2-LSTM- 83.99 LSTM-73.4 CNN- 83.72	Mel-spectrograms of the audio segments are fed to all neural networks. Results shown the combination of hybrid Convolutional-LSTM structure (C2-LSTM) is more suitable

(continued)

Table 1 (continued)

Title of paper	Dataset	Models	Overview	Accuracy%	Observations
CNN-LSTM neural network model for quantitative strategy analysis in stock markets [16]	Stock market records	CNN, LSTM	Authors have applied combined CNN-LSTM model to analyze quantitative selection and quantitative timing strategy in Stock markets	CNN- moderate LSTM – high *(as mentioned by authors)	CNN are better suited for the quantitative stock selection strategy and LSTM for the quantitative timing strategy for improving the profits
CNN and RNN mixed model for image classification [17]	CIFAR-10	CNN, CNN-RNN	Authors have combined CNN-RNN neural network where the dependency and continuity features of the hidden output of the CNN Model is calculated using RNN	CNN-77 CNN-RNN - 80	The combinations of wavelet transform method in the Fourier transform to filter the input data resulted in solving gradient exploding problem of RNN
Comparative Study of CNN and RNN for motor fault diagnosis using deep learning [18]	Motor vibration data FFT data	CNN RNN	Authors have used CNN and RNN neural networks to diagnose motor fault. The fault diagnose structure of CNN is compared with RNN	CNN- 95 RNN- 90	Backpropagation phase of RNN makes it more robust to changing environment as CNN is sensitive to variations

(continued)

Table 1 (continued)

Title of paper	Dataset	Models	Overview	Accuracy%	Observations
Automatic modulation classification using recurrent neural networks [19]	GNU radio dataset	CNN Variants of RNN	Authors have proposed novel Automatic modulation classification (AMC) method using variants RNN architecture and accuracy is compared with CNN	CNN- 80 RNN- ~ 85	RNN is uses contextual information to analyze mapping relationship. LSTM is used to overcome gradient vanishing problem
A study on a joint deep learning model for myanmar text classification [20]	Myanmar text data	CNN, CNN-LSTM, RNN	Authors have compared the performance of CNN, RNN and joint model of CNN-LSTM for Myanmar text classification	CNN- 89.6 CNN-LSTM- 89.6 RNN - 89	The use of LSTM instead of Maxpool layer reduces loss of context information I order to retain long term information
Proposed comparative study	MNIST Handwritten digits data	CNN RNN	Performance Comparison of CNN and RNN considering handwritten digits	CNN- 98 RNN- 94	RNN is better suited for temporal sequence data and CNN is for spatial data

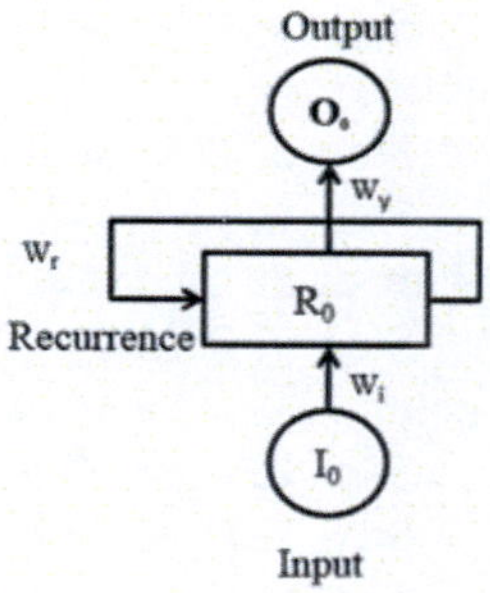

Fig. 4 Recurrent neural network with feedback

A. *Description of Digit Data*

The handwritten digit dataset contains overall 60,000 samples of handwritten digits. Among them 54,000 samples are used for training and 6,000 samples are used for testing and validation. Each digit is a binary image of size 32 × 24 pixels. The digits have been size-normalized and centered in a fixed-size image (28 × 28 pixels) with values from 0 to 1. For simplicity, each image has been flatten and converted to a 1-D Numpy array of 784 features (28 * 28).

B. *Classification Algorithms*

This section consists of working of Recurrent Neural Network and Convolutional Neural Network.

A. *Recurrent Neural Network (RNN)*

The Fig. 4 shows the basic structure of feedback neural network, consider at $t = 0$ network accepts input I_i to produce R_0 using Eq. (3) and at next timestamp i.e., $t = 1$ using Eq. (4). The output O_0 is calculated as shown in Eq. (5).

$$R^{(t)} = g_R\left(W_i I^{(t)} + b_R\right) \tag{3}$$

$$R^{(t)} = g_R\left(W_i I^{(t)} + W_y R^{(t-1)} + b_R\right) \tag{4}$$

$$O^{(t)} = g_O\left(W_y R^{(t)} + b_O\right) \tag{5}$$

Following steps describes how we trained a recurrent neural network with LSTM layer –

Step. 1. Input: A handwritten digits dataset binary image of shape (28 × 28) is given as an Input.

Step. 2. Variables like number of sequences, time steps and number of neurons for the LSTM layer are initialized.

Step. 3. The MNIST hand written digit dataset is size-normalized and centered in a fixed-size image (28 × 28 pixels) with values from 0 to 1. Each image has been flatten and converted to a 1-D NumPy array of 784 features (28*28).

Step. 4. The 60000 MNIST dataset is sliced into training (54000) and testing set (6000).

Step. 5. The dataset is shuffled and batch size is set to 5000.

Step. 6. LSTM model is created and built and forward pass is set to train input with single time step.

Step. 7. Softmax is applied to logits (a vector of non-normalized predictions) to compute cross-entropy and generate a vector of normalized probabilities.

Step. 8. Adam optimizer is used to update weights in training data.

Step. 9. Finally model ran on training set with respect to defined time steps.

The RNN model has exhibited accuracy of 94% on 10 epochs training set with loss of.38% and 89.7% validation accuracy on test set side with loss of 0.16%.

B. *Convolutional Neural Network*

The following Fig. 5 represents the constellation of Convolutional Neural Network consisting of 2 convolution layers of 32 and 64 filters respectively followed by 2 Max Pooling (down-sampling) layers with kernel size of 2 and strides of 2. The (3, 3, 64) outputs are flattened into $1-D$ vector for the fully connected layer with 1024 neurons. Followed by dropout layer with key probability of $p = 0.5$ to elude overfitting. Adam optimizer with learning rate of $\alpha = 0.001$ is used to reduce cross entropy loss. 10-way classification is performed 0–9 digits i.e., using a final layer with 10 outputs and softmax activation function.

The convolution operation

This fundamental quality is what makes it different from dense neural networks. Dense layers consider global patterns from input image, for example with a MNIST

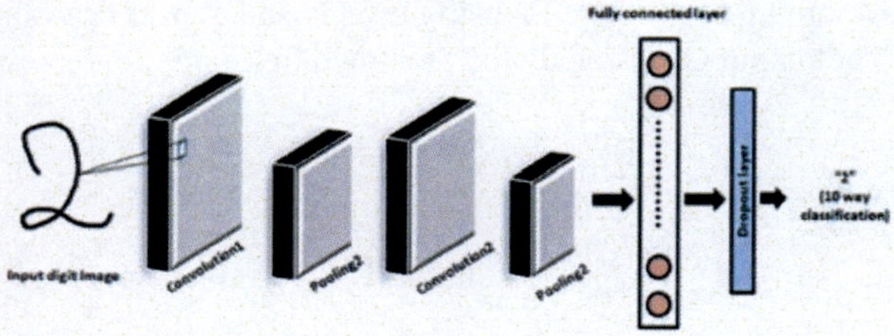

Fig. 5 Constellation of convolutional neural network

Fig. 6 Digit image fragmented into local patterns such as edges, shape of line and texture

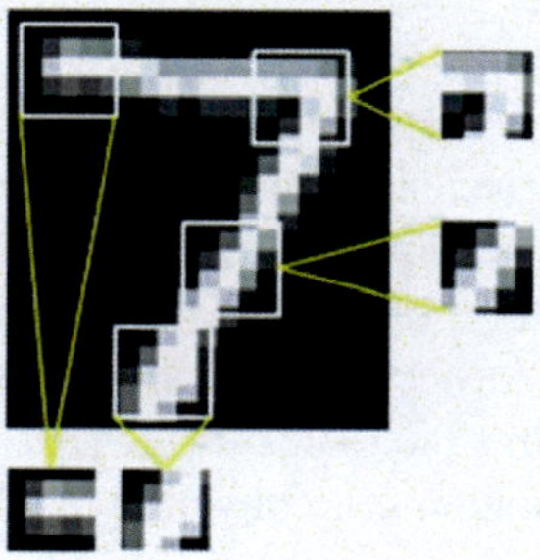

digit dataset, patterns involving all pixels. Whereas convolutional network focuses on local patterns observed in small 2D windows of the input shown in Fig. 6.

A convolution is performed by sliding 3 × 3 windows/filter over the input feature map, stopping at every possible location, and extracting the 2D patch consisting of height and width information. Each of these patches transformed to a 1D vector. All of these vectors assembled to form output feature map. This process brings difference in the shape of input image to output image because of border effects which countered by padding the input feature map and usage of strides to down sample the feature map by the factor of 2. The best part of CNN is the size of the feature maps is halved after every Max Pooling layer without compromising the information. This technique confine from having more parameters i.e. more neurons, weights and eventually avoiding overfitting.

The CNN model has exhibited persistent accuracy throughout 10 epochs of 98% on training side with loss of 9.8 and 98% validation accuracy on test set side with loss of 9.9.

The main issues with RNN are discussed below:

1. Vanishing Gradients

In Backpropagation algorithm the error term *e* is calculated between actual output and the output of model at a instance as shown in Eq. (6) to see the changes in error (*de*) with respect to the change in weight (*dw*) along with learning rate α = 0.001 given in Eq. (7).

$$e = (Actual\ Output - Model\ Output)^2 \tag{6}$$

$$\Delta w = \alpha \frac{de}{dw} \tag{7}$$

$$w = w + \Delta w \tag{8}$$

Observation noted that if the value of $\frac{de}{dw} < 1$ then multiplying this with α = 0.001 would produce value <1 in terms yielding negligible weight Δw certainly same as old weight with no effect in updating weight i.e., no learning which result in *vanishing gradient*.

2. Exploding Gradients

In contrast to the vanishing gradient, if the value of $\frac{de}{dw} > 1$ then it give raise to long term dependencies as it keeps increasing, as a result yields very different weight Δw than old weight which result in exploding the gradient value.

Solutions to these two issues with RNN comes up with many additional arrangements like adding LSTM layer, clipping gradients at threshold, ReLu activation function and so on. On the measure of complexity and supplementary arrangements, CNNs are on the lower extreme.

The following Table 2 shows the findings of this paper.

Table 2 Annotation of experimental findings

Parameters	RNN	CNN
Accuracy and loss	94% on training side with loss of 0.38% and 89.7% validation accuracy on test set side with loss of 0.16%	98% on training side with loss of 9.8 and 98% validation accuracy on test set side with loss of 9.9
Filters	RNNs reuse activation functions from previous data points in the sequence to generating the next output in line	Filters are applied on images. Convolution operation performed between image and filter to transfer data
Features	RNN includes less feature compatibility when compared to CNN	CNN focus on local patterns
Complexity	Training RNN is hard as they accept arbitrary input/output lengths	CNN accepts fixed size inputs and generates fixed size outputs
Memory	Long distance variables need to be handled so require more memory	CNN does not require additional memory
Issues	Vanishing gradient and exploding gradient are the main issues	CNN is not encountered with such issues
Time	Due to back propagation RNN consume more time for training images nearly 703 s i.e., 17 ms/sample	CNN does not build the relationship between hidden vectors of each time step so take less time. It took approximately 541 s is 9 ms/sample time to train samples

4 Conclusion

In this work we have shown the comparative study of RNN and CNN considering MNSIT handwritten digit dataset to understand the working of these neural networks. This work will act as reference for selection process. The convolutional part and usage of ReLU has surpassed the performance of RNN. The experimental results have shown the performance of CNN (98%) is more feasible for spatial data than the performance of RNN (94%). In future it is planned to extend this work to understand how these neural networks perform in geographical region wise human face classification.

References

1. Zhou H, Yu KM (2017) Imbalanced data classification for defective product prediction based on industrial wireless sensor network. In: Sixth international conference on future generation communication technologies (FGCT). IEEE

2. Sahın E, Aydos M, Orhan F (2018) Spam/ham e-mail classification using machine learning methods based on bag of words technique. In: 26th signal processing and communications applications conference (SIU). IEEE
3. Panigrahi S, Anuja N, Swarnkar T (2018) Deep learning approach for image classification. In: 2nd international conference on data science and business analytics
4. Bengio Y, Simard P, Frasconi P (1994) Learning long-term dependencies with gradient descent is difficult IEEE Trans Neural Netw 5(2):157–166
5. Chin CL, Chin MC, Tsai TC, Chen WE (2018) Facial skin image classification system using convolutional neural networks deep learning algorithm. In: 9th international conference on awareness science and technology (iCAST). IEEE
6. Wati V, Kusrini K, Al Fatta H (2019) Real time face expression classification using convolutional neural network algorithm. In: International conference on information and communications technology (ICOIACT). IEEE
7. Sultana F, Sufian A, Dutta P (2018) Advancements in image classification using convolutional neural network. In: Fourth international conference on research in computational intelligence and communication Networks (ICRCICN). IEEE
8. Zhang X, Sun Y, Jiang K, Li C, Jiao L, Zhou H (2018) Spatial sequential recurrent neural network for hyperspectral image classification J Select Top Appl Earth Obser Remote Sens 11(11):4141–4155
9. Cheng D, Liu M (2017) Combining convolutional and recurrent neural networks for Alzheimer's disease diagnosis using PET images. In: International conference on imaging systems and techniques (IST). IEEE
10. Mou L, Ghamisi P, Zhu XX (2017) Deep recurrent neural networks for hyperspectral image classification. Geosci. Remote Sens. 55(7):3639–3655
11. Chandra B, Sharma RK (2017) On improving recurrent neural network for image classification. In: 2017 International joint conference on neural networks (IJCNN). IEEE
12. Singhal N, Srishti NS (2018) Comparing CNN and RNN for prediction of judgment in video interview based on facial gestures. In: 5th International conference on signal processing and integrated networks (SPIN)
13. Yin W, Kann K, Yu M, Schütze H (2017) Comparative study of CNN and RNN for natural language processing arXiv:1702.01923
14. You SD, Liu CH, Chen WK (2018) Comparative study of singing voice detection based on deep neural networks and ensemble learning. Hum-centric Comput Inf Sci 8(34):1–18
15. de Benito-Gorron D, Lozano-Diez A, Toledano DT, Gonzalez-Rodriguez J (2019) Exploring convolutional, recurrent, and hybrid deep neural networks for speech and music detection in a large audio dataset. EURASIP J Audio Speech Music Process 9
16. Liu S, Zhang C, Ma J (2017) CNN-LSTM neural network model for quantitative strategy analysis in stock markets. In: International conference on neural information processing. Springer, Cham
17. Yin Q, Zhang R, Shao XL (2019) CNN and RNN mixed model for image classification. In: MATEC web of conferences 277:02001
18. Choi DJ, Han JH, Park UK, Hong SK (2020) Comparative study of CNN and RNN for motor fault diagnosis using deep learning. In: 7th IEEE international conference on industrial engineering and applications
19. Hong D, Zhang Z, Xu X (2017) Automatic modulation classification using recurrent neural networks. In: 3rd IEEE international conference on computer and communications
20. Phyu MS, Nwet KT (2020) A study on a joint deep learning model for Myanmar text classification. In: IEEE Conference on Computer Applications (ICCA)

Shared Buffer Crossbar Architecture for GPU-CPU

Deepika Pitliya and Namita Palecha

Abstract To achieve high-performance and energy optimized computing the GPU-CPU heterogeneous architectures are standout choice. Streaming multiprocessors (SMs) are increasing to boost throughput in GPUs. Design of on-chip interconnect for GPU-CPU diverse system is a challenge to make it scalable and efficient. Mesh network is being used in manycore CPUs but for GPU it consumes more area and power as well, due to traffic pattern of GPU. Crossbar is good fit, but it is not supporting communication among the SMs when numbers of SM are high. The motivation is to design the scalable crossbar which provide communication between SMs and SM to memory unit. The objective here to design two types of crossbar with shared buffer, crossbar local and global. Crossbar local provides communication among SMs and take all the input request which are going to the memory in coincide manner and pass it to crossbar global that divaricate these request to memory unit, Last-level cache as well as memory controllers. Sharing buffer give opportunity to all input for communication way efficient to achieve high throughput with reduce area and power. Compare to mesh network in Shared buffer crossbar network reduction in area 28% and power 32%. Scalability of design is verified by increasing the number of SMs.

Keywords Crossbar · Graphical processing unit · Buffers

1 Introduction

Designing on chip network for GPU-CPU diverse system is sustainable challenge. CPUs enhance optimization which lessen latency but hardly queuing the resources. In GPUs there is lot of congestion occurs due to massive traffic that degrade performance of CPU. High number of SM create congestion in the connection of SM to memory

D. Pitliya (✉) · N. Palecha
Department of ECE, R.V. College of Engineering, Bangalore, India

N. Palecha
e-mail: namitapalecha@rvce.edu.in

V. Nath and J. K. Mandal (eds.), *Proceeding of Fifth International Conference on Microelectronics, Computing and Communication Systems*, Lecture Notes in Electrical Engineering 748,
https://doi.org/10.1007/978-981-16-0275-7_60

controllers and last level cache [1]. In CPUs system generally mesh, butterfly and clos used which provide communication among CPUs but not suitable for GPUs, because these topologies traffic pattern is many to few and few to many [2, 3]. There is requirement of interconnect which provide communication among SMs and SMs to memory controller and last level cache.

Here motivation is to design shared buffer crossbar architecture for GPU-CPU heterogeneous system [4]. Buffers are being used on routers at input/output port to store the packet temporarily. Sharing buffer on network because every input port in the routers is not having input packets that is required to transfer at concurrently [5]. Routers, buffers and links are the important components of the Network on chip. All these elements connect the input and output ports of neighboring cores. For each router, the input buffers and the crossbar switch are the major components and consuming chip area and power. As the number of ports increases, the associated buffers, allocation logic and crossbar increase so area also increase. There is need to construct a scalable and efficient network for many core systems. which provide the connection between all the cores and all the input/output ports. Buffers occupy more area in network design, so sharing buffer gives opportunity to data packets to utilize the buffers when traffic is high by sharing them Because not all core transfer the data so if traffic at one core is high and other core is low then by sharing buffer space packet can be transfer. There is no data loss even traffic is high.

Here for crossbar designing two switching nodes, crossbar local and crossbar global rather using the conventional crossbar which is fully connected. The idea here is not all the processor wants to communicate to memory unit simultaneously for example SMs to MCs and LLCs [6, 7]. Rather than connecting all SM to memory unit, crossbar local takes the request from SM and pass it to crossbar global in converged manner then these requests are diverging to memory unit by using crossbar global.

The whole thing can be analyzed by taking one example, fully connected type crossbar in which 40 SMs that are connected to 8 LLC means all 40 is connected to 8 LLC. If taking same example in shared buffer crossbar design there are 5 crossbars local, each one connects 8 SMs input side to 3 outputs that will be input to crossbar global. 5 crossbars local converged 3 output per port all combine 15 input for crossbar global, so these 15 requests coming from SM to memory unit through crossbar local and then crossbar global and passing to 8 LLC. So observation is rather connecting all 40 requests to 8 LLC here 15 requests to 8 LLC as not all processor wants to process memory at same time. All SMs are grouped through crossbar local so communication can happen among SMs with the help of crossbar local and crossbar global.

2 Design Methology

The design methodology of network for many core systems is shown in Fig. 1 where multiple SMs grouped in crossbar local.

All crossbar local grouped and connected to crossbar global.

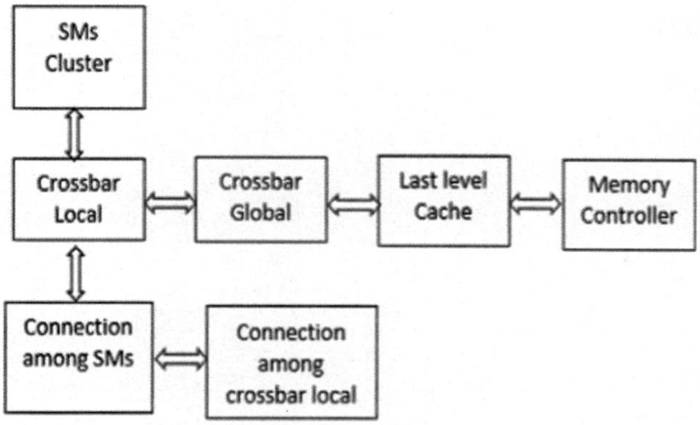

Fig. 1 Design methodology for crossbar design for GPU-GPU system

2.1 *Crossbar Local and Global*

There are two types of crossbar, combination of two crossbar gives the complete functionality, communication between SMs and SMs to memory unit. These two types of crossbars give the opportunities to utilize the networks thoroughly. All SMs are grouped to crossbar local, so there are several numbers of crossbar local that again depends on the number of SMs and how groping is done. Crossbar local provide the communication among SMs and if any request from the SM to memory then passes the request to crossbar Global in coincide manner. The key benefit here is converging the memory request from SMs to crossbar global through crossbar local.

The crossbar global provides the connection for the request coming from the crossbar local to memory unit, so crossbar global provide the communication between SMs to memory with the help of crossbar local. The crossbar global is connecting all local crossbar, MCs and LLCs. Hence gives the opportunity to all SMs to communicate each other.

2.2 *Buffers*

Buffers consume router area but to remove contention and no packet loss. There is need to store packet temporarily at input/output port. Wormhole switching technique applied here in which the credit-based flow being used, which provides the assurance for all the packets that there will be the space in buffers, so no packet loss.

Fig. 2 Round robin algorithm

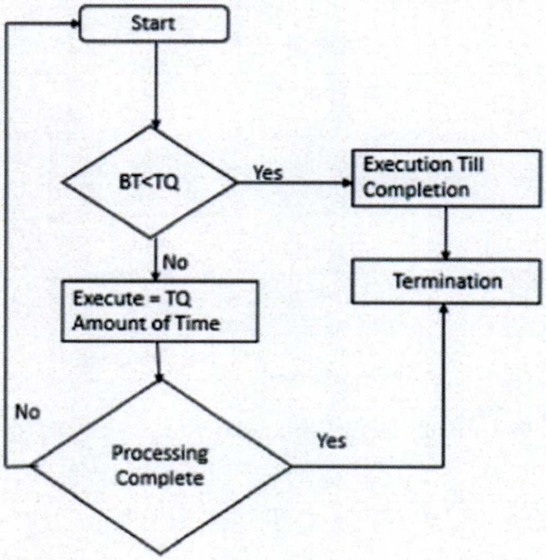

2.3 Round Robin Algorithm

In all crossbar for routing algorithm Round – robin being used, it considers all of the packets which are sending to converge port hence contention minimized. Here for example in crossbar four inputs are sending packets and at output side two converge ports. Firs algorithm chooses two out of four packets considering first come and first serve method and send to at two converged port following round robin manner. It minimizes contention of flits. Figure 2 shows flow chart of round robin microarchitecture. time quanta which is define as time amount share provide to each process in identical portion. There is cyclic order and it give opportunity to all the process considering time quantum and burst time.

2.4 LLC and MC

GPU offers enormous thread-level parallelism. Earlier GPU and CPU used to connect through PCI bus. CPU offloads data into GPU where it explicitly copy to memory, but explicit copy significantly gives performance overhead. To improve programmability and reduce communication cost integrate both on same network. Sharing the Last Level Cache and memory controllers on network on chip.

2.5 *Scalablity*

Scalability is another feature of the design. In case if the requirement is increment in the numbers of SM in the design there are two choices to scale the design. First option increase numbers of the SMs per crossbar local where output converge port remain same for crossbar local, only input number change as number of SM increases, so there is no scaling issue for crossbar global because number of crossbar local and output of crossbar local still same only input of crossbar local changes for increase in SMs. Second option if there is increase in SMs, increase number of crossbar local, in this case as increase in crossbar local which effect the crossbar global because input for crossbar global increases as number of crossbar local increases. So, the former choice is better option because in second choice scaling issue for crossbar global.

3 Implementation

3.1 *GPU-GPU Homogeneous System*

In GPU- GPU homogeneous system design with 24 SMs, 3 crossbars local with 8 SMs clustered per crossbar local. Crossbar local converging 8 inputs to 2 outputs at outputs port which are the inputs for crossbar global. Crossbar global is bridge between crossbar local and memory unit. In this design crossbar global is taking 6 inputs which are coming from crossbars local and diverging to 3 outputs to memory unit memory controller and last level cache.

In this Fig. 3 it is shown the SMs are connected through crossbar local and crossbar global being used to connect all crossbar local and memory unit. Routing algorithm round robin is giving prospect to each input for communication. Crossbar global is diverging the memory request which is coming from memory. So here rather passing all 24 request to all memory unit the design converge the 24 requests to 6 because

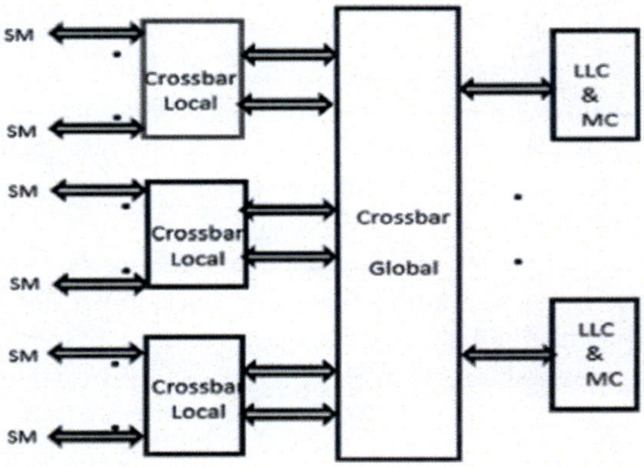

Fig. 3 Crossbar design for GPU-GPU system

each crossbar local output is 2 so all 3 crossbar local together gives 6 output that is again going through crossbar global hence output of crossbar global is 3. Hence out of 24 finally only 3 requests are going from SMs to memory unit.

Buffers are used for network interface in each network so packet can be store temporarily until the previous data transfer not complete. In this way there is minimum packet loss in the network. In routers more than 60% of area consume by buffers and if buffer less routers used in network then poor performance. So, buffer play vital role in networks. Generally, in the network not all the port having data transfer simultaneously, some ports sending more packets and while others are not sending single one, in this condition there are higher chances of packet loss, since at the port where traffic is high there is probability of packet loss.

3.2 GPU-CPU Heterogeneous System

GPU-CPU heterogeneous system is most promising design in today world. GPU does intensive computation on huge data and CPU does the task in serial manner. Here in this design there are three crossbars local in which two crossbars local is for GPUs and one crossbar local for CPU. There are 16 SMs total in two crossbars local and 8 processor equivalent (PEs) for CPU crossbar local. In GPU parallel communication is there among SMs through crossbar local and in CPU serial communication among processor.

There are three crossbars local where two crossbar local functioning is same as the previous design gives the communication among SMs and converge the request for memory to crossbar global which is coming from SMs. Third crossbar local functionality used to provide the serial communication in between PEs. This design provides parallelism of GPU and the serialization of CPUs. Here integration of GPU and CPU done with sharing memory controller and last level cache.

3.3 Mesh Topology GPU-GPU Homogeneous System

In 5 x 5 Mesh topology design shown in Fig. 4. In m x m mesh topology m^2 routers are there and per router there are 5 input/output ports. In mesh design there is unique traffic pattern, where data traverse from source node to destination in unique pattern either horizontal or vertical direction. Router design based on virtual flow base control.

In mesh network the data injected from the processing nodes and route based on deterministic routing algorithm. Each router has five input/output port and corresponding buffers, crossbars and switches. In mesh network the data transfer also in unique fashion, its work in vertical and in horizontal direction.

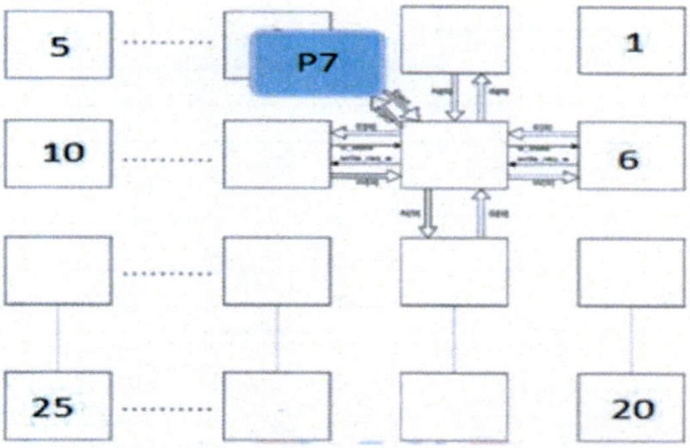

Fig. 4 Mesh 5 × 5 network for GPU-GPU system

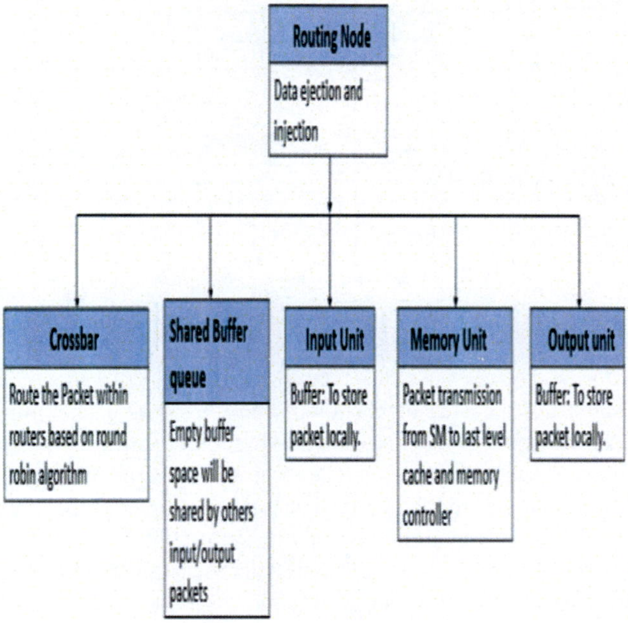

Fig. 5 Design functionality of shared buffer crossbar network

4 Result and Discussion

The shared buffer crossbar network design functionality shown in Fig. 5. There are various stages in network on chip. Data packet injects from source and crossbar routes packet based on round robin routing algorithm. At input/output units of network, buffers are placed to store data temporary.

In this section discussion on functional verification of shared buffer crossbar architecture. Crossbar local contains 8 SMs at input side and 2 converged output. The

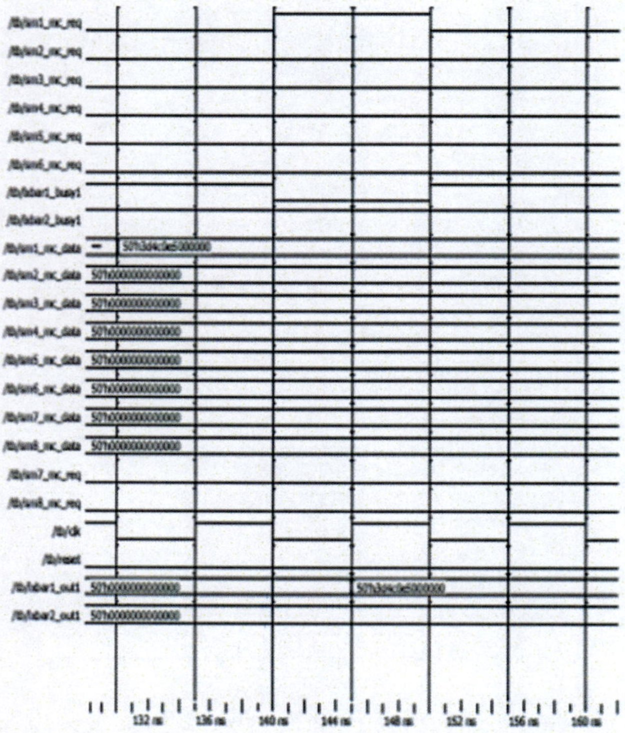

Fig. 6 Output of crossbar local

input/output signals of crossbar local shown in Fig. 6 where packet injected at SM1 in flit format. The injected packet is requesting for communication with memory. So, crossbar local according to round robin routing it checks the availability and grant the signal. The output is available for further transmission at global crossbars.

Other crossbars local function in similar fashion and converged output of all crossbar goes to crossbar global. Here 3 crossbars local designed and each crossbar local has 2 converged output so total 6 outputs. Crossbar global 6 input and these inputs are again diverging to memory unit last level cache and memory controller. Functionality of crossbar global shown in Fig. 7 there are 6 input data coming from the crossbar local and after passing various intermediate stages like buffers and routing function. Crossbar global grant the upcoming signal and pass it to memory unit. In this crossbar global connect all the crossbar local and all the memory controllers and last level cache. As not all SMs requests for memory unit. In GPU the traffic pattern is many to few so only one or two out of 8 SMs requests for memory that's the reason to use the crossbar global and crossbar local so rather doing more connections between all nodes requests are converging here.

Power Analysis of various network shown in Fig. 8. Where the shared buffer crossbar and mesh topology for 24 SMs and 8 SMs calculated. Network power for mesh network is more than shared buffer crossbar because in mesh network there

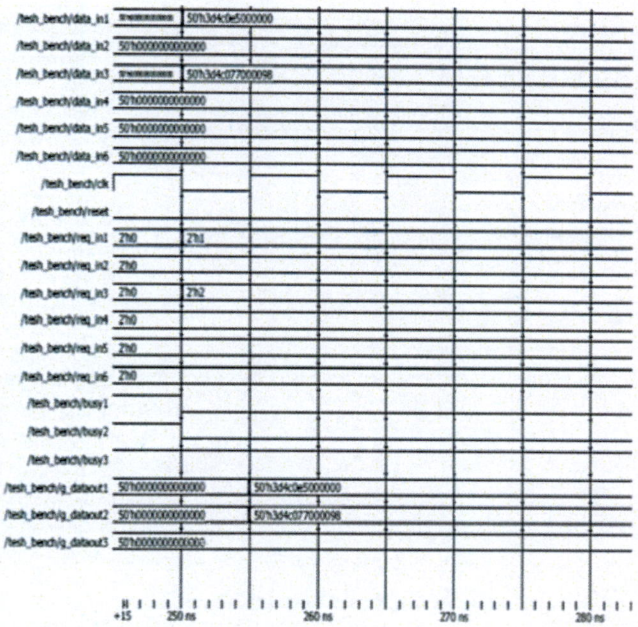

Fig. 7 Output of crossbar global

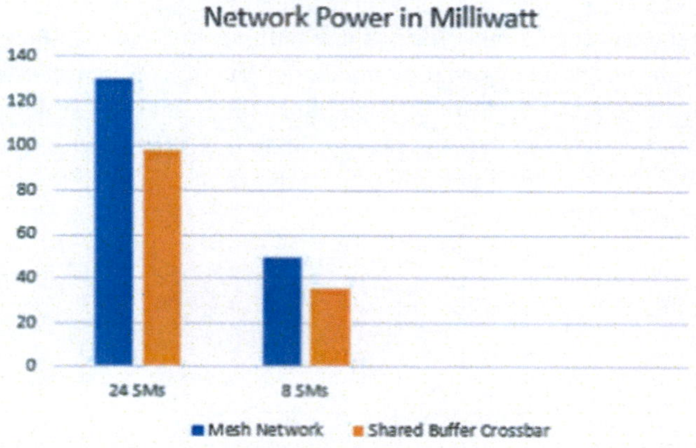

Fig. 8 Power comparison of the networks

are more input/output ports and more buffers, so numbers of links increases hence power become high compare to shared buffer crossbar.

Area report for shared buffer crossbar and mesh network shown in Table 1 and Table 2 for 24 SMs and 8 SMs respectively. Area of shared buffer crossbar network is less compared to mesh network because in shared buffer crossbar, rather using natural crossbar there are two crossbars with shared buffer. Crossbar are conversing

Table 1 Area report of shared buffer crossbar and mesh networks for 24 SMs

Network topology	No. of Slice	No. of FF.	No. of Luts
Mesh Network	634	335	1090
Shared buffer crossbar	497	252	840

Table 2 Area report of shared buffer crossbar and mesh networks for 8 SMs

Network topology	No. of Slice	No. of FF.	No. of Luts
Mesh Network	211	110	364
Shared buffer crossbar	165	84	283

and diverging the requests so there are less input/output units and less buffers which is also shared so very less area consumes in comparison with mesh network.

5 Conclusion

The work proposes shared buffers crossbar design in which for crossbar there are two switching nodes as crossbar local and crossbar global with shared buffer. Compare to natural crossbar which is fully connected the shared buffer crossbar design gives the flexibility for connection between SMs and the memory unit. Here crossbar local converging the requests coming from SMs and then diverging through crossbar global with sharing the buffers, so packets can share buffer space and maximum utilization of buffers. In shared buffer crossbar architecture reduction of area is 28% and power 32% in compare to mesh network. Mapped Scalability analysis of design in which more SMs can be added in existing design. Shared buffer crossbar architecture is highly scalable, flexible, high-performance optimized network in terms of area and power compare to others network

References

1. Zhao X, Ma S, Wang Z (2019) CD-Xbar: a converge-diverge crossbar network for high-performance GPUs. IEEE Trans Comput 68(9):1283–1296
2. Alhubail L, Bagherzadeh N (2019). Power and performance optimal NoC design for CPU-GPU architecture using formal models. In: 2019 Design, Automation & Test in Europe Conference & Exhibition
3. Garc´ıa1 V, G´omez-Luna J, Grass1 T (2016) Evaluating the effect of last-level cache sharing on integrated GPU-CPU systems with heterogeneous applications. In: IEEE International Symposium on Workload Characterization (IISWC)
4. Milic U, Villa O, Bolotin E, Jaleel A, Ramirez A, Nellans D (2017) Beyond the socket: NUMA-aware GPUs. In: Proceedings of the International Symposium on Microarchitecture (MICRO), pp 123–135

5. Mirhosseini A, Sadrosadati, M, Soltani B (2017) BiNoCHS: bimodal network-on-chip for CPU-GPU heterogeneous systems. In: Eleventh IEEE/ACM International Symposium on Networks-on-Chip (NOCS)
6. Zhao X, Ma S (2016) A heterogeneous low-cost and low-latency ring-chain network for GPGPUs. In: IEEE 34th International Conference on Computer Design (ICCD)
7. Fang J, Leng Z-Y, Liu S-T, Yao Z.-C, Sui X (2015) Exploring heterogeneous noc design space in heterogeneous gpu-cpu architectures. J Comput Sci Technol 30(1):74–83

Evaluating the Performance Indices of the Solar Photovoltaic System for Rural Area

Nimish Kumar and Nitai Pal

Abstract The output of solar photovoltaic (SPV) array mainly depends on solar radiation received on the surface of the array and the temperature of the cell. Both radiation and temperature vary throughout the day. Therefore, performance analysis of the SPV system is very essential for technical and economical point of view. This paper evaluates the performance indices of 1 kW SPV system for the rural area. A simulation model has been developed for evaluating cell temperature, power output, DC energy, AC energy, energy yield (reference yield, array yield and final yield), losses (array capture loss and system loss), performance ratio and capacity factors based on climatological data as per IEC 61724 guidelines. The annual averaged value of performance ratio and capacity factor has been found as 67.5% and 13.8% for Ramnagar block of West Champaran district, Bihar.

Keywords Solar photovoltaic system · Cell temperature · Energy yield · Losses · Performance ratio · Capacity factor

1 Introduction

Renewable energy (RE) has become the key enabler for achieving multiple goals such as to fulfil increased electricity demand, produce cleaner energy, reduce dependency on fossil fuels, improve quality of life, overcome global warming issues, enhance health benefits etc. The RE sources use the Sun's enormous energy directly or indirectly to produce electricity. The solar energy is the direct form of the Sun's energy that can be accessed using solar thermal and solar photovoltaic (SPV) technologies. The SPV systems covert solar photon energy directly into electricity [1–3].

N. Kumar (✉)
Department of Electrical and Electronics Engineering, Bakhtiyarpur College of Engineering, Patna, India
e-mail: nimish.2016dr0098@ee.ism.ac.in

N. Pal
Department of Electrical Engineering, Indian Institute of Technology (ISM), Dhanbad, India
e-mail: nitai@iitism.ac.in

V. Nath and J. K. Mandal (eds.), *Proceeding of Fifth International Conference on Microelectronics, Computing and Communication Systems*, Lecture Notes in Electrical Engineering 748,
https://doi.org/10.1007/978-981-16-0275-7_61

National Solar Mission is launched by the Government of India (GoI) for enhancing the capacity of the SPV to 100 GW by 2022 to promote the SPV systems. As per annual report 2019–20 of Ministry of New and Renewable Energy, GoI, the potential of solar energy in India is possibly of 750 GW including 11.2 GW of Bihar as on 31.12.2019 [4]. Although the solar energy has huge potential only 35.9 GW is installed as on 31.05.2020.

The solar scenario of India reveals that there is a need for the development of SPV technologies to see the country as a solar rich country. For the development of SPV technologies, it is essential to examine the performance of the system for a particular location. The information about the performance of the SPV system enables for good decision in terms of investment and for making better government policies [5].

The major features of the SPV system are photovoltaic array, inverter and tracking technologies that are enhancing their performance at a rapid rate to compete with others. Despite these features, the structured evaluation of the SPV performance is still key features for developing good features. Therefore, evaluating the comprehensive performance indices of the SPV systems to recognise potential location for electricity generation appears as an efficient tool for the development of the system on broad-scale [6].

According to report [7], some regions of West Champaran district which is at a distance of nearly 175 km from the capital of Bihar has not connected to electricity infrastructure. The Ramnagar block of this district has some villages where electricity is a distant dream. The need for electricity can be fulfilled by the use of the SPV system. In present work, the evaluation of performance indices of the SPV system has been done considering Rampur block as a potential site for effective implementation of the system as a sustainable source of electricity.

The rest of the paper is structured in the following sequence. Section 2 deals with the methodology adopted for the purpose. The evaluation results are presented in Sect. 3. In Sect. 4, the outcomes of the present work are concluded. Finally, acknowledgement and references are included in the paper.

2 Methodology

An SPV system consists of array and inverter has been considered for the present study as shown in Fig. 1. The capacity of SPV array is 1 kW and the inverter has an efficiency of 90%. The array is assumed to be mounted at a height of 2 meters from the ground in the horizontal position. The technical specifications of the SPV system have been provided in Table 1. The array produces electricity when sunlight falls on the surface of the array. The amount of electricity produced depends on two important parameters namely solar radiation and temperature. These parameters vary throughout the day/week/month/year. The detailed about solar radiation and cell temperature of the array is explained in [8]. The environmental effects on the SPV performance are detailed in [9]. The variation of electricity produced by the array for different radiation and temperature are plotted in Fig. 2.

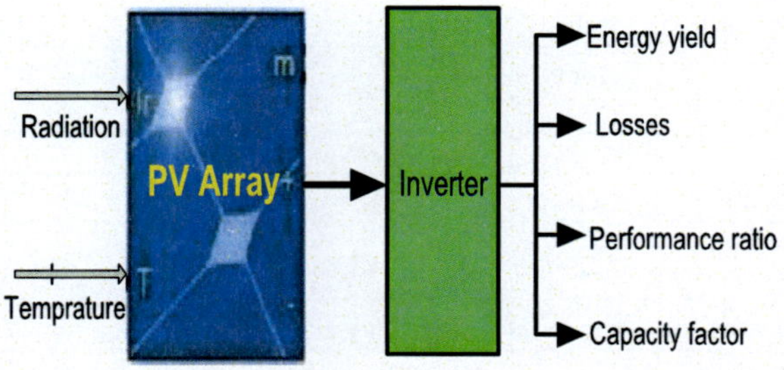

Fig. 1 Block diagram of methodology

Table 1 Specification of the SPV system

Parameters	Symbol	Value
Rated capacity	Y_{SPV}	1 kW
Derating factor	f_{SPV}	80%
Power temprature coefficient	$\propto_T$	−0.5%/ °c
Array efficiency	η_{SPV}	13%
Maximum solar radiation on earth	$G_{T,STC}$	1 kW/m^2
Product of transmittance and absorption	$\tau_{SPV}\alpha_{SPV}$	0.9
Cell temperature at standard test condition (STC)	$T_{C,STC}$	25 °c
Nominal operating cell temprature (NOCT)	$T_{c,NOCT}$	47 °c
Ambient temprature at NOCT	$T_{a,NOCT}$	20 °c
Solar radiation at NOCT	$G_{T,NOCT}$	0.8 kW/m^2
Inverter efficiency	η_{inv}	90%

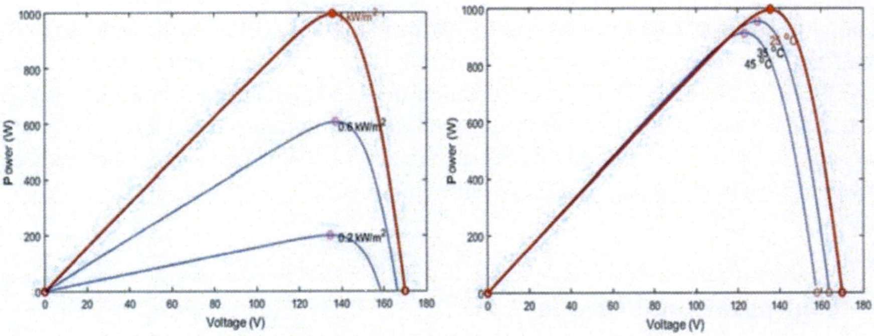

Fig. 2 PV characteristics showing variations relative to radiation and temperatures

Table 2 Modelling formulas for the SPV system

S. no.	Physical quantity	Symbol	Formulae	Unit
1.	Cell temperature	T_C	$T_a + G_T\left(\frac{T_{c,NOCT}-T_{a,NOCT}}{G_{T,NOCT}}\right)\left(1-\frac{\eta_{SPV}}{\tau_{SPV}\alpha_{SPV}}\right)$	°c
2.	Power output	P_{SPV}	$Y_{SPV}\, f_{SPV}\left(\frac{G_T}{G_{T,STC}}\right)[1+\propto_T (T_C - T_{C,STC})]$	kW
3.	DC energy	E_{DC}	$H_{sh} \times P_{SPV}$	kWh
4.	AC energy	E_{AC}	$\eta_{inv} \times H_{sh} \times P_{SPV}$	kWh
5.	Reference yield	Y_R	$\frac{G_T(KWh/m^2)}{1KWh/m^2}$	h/d
6.	Array yield	Y_A	$\frac{E_{DC}}{Y_{SPV}}$	h/d
7.	Final yield	Y_F	$\frac{E_{AC}}{Y_{SPV}}$	h/d
8.	Array capture loss	L_C	$Y_R - Y_A$	h/d
9.	System loss	L_S	$Y_A - Y_F$	h/d
10.	Performance ratio	PR	$\frac{Y_F}{Y_R}$	%
11.	Capacity factor	CF	$\frac{E_{AC}}{Y_{SPV}\times 24}$	%

The cell temperature is found with the help of ambient temperature T_a. The power output of the array can be estimated by effective radiation falling on the array surface which is denoted as G_T. The DC energy generated is formulated by multiplying power output with daylight hours H_{sh}. International Electrotechnical Commission has issued guideline namely IEC 61724 for monitoring the performance of the SPV systems [10]. It defines a detailed blueprint for analysing the productivity of the SPV systems including the evaluation of energy output, energy yield (reference, array and final), losses (array capture and system), performance ratio (PR), capacity factor (CF) etc. for effective implementation of the SPV system. All these performance indices have been tabulated in Table 2 and detailed in [11, 12]. The performance analysis of the SPV systems has been performed in various literature considering different capacity, locations, configuration, approach etc. [11–19]. The performance indices have been evaluated using MATLAB Simulink model.

3 Computational Result

The 30-year Meteorological data of solar radiation, daylight hours and temperature have been accessed from the National Aeronautics and Space Administration (NASA) website for the potential site [20]. These data are presented in Fig. 3. The

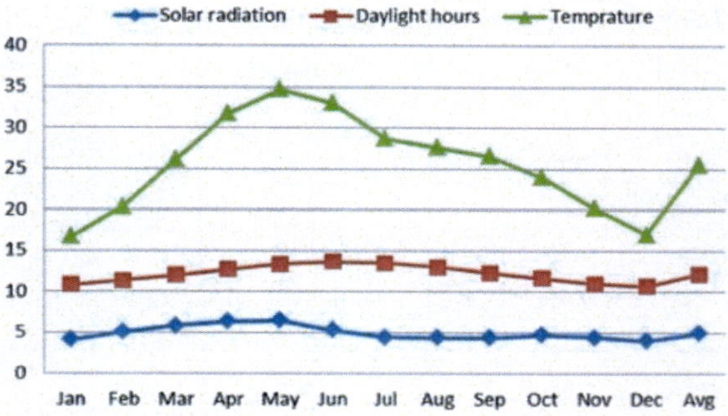

Fig. 3 Input data for the potential site

Table 3 Cell temperature, power output, DC and AC energy of the SPV system

Month	Cell temperature	Power output	DC energy	AC energy
Jan	27.69	0.3005	3.252	2.926
Feb	33.27	0.3413	3.881	3.493
Mar	40.17	0.3592	4.318	3.886
Apr	46.21	0.3576	4.548	4.094
May	48.67	0.3415	4.549	4.094
Jun	44.25	0.2808	3.825	3.442
Jul	38.1	0.2429	3.274	2.947
Aug	37.27	0.2507	3.251	2.926
Sep	36.72	0.2651	3.261	2.935
Oct	35.63	0.3051	3.545	3.191
Nov	31.72	0.3078	3.386	3.048
Dec	27.83	0.2957	3.155	2.839
Avg	37.29	0.3040	3.687	3.318

monthly averaged value of cell temperature, power output, DC and AC energy is shown in Table 3. In Table 4, monthly averaged values of energy yield and losses of the SPV system are displayed. The monthly averaged performance ration and capacity factor are sketched in Fig. 4.

The comparative study of PR and CF have been presented in Table 5 which shows that the values of PR and CF lie within the acceptable range to recommend the SPV system for supplying electricity for the considered site.

Table 4 Energy yield and losses of the system

Month	Reference yield	Array yield	Final yield	Array capture loss	System loss
Jan	4.12	3.252	2.926	0.8684	0.3252
Feb	5.06	3.881	3.493	1.179	0.3881
Mar	5.84	4.318	3.886	1.522	0.4318
Apr	6.36	4.548	4.094	1.812	0.4548
May	6.45	4.549	4.094	1.901	0.4549
Jun	5.29	3.825	3.442	1.465	0.3825
Jul	4.38	3.274	2.947	1.106	0.3274
Aug	4.33	3.251	2.926	1.079	0.3251
Sep	4.33	3.261	2.935	1.069	0.3261
Oct	4.68	3.545	3.191	1.135	0.3545
Nov	4.38	3.386	3.048	0.9937	0.3386
Dec	4	3.155	2.839	0.8454	0.3155
Avg	4.935	3.687	3.318	1.2479	0.3687

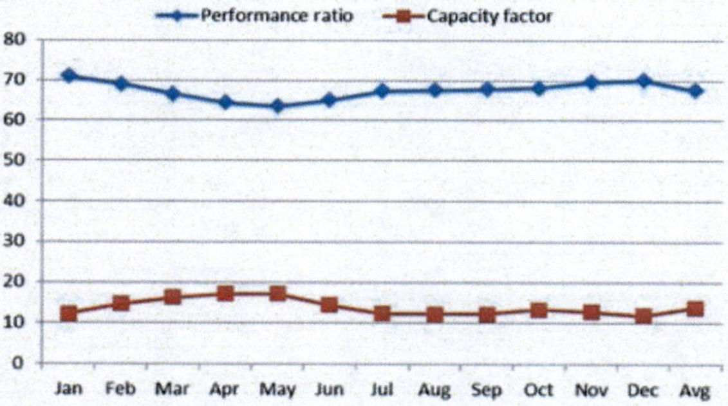

Fig. 4 PR and CF (in percentage) of the SPV system

Table 5 Comparative study of PR and CF

Location	Performance ratio (%)	Capacity factor (%)	Source
Mauritania	63.59–73.56	11.74–20.54	[11]
Lakshadweep	64.22–65.83	15.51–16.09	[12]
Port Blair	68.55–80.22	11.24–19.14	[14]
Ramagundam	86.12	17.68	[16]
Ramnagar	**67.5**	**13.83**	**Present study**

4 Conclusion

The performance indices of the SPV system tell about the suitability of the system of different configurations at different locations. The rural area, Ramnagar block, of West Champaran, Bihar has global horizontal radiation of 4–6.45 kW/m^2/day, daylight hour of 10.67–13.62 h and temperature of 16.7–34.69 °c. The annual averred value of cell temperature, power output, DC and AC energy, reference, array and final yield, performance ratio and capacity factor are examined as 37.29 °c, 0.304 kW, 3.687 kWh, 3.318 kWh, 4.935 h/d, 3.687 h/d, 3.318 h/d, 1.2479 h/d, 0.3687 h/d, 67.5% and 13.83% respectively. These performance indices are in the feasible range and hence the SPV could be a sustainable option of such rural areas where there is no sign of electricity.

Acknowledgement The authors are very grateful to the National Aeronautics and Space Administration, United States for providing access to meteorological data.

References

1. Kumar N, Pal N (2020) The existence of barriers and proposed recommendations for the development of renewable energy in Indian perspective. Environ Dev Sustain 22:2187–2205
2. Kumar CR, Majid MA (2020) Renewable energy for sustainable development in India: current status, future prospects, challenges, employment, and investment opportunities. Energ Sustain Soc 10(2):1–36
3. Das N, Pal N, Pradip SK (2015) Economic cost analysis of LED over HPS flood lights for an efficient exterior lighting design using solar PV. Build Environ 89:380–392
4. Ministry of new and renewable energy Annual report 2019–20. https://mnre.gov.in/img/documents/uploads/file_f-1585710569965.pdf
5. Woyte A, Richter M, Moser D, et al (2013) Monitoring of photovoltaic systems: good practices and systematic analysis. In: Proceedings of 28th European Photovoltaic Solar Energy Conference Paris, France, pp 3686–3694
6. Tekumalla DV, Pal D, Bajpai P (2019) Comprehensive performance evaluation of various solar PV system configurations. IET Renew Power Gen 13:1261–1270
7. https://www.news18.com/news/india/these-bihar-villages-have-electric-poles-but-they-still-depend-on-lanterns-1734217.html#:~:text=Unlike%20Madhopur%20Diara%2C%2022%20villages,solar%20panels%20on%20their%20own
8. Duffie JA, Beckman WA (2013) Solar engineering of thermal processes 4th ed. John Wiley Sons, New York
9. Chaichan MT, Kazem HA (2018) Environmental conditions and its effect on PV performance. In: Generating electricity using photovoltaic solar plants in Iraq. Springer, Cham
10. IEC Standards 61724 (1998). Photovoltaic system performance monitoring-guidelines for measurement, data exchange analysis
11. Sidi CEBE, Ndiaye ML, Bah ME et al (2016) Performance analysis of the first large-scale (15 MWp) grid-connected photovoltaic plant in Mauritania. Energy Conver Manag 119:411–421
12. Bhakta S, Mukherjee V (2016) Solar potential assessment and performance indices analysis of photovoltaic generator for isolated Lakshadweep island of India. Sustain Energy Technol Assess 17:1–10
13. Sharma SK, Palwalia DK, Shrivastava V (2019) Performance analysis of grid-connected 10.6 kW (commercial) solar PV power generation system. Appl. Sol. Energy 55:269–281

14. Bhakta S, Mukherjee V (2017) Performance indices evaluation and techno economic analysis of photovoltaic power plant for the application of isolated India's island. Sustain. Energy Technol Assess 20:9–24
15. Shukla AK, Sudhakar K, Baredar P (2016) Design, simulation and economic analysis of standalone roof top solar PV system in India. Sol Energy 136:437–449
16. Kumar BS, Sudhakar K (2015) Performance evaluation of 10 MW grid connected solar photovoltaic power plant in India. Energy Reports 1:184–192
17. Radhi MH, Mahdi EJ, Mftwol AK (2019) Design and performance analysis of solar PV System Size 2.56 kWp. In: 2019 4th scientific international conference Najaf (SICN), Al-Najef, Iraq, pp 70–73
18. Shah SFA, Khan IA, Khan HA (2019) Performance evaluation of two similar 100 MW solar PV plants located in environmentally homogeneous conditions. IEEE Access 7:161697–161707
19. Simelane SN, Ayanna, MB (2017) CSIR's single axis tracking solar PV plant performance analysis. In: 2017 IEEE PES PowerAfrica, Accra, pp 479–482
20. https://power.larc.nasa.gov/data-access-viewer

Pulse Start-up Technique Based Class C VCO Using 180 nm CMOS Technology

Dipika Simariya, R. C. Gurjar, and D. K. Mishra

Abstract This paper presents Class C VCO in which addition of kick pulse provides a way to start the oscillations without affecting output swing. In this VCO, the addition of transformer and varactor bank is an additional advantage that controls the bias current as well as self-start process with the help of kick pulse. The proposed Class C VCO oscillates at 3.85 GHz frequency with a tuning range of 900 MHz to 2.5 GHz. The design is provided by a 1.8 V supply with a difference of 0.1 V so that kick pulse is given to the oscillator. This kick pulse removes the need of giving initial condition by the Virtual tool. The design consumes 9.125 mW power which helps in achieving a good tuning range. The resulted phase noise is about −98.375 dBc/Hz at 1 MHz offset frequency and −128.33 dBc/Hz @ 3 MHz offset frequency. This design is implemented using Cadence Virtuoso Tool at 180 nm technology.

Keywords Class C oscillator · Initial conditions · Kick startup · Spectral purity · Varactor diode

1 Introduction

Voltage Controlled Oscillators are very convenient in Wireless modules and many communication systems for tuning a band of frequencies. With the rapid growth of microelectronics in wide areas of driverless automation with the use of microprocessors and clock recovery systems, the design of VCO at deep submicron is still a challenge [1, 2]. Stubby phase noise, vast frequency tuning range, adequate spectral purity, and low power consumption, are prime requirements in the efficient design of VCO. It is possible to design VCO on both CMOS and bipolar processes but CMOS

D. Simariya (✉) · R. C. Gurjar · D. K. Mishra
Department of Electronics and Instrumentation Engineering, SGSITS, Indore, MP, India

R. C. Gurjar
e-mail: rcgurjar94@gmail.com

D. K. Mishra
e-mail: dkmishra@sgsits.ac.in

V. Nath and J. K. Mandal (eds.), *Proceeding of Fifth International Conference on Microelectronics, Computing and Communication Systems*, Lecture Notes in Electrical Engineering 748,
https://doi.org/10.1007/978-981-16-0275-7_62

is preferred due to its less area coverage and low fabrication cost. The main work of VCO is to control oscillation frequency with the help of voltage input. Usually, LC VCO is preferred over ring oscillator and relaxation oscillators because of its large tuning range and high spectral purity.

The basic tuning parameters of LC oscillator are capacitance (varactor), current, and power supply. The main startup procedure of LC VCO is to apply initial conditions at the capacitor. For the start of oscillation, Capacitor needs initial voltage so that it tunes with the inductor and sustained oscillations resulted. This initial condition plays an important role in the self-start of oscillations. Usually, this is given to the oscillator by the virtual tool or need to connect an extra pin at the designed chip.

From the EDA tool design technique, it is possible to give an initial condition to the capacitor by an introduction of kick pulse. This kick pulse is generated by a voltage difference at the input supply. This will not affect voltage swing much more.

As the Varactor diode is introduced in conjunction with voltage variable capacitance that appears due to reverse biased PN junction in the early 1960s greatly helps in the development of miniature, high quality, low power, and low-cost structure of VCOs with advanced electronic tuning and highly configurable frequency [3]. Although, it is possible to design varactors, inductors, and capacitors in the form of the module due to reduction in sizes of varactors but, elimination of the extra need of pin connection for initial condition to capacitor help to design it in the more compact form [3]. By this technique, the miniature module only needs connections to supply voltage, ground, tuning voltage, and output load. Here, Class C VCO is designed in which Bias Control Circuit and the use of varactors instead of capacitor gives an extra advantage.

2 Conventional Class C Oscillator Design

A Traditional design of Class C VCO is shown in the Fig. 1. Because of nonlinear operation of M1 and M2, and increase in oscillation amplitude, average current increases highly [1, 2]. Therefore, C_{tail} is charged and V_{tail} increase to a voltage in which DC currents are balanced. The current mirror ensures robust startup while in steady state, the bias voltage drops and maximizes the output swing [1, 2].

The hybrid structure of Class C VCO is based on single pair configuration which improves the tradeoff of using a Class B pair in parallel to Class C core which ensures robust startup even with relatively low gate bias voltage [20]. If it is needed to maintain the maximum possible quality factor of the primary inductor, the capacitance of the tank is introduced in the form of MOS and varactors [4]. This design minimizes power consumption by the circuit since traditional capacitors consume a lot of power. As the bias current is increased, their sustained oscillation amplitude increases gradually. The output phase noise can be degraded when transistors enter in triode region [4]. The schematic of the Conventional Class C oscillator is shown in Fig. 1. The variable capacitors used in the design consume a high amount of power. The transistor pair provide negative resistance to the oscillator which provides noise parameters for

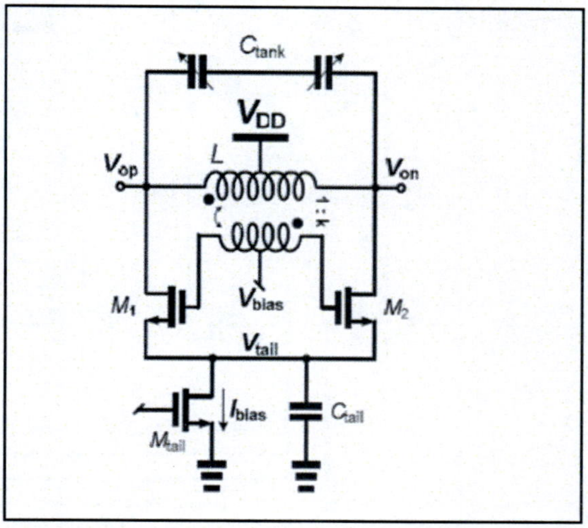

Fig. 1 Schematic of class C oscillator

startup of the oscillator. Because of these elements, VCO starts but it may take a longer time so that capacitor tunes to the inductor and a band of frequency is formed with some distortions which affect spectral purity of the transients.

3 High Swing Class C Oscillator

The high voltage technique to the output swing is provided by the tail current source. But in this design, it is removed so that outputs fed back to the switching transistors with the help of a transformer [5]. The output DC voltage allows higher output swing, a time before the transistors enter in the linear region. This Oscillator has an additional circuit to operate biasing of the control circuit so that full oscillations have been obtained. This circuit controls current consumption by the oscillator core and guarantees startup at the beginning of the oscillation. This topology greatly helps to design a circuit with a controlled current bias.

When oscillations were not started, the secondary transformer is short-circuited and $M_{3,4}$ are connected to the diode. Since $V_{g1,2}$ are equivalent to $V_{ctl,}$ I_{ref} is mirrored into M_1 and $M_{2.}$ If the current obeys oscillation condition then, it begins to oscillate [6]. On increasing amplitude of oscillations, the average current of the transistors also increases [6]. By the selection of proper sizes of the transistors, V_{ctrl} would be set up at small considerable Voltage at steady state that results in enhanced swing [6] (Fig. 2).

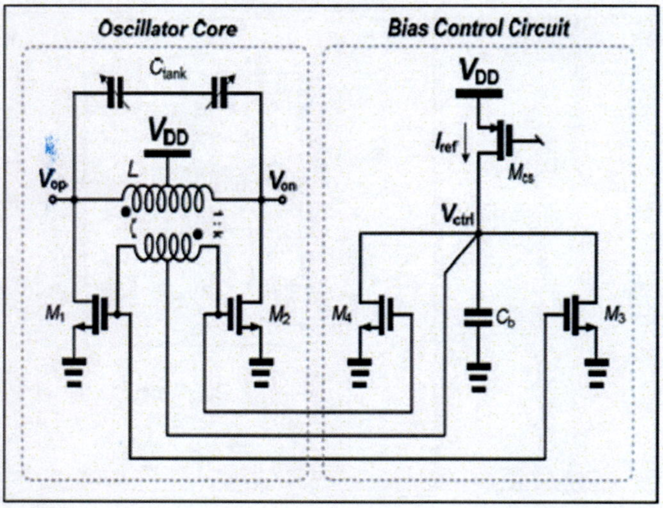

Fig. 2 HSCC Class C VCO

4 Proposed Design

The design consists of a varactor diode which works in reverse bias and stores energy in the form of charge. It is an active device that act like voltage dependent source. Therefore, it needs an initial voltage that act as a fire gun so that it provides self-start to VCO. Varactor always works in the reverse region, so, it is used in the design in place of the variable capacitor.

The swing at the output of the VCO core can be as high as $V_{dd} + V_p$ where V_p is a voltage that provides a short duration pulse to the capacitor. With this mechanism, the tuning range is increased and phase noise improved (Fig. 3).

4.1 Control of Frequency

A voltage dependent capacitance can be obtained by any reverse bias semiconductor diode which is used to change the frequency of the oscillator by varying input voltage. A varactor topology greatly helps to reduce harmonics in the oscillations and helps to produce sustained waveform which improves phase noise and tuning range.

$$Fractional\ turning\ range = (f_{max} - f_{min}) / fo \quad (1)$$

Where,

$f_o =$ *Center Oscillator Frequency* $= (fmax + fmin)/2$

$f_{max} \approx 1/(2\pi \sqrt{L_{tank} C_{min}})$

$f_{min} \approx 1/(2\pi \sqrt{L_{tank} C_{max}})$
$C_{max}/C_{min} = f^2_{max}/f^2_{min}$

At present, for CMOS processes, V_{gs} is selected such that $(V_{gs} - V_t)$ ranges from 0.4 to 0.5, where, V_t is the threshold voltage. If this value becomes too high then there is degradation in g_m due to saturation velocity effects. If the value of $(V_{gs} - V_t)$ is too small, then the width of MOS becomes very large so that the transistor will not be operated in the desirable area.

This g_m [8] is used to select suitable startup condition for VCO using (2)

$$g_m = (1/r_{ds}) + R[(C + C_{gs} + 4C_{gd0})/L] \quad (2)$$

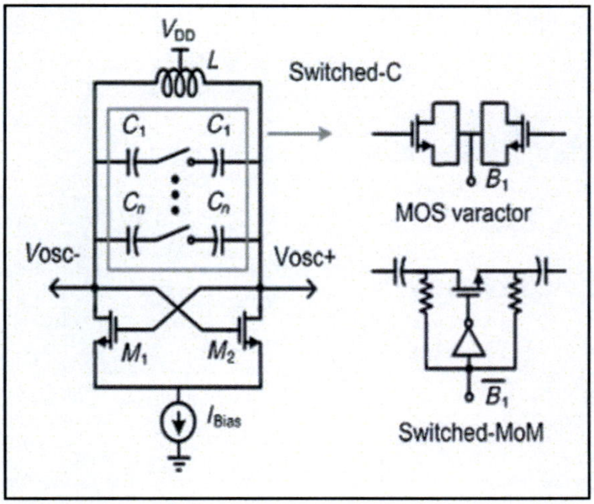

Fig. 3 Design of varactor diode

(a)

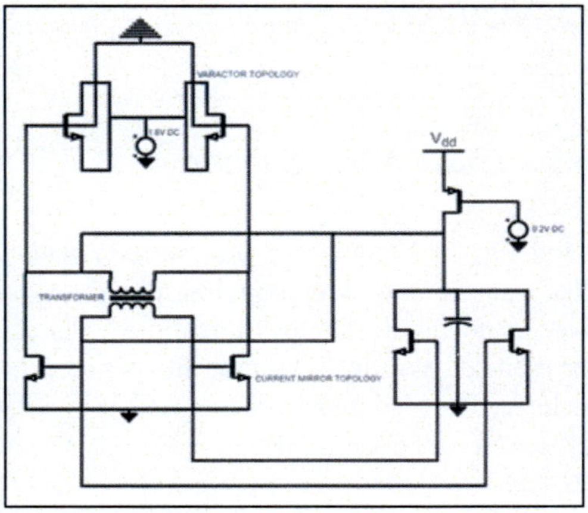

Fig. 4 (a) Proposed schematic design

Fig. 4 (b) Transient waveform

(b)

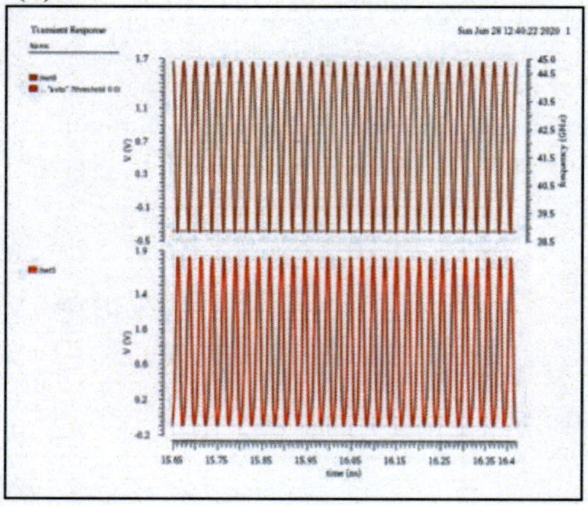

Fig. 4 (c) Layout design

(c)

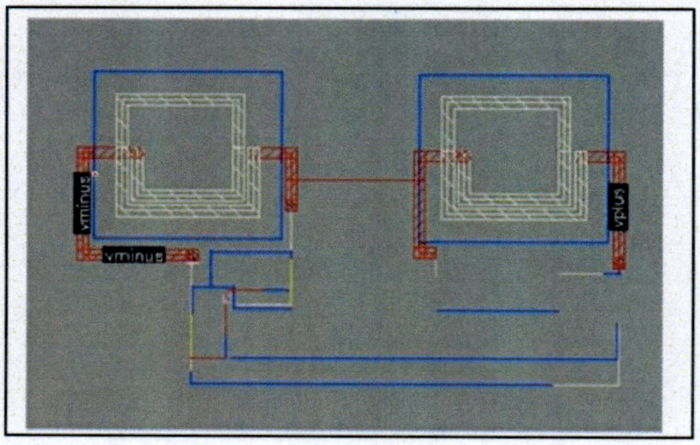

This equation defines gain margin for the oscillator in which r_{ds} is drain-source resistance and C are capacitors from ground-source and ground-drain.

4.2 Phase Noise Equations

Modeling of VCO is not done in the context of amplitude or shape of the wave-like a sinusoidal wave, triangular wave, saw tooth wave, but, phase modeling in the instantaneous form helps to read the VCO carefully. This instantaneous phase can be modeled as a linear relationship between frequency of oscillation and controlled voltage. The output phase is modeled as integral of this instantaneous frequency. The proposed equations is given by [8]

$$f(t) = f_0 + K_0.V_{in} \tag{3}$$

$$\phi(t) = \int f(t)\,dt \tag{4}$$

where,
V_{in} = *input supply voltage*
$f(t)$ = *output frequency*
$\varphi(t)$ = *phase function*

4.3 Circuit Design with Self Bias Block and Basic Startup Techniques

Basically, all time frame simulations initiate at timeslot t = 0. Initial levels or DC offsets of any source define the initial DC conditions of a circuit [12]. The initial state for the oscillators like LC VCOs which are based on tuning is defined by their DC bias conditions. So, for the oscillator to overcome its equilibrium or steady state, the presence of noise sources such as resistors, are necessary [12]. This is the reason for the design of negative resistance with the help of MOS so that the chip area can be utilized efficiently.

In many cases, this type of oscillator eventually startup due to shrouded noise which is generally due to algebraic noise generated due to defined resolution and misestimation during simulation. But this can take a longer time to run an oscillator.

To decrease the simulation time for the startup of the oscillator, an initial kick-start up condition has been introduced.

As the design of VCO depends on the design parameter defined for the capacitor. The spectral purity can be achieved by the use of a varactor diode. The capacitor charge is a function of voltage which can be defined by the Eq. (1) [12] helps to provide a kick pulse through supply. These equations define the selection of initial voltage for the capacitor

$$Q = C\, dV_{in}/dt \tag{5}$$

$$V = V_i\left(1 - exp\left(Tr/100u\right)\right) \tag{6}$$

To kick start the oscillations, the simple DC supply is replaced by a source pulsating at an initial level of desirable supply voltage and also, arranged in a way to produce a short duration pulse of the supply plus or minus with some small voltage.

This Class C oscillator with 1.8 V supply run for 500 ns consists of a PULSE source set to an initial level of 1.8 V pulsed down to 1.7 V with 1 ps rise and fall times. The Eq. (2) in which V_i (without offset supply voltage) helps to determine rise and fall times concerning supply voltage so that oscillations begin a little bit earlier than a normal startup with noise elements. The circuit can be designed in a more compact form by reducing multipliers of capacitors so that its value gets reduced and more fine transients can be achieved.

4.4 Simulation Results

The pulse startup based VCO is designed on gpdk CMOS process at 180 nm technology. The experimental results are shown in Fig. 5(a) and 5(b).

Fig. 5 (a) Spectrum of PSS

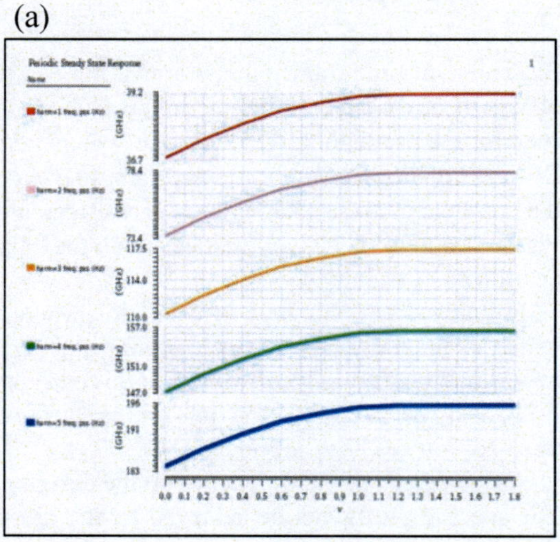

Fig. 5 (b) Spectrum of phase noise

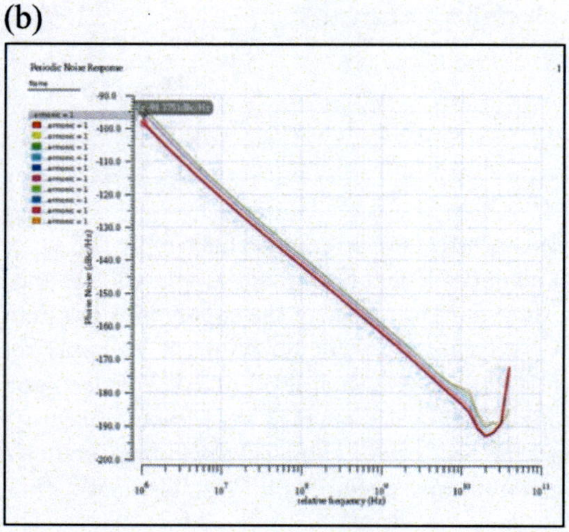

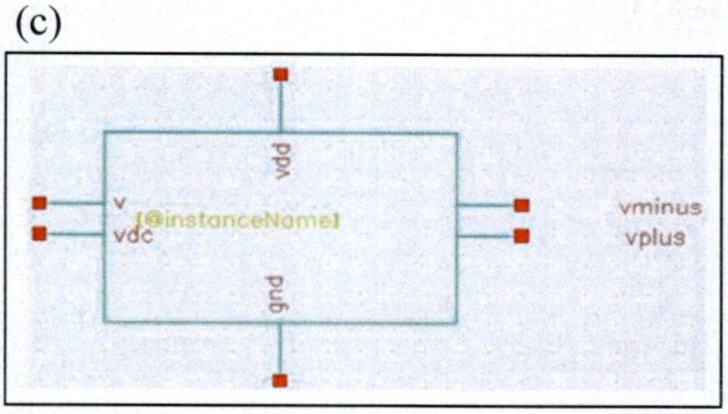

Fig. 5 (c) Generated Symbol of the design

The proposed VCO operates @ 1.8 V with a difference of 0.1 V so that a sharp kick pulse is generated which starts the capacitor so that it tunes with the inductor and oscillates at 3.85 GHz frequency with a good tuning range of 900 MHz to 2.5 GHz.

This tuning range is desirable because Class C is known for an appreciable tuning between L and C. The spectrum of output phase noise is shown in Fig. 5(b) with PSS Spectrum in Fig. 5(a). The layout of this design is shown in Fig. 4(c) with no DRC error. After proper LVS, the symbol of this design is created which is shown in Fig. 5(c). Therefore, in post layout simulation, the extra pin for application of the initial condition is eliminated.

Different parameters of the oscillator are compared in Table 1. There are excellent results of phase noise in [4] because of noise filtering of tail current and high supply voltage [1]. In general, good noise performance can be achieved with high supply voltage which can be seen in the results. As the supply voltage is reduced then noise performance of 83.32 dBc/Hz at 1 MHz is resulted in the proposed design.

These results are helpful in the design of RF filters [13] because in general, there are deferred stages that are designated as lossy integrators that exist within filter terminology. The lossy integrators usually have small signal transfer function which is given by

$$V_0/V_i = I_0/(s + p) \tag{7}$$

where,

I_0 is unity gain frequency and pole p is loss term.

On the basis of this idea, low pass filter modeled from VCO can be designed by moving pole constellation in a considerable amount towards left and by addition of exterior input to particular lossy integrators [21].

Table 1 Performance comparison of different parameters of VCO

Reference	Freq (GHz)	V_{DD} (V)	P_{DISS} (mW)	PNoise (dBc/Hz)	Freq. Offset (MHz)	Technology (um)
[1]	5.1	0.6	0.86	−127	3	90
[9]	1.2	2.5	9.25	−153	3	0.35
[13]	5.2	1	1.4	−131.5	3	0.13
[14]	1.562	0.6	1.56	−122	1	90
This work	3.85	1.8	8.13	−128.3	3	0.18

5 Conclusion

The introduction of kick pulse at the supply source eliminates the need for convergence aids (a tool used for application of initial conditions in Cadence) for the startup of VCO by less affecting the output swing of the designed oscillator. The area is also reduced by the use of transistor based active capacitors. Power is also minimized by the removal of variable capacitors. This technique can be applied to different types of VCOs. The basic VCO gives excellent performance by applying short duration pulse through PULSE supply instead of a simple DC supply. Overall, it can achieve good noise performance with the lesser area and low power consumption.

References

1. Tohidian M (2011) High swing class C VCO. In: Proceedings of IEEE Applied Electromagnetic Systems, University of Tehran, pp 495–498
2. Chen Z, Wang W, Liao Y, Ye J, Zheng W, Liu R, Chi B (2008) A Low phase noise and high linearity class-C VCO for FMCW radar system. In: Institute of Microelectronics, Tsinghua University, Beijing, China, 100084, China Academy of Information and Communications Technology, MIIT, Beijing, China, p 100804
3. Hung C, Kenneth KO (2000) A packaged 1.1 GHz CMOS VCO with phase noise of −126 dBc/Hz at a 600 kHz offset. IEEE J Solid-State Circ 35:100–103
4. Fanori L, Andreani P (2013) A high-swing complementary class-C VCO" Department of electrical and information technology, Lund University, Sweden, IEEE Proceeding. 978-1-4799-0645-1/13
5. Leeson DB (1966) A simple model of feedback oscillator noise spectrum In: Proceedings IEEE, pp 329–330
6. Mazzanti A, Andreani P (2008) Class-C harmonic CMOS VCOs, with a general result on phase noise. IEEE J Solid-State Circ 43(12):2716–2729
7. Andreani P, Wang X, Vandi L, Fard A (2005) A study of phase noise in colpitts and LC-tank CMOS oscillators. IEEE J Solid-State Circ 40(5):1107–1118
8. Razavi B (1997) A 1.8 GHz CMOS voltage-controlled oscillator In: ISSCC digest of technical papers, pp 390–391
9. Hegazi E, Sjöland H, Abidi AA (2001) A filtering technique to lower LC oscillator phase noise. IEEE J Solid-State Circ 36(12):1921–1930
10. Kinget P (1999) Integrated gigahertz voltage controlled oscillators In: Analog circuit design, Kluwer, Norwell, MA, pp 353–381

11. Cheung TSD, Long JR (2006) Shielded passive devices for silicon-based monolithic microwave and millimeter-wave integrated circuits. IEEE J Solid-State Circ 41(5):1183–1200
12. Lee TH (2004) The design of CMOS radio-frequency integrated circuits, 2nd edn. Cambridge Univ. Press, Cambridge, UK
13. Kwok K, Luong HC (2005) Ultra-Low-voltage high-performance CMOS VCOs using transformer feedback. IEEE J Solid-State Circ 40(3):652–660
14. Aspemyr L, Linten D (2006) An ultra-low voltage, low power, fully integrated VCO for GPS in 90 nm RF-CMOS. Analog Integr Circ Sign Process 40:57–63
15. Hsieh HH, LH Lu (2007) A high-performance CMOS voltage-controlled oscillator for ultra-low-voltage operations. IEEE Trans Microw Theory Tech 55(3):467–473
16. Piper SO, Alex-andria VA (1995) Homodyne FMCW radar range resolution effects with sinusoidal nonlinearities in the frequency sweep. In: Proceedings international radar conference, pp 563–567
17. Ayhan S, Scherr S, Bhutani A, Fischbach B, Pauli M, Zwick T (2016) Impact of frequency ramp nonlinearity, phase noise, and SNR on FMCW radar accuracy. IEEE Trans Microw Theory Tech 64(10):3290–3301
18. Dhakad NS, Gurjar RC (2017) A low phase noise VCO using injection locking. In: IEEE, international conference on information communication, instrumentation and control (ICICIC)
19. Tiwari AA, Mishra DK (2017) A DC-DC boost converter with soft start function and DPEA technique using 180 nm technology. In: IEEE, international conference on information communication, instrumentation and control (ICICIC)
20. Fanori L, Andreani P (2012) Low-phase-noise 3.4–4.5 GHz dynamic-bias class-C CMOS VCOs with a FoM of 191 dBc/Hz. In: Proceeding of IEEE, ESSCIRC, Lund University, Sweden, pp 406-409
21. Geiger RL (2000) High frequency VCO-derived filters. In: Proceeding IEEE, midwest symposium on circuits and systems (Cat No CH37144) MWSCAS-00, pp 208–211

A Message Forwarding Scheme with Some Incentive and Minimum Number of Hops in Delay Tolerant Networks

Nidhi Sonkar, Sudhakar Pandey, and Sanjay Kumar

Abstract Delay tolerant networks got more attraction due to its property of enabling communication where no end to end connectivity exist between nodes, by storing data to the storage till the connectivity does not exist. Now a days we can not survive without mobile devices in our life for sharing data and we cannot compromise with the forwarding data. İf connection not exist between nodes there will be no communication between them. DTNs enable communication in this type of networks. Traditional protocols for routing can not be applied in DTNs, many protocols have been proposed for DTN but performance is not good in terms of delivery ratio with minimum delay. In this study, we proposed a message forwarding scheme in DTNs to improve the message delivery ratio and minimum hop count with minimum delay by selecting node that have higher encounters and farthest from source with giving incentive to the forwarder node. We simulate the environment in ONE simulator and compare the algorithm with existing routing protocols and find that proposed algorithm improves the performance in terms of delivery ratio and latency.

Keywords Delay Tolerant Networks · Routing · Incentive · Delivery probability · Number of hops · Delay

1 Introduction

Delay Tolerant Networks, a type of network that enable communicaton where communication issues like intermittent connectivity, high error rate or even no end to end communication exist. DTNs enable communication by using store carry forwaed method that is, if connectivity does not exist or connection lost due to any natural disasters then node stores data in the storage device till the connectivity does not exist [1]. Now a days mobile devices are the most important for people for data sharing. The data will be of any format like audio, video, image etc. Delay or losing of data is intolerable for people. So Delay Tolerant Networks protocols are preferable in that

N. Sonkar (✉) · S. Pandey · S. Kumar
Department of Information Technology, National Institute of Technology Raipur, Raipur, India

V. Nath and J. K. Mandal (eds.), *Proceeding of Fifth International Conference on Microelectronics, Computing and Communication Systems*, Lecture Notes in Electrical Engineering 748,
https://doi.org/10.1007/978-981-16-0275-7_63

type of network to satisfy peolpe by sending data with minimum delay and no loss of data [2]. There are many aplications of DTNs in the area of Wildlife tracking [3], Vehicular Networks [4], Mritime Networks [5], Space communication [6], rural area network communication [7] etc.

Routing is important issue is Delay Tolerant Networks, since it has intermittent or no connectivity exist. Many researchers have proposed routing scheme but their performane in terms of message delivery ratio with minimum number of hops is still missing in them [8]. Therefore in this paper, we proposed the routing scheme by giving some incentive to the node that forward message to other node for getting the interest on the node to take part in the transmission so that delivery ratio can be increased. For getting minimum latency we have to try minimize number of hops to transmit the messages. We use number of encounter and distance between nodes to select node for data forwarding to minimize number of hops. We select the node that have more number of encounters in history and farthest from the sender node. So that we can transmit data with minimum latency and by best suitable. In this work, we combine both scheme to create a new forwarding scheme. We select node which have more number of encounters and farthest from node, when receiver node forward data to other node we give some incentive to the node so that we can create interest to the node for sending data. In this way, the performance of Delay Tolerant Networks can be improve.

The organization of paper is in the way that, Sect. 2 describes the study existing works, proposed methodology and description about that is explained in Sect. 3. The simulation and discussion of result is assembled in Sect. 4. The paper is concluded in Sect. 5 by conclusions and future work.

2 Related Works

In Delay Tolerant Networks, many researchers are proposing different routing algorithm. Some are single copy based routing, some are multiple copy based routing and social based routing that describes that all node behaves like human [9]. Here we only discuss main algorithms that are used in real environment which are as follows:

Epidemic [10] routing scheme is muli copy routing algorithm that simply forward to all the neighbour node blindly. So the performance of Epidemic is poor in DTNs. Spray and Wait [11] routing uses the multiple copy scheme and forward messages to the node of L numbers and wait for the delivery if message delivered we can save remaining 1-L number of nodes. PROPHET [12], Probabilistic rouoting protocols using history of encounters and transitivity routing protocol set the probility of delivery based on encounter history of a node. MaxProp [13], routing protocol detects the packet when buffer space is very low to deliver the message in minimum number of hops. Minvisit [14] algorithm minimize the number of hops by using the node farthest from the source. In Social based DTN routing Schemes many recent researches have focused on social based routing algorithms (mainly a human with mobile phones) to get the effective routing algorithm in Delay Tolerant Networks.

Authors in [15], used both similarity and betweenness centrality to form a new routing algorithm which have a good performance. In [16], authors proposed Bubble Rap a routing scheme in which every node has two ranking global and local. This algorithm got a high attention due to high delivery ratio. Epidemic with community based routing is proposed in [17]. The effect of socially selfish routing algorithm is presented in [18]. MobiCent [19] routing protocol is the credit based routing that find the optimistic path to forward data to the destination. İncentive based routing is proposed in [20] that uses the node that is in minimum distance from the source node and give some incentive to the node to increase delivery ratio.

All the algorithms are assume that uses are interested and cooperative to forward the data to other node and dont refuse for transmission. But mobile nodes behaves like human they find the profit to do some work. Therefore we usen the concept of reward in this work. Existing routing scheme does not preform good in terms of delay and message delivery ratio. Therefore, in this article we propose the routing algorithm that uses the concept of number of encounters in history and farthest neighbour node to choose node for data forwarding with giving then some incentive so that we can improve the performance of DTNs by improving message delivery ratio, minimum delay with minimum number of hops.

3 Proposed Methodology

In Delay Tolerant Networks, to enable communication in intermittent connection Store-carry-forward scheme is used. Node stores the message till the connectivity does not exist. So routing and storage management is a big issue is DTNs. In this article we only propose the routing scheme. İn extended version of this paper we are working on the buffer management with routing scheme. Therefore, we can improve the performance of DTNs by improvement in message delivery ratio, delay and buffer utilization also.

Figure 1 shows the model for proposed routing scheme, when source node arises in the range of any node they exchange information of each other. The information like contact history, transmission time, encounter time etc. Proposed methodology chooses the node that is farthest neighbor of the source node and had highest number of encounters in history. İf source node find that node it forward message to that node and give some incentive to the node by increasing reward point by one. We assume that all node are selfish node if they gain something that is profitable for them then only they forward data to other node. Therefore, we are adding the incentive method in our proposed scheme so that node get some interest to forward data. When the forwarding node fond the destination it stops unless forward the data to another node.

The steps to perform routing scheme is as follows.

A. *Node Selection*

In proposed routing scheme, we select node that have the highest number of encounters in history of encoounters and node is farthest among all neighbour node.

Fig. 1 Model for proposed routing technique

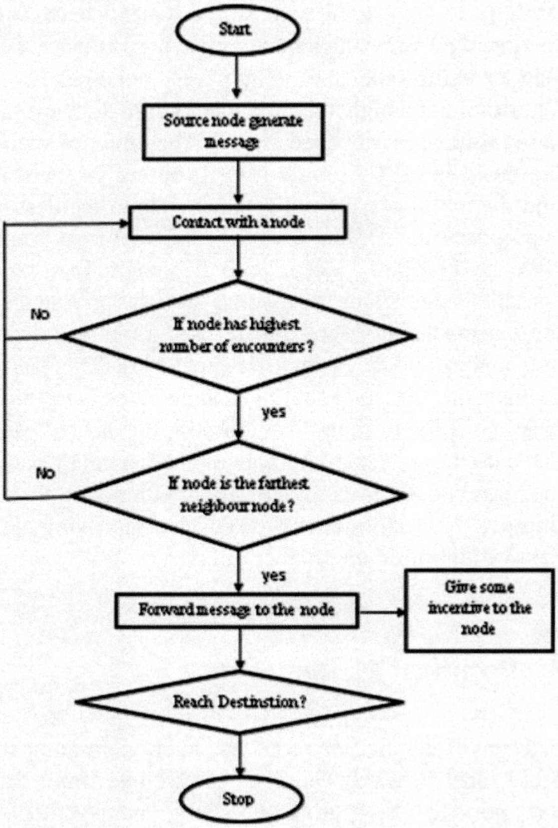

For calculating distant between node we used Ecludien distance theorem. According to Ecludien the distance between point with (x1, y1) and (x2, y2) cordinates is given by the formulae:

$$d = \sqrt{(x2 - x1)^2 + (y2 - y1)^2} \quad (1)$$

That means we can calculate the deistance between two nodes by their coordinates. The square root of sum of distance between two nodes are refered as the destance between two nodes.

B. *Inncentive Calculation*

Here we consider all nodes are selfish node like human behaviour. So we have to give them some reward that why they encourage to forward the data. The incentive can be calculated as one reward per hop. İf a node forward data to one node their reward is increased by one. So that, node can be get some interest to forward data to other nodes.

C. *Forwarding Strategy*

In proposed routing scheme we are choosing node that has the higher number of encounters in their history of transaction and farthest node among all the neighbour so that we can send the data to the destination with higher delivey ratio and minimum number of nodes. Here we suppose that, all the nodes are selfish node like human behaviour. Therefore we have to send them some reward so they encourage to forward the messages to other nodes. The algorithm for proposed routing scheme is as follows:

Algorithm for proposed methodology

[Input: History of all nodes, Encounter Time, Destination node.]
[Output: Data transfered to the destination.]

1. Get information about nodes that comes in the range of contact.
2. Find the distanve between nodes by equation (1).
3. Find the number of encounters by their contact history.
4. if node is farthest among all neighbours
5. if node had higher number of encounters in contact history
6. transfer data to that node
7. else
8. go to 1.
9. Deliver data to the destination.
10. End.

4 Simulations and Results

We simulate the surroundings in ONE simulator [21], that is actually created for simulation for Delay Tolerant Networks and its applications. Table 1 shows the parameter we have used to simulate the environments.

In Fig. 2, we can see that the delivery ratio has been increased when we increase the number of nodes. The delivery ratio increases as compare to spray and wait because we choose the node that has higher number of encounters compare to all neighbor nodes. The opportunity to forward message increases when number of nodes increases. We used the concept of incentive when node transferred the message to other node we send the reward so that they can get interest to send the messages therefore we can increase the delivery ratio by giving incentive to the nodes.

In Fig. 3, we can see that when number of hops is decreases as compare to spray and wait routing algorithm because we select the node farthest from the source node

Table 1 Simulation parameters

Parameter	Value
Message size	100 b
Buffer size	50 MB
Transmit range	100 m
Message TTL	300 min
Data rate	2048 bytes per second
Simulation time	5000 s
Number of nodes	20–100

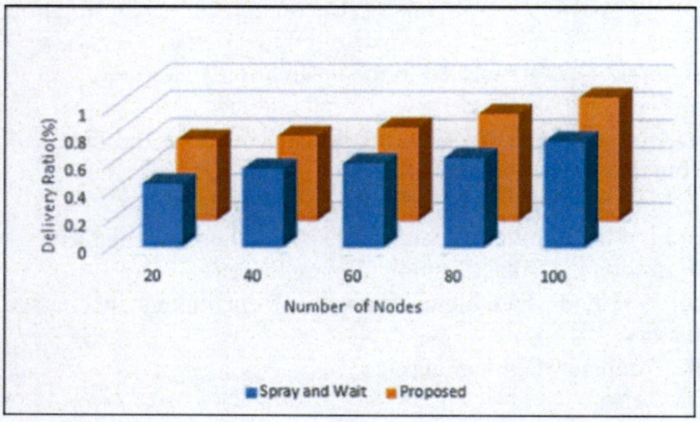

Fig. 2 Number of nodes vs delivery ratio

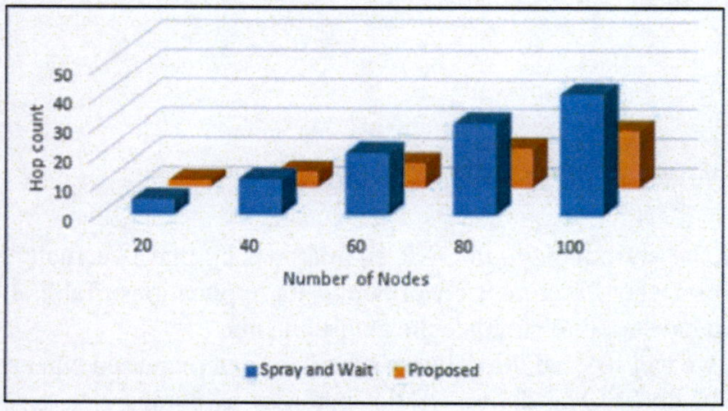

Fig. 3 Number of nodes vs number of hops

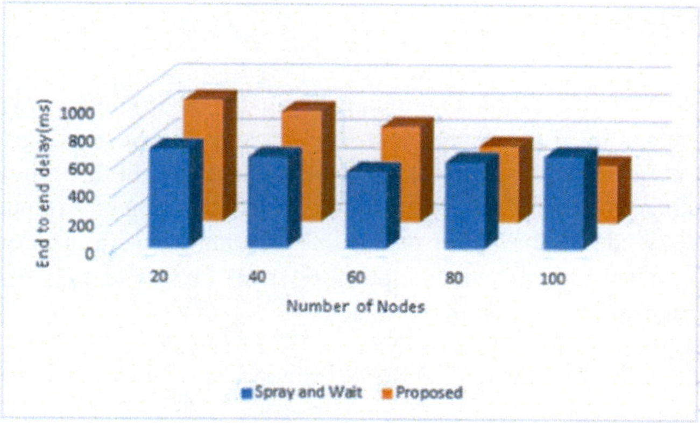

Fig. 4 Number of nodes vs end to end delay

and is neighbor of source node so that they can transfer the message in the minimum number of hops.

In Fig. 4, we can see that end to end delay has been decreased as compare to spray and wait because we choose the node that is farthest from the source so they can transfer message quicker than others. And we give rewards to the node after they transmit data to other node so that for getting reward they transfer data with minimum delay.

5 Conclusion

In this article, we proposed a message forwarding scheme that choose the node that has a higher number of encounters in contact history for increasing the delivery ratio and node that is farthest among all neighbors so that in minimum number of hops we can transfer the data. Our assumption is all nodes are selfish node like human behavior so that we have to give them some incentive to encourage node for forwarding message so all the nodes can participate in message forwarding. Therefore we can decrease the delivery delay. We used ONE simulator to simulate the environment and compare the results with existing routing algorithm and find that proposed methodology can transfer the data with less number of hops in minimum delay with higher delivery ratio. So that we can say that proposed routing algorithm increase the performance of Delay Tolerant Networks.

6 Future Work

Delay Tolerant Networks enable communication in the environment where connectivity issues arises by using Store-carry-forward strategy. So routing and buffer management is important issue in intermittent connectivity. In this article we only proposed routing scheme for data forwarding. We are working on extended version of this article by adding the buffer management technique with proposed routing scheme so that our work will be more optimistic in all the way to use in the physical real world scenario.

References

1. Warthman F, et al 2012 Delay-and disruption-tolerant networks (DTNs), a tutorial V 0, interplanet. internet spec. interes. Gr 5–9. https://www.bit-netlab.org/songtian/courses/docs/dtn_tutorial_v2.03.pdf
2. Benhamida FZ, Bouabdellah A, Challal Y (2017) Using delay tolerant network for the internet of things: opportunities and challenges, pp 252–257. https://doi.org/10.1109/IACS.2017.7921980.
3. Kaviani M, Kusy B, Jurdak R, Bergmann N, Liu V (2016) Energy-aware forwarding strategies for delay tolerant network routing protocols. J Sens Actuator Netw 5(4):18. https://doi.org/10.3390/jsan5040018
4. Soares VNGJ, Farahmand F, Rodrigues JJPC (2009) A layered architecture for vehicular delay-tolerant networks. In: Proceeding - IEEE symposium computer communication, pp 122–127. https://doi.org/10.1109/ISCC.2009.5202332.
5. Geng X (2017) Lanepost : lane-based optimal routing protocol for delay-tolerant maritime networks
6. Sun W, Liu C, Wang D, (2011) On delay-tolerant networking and its application, delay-tolerant Netw Appl Wei 51:238–244. https://doi.org/10.7763/IPCSIT.2012.V51.42.
7. Zhang M, Wolff RS (2010) A border node based routing protocol for partially connected vehicular Ad Hoc networks. J Commun 5(2):130–143. https://doi.org/10.4304/jcm.5.2.130-143
8. Puri P, Singh MP (2013) A survey paper on routing in delay-tolerant networks. In: Proceeding 2013 International conference information system computer networks ISCON, pp 215–220, 2013. https://doi.org/10.1109/ICISCON.2013.6524206
9. Ali S, Qadir J, Baig A (2010) Routing protocols in delay tolerant networks - a survey. In: 2010 6th international. conference emerging technologies ICET, pp 70–75. https://doi.org/10.1109/ICET.2010.5638377
10. Vahdat A, Becker D (2000) Epidemic routing for partially-connected Ad Hoc networks. technical report CS-200006. department of computer science, Duke university. Recuperado de: https://issg.cs.duke.edu/epidemic/epidemic.pdf
11. Spyropoulos T, Psounis K, Raghavendra CS (2005) Spray and wait: an efficient routing scheme for intermittently connected mobile networks. In: ACM SIGCOMM work. delay-tolerant networking, pp 252–259. http://doi.acm.org/https://doi.org/10.1145/1080139.1080143
12. Pathak S, Gondaliya N, Raja N (2017) A survey on PROPHET based routing protocol in delay tolerant network. In: 2017 International conference emerging trends innovation. ICT, ICEI 2017, pp. 110–115. https://doi.org/10.1109/ETIICT.2017.7977020
13. Abdelkader T, Naik K, Nayak A, Goel N, Srivastava V (2016) A performance comparison of delay-tolerant network routing protocols. IEEE Netw 30(2):46–53. https://doi.org/10.1109/MNET.2016.7437024

14. Veas-Castillo L, Ovando-Leon G, Gil-Costa V, Marin M (2018) minvisited: a message routing protocol for delay tolerant network. In: Proceedings - 26th euromicro international conference parallel, distributed network-based process. PDP 2018, pp 325–328. https://doi.org/10.1109/PDP2018.2018.00057.
15. Daly EM, Haahr M, (2007) Social network analysis for routing in disconnected delay-tolerant MANETs. In: Proceeding of 8th ACM intertnationl symposium on mobile ad hoc networking and computing. - MobiHoc 2007, p 32. https://doi.org/10.1145/1288107.1288113.
16. Hui P, Crowcroft J, Yoneki E (2011) Bubble rap: social-based forwarding in delay-tolerant networks. IEEE Trans Mob Comput. https://doi.org/10.1109/TMC.2010.246
17. Li F, Wu, J (2009) LocalCom : a community-based epidemic forwarding scheme in disruption-tolerant networks. In: 2009 6th annual IEEE communication society conference. sensor, mesh Ad Hoc communication networks, pp 1–9. https://doi.org/10.1109/SAHCN.2009.5168942.
18. Li Q, Zhu S, Cao G (2010) Routing in socially selfish delay tolerant networks. In: 2010 Proceeding of IEEE INFOCOM, pp 1–9. https://doi.org/10.1109/INFCOM.2010.5462138.
19. Chen BB, Chan MC (2010) "MobiCent: a credit-based incentive system for disruption tolerant network. https://doi.org/10.1109/INFCOM.2010.5462136.
20. Gholap M, Shaikh S (2016) Information sharing in delay tolerant mobile networks with some incentive and fewest transmissions. In: Proceeding of - 2016 8th international conference computational intelligence communication networks, CICN 2016, pp 119–123 https://doi.org/10.1109/CICN.2016.172.
21. Keränen A, Kärkkäinen T, Ott J (2010) Simulating mobility and DTNs with the One. J. Commun 5(2):92–105. https://doi.org/10.4304/jcm.5.2.92-105

Design of a Novel Wide Band Antenna with Defected Ground Structure for Mm-Wave System on Chip Applications

P. Loktongbam, Chaitali Koley, Debasish Pal, and Ayan Kumar Bandyopadhyay

Abstract This paper represents a novel fish hooked E shaped dipole antenna for the wideband operation of the system on chip applications in the mm-frequency band. The antenna is printed on a corrugated silicon substrate and also provides good radiation efficiency. Twin concentric square-shaped defected ground is introduced to enhance the gain. This antenna reports of simulated gain of 6.04 dBi and 3 dB beam-width of 91.2°. The proposed antenna has a radiation efficiency of 91.7%. Antenna resonates at 144 GHz with an approximate −10 dB bandwidth of 46 GHz with impedance bandwidth of almost 32% which is yet to be reported, to the best of the author's knowledge.

Index Terms Dipole · Corrugated substrate · Defected ground

1 Introduction

Commercial cellular handset manufacturers are putting gigantic efforts in implementing a fifth-generation (5G) communication system to provide amplified broadband data bandwidth-based communication services compared to the 4G system [1]. To counter this ever-increasing data traffic, the use of millimeter-wave (mm-wave) bands beyond 20 GHz have been proposed [2]. During the WRC-2015 (World Radiocommunication conference), many mm-wave frequency bands were proposed for 5G communications. It was supposed to be finalized at WRC-2019 [3, 4]. Till date, mm-wave frequency bands are being used for military, satellites, and internal networks. They are assigned only for designated wireless links or inter-facility networks and the assigned frequency bands vary across nations. The 5G bands are already congested due to the spectrum license owned by different nations of the world. Technically the 5G systems will operate up to 60 GHz [5], so the future of data communications

P. Loktongbam (✉) · C. Koley
Department of ECE, NIT Mizoram, Aizawl, India

D. Pal · A. K. Bandyopadhyay
CSIR CEERI, Pilani, Rajasthan, India

V. Nath and J. K. Mandal (eds.), *Proceeding of Fifth International Conference on Microelectronics, Computing and Communication Systems*, Lecture Notes in Electrical Engineering 748,
https://doi.org/10.1007/978-981-16-0275-7_64

will rely on mm-wave frequencies beyond 60 GHz. Now as mm-wave antennas are part and parcel of the present-day data-based system on chip applications (SOC), it is an inevitable requirement to design simplified antennas that can fit the size and compactness required in on-chip systems. In this paper, we report a novel E shaped fish hooked ended wideband dipole antenna with the defected ground. It is expected that this simple yet robust reported antenna design might cater to the ever-demanding enhanced data traffic beyond 5G in the near future. The antenna is operational in the mm-wave minimum loss propagation window of 140 GHz [6–8].

2 Antenna Structure

The proposed E shaped dipole antenna is shown in Fig. 1. Figure 1(a) displays the top view of the corrugated substrate and Fig. 1(b) the top view of the printed antenna. Figure 2 shows the front view of the antenna while Fig. 3 shows the top view of the ground plane. The whole antenna consists of five layers. The top and bottom are

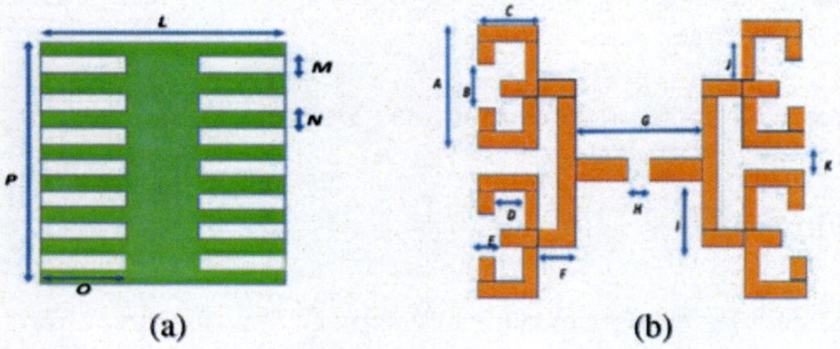

Fig. 1 (a) Substrate top view (b) Antenna top view

Fig. 2 Front view of the antenna

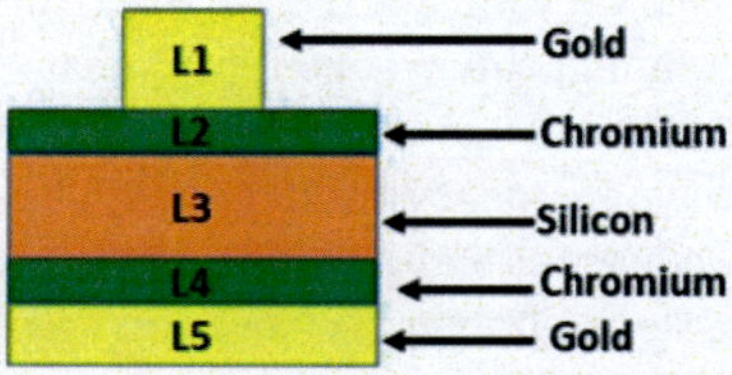

Fig. 3 Ground plane top view

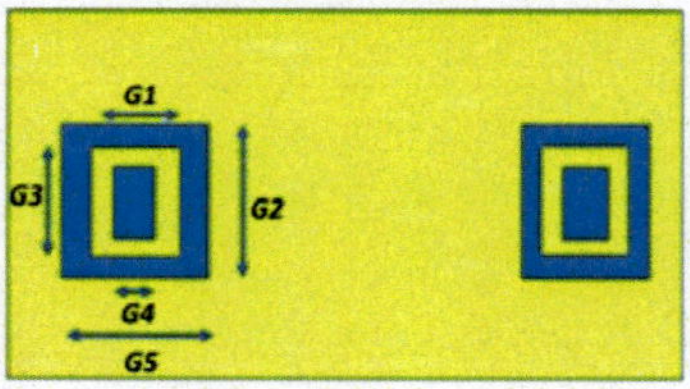

metallic layers. A silicon substrate layer is sandwiched between the top and bottom layers. Two chromium layers have been used as adhesives between substrate and gold layers. In some works aluminum oxide [9] and silicon dioxide [10] and has also been used as adhesives. The Dipole antenna consists of four E shaped units with fish hooked ending. Each side of the dipole consists of two E shaped units connected in parallel through a feed line. Dimensions of all the E units are the same. The substrate has a corrugated geometry. In the case of mm-wave printed antennas, thick substrates have the disadvantage of power loss due to the excitation of higher-order modes [11, 12] while thin substrates are challenging to implement. In this work, substrate thickness was optimized in such a manner that wide bandwidth and high gain achieved simultaneously. The substrate has corrugation which helps in reducing the surface wave propagation to a large extent. Defected ground structures of 2 concentric square shapes have also been used to further enhance the gain. The ground plane is made of gold dimensional details of the antenna is provided in Table 1. A discrete port feed was used to excite the antenna.

Table 1 Antenna dimensions

Serial no.	Notations	Dimensions (μm)
1	A	280
2	B	174
3	C	170
4	D	130
5	E	20
6	F	55
7	G	40
8	H	15
9	I	170
10	J	70
11	K	71
12	L	1090
13	M	20
14	N	80
15	O	110
16	P	840
17	G1	100
18	G2	130
19	G3	96
20	G4	60
21	G5	100

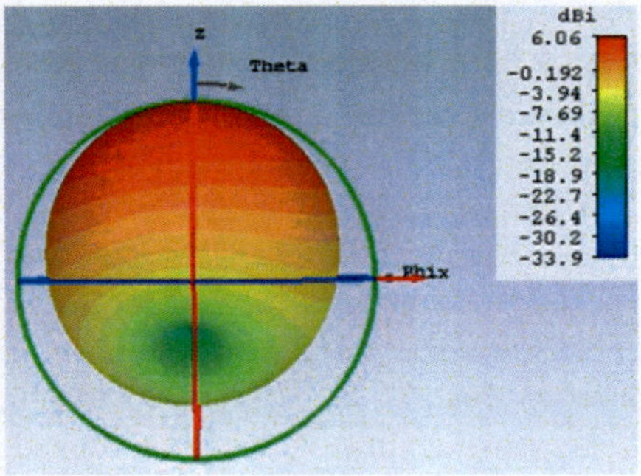

Fig. 4 3D plot of radiation pattern showing directivity

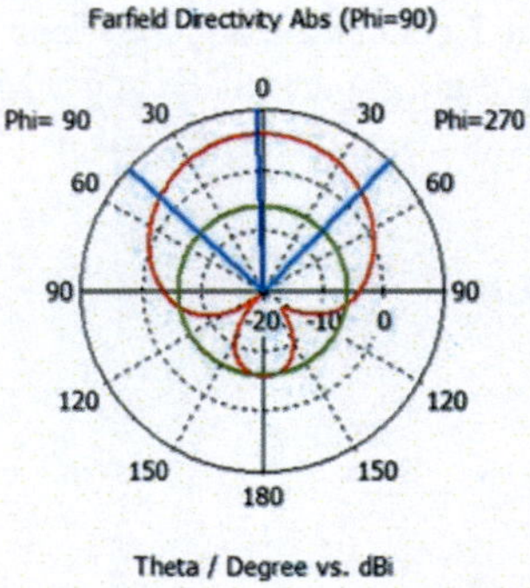

Fig. 5 2D plot of radiation pattern

3 Simulated Results

Industry standard CST software was used to perform the simulation. The S parameter plot of the antenna is displayed in Fig. 6. As shown in Fig. 6, bandwidth (−10 dB) of the proposed antenna is approximately 46 GHz with an impedance bandwidth of almost 32%. The simulated 3D radiation pattern plot is shown in Fig. 4. The simulated gain of the antenna is around 6.04 dBi. The radiation efficiency of the proposed antenna model is high. The value of calculated radiation efficiency is 91.7%. The simulated 3 dB beam-width of the proposed antenna was found to be around 91.2

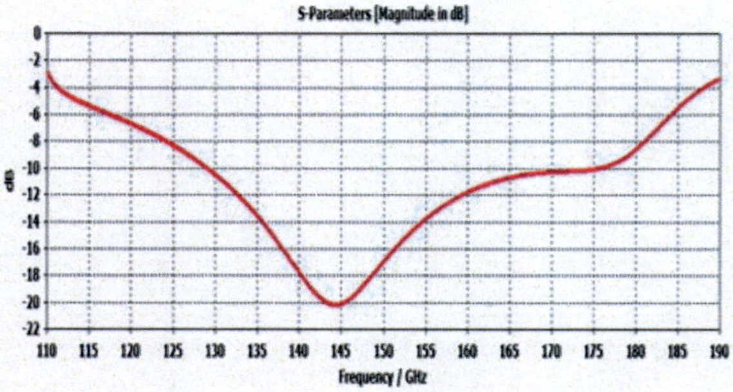

Fig. 6 S_{11} parameter of the antenna model

Table 2 State of the art comparison

Ref no	Parameter		
	Fractional bandwidth (%)	Operating frequency (GHz)	Gain (dBi)
[13]	13.9	143	7
[14]	14.2	135	15.5
[15]	6.25	160	7
[16]	20.28	138	25
[17]	6.06	165	5
[18]	23.54	140	33
[19]	7.77	120	13.36
[20]	15.1	149	31.7
[21]	14	143	6
Our work	32	144	6.04

degrees as measured from the polar plot data of Fig. 5. A comparison of contemporary antennas (Table 2) shows that the proposed antenna has high gain with high fractional bandwidth to the best of our knowledge. Also, it is compact in size and has a low loss. At the same time, it is also compatible with on-chip applications.

4 Conclusion

A customized dipole antenna with defected ground structure with wide operational bandwidth and high gain has been reported. This antenna has a simulated −10 dB bandwidth of 46 GHz with an impedance bandwidth of 32%. A directive gain of 6.04 dBi is reported. This design is expected to work with minimum atmospheric attenuation as it is operational in the 140 GHz transmission window. The antenna also has high efficiency.

References

1. Framework and overall objectives of the future development of IMT for 2020 and beyond, document ITU-R M.2083-0, international telecommunication union-radio communication sector, geneva, Switzerland, Sep. 2015
2. Rappaport TS et al (2013) Millimeter wave mobile communications for 5G cellular: It will work! IEEE Access 1:335–349
3. Marcus MJ (2017) WRC-19 issues: A survey. IEEE Wireless Commun 24:2–3
4. 5G spectrum recommendations (2017). 5G Americas, White Paper
5. Rahimian A, Alomainy A, Alfadhl Y (2016) A flexible printed millimetre-wave beamforming network for WiGig and 5G wireless subsystems. In: 2016 loughborough antennas & propagation conference (LAPC), loughborough, pp 1–5

6. https://transition.fcc.gov/Bureaus/Engineering_Technology/Documents/bulletins/oet70/oet70a.pdf
7. CCIR Doc. Rep. 719-3, 'Attenuation by Atmospheric Gases,' ITU 1990
8. Xing Y, Rappaport TS (2018) Propagation measurement system and approach at 140 GHz–moving to 6G and above 100 GHz. In: 2018 IEEE global communications conference (GLOBECOM), Abu Dhabi, UAE, pp 1–6
9. Correas-Serrano D, Gomez-Diaz JS (2017) Graphene-based antennas for terahertz systems: A review. arXiv preprint arXiv:1704.00371
10. Khan S, Tahir FA, Cheema HM (2017) A multiband on-chip antenna for 94 and 140 GHz applications. In: 2017 11th european conference on antennas and propagation (EUCAP), Paris, pp 1992–1994
11. Filipovic DF et al (1993) Double-slot antennas on extended hemispherical and elliptical silicon dielectric lenses. IEEE Trans Microw Theor Tech 41:1738–1749
12. Filipovic DF et al (1992) Millimeter-wave double-dipole antennas for high gain integrated reflector illumination. IEEE Trans Microw Theor Tech 40(5):962–967
13. Lamminen A, Säily J, Ala-Laurinaho J, de Cos J, Ermolov V (2020) Patch antenna and antenna array on multilayer high-frequency PCB for D-band. IEEE Open J Antennas Propag https://doi.org/10.1109/ojap.2020.3004533
14. Chao G et al (2020) A D-band 3D-printed antenna. IEEE Trans Terahertz Sci Technol https://doi.org/10.1109/tthz.2020.2986650
15. Ahmad M, Kucharski M, Ng H, Kissinger D (2019) A compact efficient d-band micromachined on-chip differential patch antenna for radar applications. In: 2019 IEEE international symposium on antennas and propagation and USNC-URSI radio science meeting, Atlanta, GA, USA, pp 2201–2202
16. Manzillo FF, Gonzalez-Jimenez, JL, Clemente A, Siligaris A, Blampey B, Dehos C (2019) low-cost, high-gain antenna module integrating a cmos frequency multiplier driver for communications at d-band. In: 2019 IEEE radio frequency integrated circuits symposium (RFIC), Boston, MA, USA, pp. 19–22 https://doi.org/10.1109/RFIC.2019.8701772
17. Ahmad WA, Kucharski M, Di Serio A, Ng HJ, Waldschmidt C, Kissinger D (2019) Planar highly efficient high-gain 165 GHz on-chip antennas for integrated radar sensors. IEEE Antennas Wireless Propag Lett 18(11):2429–2433
18. Yao SS, Cheng YJ, Zhou MM, Wu YF, Fan Y (2020) D-band wideband air-filled plate array antenna with multistage impedance matching based on MEMS micromachining technology. IEEE Trans Antennas Propag 68(6):4502–4511
19. Qian J, Tang M, Chen Q, Zhang Y, Mao J (2019) Integration of S/Ka/D-band antennas in LTCC with a cylindrical radome for triband applications. IEEE Tran Antennas Propag 67(9):5781–5789
20. Zhou MM, Cheng YJ (2018) D-band high-gain circular polarized plate array antenna. IEEE Trans Antennas Propag 66(3):1280–1287
21. Khan WT et al (2015) A D-band micromachined end-fire antenna in 130-nm SiGe BiCMOS technology. IEEE Trans Antennas Propag 63(6):2449–2459

Missing Person Tracking System

Arjun Besra, Amir Junaid Ahmed, and Sabina Priyadarshini

Abstract In this paper, we propose a Missing Person Tracking System. It's a web-based application for the police investigation team to track the missing person. This system, therefore, was to investigate weaknesses in the current system of searching for missing persons in India, with the interest to offer a more significant solution. The results indicate that most respondents do not have much confidence in the current process of searching for the missing persons. From the results, we can see that the current system is not efficient enough, because many people reported as missing are either found after a long search or not found at all. The paper concludes that we need a system and amore efficient method of finding missing persons more easily and faster. A new approach has been proposed by us for the police and government to trace out missing people easily.

Keywords Biometric identification · Finding missing person · Aadhar data · Missing person data · Web application · Web server

1 Introduction

As per Google Shows data on missing persons in India, a total of 2,90,439 people in year 2016, 3,05,267 people in 2017, and 3,47,524 people in 2018 have been reported as missing. Each year, police stations across the country receive thousands of reports of missing persons. Fortunately, many of those who are reported to police as missing are located within a short span of time. There are however, others who are never found or who are, eventually, identified as victims of crime or misfortune. Besides, there are persons missing but not reported to police or inquired into. To make task of track missing person easy we designed missing person tracking system to find lost person within minimum time period.

A. Besra (✉) · A. J. Ahmed · S. Priyadarshini
Department of CSE, BIT Mesra, Ranchi, India

V. Nath and J. K. Mandal (eds.), *Proceeding of Fifth International Conference on Microelectronics, Computing and Communication Systems*, Lecture Notes in Electrical Engineering 748,
https://doi.org/10.1007/978-981-16-0275-7_65

2 Problem Statement

There has been concern about the alarming number of cases of missing people that end up not being identified. This study undertook to investigate the current process of searching and tracking for the missing people, to identify its weaknesses and device a more efficient method of ensuring more recovery of missing people within the shortest time possible.

3 Existing System

In the existing system, all the police station uses the manual process. Manual System for finding missing person First Each missing person case is allocated to a police officer to obtain initial details and that officer and their supervisors do conduct a risk assessment to decide the correct response to every case. This will depend on various factors including age, vulnerability, physical or mental health conditions, the time elapsed since last seen, circumstances of disappearance, whether the person has been missing before and previous behavior whilst missing. And then police start physically searching the missing person. Police searchplaces or locations where the missing person may have been or may currently be. This could be their home address, workplaces, if someone seen or have some information, the police use this information and predict where the person may be. The police have specialist officers to manage these types of searches and will draw upon resources such as police dogs, police helicopters, police horses, and specialist search teams to conduct these. This process has a very long procedure and may take more time. Also, more times required for finding lost a person. Also, during the manual process number of police officers required for searching the lost person. There has some existing system which some system shows their information in the social media and missing person related website. And some existing system does not show the proper information which is difficult to find out the missing person.

4 Literature Review

In [1, 4], the authors have created an android based system with face detection technique. They have created four modules- police, user, complaint holder and admin. Complaint is uploaded on web server which is seen by what they call as trust members who have the application.

In [2], they have used a microcontroller based system. It has been programmed. They have used google map and GPS. [3] describes a system for jeevan vikas charitable trust. It is android based.

[5] talks about the fingerprint matching technique. They first take the fingerprint. Then, they clear the unwanted data from the picture. Then, they extract features from the fingerprint and match it with the features already present. This work is important to match a fingerprint with fingerprints in the database.

[5–15] describe related techniques and methods to find a missing person. Some focus on video surveillance, while some on missing children, while some others focus on more efficient face recognition.

5 Proposed System

Proposed approach to find a missing person is given as follows:

1. We propose that the biometric machines should be made mandatory by the government in every place. We also propose that every person will have to give an entry in biometric machine wherever he goes and come out from. The biometric machine is programmed to take out his aadhar card number by matching his fingerprint with a centralized database maintained and governed by the government including his name, age and gender.
2. Related to every person, there will be a separate database table maintained by the police. The name of the table would be his aadhar card number. This table would contain fields/columns:-(name, age, from location, to location, in-time, out-time, date).
3. Whenever a person goes to some place and give his fingerprint the biometric machine would search his database by the name of his aadhar card number
4. The biometric machine would update all fields of his database table and add a new record as soon as the person gives his fingerprint.
5. So, every person's movement from one place to other gets recorded in database for lifetime.
6. We propose that biometric entry must be mandatory for each person wherever he goes, be it a college, a bus, a train, a hospital, an office, a shop or any other place mandated by government.
7. Whenever a person goes missing, the database by the name of his aadhar card number would be searched. Then the records would be fetched using the date and time as indexes. The database of people can be accessed only by the police or the government and nobody else. In this way a missing person's wherever about can be easily known by the police.

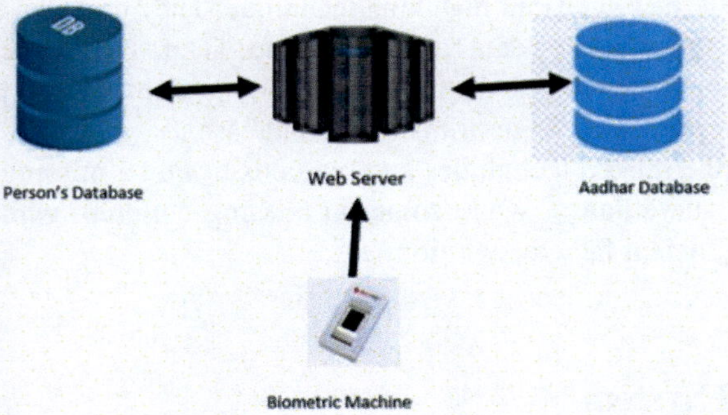

6 Requirements

We are going to perform the project on windows platform so we need the os as windows. Any version of windows as windows xp, windows 7 or windows 8 or windows 10. The system should have minimum ram of 500 MB as well as minimum storage capacity of 500 GB.

7 Modules

User Login: Each Police station there will be a user. Each user will get the user id and password by which police officer can sign-in to the website.

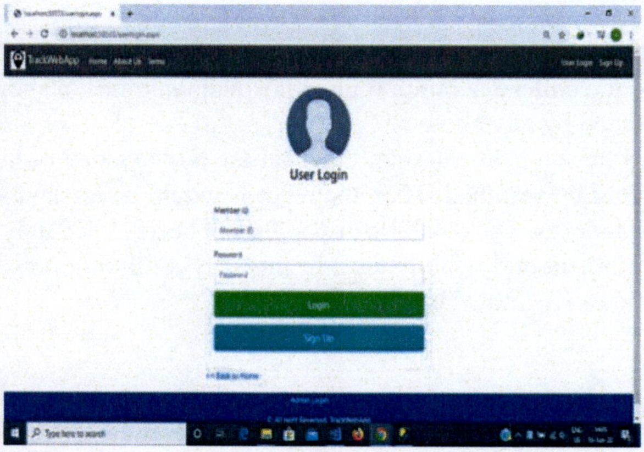

Search Module: After sign In User or officer UI for person profile in this page user can see all the details in which the person punches his finger and it will show the details of visited places, date and time.

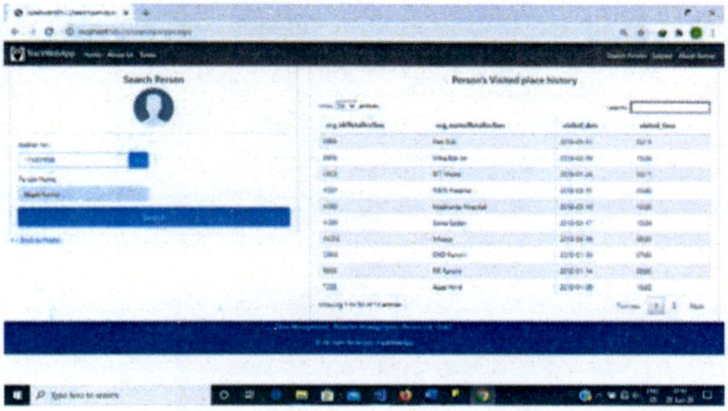

Missing Person Management: In this module user can manage the missing person like when the complaint come the user can add the missing person some basic details and when person found then user can delete.

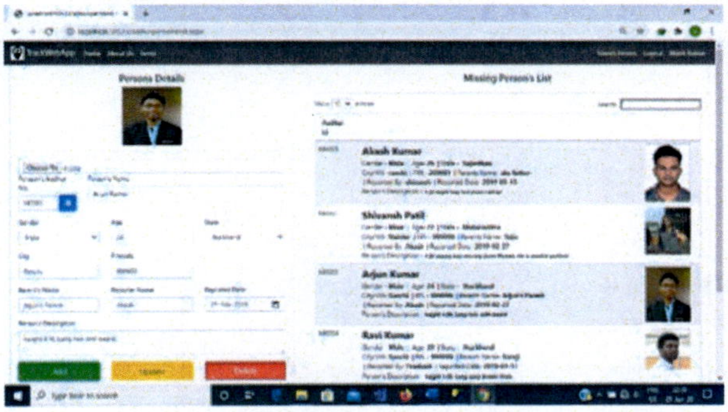

Administrators: The module will be protected by user ID and password. Ordinary users of the application will not be permitted to enter into this area of the software. The module will be focusing on the maintenance like Master Data Maintenance add, update, delete data from the web server etc.

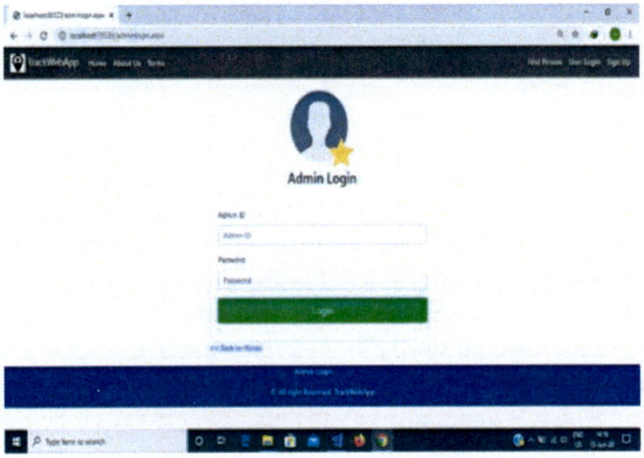

Admin User Management: In this module admin can manage the user and give authentication to user. And admin can add and delete the user.

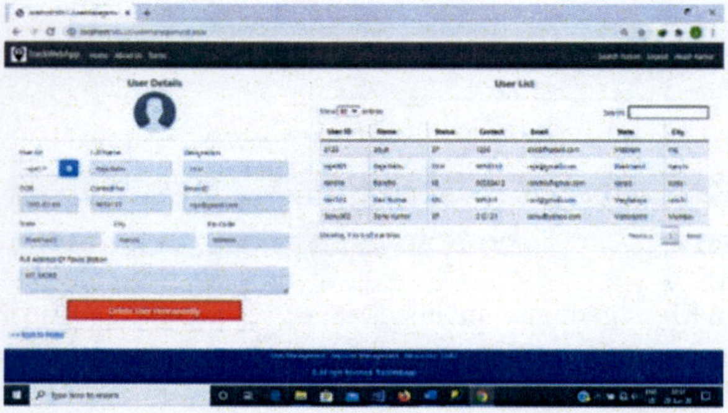

8 Technology Stack

Front end

HTML
CSS & Bootstrap 4.0
JavaScript with jQuery

Back End

ASP.NET

Database

MS SQL Server

9 Comparison Study

Previously, [9] Ankush Pandita et al. presented a paper which deals with a little bit similar problem statement and objective. The system proposed by them makes use of face detection, they use the CCTV video streaming. When the missing person found in the CCTV video streaming, they track this location the missing person.

The main difference between their work and ours is that in their system they require the image that must be stored in their database and from that database they compare with the video streaming. In our system we use Aadhar database and separate person's database which is maintained by the police for every citizen. In the persons database all visited places data will be updated automatically and from that data police can track the missing person.

Our system doesn't rely on face recognition and is better because there are many similar faces in the world. Video surveillance is also complicated. Our method is simpler and more accurate.

10 Conclusions

The study ended successfully with the development of a web-based application prototype as was desired in the main objective. The application has the capability to reunite many families and friends. The project was the ultimate objective is to reunite friends and family who have been separated by reason. And this application is also useful for the CBI, CID and other agencies of our country for investigation based on stored data.

References

1. Hetal B, Rakesh S, Porjerohan, Harish S (2018) Android based application-missing person finder. Iconic Res Eng J 1(12). ISSN: 2456-8880
2. Dzulkifli MA, Sulaiman S, Saparon A (2017) Tracking system for missing person. In: 2017 international conference on electrical, electronics and system engineering (ICEESE)
3. Thali P, Mujmule D, Bhingarde A, Salve Y (2018) Crime reporter and missing person finder. Int J Sci Eng Res 9(4):2229-5518
4. Mote R, Rakate P, Thavare D, Patil R (2015) Android Based Application on "Missing Person Finder". IJSRD Int. J Sci Res Dev 3(01): 2321–0613
5. Ali MMH, Mahale VH, Yannawar P, Gaikwad A.T (2016) Overview of fingerprint recognition system. In: 2016 international conference on electrical, electronics, and optimization techniques (ICEEOT)
6. Priya BS, Rajesh R (2011) A note on fingerprint recognition systems. In: 2011 3rd international conference on electronics computer technology
7. Singh R, Shah U, Gupta V (2009) Fingerprint Recognition- Department of Computer Science and Engineering, Indian Institute of Technomogy, Kanpur
8. Thai LH, Tam HN (2010) Fingerprint recognition using standardized fingerprint model. IJCSI Int J Comput Sci Issues 7(3):1694-0784
9. Pandita A, Dahenkar R, Dalve J, Kumar S, Mulay SA (2017) Missing person identification and tracking for intelligent video surveillance. Int J Innovative Res Comput Commun Eng-5(3):2320-9801
10. Balar BD, Kavya DS, Chandana M, Anush E, Vishwanath RH (2019) Efficient face recognition system for identifying lost people. Int J Eng Adv Technol (IJEAT) 8(5S): 2249–8958
11. Hou L, Wan W, Hwang JN, Muhammad R, Yang M, Han K (2017) Human tracking over camera networks: a review. EURASIP J Adv Signal Process 2017(1): 1–20
12. Desale H, Tavasalkar P, Vare S, Shintre R (2020) Android crime reporter and missing person finder. Int Res J Eng Technol (IRJET) 07(03):2395-0056
13. Saeed M, Malallah FL (2016) Children tracking system using fingerprint verification. Int J Adv Res 4(12): 1381–1391

14. Ali MMH, Mahale VH, Yannawar P, Gaikwad A (2016) Overview of fingerprint recognition system. In: 2016 international conference on electrical, electronics, and optimization techniques (ICEEOT)
15. Nath D, Ray S (2011) Fingerprint recognition system: design & analysis. In: conference international conference on scientific paradigm shift in information technology & management, SPSITM 2011

Delay and Crosstalk Aware Analysis for High Speed On-Chip Global RLC VLSI Interconnects

Apoorva Gupta, Vikas Maheshwari, Somashekhar Malipatil, and Rajib Kar

Abstract At the advanced stage of technologies, when feature size is being reduced, it is compulsory to introduce other parameters for more accurate modelling of transmission line interconnects. So mutual inductance and coupling capacitance how have become more important role for analysis of high-speed on-line VLSI interconnects. This paper introduces a mathematical aware analysis result for crosstalk noise of 'L' type RLC interconnections using mutual inductance. Two RLC interconnect lines of 'L' type, are equidistant to each other and used as 'Aggressor line' and 'Victim line' respectively, whereas, a step signal voltage is employed as input to aggressor line. Other calculative results for Delay and peak noise voltage between these two RLC electrical lines with using mutual inductance, are also introduced in this paper. This paper also shows a comparative result between our derived expression values and BKM values for simulation purpose.

Keywords Integrated circuits · RLC VLSI interconnects · Mutual inductance · Coupling capacitance · Delay · Crosstalk noise

1 Introduction

Presently, VLSI is the present level of designing and fabrication of ICs and microchips which consist of lacs of transistors on a single chip [1, 2]. In DSM region [3], now it is considered to study of inductive effect as well as capacitive-coupling effects to

A. Gupta (✉) · V. Maheshwari
Department of ECE, Guru Nanak Institutions Technical Campus, Ranga Reddy, Hyderabad, Telangana, India

S. Malipatil
Department of ECE, Malla Reddy Engineering College and Management Sciences, Medchal, Hyaderabad, Telangana, India

R. Kar
Department of ECE, National Institute of Technology, Durgapur, West-Bengal, India

V. Nath and J. K. Mandal (eds.), *Proceeding of Fifth International Conference on Microelectronics, Computing and Communication Systems*, Lecture Notes in Electrical Engineering 748,
https://doi.org/10.1007/978-981-16-0275-7_66

develop and explain the more accurate and real behaviour of on-chip VLSI interconnects. Delay and Crosstalk noise between VLSI high speed interconnected networks can have occurred due to self and mutual inductance. There are so many approaches presented [4–18] for the modelling of interconnect structures. This paper introduces use of closed loop of 'L' type RLC interconnect network. Two RLC parallel interconnects and a mutual inductance and coupling capacitance are occurred automatically. These two RLC networks are named as 'aggressor line' and 'victim line' respectively. The proposed work is much improved work of the BKM [19] model. This paper establishes a mathematical equation of crosstalk voltage of mutually inductively coupled interconnections of RLC type. This paper also introduces expressions for delay and peak noise voltage between adjacent RLC network. This paper is organized remaining follows: Sect. 2 describes Proposed models and of crosstalk voltage and delay analysis. Detailed results of simulation are discussed in portion 3 and portion 4 conclude the paper.

2 Proposed Model

Mathematical and analytical expressions for the crosstalk voltage, delay and peak crosstalk voltage are derived in the case only when victim lines are grounded and excitation is connected to aggressor line. Figure 1 shows lumped RLC model of 'L' shaped interconnection system considering Mutual Inductance coupling between the parallel lines. Step input voltage is used for the analysis of the interconnection system.

A step input voltage supply is given to the input of aggressor line which is equidistant to the victim lines. Coupling capacitance is generated because those two RLC networks are proximate to each other and mutual inductance is induced due to using inductor coil.in this paper we use 90 nm technology.

As per Moore's law, in the process of designing the ICs, the number of transistors will continue to double in every 18 months [1]. That means the same silicon area would accommodate a greater number of transistors. Transistors size is gradually getting reduced for achieving this or we can say that transistor size is shifting from one technology node to smaller technology node by using scaling process. A specific technology gets used by the industries for the period of time till the time when

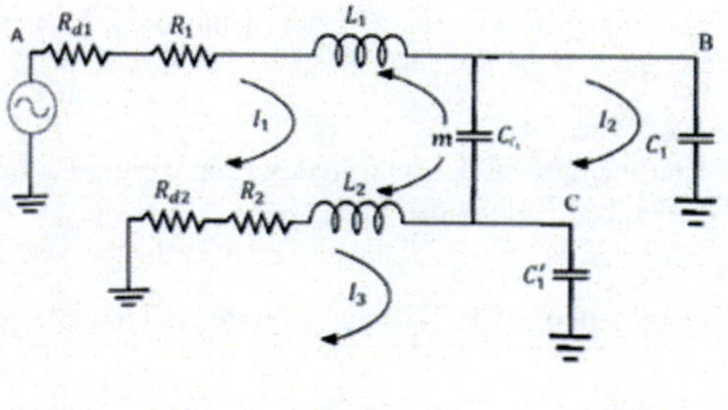

Fig. 1 Equivalent Circuit of RLC interconnects using 'L' shaped interconnect model with effect of Mutual Inductance m

the next feasible smaller technology node would be ready for implantation. For example, 180 nm technology was used mostly in 1999–2000-time period whereas 90 nm technology was used in 2004–2005. The technology's numbers represent the minimum feature size of transistor or CMOS. Minimum channel length that can be used in fabrication of CMOS or transistor is known as Feature size of transistor. These numbers are decided by dividing the previous number (technology) by square root of two ($\sqrt{2}$).

In the circuit shown in Fig. 1, at node C, we develop the mathematical expression for the voltage for RLC victim line.

On employing KVL in 1st loop:

$$V_{S1} = I_1 R_{d1} + I_1 R_1 + L_1 \frac{dI_1}{dt} - m\frac{dI_3}{dt} + \frac{1}{C_2}\int (I_1 - I_2)dt \tag{1}$$

On taking Laplace,

$$V_{S1}(s) = I_1(s)R_{d1} + I_1(s)R_1 + L_1 S I_1(s) - mSI_3(s) + \frac{1}{sC_1}[I_1(s) - I_2(s)] \tag{2}$$

$$V_{S1}(s) = \frac{(M_1 SC_1)I_1(s) - I_2(s)}{SC_1} - mSI_3(s) \tag{3}$$

Where,

$$M_1 = R_{d1} + R_1 + L_1 S \tag{4}$$

Similarly, on applying KVL in 2nd mesh,

$$\frac{1}{sC_1}[I_1(s) - I_2(s)] = \frac{1}{SC_{C1}} I_2(s) + \frac{1}{SC_1'}[I_2(s) + I_3(s)]$$

$$\frac{1}{sC_1} I_1(s) - \left[\frac{1}{sC_1} + \frac{1}{SC_1'} + \frac{1}{SC_{C1}}\right] I_2(s) - \frac{1}{SC_1'} I_3(s) = 0 \tag{5}$$

Similarly, on applying KVL in 3rd mesh,

$$0 = I_3(s)R_{d2} + I_3(s)R_1' + SL_1' I_3(s) - SmI_1(s) + \frac{1}{SC_1'}[I_2(s) + I_3(s)]$$

$$I_1(s)\left[-s^2 mC_1'\right] + I_2(s) + \left(SM_2 C_1' + 1\right) = 0 \tag{6}$$

Where

$$M_2 = R_{d1} + R_1' + sL_1'$$

From Eqs. (3), (5) and (6), we get a matrix:

$$\begin{bmatrix} \frac{M_1SC_1+1}{SC_1} & -\frac{1}{sC_1} & -ms \\ \frac{1}{sC_1} & -\left[\frac{1}{sC_1}+\frac{1}{SC_1'}+\frac{1}{SC_{C1}}\right] & -\frac{1}{sC_1} \\ -s^2mC_1' & 1 & (SM_2C_1'+1) \end{bmatrix}\begin{bmatrix} I_1(s) \\ I_2(s) \\ I_3(s) \end{bmatrix} = \begin{bmatrix} V_{s1}(s) \\ 0 \\ 0 \end{bmatrix}$$

Let,

$$M_1SC_1+1=A,\ B=-\left[\frac{1}{sC_1}+\frac{1}{SC_1'}+\frac{1}{SC_{C1}}\right],$$

$$\left(SM_2C_1'+1\right)=C$$

Then required matrix is,

$$\begin{bmatrix} \frac{A}{sC_1} & \frac{-1}{sC_1} & -mS \\ \frac{1}{sC_1} & -B & \frac{-1}{sC_1} \\ -s^2mC_1' & 1 & C \end{bmatrix}\begin{bmatrix} I_1(s) \\ I_2(s) \\ I_3(s) \end{bmatrix} = \begin{bmatrix} V_{s1}(s) \\ 0 \\ 0 \end{bmatrix}$$

After solving by Cramer's rule [20],

$$I_1(s)=\frac{sC_1^2\left[1-CC_1's\right]V_{s1}}{-sABCC_1C_1'-2s^2mC_1C_1'+Bs^5m^2C_1^2C_1'^2+AC_1+CC_1} \tag{7}$$

$$I_2(s)=\frac{sC_1C_1'\left[ms^2C_1-C\right]V_{s1}}{-sABCC_1C_1'-2s^2mC_1C_1'+Bs^5m^2C_1^2C_1'^2+AC_1+CC_1} \tag{8}$$

$$I_3(s)=\frac{sC_1C_1'\left[1-3s^3mC_1C_1'\right]V_{s1}}{-sABCC_1C_1'-2s^2mC_1C_1'+Bs^5m^2C_1^2C_1'^2+AC_1+CC_1} \tag{9}$$

Now, at node 'C':

$$V_c=\frac{1}{sC_1'}[I_2(s)+I_3(s)] \tag{10}$$

So,

$$V_c=\frac{sC_1C_1'V_{s1}}{sC_1'}\left[\frac{ms^2C_1-C+1-Bs^3mC_1C_1'}{-sABCC_1C_1'-2s^2mC_1C_1'+Bs^5m^2C_1^2C_1'^2+AC_1+CC_1}\right]$$

$$V_c=\frac{C_1V_{s1}\left[ms^2C_1-C+1-Bs^3mC_1C_1'\right]}{P} \tag{11}$$

Where,
$P = -sABCC_1C_1' - 2s^2mC_1C_1' + Bs^5m^2C_1^2C_1'^2 + AC_1 + CC_1$ After substituting the values of A, B and C:

$$P = -sC_1C_1'(M_1sC_1+1)\left(\frac{\propto}{sC_1C_1'C_{C1}}\right)\left(sM_2C_1'+1\right) - 2s^2mC_1C_1' + \left(\frac{\propto}{sC_1C_1'C_{C1}}\right)s^5m^2C_1^2C_1^2 + (M_1sC_1+1)C_1 + \left(sM_2C_1'+1\right)C_1'$$

Where,

$$B = \frac{\propto}{sC_{C1}C_1C_1'}$$

Where,

$$\propto = C_1'C_{C1} + C_{C1}C_1 + C_1C_1'$$

Now,

$$P = -\propto (M_1sC_1+1)\left(sM_2C_1'+1\right) - 2s^2mC_1C_1' + \frac{\propto}{C_1'}s^4m^2C_1C_1' + (M_1sC_1+1)C_1 + \left(sM_2C_1'+1\right)C_1'$$

After neglecting all high-power terms:

$$P = s^2C_1C_1'[\propto M_1M_2 + 2m] + s\left[-\propto M_1C_1 - \propto M_1C_1' + M_1C_1 + M_2C_1'\right] + C_1 + C_1' - \propto \tag{12}$$

Now after substituting the value of $V_{s1} = \frac{1}{s}$ (for step input voltage) & P in Eq. (11),

$$V_c = \frac{C_1\left(smC_1 - M_2C_1'\right)}{s^2C_1C_1'[\propto M_1M_2 + 2m] + s\left[-\propto M_1C_1 - \propto M_1C_1' + M_1C_1 + M_2C_1'\right] + C_1' + C_1' - \propto}$$

Now, let us assume:

$$X = C_1C_1'[\propto M_1M_2 + 2m]$$

$$Y = -\propto M_1C_1 - \propto M_1C_1' + M_1C_1 + M_2C_1'$$

$$Z = C_1' + C_1' - \propto$$

$$P = -X^2 s + Ys + Z \tag{13}$$

$$V_c = \frac{C_1\left(smC_1 - M_2C_1^{'}\right)}{-Xs^2 + Ys + Z}$$

$$V_c = \frac{-mC_1^2 s}{Xs^2 - Ys - Z} + \frac{M_2C_1^{'}}{Xs^2 - Ys - Z}$$

$$V_c = \frac{-mC_1^2 s/X}{s^2 - \frac{Y}{X}s - \frac{Z}{X}} + \frac{M_2C_1^{'}/X}{s^2 - \frac{Y}{X}s - \frac{Z}{X}}$$

$$V_c = \frac{-mC_1^2 s/X}{\left[s - \frac{Y}{2X}\right]^2 + \left[\sqrt{\frac{-Z}{X} - \frac{Y^2}{4X^2}}\right]^2} + \frac{M_2C_1^{'}/X}{\left[s - \frac{Y}{2X}\right]^2 + \left[\sqrt{\frac{-Z}{X} - \frac{Y^2}{4X^2}}\right]^2}$$

If, $R = \left[\sqrt{\frac{-Z}{X} - \frac{Y^2}{4X^2}}\right]^2$

After taking Inverse Laplace transform:

$$V_c(t) = -\left[e^{\frac{Y}{2X}t}\cos Rt\right]\frac{mC_1^2}{X} + \left[e^{\frac{Y}{2X}t}\sin Rt\right]\frac{M_2C_1^{'}}{X}$$

$$V_c(t) = \frac{e^{\frac{Y}{2X}t}}{X}\left[M_2C_1^{'}\sin Rt - mC_1^2\cos Rt\right] \tag{14}$$

Peak time value t_{p_c} is calculated by equating first derivative of $V_c(t)$ to zero,

$$\frac{dV_c(t)}{dt} = 0$$

$$M_2C_1^{'}\frac{d}{dt}\left[e^{\frac{Y}{2X}t}\sin Rt\right] - mC_1^2\frac{d}{dt}\left[e^{\frac{Y}{2X}t}\cos Rt\right] = 0$$

After simplification we get

$$t_{p_c} = t = \frac{1}{R}\tan^{-1}\left[\frac{-RM_2C_1^{'} + \left(\frac{mC_1^2}{2}\right)\left(\frac{Y}{X}\right)}{RmC_1^2 + \left(\frac{M_2C_1^{'}}{2}\right)\left(\frac{Y}{X}\right)}\right] \tag{15}$$

Let's put the value t = t_{p_c} in Eq. (14) so that,

$$V_c(t) = \frac{t_{c_{(t)max}}}{t_{=t_{pc}}}$$

$$V_c(t)_{max} = e^{\frac{\gamma}{2X}t_{Pc}}\left[M_2C_1^{'}\sin Rt_{p_c} - mC_1^2\cos Rt_{p_c}\right] \tag{16}$$

Now calculate the value of V_B:

$$V_B = \frac{1}{sC_1}[I_1(s) - I_2(s)]$$

$$V_B = \frac{V_{s1}}{sC_1}\left[\frac{sC_1^2[1 - BCC_1^{'}s] - sC_1C_1^{'}[ms^2C_1 - C]}{-sABCC_1C_1^{'} - 2s^2mC_1C_1^{'} + Bs^5m^2C_1^2C_1^{'2} + AC_1 + CC_1}\right]$$

For unit step input, $V_{s1} = \frac{1}{s}$:

$$V_B = \frac{\frac{1}{s}\left[C_1 - \frac{s\propto C_1C_1^{'}}{sC_1C_1^{'}C_{C1}}(1 + sM_2C_1^{'}) - mC_1C_1^{'}s^2 - C_1^{'}(1 + sM_2C_1^{'})\right]}{P}$$

$$V_B = \frac{\frac{1}{s}\left[C_1 - \frac{\propto}{C_{C1}} - \frac{sM_2C_1^{'}}{C_{C1}} - mC_1C_1^{'}s^2 - M_2s\right]}{P}$$

Now let's put the value of P from expression (12) and have,

$$V_B = \frac{C_1C_{C1} - \propto - s\left(M_2C_1^{'} + M_2C_1^{'2}C_{C1}\right) - mC_1C_1^{'}C_{C1}s^2 - C_1^{'}C_{C1}}{C_{C1}s\left[-s^2C_1C_1^{'}(\propto M_1M_2 + 2m) + s\left(-\propto M_1C_1 - \propto M_2C_1^{'} + M_1C_1 + M_2C_1^{'}\right) + \left(C_1 + C_1^{'} - \propto\right)\right]}$$

$$V_B = \frac{C_1C_{C1} - \propto - C_1^{'}C_{C1}}{sPC_{C1}} - \frac{msC_1C_1^{'}C_{C1}}{PC_{C1}} - \frac{\left(M_2C_1^{'} + M_2C_1^{'2}C_{C1}\right)}{PC_{C1}}$$

$$V_B = \frac{C_1C_{C1} - \propto - C_1^{'}C_{C1}/C_{C1}}{sP} - \frac{mC_1C_1^{'}s}{P} - \frac{\left(M_2C_1^{'} + M_2C_1^{'}C_{C1}\right)/C_{C1}}{P}$$

$$V_B = V_{B1} - V_{B2} - V_{B3} \tag{17}$$

Where,

$$V_{B1} = \frac{C_1C_{C1} - \propto - C_1^{'}C_{C1}/C_{C1}}{sP}$$

$$V_{B2} = \frac{msC_1C_1^{'}}{P}$$

$$V_{B3} = \frac{\left(M_2C_1^{'} + M_2C_1^{'}C_{C1}\right)/C_{C1}}{P}$$

Putting the result of P in the equation of V_{B1} from Eq. (13)

$$V_{B1} = \frac{C_1 C_{C1} - \propto - C_1^{'} C_{C1} / C_{C1}}{s\left(-Xs^2 + Ys + Z\right)}$$

after simplification, we get,

$$V_{B1} = \frac{C_1 C_{C1} - \propto - C_1^{'} C_{C1}}{C_{C1}} \left[\frac{1}{Zs} - \frac{1}{Z\left(s + \frac{Z}{Y}\right)} \right]$$

In similar way, the expressions of V_{B2} and V_{B3} after substituting the expression of P from Eq. (13),

$$V_{B2} = \frac{\frac{-mC_1 C_1^{'}}{X} s}{\left[s - \frac{Y}{2X}\right]^2 + R^2}, \quad V_{B3} = \frac{\left(M_2 C_1^{'} + M_2 C_1^{'2} C_{C1}\right) / XC_{C1}}{\left[s - \frac{Y}{2X}\right]^2 + R^2}$$

Where,

$$R = \left[\sqrt{\frac{-Z}{X} - \frac{Y^2}{4X^2}} \right]^2$$

substituting the values of V_{B1}, V_{B2}, V_{B3} in Eq. (17)

$$V_B(s) = \frac{C_1 C_{C1} - \propto - C_1^{'} C_{C1}}{C_{C1}} \left[\frac{1}{Z} \left(\frac{1}{s} - \frac{1}{\left(s + \frac{Z}{Y}\right)} \right) \right] + \frac{\frac{mC_1 C_1^{'}}{X} s}{\left[s - \frac{Y}{2X}\right]^2 + R^2} + \frac{\left(M_2 C_1^{'} + M_2 C_1^{'2} C_{C1}\right) / XC_{C1}}{\left[s - \frac{Y}{2X}\right]^2 + R^2}$$

After using inverse Laplace Transform in the above expression

$$V_B(t) = \frac{\left(C_1 C_{C1} - \propto - C_1^{'} C_{C1}\right) \left(1 - e^{-\left(Z/Y\right) t} \right)}{C_{C1} Z} + \left[e^{\frac{Y}{2X} t} \cos Rt \right] \frac{mC_1 C_1^{'}}{X} + \left[e^{\frac{Y}{2X} t} \sin Rt \right] \frac{\left(M_2 C_1^{'} + M_2 C_1^{'2} C_{C1}\right)}{C_{C1} X} \quad (18)$$

On differentiating with respect to t

$$\frac{dV_B(t)}{dt} = \frac{\left(C_1 C_{C1} - \propto - C_1^{'} C_{C1}\right)}{C_{C1} Y} e^{-\left(Z/Y\right) t} + \frac{mC_1 C_1^{'}}{X} \left[\frac{Y}{2X} e^{\frac{Y}{2X} t} \cos Rt + (-R \sin Rt) e^{\frac{Y}{2X} t} \right] + \frac{\left(M_2 C_1^{'} + M_2 C_1^{'2} C_{C1}\right)}{C_{C1} X} \left[\frac{Y}{2X} e^{\frac{Y}{2X} t} \sin Rt + R \cos Rt e^{\frac{Y}{2X} t} \right]$$

Peak time value is calculated by equating first derivative to zero,

$$\frac{\left(C_1C_{C1}-\propto-C_1^{'}C_{C1}\right)}{C_{C1}Y}e^{-\left(Z/Y\right)t_p}+\cos Rt_pe^{\frac{Y}{2X}t_p}\left[\frac{Y}{2X^2}mC_1C_1^{'}+\frac{R\left(M_2C_1^{'}+M_2C_1^{'2}C_{C1}\right)}{C_{C1}X}\right]+$$
$$\sin Rt_pe^{\frac{Y}{2X}t_p}\left[\frac{Y}{2X^2C_{C1}}\left(M_2C_1^{'}+M_2C_1^{'2}C_{C1}\right)-\frac{R}{X}mC_1C_1^{'}\right]=0$$

Let us assume for simplicity,

$$q=\frac{\left(C_1C_{C1}-\propto-C_1^{'}C_{C1}\right)}{C_{C1}Y}$$

$$u=\frac{Y}{2X^2}mC_1C_1^{'}+\frac{R\left(M_2C_1^{'}+M_2C_1^{'2}C_{C1}\right)}{C_{C1}X}$$

$$v=\frac{Y}{2X^2C_{C1}}\left(M_2C_1^{'}+M_2C_1^{'2}C_{C1}\right)-\frac{R}{X}mC_1C_1^{'}$$

After simplification above equation becomes,

$$q.e^{-\left(Z/Y\right)t_p}+\cos Rt_pe^{\frac{Y}{2X}t_p}.u+\sin Rt_pe^{\frac{Y}{2X}t_p}.v=0$$

$$q.e^{-\left(Z/Y\right)t_p}+e^{\frac{Y}{2X}t_p}\left(u.\cos Rt_p+v.\sin Rt_p\right)=0 \tag{19}$$

After simplification and approximation to lower degree terms of above Eq. (19), we get

$$t_p=\frac{2XY(u+q)}{qZ-2XYvR-uY^2} \tag{20}$$

$$t_p=\frac{H}{I}$$

where,

$$H=2XY(u+q),\ I=qZ-2XYvR-uY^2$$

After substituting the values of t_p in equation, so now

$$V_{Bmax}(t)=\left(\frac{C_1C_{C1}-\propto-C'_1C_{C1}}{C_{C1}Z}\right)\left(1-e^{\left(Z/Y\right)\left(H/I\right)}\right)$$
$$+\frac{mC'_1C_{C1}}{X}\left[e^{\left(\frac{Y}{2X}\right)\left(\frac{H}{I}\right)}\cdot\cos\left(\frac{RH}{I}\right)\right]+\frac{M_2C'_1+M_2C_1^{'2}C_{C1}}{C_{C1}X}\left[e^{\left(\frac{Y}{2X}\right)\left(\frac{H}{I}\right)}\cdot\sin\left(\frac{RH}{I}\right)\right] \tag{21}$$

The expressions for peak delay time and peak voltages at node C and B are discussed by Eqs. (15), (16), (20) and (21) respectively. The essential proposed crosstalk voltage and peak crosstalk noise voltage respectively at node C are discussed by Eqs. (14) and (16).

3 Simulation Result and Discussion

Figure-1 shows simulation set-up of two L type High speed RLC mutually coupled interconnection system having 1000 μm of length. High performance CPU system designs typically consist of such type of bus structures. Symmetrical Step signal having finite and equal rise/fall time of 10 ps is used to excite the aggressor line. It is assumed that the interconnection system is identical and symmetrically distributed by considering that the system is connected with identical size of inverters for drivers and loads. Variations in the input slew times values up to 200 ps are used for the simulation of Mutually coupled interconnection system connected with identical driver size.

For the testing and verification purpose, we have compared our proposed model values with BKM [19] model values to show the novelty of our proposed work. This comparison was done on the same set of circuit parameters. Our work is much-improved version of BKM model [19] for the same L-interconnect model with the consideration of mutual inductance for high operating frequencies. Comparison of simulated results at node C for the expression given by Eq. (16) for proposed model and BKM model are demonstrated in Table 1 for various input slew times. Table 2 discusses the comparison of aggressor line voltage described by Eq. (21) with BKM model values and our proposed model for the various input slew values. Comparative results for the peak times t_{pc} and t_{pb} at node C and node B of the victim line and aggressor line described by Eqs. (15) and (20) respectively is discussed in Table 3 and Table 4. Comparative results for proposed model Aggressor voltage, BKM and SPICE for different values of T_s are shown in Figs. 2, 3 and 4 respectively. Similarly, comparative results for aggressor line voltage, victim line peak time and aggressor line peak time values from proposed model and the values from SPICE simulations are shown in Figs. 5, 6, 7, 8, 9, 10, 11, 12 and 13 respectively.

Table 1 Comparative analysis for peak crosstalk noise received from proposed model and BKM [19]

R_{D1} (Ω)	R_{D2} (Ω)	C_L (fF)	$T_s = 0$		$T_s = 100$		$T_s = 200$	
			BKM values (mV)	Our model values (mV)	BKM values (mV)	Our model values (mV)	BKM values (mV)	Our model values (mV)
10	10	1.2	121	101	161	147	210	181
10	15	1.2	139	118	179	167	231	197
15	20	1.2	147	143	202	189	245	231
15	25	2.4	182	167	238	216	279	268
15	50	2.4	221	192	271	243	291	294

Table 2 Comparative analysis for aggressor line voltage received from proposed model and BKM [19]

R_{D1} (Ω)	R_{D2} (Ω)	C_L (fF)	$T_s = 0$		$T_s = 100$		$T_s = 200$	
			BKM values (mV)	Our model values (mV)	BKM values (mV)	Our model values (mV)	BKM values (mV)	Our model values (mV)
10	10	1.2	0.761	0.551	0.951	0.745	1.17	1.002
10	15	1.2	0.872	0.649	1.02	0.812	1.26	1.012
15	20	1.2	1.01	0.998	1.35	0.903	1.41	1.213
15	25	2.4	1.29	1.123	1.48	1.212	1.68	1.534
15	50	2.4	1.43	1.324	1.71	2.012	1.74	2.121

Table 3 Comparative analysis for victim line peak time got from proposed model and BKM [19]

R_{D1} (Ω)	R_{D2} (Ω)	C_L (fF)	$T_s = 0$		$T_s = 100$		$T_s = 200$	
			BKM values (nS)	Our model values (nS)	BKM values (nS)	Our model values (nS)	BKM values (nS)	Our model values (nS)
10	10	1.2	23.12	17.13	37.73	23.87	59.29	32.73
10	15	1.2	27.34	18.95	48.24	26.12	77.92	37.49
15	20	1.2	39.71	20.32	73.84	28.78	98.28	39.15
15	25	2.4	45.64	21.78	91.27	30.19	103.37	41.29
15	50	2.4	59.59	22.12	103.12	32.19	118.23	44.29

Table 4 Comparative analysis for aggressor line peak time vn ‘received from proposed model and BKM [19]

R_{D1} (Ω)	R_{D2} (Ω)	C_L (fF)	$T_s = 0$		$T_s = 100$		$T_s = 200$	
			BKM values (nS)	Our model values (nS)	BKM values (nS)	Our model values (nS)	BKM values (nS)	Our model values (nS)
10	10	1.2	10.54	7.13	28.43	10.2	46.54	15.17
10	15	1.2	25.32	7.98	36.21	12.28	59.32	16.29
15	20	1.2	38.65	8.54	47.32	15.19	73.72	18.54
15	25	2.4	55.23	10.19	65.81	18.39	82.82	20.18
15	50	2.4	68.84	10.36	75.32	21.26	99.32	22.87

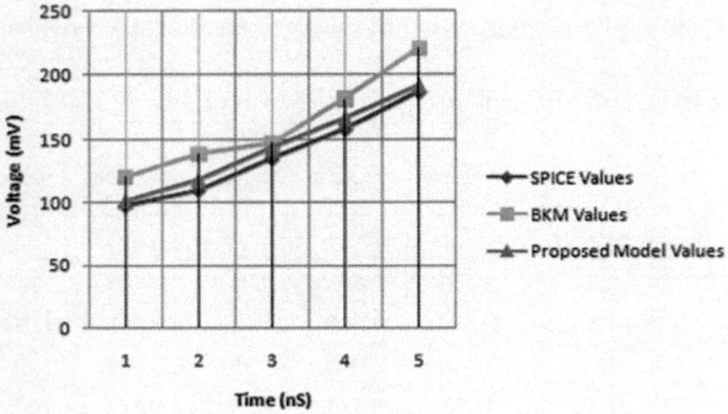

Fig. 2 Comparative graph of peak crosstalk noise obtained from proposed model, SPICE and BKM with $T_s = 0$

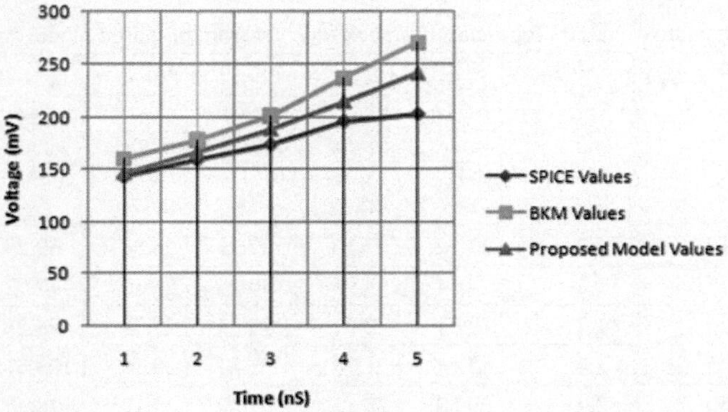

Fig. 3 Comparative graph of peak crosstalk noise obtained from proposed model, SPICE and BKM with $T_s = 100$

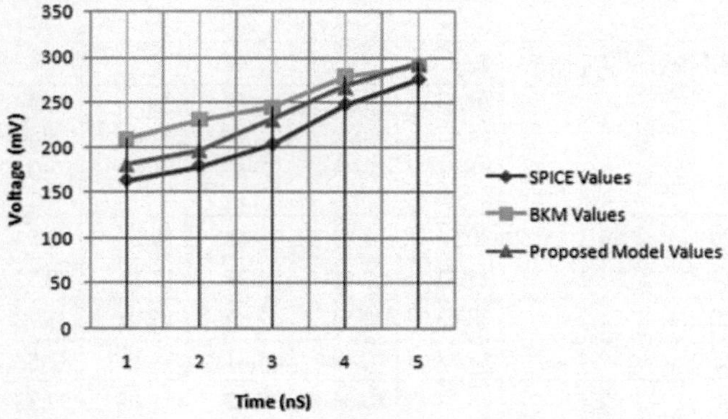

Fig. 4 Comparative graph of peak crosstalk noise obtained from proposed model, SPICE and BKM with $T_s = 200$

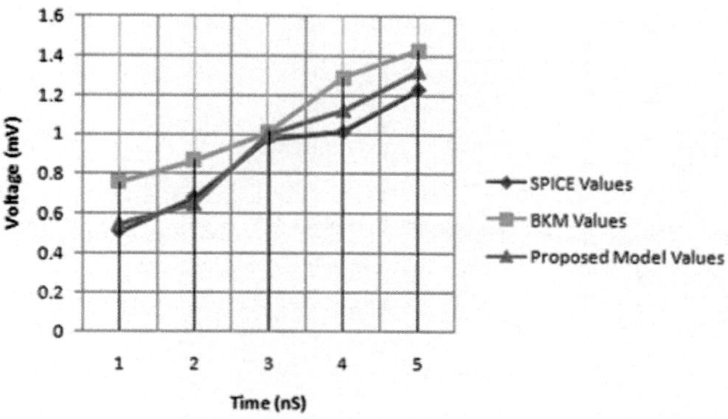

Fig. 5 Comparison of aggressor line voltage obtained from proposed model, SPICE and BKM with $T_s = 0$

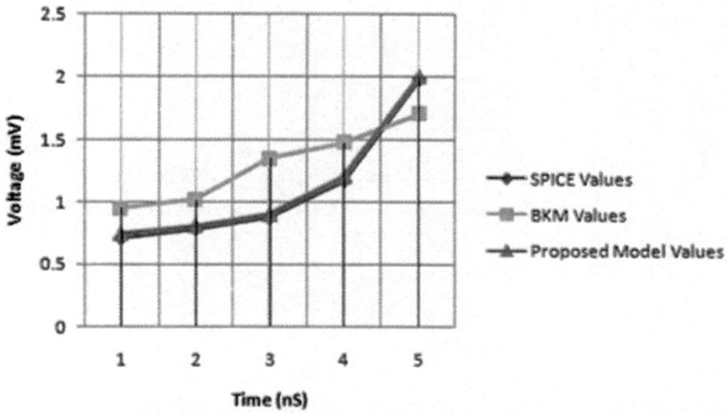

Fig. 6 Comparison of aggressor line voltage obtained from proposed model, SPICE and BKM with $T_s = 100$

Fig. 7 Comparison of aggressor line voltage obtained from proposed model, SPICE and BKM with $T_s = 200$

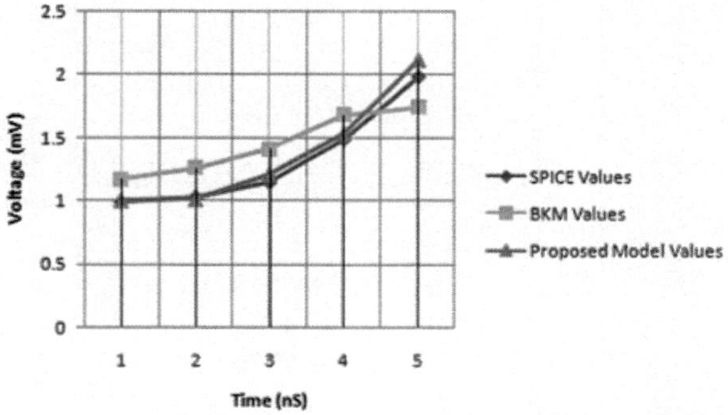

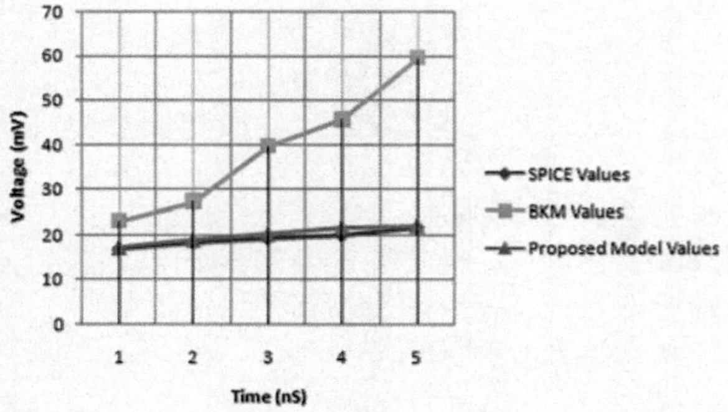

Fig. 8 Comparison victim line peak time obtained from proposed model, SPICE and BKM with $T_s = 0$

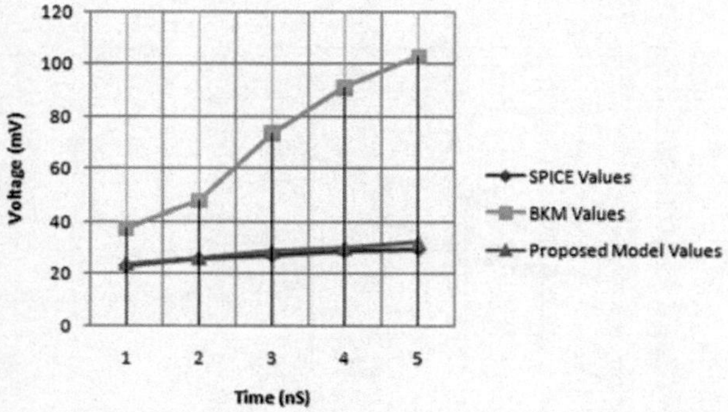

Fig. 9 Comparison victim line peak time obtained from proposed model, SPICE and BKM with $T_s = 100$

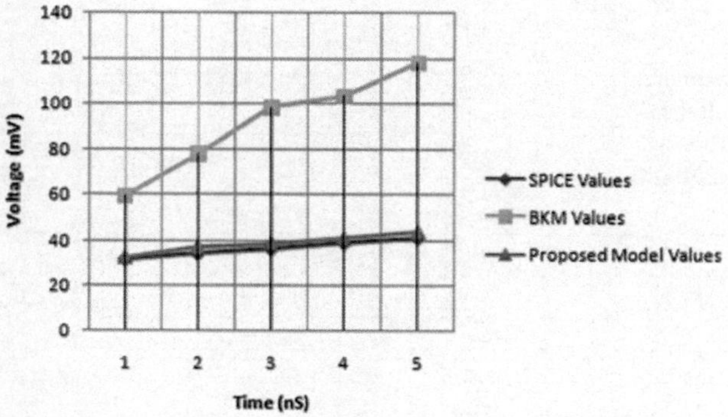

Fig. 10 Comparison victim line peak time obtained from proposed model, SPICE and BKM with $T_s = 200$

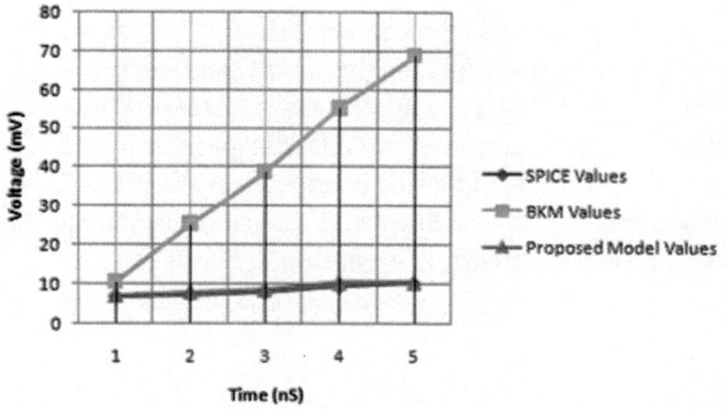

Fig. 11 Comparison of aggressor line peak time obtained from proposed model, SPICE and BKM with $T_s = 0$

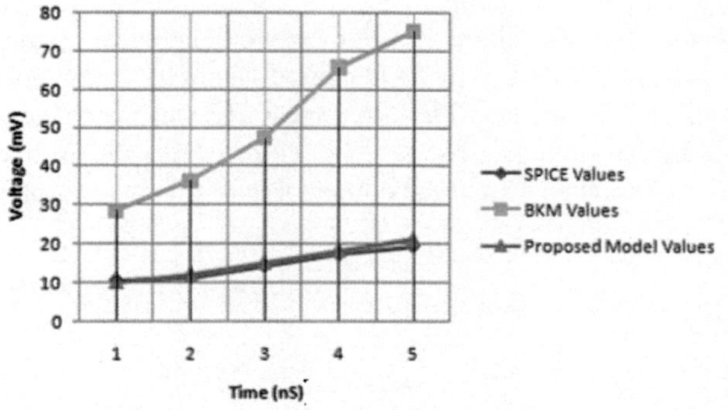

Fig. 12 Comparison of aggressor line peak time obtained from proposed model, SPICE and BKM with $T_s = 100$

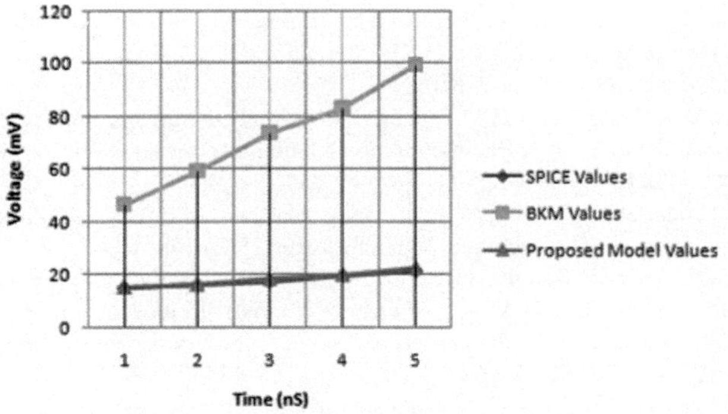

Fig. 13 Comparative graph of aggressor line peak time obtained from proposed model, SPICE and BKM with $T_s = 200$

After analysing the simulated result related Figs. 2, 3, 4, 5, 6, 7, 8, 9 and 10 and Tables 1, 2, 3 and 4, we can easily find out the novelty and importance of our proposed model in comparison to BKM [19] model. By considering mutual inductance in between two coupled interconnection models, our proposed model becomes more realistic and generic as it follows SPICE results better than BKM model. Deviation in between BKM model values with SPICE values is very large therefore; BKM model becomes appropriate in current scenario.

4 Conclusion

Proposed research work discussed about the mathematical analysis of delay and crosstalk voltage in mutually coupled RLC VLSI interconnection structures. The derived models for crosstalk voltage and delay are found precise as simulation results are very close to SPICE. The L-type RLC mutually coupled interconnection system is proposed in this research work. The correctness and validity of the research work is demonstrated by the simulation results. Simulation results shows that the proposed models are having less than 10% error comparable to the results obtained from the SPICE.

References

1. Wu B (2017) High bandwidth IC interconnects with silicon interposers and bridges for 3D multi-chip integration and packaging. In IEEE Semiconductor Technology International Conference (CSTIC) China, 12–13 April 2017, pp 1–3
2. Zhang XJ, Jiang W, Gao L, li H (2016) Impact of cross talk on signal integrity of high-speed density ceramic package for IC. In: IEEE international conference on electronic packaging technology Wuhan, 16–19 August 2016
3. Wu SY, Liew BK, Young KL, Yu CH, Sun SC (1999) Analysis of interconnect delay for 0.18 μm technology and beyond. In: IEEE International Conference on Interconnect Technology, San Francisco, CA, USA May 1999, pp 68–70
4. Agrawal Y, Chandel R (2016) Crosstalk analysis of current-mode signalling-coupled RLC interconnects using FDTD technique. IETE Tech Rev 33(2):148–159
5. Gupta A, Maheshwari V, Sharma S, Kar R (2015) Crosstalk noise and delay analysis for high speed on-chip global RLC VLSI interconnects with mutual inductance using 90 nm process technology. In: International Conference on computing on computing, communication and automation (ICCCA-2015), 15–16 May 2015, pp 1215–1219
6. Fattah G, Masoumi N (2018) Comprehensive evaluation of crosstalk and delay profiles in VLSI interconnect structure with partially coupled lines. J Iran Assoc Electric Electron Eng 14(4):41–54
7. Rebelli S, Nestala BR (2018) An efficient MRTD model for analysis of crosstalk in CMOS-driven coupled cu interconnects. Radio Eng 27(2):532–540
8. Rebelli S, Nistala BR (2020) A multiresolution time domain (MRTD) method for crosstalk noise modeling of CMOS-gate-driven coupled MWCNT interconnects. IEEE Trans Electromagn Compat 62(2):521–531

9. Bhattacharya S, Rahaman DDH (2017) Analysis of temperature- dependent crosstalk for graphene nanoribbon and copper interconnects. IETE J Res 10(20):187–205
10. Maheshwari V, Gupta S, Khare K, Yadav V, Kar R, Mandal D, Bhattacharjee AK (2012) Efficient coupled noise estimation for RLC on-chip interconnect. In: IEEE Symposium on Humanities, Science and Engineering Research (SHUSER-2012), Kuala Lumpur, Malaysia, 24–27 June 2012, pp 1125–1129
11. Hunagund PV, Kalpana AB (2010) Crosstalk noise modeling for RC and RLC interconnects in deep submicron VLSI circuits. J Comput 2:60–65
12. Kar R, Maheshwari V, Mal AK, Bhattacharjee AK (2010) Delay analysis for on-chip VLSI interconnect using gamma distribution function. Int J Comput Appl 1(3):65–68. Article 11
13. Peng X, Pan Z (2017) The analytical model for crosstalk noise of current-mode signaling in coupled RLC interconnects of VLSI circuits. J Semiconductors 38(9):095003
14. Mudavath R, Naik BR (2018) Estimation of far end crosstalk and near end crosstalk noise with mutually coupled RLC interconnect models. In: IEEE International Conference on Communication and Signal Processing India, 3–5 April 2018, pp 0198–0201
15. Maheshwari V, Suvra R, Kar D, Mandal AKB (2012) Analytical crosstalk modelling of on-chip RLC global interconnects with skin effect for ramp input. Procedia Technol 6:814–821
16. Mudavath R, Naik BR, Gugulothu B (2019) Analysis of crosstalk noise for coupled microstrip interconnect models in high-speed PCB design. In: International Conference on Electronics, Information, and Communication (ICEIC) New Zealand, 22–29 January 2019
17. Chhotray SK, Mohapatra DD, Dash SS, Swain S (2013) An analytical delay expression for deep sub-micron RLC interconnect. Int J Eng Res Appl 3(5):1727–1730
18. Chowdhury MMH (2004) Noise Analysis and Design Methodologies in Deep Sub-Micron VLSI circuits. PhD Dissertation, North-Western university
19. Kumar B, Maheshwari V, Dhubkariya DC, Kar R (2013) Analytical modeling of crosstalk noise and delay for high speed on-chip global RLC VLSI interconnects. J Electron Dev 17:1452–1456
20. Strang G (1998) Linear Algebra and its Applications, 3rd ed. Harcourt Brace Jovanich, Inc.

Hardware Software Interfacing for FPGA Configuration

Yashwarya Khandait, R. S. Gamad, and Trinath Somarouthu

Abstract In this paper, FPGA Configuration is implemented in which programmable circuit of a FPGA and a USB-PHY transceiver are bring together to form new programming interface. Purpose of FPGA configurartion by Flash programmer is to minimize software and communication overhead to achieve close to theoretical programming time possible on polarfire devices. Although the focus is to speed up polarfire configuration, previous generation of flash devices could take advantage of the hardware architecture and achieve faster programming time as well. FPGA offers a variety of programming Option; FPGA device usage JTAG or SPI interface to configure on-chip system controller through its dedicated USB port. To achieve that we used java native Interface which creates libraries to be used with a proprietary interpreter tool, which provide us flexibility to configure FPGA devices with JTAG Instructions. Visual studio is used to inherit product specific libraries along with JNI. This platform can also be used by broad range of applications wherever high-speed data transfer is required. USB 3.0 capable host is used for high speed communication.

1 Introduction

Implementing new interface is inspirational and challenging at the same time such as software and Hardware interface development. To effectuate that author develop a new Interface efficiently, using Visual Studio for generating framework for the interface flow and to authenticate the new native method implementations. In this work, native interface is presented that is used to interface a Flash programmer that contains USB-3.0 transceiver module, which acknowledge FPGA's configuration

Y. Khandait (✉) · R. S. Gamad
Department of Electronics and Instrumentation, Shri G. S. Institute of Technology and Science, Indore, India

T. Somarouthu
Design Engineering, R&D India, Microchip, Hyderabad, India
e-mail: Trinath.Somarouthu@microchip.com

V. Nath and J. K. Mandal (eds.), *Proceeding of Fifth International Conference on Microelectronics, Computing and Communication Systems*, Lecture Notes in Electrical Engineering 748,
https://doi.org/10.1007/978-981-16-0275-7_67

to transfer data with a processing unit over USB-3.0 high Speed via High level programming Language [1]. In this configuration, java platform is used in a central processing unit for code compilation and decoding data for FPGA configuration during initial phase.

FPGAs offer variety of configuration to cater to diverse end-user applications. Semi-custom device usage JTAG or SPI interface to program on-chip controller of device by its software. Three programming modes are supported based on the interface used; JTAG, SPI master, and SPI slave.

In JTAG and SPI slave programming modes, the device can be programmed either using an external programmer such as a microprocessor or use microchip FlashPro programmer (version 5 or later). The external master fetches the configured data (in the form of bitstream) from an external memory.

In SPI master programming mode, the system controller itself acts as the master and fetches the bitstream from an external flash memory to program the device. This mode supports two programming features: auto update and in-application programming (IAP). In auto update, the device reprograms itself on power-up, and in IAP, the device is programmed when the user application actuate programming [1].

Motivation behind this paper work is discussed in the next section, also a brief introduction of the Flashpro Programming, along with a demonstration of the previous literature in Sect. 3. System architecture is discussed briefly in Sect. 4. Subsect. 4.1 presents the JNI native libraries implementation. Idea of interfacing Flashpro6 with host and controller logic is describe in Subsect. 4.2. Section 5 presents achieved implementation results with the advantages of the interface architecture and communication speed over different devices with future expansion of this Configuration.

2 Motivation

The idea of this paper is taken by previously developed Flash programmers which is used for same purpose with moderate speed, so to achieve wide bandwidth and high baud rate output Flashpro6 is developed. Configuration and architecture of Flashpro are discussed in the next section of this paper. Interface presented over here should help to develop an equivalent software Interface by eliminating the Software need and generated delay for debugging and output analysis. Idea behind this paper was to set up a platform that can replace the need of specific licensed software to program FPGA for various applications. For the reason number of use cases become visible now days. Future applications are also deliberate in end of this paper. Logical thought for selecting Flash programmer6 with USB-3.0 port is because of its immortality, popularity and supporting systems to work as main unit for configuration, visual representation, validation and so on. In Addition to that USB-3.0 consist of error rectification methods and shorten the efforts for communication overhead [2].

Field programmable gate array is a semiconductor device that are based around a matrix of configurable logic blocks connected via programmable interconnects. It fall under the semi-custom devices which means it can be programmed or reprogrammed to the required functionality after manufacturing. FPGA is an integrated circuit consisting of an arrangement of uncommitteds element. The connection between those element is user-programmabled. Logic blocks are programmed to implement the desired function Using RAM, high-density logic is provided. The design flows is follow until the timing simulations and then the generate file is download into the target device (FPGA) [11].

To use Microchip Flashpro6, This hardware software interfacing is presented. For Hardware interfacing JTAG, USB-PHY is used. Debug port is allocated for implementing a serial communications interface between FPGA devices and USB 3.0 which provides minimal overhead access without a need of direct external access to the system memory data and address buses. Dynamic protocol is implemented in on-chip Test Aceess port (TAP) which used to access component specification and test registers of chip logic levels.

Flashpro6 can operate in two modes: Flashpro5 and Turbo mode. M2S device is used to use existing CM3 and SPI IP core already available and to implement JTAG IP block in fabric [16].

2.1 Flashpro5 mode

In this mode, Flashpro6 operates exactly like Flashpro5 programmers with respect to JTAG and SPI programming. The same software libraries already implemented for Flashpro5 could be used for Flashpro6 support in this mode with minor modification. Flashpro5 mode is used to do the following:

- Run scan chain.
- Detect vjtag voltage.
- Detect and apply vpump voltage.
- Run STAPL player for polarfire and legacy flash devices.
- Product Specific API support.

2.2 Turbo Mode

In this mode, USB-PHY is used in FIFO mode to send chain configuration, JTAG/SPI clock frequency value, dat file, compiled code and the action to be performed to fabric. CM3 is then released from reset mode to start executing the requested action. In this mode use, the existing Direct C solution and dat files should be used with minor modification if required.

For Software Interfacing JNI creates libraries is developed, JNI creates libraries provides interface where we are creating Implementation of each instruction, parameters and components in High Level Language(java). link the source code with product libraries In visual studio by using java native Interface. Create Libraries and.dll files according to system dependencies, Create a jar file out of it so that we can place it anywhere and call from server location through Batch files.

3 Related Work

Various journals and research paper work have been published in the field of FPGA programming, in which authors are using Flash programmers to load bitstream generated with license software into the FPGA. many other Interface with different software like Intel® Quartus ® Prime software, MAX + PLUS ® II, FlashPro_v11_5 and Xilinx have been introduced in the field of FPGA programming.

An USB–DAQ interface unit that allows connecting data acquisition (DAQ) application to USB is presented. The unit contains two main components: USB to FIFO IC controller and application controller, designed as VHDL core for FPGA. DAQ logic can be connected to USB through simple I/O and streaming interfaces, thus development time of user application can be reduced. The design was tested with high-speed and SuperSpeed FTDI and USB – FIFO controllers [10]. A radar transceiver chip that usage USB Interface for frequency modulation is described In [3] which operates on a frequency of around 24 GHz with 8 GHz bandwidth. USB interface is used by the system to power all components and also for processing the data [4]. In [5, 15] USB interface for radar data transmission to the host via MATLAB is used.

Analysis shows that software and hardware overhead of previous version of Flash programmer is caused by two major factors: USB bidirectional communication overhead and JTAG/SPI clock speed. To eliminate all issues causing the overhead seen with the current Flashpro3/4/5 programmers, an embedded solution could be used with a fast memory to fetch the data as well as SPI and JTAG IP blocks implemented in fabric to run the JTAG and SPI clocks at their maximum frequencies allowed by the FPGA devices.

- JTAG - 50 MHz
- SPI - 80 MHz

Fig. 1 FPGA configuration modes and interface.

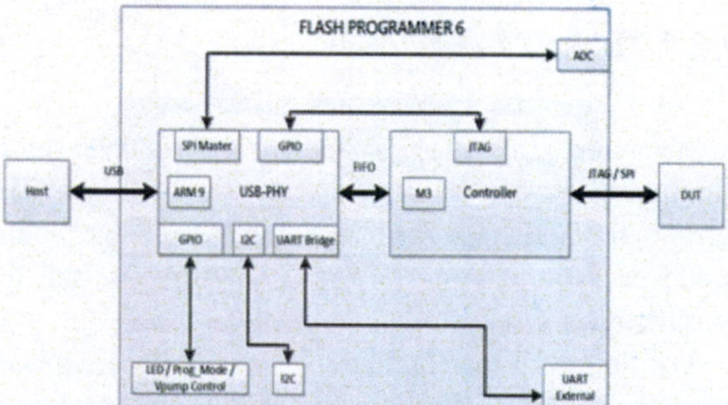

4 System Architecture

Overall System's architecture of Flash programmer is presented in Fig. 1. The first shown module in the Left side of Fig. 1 represents the host computer, in which visual studio is running for the evaluation of the native code followed by communication chain which contains the USB-PHY 3.0, FPGA and DDR SRAM. On the right side of Fig. 1 analog-to-digital converter unit (ADC) is used for data conversion. DUT is connected to Flash programmer with 10-pin JTAG connector on the right side [18].

4.1 JNI Creates Libraries

Wrapper libraries for flash programmers is implemented few years back, each time it need modification in source code whenever programmer is updated [12].

Product specific API which is allow vendors to use Flash-programmer supported programming hardware in their own application without the need to worry about programmer specific drivers and the need to update the source code when a new programmer is introduced. JTAG protocol is the only protocol supported with this API. All programmers are supported on Windows based systems with the exception of Flashpro5 which is also supported on Linux platforms as well. "C" API library that provides JTAG level access to Flash programmers. To use product specific API, (.h) header file, (.lib) library file and (.dll) linker file is created out of API source code [12, 17].

To create a graphical user interface optimize with C visual studio is used. To declare native methods along with main method java file is created. Implementation of all these native methods are available in C language. This native methods are basically implementation of Tap controller states, which is used for JTAG programming [13, 14].

To call API, need to load.dll file in our program first for that we created static block and usage default java function System.Loadlirary to load.dll file and created Main method To call Native method at the Time of Compilation.

4.2 Flashpro6 Interface

As mention in previous sections of this paper, user need to update the source code when a new programmer is introduced Because every programmer are different in terms of architecture schematic and components. Flashpro6 has USB-PHY Which other programmers did not have. USB-PHY is the High-Speed peripheral controller USB 3.0 which provides developers an option to add USB 3.0 for enhance device functionality to any methodology. USB-PHY has a fully configurable, General

Programmable Interface that can interface with any processor, ASIC, image sensor, or FPGA [20].

Before importing Flashpro6 specific functions into native libraries, we have to use USB libraries, Header File to Create Native Interface so that we can import that function into pre created native libraries and call each function independently which improve speed of validation Test. For Flashpro6 Interface all the java were implemented on a java development kit jdk1.8.0_111. jdk Platform is a 32-bit, creating Interface in java provides us Platform Independency, Automatic Storage Management, Avoids Unsafe Constructs, Exception Handling Support, Multi-threading Support and we can easily Avoids Unsafe Constructs.

5 Result

In this paper author presented an interfaces created between USB-3.0 transceiver port and enables FPGA Configuration logic to improve communication to a host via USB-3.0 with java programming (Fig. 2).

Although development of FPGA configuration is taken by previous versions to improve data transfer rate, which make it useful for broad variety of System application. It is beneficial to use it in many other applications that require a high-speed communication to a USB-3.0 capable host.

The data transfer tests have been performed on a host computer which is based on an Intel Core i7-6500U processor, containing 32.0 GB of RAM, 64-bit Operating System running Microsoft Windows 10 and providing native USB-3.0 high speed connectivity. In order to conduct the tests a command line demonstration application has been written which transfers data in both directions [6].

Speed achieved with Flashpro6 programmer for JTAG-23 Mhz for Polarfire devices, 20 Mhz for Smartfusion devices and 17 Mhz for RTG4 and SPI—13.33 MHz for Polarfire devices. Supported devices for Flashpro6 are Smartfusion, RTG4 and Polarfire for JTAG Programming and Smartfusion and Polarfire devices for SPI Programming.

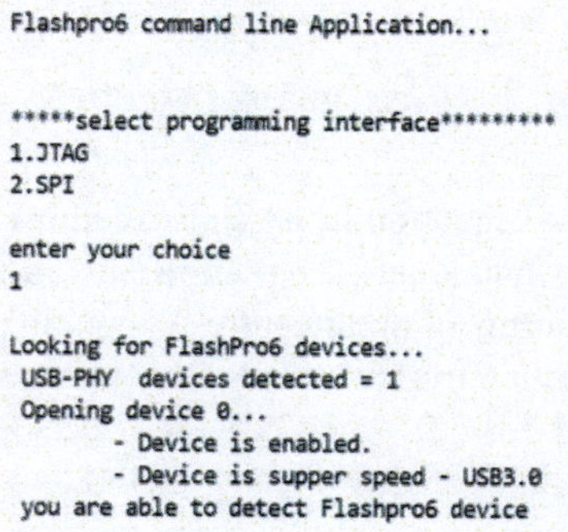

Fig. 2 Output window

In FPGA configuration, depending upon the type of application, design and practicality many different strategies other than High Speed USB port and DDR, can be performed for magnify the system performance by implementing it on hardware. However, the structure of algorithm includes a multiple optimization process and components; practical application of this scheme may demand high computational integrity. Hence, the hardware implementation of proposed Interface configurations is a challenging task. Contrarily, many other Interface with different software like Intel® Quartus ® Prime software, MAX + PLUS ® II, FlashPro_v11_5 and Xilinx have been introduced in the field of FPGA configuration. They can provide good results with various combinations in real-time applications by generating bitstream. Hence, there is a large scope to implement the suggested Flashpro6 native interface in real time application. It would be interesting to make experiments or simulations on validation & verification that include test usecases, and usermodels along with the fabric, Secure non-volatile memory (snvm) components and magnetic storage elements to find possible optimization configuration to implement the proposed interface. Therefore this Software Hardware Interface can be used various applications for high-speed communication.

References

1. Andrews D, Niehaus D, Ashenden P (2004) Programming models for hybrid CPU/FPGA chips. Computer 37(1):118–120
2. Hewlett-Packard. Intel, Microsoft, NEC, ST-Ericsson and T. Instruments, "Universal Serial Bus 3.0 Specification" ,USB Implementers Forum (2011)
3. Bredendiek C, Pohl N, Jaeschke T, Thomas S, Aufinger K, Bilgic A (2012) A 24 GHz wideband single-channel SiGe bipolar transceiver chip for monostatic FMCW radar systems.In: The 7th European Microwave Integrated Circuits Conference, Amsterdam, The Netherlands
4. ARM (2011). AMBA AXI and ACE Protocol Specification, ARM
5. HuX, Jin Y, Li Z (2018) A parallel JTAG-based debugging and selection scheme for multi-core digital signal processors. In: 2018 IEEE International Conference of Safety Produce Informatization (IICSPI)
6. Janssen B, Hubner M, Jaeschke T (2014) An AXI compatible cypress EZ-USB FX3 interface for USB-3.0 SuperSpeed. In: 2014 International Conference on ReConFigurable Computing and FPGAs (ReConFig14)
7. BakshiA, Pandey S, Pradhan T, Dey R (2013) ASIC implementation of DDR SDRAM memory controller. In: 2013 IEEE International Conference ON Emerging Trends in Computing, Communication and Nanotechnology (ICECCN)
8. Li Ling (2013) Design of FPGA debugging interface based on EZ-USB slave FIFO. Appl Mech Mater 367:240–244. https://doi.org/10.4028/www.scientific.net/AMM.367.240
9. Jiang X, Cui X, Yu D (2011) A JTAG-based configuration circuit applied in SerDes chip. In: 2011 9th IEEE International Conference on ASIC
10. ChenX, Zhang D, Yang H (2008) Design and implementation of a single-chip ARM-based USB interface JTAG emulator. In: 2008 Fifth IEEE International Symposium on Embedded Computing
11. Zou H, Huang J, Gao M (2011) The application of Virtual JTAG technology in FPGA design and debugging. In: 2011 International Conference on Electrical and Control Engineering (2011)
12. Vairale VS, Honwadkar KN (2010) Wrapper Generator using Java Native Interface. Int J Comput Sci Inf Technol 2(2):127–141

13. Eschermann B (1992) An implicitly testable boundary scan TAP controller. J Electron Test 3(2):159–169
14. Tamara TT, Affandi A, Imam Arifin M, Adityo F (2019) Graphical user interface optimalization based C# with visual studio to operator control unit rhino robot. IPTEK J Proc Ser 0(3):54
15. Reddy NK, Suresh N (2015) An efficient approach for design and testing of FPGA programming using LabVIEW. Int J Reconfigurable Embed Syst (IJRES) 4(3):192
16. Sutikno T, Idris NRN, Jidin AZ (2014) Overview on strategies and approaches for FPGA programming. TELKOMNIKA (Telecommun Comput Electron Control) 12(2):273
17. Sotiriadis S, Omosebi O, Ayapbergenova A, Saparkhojayev NP (2018) Evaluating the java native interface (JNI): leveraging existing native code, libraries and threads to a running java virtual machine. Int J Distrib Syst Technol 9(2):39–61
18. Abdullah D, Mahmood R (2006) Design flash memory programmer device. AL-Rafidain J Comput Sci Math 3(1):55–83. https://doi.org/10.33899/csmj.2006.164045
19. Raju A (2012) High speed data acquisition system with ethernet interface. IOSR J Electron Commun Eng 2(5):12–17
20. Jithendar C (2012) Landela P (2012) FPGA implementation of real time ethernet communication using RGMII interface. Int J Sci Res 2(3):175–177

OpenCLTM Implementation of Rapid Image Restoration Kernels Based on Blind/Non-blind Deconvolution Techniques for Heterogeneous Parallel Systems

Ashutosh Satapathy and L. M. Jenila Livingston

Abstract Over the years, image enhancement is one of the research areas in the field of image processing and computer vision. In the real world, images or videos captured and transmitted by different sensors are not fit for further processing. Image restoration is the sub-branch of image enhancement used to reconstruct the original image from blur images using restoration models. It plays an immense role in many real-time applications such as traffic monitoring, vigilance system, human identification, and recognition, while it also requires faster implementation. In this paper, we have discussed different image degradation models and the most well-known restoration techniques such as Inverse, Pseudo Inverse filter, Wiener filter, Constraint Least Square Error filter, and Lucy Richardson filter are used to reestablish an original image from a distorted one. Apart from that, we have developed OpenCL kernels for those above filters for faster image reconstruction, which are also easily deployed on multiple heterogeneous parallel systems. At last, we have done a performance analysis of both OpenCL CPU and GPU implementation of the above restoration techniques in terms of accuracy and time consumption. Different performance metrics like Root Mean Square Error, Root Mean Square Error, Peak Signal to Noise Ratio, and Structural Similarity Index have used to evaluate the accuracy of the above techniques.

Keywords Point spread function · OpenCL architecture · Blur models · Restoration models · Restoration techniques

---

A. Satapathy (✉) · L. M. J. Livingston
SCOPE, Vellore Institute of Technology, Chennai 600127, India
e-mail: ashutosh.satapathy2013@vit.ac.in

L. M. J. Livingston
e-mail: jenila.lm@vit.ac.in

V. Nath and J. K. Mandal (eds.), *Proceeding of Fifth International Conference on Microelectronics, Computing and Communication Systems*, Lecture Notes in Electrical Engineering 748,
https://doi.org/10.1007/978-981-16-0275-7_68

1 Introduction

An image speaks millions of words, and its quality is affected due to various circumstances such as bad weather, motion, sensor noise, and noise in the transmission medium. It causes degradation to image or video data, which recover through a restoration process. Image restoration is one of the research areas in image processing, and fast image restoration is a challenging task too. It can be useful for lots of real-time applications such as traffic monitoring, vigilance system, human identification, and recognition to provide better results [1, 2]. An image restoration operation must require a greater understanding of its quality degradation. The main idea behind an image restoration is to regain the original image from its blurred picture, which fades due to some degradation functions [3, 4]. The degradation function is also called the Transfer function and Point Spread Function (PSF). Image degradation and its restoration, are commonly described by two mathematical functions known as degradation model and restoration model. A degraded image and its recovered image using Constraint Least Square Error (LSE) filter; are shown in Fig. 1.

At the time of image acquisition, the final image degrades due to fault in the capturing process, and it has an impact on the consequence image processing operations, which made the whole method a complete disaster. There are broad areas of image deterioration, such as additive and multiplicative noises, spatial deformation, color, contrast, and illumination deformation cause an image quality degradation that happen due to the sensor motion or the involvement of electromechanical devices [5]. Some of the most occurred degradations happen during the acquisition of the original scene while the camera is moving, improper focus by an optical system, or by remote sensing devices while capturing an aerial image. Atmospheric turbulence can also be a driving factor for inducing blur during the image acquisition process [6]. Apart from that, CT scan image and electron micrographs degradation depend on focal points' position of electron lenses and their positional deviation [7]. However, real-time image degradation happens not only by various blurring effects but also there are some noises get added during this process from acquisition to transmitting the data to its endmost terminal [8]. Image restoration is a method of recovering the original image from its corrupted one that tries to inverse the deformation operation that occurs in the image formation system. It estimates the whole blur identification

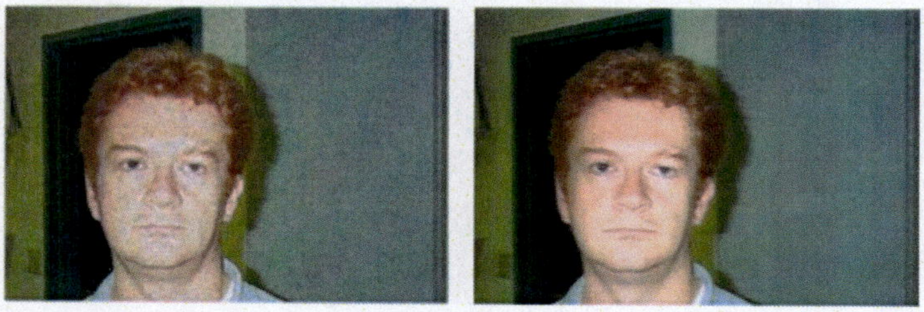

Fig. 1 Degraded image and its restored image using CLSE filter

plus its restoration operations by observing the degenerated image in the form of a single function called PSF [9].

To make the restoration process effective, it requires an efficient mathematical model to represent the degradation process and implementation of a restoration filter that corrects that degradation. The process of getting blurred is very complex, nonlinear, and mostly unknown. So, it is often modeled as a linear function to make the calculation simple. The degradation process is generalized as a linear system, as shown in Fig. 2. Here, f(x,y) is an ideal input image that is degraded by the transfer function H(x, y), and it is also added with noise n(x,y) to produce the degraded image g(x, y) [10]. The transfer function H(x,y); is also called as Point Spread Function (PSF), and it satisfies linearity, homogeneous and shift-invariant properties. The mathematical expression of such a system; is provided in Eq. (1).

$$g(x, y) = H(x, y) * f(x, y) + n(x, y) \tag{1}$$

As shown in Fig. 3, the restoration model takes a degraded image as input and uses filter function h(x, y) to produce its restored image f′(x, y). In the spatial domain, an image has recreated by taking the convolution of its noisy image with a filter function. Here, filter function h (x, y) must satisfy the properties of inverse transfer function H(x, y). Mathematical expression for image restoration; is given in Eq. (2).

$$f'(x, y) = g(x, y) * h(x, y) \tag{2}$$

Both g(x, y) and f′(x, y) functions are of size M × N pixels; i.e., pixels have distributed across M columns and N rows. f′(x, y) has constructed in such a way that error between a restored image and its actual image has minimized. The objective function of the restoration method has given below.

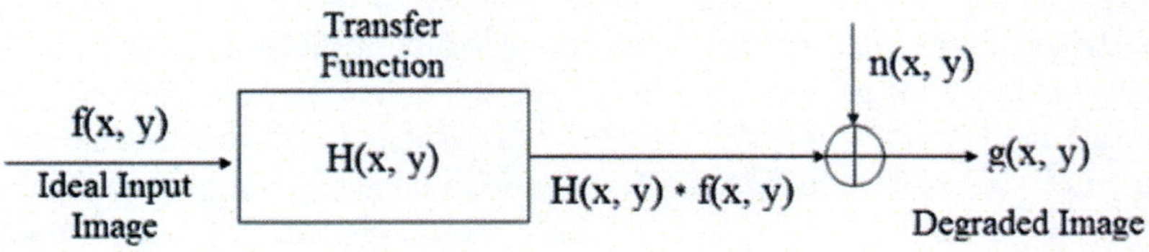

Fig. 2 Degradation model

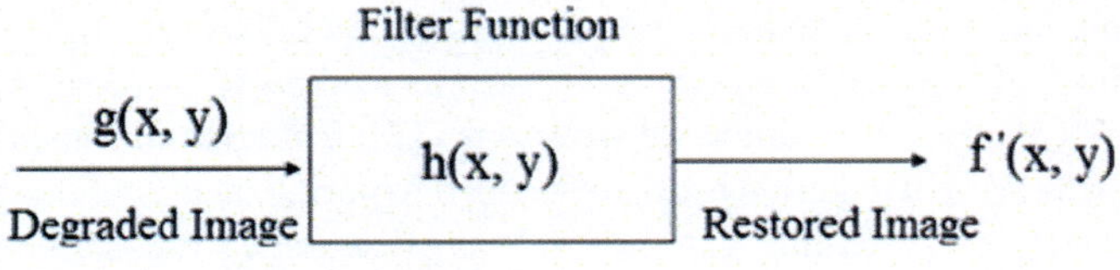

Fig. 3 Restoration model

$$\begin{aligned} &\text{Minimize } z = \left| f'(x, y) - f(x, y) \right| \\ &0 \le x \le M - 1 \text{ and } 0 \le y \le N - 1 \end{aligned} \quad (3)$$

Mainly, image restoration techniques are of two types, the first one is the blind deconvolution techniques, and the other one is the non-blind deconvolution techniques. Blind convolution techniques try to recover an image with a little knowledge or without any knowledge of the PSF, whereas the non-blind deconvolution techniques use prior knowledge of the PSF to restore the image quality [11]. In the upcoming sections, we have discussed OpenCL architecture with well-known blind and non-blind deconvolution filters and implemented OpenCL kernels for their fast computation.

2 Related Work

L. Xu, X. Tao and J. Jia from the Image and Visual Computing Lab, University of Hong Kong, came with a spatial deconvolution method that includes sparse prior to deal with noise and ocular artifact. Using this method, an inverse filter was successfully estimated using a set of 1D kernels, whose Computation time around 0.57 s for a 325 × 365 image on a multicore CPU [12]. In the year 2017, Zon, N. et al. developed a recursive framework, so-called Progressive Removal of Blur Residual (PROBE) that used an Inverse filter for fast and easy deburring and consumed 5 s to deblur a 512 × 512 image on a 3.4 GHz CPU [13]. Z. Mbarki, H. Seddik, and E. B. Braiek created a parametric Wiener filter to perform the deconvolution in the Fourier space and subsequently smoothed an image in the wave atom transformation by adjusting the threshold in between 0 and 1, which required approximately 8 s to restore a 256 × 256 image [14]. F. Baselice et al. from Università Degli Studi di Napoli Parthenope, Italy in 2018, proposed an enhanced Wiener filter whose kernels can be dynamically tuned to achieve better noise reduction including edge and details preservation in an image at the same time [15]. It required 21.2 s to process 1556 × 360 pixels on a Core i7 workstation, which is not suitable for any real-time image and video processing applications. Apart from these, a blind restoration technique was proposed by T. Goto, S. Otake, and S. Hirano, considers local patches for optimal PSF estimation for rapid image restoration, and it accomplished such task in 3.6 s and 3.1 s on the Intel Xeon E5-2360 CPU and 2688 cores NVIDIA GTX Titan GPU for a 482 × 482 RGB image [16]. In 2017, W. Liu et al. created a Semi-Global Weighted Least Squares filter that solved a large linear system by breaking it into a sequence of subsystems and calls them iteratively. It consumes 0.35 s for a 1 MB RGB image on an Intel i7 CPU for four number of iterations [17]. Unlike traditional restoration techniques, which involve kernel estimation, deconvolution, and final image generation; in the same year, L. Wang, Y. Li and S. Wang discussed a one-step rapid image restoration technique using a Convolutional Neural Network, whose consumption time for a 112 × 96 pixels sized face image is around 0.0148 s [18]. Fast space-variant image deconvolution using the Lucy Richardson filter for a 15 number of iterations, consumed 40 ms to enhance a 640 × 480 pixels grayscale image on an Altera Stratix V family FPGA [19]. HW/SW codesign of Lucy-Richardson deconvolution algorithm, accelerated on an FPGA, took around 180 ms for one iteration on a 610 × 340 grayscale image [20]. However, R. Das, A. Bajpai, and S. M. Venkatesan showed that the fast non-blind image deburring using the Lucy Richardson filter required 1.24 s, 1.7 s, and 6.63 s while restoring the RGB images of size 129 × 129, 267 × 267, and 512 × 512 pixels respectively [21].

3 OpenCL Architecture

Parallel programming gains lots of interest over the last few years in various institutions from universities to corporates as its users have been seen in multiple fields of computer science like mathematical modeling, Artificial intelligence, speech, image, and video processing. It uses a synchronous mode of communication between multiple working units present inside devices to achieve a highly complex task most effectively and reliably at the expense of a few times. Shortage of computing elements like processing power, data transmission power, memory, a computing device with a lack of parallelization, introduces different parallel devices so-called multi-threaded CPU, Graphics Processing Unit (GPU), and Free Programmable Gateway Array (FPGA), etc. Subsequently, various proprietary or open-source software packages like CUDA, OpenCL, MPI, OpenMP, OpenACC, Renderscripts, and Pthread came with their built-in functions to support data flow customization while utilizing these devices' memories and cores present inside them. These programming interfaces and APIs are broadly classified into two types, some of them are mapped to specific vendors' devices while others are platform-independent [22].

OpenCL is one of the platform-independent programming packages, developed by Apple Inc. that provides a platform for computing devices from various vendors like NVIDIA, RADEON, ARM, and FPGA. Later, its license had been transferred to the Khronos Group, a nonprofit organization is responsible for the creation of various open-source programming packages related to computer vision, image processing, high-performance computing, and sensor computing, meant for heterogeneous platforms and devices. The low-level abstraction of this language has been specifically designed for C, and Python languages with JACKET, MAGMA, BOLT C++, clAMDFFT, and clAMDBLAS accelerated library files supply additional benefits from the creation of a kernel to call and manage it by the host program. To provide a better understanding of the data flow between the host and the computing device until its kernel execution to produce the final output. OpenCL architecture has mainly classified into four atomic models, and the detailed information regarding these models is described in the upcoming subsections [23].

3.1 OpenCL Platform Model

Fundamental information about the underlying architecture of mounted devices is quiet necessary to run kernels by the host program, and platform models help to provide that information at the initial stage during program execution. Heterogeneous parallel devices from different manufacturers like Intel, NVIDIA, AMD, and ARM provide their OpenCL platforms; to be installed after connecting those devices to the host machine. Each parallel device has its own set of computation units, and those computation units have split into a set of work-items that indirectly reflect total processing elements inside the instruction engine. NVIDIA GTX 1050 Ti uses proprietary OpenCL 1.2 platform that divides twenty-four streaming processors consists of thirty two processing elements each into six computation units having 1024 work items each [24], whereas the Xeon E3 1225 V5 uses OpenCL 2.0 to divide its processing elements into four cores composed of 8192 work items individually [25]. Apart from that, those work-items can be distributed into three-dimensional index space for better program implementation, and details about these are discussed below in the OpenCL execution model.

3.2 OpenCL Execution Model

The OpenCL execution model consists a set of two execution process, the first one is the host executing a program that runs on the host computer, and the other one are multiple kernels execution run on the targeted parallel device. From the creation of host variables, platform, device attributes, context, command queue, programs and kernels to the building of programs until releasing them are handled by the host program, while the kernel execution model utilizes the work items present inside index space to run the application in quite a least amount of time. The host program divides the processing elements; present inside the device into N-dimensional NDrange or index space, which can be arranged in 1D, 2D, or 3D to provide benefits to varieties of applications like speech processing, image, and video processing. Figure 4 shows sixty-four processing elements of a device that are structured into a 1D, 2D, or 3D index space where each coordinate represents a single work item, and each work item is responsible for running a single thread of kernel [23]. In image processing applications, a total number of picture elements in an image is considered as the size of the ND-Range, and OpenCL clEnqueueNDRange() function is responsible for the creation of N-dimensional index space during its execution by host program at runtime.

In OpenCL, terms like work_item, work_group, global_size, local_size, global_id, and local_id have coined for better understanding about the NDRange whose requirement provides an extra bit of smoothness to a program. global_size in OpenCL represents the total size of an index space along each direction, and its size must be equal to the size of an input space organized in a 1, 2, or 3 dimensions. As

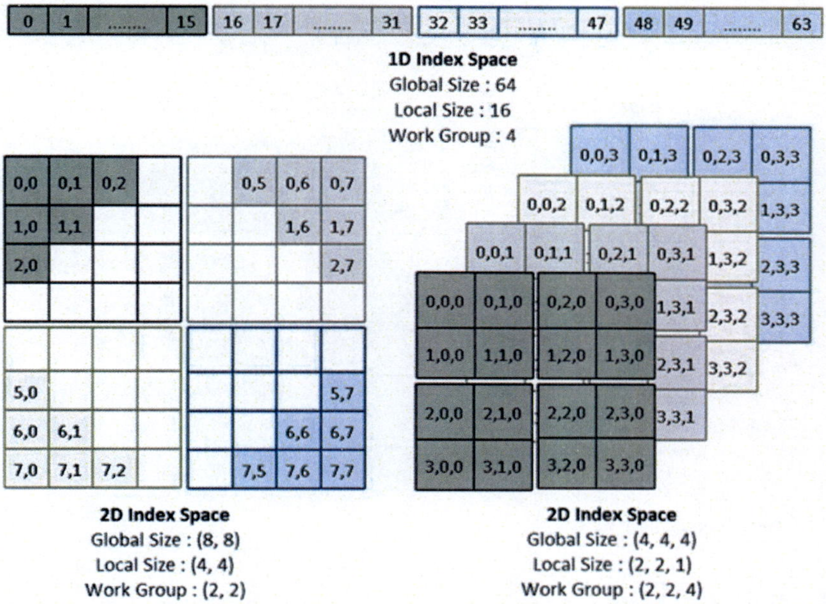

Fig. 4 OpenCL index space

we discussed above, a work_item denotes a processing element present inside a computing device, where it has identified by its global_id inside the index space. According to the application requirements, a set of work-items are logically well connected to form a work_group to achieve a particular task, and each work_item inside a group has a unique local_id that expresses its position inside a work_group. At last, work_group_id and work_group_size stand for the position of a work_group inside the index space and the total number of work_items present inside a work_group respectively. Figure 5 shows a brief description of those attributes in a three-dimensional index space made of 64 processing elements and assignment of those values have specified through the host program during its kernels' execution.

From the above figure, we have concluded that global size (Gx, Gy, Gz) of the 3D ND Range is (4, 4, 4) and local Size (Sx = 2, Sy = 2, Sz = 1) indicates the total number of work items are organized along each dimension in a 3D index space to form a group. (sx, sy, sz) illustrates local_id of a processing element inside a group whose global_id is computed with the help of its work_group_id (wx, wy, wz) and local_id (sx, sy, sz) as given below.

$$\text{global\_id}(gx, gy, gz) = (wx.Sx + sx, wy.Sy + sy, wz.Sz + sz) \quad (4)$$

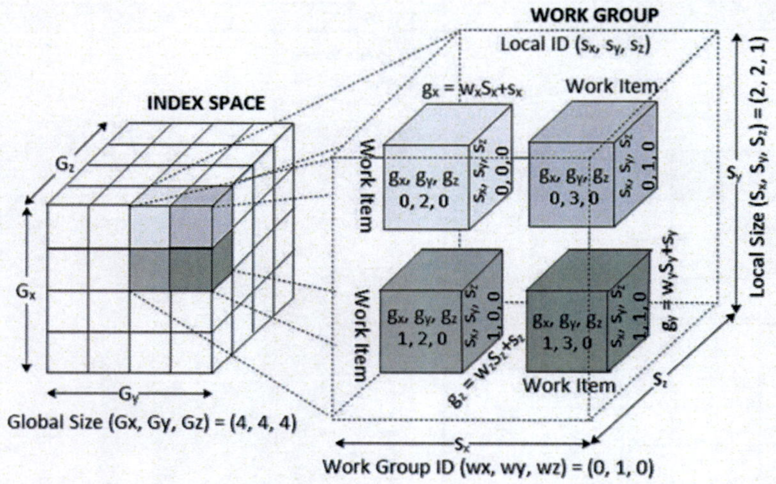

Fig. 5 OpenCL execution model

3.3 *OpenCL Memory Model*

The OpenCL memory model is responsible for the creation and distribution of information across the device memory to maintain data consistency between the host and devices by synchronizing information between work-items present inside the computing devices. That requires proper knowledge on the partition of data across the private, local, and global memory sections plus their working principles to gain efficiency in enormously parallel divergent devices [23]. Each computing device has a large part of the section utilized as global memory, which is shared by all work items inside all computation units to allow them to read and write into it. __global or global keyword specifies the attributes that have created in this memory, and initialization of those variables happens by the host program; are dynamically passed to the kernel at run time. Mapping of information from host memory to this memory makes it quite slower than other memory sectors, but information access cost of it is relatively lower than the host memory. Constant memory section behaves just like the global memory, which has also shared between all the processing elements from different units, but data present inside this memory will not change throughout kernel execution. The read-only nature of the constant memory section makes it slightly faster than the global memory section. Apart from those two universal memory sections, there are multiple local memory sections unique to their respective workgroups, and work items inside a workgroup share a local memory defined by a keyword called __local or local. In addition to that, each work-item has its own set of registers act like its private memory section, but the creation and allocation of data in these regions happen during kernel execution, unlike the constant or global memory regions. It provides a proper enclosure to data items of any processing elements

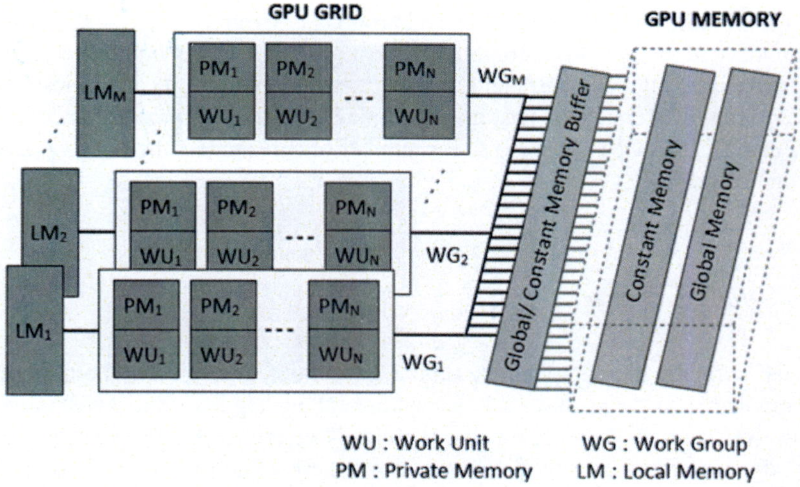

Fig. 6 OpenCL memory model

because modification of information in one memory region has not replicated to other processing elements' private memory sections (Fig. 6).

In the OpenCL memory model, the size of each memory region has indirectly related to the capacity of that region, i.e., as the size of the section increases, the speed of this region gradually decreases. Within those memory regions, the private memory region provides the fastest information access to GPU cores that made from a set of registers with hardware scheduler limit the size of each private section to 1 KB.

3.4 OpenCL Programming Model

The creation of an OpenCL program object requires a compilation of the source code using clCreateProgramWithSource commands to generate binary and then clBuild-Program helps to build this program by linking the binary with the main program during execution. On the other hand, building the program, the object is directly done by linking the precompiled binary file with the help of clCreateProgramWith-Binary and the linking process is one to one for a context in a device [26]. A single binary file is responsible for creating a single program object whose performance will be stimulated by creating multiple kernel objects from a program object at a later stage of host program execution. OpenCL Installable Client Driver allows to keep multiple platforms on the host system in passive mode and selection of the platform is done at execution time by the host program. As the creation of a single context is possible for all devices present in a platform, transferring the information between two contexts present in two platforms is not possible which is done through the host program itself in a synchronous environment. After the creation and building of the OpenCL program, the host program creates multiple threads of that program so-called kernels and distribute them across multiple cores present in a device to run independently. That makes the device with more processing cores run faster than CPU or GPU having a fewer number of cores unless the running threads are cooperative.

3.5 OpenCL Kernel

OpenCL kernel is the compiled executable kernel object created by both program and kernel objects to run on work items, whose calling happens from a host program during its execution. Application Programming Interface (API) in OpenCL framework supports across diverse platforms, allows these kernels to easily transportable between heterogeneous parallel devices. Here, OpenCL kernel has written in C based on the C99 compiler where it uses data types are from C built-in data type, and some of the composite data types like int2, float3, and double4 have derived from those basic data types [27]. Each work item inside the index space creates its instance of this kernel and execute them in parallel. A bunch of work items together form a workgroup during kernel execution to maintain a relation between them to achieve a common interest. Instances of one work item make use of local or global memory regions to communicate with the instances of other work-items to not only brings reliability but also diminishes the time utilization by a remarkable amount. Several OpenCL kernels can be created from a program object, while multiple program objects can be built from each OpenCL context, dedicated to a computing device in the OpenCL platform [26]. Proper distribution of work items to form a workgroup in the index space optimizes kernel execution that ultimately makes the application performs quickly. As the compilation of a kernel happens during host program execution, it requires proper care while writing this function because small mistakes create trouble to identify errors present inside it.

As shown in Fig. 7, the I/O device transfers an image to system memory during the image acquisition phase, where an image Red, Green, and Blue channels are extracted and stored in separate cv::Mat variables. Using the OpenCL clEnqueueWriteBuffer() command channels' information is transferred from Host DDR4 memory to GPU DDR5 memory, and that information is utilized as inputs to OpenCL kernels for image restoration [28]. DFT Kernels take the Red, Green, and Blue frames separately as their inputs and generate the transformed frequency maps using two dimensional Discrete Fourier Transformation. The filter and Filtration kernels are responsible for the filter creation and do the pointwise multiplication with those DFT transformed frames to generate the restored image in the transformed domain. After that, IDFT kernels take these output frequency map and apply the 2D inverse Fourier Transformation to create restored pixels in the spatial domain, which are then transferred back to the system host memory from the parallel device memory using clEnqueueReadBuffer() OpenCL Command [28]. At last, all three channels

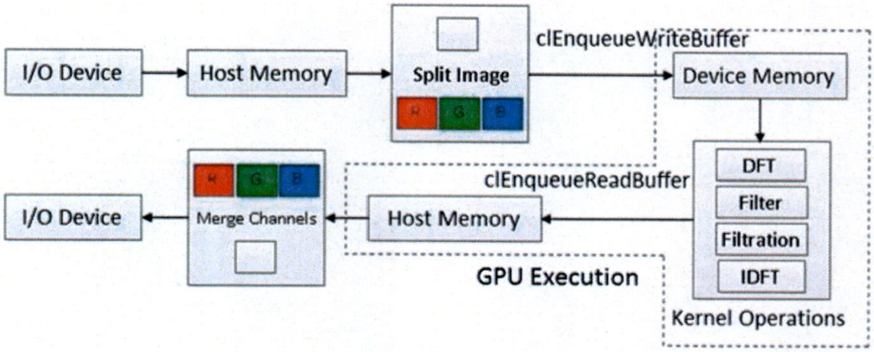

Fig. 7 OpenCL Kernels' GPU execution flow diagram

have merged using cv::merge() to generate the final RGB image that will move to display devices for display purposes. The total computation time of an image restoration process includes splitting an RGB image to joining its enhanced frames to produce the final one, while the GPU execution time is the time gap between transferring the data from host memory to device memory and transfer the modified data back to the host memory (Fig. 7).

4 Degradation Model

As we discussed earlier, blurring happens due to various reasons. Bad weather, poor illumination, improper focusing, and atmospheric turbulence are some of the major factors behind image degradation. It may occur due to considerable movement between the visual sensor and sensing environment. Blurring causes a reduction of high-frequency components in images and produces low quality blurred images. The types of image blurring are mainly classified into five types; are discussed below.

4.1 Average Blur

In Image enhancement, the average filter has used for removing noise from the image [29]. If noise is spread throughout an image, then the average filter is one of the suitable choices for image smoothing, which is not only computationally much faster but also creates a blurred image that needs to be enhanced by a high-boost filter. The mathematical equation for this degradation has given below.

$$\mathrm{d} = 2 * \sqrt{\sqrt{g^2} + f^2} \tag{5}$$

a. Average blur, 7 x 7 Average Filter.

b. Defocus blur, Focus Point = 0.5, Focal Length = 7.

c. Gaussian blur, Mean (μ) = 0, SD (σ) = 3.

d. Motion blur, Length (S) = 15, Angle (Θ) = 35

e. Turbulence blur, 5 x 5 Gaussian Filter, SD = 2, Angle (Θ) = 45.

f. Inverse filter, Cutoff frequency (D_f) = 80.

Fig. 8 Restored images (8.f, …, 8.j) generated from their blurred images (8.a, …, 8.e) using Inverse, Pseudo Inverse, Wiener, Least Square Error and Lucy Richardson filters

g. Pseudo Inverse filter, Cutoff frequency (D_f) = 25.

h. Wiener filter, Cutoff frequency (D_f) = 30.

i. Lucy Richardson filter, Iteration = 15.

j. CLSE filter, Cutoff frequency (D_f) = 35.

Fig. 8 (continued)

Here, 'g' is blurring along the horizontal direction, and 'f' is blurring along the vertical direction. 'd' is the diameter of the circular-average blurring.

4.2 Gaussian Blur

It has also called the normal distribution blurring. Images taken in the outer environment are mostly affected by various forms of Gaussian blur [30]. Point Spread Function of simple Gaussian blurring has given in Eq. (6). 'σ' represents the standard deviation of this blur.

$$H(x, y) = \frac{1}{\sigma\sqrt{2\pi}} e^{-\frac{x^2+y^2}{2\sigma^2}} \tag{6}$$

4.3 Defocus Blur

The improper focus of the camera on an image creates out of focus blur. Quality of image has degraded due to improper adjustment of the lens. Focal length and displacement between objects and cameras have a high impact on image blurring. It spreads uniformly across the center of out of focus point spread function [31]. The point spread function of the defocus blur has given in Eq. (7).

$$H(x, y) = c\begin{cases} 1, \sqrt{(x - c_x)^2 + (y - c_y)^2} \le r \\ 0, \qquad otherwise \end{cases} \tag{7}$$

(cx, cy) is the center and ‘r’ is the radius of point spread function. Scaling constant ‘c’ must satisfy $\iint H\,(x, y)dxdy = 1$.

4.4 Motion Blur

Considerable movement between objects and camera adds motion blur to the captured image. Various form of transformations like translation, rotation, scaling, or any of these combinations creates motion blur [32]. An object is approaching the camera at a velocity υ, making an angle θ with a horizontal axis during an interval (0, t). Here, the transfer function is depending on length S and angle θ; has given below in Eq. (8) and (9).

$$S = \nu t \tag{8}$$

$$H(x, y, S, \theta) = \begin{cases} \frac{1}{S}\ if \sqrt{x^2 + y^2} \le \frac{S}{2} and \frac{x}{y} = -tan\theta \\ 0 \qquad otherwise \end{cases} \tag{9}$$

4.5 Atmospheric Turbulence Blur

An atmospheric turbulence blur is a special form of Gaussian blur [33]. It occurs due to various factors such as temperature variation, wind speed, long exposure time, and noise in the transmission medium. It has a serious limitation; most remote sensing images are affected by this degradation. The transfer function has given below.

$$H(x, y, \sigma) = Ce^{-\frac{x^2+y^2}{2\sigma^2}} \tag{10}$$

Here, σ^2 is the variance of Gaussian blur. Scaling factor 'C' depends on the environment, and it must satisfy the point spread function.

5 Image Restoration

As discussed earlier, image restoration techniques are classified into two types based on the involvement of any prior knowledge regarding various degradation functions [34, 35]. One is blind deconvolution techniques, and the other one is non-blind deconvolution techniques. Blind deconvolution techniques have little knowledge or without any knowledge of transfer function, which makes them less accurate. Non-blind deconvolution techniques use prior knowledge of transfer function to design filter function and make them suitable for restoration. The most familiar blind and non-blind deconvolution filters are discussed below.

$$f'(k, l) = h(k, l).g(k, l) \tag{11}$$

$$f'(x, y) = f^{-1}\left[f'(k, l)\right] \tag{12}$$

g(k, l) and f′(k, l) are corresponding Fourier transformation of g(x, y) and f′(x, y). f′(x, y) is constructed by taking inverse Fourier transformation of f′(k, l). OpenCL kernel for 2D DFT and 2D IDFT transformation of an image is given below.

```
#pragma OPENCL EXTENSION cl_khr_fp32 : enable
__kernel void DFT_Kernel (__global float2* data, __global float2* out,__local float2*
sharedMemory, int width, float norm) { //Get the index of work items
int globalID = get_global_id(0); int localID = get_local_id(0);
int groupID = get_group_id(0); int groupSize = get_local_size(0);
float2* sharedArray = sharedMemory;
int v = globalID % width; float param = (-2.0*v)/width;
//-2.0 is replaced by 2.0 in case of IDFT
float c, s; float2 valueH, valueV; sharedArray[localID] = (0.0, 0.0);
// Horizontal DFT Transformation
for(int i = 0; i<groupSize; i++){
valueH = data[groupSize*groupID+i];s = sinpi(param * i); c = cospi(param * i);
sharedArray[localID] += (float2)(valueH.x*c-valueH.y*s,valueH.x*s+valueH.y*c);
} out[groupSize*groupID+localID]=dot(sharedArray[localID],float2(norm, norm));
sharedArray[localID] = (0.0, 0.0);
// Vertical DFT Transformation
for(int i = 0; i<groupSize; i++){
valueV = out[groupSize*i+groupID];s = sinpi(param * i); c = cospi(param * i);
sharedArray[localID] += (float2)(valueH.x*c-valueH.y*s,valueH.x*s+valueH.y*c);
} out[groupSize*localID+groupID]=dot(sharedArray[localID],float2(norm, norm));
//* Replace 'data' buffer with 'out' buffer and vice versa in IDFT
```

5.1 Inverse Filter

Inverse filter is one of the simple and straightforward blind deconvolution methods that restore images by taking the inverse of the PSF, normally represented by an inverse low pass Gaussian distribution function [36]. As it neglects the noise effect on degradation, the final restoration results are not perfect. In the absence of noise, it produces better results compared to other filters. The inverse filter cannot be applied everywhere because it requires Point Spread Matrix (PSM) to be nonsingular in nature. The filter function and OpenCL kernel of the inverse filter is given below.

$$\mathrm{h(k, l)} = \frac{1}{H(k,l)} \tag{13}$$

```
#define EXP 2.72
#pragma OPENCL EXTENSION cl_khr fp32 : enable
__kernel void inverse_kernel (__global float2* data, int height, int width, int CUTOFF)
{       //Get the index of work items
        uint index = get_global_id(0); int U = index / width; int V = index % width;
        float imgx = data[index].x;     float imgy = data[index].y;
        float D = pow(height/2-abs(U-height/2),2.0) + pow(width/2-abs(V-width/2),2.0);
        float H = pow(EXP, (D / (2.0 * pow(CUTOFF, 2.0))));
        data[index].x = imgx * H; data[index].y = imgy * H;       }
```

5.2 Pseudo-Inverse Filter

The difficulty of non-singularity in the inverse filter; is recovered by using pseudo-inverse filters [37]. Like an inverse filter, negligence towards noise makes the results imperfect. If H(k,l) is less than Є, it replaces h(k,l) with Є where Є is a constant. The selection of Є is not clear, and it will affect the quality of an image. Generally, the pseudo inverse filter produces noisy images and makes the images not worthy for further processing. The PSF and OpenCL kernel implementation of the Pseudo Inverse filter, are given below.

$$\mathrm{h(k, l)} = \begin{cases} \frac{1}{H(k,l)} & if H > \in \\ \in & if H \leq \in \end{cases} \tag{14}$$

```
#define EXP 2.72
#pragma OPENCL EXTENSION cl_khr_fp32 : enable
__kernel void inverse_kernel (__global float2* data, int height, int width, int CUTOFF)
{       //Get the index of work items
        uint index = get_global_id(0); int U = index / width; int V = index % width;
        float imgx = data[index].x;     float imgy = data[index].y;
        float D = pow(height/2-abs(U-height/2),2.0) + pow(width/2-abs(V-width/2),2.0);
        float H = pow(EXP, (-1.0 * D / (2.0 * pow(CUTOFF, 2.0))));
        H = (H > 0.05)? 1.0/H : 0.05; data[index].x = imgx * H; data[index].y = imgy * H;}
```

5.3 Wiener Filter

Unlike inverse and pseudo-inverse filter, Wiener filter considers noise artifact and try to minimize the mean square error between restored image and actual-ideal image. It has characteristics of both high pass filter and low pass filter. A high pass filter is used for deconvolution, whereas a low pass filter is used to remove additive noise from the image. An optimal filter function based on a minimum mean square error, is given in Eq. (13). Here, Sn(k,l) and Sf(k,l) represent noise spectral density and signal spectral density respectively. Without any prior knowledge makes SNR difficult to select. If $\gamma = 1$ and $\frac{1}{SNR} =$ Constant, then the above filter is called Wiener filter [38]. The PSF of Wiener filter with $\frac{1}{SNR} = 0.005$ performs quite well than Pseudo Inverse filter and the corresponding OpenCL kernel is presented below.

$$\text{h(k, l)} = \frac{1}{H(k,l)}\left[\frac{|H(k,l)|^2}{|H(k,l)|^2 + \gamma\frac{S_n(k,l)}{S_f(k,l)}}\right] \tag{15}$$

$$\text{h(k, l)} = \frac{1}{H(k,l)}\left[\frac{|H(k,l)|^2}{|H(k,l)|^2 + \gamma\frac{1}{SNR}}\right] \tag{16}$$

```
#define EXP 2.72
#pragma OPENCL EXTENSION cl_khr_fp32 : enable
__kernel void Wiener_kernel (__global float2* data, int height, int width, int CUTOFF)
{       //Get the index of work items
        uint index = get_global_id(0); int U = index / width; int V = index % width;
        float imgx = data[index].x;      float imgy = data[index].y;
        float D = pow(height/2-abs(U-height/2),2.0) + pow(width/2-abs(V-width/2),2.0);
        float H = pow(EXP, (-1.0 * D / (2.0 * pow(CUTOFF, 2.0))));
        H = H / (H * H + 0.005); data[index].x = imgx * H; data[index].y = imgy * H; }
```

5.4 Constraint Least Squared Error Filter

Like Wiener filter, Constraint Least Square Error (CLSE) filter minimizes mean square error between the ideal image and the reconstructed image by taking additive noise into consideration [39, 40]. The addition of noise to image pixels, changes frequencies in both directions (vertical and horizontal) and the changes can be captured by a high pass filter like Laplacian. So, the Laplacian filter is a suitable choice for SNR to remove noise artifact. Here, L(k,l) is the Fourier transform of Laplacian filter and γ is the control parameter between noise and signal data which is assumed to be constant. In most cases, constraint least square error filter outperforms all the above filters.

$$h(k,l) = \frac{1}{H(k,l)}\left[\frac{|H(k,l)|^2}{|H(k,l)|^2 + \gamma|L(k,l)|^2}\right] \tag{17}$$

```
#define EXP 2.72
#pragma OPENCL EXTENSION cl_khr_fp32 : enable
__kernel void LSE_kernel (__global float2* data,__global float* Lap,int height, int
width, int CUTOFF)
{       //Get the index of work items
        uint index = get_global_id(0); int U = index / width; int V = index % width;
        float imgx = data[index].x;      float imgy = data[index].y;

        float D = pow(height/2-abs(U-height/2),2.0)+pow(width/2-abs(V-width/2),2.0);
        float H = pow(EXP, (-1.0 * D / (2.0 * pow(CUTOFF, 2.0))));
        float L = Lap[index]; H = H / (H * H + 0.005 * L * L);
        data[index].x = imgx * H; data[index].y = imgy * H; }
```

5.5 Lucy Richardson Filter

Unlike the above non-blind deconvolution techniques, it is a nonlinear filter, and it works faster in the special domain compare to the frequency domain. It provides better image quality in the presence of more noise, but it is slower than other restoration techniques as it uses maximum-likelihood optimization to find the best solution [41]. As it works on expectation–maximization, each iteration provides a better result compare to the previous one. So, it requires a successful trade-off between the number of filter iterations and the output image quality, which is observed by checking the difference between MSE of two successive constructed images. Here, H(x, y), g(x, y), and fi′(x,y) is the PSF, blurred image, and ith latent image respectively. H (−x, −y) is the Flipped Point Spread Function (FPSF).

$$\text{fi} + 1'(x, y) = \text{fi}'(x, y)[\text{FPSF} * \frac{g(x, y)}{H(x, y) * f_i'(x, y)}] \tag{18}$$

```
__constant sampler_t image_sampler = CLK_NORMALIZED_COORDS_FALSE |
CLK_ADDRESS_CLAMP_TO_EDGE | CLK_FILTER_NEAREST;
__kernel void Lucy_Richard_St1 (__read_only image2d_t iimage,  __read_only image2d_t
kimage,__write_only image2d_t timage,__write_only image2d_t oimage,__constant float*
Filter, int windowSize)
{     unsigned int x = get_global_id(0);    unsigned int y = get_global_id(1);

      int halfWindow = windowSize/2; float4 kpixelValue, ipixelValue, tpixelValue;
      float4 computedFilter = 0.0f; int i, j, ifilter, jfilter;
      tpixelValue = read_imagef(kimage, image_sampler, (int2)(x, y));
      ipixelValue = read_imagef(iimage, image_sampler, (int2)(x, y));
      for(i = -halfWindow, ifilter=0; i <= halfWindow; i++, ifilter++)
      {     for(j = -halfWindow, jfilter = 0; j <= halfWindow; j++, jfilter++) {
            kpixelValue = read_imagef(kimage, image_sampler, (int2)(x+i, y+j));
            computedFilter += Filter[ifilter * windowSize + jfilter] * kpixelValue;}}
            computedFilter = ipixelValue / computedFilter;
            write_imagef(timage, (int2)(x, y), tpixelValue);
            write_imagef(oimage, (int2)(x, y), computedFilter); }

__kernel void Lucy_Richard_St2 (__read_only image2d_t iimage,  __read_only image2d_t
timage,__write_only image2d_t oimage,__constant float* Filter, int windowSize)
{     unsigned int x = get_global_id(0);    unsigned int y = get_global_id(1);

      int halfWindow = windowSize/2; float4 ipixelValue, tpixelValue;
      float4 computedFilter = 0.0f; int i, j, ifilter, jfilter;
      tpixelValue = read_imagef(timage, image_sampler, (int2)(x, y));
      for(i = -halfWindow, ifilter=0; i <= halfWindow; i++, ifilter++){
      for(j = -halfWindow, jfilter = 0; j <= halfWindow; j++, jfilter++) {
            ipixelValue = read_imagef(iimage, image_sampler, (int2)(x+i, y+j));
            computedFilter += Filter[ifilter * windowSize + jfilter] * ipixelValue;}}
            computedFilter = tpixelValue * computedFilter;
            write_imagef(oimage, (int2)(x, y), computedFilter); }
```

5.6 *Global and Local Size*

For 2D-DFT and IDFT transformation, the global_size of index space is the total number of pixels that constitute a single channel of an RGB image, where its group_size is the total number of pixels present in each row of an input channel, and the number is 512 for a 512 × 512 RGB image. However, the same global_size is followed for the filtering operations, but its work_group_size increases to 1024 and 8192 for the GPU, and CPU operations respectively. At last, all the OpenCL kernels run for red, green, and blue channels separately to not only reduce the complexity for work items but also execute each channel independently to make the whole operation faster.

```
size_t global_size = Image.rows * Image.cols, size_t local_size = 512; /*DFT & IDFT*/
size_t global_size = Image.rows * Image.cols, size_t local_size = 1024;
/*Filtering*/clStatus = clEnqueueNDRangeKernel(command_queue, kernel, 1, NULL,
&global_size, &local_size, 0, NULL, NULL);
```

6 Results

On the above, we have discussed the blind and non-blind deconvolution image restoration filters try to reproduce original images from the blurred images which are

Table 1 Environment specification

Components	Specification
Processor	Intel® Xeon® E3-1225 v5
Memory	DDR4 8 GB 2133 MHz
Storage	HDD 1 TB
Graphics processor	NVIDIA GTX 1050 Ti
OS	Windows Server 2012 R2
Packages	C, OpenCV, OpenCL
Image	Size = 512 × 512, Bit Depth = 24

normally affected due to atmospheric turbulence, the improper focus of camera lens, or movement of the target object. For the frequency domain filters, the transformation of a distorted image was taken place using 2D DFT and after refining, 2D IDFT transforms it back to produce the reconstructed image, while Lucy Richardson's spatial domain filter calls the kernel repeatedly over the time to improve the image quality gradually. Those filters were implemented using C, OpenCV for Xeon E3 1225 V5 processor, and their computation time was compared with respect to C-OpenCL implementation of filter kernels for both Xeon E3 1225 V5 and NVIDIA GTX 1050 Ti without altering their precision. V. P. S. Naidu and J. R. Raol explain about various performance metrics like entropy, Root Mean Square Error, Peak Signal to Noise Ratio and Structural Symmetric Index is used to examine the rightness of any image filters with the help of its produced image and the original one [42]. The environment along with various uncleansed images are used for the purpose of enhancement to produce their restored images are labeled in Table 1 and shown in Fig. 10 respectively.

Table 2 shows, as the cutoff frequencies of inverse filter increases, there is an increase in Entropy SNR, PSNR, SSIM, and correlation values of the enhanced images while there is a significant reduction in MAE, RMSE, SD and PFE values. Apart from that, the overall structural similarity index values of an inverse filter are absolutely lowest compare to other filters which mean distortion of structures inside the image is largely present even after the restoration process and that makes the inverse filter unfit for most of the real-time applications. To overcome the limitations present inside the inverse filter, Pseudo Inverse filter comes with an idea of checkpoint filtrations i.e. filter coefficient above the checkpoint behaves like an inverse filter, otherwise, checkpoint value is used to improve the image quality. Table 3 shows that a pseudo-inverse filter with 0.05 as its checkpoints increases overall SSIM, correlation, and PSNR to 0.6793, 0.8713, and 19.22 from 0.2926, 0.554, and 12.52 of the inverse filter. On the other hand, MAE, RMSE, SD, and PFE values are dropped to 20.15, 29.03, 59.15, and 9.84 from 51.68, 64.98, 77.13, and 15.82.

A better understanding of change in signal value with respect to noise in Wiener and LSE filters made those two filters quite better than the previous filters as shown in Fig. 9. Average SSIM, correlation, and PSNR are (0.798, 0.9226, 21.6) and (81.53, 93.66, 22.6) for Wiener and LSE filters which are much higher than inverse and

Table 2 Performance metrics of an Inverse Filter in C-OpenCV and C-OpenCL implementation

C. Freq.	Images	Entropy	MAE	RMSE	SD	SNR	PSNR	PFE	SSIM	Correlation	Time (s) OpenCV-CPU	Time (s) OpenCL-GPU	Time (s) OpenCL-CPU
(Df = 85)	8.a	6.05	79.0	95.0	96.3	1.92	8.57	22.9	0.11	0.30	0.922	0.109	0.849
	8.b	5.89	79.5	96.4	95.8	1.04	8.45	27.4	0.14	0.25	0.965	0.107	0.877
	8.c	6.14	78.9	95.0	96.3	2.36	8.57	19.2	0.08	0.31	0.939	0.108	0.865
	8.d	5.64	81.2	101	99.0	1.30	8.02	31.5	0.14	0.28	0.963	0.109	0.877
	8.e	5.77	78.7	96.7	100	4.05	8.42	16.4	0.06	0.37	0.944	0.111	0.858
(Df = 90)	8.a	7.48	45.9	58.0	71.4	6.20	12.9	12.8	0.29	0.59	0.964	0.109	0.871
	8.b	7.32	47.1	59.9	69.6	5.18	12.6	16.4	0.36	0.51	0.977	0.109	0.884
	8.c	7.59	47.0	59.5	71.9	6.42	12.6	12.5	0.22	0.58	0.963	0.108	0.882
	8.d	7.15	49.1	62.9	74.7	5.43	12.1	15.8	0.36	0.57	0.991	0.110	0.885
	8.e	7.28	48.3	62.8	80.8	7.79	12.2	12.4	0.18	0.64	0.960	0.112	0.880
(Df = 95)	8.a	7.72	26.0	34.3	57.6	10.8	17.4	10.5	0.52	0.80	0.984	0.110	0.872
	8.b	7.66	27.2	36.8	54.6	9.40	16.8	10.4	0.59	0.73	0.979	0.111	0.892
	8.c	7.79	28.1	36.9	57.8	10.9	16.8	9.23	0.39	0.78	0.957	0.109	0.890
	8.d	7.58	29.6	38.1	61.0	9.78	16.5	9.29	0.60	0.79	1.014	0.111	0.905
	8.e	7.65	29.6	40.4	70.1	11.6	16.0	10.6	0.35	0.82	0.975	0.112	0.890

Table 3 Performance metrics of a Pseudo Inverse Filter in C-OpenCV and C-OpenCL implementation

C. Freq.	Images	Entropy	MAE	RMSE	SD	SNR	PSNR	PFE	SSIM	Correlation	Time (s) OpenCV-CPU	Time (s) OpenCL-GPU	Time (s) OpenCL-CPU
(Df = 25)	8.a	7.74	17.4	25.2	56.3	13.4	20.1	12.2	0.75	0.89	0.967	0.108	0.863
	8.b	7.64	22.8	31.4	55.6	10.8	18.2	9.23	0.70	0.83	0.958	0.109	0.850
	8.c	7.77	16.1	23.2	55.1	14.6	20.8	9.32	0.64	0.91	0.930	0.111	0.842
	8.d	7.69	24.5	34.7	62.6	10.6	17.3	11.4	0.72	0.83	0.951	0.108	0.848
	8.e	7.20	37.3	50.9	78.3	9.61	14.0	12.6	0.36	0.76	0.982	0.109	0.876
(Df = 30)	8.a	7.74	13.5	20.5	53.6	15.2	21.9	11.1	0.81	0.92	0.981	0.109	0.871
	8.b	7.66	18.2	26.5	52.5	12.2	19.6	8.28	0.76	0.86	0.953	0.109	0.848
	8.c	7.72	14.4	21.5	54.7	15.3	21.5	8.56	0.67	0.92	0.937	0.111	0.844
	8.d	7.76	21.1	30.2	59.5	11.8	18.5	9.37	0.76	0.86	0.962	0.109	0.862
	8.e	7.37	30.9	43.6	74.6	10.9	15.3	11.0	0.43	0.81	0.990	0.110	0.879
(Df = 35)	8.a	7.73	12.2	18.8	52.3	16.0	22.6	10.3	0.83	0.93	0.992	0.109	0.884
	8.b	7.66	14.9	23.0	50.4	13.5	20.9	7.50	0.81	0.89	0.960	0.110	0.856
	8.c	7.70	14.2	21.2	52.0	15.4	21.6	7.95	0.66	0.92	1.014	0.110	0.901
	8.d	7.77	18.9	27.1	57.8	12.7	19.5	8.16	0.80	0.89	0.967	0.109	0.866
	8.e	7.51	25.9	37.7	71.9	12.2	16.6	10.7	0.49	0.85	1.010	0.111	0.898

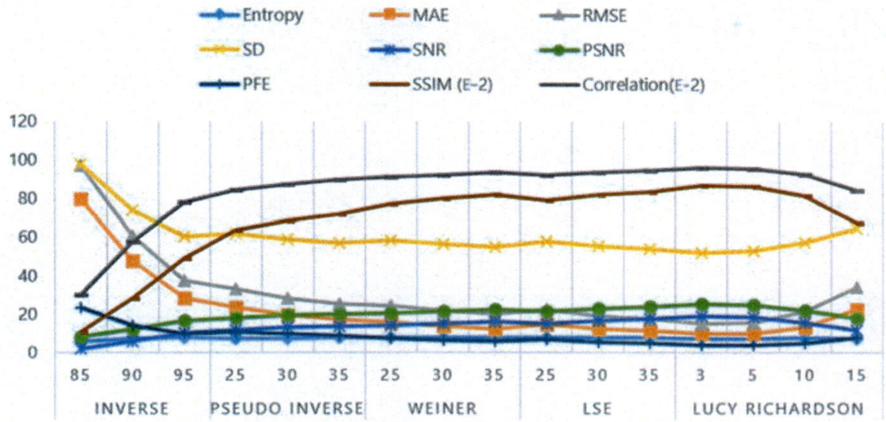

Fig. 9 Performance metrics of Inverse, Pseudo Inverse, Wiener, LSE and Lucy Richardson filters with reference to their cutoff frequencies and Iterations

Pseudo Inverse filters as displayed in Table 4 and 5. Although Wiener and LSE filters perform relatively well, the introduction of the Laplacian high pass filter in LSE's PSF catches change in signal effectively contrast to Wiener filter which is equal to 0.005 in our experiment. Table 6 exhibits that the Lucy Richardson filter called for three iterations outperformed than its higher iterations and its SSIM, correlation and PSNR is 0.88, 0.962, and 25.12 sufficiently larger than its 5, 10, 15 iterations. Generally, Lucy Richardson filter ran for three iterations provides the best results among those above filters whereas LSE filter achieves the highest image enhancement between those frequency domain restoration filters as shown in Fig. 9.

As shown in Fig. 10, the computation time of the inverse filter are least among the four frequency domain restoration filters' OpenCV-CPU, OpenCL-CPU, and OpenCL-GPU implementation, but in term of enhancement, CLSE filter provides better restoration than the other three frequency domain filters. As we have discussed earlier, estimated inverse filter operations using a set of 1D kernels require 0.57 s to enhance an image of size 325 × 365 pixels on a multicore CPU [12]. On the other hand, fast and easy deburring using the Progressive Removal of Blur Residual framework (PROBE) that used an Inverse filter, needed 5 s to restore an in 512 × 512 image on a 3.4 GHz CPU [13], whereas the OpenCL-CPU and OpenCL-GPU implementations of an Inverse filter kernel demand 0.878 s and 0.1096 s respectively to deburr a 512 × 512 RGB image. A parametric Wiener filter in Fourier space performs the deconvolution on a 256 × 256 distorted image and subsequent smoothing of the image happens in the wave atom transformation by tuning its threshold between 0 and 1, whose execution time stands at 8 s. approximately [14]. However, a dynamically tuned Wiener filter kernel for fine noise reduction as well as edge and details preservation processes 1556 × 360 pixels just in 21.2 s on a core i7 workstation [15]. Here, OpenCL implementation of the Wiener filter needs around 0.8906 s and 0.1106 s to enhance a 512 × 512 pixels RGB image on Xeon E3-1225 CPU and GTX

Table 4 Performance metrics of a Wiener Filter in C-OpenCV and C-OpenCL implementation

C. Freq.	Images	Entropy	MAE	RMSE	SD	SNR	PSNR	PFE	SSIM	Correlation	Time (s) OpenCV-CPU	Time (s) OpenCL-GPU	Time (s) OpenCL-CPU
(Df = 25)	8.a	7.77	11.4	18.3	53.8	16.2	22.9	7.34	0.87	0.94	1.006	0.110	0.902
	8.b	7.65	12.7	20.3	50.7	14.6	22.0	5.68	0.85	0.92	0.971	0.109	0.868
	8.c	7.70	12.3	19.3	53.4	16.2	22.4	5.84	0.71	0.93	0.969	0.110	0.870
	8.d	7.76	17.8	27.0	59.1	12.7	19.5	9.83	0.82	0.89	0.974	0.111	0.879
	8.e	7.53	24.3	36.3	74.2	12.5	16.9	7.95	0.62	0.87	0.994	0.109	0.892
(Df = 30)	8.a	7.75	9.49	15.7	52.1	17.5	24.2	6.70	0.90	0.95	1.010	0.110	0.896
	8.b	7.63	10.2	17.2	48.7	16.0	23.4	5.07	0.88	0.93	0.979	0.110	0.881
	8.c	7.66	11.6	18.7	51.7	16.5	22.7	5.36	0.72	0.94	0.999	0.111	0.888
	8.d	7.77	16.1	24.5	57.1	13.6	20.3	8.23	0.84	0.90	1.001	0.112	0.890
	8.e	7.62	21.1	32.6	71.9	13.5	17.8	7.36	0.66	0.89	1.003	0.110	0.892
(Df = 35)	8.a	7.74	8.74	14.6	51.2	18.2	24.8	6.24	0.91	0.96	1.012	0.111	0.907
	8.b	7.60	8.52	15.0	47.2	17.2	24.6	4.65	0.91	0.95	0.992	0.111	0.889
	8.c	7.64	11.8	18.7	50.8	16.4	22.7	5.00	0.72	0.94	1.015	0.111	0.908
	8.d	7.77	14.7	22.3	55.9	14.4	21.1	7.31	0.86	0.92	1.006	0.112	0.895
	8.e	7.69	18.4	29.3	70.1	14.4	18.8	7.05	0.70	0.91	1.011	0.112	0.903

Table 5 Performance metrics of a LSE Filter in C-OpenCV and C-OpenCL implementation

C. Freq.	Images	Entropy	MAE	RMSE	SD	SNR	PSNR	PFE	SSIM	Correlation	Time (s) OpenCV-CPU	Time (s) OpenCL-GPU	Time (s) OpenCL-CPU
(Df = 20)	8.a	7.77	11.1	17.2	53.7	16.7	23.4	6.53	0.88	0.95	1.013	0.112	0.915
	8.b	7.65	11.7	18.2	49.5	15.5	22.9	4.98	0.86	0.93	1.010	0.110	0.914
	8.c	7.71	12.3	19.0	53.4	16.3	22.5	5.15	0.71	0.94	1.001	0.111	0.877
	8.d	7.78	16.3	24.7	58.4	13.5	20.3	9.65	0.85	0.91	1.003	0.113	0.889
	8.e	7.57	22.7	33.5	73.3	13.3	17.6	7.68	0.65	0.89	1.008	0.112	0.893
(Df = 25)	8.a	7.75	8.90	14.1	51.9	18.5	25.2	5.22	0.91	0.96	1.019	0.113	0.923
	8.b	7.62	8.68	13.7	46.9	17.9	25.3	3.65	0.90	0.95	1.018	0.112	0.920
	8.c	7.66	11.4	18.0	51.5	16.8	23.0	4.07	0.72	0.94	1.006	0.112	0.892
	8.d	7.78	14.4	21.8	56.2	14.6	21.3	8.12	0.87	0.92	1.015	0.113	0.918
	8.e	7.67	18.6	28.2	70.4	14.7	19.1	6.72	0.70	0.92	1.013	0.112	0.916
(Df = 30)	8.a	7.74	7.99	12.6	50.9	19.4	26.1	4.38	0.92	0.97	1.027	0.113	0.931
	8.b	7.57	7.01	11.2	45.4	19.7	27.1	2.90	0.92	0.97	1.022	0.113	0.934
	8.c	7.64	11.4	17.9	50.6	16.8	23.1	3.43	0.72	0.94	1.010	0.112	0.914
	8.d	7.77	13.3	20.2	54.9	15.3	22.0	7.38	0.88	0.93	1.042	0.114	0.949
	8.e	7.72	16.2	25.1	68.6	15.7	20.1	6.21	0.74	0.93	1.018	0.113	0.921

Table 6 Performance metrics of [5 × 5] Lucy Richardson Filter in C-OpenCV and C-OpenCL implementation

Iteration	Images	Entropy	MAE	RMSE	SD	SNR	PSNR	PFE	SSIM	Correlation	Time (s) OpenCV- CPU	Time (s) OpenCL-GPU	Time (s) OpenCL-CPU
(Itr = 3)	8.a	7.70	7.04	10.9	49.6	20.7	27.4	2.40	0.94	0.98	1.232	0.078	0.125
	8.b	7.44	4.84	7.87	43.1	22.8	30.2	1.34	0.95	0.98	1.286	0.083	0.154
	8.c	7.60	11.6	18.0	49.4	16.8	23.0	3.12	0.72	0.94	1.269	0.082	0.147
	8.d	7.75	11.8	18.1	53.4	16.3	23.0	6.83	0.91	0.95	1.210	0.080	0.120
	8.e	7.73	12.8	20.3	65.4	17.6	22.0	5.35	0.82	0.96	1.253	0.081	0.133
(Itr = 5)	8.a	7.71	7.13	11.1	50.1	20.6	27.2	2.43	0.94	0.97	2.016	0.085	0.209
	8.b	7.49	5.13	8.43	44.3	22.2	29.6	1.44	0.95	0.98	2.022	0.086	0.212
	8.c	7.61	11.1	17.6	49.9	17.0	23.2	3.12	0.73	0.94	2.023	0.087	0.210
	8.d	7.77	12.2	18.9	54.3	15.9	22.6	7.35	0.90	0.94	2.008	0.084	0.203
	8.e	7.75	13.9	22.1	67.2	16.9	21.2	5.38	0.81	0.95	2.032	0.088	0.223
((Itr = 10)	8.a	7.76	8.82	13.7	52.0	18.7	25.3	2.97	0.92	0.96	3.565	0.091	0.328
	8.b	7.64	9.43	17.3	49.9	15.9	23.4	3.51	0.91	0.94	3.598	0.092	0.345
	8.c	7.64	10.5	16.8	51.7	17.4	23.6	3.12	0.74	0.95	3.632	0.093	0.367
	8.d	7.80	16.7	26.5	59.2	12.9	19.7	9.67	0.83	0.89	3.567	0.089	0.332
	8.e	7.72	20.6	32.5	73.7	13.5	17.9	5.71	0.68	0.90	3.629	0.092	0.358
(Itr = 15)	8.a	7.81	14.8	22.9	56.5	14.2	20.9	4.65	0.81	0.91	5.185	0.096	0.415
	8.b	7.66	20.0	33.2	58.6	10.3	17.7	7.03	0.76	0.82	5.221	0.099	0.442
	8.c	7.72	12.2	18.3	54.9	16.6	22.8	3.22	0.74	0.94	5.250	0.099	0.453
	8.d	7.57	30.0	44.8	69.8	8.38	15.1	14.1	0.64	0.76	5.170	0.097	0.414
	8.e	7.44	35.4	51.2	83.4	9.57	13.9	9.40	0.42	0.79	5.208	0.098	0.430

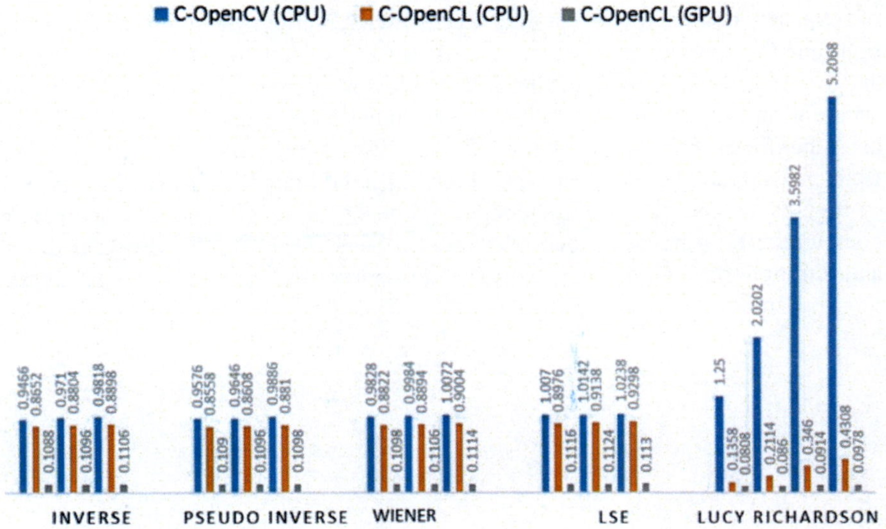

Fig. 10 Time consumption in seconds by C-OpenCV and C-OpenCL implementations of blind and non-blind deconvolution image filters in reference to cutoff frequencies and iterations

1050Ti GPU respectively. Image dependent restoration filter whose PSF depends on the local patches of an input image, strengthen a 482 × 482 RGB image in 3.6 s, and 3.1 s using the Intel Xeon E5-2360 CPU and 2688 cores NVIDIA GTX Titan GPU [16]. Besides that, a Semi-Global Weighted Least Squares filter is a linear system made of a sequence of subsystems that are called iteratively to restore an image, and four number of the successful calling of these subsystems revamp a 1 MB RGB image in 0.35 s using an Intel i7 CPU [17]. On the other side, Using the Xeon E3 1225 CPU and GTX 1050Ti platforms, OpenCL CLSE filter kernel restores a 512 × 512 RGB image in 0.9137 s and 0.1123 s as shown in Fig. 10.

From the literature, fast space-variant image deconvolution using the Lucy Richardson filter for a 15 number of iterations, consumed 40 ms to enhance a 640 × 480 pixels grayscale image on an Altera Stratix V family FPGA [19]. HW/SW codesign of the Lucy-Richardson deconvolution algorithm, accelerated on an FPGA, took around 180 ms for its first iteration to enhance a 610 × 340 grayscale image [20]. The fast non-blind image deburring using the Lucy Richardson filter requires 1.24 s, 1.7 s, and 6.63 s while restoring the RGB images of size 129 × 129, 267 × 267, and 512 × 512 pixels respectively [21]. As shown in Figs. 9 and 10, the Lucy Richardson filter provides the best image restoration, but the OpenCV CPU implementations of it perform worse than the other filters, whose time consumption exponentially increases with an increase in the number of iterations. The OpenCL CPU and GPU implementations drastically reduces the overall computation time to (0.1358 s, 0.2114 s, 0.346 s, 0.4308 s) and (0.0808 s, 0.086 s, 0.0914 s, 0.0978 s) for 3, 5, 10 and 15 iterations respectively. On the other hand, there is not much notable difference between OpenCV and OpenCL implementations for CPU In case of frequency domain filters, but their OpenCL GPU kernels minimize the computation time to approximately one-ninth of their CPU computation time. From Figs. 9 and 10, we have concluded that the GPU implementation of Lucy Richardson filters accomplishes the whole restoration process quite faster with better accuracy than any other filters, but optimizing the DFT and IDFT kernels further lower the above frequency domain filters' consumption time than the Lucy Richardson filter.

Figures 11 and 12 open up modularization of the whole OpenCL restoration computation time in case various filters into 2D-DFT, 2D-IDFT, filter matrix creation, and filtration time. As we have seen, the most part of the frequency domain filters'

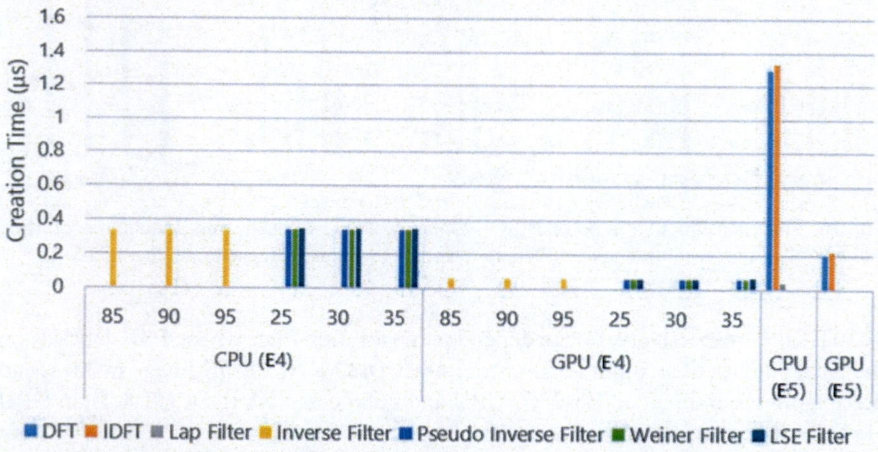

Fig. 11 Time consumption by 2D-DFT, 2D-IDFT image transformation and filters' matrices creation with respect to multiple cutoff frequencies in microseconds

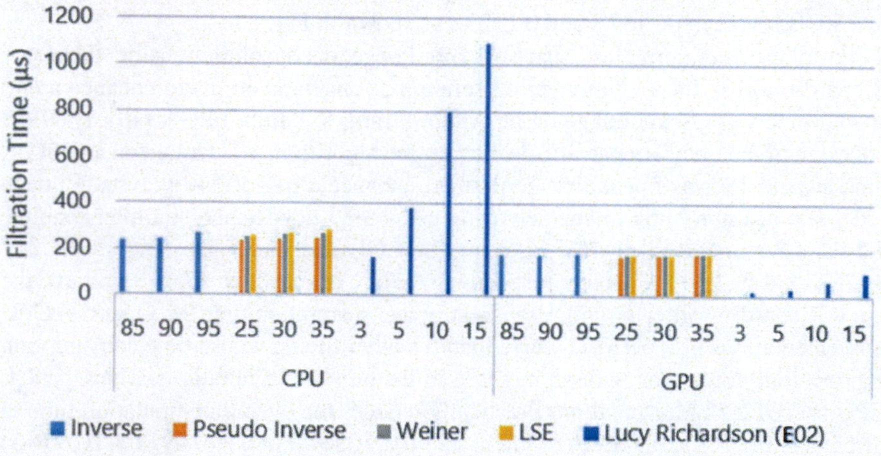

Fig. 12 Time consumption in microseconds by the blind and non-blind deconvolution image restoration filters with respect to the cutoff frequencies or number of iterations

computation time is consumed during the DFT/IDFT transformation of image matrix and creation of the filter matrix, which will be reduced by replacing DFT/IDFT transformation by 8 × 8 or 16 × 16 FFT/IFFT transformation [43]. As the creation of the filter matrix happens one time, it will not affect the performance of those filters in practical applications. If we see the overall filtration time of those frequency domain filters on various computing devices, they are utmost lower comparable to Lucy Richardson filter's computation time, expressed in terms of microseconds. At last, from our experiments, we have concluded that the utilization of FFT and IFFT kernels instead of DFT and IDFT kernels will make the LSE filter efficient and reliable one in all possible situations.

7 Conclusion

Image restoration is the image processing technique that allows the reconstruction of original images from faded images, which are affected by various factors like out of focus, motion, and atmospheric turbulence. In this paper, we have discussed the degradation model and the multiple factors that are affecting image degradation. It has also explained the restoration model and the well-known blind and non-blind deconvolution restoration techniques. In addition to that, a brief introduction about OpenCL architecture has presented, and OpenCL kernels of the above filters are designed for faster restoration, which could be used in a wide range of image and video processing applications. From the past work and performance analysis, we have found that Lucy Richardson's filter OpenCL kernel on an average is better than other filters in terms of time consumption and accuracy. In some of the practical applications where restoration, noise reduction, and high boosting all together play a major role, by Introducing 8 × 8 or 16 × 16 FFT/IFFT kernels instead of DFT/IDFT kernels, will make OpenCL CLSE filter kernel preferable one, as not only its filter creation and filtration time of are quite lower than the filtration time of a Lucy Richardson filter but also cascading the CLSE filter with any noise reduction and high boosting filters in frequency domain makes the overall operation much faster. At last, in the future, more research has to be done to increase the level of efficiency and reduce time consumption.

References

1. Gonzalez RC, Woods RE (2002) Introduction. In: Horton MJ (ed) Digital Image Processing. Prentice-Hall, Upper Saddle River, pp 15–45
2. Jayaraman S, Esakkirajan S, Veerakumar T (2015) Introduction to image-processing system. In: Jha S (ed) Digital Image Processing. Mc Graw Hill Education, New Delhi, pp 1–46
3. Singh T (2014) Novel restoration process for degraded image. In: 2014 Fifth International Conference Signal and Image Processing (ICSIP), pp 81–86. IEEE, Bangalore
4. Kaur R, Singh EN (2014) Image restoration - a survey. IOSR J Comput Eng 16(4):107–111

5. Lagendijk RL, Biemond J (1999). Basic methods for image restoration and identification. https://www-inst.eecs.berkeley.edu/~ee225b/sp07/handouts/Image_Restoration_99.pdf
6. Kundur D, Hatzinakos D (1996) Blind image deconvolution. IEEE Signal Process Mag 13(3):43–64. https://doi.org/10.1109/79.489268
7. Bhatele KR, Tiwari D (2018) The fundamentals of medical image restoration. In: Patnaik S, Das P (eds) Medical Image Processing for Improved Clinical Diagnosis. IGI Global, Hershey, pp 81–99
8. Bhattacharya A, Palit S (2013) Determination of the cause and amount of image degradation using a reduced reference approach. In: Twenty Eighth International Conference Image and Vision Computing, pp 418–423. IEEE, Wellington
9. Kaufman H, Tekalp AM (1991) Survey of estimation techniques in image restoration. IEEE Control Syst 11(1):16–24. https://doi.org/10.1109/37.103345
10. Boyat AK, Joshi BK (2015) A review paper: noise models in digital image processing. Signal Image Process Int J (SIPIJ) 6(2):63–75. https://doi.org/10.5121/sipij.2015.6206
11. Samarasinghe PD, Kennedy RA, Li H (2009) On non-blind image restoration. In: Third International ConferenceSignal Processing and Communication Systems, pp 1–7. IEEE, Omaha
12. Xu L, Tao X, Jia J (2014) Inverse kernels for fast spatial deconvolution. In: European Conference on Computer Vision, pp 33–48. Springer, Cham
13. Zon N, Hanocka R, Kiryati N (2017) Fast and easy blind deblurring using an inverse filter and PROBE. In: International Conference on Computer Analysis of Images and Patterns, pp 269–281. Springer, Cham
14. Mbarki Z, Seddik H, Braiek EB (2016) A rapid hybrid algorithm for image restoration combining parametric Wiener filtering and wave atom transform. J Vis Commun Image Represent 40:694–707
15. Baselice F, Ferraioli G, Ambrosanio M, Pascazio V, Schirinzi G (2018) Enhanced Wiener filter for ultrasound image restoration. Comput Methods Programs Biomed 153:71–81
16. Goto T, Otake S, Hirano S (2017) Blind image restoration for blurred images implemented on GPU. In: 2017 IEEE International Conference on Consumer Electronics-Taiwan (ICCE-TW), pp 213–214. IEEE
17. Liu W, Chen X, Shen C, Liu Z, Yang J (2017) Semi-global weighted least squares in image filtering. In: Proceedings of the IEEE International Conference on Computer Vision, pp 5861–5869
18. Wang L, Li Y, Wang S (2017) DeepDeblur: fast one-step blurry face images restoration. arXiv preprint arXiv:1711.09515.
19. Carrato S, Ramponi G, Marsi S, Jerian M, Tenze L (2015) FPGA implementation of the lucy-richardson algorithm for fast space-variant image deconvolution. In: 2015 9th International Symposium on Image and Signal Processing and Analysis (ISPA), pp 137–142. IEEE
20. Avagian K, Orlandić M, Johansen TA (2019) An FPGA-oriented HW/SW codesign of lucy-richardson deconvolution algorithm for hyperspectral images. In: 2019 8th Mediterranean Conference on Embedded Computing (MECO), pp 1–6. IEEE
21. Das R, Bajpai A, Venkatesan SM (2017) Fast non-blind image deblurring with sparse priors. In: Proceedings of International Conference on Computer Vision and Image Processing, pp 629–641. Springer, Singapore
22. Banger R, Bhattacharyya K (2013) Hello OpenCL. In: D'souza W, Pandey K, Colaco K (eds.) OpenCL Programming by Example, pp 1–34. Packt Publishing, Birmingham
23. Banger R, Bhattacharyya K (2013) OpenCL architecture. In: D'souza W, Pandey K, Colaco K (eds.) OpenCL Programming by Example, pp 35–58. Packt Publishing, Birmingham
24. Mann P (2016) Review: EVGA GeForce GTX 1050 Ti SC gaming. https://hexus.net/tech/reviews/graphics/98329-evga-geforce-gtx-1050-ti-sc-gaming/
25. Intel (2016) Intel Xeon processor-based platforms for Internet of Things (IoT) solutions [pdf]. https://www.intel.com/content/dam/www/public/us/en/documents/platform-briefs/xeon-processor-e3–1200-v5-workstation-platform-brief.pdf

26. Banger R, Bhattacharyya K (2013) OpenCL program and kernel objects. In: D'souza W, Pandey K, Colaco K (eds.) OpenCL Programming by Example, pp 109–136. Packt Publishing, Birmingham
27. Banger R, Bhattacharyya K (2013) OpenCL C programming. In: D'souza W, Pandey K, Colaco K (eds.) OpenCL Programming by Example, pp 155–178. Packt Publishing, Birmingham
28. Banger R, Bhattacharyya K (2013) OpenCL buffer objects. In: D'souza W, Pandey K, Colaco K (eds.) OpenCL Programming by Example, pp 59–86. Packt Publishing, Birmingham
29. Maurya A, Tiwari R (2014) A novel method of image restoration by using different types of filtering techniques. Int J Eng Sci Innov Technol 3(4): 124–129. https://www.ijesit.com/Volume%203/Issue%204
30. Damon J (1998) Generic structure of two-dimensional images under Gaussian blurring. Soc Ind Appl Math 59(1):97–138. https://doi.org/10.1137/S0036139997318032
31. Dhole P, Chopde N (2015) A comprehsive approach for analysis of image restoration using image deburring techniques. Int J Current Eng Technol 5(2):1046–1049. https://inpressco.com/category/ijcet
32. Gupta A, Joshi N, Zitnick L, Cohen M, Curless B (2010) Single image deblurring using motion density functions. In: Daniilidis K, Maragos P, Paragios N (eds) Computer Vision: Eleventh European Conference. Springer, Heraklion, pp 171–184
33. Singh A, Singh RK (2015) A survey on restoration techniques of atmospheric turbulence blur image. J Innov Comput Sci Eng, 4(2):1–9. https://pdfs.semanticscholar.org/d8cb
34. Gonzalez RC, Woods RE (2002) Image restoration. In: Horton MJ (ed) Digital Image Processing. Prentice-Hall, Upper Saddle River, pp 220–278
35. Jayaraman S, Esakkirajan S, Veerakumar T (2015) Image restoration and denoising. In: Jha S (ed) Digital Image Processing. Mc Graw Hill Education, New Delhi, pp 1–46
36. Stathaki DT (2014) Digital image processing: image restoration-inverse filtering. https://www.commsp.ee.ic.ac.uk/~tania/teaching/DIP%202014/Image%20Restoration%20Pres.pdf
37. Miljkovic S, Miladinovic M, Stanimirovic P, Stojanovic I (2012) Application of the pseudoinverse computation in reconstruction of blurred images. Fac Sci Math Univ Nis Serbia 26(3):453–465. https://doi.org/10.2298/FIL1203453M
38. Kumar N, Singh KK (2011) Wiener filter using digital image restoration. Int J Electron Eng 3(2):345–348
39. Galatsanos NP, Katsaggelos AK, Chin RT (1991) Least squares restoration of multichannel images. IEEE Trans Signal Process 39(10):2222–2236
40. Mesarovic VZ, Galatsanos NP, Katsaggelos AK (1995) Regularized constrained total least squares image restoration. IEEE Trans Image Process 4(8):1096–1108
41. Sharma S, Sharma S, Mehra R (2013) 'Image restoration using modified Lucy Richardson algorithm in the presence of Gaussian and motion blur'. Adv Electron Electric Eng 3(8):1063–1070
42. Naidu VPS, Raol JR (2008) Pixel-level image fusion using wavelets and principal component analysis. Def Sci J 58(3):338–352
43. Zhang F, et al (2017) A GPU based memory optimized paral-lel method for FFT implementation. arXiv:1707.07263 [cs.DC], Accessed 15 Dec 2018

Design of Area Efficient Multiply Accumulator Unit in Quantom Dot Cellular Automata (QCA)

Mohammed Arshad Abrar and Arun Kumar P. Chavan

Abstract The Quantum Dot Cellular Automata (QCA) is a latest technology that has gained the curiosity of researchers due to its small size and low power consumption, so it is being researched as a suitable alternate to the CMOS technology. The QCA technology provides a new computational platform in designing digital circuits with the help of quantum-dots. Quantum dots have wide range of application like LED, medical imaging and quantum computing etc. The quantum-dots are Nano crystals that transfer information at Nano scale. This paper proposed an area efficient design and implementation of a 16-bit Multiply-Accumulate unit (MAC). It's been designed by designing an area efficient novel 1-bit Full adder using QCA it is further used in a 16-bit Parallel binary adder and 8-bit multiplier modules. The design put forth for proposed 1-bit adder has brought down the cell count by 9.09% and area by 33.33%.The proposed design of multiplier has decreased the cell count by 18.7% and area by 9.92% in contrast to other previous designs. The proposed design of MAC has 27110 cells and its area is 77.89 μm^2. The simulation results have been verified using the QCADesigner Simulation tool.

Keywords CMOS · QCA design · Adder · Multiplier · MAC unit

1 Introduction

The most impactful technology being used which is known for its high speed processing and small size is the CMOS technology. But as the technological aspects are changing at higher speed, it's a need to scale down the dimensions of the technology and increase its efficiency. The notable shortcomings of CMOS innovations

M. A. Abrar (✉) · A. K. P. Chavan
Department of Electronics and Communication, R.V. College of Engineering, Bengaluru, India

A. K. P. Chavan
e-mail: arunkumarpc@rvce.edu.in

V. Nath and J. K. Mandal (eds.), *Proceeding of Fifth International Conference on Microelectronics, Computing and Communication Systems*, Lecture Notes in Electrical Engineering 748,
https://doi.org/10.1007/978-981-16-0275-7_69

are thin gate oxide doping fluctuations,short channel effects and expensive lithography at Nano scale level [1]. In order to control the shortcomings of CMOS innovations, QCA frameworks has come up with many good features like reducing the area, lower power consumption and higher speed.

The cellular automaton (CA) is a framework which is discrete and dynamic in nature. It comprises of a uniform network of cells. At a discrete interval each cell can be in at most one of a limited number of states. As time continues to increase, the state of the cells in the network is ascertained by a rule of transformation. This rule takes under consideration the previous state and the states of the cells which are right next to it. Lent et al. suggested the use of quantum-dot cells in the physical implementation of the automaton. In 1997 it was fabricated for the first time and it quickly gained popularity.

QCA is a new innovation which promises and performs transfer and computation of information at Nano scale. In QCA the cell is used to represent the logic states. A cell is a small Nano device which can transfer information by two state electron arrangements. The QCA has advantages over CMOS in terms of size, consumption of power and less delay.

Many endeavors were made in the designing of logic circuits in QCA such as adder [2–10], multipliers [11–13], divider, decoders, and memory circuits. Many of these circuits use large number of cells in the implementation and consume more area which is not efficient and results in more power dissipation. The proposed design overcomes these problems by decreasing the number of cell count and decreasing the area of the adder, multiplier and MAC unit. QCA circuit dissipates ultra-low power compared to CMOS circuits.

The paper presents design and implementation of an area efficient MAC unit by designing an area efficient novel 1-bit full adder using QCA and area efficient novel 8-bit multiplier using QCA. The structure of the paper is: Sect. 2 speaks about the reviews of QCA technology. Section 3 states the design structure of the proposed MAC unit. Section 4 gives the detail about the design and implementation of MAC unit. The design is concluded in Sect. 5.

2 Review of Quantum-Dot-Cellular Automata

2.1 QCA Cell

The QCA Cell is main block of any QCA design. QCA cell stores logic states based on the electron position. Each cell consists of four Quantum dots. There exists a tunneling junction between the Quantum-dots. The electrons occupy the Quantum-dots by tunneling to them. There is a Coulombic repulsion between these electrons, so they occupy sites in diagonal to one another in a cell. A QCA cell has two polarization states which constitute the value ‘0’ (FALSE) and ‘1’ (TRUE) respectively as can be seen in Fig. 1.

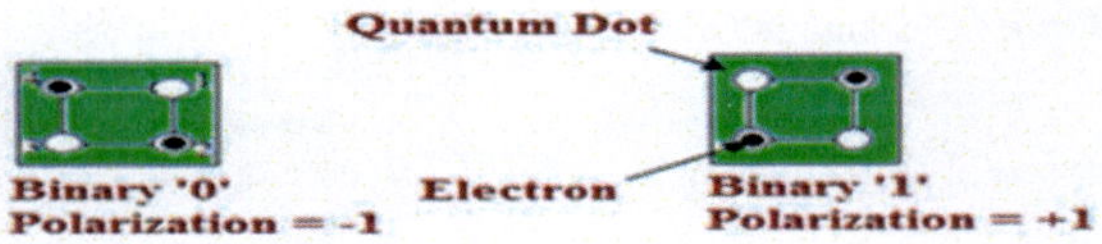

Fig. 1 QCA cell with two polarization state

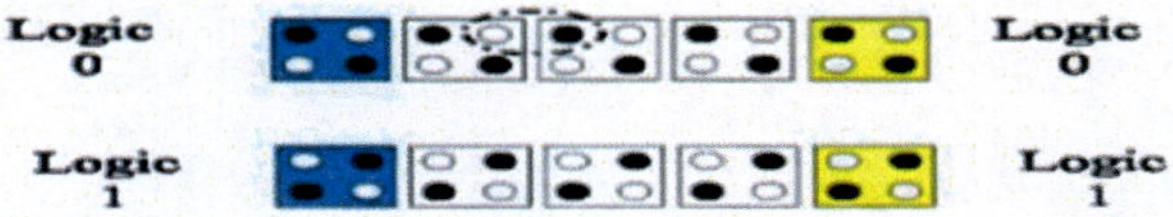

Fig. 2 QCA binary wire

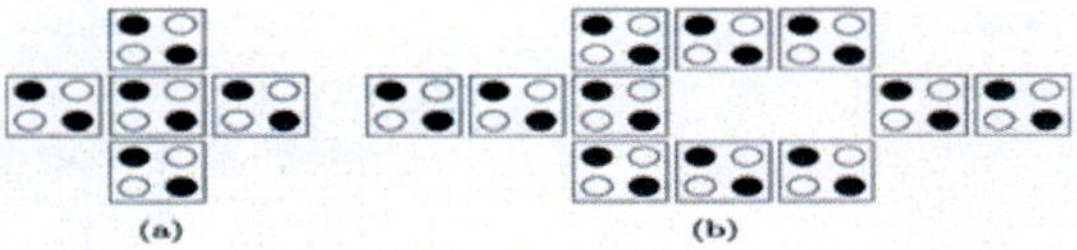

Fig. 3 (a) Majority gate, (b) Inverter

2.2 QCA Wire

A group of cell representing in the form of array is called as QCA wire. QCA wire is used to transfer data from input to output. The transfer of data between QCA wires is due to Coulombic interaction between QCA cells in Fig. 2.

2.3 QCA Gates

The basic gate of QCA is majority voters. To design a majority voter function of Eq. 1 five QCA cells are used in a majority gate which consists of 3 inputs, 1 output and 1 driving cell. The majority gate can function either as an AND gate or an OR gate by permanently fixing the input of one cell to '0' or '1'.

$$M\ (X,\ Y,\ Z) = XY + YZ + XZ \tag{1}$$

The basic structure of the Majority gate and the inverter using QCA cell is shown in Figs. 3 (a) and (b).

2.4 Clocking in QCA

QCA clocking strategy includes four clock phases. These phases are switch-phase, hold-phase, release-phase, and relax-phase shown in Fig. 4. There is a 90° shift in each phase with respect to the previous phase. Initially in the switching phase the cells have no polarization and the barriers in potential are low. The barriers are kept

Fig. 4 Clocking in QCA

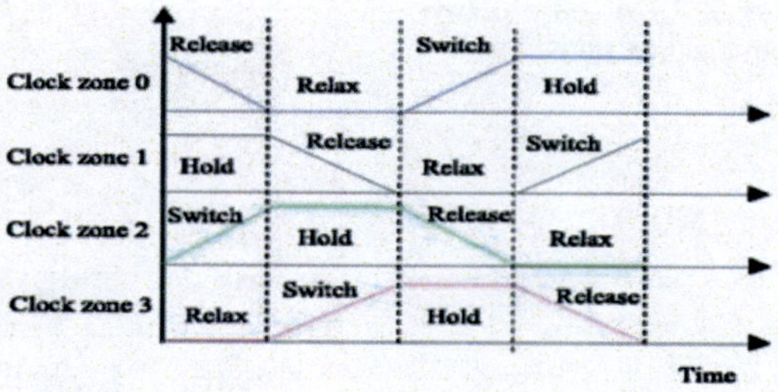

Fig. 5 (a) Coplanar crossover, (b) Multilayer crossover

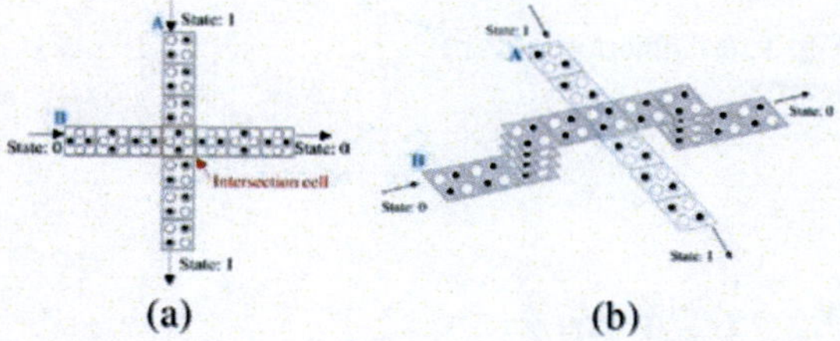

high in the hold phase, so that the data can propagate. The barriers are decreased in the release phase, and in relax phase the barriers rests low as in the previous phase. But the cell will be in un-polarized state.

2.5 *Crossover in QCA*

The two different methods of crossing of wire in QCA are, 1) Co-planar Wire Crossing and 2) Multi-layer Crossing. Multilayer intersections utilize just 90° wire to frame an extension type structure as appeared in the Fig. 5. (b). A coplanar crossing is a type of crossing in which two wires intertwine each other in a single layer by rotating the cells of one wire. Figure 5 (a) depicts the coplanar crossing in which both 90° and 45° wires are used in a single layer.

3 Multiply-Acumulate (Mac) Unit

The fundamental building unit in Arithmetic and Logical unit is the MAC unit. It is incorporated in ALU to achieve high performance. The performance of MAC unit majorly affects the speed of overall system.

The MAC unit incorporates three sub modules Multiplier, Adder and Accumulator respectively shown in Fig. 6. The inputs to the MAC unit are first multiplied, and then the multiplied product is summed with the value from the accumulator unit in the adder. The added sum is then passed on to the accumulator unit. This value from

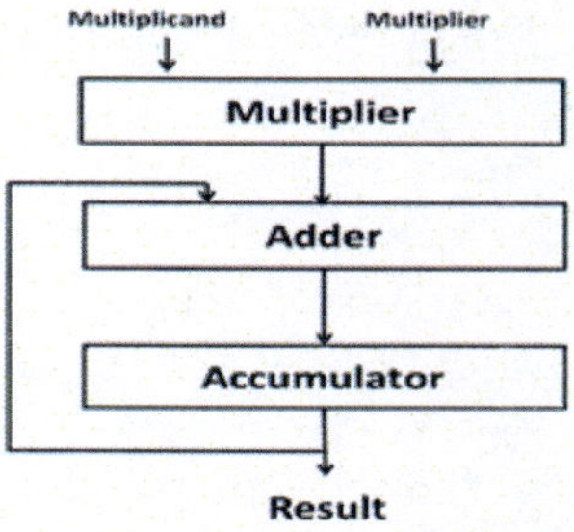

Fig. 6 Block diagram of MAC unit

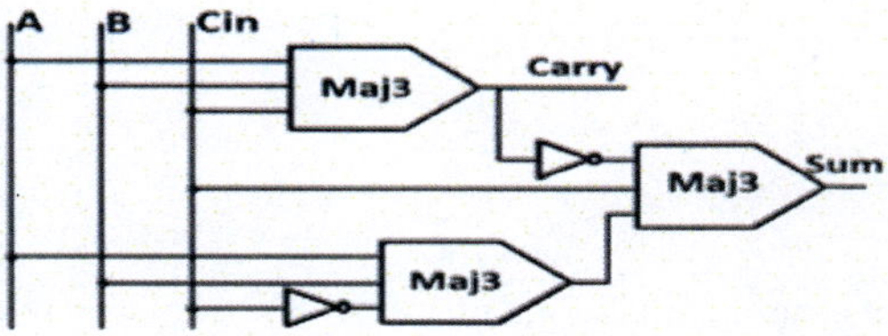

Fig. 7 Circuit Diagram of full adder in QCA

the adder unit is stored in the accumulator unit. There is a feedback to the adder unit from the accumulator unit. Then finally the result is obtained from the accumulator region of the MAC unit.

3.1 Adder Unit

In digital circuits, full adder plays a crucial role. Thus, great performance execution of adder circuit is an interesting research area. The logical equation for the output carry and the sum for 1-bit full adder are given as follows.

$$C_{out} = AB + BC + AC = Maj_3(A, B, C) \tag{2}$$

$$Sum = A \oplus B \oplus C = Maj_3\left(C_{in}, Maj_3\left(A, B, C'_{in}\right), C'_{out}\right) \tag{3}$$

In (2) and (3), the inputs to the adder design are A and B. The carry input is given as C_{in} and the carry output is given as C_{out} respectively. The output of the circuit is given as Sum. The three input majority function is denoted by Maj_3. This function can be used with the help of a 3-input majority gate in QCA. Figure 7 depicts the circuit diagram of 1-bit full adder in QCA.

The four bit RCA is constructed using four 1-bit full adders. Figure 8 shows the circuit of 4-bit RCA in QCA.

In this circuit, A (A_3, A_2, A_1, A_0), B (B_3, B_2, B_1, B_0) are two four-bit inputs. The carry input and the carry output of the ripple carry adder are denoted by Cin and Cout respectively. The Sum (Sum_3, Sum_2, Sum_1, and Sum_0) is four-bit output. This

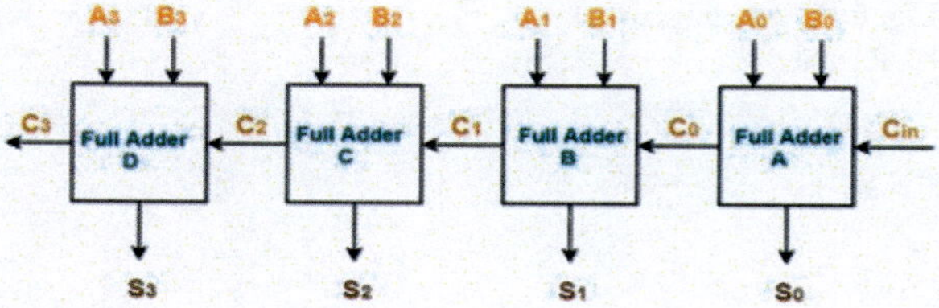

Fig. 8 4-bit RCA circuit

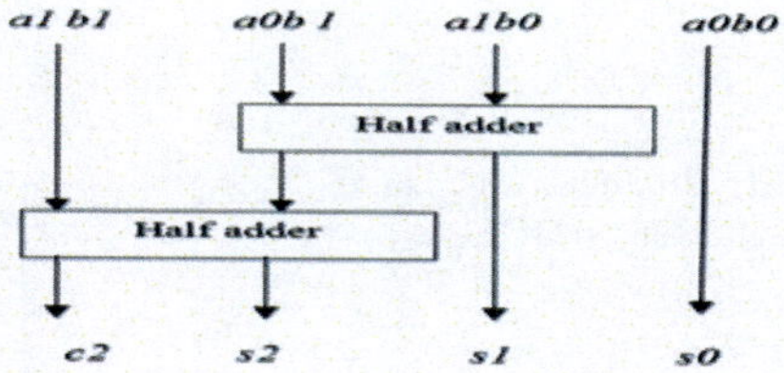

Fig. 9 Block diagram of a 2 × 2 Multiplier

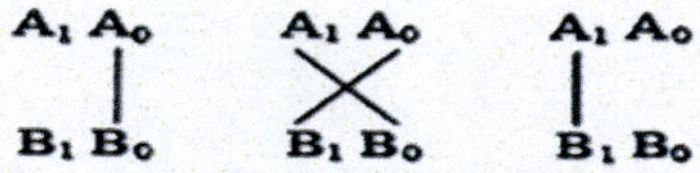

Fig. 10 Line diagram of 2 × 2 Vedic Multiplier

RCA can further be used in the design of 8-bit RCA, which is the adder unit in the MAC unit. It is used to add the multiplied product and the pre-existing value in the accumulator. This can also be used along with half adders in the design of multiplier unit.

3.2 *Multiplier Unit*

Multiplier is the main significant blocks in DSP frameworks. Different algorithms are developed to build a multiplier; the major drawback of those algorithms is large propagation delay. Large propagation delay issue is defeated by designing multiplier using Vedic algorithm. The advantage of Vedic algorithm is least propagation delay.

Urdhwa and Triyakbhyam multiplication algorithm is used in the designing of the multiplier. Figure 10 shows the line diagram of a 2 × 2 multiplier. The two bit inputs are A and B. First the LSB of A and LSB of B are multiplied; the obtained product of the LSBs is taken as the LSB of final product denoted by s_0. Then the LSB of A_0 and the MSB of B_1 are multiplied. The MSB of A_1 and LSB of A_0 are multiplied. The obtained products $A_0 \times B_1$ and $B_0 \times B_1$are added with the help of the half adder 1. The sum obtained from the half adder 1, s_1 is taken as the second bit of the final product and the carry c_1 is taken as an input to the second half adder. The MSB of B_1 and MSB of A_1 are multiplied, the obtained product $A_1 \times B_1$ is taken as the second input to the half adder 2. The sum s_2 and the carry c_2 of the half adder 2 are taken as the third and fourth bit of the final product. Thus the product of the 2 × 2 multiplier

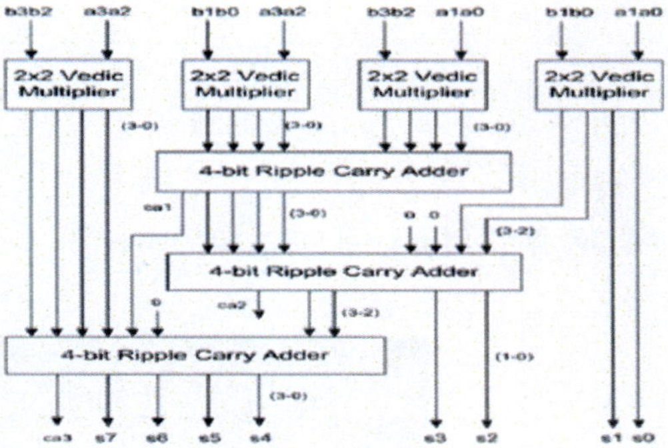

Fig. 11 4 × 4 Multiplier block diagram

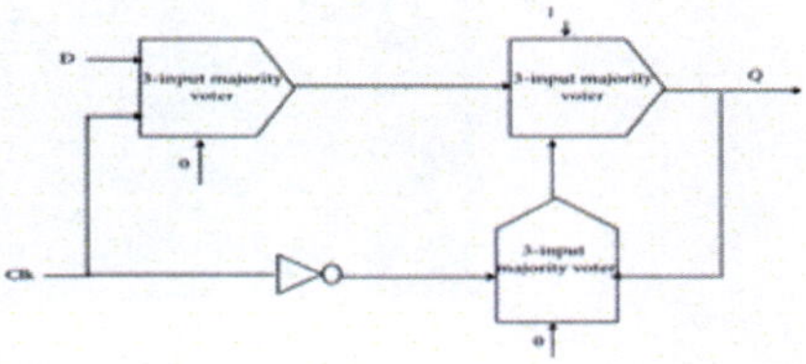

Fig. 12 D-flip-flop logic diagram

obtained is c_2 s_2 s_1 s_0. The Block diagram of a 2 × 2 Vedic multiplier is depicted in Fig. 9.

Four 2 × 2 modules with 3-RCA of 4-bit each is used to design a 4 × 4 multiplier. The inputs to the 4-bit multiplier are a (a_3, a_2, a_1, a_0) and b (b_3, b_2, b_1, b_0). The output of the multiplier after multiplication process is carried out is – s_7, s_6, s_5, s_4, s_3, s_2, s_1, s_0. The proposed design of the multiplier is area efficient. Figure 11 shows the block diagram of a 4 × 4 multiplier.

3.3 Accumulator Unit

The Accumulator unit is the final section of MAC unit. The Accumulator is a set of register that stores the information. The register is designed using n-D Flip-flops. The design of D Flip-flop is done using QCA is shown in Fig. 12. The design of the Flip-flop consists of 3 three input majority gates and inverter. The expression of D-FF is given in Eq. 4.

$$Q_{(t)} = (\text{Clk X D}) + \left(\text{Clk X } Q_{(t-1)}\right) \tag{4}$$

The Accumulator unit is designed using n-QCA D flip-flops that are linked to one another using a clock signal as illustrated in Fig. 13. In accordance to the shown figure, the input stream of data are D_0, D_1,.. D_{n-1} which is loaded in parallel to each register unit. The output of the register unit is Q_0, Q_1,..., Q_{n-1} that is obtained parallel at the output of every D Flip-flop.

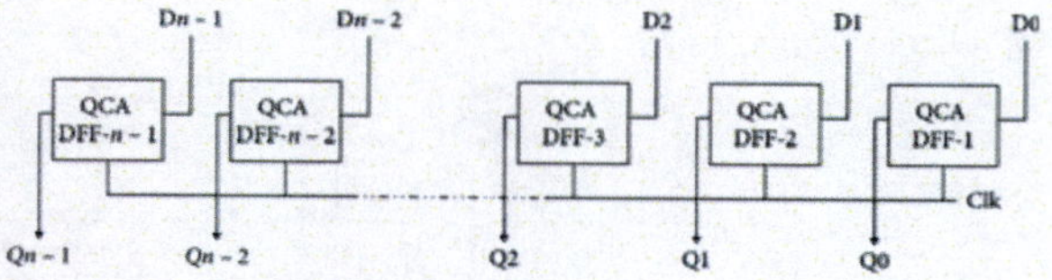

Fig. 13 Block Diagram of PIPO shift register

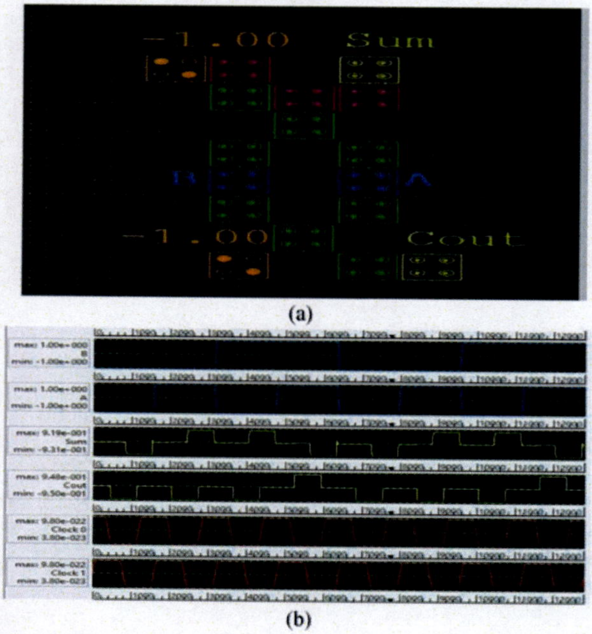

Fig. 14 (a) Layout design, (b) Functional result of Half adder

4 Design and Implementation of the Modules

The designing and implementation of the proposed sub module hardware design are carried out in QCADesigner 2.0.3 simulation tool. The Bistable Approximation is used as a simulation engine for the simulation of the designs in the QCADesigner software.

4.1 *Half Adder Design*

The simulation results obtained after simulating of half adder design in QCADesigner software is depicted in Fig. 14. The result obtained is verified using the truth table of the half adder. Thus it can be seen that the simulation results obtained are true with respect to the truth table.

4.2 Proposed 1-Bit Full Adder Design

A, B and Cin are the three inputs to the 1-bit full adder circuit, where A and B are the two inputs that are being summed and Cin is the input carry. Sum and Cout are the outputs of 1-bit full adder. The proposed Full adder has cell count as 30, which is lesser compared to other design.

Figure 15 (a) depicts the QCA layout design of a 1-bit Adder which is designed using QCADesigner version 2.0.3. The design consists of three majority voter gates and two inverter gates. A suitable clocking scheme is used in the design. In Fig. 15 (b), the simulation results of 1-bit Adder are shown. The latency in the output is 0.5, since the output is delayed by two clock phase, each accounting to a latency of 0.25

Figure 16 depicts the QCA layout design of a 16-bit Parallel Binary Adder which is designed using QCADesigner version 2.0.3.The one bit adders are cascaded to one another to using QCA wires in the QCADesigner software to design the PBA.

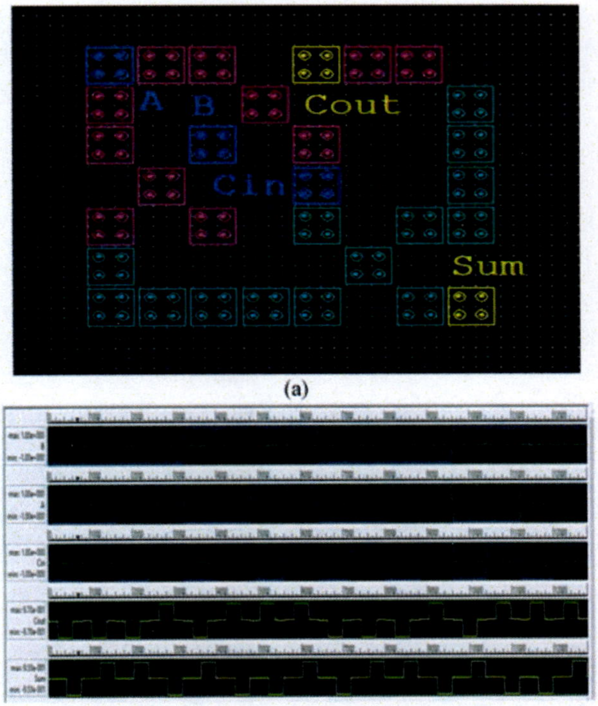

(a)

(b)

Fig. 15 (a) Layout design, (b) Functional result of Full adder

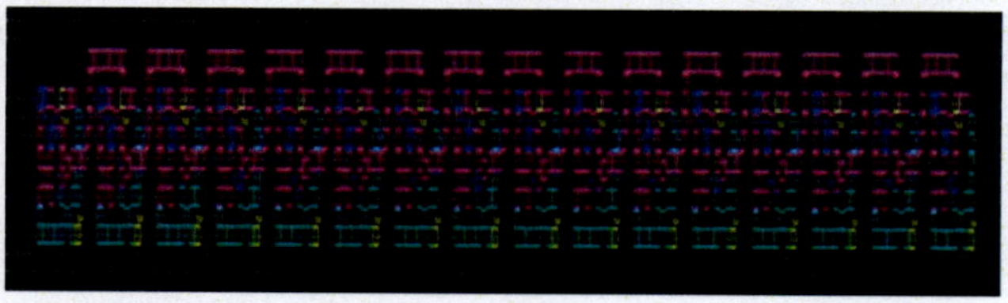

Fig. 16 QCA design layout of a 16 bit parallel binary adder

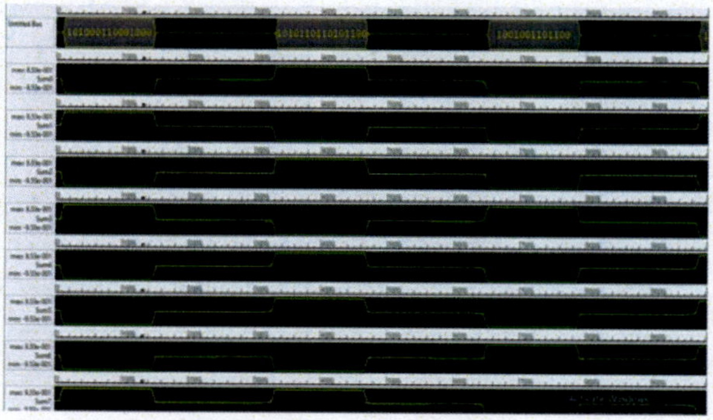

Fig. 17 Functional result of 16-bit adder design

The simulation results of 16-bit Adder are shown in Fig. 17. The number of used cells in the design of the adder is 769 and the area covered by the design is 0.69 μm^2.

4.3 *Proposed Multiplier Design*

Figure 18 (a) depicts the QCA layout design of a 2 × 2 multiplier which is designed using QCADesigner version 2.0.3. The Urdhwa and Triyakbhyam multiplication algorithm is used in the design of the multiplier. The simulation results obtained after simulation of 2 × 2 Multiplier in QCADesigner software is depicted in Fig. 18 (b). The result obtained is verified using the truth table of the multiplier.

Figure 19 (a) depicts the QCA layout design of an 8-bit multiplier which is designed using QCADesigner version 2.0.3. The Urdhwa and Triyakbhyam multiplication algorithm is used in the design of the multiplier. The output obtained is given to Ripple carry adder. The un-used inputs are given an input 0. The carry of the first RC adder is given to the second RC adder together with other inputs from the 2-bit multiplier. The resulting outputs from the RC adder are the output of the Multiplier design. The results obtained after simulation of 8-bit Multiplier in QCADesigner software is shown in Fig. 19 (b).

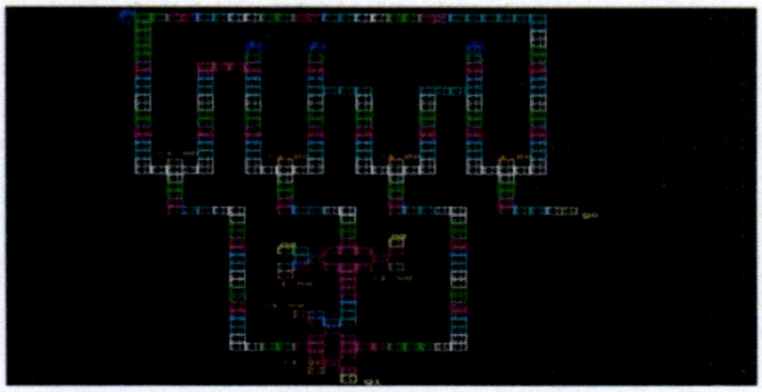

(a)

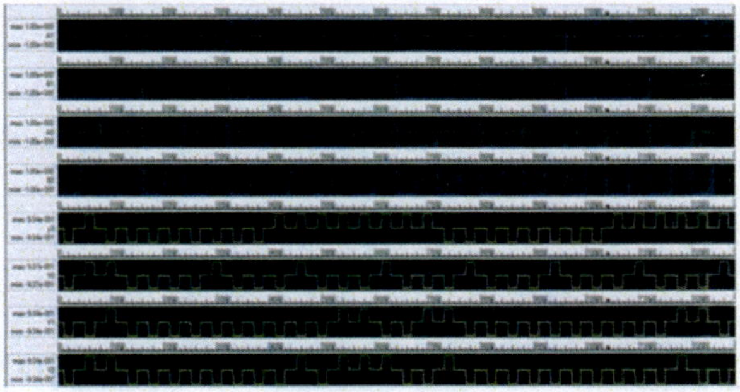

(b)

Fig. 18 (a) Layout design, (b) Functional result of 2 × 2 Multiplier

4.4 PIPO Shift Register Design

The QCA layout design of a D-Flip-flop which is designed using QCADesigner version 2.0.3 is shown in Fig. 20. The simulation result is depicted in Fig. 21. The result obtained is verified using the truth table of the D-Flip-flop.

The QCA layout design of a PIPO shift register which is designed using QCADesigner version 2.0.3 is depicted in Fig. 22 (a). The results obtained after simulation of the PIPO shift register in QCADesigner software is depicted in Fig. 22 (b). The result obtained is verified using the truth table of the D-Flip-flop.

4.5 MAC Unit Design

Figure 23 depicts the QCA layout design of a MAC unit which is designed using QCADesigner version 2.0.3. The Multiplier unit previously designed is integrated with the Adder unit. The inputs undergo multiplication then the resulting output is given as input to the Adder unit. The output is given to the shift register which feedbacks the output to the adder resulting in the operation of addition and shifting.

The simulation results obtained after simulation of MAC Unit in QCA Designer software is depicted in Fig. 24. Thus it can be seen that the simulation results obtained are true with respect to the truth table. This work provides has high performance

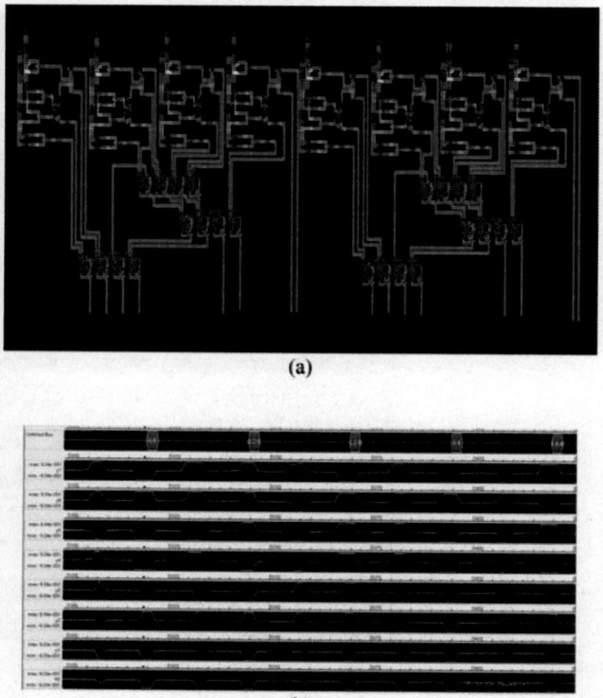

(a)

(b)

Fig. 19 (a) Layout design, (b) Functional result of 8 bit Multiplier

Fig. 20 QCA design layout of D-Flipflop

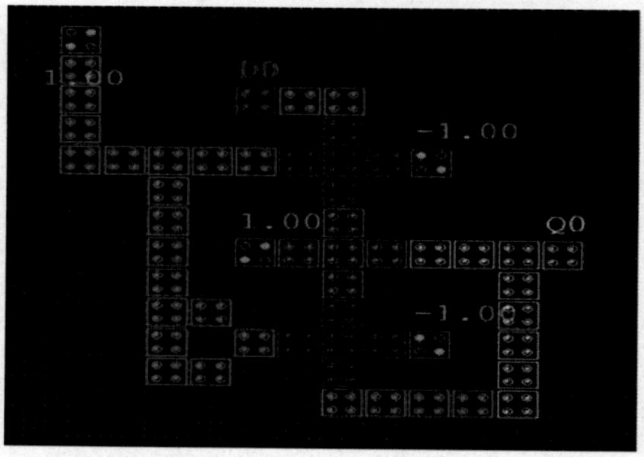

circuits with very small dimensions and lesser power consumption compared to standard and typical VLSI technology. The number of used cells in the design is 27110 and the area of the MAC unit design is 77.89 μm^2. This is very much less compared to previous works.

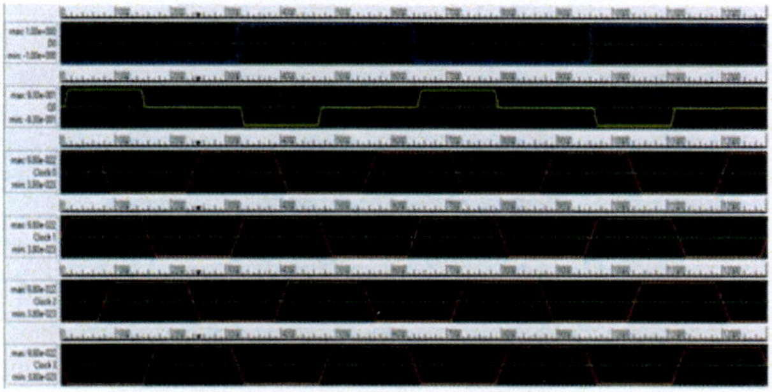

Fig. 21 Functional result of D-Flip-Flop

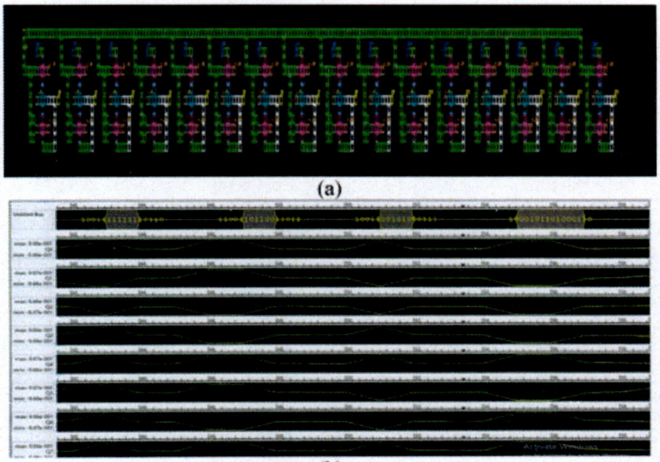

(a)

(b)

Fig. 22 (a) Layout design, (b) Functional result of PIPO shift register

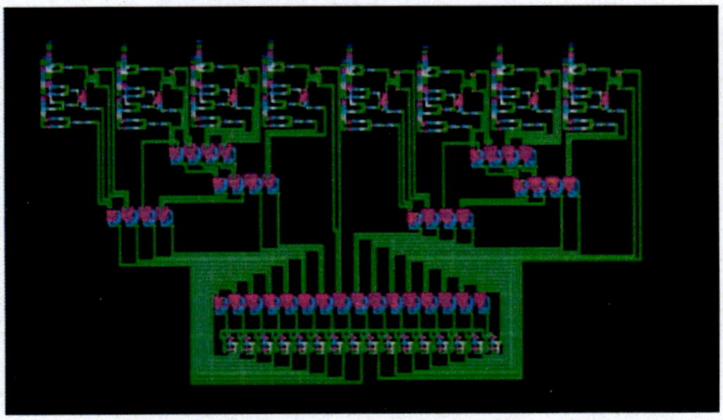

Fig. 23 QCA design layout of MAC Unit

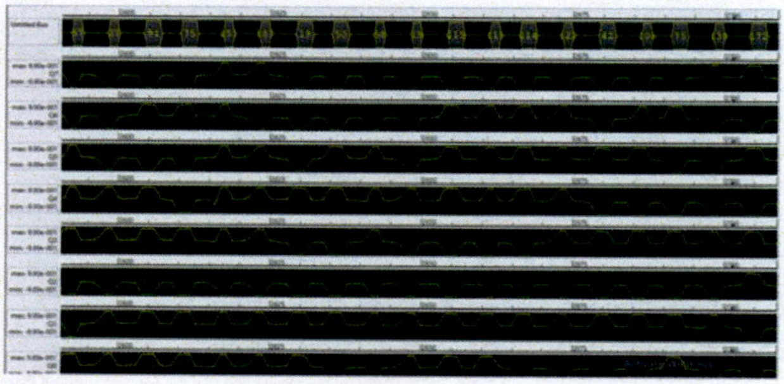

Fig. 24 Functional result of MAC Unit

Table 1 Comparison of adder design

Sl.No.	Design of adder	Cell count	Area (μm^2)	Clock phases
1.	In [2]	46	0.04	1
2.	In [3]	33	0.03	0.5
3.	In [4]	63	0.06	0.75
4.	In [5]	69	0.10	1
5.	In [6]	58	0.04	0.75
6.	In [7]	105	0.14	0.75
7.	In [8]	86	0.08	1
8.	In [9]	71	0.06	1.5
9.	In [10]	59	0.043	1
10.	Proposed design	30	0.02	0.5

Table 2 Comparison of multiplier design

Sl.No.	Design of multiplier	Cell count	Area (μm^2)	Clock phases
1.	In [11]	26499	82.18	38
2.	In [11]	26973	82.19	38
3.	In [12]	33894	87.47	44
4.	In [12]	–	28.16	31
5.	In [12]	34904	92.60	47
6.	In [11]	13839	22.57	30
7.	In [11]	15106	21.39	30
8.	In [13]	13533	18.44	10.75
9.	Proposed design	10997	16.61	11

The Tables 1 and 2 show the comparison of one bit full adder and multiplier designs with that of existing designs. It is noted from the Table 1 and 2 the proposed design of full adder circuit has least cell count and has minimum area when compared to other designs thus making the design more efficient. The proposed 8-bit multiplier significantly improves the design over the other multiplier with regards to number of cell count and area. In the above table, area is conveyed in the units of μm^2, delay is conveyed in regard to the number of needed clock cycles ($\times 10^{-12}$ s), and complexity is conveyed in regards to the number of required cells.

The graphs in Figs. 25 and 26 show the comparison of the parameters of the proposed adder and multiplier design with previous designs respectively. The parameters under comparison are the number of cells in the design, the area of the design and the number of clock phases. It is verified proposed design has reduced area, lesser number of cell counts and less number of clock phase delay. This also shows that since less number of cells is utilized, the power dissipation of the design also gets reduced, which leads to better performance.

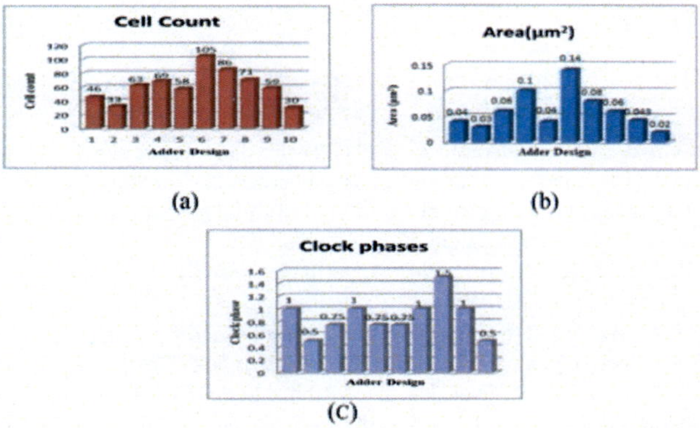

Fig. 25 Comparison graph of (a) Cell count, (b) Area, (c) Cock phases of Adder

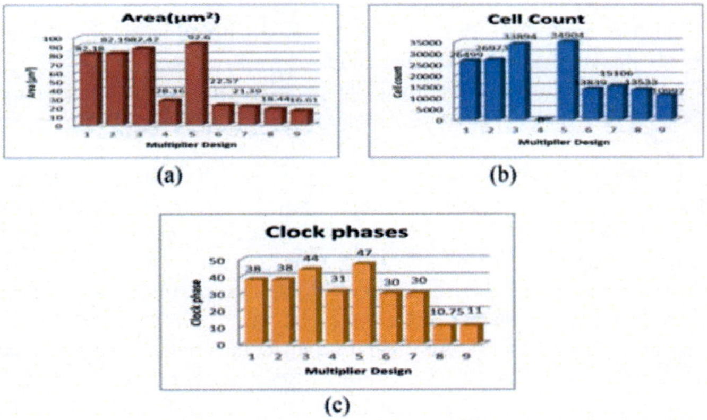

Fig. 26 Comparison graph of (a) Area, (b) Cell count, (c) Cock phases of Multiplier

5 Conclusion

This Paper presents the design of an area efficient MAC unit using QCA. The proposed design has three sub modules: An 8-bit multiplier, parallel binary adder unit, and accumulator unit. Not much research has been made in regards to the MAC unit in this field. The proposed design of adder has reduced cell count by 9.09% and area by 33.33%.The proposed design of multiplier has narrowed down the cell count by 18.7% and area by 9.92% in contrast to preceding designs. The design of Proposed MAC unit has 27110 cells and its area is 77.89 μm^2. In this way it make the proposed model valuable for digital signal and image processing frameworks implemented in gadgets where power utilization is low and constant handling are required.

References

1. Paul DJ, Compano R, Molenkamp L (2001) Technology Roadmap for nanoelctronics. In: European commission IST programme-future and emerging Technologies
2. Mokhtari D et al (2018) Design of novel efficient full adder circuit for quantum-dot cellular automata technology. Electron Energ 31(2):279–285
3. Rezai A (2017) Rashidi H (2017) High-performance full adder architecture in quantum-dot cellular automata. J Eng 7:394–402
4. Labrado C, Thapliyal H (2016) Design of adder and subtractor circuits in majority logic-based field-coupled QCA nanocomputing. Electron Lett 52(6):464–466
5. Kianpour M et al (2014) A novel design of 8-bit adder/subtractor by quantum-dot cellular automata. J Comput Syst Sci 80(7):1404–1414
6. Sayedsalehi S, Khademolhosseini H, Roohi A et al (2014) A symmetric quantum- dot cellular automata design for 5-input majority gate. J Comput Electron 13(3):701–708
7. Azghadi MR, Sayedsalehi S et al (2015) Restoring and nonrestoring array divider designs in quantum-dot cellular automata. Inf Sci 311:86–101
8. Sen B, Mukherjee R, Sahu Y et al (2016) On the reliability of majority logic structure in quantum-dot cellular automata. Microelectron J 47:7–18
9. Hashemi S, Navi K (2015) A novel robust QCA full-adder. In: Fifth International Biennial Conference Ultrafine Grained and Nanostructured Materials, Procedia Materials Science 11: 376–380
10. Jaberipur G, Abedi D, Sangsefidi M (2015) Coplanar full adder in quantum-dot cellular automata via clock-zone based crossover. IEEE Trans Nanotechnol 14(3):497–504
11. Kim SW (2011) Design of parallel multipliers and dividers in quantum-dot cellular automata. The University of Texas at Austin
12. Kim SW, Swartzlander EE (2010) Multipliers with coplanar crossings for quantum-dot cellular automata. In: 10th IEEE Conference on Nanotechnology, Seoul, pp. 953–957
13. Yadav J, Chudasam A (2018) Trailokya nath sasamal, an efficient design of vedic multiplier using ripple carry adder in quantum dot cellular automata. Elsevier Comput Electr Eng 65:527–542

Stress Enhancement in the Piezoelectric Cantilever with Tapered Substrate and Piezoelectric Layer Thickness

Ashutosh Anand, Srikanta Pal, and Sudip Kundu

Abstract In this paper, the modification of the thickness profile of the uimorph piezoelectric cantilever-based energy harvester is presented to achieve the uniform and improved stress distribution profile along the length of the cantilever. The thickness of both substrate and the piezoelectric layer is non uniform and tapered thickness profile is achieved in case of the proposed cantilever structure. Non uniform thickness profile not only improved the stress distribution along the length but also reduces the resonant frequency of the cantilever. The proposed cantilever has frequency 14.32% less resonant than the normal rectangular cantilever. The proposed rectangular cantilever with tapered thickness generates the peak output voltage of 10 V at the resonant frequency of 102.52 Hz. The proposed cantilever generates 17.65% more output voltage than the normal rectangular cantilever.

Keywords Piezoelectric · Energy harvester · Tapered · Stress · Euler bernoulli beam

1 Introduction

The advancement in the field of the VLSI technology has drastically reduced the size and power of the electronic devices. The low power micro devices have been used in numerous application such as wireless communication sensors, military equipment and consumers electronic devices biomedical devices etc. All these devices have been powered by the conventional batteries, which has the limited life span which needs

A. Anand (✉) · S. Pal
Electronics and Communication Engineering, Birla Institute of Technology Mesra, Ranchi, India
e-mail: phdec10052.15@bitmesra.ac.in

S. Pal
e-mail: spal@bitmesra.ac.in

S. Kundu
Electronics and Communication Engineering, National Institute of Technology Rourkela, Rourkela, India

V. Nath and J. K. Mandal (eds.), *Proceeding of Fifth International Conference on Microelectronics, Computing and Communication Systems*, Lecture Notes in Electrical Engineering 748,
https://doi.org/10.1007/978-981-16-0275-7_70

to be replaced after some year. The device with the integrated energy harvester can reduce the dependency on the battery. The vibration-based energy harvester (EH) can be power these devices. Piezoelectric [1, 2] based vibration-based energy harvester is the most suited than electromagnetic [2] and electrostatic [2] energy harvester because of high power density and non-dependence on external power source. It can also be easily integrated in the MEMS technology [3, 4].

The cantilever structure is most preferred for vibration-based EH as the average strain produced in the cantilever structure for a given input force is more than any other structure such as cymbal type structure, stack type structure, shell type structure [3]. The rectangular cantilever structures are easy to fabricate and implement [5].

The vibration of the PZEH results in the strain developed in the piezoelectric layer, which generates electrical energy. The maximum output voltage can be generated by the piezoelectric energy harvester (PEH) when the resonant frequency of the cantilever matches with the ambient vibration frequency. The slight deviation in the frequency cause drastic fall in the output voltage of the PEH [4, 6].

The conventional cantilever with the uniform thickness has been used to generate the electricity [5]. In this form, the stress distribution is maximum at the fixed end and gradually decreases towards free end. This results in the under-utilization of the piezoelectric layer at the free end. Several researchers try to improve the efficiency of PEH by modifying the structure of the cantilever.

Baker et al. [7] shows that the tapering of the rectangular cantilever at the free end increases the power density of the PEH. Muthalif and Nordin [8] shows that the triangular cantilever has double average strain along the length than the normal rectangular cantilever of same fabrication area. Anand and Kundu [1] designed the spiral cantilever and shows the increase in the stress distribution at the corner point of the cantilever. Anand and Kundu [9] show that the spiral cantilever can increase the effective length of the cantilever while maintaining the compactness of the cantilever. It can be noted that all the structures discussed above has the uniform thickness of the substrate and piezoelectric layer in the cantilever beam, but as per authors knowledge, there is no such result which shows the use of tapered thickness in both substrate and piezoelectric layer in the micro-cantilever structure.

This paper tries to investigate the thickness profile of the unimorph cantilever and compare the performance of the EH with uniform thickness. The proposed PEH is unimorph cantilever with tapered substrate thickness and tapered piezoelectric thickness layer. All the structures have been designed and simulated in the COMSOL Multiphysics which is finite element analysis software.

The paper has been organized as follows. The design and analysis of the structure are discussed in Sect. 2. This section also discusses the different properties of the materials used. Section 3 discusses about the mechanical and electrical output of the PEH. Finally, the conclusion has been drawn in Sect. 4.

2 Design and Analysis

In this paper a cantilever beam is designed for the PEH with the varying thickness. The thickness of the substrate layer and piezoelectric layer varied along the length. The silicon is used as the substrate layer the Zinc oxide is used as the piezoelectric layer in the design of the cantilever. The proof mass is also of Si materials. The rectangular cantilever with uniform thickness is also designed with the same layer configuration. The different properties of the materials used in the cantilever is tabulated in Table 1.

The resonant frequency of the rectangular cantilever according to the Euler Bernoulli beam theory is given by Eq. 1 [10].

$$f_n = \frac{v_n^2}{2\pi}\frac{1}{L^2}\sqrt{\frac{YI}{m}} \tag{1}$$

where f_n is the nth mode resonant frequency, v_n is the nth mode eigen value, L is the length of the beam, Y is the young's modulus, I is the moment of inertia and m is mass per unit length. The stiffness of the tapered thickness composite beam (*EI(x)*) is given by Eq. (2) [11].

$$YI(x) = \frac{Y_P W h_p(x)^3}{12} + \left(Y_s - Y_p\right)\frac{W h_s(x)^3}{12} \tag{2}$$

where Yp and Ys is the young's modulus of piezoelectric layer and substrate layer. The value of hp(x) and hs(x) is given by Eq. (3) and (4).

$$h_P(x) = h_{P0}\left(1 - \alpha\frac{x}{L}\right) \tag{3}$$

$$h_s(x) = h_{s0}\left(1 - \beta\frac{x}{L}\right) \tag{4}$$

Table 1 Properties of the materials

Properties	Si	ZnO
Young's modulus	170 GPa	210 GPa
Poisson ratio	0.29	0.33
Density	2329 (Kg/m^3)	5680 (Kg/m^3)
Charge coefficient, d_{31}: (pC/N)	–	11.34
Dielectric constant, ε_r	–	12.64
Voltage coefficient, g_{31}: (Vm/N)	–	0.217

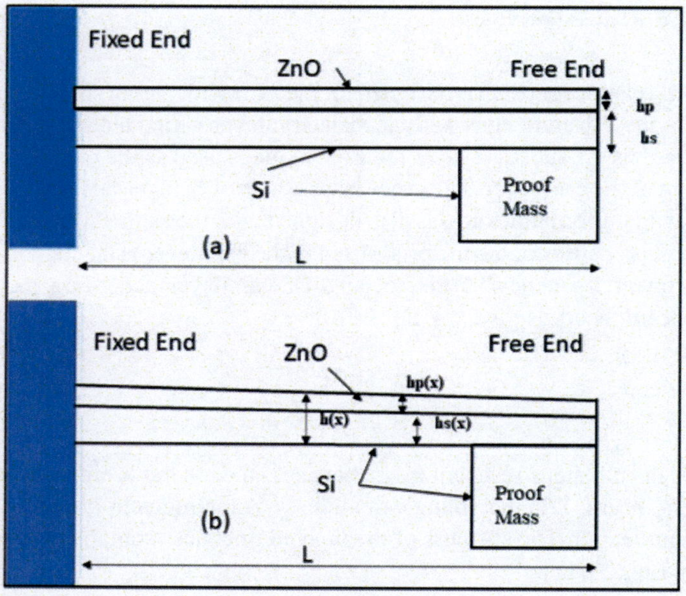

Fig. 1 Side view of the energy harvester geometry (a) Normal Rectangular (b) Rectangular with tapered thickness

where $\alpha = \frac{h_{P0}-h_{PL}}{h_{P0}}$ and $\beta = \frac{h_{s0}-h_{sL}}{h_{s0}}$. Here h_{P0} and h_{PL} is the thickness of piezoelectric layer at fixed end and free end respectively. h_{s0} and h_{sL} is the thickness of substrate layer at fixed end and free end respectively.

Using the above equations, the rectangular cantilever beam has been designed. The side view of the rectangular cantilever with tapered thickness and normal rectangular cantilever is shown in the Fig. 1.

The total thickness of the beam is h(x) which is the summation of piezoelectric layers of thickness hp(x) and a substrate layer of thickness hs(x), and the length of the piezoelectric layer (same as that of the cantilever beam) is L. The width of both piezoelectric layer and substrate layer is W. The dimension of the proof mass attached at the free end of the cantilever is given in the Table 2.

The charge generated by the piezoelectric layer under transverse motion of cantilever is given by

$$q(t) = \int_{0}^{L}(d_{31}Y_{P}h(x)\left(-\frac{\partial^2 w(x,t)}{\partial x^2}\right)Wdx) \tag{5}$$

where $w(x,t)$ transverse bending of the cantilever beam.

After discussing the brief theory related to the piezoelectric cantilever. Next section will discuss the simulation and results.

Table 2 Geometric parameters of the simulated cantilevers

Parameter	Description	Value
L	Beam length	10 mm
W	Beam width	1 mm
hs	Thickness of substrate layer of normal rectangular cantilever	25 μm
hp	Thickness of piezoelectric layer of normal rectangular cantilever	5 μm
hp(x)	Thickness of the piezoelectric layer of rectangular cantilever with tapered thickness	hp(x = 0) = 5 μm hp(x = L) = 3 μm
hs(x)	Thickness of the substrate layer of rectangular cantilever with tapered thickness	hs(x = 0) = 25 μm hs(x = L) = 15 μm
Proof mass	Dimension of the proof mass	$1 \times 1 \times 0.85$ mm^3

3 Simulation and Results

3.1 Mechanical Analysis

After designing both cantilevers, eigen frequency analysis is carried out to find the resonant frequency of the cantilever. The normal meshing technique is used to mesh the cantilever. The resonant frequency of the normal rectangular cantilever is 119.65 Hz while the resonant frequency of the rectangular cantilever with tapered thickness has 102.52 Hz.

It can be observed from the Eq. (5) that the charge generated in the piezoelectric layer depends on the deflection of the cantilever. The Fig. 2 shows the comparison of the deflection between the normal rectangular cantilever and rectangular cantilever with tapered thickness.

In case of the piezoelectric layer of the beam, the applied stress (T_P) not only induces a corresponding strain (S_P) along the piezoelectric layer, but also generates an electric field (E). The total stress in the middle of the piezoelectric layer is given by

$$T_P = Y_P\left(-h(x)\frac{\partial^2 w(x,t)}{\partial x^2}\right) - d_{31} Y_P E \quad (6)$$

The output voltage (V) of the PEH depend on the average stress developed in the cantilever beam [12], as shown in Eq. 6.

$$V = T_P \times g_{31} \times h_P \quad (7)$$

So, the better stress distribution in the cantilever will results in the better output voltage of PEH. The Fig. 2 shows the comparison between the normal rectangular cantilever and rectangular cantilever with tapered thickness (Fig. 3).

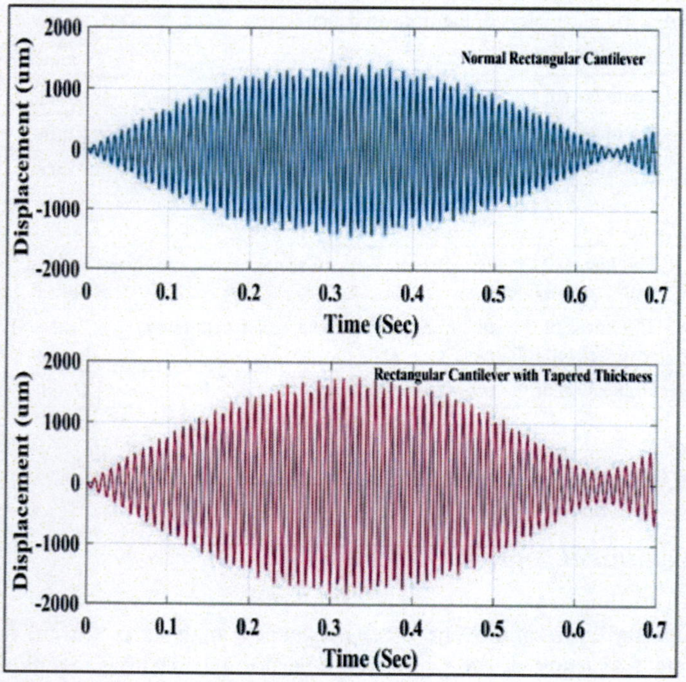

Fig. 2 Transient displacement of the tip point of cantilever structure

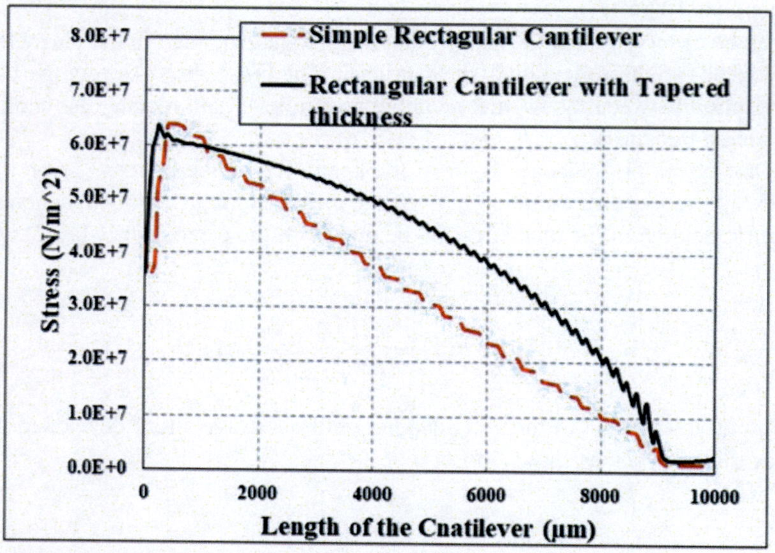

Fig. 3 Comparision between the variation in stress distribution along the arc length of all cantilever structures

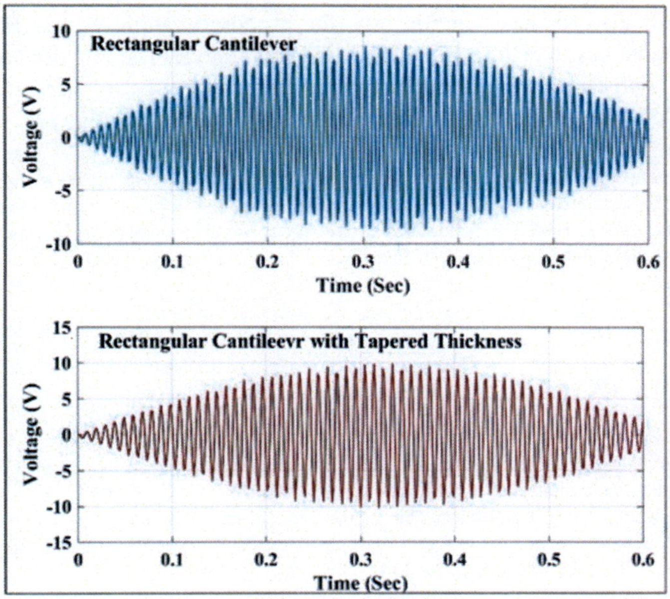

Fig. 4 Transient output voltage cantilever structure

3.2 Voltage Output of Simulated Structures

It is assumed that the whole system is vibrating by the external sinusoidal load, so that the proper time dependent simulation of the cantilever EH is carried out. The time domain analysis is carried out in the COMSOL Multiphysics for each cantilever EH and peak voltage is observed. The normal rectangular cantilever beam generates the peak output voltage of 8.5 V at the resonant frequency of 119.65 Hz. The rectangular cantilever with tapered thickness generates the peak output voltage of 10 V at the resonant frequency of 102.52 Hz. The transient output voltage of both cantilever structures is given in the Fig. 4. It can be observed that there is a improvement of 17.65% in the output voltage in case rectangular cantilever with tapered thickness than the normal rectangular cantilever.

4 Conclusion

In order to improve the stress distribution and minimize the non-uniformity in the normal rectangular cantilever structure, a tapered substrate thickness and tapered piezoelectric thickness is proposed. The FEM analysis is done in the COMSOL Multiphysics. The proposed rectangular cantilever with tapered thickness generates 17.65% more output voltage than the normal rectangular cantilever. The resonant frequency is also reduced significantly, the proposed cantilever has 14.32% less resonant frequency than the normal rectangular cantilever. The proposed rectangular cantilever with tapered thickness has more uniform and better stress distribution along the length in comparison with the normal rectangular cantilever.

Acknowledgements We, all the authors, would like to express our special thanks to the Science and Engineering Research Board (SERB), Govt. of India for funding to execute this research through project ref. no. ECR/2016/001279. This research was not possible without the help and support of BIT Mesra, Ranchi and SERB, Govt. of India.

References

1. Anand A, Kundu S (2019) Design of a spiral-shaped piezoelectric energy harvester for powering pacemakers. Nanomaterials and Energy 8(2):139–150
2. Priya S, Inman, DJ (2009) Energy Harvesting Technologies, New York
3. Kim HS, Kim JH, Kim J (2011) A review of piezoelectric energy harvesting based on vibration. Int J Precis Eng Manuf 12(6):1129–1141
4. Anand A, Naval S, Sinha PK, Das NK, Kundu S (2020) Effects of coupling in piezoelectric multi-beam structure. Microsyst Technol 26(4):1235–1252
5. Roundy S, Wright PK, Rabaey J (2003) A study of low level vibrations as a power source for wireless sensor nodes. Comput Commun 26(11):1131–1144
6. Naval S, Sinha PK., Das NK., Anand A, Kundu S (2019) Bandwidth increment of piezoelectric energy harvester using multi-beam structure. In: 2019 devices for integrated circuit (DevIC), pp 370–373. IEEE
7. Baker J, Roundy S, Wright P (2005) Alternative geometries for increasing power density in vibration energy scavenging for wireless sensor networks. In: collection of technical papers - 3rd international energy conversion engineering conference, vol 2, pp 959–70
8. Muthalif AGA, Nordin NHD (2015) Optimal piezoelectric beam shape for single and broadband vibration energy harvesting: Modeling, simulation and experimental results. Mech Syst Signal Process 54–55:417–426. https://doi.org/10.1016/j.ymssp.2014.07.014
9. Anand A, Kundu S (2019) Design of mems based piezoelectric energy harvester for pacemaker. In: 2019 devices for integrated circuit (DevIC), pp 465–469. IEEE
10. Shen D, Park JH, Ajitsaria J, Choe SY, Wikle HC, Kim DJ (2008) The design, fabrication and evaluation of a MEMS PZT cantilever with an integrated Si proof mass for vibration energy harvesting. J Micromech Microeng 18(5): 055017
11. Erturk A, Inman DJ (2009) An experimentally validated bimorph cantilever model for piezoelectric energy harvesting from base excitations. Smart Mater Struct 18:025009. https://doi.org/10/0964-1726/18/2/025009
12. Anand A, Kundu S (2019) Improvement of output power in piezoelectric energy harvester under magnetic influence. In: 2019 devices for integrated circuit (DevIC), pp 382–385. IEEE

Author Index

V. Nath and J. K. Mandal (eds.), *Proceeding of Fifth International Conference on Microelectronics, Computing and Communication Systems*, Lecture Notes in Electrical Engineering 748,
https://doi.org/10.1007/978-981-16-0275-7